90 0591032 3

Enantioselective Synthesis of β-Amino Acids

Enantioselective Synthesis of β-Amino Acids

EDITED BY

Eusebio Juaristi

WILEY-VCH

New York • Chichester • Weinheim • Brisbane • Singapore • Toronto

Eusebio Juaristi
Department of Chemistry
Centro de Investigación y de Estudios Avanzados
del Instituto Politécnico Nacional
Apartado Postal 14-740
07000 México, D.F., México

This book is printed on acid-free paper. ♾

Library of Congress Cataloging-in-Publication Data

Enantioselective synthesis of [beta]-amino acids / edited by Eusebio Juaristi.
p. cm.
Includes bibliographical references and index.
ISBN 0-471-18627-9
1. Beta lactam antibiotics—Synthesis. 2. Amino acids—Synthesis. 3. Chiral drugs—Synthesis. 4. Enantiomers—Synthesis.
I. Juaristi, Eusebio.
RS431.B48E52 1996
615′.19—dc20 96-5913
CIP

Printed in the United States of America

ISBN 0-471-18627-9 Wiley-VCH, Inc.

10 9 8 7 6 5 4 3 2 1

Preface

β-Amino acids, although less abundant than their α-analogues, are also present in peptides and in other natural products, and in free form they show interesting pharmacological effects. Furthermore, β-amino acids are synthetic precursors of β-lactams, which are potentially biologically active and of current interest. Although several methods for the synthesis of racemic β-amino acids have been developed, only recently has the preparation of enantiomerically pure compounds emerged as an important and challenging synthetic endeavor.

Following introductory overviews of the relevance of β-amino acids in pharmaceutical sciences, this book presents a discussion of the most important methods that have been developed for the asymmetric synthesis of β-amino acids. Each important method is described by the original leader chemist who developed it, and thus each chapter is written with authority and firsthand knowledge.

I hope that the relevance and timeliness of the topics discussed in this book—the first comprehensive monograph in the area—will render it of interest to a broad group of chemists in universities and in the pharmaceutical and related industries.

Eusebio Juaristi

Contents

Contributors

Jesús M. Aizpurua
Departamento de Química Orgánica
Facultad de Química
Universidad del País Vasco
Apdo. 1072
20080 San Sebastián
Spain

Amit Basak
Department of Chemistry
Indian Institute of Technology
Kharagpur 721302
India

Wolfgang Bettray
Institut für Organische Chemie
Rheinisch-Westfälische Technische Hochschule
Professor-Pirlet Strasse 1
D-52074 Aachen
Germany

Thomas C. Boge
Department of Medicinal Chemistry
University of Kansas
Lawrence, KS 66045
U.S.A.

Christine Braun
Laboratorium für Organische Chemie
Eidgenössischen Technischen Hochschule
ETH-Zentrum
Universitätstrasse 16
CH-8092 Zürich
Switzerland

Andreas Burgard
Institut für Organische Chemie
Johannes Gutenberg-Universität Mainz
Becher-Weg 18-20
55099 Mainz
Germany

Giuliana Cardillo
Departimento di Chimica "G. Ciamician"
Università di Bologna
Via Selmi, 2
40126 Bologna
Italy

Franklin A. Davis
Department of Chemistry
Temple University
Philadelphia, PA 19122-2585
U.S.A.

Dieter Enders
Institut für Organische Chemie
Rheinisch-Westfälische Technische Hochschule
Professor-Pirlet Strasse 1
D-52074 Aachen
Germany

Iñaki Ganboa
Departamento de Química Orgánica
Facultad de Química
Universidad del País Vasco
Apdo. 1072
20080 San Sebastián
Spain

Oscar García-Barradas
Departamento de Química
Centro de Investigación y de Estudios Avanzados
del Instituto Politécnico Nacional
Apartado Postal 14-740
07000 México, D.F.
Mexico

Cesare Gennari
Departimento di Chimica Organica e Industriale
II Facoltà di Scienze Matematiche Fisiche e Naturali
Università di Milano
via G. Venezian 21
20133 Milano
Italy

Gunda I. Georg
Department of Medicinal Chemistry
University of Kansas
Lawrence, KS 66045
U.S.A.

Peter Gmeiner
Universität Erlangen–Nürnberg
Institut für Pharmazie und Lebensmittelchemie
Schuhstraße 19
91052 Erlangen
Germany

Christel Guibourdenche
Laboratorium für Organische Chemie
Eidgenössischen Technischen Hochschule
ETH-Zentrum
Universitätstrasse 16
CH-8092 Zürich
Switzerland

Kouji Hattori
School of Engineering
Nagoya University
Chikusa, Nagoya 464-01
Japan

Kazuaki Ishihara
School of Engineering
Nagoya University
Chikusa, Nagoya 464-01
Japan

Peter A. Jacobi
Department of Chemistry
Wesleyan University
Middletown, CT 06459-0180
U.S.A.

Charles W. Jefford
Department of Organic Chemistry
University of Geneva
CH-1211 Geneva 4
Switzerland

Eusebio Juaristi
Departamento de Química
Centro de Investigación y de Estudios Avanzados
del Instituto Politécnico Nacional
Apartado Postal 14-740
07000 México, D.F.
Mexico

Esko K. Karvinen
Department of Chemistry
University of Oulu
Linnanmaa
SF 90570 Oulu
Finland

Joseph P. Konopelski
Department of Chemistry and Biochemistry
University of California
Santa Cruz, CA 95064
U.S.A.

Ari M.P. Koskinen
Department of Chemistry
University of Oulu
Linnanmaa
SF 90570 Oulu
Finland

Horst Kunz
Institut für Organische Chemie
Johannes Gutenberg-Universität Mainz
Becher-Weg 18-20
55099 Mainz
Germany

Jennifer L. Matthews
Laboratorium für Organische Chemie
Eidgenössischen Technischen Hochschule
ETH-Zentrum
Universitätstrasse 16
CH-8092 Zürich
Switzerland

James McNulty
Department of Organic Chemistry
University of Geneva
CH-1211 Geneva 4
Switzerland

Albert Moyano
Unitat de Recerca en Síntesi Asimètrica
Departament de Química Orgànica
Universitat de Barcelona
c/Martí i Franquès, 1-11
E-08028 Barcelona
Spain

Mark Overhand
Laboratorium für Organische Chemie
Eidgenössischen Technischen Hochschule
ETH-Zentrum
Universitätstrasse 16
CH-8092 Zürich
Switzerland

Claudio Palomo
Departamento de Química Orgánica
Facultad de Química
Universidad del País Vasco
Apdo. 1072
20080 San Sebastián
Spain

Rafael Pedrosa
Departamento de Química Orgánica
Universidad de Valladolid
Dr. Mergelina s/n
47011-Valladolid
Spain

Miquel A. Pericàs
Unitat de Recerca en Síntesi Asimètrica
Departament de Química Orgànica
Universitat de Barcelona
c/Martí i Franquès, 1-11
E-08028 Barcelona
Spain

Rajarathnam E. Reddy
Department 9NM, Bldg. AP20
Diagnostics Division
Abbott Laboratories
100 Abbott Park Road
Abbott Park, IL 60064-3500
U.S.A.

Antoni Riera
Unitat de Recerca en Sintesi Asimètrica
Departament de Química Orgànica
Universitat de Barcelona
c/Martí i Franquès, 1-11
E-08028 Barcelona
Spain

Jürgen Schankat
Institut für Organische Chemie
Rheinisch-Westfälische Technische Hochschule
Professor-Pirlet Strasse 1
D-52074 Aachen
Germany

Dieter Seebach
Laboratorium für Organische Chemie
Eidgenössischen Technischen Hochschule
ETH-Zentrum
Universitätstrasse 16
CH-8092 Zürich
Switzerland

Charles J. Sih
School of Pharmacy
University of Wisconsin-Madison
425 North Charter Street
Madison, Wisconsin 53706
U.S.A.

Vadim A. Soloshonok
Institute of Bioorganic and Petrochemistry
Ukrainian Academy of Sciences
Kiev 253160
Ukraine

Joaquín Tamariz
Departamento de Química Orgánica
Escuela Nacional de Ciencias Biológicas
Instituto Politécnico Nacional
Prol. Carpio y Plan de Ayala
Col. Santo Tomás
11340 México, D.F.
Mexico

Claudia Tomasini
Dipartimento di Chimica "G. Ciamician"
Università di Bologna
Via Selmi, 2
40126 Bologna
Italy

Anna Vulpetti
Departimento di Chimica Organica e Industriale
II Facoltà di Scienze Matematiche Fisiche e Naturali
Università di Milano
via G. Venezian 21
20133 Milano
Italy

Markus Weymann
Institut für Organische Chemie
Johannes Gutenberg-Universität Mainz
Becher-Weg 18-20
55099 Mainz
Germany

Jürgen Wiedemann
Institut für Organische Chemie
Rheinisch-Westfälische Technische Hochschule
Professor-Pirlet Strasse 1
D-52074 Aachen
Germany

Hisashi Yamamoto
School of Engineering
Nagoya University
Chikusa, Nagoya 464-01
Japan

Wanjun Zheng
Department of Chemistry
Wesleyan University
Middletown, CT 06459-0180
U.S.A.

CHAPTER

1

The Medicinal Chemistry of β-Amino Acids: Paclitaxel as an Illustrative Example

Thomas C. Boge and Gunda I. Georg

1.1 Introduction

β-Amino acids are emerging as an interesting class of compounds for the medicinal chemist. They are produced in humans, animals, microorganisms, marine organisms, and plants either in free form or as part of a peptide or depsipeptide.[1–3] They are found as components of peptidic natural products with antibiotic,[2] antifungal,[4,5] cytotoxic,[4–6] and other pharmacological properties.[3,7,8] The most well-known, medicinally important class of nonpeptidic β-amino acids are found in β-lactams. These include antibiotics,[9,10] β-lactamase inhibitors,[11] human leukocyte elastase inhibitors,[12] and cholesterol uptake inhibitors.[13] Other potent enzyme inhibitors such as statin, pepstatin, and amastatin contain β-amino acids as key structural components.[1] The chemical synthesis of β-amino acids has been the subject of several reviews.[3,4,14]

Peptidic molecules containing β-amino acids are generally more stable to enzymatic hydrolysis due to the inability of peptidases to cleave the amide bonds adjacent to the β-amino acid.[1,15] Unfortunately, attempts to exploit this increased stability by substituting α-amino acids with β-amino acids usually results in reduced biological activity, which is at least partly due to a loss of conformational integrity.[1,15]

Arguably the most extensively studied β-amino acid from the medicinal chemistry viewpoint is (2*R*,3*S*)-N-benzoyl-3-phenylisoserine, a component of the medicinally important natural product paclitaxel (**1**, Taxol®).[16,17] As an illustration of the potential role of β-amino acids in medicinal agents this review covers: (1) the asymmetric synthesis of 3-substituted isoserines and their potential for modification, (2) semisynthesis of paclitaxel and paclitaxel analogues from 3-phenylisoserine equivalents and 10-deacetylbaccatin III (**2**, 10-DAB), and (3) the structure–activity relationships (SARs) obtained from these studies. The creative synthetic medicinal chemistry presented herein should be applicable to any β-amino acid being considered for elaboration.

In the mid-1960s extracts of the stem and bark tissues of the Pacific yew (*Taxus brevifolia*) were found to be cytotoxic toward KB cells.[18] The isolation of the active components, assisted by bioassay-directed fractionation, resulted in the identification of paclitaxel (**1**).[19]

The initial investigations also showed that the intact paclitaxel molecule was required for optimal cytotoxicity. Neither of the methanolysis products, 10-deacetylbaccatain III (**2**) nor methyl (2*R*,3*S*)-N-benzoyl-3-phenylisoserine (**3**), displays significant activity.[19] Although the conditions used cleaved the 10-O-acetate of the diterpene, it was later found to be inconsequential as baccatin III (**4**) is also inactive.[20]

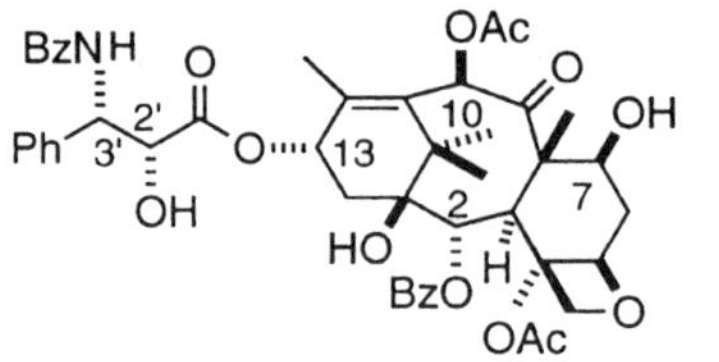

1 paclitaxel

Subsequently, paclitaxel was identified as a "spindle poison," causing cell death by disturbing the tubulin–microtubule dynamic equilibrium required for cell division.[21] However, instead of acting as a depolymerizing agent as the previously known spindle poisons (e.g., colchicine, vinca alkaloids), paclitaxel induces rapid polymerization of tubulin and the formation of hyperstable microtubules.[22] This new and unique mechanism of action[23,24] generated great interest in paclitaxel as a chemotherapeutic agent. Paclitaxel has been approved by the FDA for the treatment of drug refractory ovarian and metastatic breast cancer. Furthermore, encouraging results in lung, head, neck, and esophageal cancer are emerging from clinical trials.[25] The antitumor activity of paclitaxel may be influenced by other factors. It was recently reported that it blocks processes essential for invasion and metastases in prostate tumor cells.[26] Because microtubules are involved in a variety of cellular functions, paclitaxel may be useful in other therapeutic areas. For example, the potential for paclitaxel to stem the progression of fibrotic disease has been demonstrated in mice.[27]

The earliest chemical problem associated with paclitaxel was one of supply. It was feared that should paclitaxel become a useful chemotherapeutic agent, the slow-growing yew trees could not provide paclitaxel in sufficient amounts, and the survival of the species might be in jeopardy.[18]

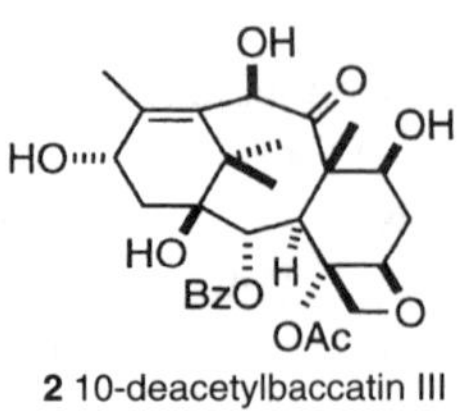

2 10-deacetylbaccatin III

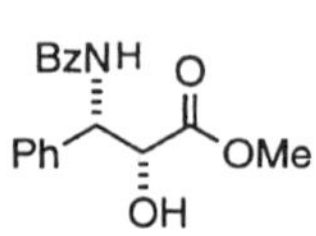

3 methyl (2*R*,3*S*)-N-benzoyl-3-phenylisoserine

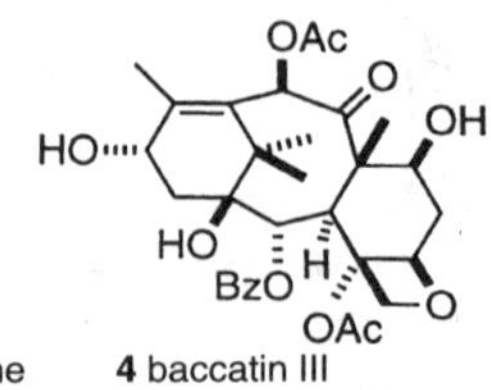

4 baccatin III

Total synthesis of the highly complex diterpene core structure was considered unfeasible for the short-term supply problem. In fact, only recently has paclitaxel succumbed to total synthesis, 23 years after the original publication of its structure![28–31]

Semisynthesis was considered the most likely answer to the supply problem. The breakthrough for a semisynthetic approach was the isolation of relatively large amounts of 10-deacetylbaccatin III (**2**) from the needles of the European yew (*Taxus baccata*), a regenerable source.[32] The FDA has recently approved a semisynthetic process for the industrial production of paclitaxel from 10-deacetylbaccatin III.

1.2 Synthesis of (2*R*,3*S*)-N-Acyl-3-phenylisoserines[33–39]

Acetylation studies suggested that O13 was the least reactive of the secondary alcohols (O7 ≥ O10 > O13 > O1) present in **2**.[39,40] This poor reactivity can be understood by examining models that clearly show that O13 is directed into the concavity of the cup-shaped diterpene.[34,41] Additionally, H bonding to the 4-acetyl carbonyl may further decrease reactivity.[42] An approach that circumvents the difficulties associated with the esterification of O13 with a sterically demanding, protected form of N-benzoyl-3-phenylisoserine uses suitably protected 13-O-[(*E*)-cinnamoyl]10-deacetylbaccatin III **5a** or **5b** as the starting material.[43] As can be seen in Scheme 1.1, stoichiometric hydroxyamination of **5a** led to a mixture of four isomers with virtually no regio- or diastereoselectivity. Increased facial selectivity for the hydroxyamination was observed with the chiral additives dihydroquinine acetate (4:1 favoring 2′*R*,3′*S* isomers) and dihydroquinidine acetate (4:1 favoring the 2′*S*,3′*R* isomers). Fortunately, the diastereomers could be separated, leading to the first successful semisynthesis of paclitaxel from **6b**. Although this method is not suitable for the large-scale synthesis of paclitaxel, the analogues of the undesired isomers provided a wealth of information on the regio- and stereochemical requirements of the paclitaxel C13 side chain for bioactivity (see Section 1.4, Table 1.2).[44] In fact, deprotection of intermediate **6a** led to docetaxel (**7**; Taxotere®), which became the lead compound for a number of research groups since it was twice as active as paclitaxel in a microtubule assay[44] and its in vitro cytotoxicity[44,45] was also higher than that of paclitaxel. It is of note that paclitaxel and docetaxel also differ at C10. A comparison of the biological data for 10-deacetylpaclitaxel (**9**),[46] which is slightly less active, and 10-acetyldocetaxel (**8**),[44] which is twice as active as paclitaxel in the tubulin assay, clearly shows that docetaxel's improved activity is due to the N3′ substituent. These results left the chemist with the challenge of preparing suitable (2*R*,3*S*)-N-benzoyl- and N-(*t*-butoxycarbonyl)-3-phenylisoserine equivalents for semisynthesis.

1.2.1 Synthesis via Asymmetric Epoxidation and Dihydroxylation

(2*S*,3*R*)-3-Phenylglycidic esters are the most popular intermediates for the synthesis of the 3-phenylisoserine side chains of paclitaxel and docetaxel. Greene and co-workers reported the first asymmetric synthesis of methyl (2*R*,3*S*)-N-benzoyl-3-phenylisoserine (**3**).[47] Based on Sharpless' asymmetric epoxidation methodology, the

Scheme 1.1

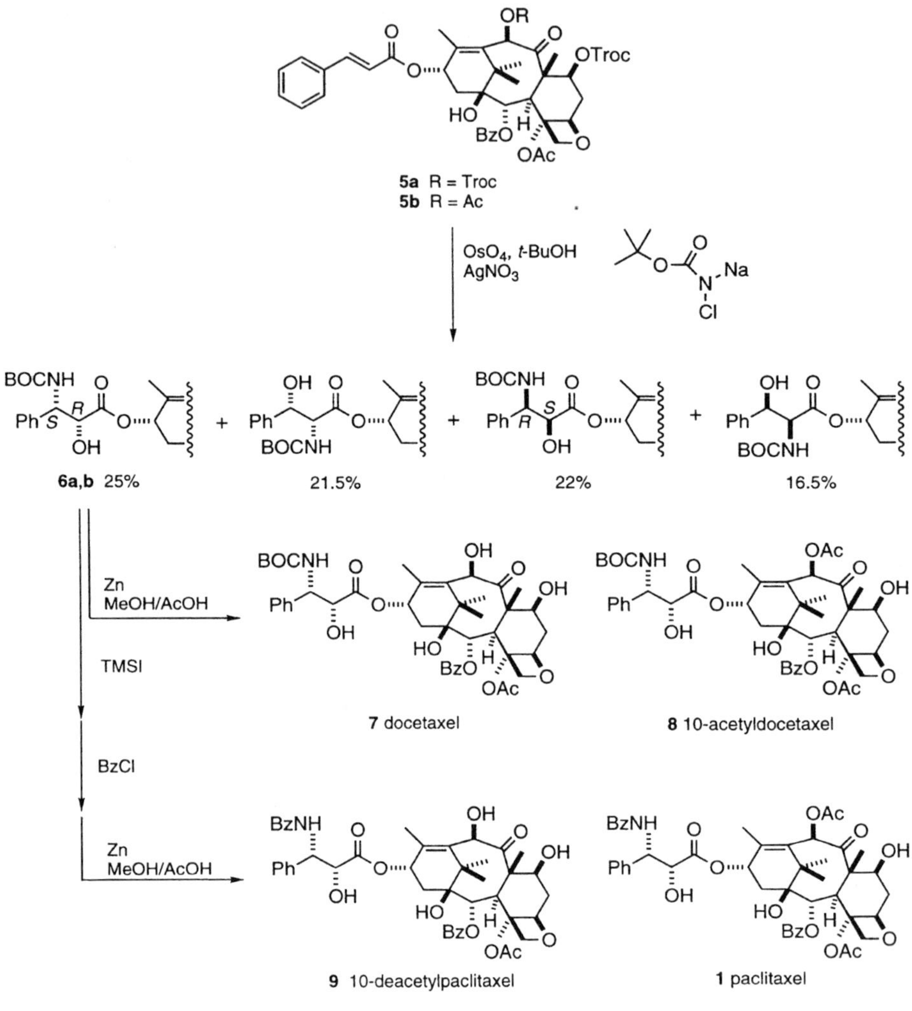

desired stereochemistry is introduced at C2, and a subsequent regioselective epoxide cleavage with a nitrogen nucleophile sets the C3 stereochemistry. This approach (Scheme 1.2) requires epoxidation of (*Z*)-cinnamyl alcohol (**10**), with the L-(+)-diethyl tartrate titanium complex [L-(+)-DET] as the chiral director, to give the desired (2*S*,3*R*)-epoxy alcohol **11** with modest enantioselectivity (76–80% ee). The alcohol oxidation was performed in the presence of base, and the resulting acid was methylated without isolation to minimize product loss due to the instability of the intermediate glycidic acid. The authors' concerns of poor regioselectivity in the epoxide cleavage of **12** turned out to be unfounded, as trimethylsilyl azide and zinc chloride produced the *syn*-azido alcohol **13** exclusively. In fact this regiospecificity is general

Scheme 1.2

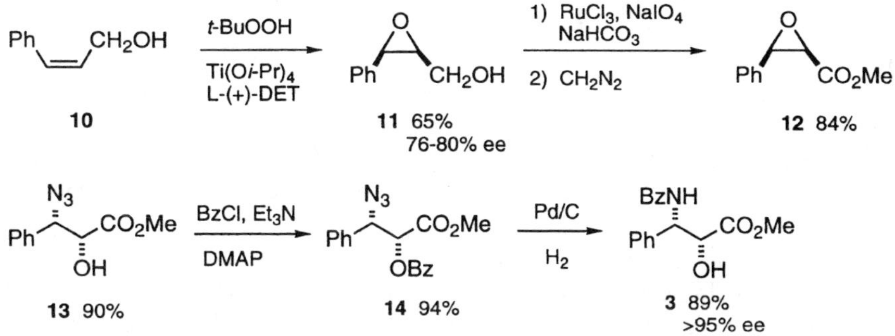

and occurs as such under a variety of conditions (vide infra). The synthesis was concluded by reduction of the azido group of **14** with concomitant O-to-N acyl migration to furnish **3** in 23% overall yield and 95% ee after recrystallization.

An improvement of the epoxidation strategy was reported by Deng and Jacobsen (Scheme 1.3).[48] Epoxidation of (*Z*)-ethyl cinnamate (**15**) with the salen-manganese catalyst **16**, which does not require an allylic alcohol for high stereoselectivity, resulted in a diastereomeric mixture (3.5:1 cis–trans) of products. NMR analysis of the mixture with $Eu(hfc)_3$ revealed that glycidic ester **17** was formed with excellent enantioselectivity (95–97% ee). The authors make special mention that rigorous purification is not required for the epoxidation step since byproducts can be removed by recrystallization at a later stage. Thus the partially purified mixture containing 56% **17** was carried through as shown in Scheme 1.3. This operationally simple synthesis requires no chromatography and produces the side chain acid **18** in 25% overall yield.

Subsequently, Denis et al.[49] reported the synthesis of glycidic ester **12** from commercially available methyl (*E*)-cinnamate via a C2,C3 double-inversion strategy

Scheme 1.3

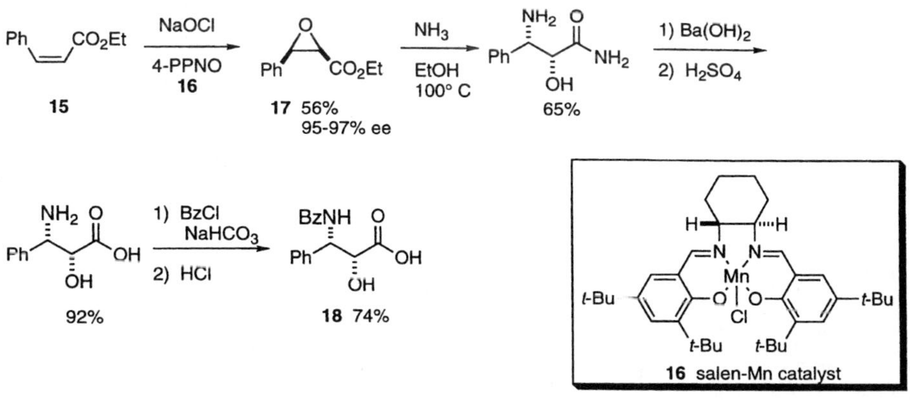

(Scheme 1.4). Asymmetric dihydroxylation, mediated by dihyroquinidine 4-chlorobenzoate (DQCB), gave the (2*S*,3*R*)-diol **19** (82% ee) with excellent enantioselectivity after recrystallization (>98% ee). Surprisingly, only the 2-O-tosylate **20** was formed upon treatment of **19** with *p*-toluenesulfonyl chloride (TsCl). The authors suggest that hydrogen bonding between the 3-hydroxyl and carbonyl oxygen may play a role in this regiospecificity. Intramolecular epoxidation to give **12** followed by ring cleavage with NaN_3 (**23**) comprises the double-inversion process. Both methyl N-benzoyl- and N-(*t*-BOC)-phenylisoserinates **3** and **24** (R^2 = Me) were produced in 36% overall yield.

A stereoselective aldol reaction directed by Evans' chiral auxiliary (Scheme 1.4) has also been used to generate the ethyl glycidate **17**.[50] The boron enolate of **21** was condensed with benzaldehyde to give 2-bromo-3-hydroxyimide **22**, the stereochemistry of which requires a similar vicinal double inversion as described above to give **23b**. Docetaxel side chain **24** was prepared in 30% overall yield by this approach. An advantage of this procedure is the ability to introduce phenyl substitution from readily available substituted benzaldehydes.

Examples of the geminal double-inversion concept have been reported (Scheme 1.5). Asymmetric dihydroxylation of methyl (*E*)-cinnamate with the osmium-$(DHQ)_2$PHAL catalyst produced the (2*R*,3*S*)-diol **25** with good yields and enantioselectivity (92% ee and recrystallized 99% ee).[51,52] Protection and activation of the diol was achieved by formation of the cyclic ortho ester followed by regioselective bromination (6:1) favoring **26**.[51] Treatment of the isolated bromo ester **26** with NaN_3

Scheme 1.4

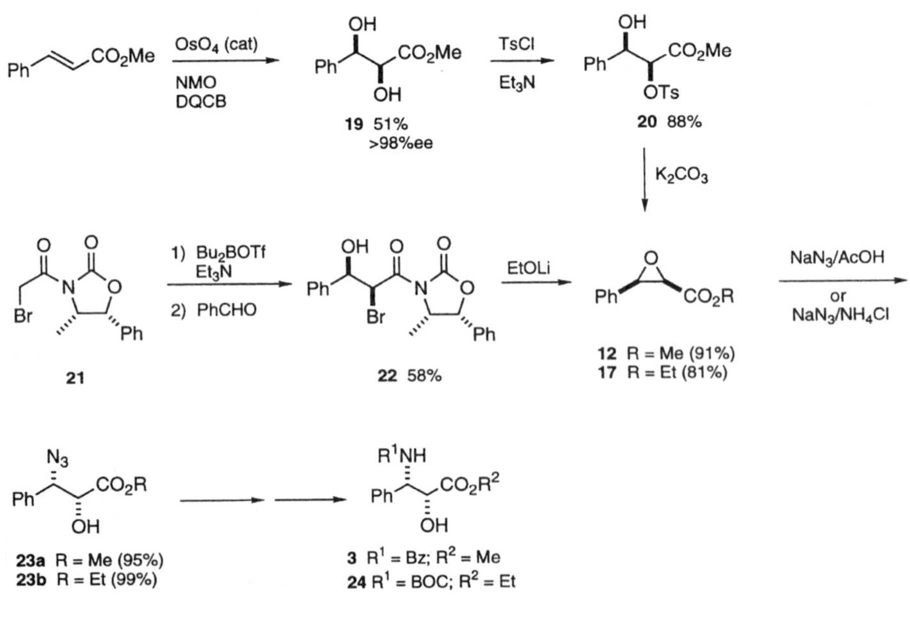

Scheme 1.5

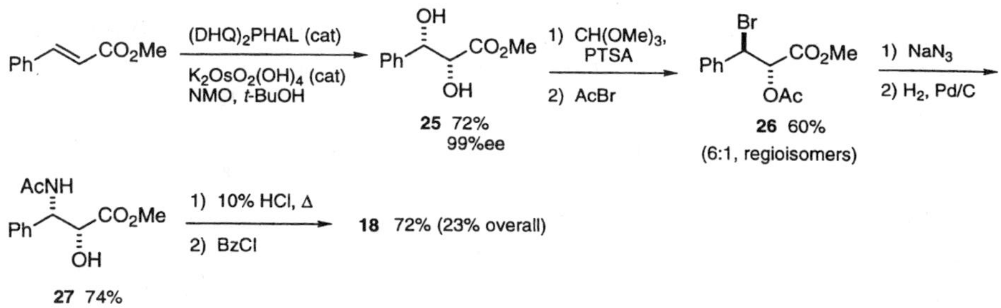

followed by hydrogenolysis introduces the C3 nitrogen into **27** with an overall retention of stereochemistry from **25**.

Better bromination regioselectivity (9:1) has been obtained in the conversion of cyclic sulfate **28** to **29** (eq. 1.1).[52] However, the azido alcohol **30** obtained from the crude bromo alcohol was contaminated by its *anti*-diastereomer (20%) which required preparative HPLC for isolation.

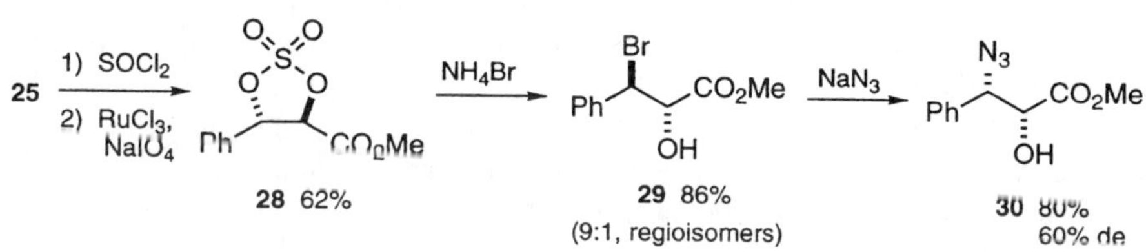

(eq. 1.1)

1.2.2 α-Amino Acids As Chiral Templates

(*S*)-α-Amino acids, used as chiral templates, have led to efficient syntheses of (2*R*,3*S*)-3-phenylisoserines and 3-alklyisoserines. Syntheses of the paclitaxel and docetaxel side chains starting with (*S*)-phenylglycine are shown in Scheme 1.6.[53,54] The C2 stereochemistry is installed via a chelation-controlled diastereoselective Grignard reaction [see transition state (**TS**) in Scheme 1.6]. To protect the configuration of the racemization-prone α-amino aldehyde intermediate, phenylglycinol **31** was oxidized and reacted in situ with vinylmagnesium bromide to yield amido alcohol **32** (80% de). The C2 hydroxyl group was protected as the ethoxyethyl

Scheme 1.6

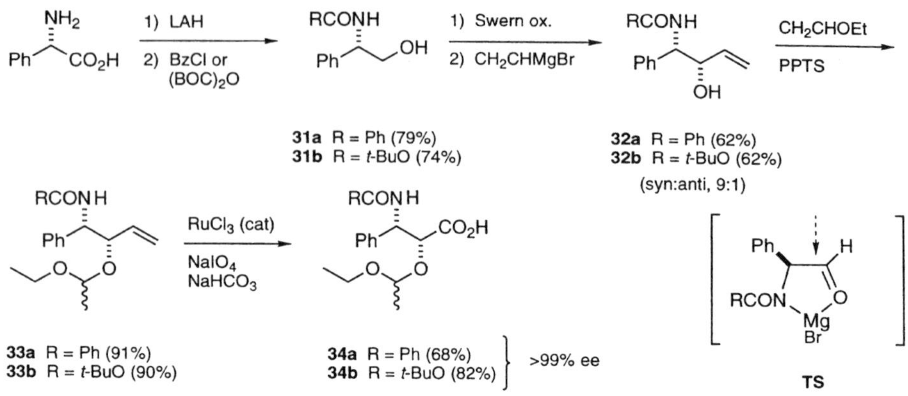

ether **33**,[32] and the terminal vinyl group oxidatively cleaved to give **34a** and **34b** in 30 and 34% overall yields, respectively.[55] This approach is not amenable to modified 3′-phenyl paclitaxel analogue synthesis since substituted phenylglycines are not readily available.

However, the use of other α-amino acids, such as L-phenylalanine,[56] and especially L-*t*-leucine (**35**, Scheme 1.7)[57] has led to numerous 3′-alkyl-3′-dephenylpaclitaxel derivatives. Reduction of L-*t*-leucine (**35**, Scheme 1.7) to alcohol **36** was followed by Swern oxidation and in situ vinylation of the resulting aldehyde. Due to the steric demands of the *t*-butyl group, intermediate **37** was formed as a single diastereoisomer. After formation of oxazolidine **38**, the double bond was oxidized to provide oxazolidinecarboxylic acid **39** in 33% overall yield.[57]

Scheme 1.7

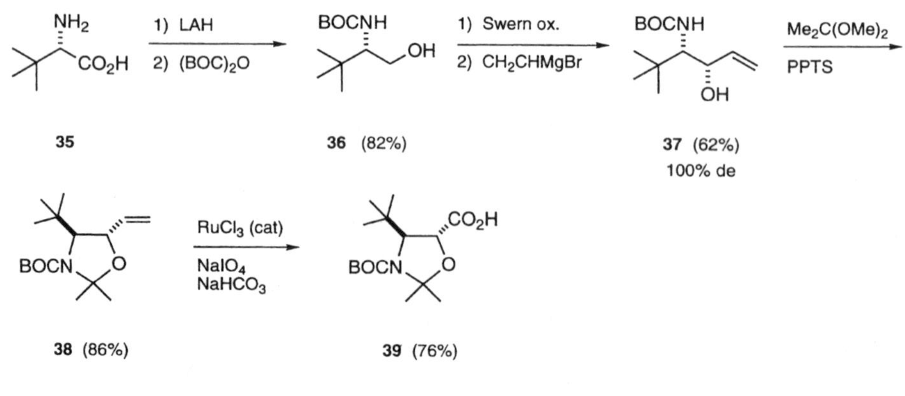

1.2.3 Amines As Chiral Templates

Davis and co-workers have reported a method in which chirality was transferred from a chiral template bound to what eventually becomes N3 of the phenylisoserine framework.[58] In Scheme 1.8,[58] the aldol-like reaction of chiral sulfinimine **40** (>97% ee) and the lithium enolate of methyl acetate afforded **41** (>95% ee). Following hydrolysis and benzoylation, asymmetric hydroxylation of the enolate dianion of **42** with (+)-(camphorylsulfonyl)oxaziridine (**43**) resulted in a *syn*-enriched product (72% de) of which **3** could be isolated with >93% ee. An eight-membered chelate for the dianion of **42** (Scheme 1.8) was proposed to rationalize the unusual *syn*-selectivity of the hydroxylation.

An interesting approach that allows access to all four stereoisomers of the paclitaxel side chain has been reported based on a single-step asymmetric Michael reaction–hydroxylation sequence (Scheme 1.9).[59] The conjugate addition of lithium

Scheme 1.8

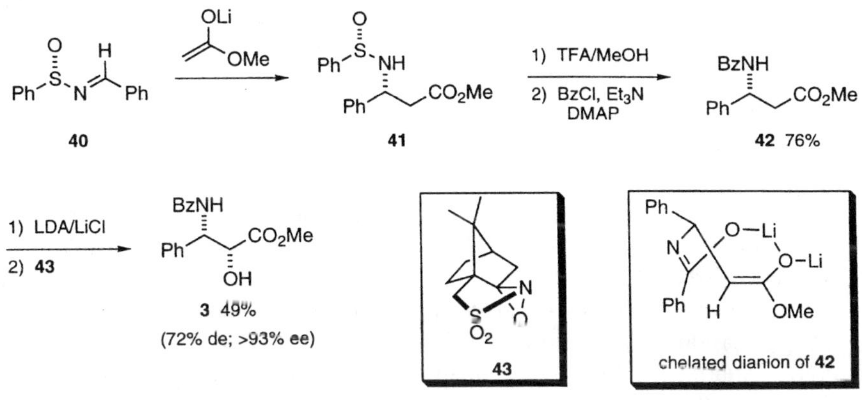

Scheme 1.9

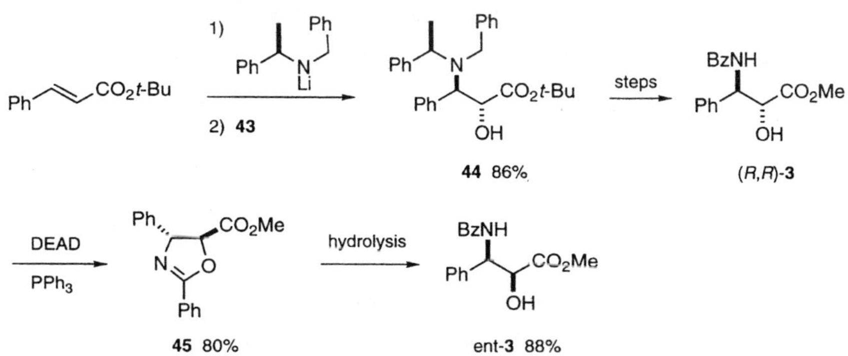

(*R*)-(α-methylbenzyl)benzylamide to *t*-butyl (*E*)-cinnamate followed by electrophilic hydroxylation with oxaziridine **43**, of the resulting enolate, gave **44** (92% de) as a single isomer after chromatography. Following standard deprotection and functionalization, (*R*,*R*)-**3** was subjected to an intramolecular Mitsunobu inversion producing oxazoline **45**, which was subsequently hydrolyzed to ent-**3**. The overall yield for the conversion of **44** to ent-**3** was 56%. Although these experiments were conducted in the wrong stereochemical series, (*S*)-(α-methylbenzyl)benzylamine is available for synthesis of the paclitaxel side chain.

Taking advantage of the availability of **46** (de >95%), the conjugate addition product of (*R*)-(α-methylbenzyl)benzylamide and *t*-butyl (*E*)-cinnamate, perhydropyrimidine **47** was prepared (eq. 1.2).[60] Due to the conformational constrains of the heterocyclic ring system and the presence of the stereodirecting phenyl group, a single diastereoisomer **48** was obtained in the electrophilic hydroxylation of the lithium enolate of **47** with oxaziridine **43**. The synthesis of (*R*,*R*)-**3** was achieved in 17% overall yield from **46**.

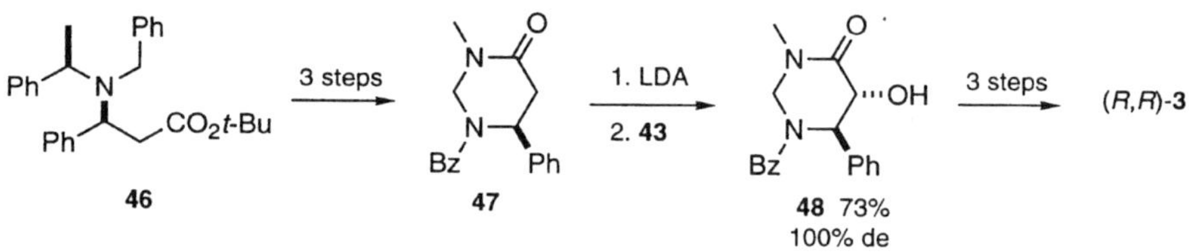

(eq. 1.2)

Phenylisoserine has also been prepared by an asymmetric Mukaiyama condensation of a chiral benzaldimine and silyl ketene acetals in the presence of chiral borate ester Lewis acids (eq. 1.3).[61] Reaction of N-(*S*)-α-methylbenzylbenzaldimine (**49**), silyl ketene acetal **51** (>99% *Z*), and (*S*)-binaphthol-derived borate ester **50** gave essentially pure product **52**(>99:1 syn:anti; >99:1 de) as a result of matched double diastereodifferentiation. Paclitaxel side-chain ester **3** was obtained from **52** via standard deprotection and benzoylation (68%). The anti product (*S*,*S*)-**52** can be prepared in a similar manner from an *E*-silyl ketene acetal.

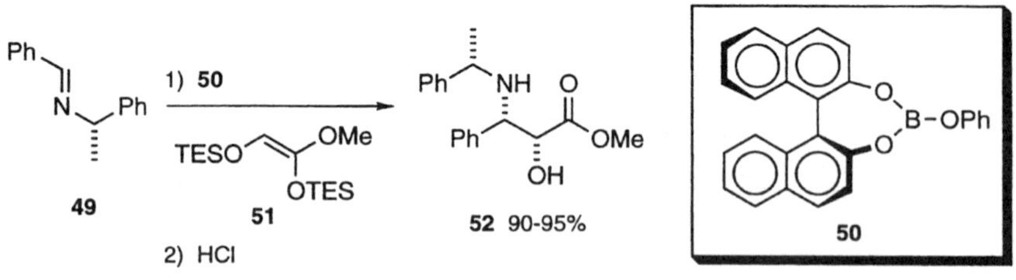

(eq. 1.3)

1.2.4 Amides and Esters As Chiral Auxiliaries

Aldol-type reactions of chiral amide enolates with N-acylimines have provided products with excellent diastereo- and enantioselectivity (eq. 1.4).[62,63] Addition of the

L-(+)-camphorsultam-derived lithium enolate of **53** to N-(*t*-butyloxycarbonyl) benzaldimine gave **54** as the only product. Lithium hydroperoxide cleavage of the chiral auxiliary gave "esterification ready" **55** in 45% overall yield [from L-(+)-camphorsultam].

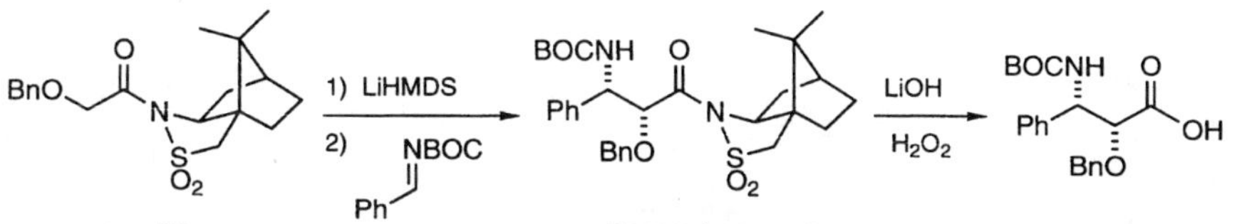

53 **54** 66% (>99% ee) **55** 70%

(eq. 1.4)

A surprisingly simple synthesis of the paclitaxel side chain via hydrolysis of an intermediate dihydrooxazine has been reported (eq. 1.5).[64] The proposed dihydrooxazine intermediate **58** was prepared by an inverse electron demand hetero-Diels–Alder reaction of the chiral (*Z*)-silyl ketene acetal **56** and N-benzoylbenzaldimine **57**. Various chiral auxiliaries (X_c) were examined with (1*S*,2*R*)-(+)-*trans*-2-(1-methyl-1-phenylethyl)-1-cyclohexanol, resulting in the best selectivity for **59** (86% de; 100% ee).

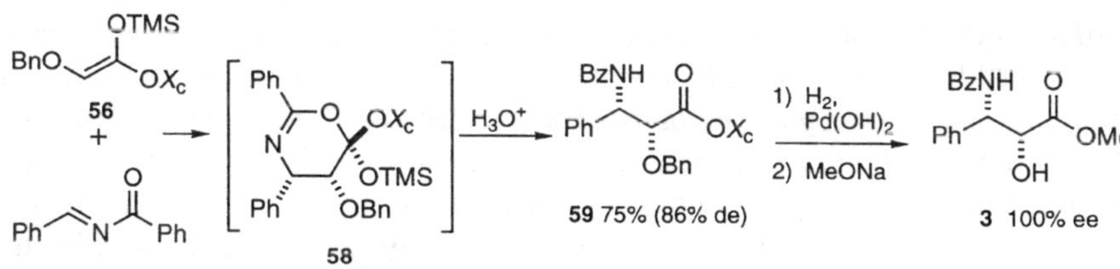

56 **57** **58** **59** 75% (86% de) **3** 100% ee

(eq. 1.5)

1.2.5 Chiral Aldehyde As the Template

In an interesting approach to the synthesis of the paclitaxel and docetaxel side chains,[65,66] the chiral benzaldehyde–chromium complex **60** was subjected to the titanium thioenolate of *t*-butyl-2-benzyloxythioacetate to give predominantly the *anti*-aldol product **61** (90% de). Interestingly, the Ti-thioenolate is necessary for *anti*-selectivity as the Li-thioenolate and corresponding Ti-enolate were *syn*-selective. Following decomplexation and disilylation, the nitrogen functionality was installed with inversion under Mitsunobu conditions (**61–63**). The paclitaxel side-chain methyl ester **3** was obtained in 29% overall yield essentially enantiopure (>98% ee) (Scheme 1.10).

Scheme 1.10

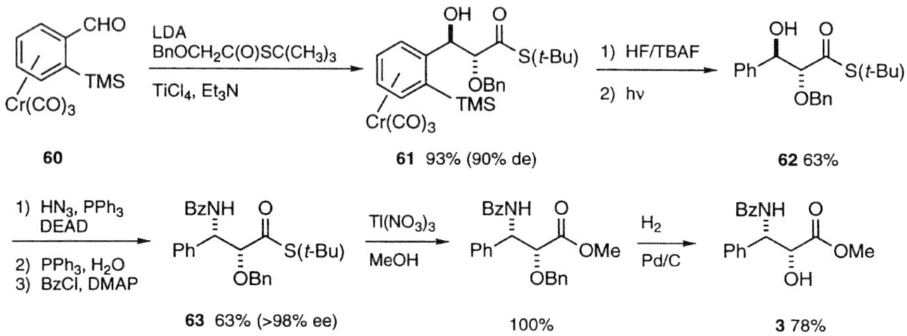

1.2.6 β-Lactams

β-Lactams are versatile intermediates since substituents can be readily introduced with outstanding *syn*-diastereoselectivity required for the paclitaxel side chain.[10,67]

1.2.6.1 *Staudinger Reaction*

The Staudinger reaction[68] is a common method for the preparation of β-lactams and has been used for the synthesis of 3-hydroxy-4-phenyl-2-azetidinones related to paclitaxel.[69–76] The Staudinger reaction is attractive because of its operational simplicity. Cycloaddition of a ketene, often generated in situ from an acid chloride, and an imine (Figure 1.1) typically proceeds with excellent cis-selectivity.[68] Selectivity is dependent on substituents, order of addition of reagents, and solvent effects.[68] Asymmetric induction has generally been modest with imine-based chiral auxiliaries, as shown in Figure 1.1 for β-lactams **64**, **65**, and **67**.[70,71,73] An improved diastereomeric excess (70% de) was obtained using a novel α-methylbenzylamine-derived chiral template, 1-(2,6-dichlorophenyl)ethylamine, for the formation of β-lactam **66**.[72] Due to the precedence in the literature,[77,78] it is not surprising that the best results (83% de) were obtained, when a sterically demanding L-threonine-derived chiral template was used in the synthesis of β-lactam **68**.[69] Deprotection and hydrolysis of the β-lactams provides phenylisoserines.

It is well known that imines derived from glyceraldehyde acetonide[79,80] and related α-substituted aldehydes[68] provide cis-β-lactams such as **69** as single diastereoisomers in the Staudinger reaction. β-Lactam **69** was used to prepare multiple 3-alkyl and 3-heteroaromatic isoserines.[81]

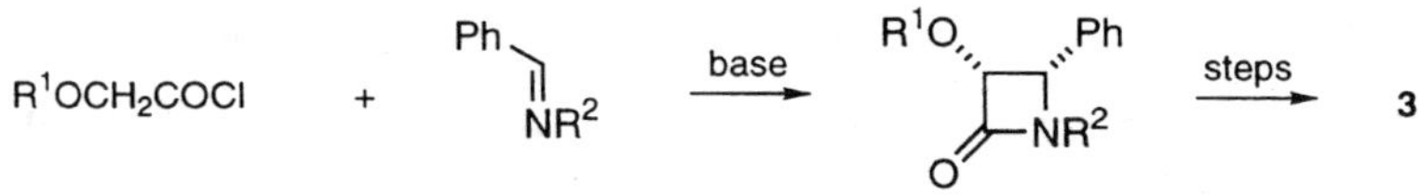

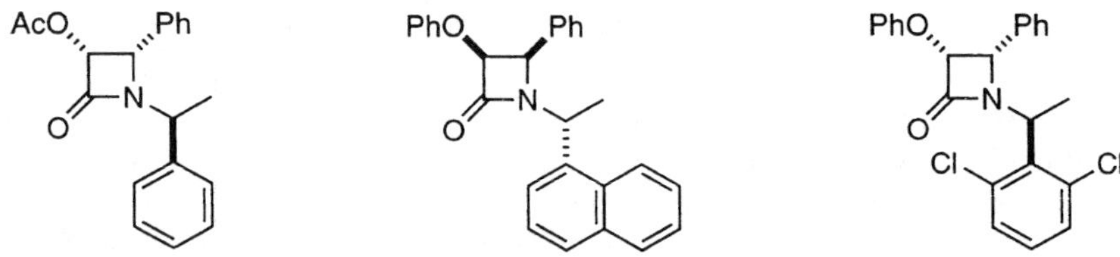

64 50% de **65** 40% de **66** 70% de

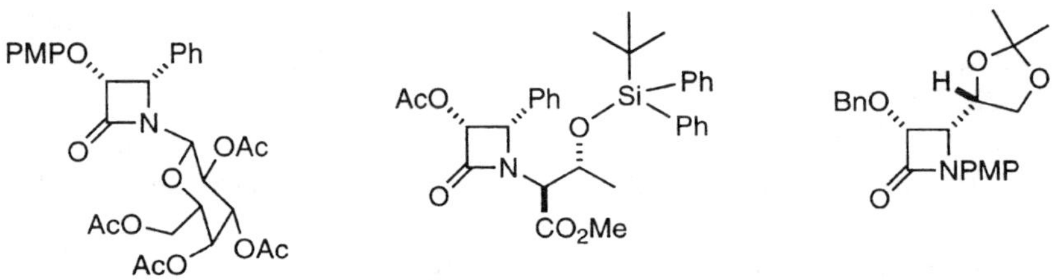

67 20% de **68** 83% de **69** 100% de

Figure 1.1 Staudinger reaction with chiral imines.

Outstanding diastereoselection has been reported for an acid-based chiral auxiliary[82] in the Staudinger reaction (eq. 1.6).[75,76] The ketene of chiral acid chloride **70** reacted with N-(*p*-methoxyphenyl)benzaldimine to give β-lactam **71**, in which the auxiliary was oxidatively cleaved to the α-keto β-lactam **72**. Borohydride reduction produces **73** with the desired *syn*-selectivity. Removal of the *p*-methoxyphenyl (PMP) protecting group can be accomplished with ceric ammonium nitrate (CAN).[83]

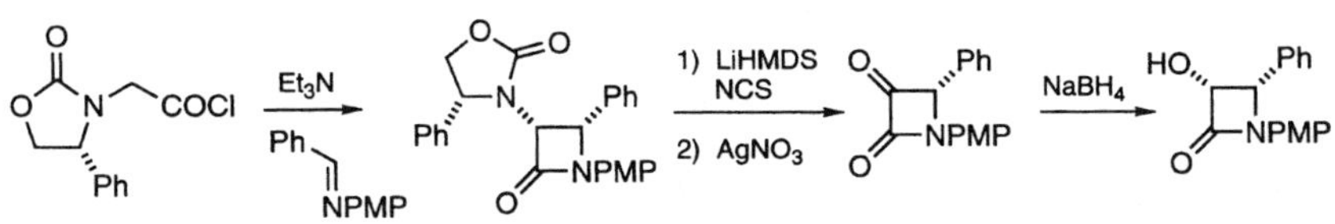

70 **71** 93% (>99% ee) **72** 95% **73** 100%

(eq. 1.6)

1.2.6.2 The Ester Enolate–Imine Cyclocondensation Reaction

The ester enolate–imine cyclocondensation[84,85] was used in a highly enantioselective asymmetric synthesis of **18** (Scheme 1.11).[86] In this reaction, an enolate is added to an imine, followed by intramolecular ring closure of the resultant lithiated amide. Asymmetric induction was investigated using several chiral esters, with the triisopropylsilyloxyacetyl ester of *trans*-2-phenylcyclohexanol[87] (**74**) giving the best combination of yield and enantioselectivity for β-lactam **75**.[86] In this series the triisopropylsilyl (TIPS) ether was necessary for high diastereoselectivity, as the *t*-butyldimethylsilyl (TBS) ether resulted in decreased enantiopurity (76% ee). Conversion of β-lactam **75** to **18** was achieved in 58% overall yield, as shown in Scheme 1.11. This method allows the introduction of 3-arylisoserines from readily available aromatic and heteroaromatic aldehydes.[88,89]

Two advantages of this approach are readily apparent. First, removal of the bulky chiral auxiliary, which can often be difficult, is not required since it is displaced during cyclization. The preparation of β-lactams from N-TMS-aldimines, derived from enolizable aldehydes, often proceeds in low yields.[90,91] This limitation can be overcome by using N-(*p*-methoxyphenyl)aldimines in the ester enolate–imine cyclocondensation described in Scheme 1.11.[92] β-Lactams with excellent enantioselectivity can be obtained by reacting aromatic, heteroaromatic, and alkyl PMP-aldimines with **74**.[92]

The protecting groups used for the chiral hydroxyacetate in the ester enolate–imine cyclocondensation can have profound effects on the enantioselectivity of the reaction. Generally bulky silyl ether-protecting groups (TIPS,[86] TBS[93]) provide excellent results. In preliminary studies on the asymmetric synthesis of 3-hydroxy-4-phenyl-β-lactam (**76**), the importance of steric bulk in the silyl-protecting group of the hydroxyacetate derived from Helmchen's chiral auxiliary[94] (**77b**–**77d**) became apparent (eq. 1.7).[95] Most interestingly, substituting a benzyl (**77a**) for the silyl protecting group also results in the formation of *cis*-β-lactams exclusively, however, with the opposite absolute configuration. The geometry of the lithium enolate was considered as the likely cause of this enantiomeric switch. Lithium chelation of the benzyl ether leads to the ester enolate with predominantly

Scheme 1.11

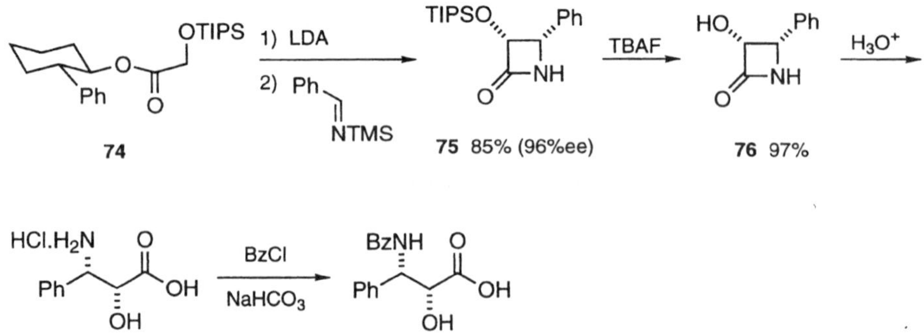

the *Z* configuration (97:3) for **77a**, and the *E* configuration of the ester enolate is obtained exclusively with TIPS-protected **77d**, as determined by analysis of the respective trimethylsilyl ketene acetals.[95]

(eq. 1.7)

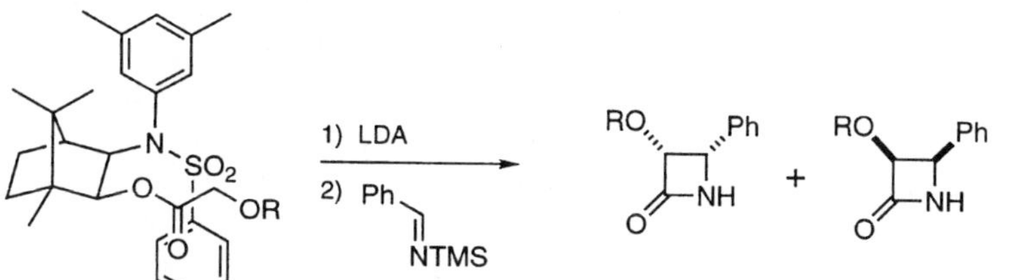

77a R = Bn	**78a** R = Bn	**79a** R = Bn	Ratio **78:79**
77b R = TES	**78b** R = TES	**79b** R = TES	99:1
77c R = TBS	**78c** R = TBS	**79c** R = TBS	45:55
77d R = TIPS	**75** R = TIPS	**79d** R = TIPS	13:87
			3:97

Thus the diastereoselectivity of the β-lactam products (**79b–d**) from the silyl-protected enolates can be rationalized on the basis of the six-membered Zimmerman–Traxler transition state[96] for this reaction (**TS-1**). On the other hand, the *Z*-Bn enolate must proceed through a different transition state since it would be expected to produce trans β-lactams if transition state **TS-1** were invoked. In the formation of product **78a** from the *Z*-Bn enolate, cis-diastereoselectivity can arise from a boat-like six-membered transition state[97] (**TS-2**) involving benzyl ether oxygen chelation (Fig. 1.2).

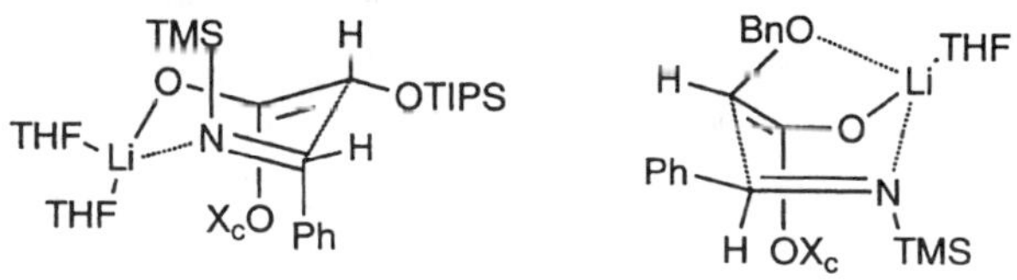

TS-1 **TS-2**

Figure 1.2 Chair-like (TS-1) and boat-like (TS-2) transition states in the ester enolate–imine cyclocondensation reaction.

1.2.7 Enzymes

Chemoenzymatic methods have been used in the synthesis of the paclitaxel side chain, such as lipase-mediated resolution of β-lactams,[98,99] phenyl glycidic esters,[100] 3-azido-2-hydroxy-3-phenylpropionates,[101] and microbial reduction of α-keto esters.[102,103]

1.3 Attachment of Phenylisoserines to Baccatin III Derivatives

As noted previously, the C7 and C10 hydroxyl groups of **2** are relatively reactive and require acetylation and/or protection prior to coupling between phenylisoserine and baccatin III derivatives. For the preparation of paclitaxel, 10-DAB (**2**) is selectively silylated and subsequently acetylated to give 7-O-triethylsilylbaccatin III (**80**).[32,104] Docetaxel derivates are prepared from 10-deacetyl-7,10-di-O-(trichloroethoxycarbonyl)baccatin III (**83a**).[44] The esterification of protected baccatin III was expected to be difficult based on models that showed that O13 was directed into the concavity of the cup-shaved molecule and the additional bulk of an alcohol-protecting group on phenylisoserine would not improve this situation. Indeed this was the case, as very specific conditions are required to effect this conversion.

1.3.1 Esterification Reactions

After extensive investigation of protection strategies and acylation methods, coupling of the two components **34a** and **80** was accomplished as shown in Scheme 1.12.[32] Note that the 80% yield of **81** is based on the recovery of 50% baccatin **80**. Pushing the reaction toward completion gave a lower yield (60% at 85% conversion). Acid-mediated removal of the protecting groups resulted in the first paclitaxel semisynthesis from a side-chain precursor. The ethoxyethyl ether (EE), with the unfortunate addition of another chiral center, was necessary because the simpler mixed acetal, 2′-O-methoxymethyl ether, could not be removed following esterification.[32] Apparently C2′ epimerization of the activated acid during esterification can be a significant problem.[50,63,105] For example, baccatin III derivatives were efficiently esterified with 2-O-benzyl-3-phenylisoserine **55** (for the structure of **55**, see eq. 1.4)

Scheme 1.12

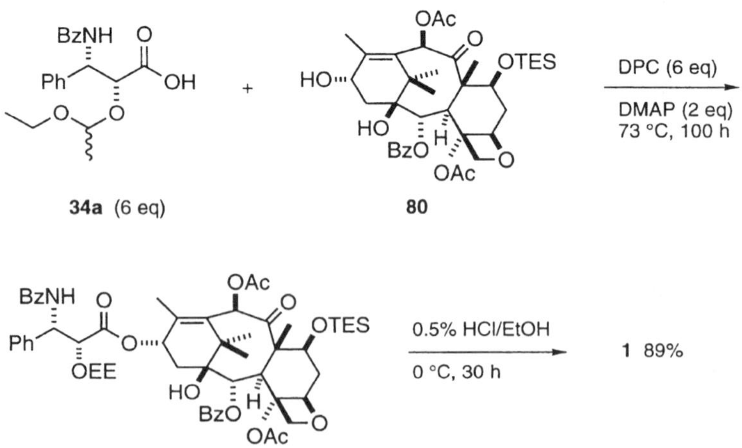

34a (6 eq)

80

81 80% (50% conversion)

mediated by DCC/DMAP at room temperature (93%); however, a significant amount (15%) of 2′-epimer contaminated the products. Subsequently, better methods were discovered that allowed milder conditions to be used, provided better yields, and suppressed epimerization.

A new protection strategy resulted in the first breakthrough for an improved coupling procedure (Scheme 1.13). Coupling of the oxazolidine **82** (prepared in quantitative yield by ketalization and hydrolysis from **24**) and **83** resulted in the formation of protected intermediates **84** in good yields without the need for large excess of reagents, and no epimerization was observed.[50] This cyclic protection effectively reduces steric bulk of the acid by eliminating rotation and epimerization is disfavored due to the thermodynamically stable trans disposition of substituents on the oxazolidine ring.

Scheme 1.13

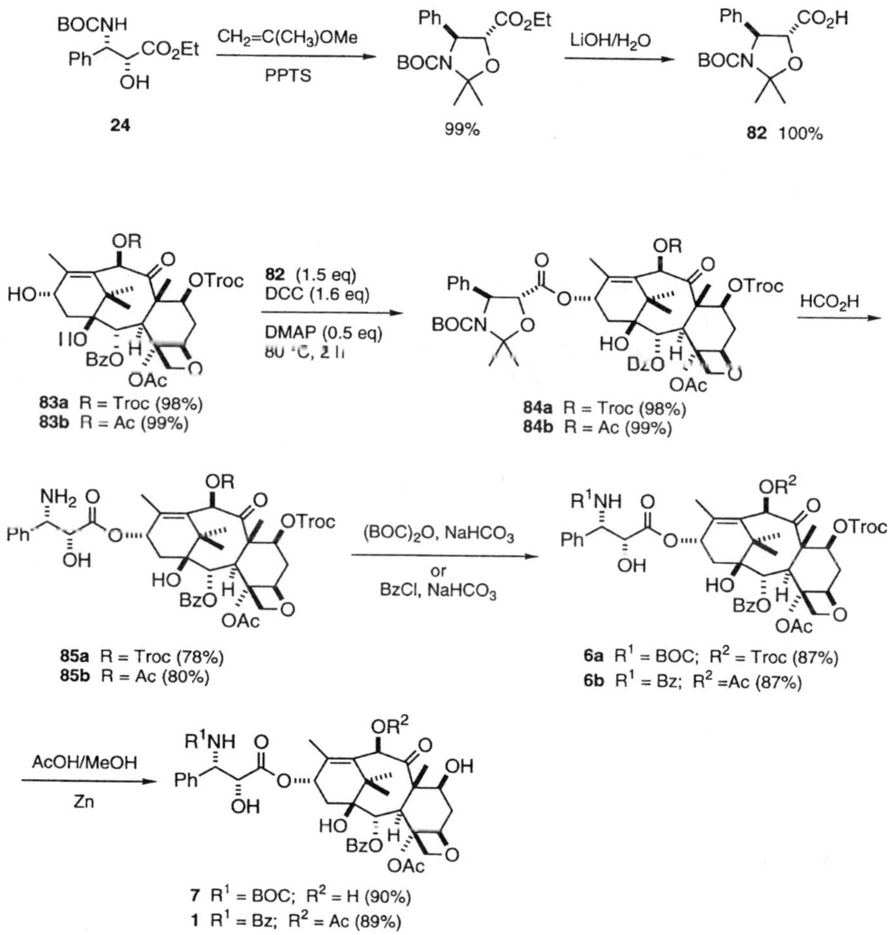

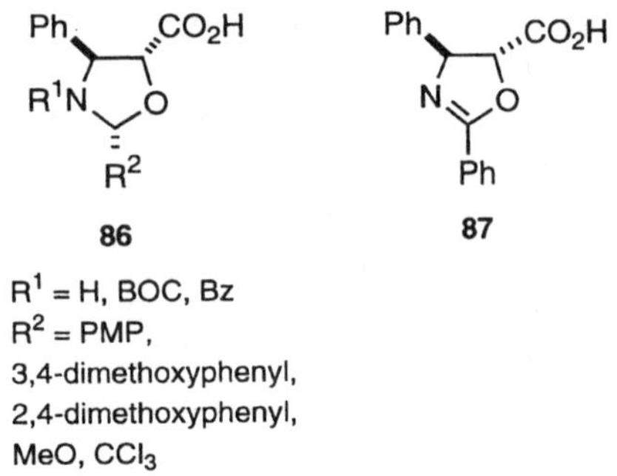

Figure 1.3. Oxazolidines and oxazolines.

Although the oxazolidinone procedure is a marked improvement for esterification, deprotection of the N,O-ketal with formic acid also removes the N-BOC group (**84–85**) decreasing its utility in the preparation of docetaxel. Resterificaiton and deprotection gives paclitaxel (**1**) and docetaxel (**7**). However, **85** is a versatile intermediate, which allows investigation of N-acyl substitution.

N,O-Acetals (**86**, Figure 1.3), which can be formed with high stereoselectivity by intra-[62,63] or intermolecular[106,107] routes, were subsequently investigated.

An example[63] of the intramolecular approach (eq. 1.8) involves trapping of the intermediate aldehyde from oxidation of **88** to give oxazolidine **89**. Hydrolysis results in the side-chain equivalent **86**, which could be coupled to **83a** or **83** very efficiently (DCC/DMAP/25°C/toluene). The N,O-acetal moiety can be removed with *p*-toluenesulfonic acid from the coupling products without concomitant N-BOC cleavage. In addition, the acetal group of the related 3,4-dimethoxy analogue was also easily cleaved while retaining the BOC group.[107]

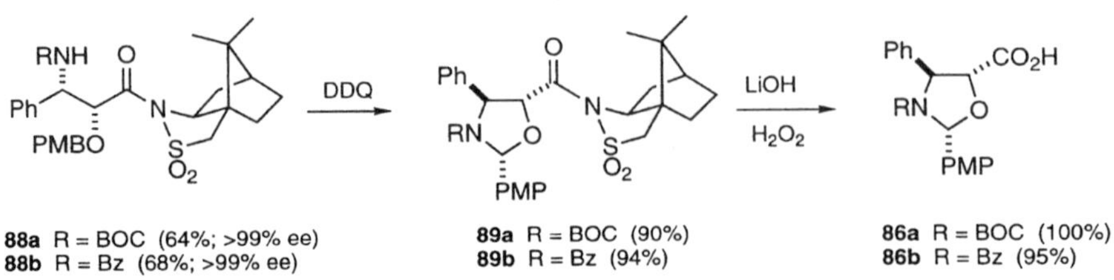

(eq. 1.8)

In a very interesting report,[105] it was hypothesized that *cis*-oxazolidines such as 2-epi-**82** would undergo ready epimerization under the acylation conditions to afford intermediates with the proper (2′*R*) stereochemistry. The reaction of 2-epi-**82** (structure not shown) with **83a** gave docetaxel intermediate **84a** (86% yield) exclusively, confirming the hypothesis (C2′ of **84a** is configurationally stable under the conditions).

In a related approach,[108] oxazoline carboxylic acid **87** (Figure 1.3; for its synthesis[59] see Scheme 1.9) was coupled (DCC/pyrrolidinopyridine) to 7-triethylsilylbaccatin III (**80**) in 95% yield. Hydrolysis of the oxazolone moiety provided paclitaxel in 75% yield.

1.3.2 Baccatin III Acylations with N-Acyl β-Lactams

Initially β-lactams were cleaved to form the open-chain phenylisoserine, then acylated and protected for coupling to the taxane ring system (see Scheme 1.11). The synthetic utility of β-lactams was further increased upon the finding that N-acyl-β-lactams are powerful acylating agents (eq. 1.9). β-Lactam **90** reacts with O13 of **80** to give paclitaxel intermediate **81** in excellent yield.[34] N-Acyl-β-lactams can be synthesized conveniently from the N-unsubstituted β-lactam with an acid chloride in the presence of triethylamine and catalytic DMAP at 0°C.[88] A large excess of β-lactam and very concentrated reaction conditions are required for good results. More sterically demanding O-protecting groups such as triethylsilyl[34] and triisopropylsilyl[88] failed to react under these conditions.

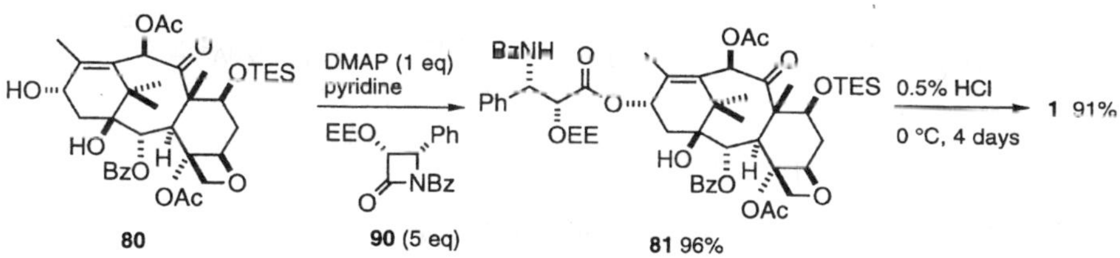

80 — **90** (5 eq) — **81** 96%

(eq. 1.9)

Subsequent investigation led to the use of O13 baccatin III alkoxides (Na, K, Li) as nucleophiles in the β-lactam coupling procedure.[34,93,109,110] This modification allows the use of 1.1 to 1.5 equivalents of β-lactam and gives excellent yields of protected paclitaxel and docetaxel intermediates. Moreover, bulky O-protecting groups can be esterified according to this procedure. These improved conditions make the β-lactam semisynthesis approach very efficient for paclitaxel (Scheme 1.14)[93] and docetaxel,[110] and very versatile for the preparation of analogues. Interestingly, racemic

Scheme 1.14

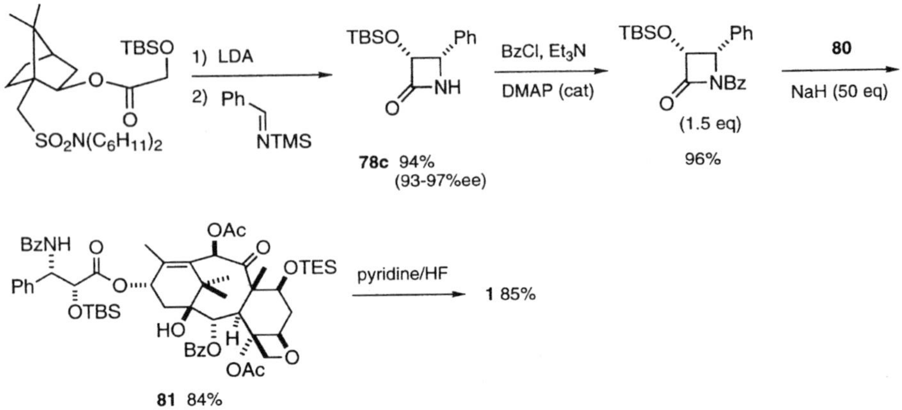

cis-β-lactams exhibit good to excellent kinetic resolution (3:1 to 100:1) for the 2′*R*,3′*S* isomers under these conditions.[34]

Initially isolated as a byproduct in pyridine/DMAP-catalyzed acylations of baccatin III with β-lactams, oxazinine **91** can also be used for the acylation of 7-triethylsilylbaccatin III (eq. 1.10).[34] However, due to the limited stability of **91**, this method has not found widespread use.

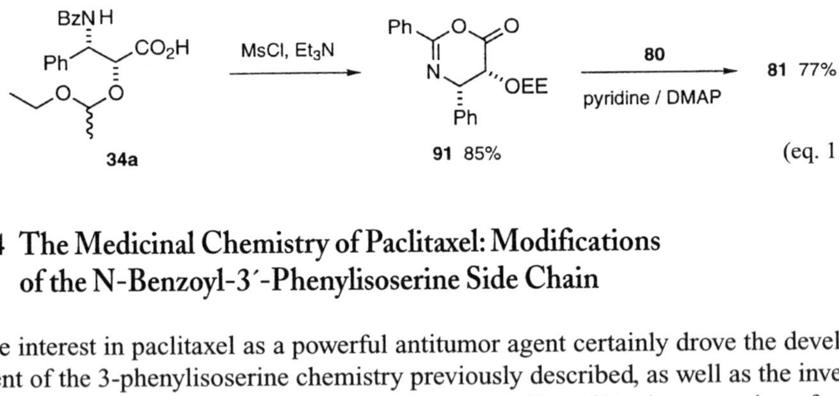

(eq. 1.10)

1.4 The Medicinal Chemistry of Paclitaxel: Modifications of the N-Benzoyl-3′-Phenylisoserine Side Chain

The interest in paclitaxel as a powerful antitumor agent certainly drove the development of the 3-phenylisoserine chemistry previously described, as well as the investigation of analogues that might develop an understanding of its pharmacophore for the design of future generation taxane anticancer agents. In the following, structure–activity results are discussed as they relate to modifications at the C13 side chain of paclitaxel. Information on the influence of structural modifications of the diterpene moiety of paclitaxel on bioactivity has been reviewed recently.[40]

Bioactivity profiles for paclitaxel analogues are best determined through analysis of in vitro microtubule assays[41] and cytotoxicity toward cultured tumor cell lines.[111] The use of two assay systems is primarily due to two major points. First, to develop meaningful structure–activity relationships it is important to gauge each new analogue's activity toward the molecular target, in this case microtubules. SARs formed from cytotoxicity data alone may be unreliable due to extraneous pharmacodynamic factors (i.e., cellular permeability, metabolism). Second, it is desirable to link the molecular mechanism of action with the expected "clinical" result—tumor cell death. From the following tables, the general observation can be made that microtubule and cytotoxicity assays correlate well. However, as the microtubule assay EC_{50}^{rel} values increase beyond ~fivefold, the magnitude of the cytotoxicity EC_{50}^{rel} values increases disproportionately. In cases where the microtubule assay predicts cytotoxicity and this is not realized in the cell line data, one might expect the poor correlation to be a result of poor cellular permeability, metabolic inactivation, or poor solubility. In some cases, the explanation is relatively apparent; in others a rationalization based on "extramechanistic factors" is more difficult.

The ability of paclitaxel to hyperstabilize microtubules is the basis of the two complementary in vitro microtubule assays most frequently used. In the microtubule disassembly method (MTD), microtubules at 30°C, in the presence of the analogue being tested, are subjected to a rapid temperature decrease to 4°C, conditions that normally rapidly depolymerize microtubules.[20,41] The velocity of depolymerization for the analogue is compared with a control, and the concentration of drug decreasing this velocity by 50% is the IC_{50}. The microtubule assembly method (MTA) measures the extent to which tubulin polymerizes to microtubules under conditions that normally do not cause polymerization.[41,57] An ED_{50} is calculated for the analogue concentration polymerizing half of the tubulin. The MTA and MTD assays give essentially similar results when compared with paclitaxel.[41] Thus data are generally presented as a relative activity: $IC_{50}/IC_{50(paclitaxel)}$ or $ED_{50}/ED_{50(paclitaxel)}$. A less frequently used assay, the initial slope of polymerization (MTI), takes advantage of paclitaxel's ability to enhance the nucleation phase of microtubule assembly.[41,112] In this method, the initial slope of polymerization for an analogue is measured and compared with paclitaxel, which is given a value of 100. Thus the data for paclitaxel analogues are essentially a percentage of tubulin nucleation compared to paclitaxel. The reader should take special note that the data for the MTI assay have been converted from the percentage value, found in the literature, to a relative value (100/relative $slope_{(analogue)}$) to facilitate comparison with the more prevalent tubulin assays. In this review, the term EC_{50}^{rel}, effective concentration, which collects the various terminologies for bioactivity (ID_{50}, ED_{50}, relative slope) has been used exclusively.

1.4.1 Deletion and Transposition Analogues and the Effects of Stereochemistry

When the initial SAR data were investigated, methods for the synthesis of 3-phenylisoserine, let alone specific analogues, and coupling methods had not been developed. Thus the first SAR data available were derived from O13 side chains, which could be attached to baccatin III or 10-DAB and would allow the desired functionalization (e.g., crotonate and cinnamate).[43] Unfortunately, molecular simplification of the paclitaxel side chain has not produced analogues (**92–106**) with any advantages over paclitaxel (Table 1.1). Analogues of the simple unfunctionalized alkanoic acids, acetate [7,13-diacetylbaccatin III, ED_{50}^{rel} = 670 (MTD)],[20] cinnamate (**92**), crotonate (**93**), propionic acid (**94**), and 3-phenylpropionate (**95**) show considerably reduced ability to stabilize microtubules.

Analogues **94**–**103** represent the systematic removal of functional groups from the 3′-phenylisoserine core. In considering the activity of the series, **1** > **96, 97, 100** > **95, 98, 99** > **94**, several trends appear. Four clusters of activity, differing by approximately two- to fourfold, are observed in which none (**94**), one (**95, 98, 99**), two (**96, 97, 100**), or all (**1**) of the side-chain functionality (2′-hydroxy, 3′-phenyl, and 3′-amide) is

Table 1.1 DELETION ANALOGUES

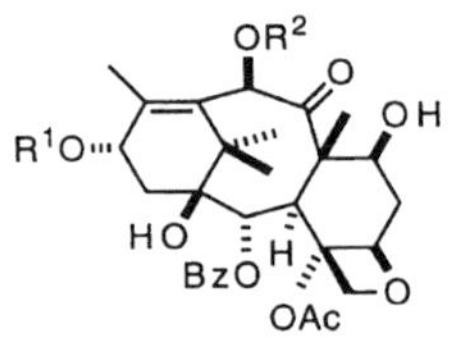

Compound	R^1	R^2	EC_{50}^{rel}
92[43,44,113]	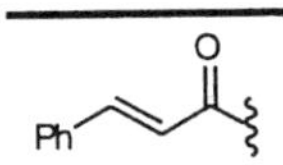	H	MTD- 23
93[44,113]	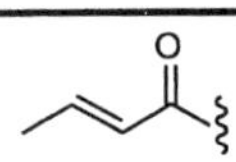	H	MTD- 100
94[114]	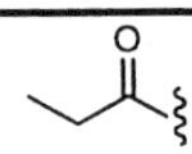	H	MTD- 41
95[44]	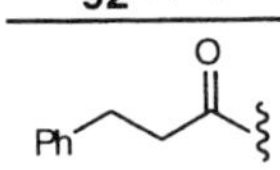	H	MTD- 17
96[44,112]	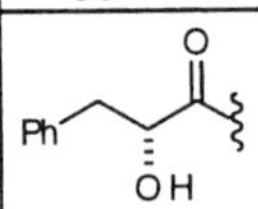	H Ac	MTD- 4.5 MTI- 7.6 J774.2- 39
97[44,112]	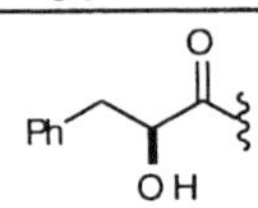	H Ac	MTD- 3.5 MTI- 6.40 J774.2- 38
98[112]	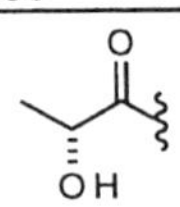	Ac	MTI- 18 J774.2- 300
99[112]	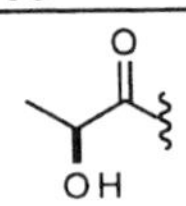	Ac	MTI- 16 J774.2- 322
100[44]	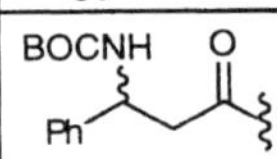	H	MTD- 2.3[a] 4.1[a]
101[112]	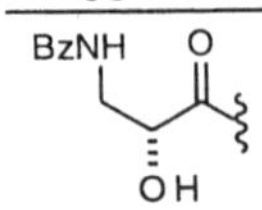	Ac	MTI- 18 J774.2- 222
102[112]	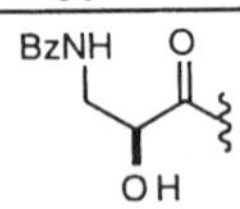	Ac	MTI- n.d. J774.2- >500
103[44]	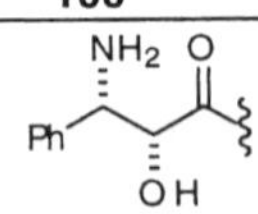	Ac	MTD- 44
104[44]	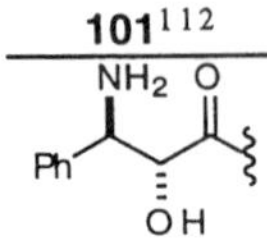	H	MTD-30[a] MTD-30[a]
105[44]	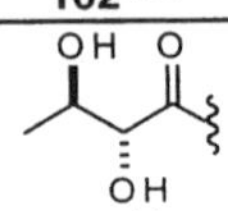	Ac	MTD- 60[b]
106[44]	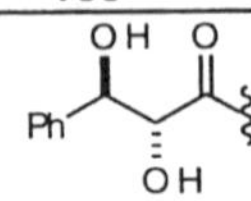	H	MTD- 3.0[b]

[a] Stereoisomers tested separately, but absolute stereochemistry unknown.
[b] Mixture of anti isomers (2′*R*,3′*R* or 2′*S*,3′*S*).

present.[114] Interestingly, stereochemistry has little effect on activity of the lactate (**98** and **99**), phenyllactate (**96** and **97**), and N-BOC-β-phenylalanine (**100**) isomers. Unfortunately the N-benzoylisoserines (**101** and **102**) do not fit nicely into this simple additive analysis, and there is some cooperativity between the side-chain functionalities. Apparently, the 3′-phenyl moiety exerts more of an influence on activity than the 2′-hydroxy and 3′-amide groups; otherwise one would expect **101** to have activity comparable to **96**, **97**, and **100**. Also, stereochemistry indeed plays a large role for the N-benzoylisoserines, where **102** exhibits considerably less activity than its epimer **101**. The cytotoxicity of these deletion analogues (**96–99**, **101**, **102**) are uniformly poor. Reduced cellular permeability could rationalize the cell line data, as one would expect these analogues to be more polar than paclitaxel. The poor activity of the phenylisoserine analogues **103** and **104** in the MTD assay is likely due to a combination of a removed binding interaction and an unfavorable interaction with the charged amine group, since diol **106** possesses only slightly reduced bioactivity. The importance of the 3-phenyl moiety is reinforced in the diols **105** and **106**.

As discussed earlier, the difficulty in preparing paclitaxel from the 13-O-cinnamate derivative (see Scheme 1.1) turned out to be quite productive with the discovery of docetaxel, a compound with slightly improved activity over paclitaxel, and a number of analogues with altered stereochemistry and/or regiochemistry (Table 1.2). Two points are readily apparent from analysis of these analogues. First, the 3′-phenylserine regioisomers **108**, **110**, **111**, **114**, and **115** have poor microtubule activity. This is likely due primarily to the bulky 2′-amide substituent (compared with diol **106**). More intriguing, the (2′*S*,3′*R*)-phenylisoserine analogues **109** and **113** are less active than their respective counterparts docetaxel and paclitaxel with the natural (2′*R*,3′*S*) stereochemistry, while the anti-analogues **112** and **116** are only slightly less active. Thus the stereochemical promiscuity observed for the simpler deletion analogues is still operating in the full 3′-phenylisoserine series. However, it is clear that the natural stereochemistry results in optimal activity.

Molecular modeling and NMR studies of some of the deletion analogues and docetaxel analogues have been undertaken to rationalize the interesting results found in Tables 1.1 and 1.2.[114,117] In general, active analogues can adopt relatively similar conformations to the X-ray structure[118] of docetaxel irrespective of stereochemistry. NMR studies of the deletion analogues (**96**, **97**, **101**, **102**) suggest a correlation between side-chain flexibility and activity. Thus the most rotationally restricted analogue **97** is also the most active of the compounds studied.[117] It has been claimed through comparison of NMR and molecular modeling data of the docetaxel stereoisomers **7**, **109** ($R = H$) and **112** (2′*S*,3′*S* and 2′*R*, 3′*R*) that the MTD activity trend is related to the conformational deviation from the optimum represented by docetaxel (**7**).[114] The difference in MTD activity between (2′*S*,3′*R*)-docetaxel analogues (**109**) and (2′*S*,3′*R*)-paclitaxel analogues (**113**) may be conformational in nature.[114]

1.4.2 Modifications at the 2′-Hydroxy Group

Modification of the 2′-hydroxy group results in analogues with greatly reduced cytotoxicity (Table 1.3), while removal altogether (**100**, Table 1.1) or inversion (**112** and

Table 1.2 INVERSION AND TRANSPOSITION ANALOGUE

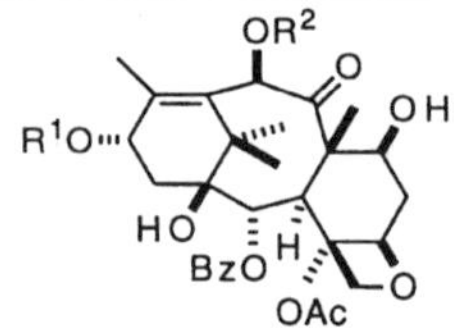

R^1	R^2	EC_{50}^{rel}	R^1	R^2	EC_{50}^{rel}	R^1	R^2	EC_{50}^{rel}
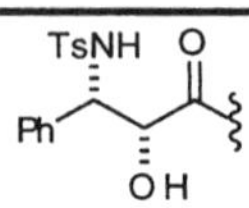 **107**[44,115]	H	MTD- 5	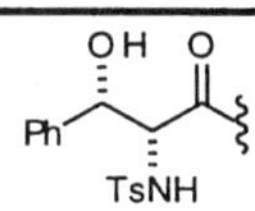 **108**[44,115]	H	MTD- 15	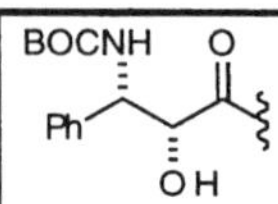 **7**[43,44,45,116] H **8**[43,44,45,116] Ac	H Ac	MTD- 0.5 P388- 0.5 MTD- 0.5 MTA- 0.55 P388- 0.6 B16- 0.23 J774.2- 0.4
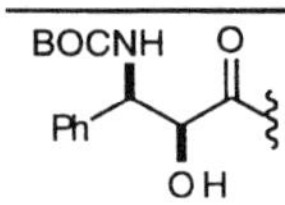 **109**[43,44]	H Ac	MTD- 30 MTD- 30	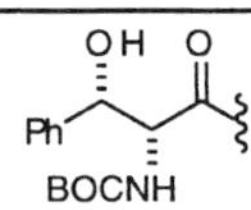 **110**[43,44]	H Ac	MTD- 10 MTD- 10	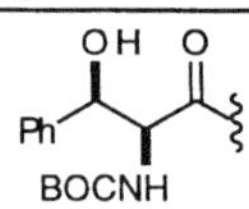 **111**[43,44]	H Ac	MTD- 160 MTD- 110
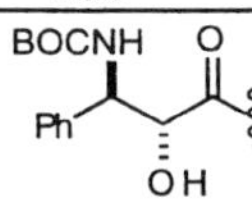 **112**[44]	H	MTD- 1.8[a] 4.3[a]	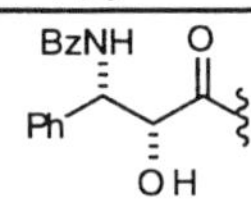 **9**[43,44]	H	MTD- 1.3	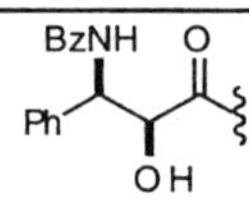 **113**[43,44]	H Ac	MTD- 4 MTD- 4.5
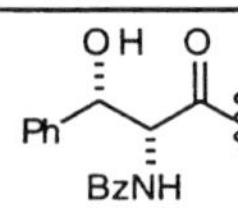 **114**[43,44]	H Ac	MTD- 10 MTD- 10	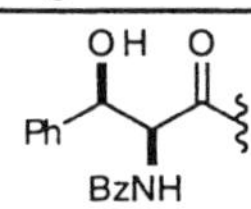 **115**[43,44]	H Ac	MTD- 170 MTD- 110	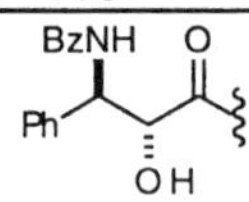 **116**[44]	H	MTD- 1.3[a] 1.3[a]

[a] Anti isomers tested separately, absolute stereochemistry undetermined.

116, Table 1.2) reduces MTD activity only slightly. The ethers (**117** and **118**), the bioisosteric 2′-deoxy-2′-fluoro (**119**), and the 2′-deoxy (**121**) analogues virtually lack cytotoxicity. The 2′-O-acetyl analogue **120** has significantly reduced activity in the MTD assay but is cytotoxic in the KB tumor cell assay. This has been attributed to cleavage of the acetate by cellular esterases. Exploitation of this metabolic cleavage has led to various 2′-O-ester prodrugs.[41]

Since microtubule assays were not performed with these analogues, strict SARs cannot be established. Nevertheless the poor cytotoxicity of **121** is somewhat surprising in light of the slightly reduced MTD activity of the 2′-deoxydocetaxel analogues **100**.

Table 1.3 2′-HYDROXY MODIFICATIONS

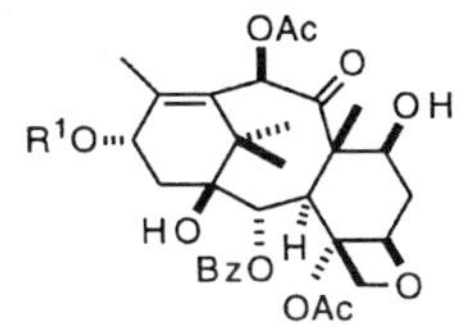

R^1	EC_{50}^{rel}	R^1	EC_{50}^{rel}	R^1	EC_{50}^{rel}
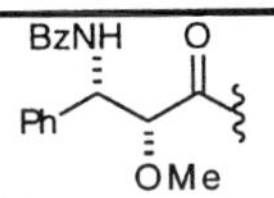 **117**[119]	HCT116-216	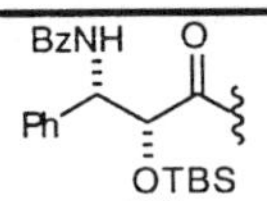 **118**[120]	KB- 3 x 10^4	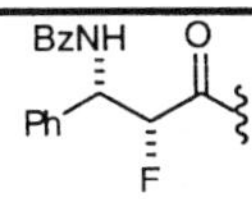 **119**[119]	HCT116-119
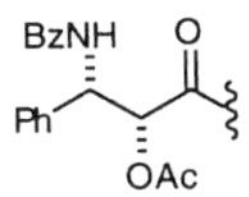 **120**[20,120,121]	MTD- 30 KB- 2	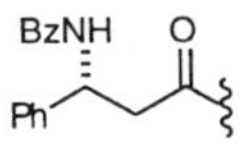 **121**[119]	HCT116-74	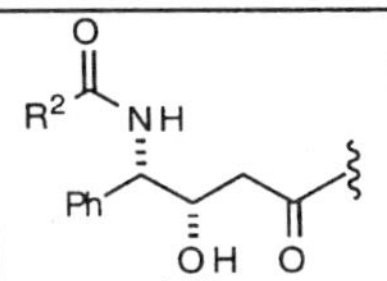 **122**[122]	MTA > 27[a]

[a] 10-Deacetyl; R^2 = *t*-BuO or Ph.

These results suggest that the (*R*)-2′-hydroxy group is involved in a binding interaction and/or the 2′-hydroxy acts to increase the conformational rigidity of the side chain. Consistent with this interpretation is that alkylation, as in 2′-O-methyl ether **117**, would disrupt both an intermolecular hydrogen-bonding paclitaxel–protein interaction or a conformational contribution from intramolecular hydrogen bonding, resulting in diminished activity if the 2′-hydroxy acts a hydrogen-bond donor. But the drastic drop in cytotoxicity for **117** is inconsistent, considering the MTD activity of 2′-deoxydocetaxel (**100**), suggesting an unfavorable steric interaction in the paclitaxel binding site as well. The reduced, but surprisingly decent, MTD activity of the transposition analogues **110** and **114** may be rationalized along these lines. The (2′*R*)-amido function can participate in a favorable hydrogen-bonding interaction with tubulin but suffers from an unfavorable steric interaction. The lack of the intermolecular hydrogen-bonding interaction with tubulin, in addition to steric factors, would then rationalize the MTD results of the virtually inactive (2′*S*,3′*R*) transposition analogues **111** and **115**.

The novel homologue **122** was inactive in the MTA assay.[122] The O13 side chain of **122** was synthesized from phenylglycine according to Scheme 1.6 by substitution of vinylmagnesium bromide with allylmagnesium bromide.

1.4.3 N3′ Substitution Analogues

Significant interest was generated in the modification of the N-benzoyl moiety of paclitaxel due to the cytotoxic properties of the natural product cephalomannine (**143**,

Table 1.5) and the early identification of docetaxel as a potent microtubule stabilizing agent. The results of these investigations are presented in Tables 1.4 and 1.5. An attempt to optimize the binding characteristics of the benzamide group by a systematic sampling of substituted N-benzoyl amides was undertaken according to the Topliss Operational Scheme.[123–125] Analogues **122–131**, containing various functionalities to probe the steric, electronic, and lipophilic parameters for activity, were synthesized either from N-aroyl β-lactams of type **90** (eq. 1.9) or by amidation of N-debenzoylpaclitaxel (**103**).[88,116] The remarkably flat MTA response of **122–130** suggests that little improvement would be gained by further modification of steric, electronic, and lipophilic parameters. However, the trend of increasing microtubule assembly with decreasing Hansch π and Hammet σ constants, observed in the 4-substituted N-benzoyl analogues (**123** > **1** > **126** > **122** > **131**; **130** is an outlier in this analysis) might warrant further attention.[126] Photoaffinity labeled analogues such as **132–134** and other miscellaneous substituted N-benzoyl analogues (**135–137**) also follow this trend, being of similar or only slightly reduced activity.

Worthy of note is the binding-site information gained from microtubule photolabeling experiments with **132**. Briefly, **132** specifically labels the β-tubulin subunit in the presence of microtubule-associated proteins, and following protein digestion the label is found in the N-terminal 31 amino acid fragment.[127,128] Labeling of both tubulin subunits (α:β 1:4) has also been noted for **132**[129] and **159**[130] (Table 1.5), suggesting that the paclitaxel binding site overlaps to some extent the α- and β-tubulin subunits.

A series of bioisosteric pyridyl and furyl analogues (**138–142**) have been prepared and evaluated.[131] Generally these analogues are of similar activity as paclitaxel, with **142** the best compound of the series. Interestingly, a poor correlation between MTA activity and cytotoxicity exists for the pyridyl bioisosteres, while this effect is not observed with the furyl analogues. Replacement of the N-benzoyl moiety with aliphatic amides and carbamates is tolerated well (Table 1.5). Synthesis of these compounds was achieved primarily by acylation of N-debenzoylpaclitaxel (**103**) or its 7,10-di-Troc analogue. Generally the amides (**143–150**) are slightly less active and the carbamates (**151–158**) are slightly more active than paclitaxel based on microtubule and cytotoxicity assays. While most of the synthetic amide and carbamate analogues contain branched alkyl groups (apparently driven by the success of docetaxel), the linear alkyl amide natural products (**144**, **148**, and **149**, Table 1.5) are quite competent microtubule-stabilizing agents. Interestingly, the tertiary amide **149** retains microtubule activity indicating that the amide proton has little influence on microtubule binding. The hemisuccinate **150**, which is likely to be charged at assay pH, is also tolerated (see also **135**, Table 1.4). The N-pivaloyl analogue **145** and methyl carbamate **151** indicate that branching adjacent to the carbonyl and small substituents are detrimental to good cytotoxicity. The urea photoaffinity label **159** also exhibits good MTD activity.

Although the carbamate analogues tend to be more active than the amide analogues described in Table 1.5, no general statement can be made about methylene–oxygen interchange by direct comparison of the bioisosteres **7** (Table 1.2), **8** (Table 1.2), and **153** (Table 1.5) with **147** (R^2 = H), **147** (R^2 = Ac), and **148**, respectively. Microtubule

Table 1.4 N3′ AROMATIC AND HETEROAROMATIC AMIDE SUBSTITUTION ANALOGUES

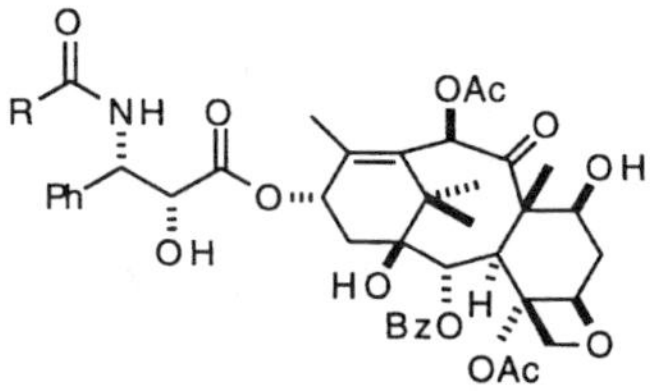

R	EC_{50}^{rel}	R	EC_{50}^{rel}	R	EC_{50}^{rel}
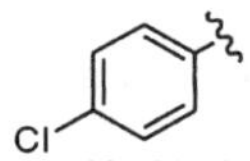 **122**[88,132,133]	MTA- 2.4 B16- 1.5	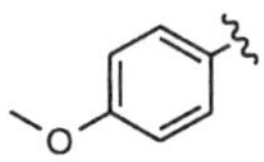 **123**[133]	MTA- 0.55 B16- 1.3	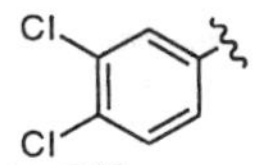 **124**[133]	MTA- 2.1 B16- 11
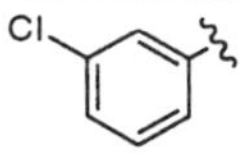 **125**[133]	MTA- 2 B16- 1.8	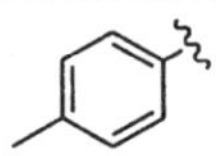 **126**[133,134]	MTA- 1.6 B16- 1.4	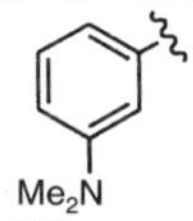 **127**[133]	MTA- 1.4 B16- 1.6
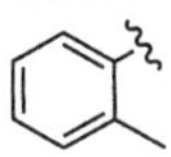 **128**[133]	MTA- 1.9 B16- 7.5	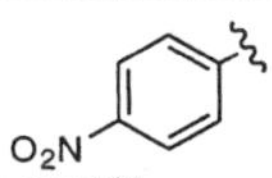 **129**[133]	MTA- 2.0 B16- 21	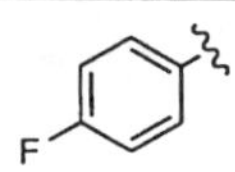 **130**[133]	MTA- 1.2 B16- 4.3
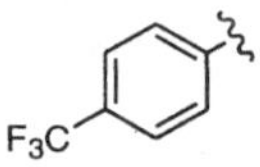 **131**[133,134]	MTA- 6.0 B16- 18 J774.2- 10	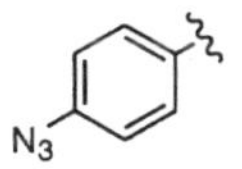 **132**[129,135,136]	MTI- 1.2 MTA- 2.0 J744.2- 3.1 B16- 3.1	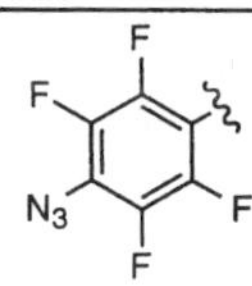 **133**[116,136]	MTA- 4.6 B16- 15
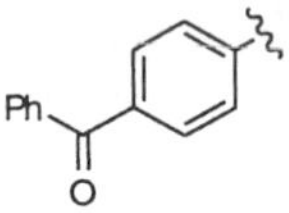 **134**[135]	MTI- 6.7 J744.2- 17	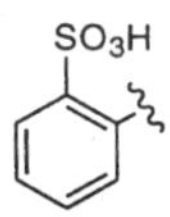 **135**[44]	MTD- 5.5[a]	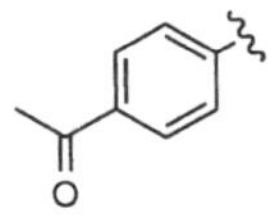 **136**[135]	MTI- 2.3 J744.2- 18
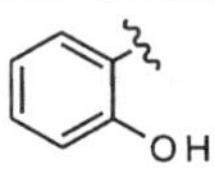 **137**[135]	MTI- 1 J744.2- 6.2	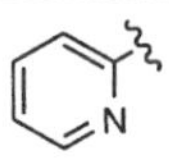 **138**[131]	MTA- 3.5 B16- 4.6	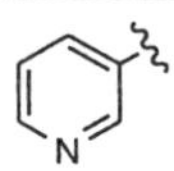 **139**[131]	MTA- 0.63 B16- 26
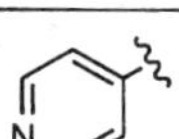 **140**[131]	MTA- 2.0 B16- >36	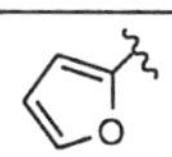 **141**[131]	MTA- 0.78 B16- 1.0	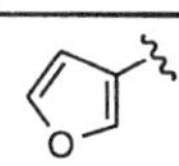 **142**[131]	MTA- 0.36 B16- 0.5

[a] 10-Deacetyl.

Table 1.5 N3′ ALIPHATIC AMIDE AND CARBAMATE SUBSTITUTION ANALOGUES

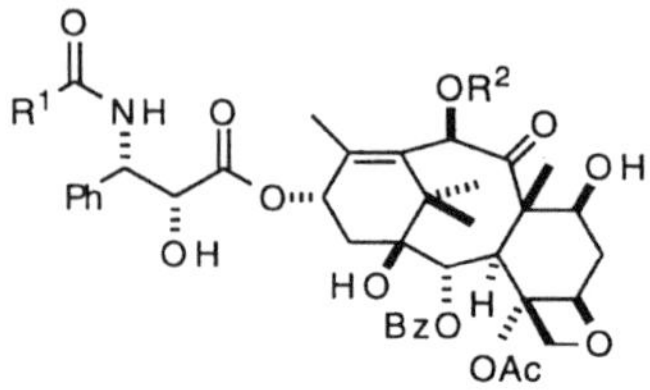

R^1	R^2	EC_{50}^{rel}	R^1	R^2	EC_{50}^{rel}	R^1	R^2	EC_{50}^{rel}
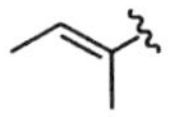 **143**[137,138]	Ac	MTD- 1.5 J774.2- 1.4 KB- 4	*n*-C_3H_7 **144**[139]	Ac	MTA- 2.0[a]	 **145**[116]	Ac	MTA- 2.6 B16- 22
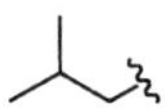 **146**[116]	Ac	MTA- 1.2 B16- 2.7	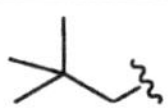 **147**[56,116]	H Ac	MTD- 1.3 P388-13[c] MTA- 0.74 B16- 3.1	*n*-C_5H_{11} **148**[116,139]	H Ac	MTA- 2.6[a] MTA- 0.32[b] 1.1[a] B16- 0.51[b]
n-C_5H_{11} (*N*-methyl) **149**[139]	Ac	MTA- 1.7	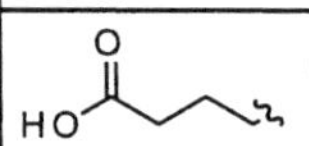 **150**[44]	H	MTD- 1	MeO_3O **151**[56]	H	MTD- 0.7 P388- 150[c]
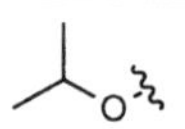 **152**[56]	H	MTD- 1.2 P388- 1.5[c]	*n*-C_4H_9O **153**[56,116]	H Ac	MTD- 1.1 P388- 7.5[c] MTA- 0.8 B16- 0.6	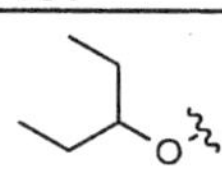 **154**[56]	H	MTD- 0.5 P388- 2[c]
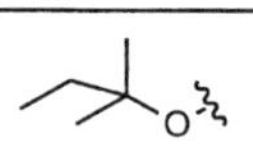 **155**[56]	H	MTD- 0.6 P388- 1[c]	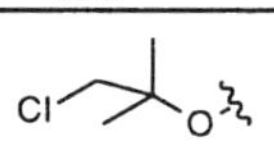 **156**[56]	H	MTD- 0.6 P388- 1.8[c]	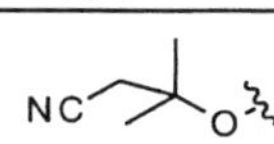 **157**[56]	H	MTD- 0.65 P388- 11[c]
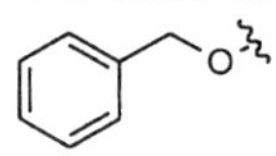 **158**[116,135]	Ac	MTA- 3.3 MTI- 1.2 B16- 4.0 J744.2- 2.1	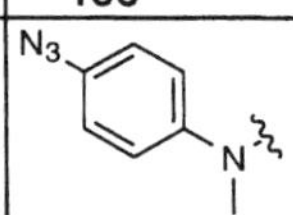 **159**[130]	H	MTD- 1.8			

[a] Natural product.
[b] Synthetic compound.
[c] $EC_{50}/EC_{50(docetaxel)}$; docetaxel (**7**, Table 1.2) is twofold more active than paclitaxel in P388 cells.

data differ little; however, there is a distinct cytotoxicity advantage favoring *t*-butyl carbamates **7** and **8** compared to their corresponding N-(3,3-dimethyl)butanoyl analogues **147**. This correlation is not observed for bioisosteres **148** and **153**, nor is it obvious for potential bioisosteres **146** and **152** (caution must be exercised since these analogues differ in O10 substitution and different microtubule and cytotoxicity assays were used).

In summary, with respect to the microtubule assays, the large variety of N3′-substituted analogues (Tables 1.4 and 1.5) show comparable activity to paclitaxel (EC_{50}^{rel} = 0.32–6.7). Substituted benzoyl and alkyl amides, pyridyl, and the urea analogues are slightly less active, and the furyl and carbamate analogues are slightly more active than paclitaxel. Unfavorable steric interaction may become important with increasing substituent size (**134** and **158**). The cytotoxicity assays correlate quite well; however, for some analogues (e.g., **129**, **136**, **137**, **139**, **140**, **145**, **151**, and **157**) cell-related effects (e.g., absorption, metabolism, etc.) probably contribute to diminished bioactivity. The best analogues of this series based on microtubule and cytotoxicity assays are the structurally diverse, taxol C (**148**), docetaxel (**7**), and *N*-3-furoyl analogues **142**.

1.4.3 Analogues with Modifications at C3′

Since the 3′-phenyl moiety of the paclitaxel side chain was shown to be an important element, it could be envisioned that modifications to this position might lead to interesting compounds. To this end, SAR of 3′-dephenylpaclitaxel (Table 1.6) and docetaxel (Table 1.7) analogues have been studied. An investigation of substituted phenyl substituents synthesized by the β-lactam method (Scheme 1.14) based on the Topliss Operational Scheme resulted in analogues **160–168** (Table 1.6) with slightly reduced activity. No systematic change in bioactivity as a result of Hansch parameters was observed, but the best analogues in both the MTA and cytotoxicity assays were *para*-substituted compared to the *ortho*- and *meta*-substituted analogues. The 3′-dephenyldocetaxel analogues **178–188** (Table 1.7) prepared from the oxazoline-protected side-chain equivalents of type **82** (Scheme 1.13) lead to essentially the same conclusion. Where applicable, comparison of the paclitaxel and docetaxel analogues demonstrate the "docetaxel effect," where a twofold increase in activity is observed for the docetaxel analogues over the paclitaxel analogues, suggesting modification at C3′ and N3′ are additive to some extent. Interestingly, the three fluorine positional isomers **181–183** (Table 1.7) exhibit similar activity, and the 4-halophenyl analogues **183**, **186**, and **187** show a trend related to size and/or electronegativity.

A large number of substituted 4-phenyl analogues (**189–196**, Table 1.7) have been synthesized from the versatile oxazoline **204** via palladium-catalyzed Suzuki reactions. Most of the analogues prepared are two-atom homologues that exhibit excellent performance in MTD and P388 cell assays. Larger para substitution (**189**, **191**, and **193**) and replacement of the 3′-phenyl group with β-naphthyl as in **170** (Table 1.6) result in analogues with diminished activity.

Bioisosteric replacement of the 3′-phenyl with heterocyclic pyridyl (**171–173**) and furyl (**174** and **175**) moieties (Scheme 1.6) provided compounds with better activity than paclitaxel. A limitation of the β-lactam method is evident, as some of the heterocyclic β-lactams, corresponding to **205–207** (*R* = H, Figure 1.4), synthesized by the ester enolate–imine cyclocondensation, exhibit poor enantiomeric excesses (31–44% ee). However, excellent kinetic resolution was observed during the coupling of **205**

Table 1.6 C3′ AROMATIC, HETEROAROMATIC, AND ALIPHATIC SUBSTITUTION ANALOGUES OF PACLITAXEL

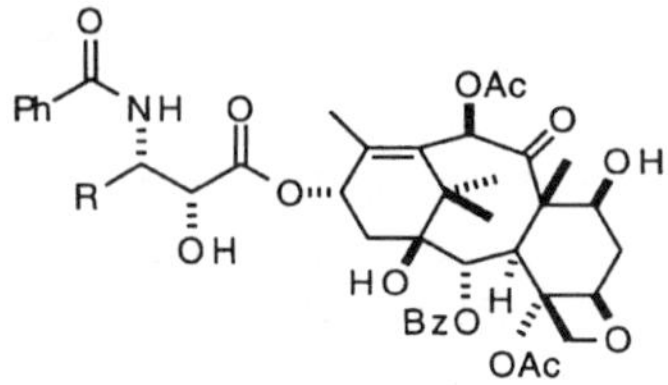

R	EC_{50}^{rel}	R	EC_{50}^{rel}	R	EC_{50}^{rel}
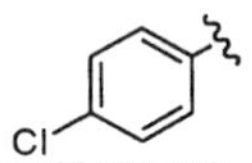 **160**[88,132,140]	MTA- 1.9 B16- 2.2	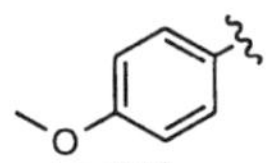 **161**[140]	MTA- 0.51 B16- 1.0	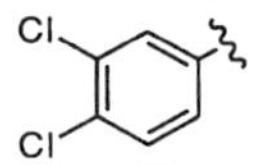 **162**[140]	MTA- 7.1 B16- 40
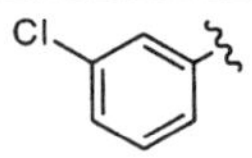 **163**[128,132,140]	MTA- 4.4 B16- 6.7	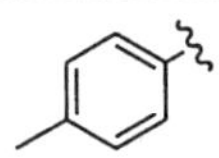 **164**[134,140]	MTA- 2.4 B16- 3.0	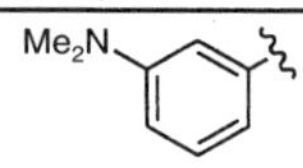 **165**[140]	MTA- 4.6 B16- 5.8
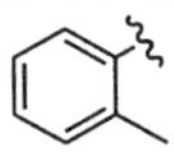 **166**[140]	MTA- 5.1 B16- >33	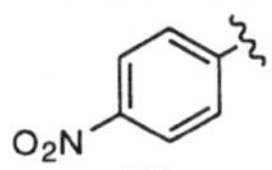 **167**[133]	MTA- 2.0 B16- 21	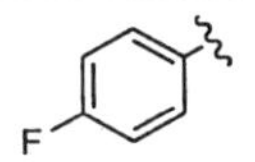 **168**[140]	MTA- 1.1 B16- 1.2
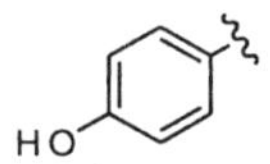 **169**[141]	MTD- 0.8 L1210- 10	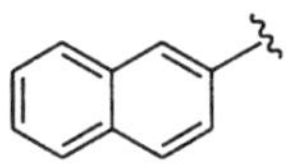 **170**[37]	MTA- 7.1 B16- 6.7	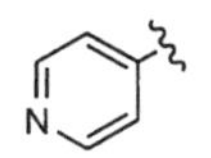 **171**[89]	MTA- 0.42 B16- 1.3
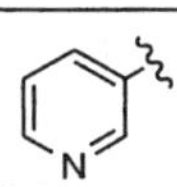 **172**[89]	MTA- 0.5[a] B16- 27[a]	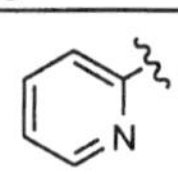 **173**[89]	MTA- 0.69 B16- 0.78	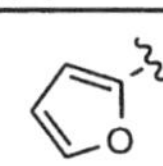 **174**[89]	MTA- 0.85 B16- 0.31
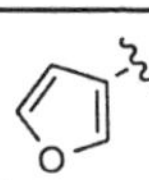 **175**[89]	MTA- 0.9[b] B16- 3.3[b]	 **176**[142]	MTA- 0.29 B16- 0.91	 **177**[57]	MTA- 1.8[c] B16- 7.5[c]

a 2:1 Diastereomeric mixture.
b 3:1 Diastereomeric mixture.
c 10-Deacetyl derivative.

(*R* = Bz) with 7-TES-baccatin III (**80**), resulting in **173** (Table 1.6) as a single isomer. Less kinetic resolution was observed with **206** (*R* = Bz) and **207** (*R* = Bz), leading to inseparable diastereomeric mixtures for **172** and **175**, which may explain the decreased cytotoxicity of these analogues compared to their heterocyclic cousins.

3′-Alkyl-3′-dephenylpaclitaxel and docetaxel analogues have also been prepared. Reduction of the 4-phenyl group of β-lactam **78c** (see Scheme 1.14) leads to 3′-cyclohexyl analogues.[142] In the paclitaxel series, **176** is a most potent analogue, while the docetaxel derivative **197** exhibits good but slightly reduced cytotoxicity.

Table 1.7 C3′ AROMATIC, HETEROAROMATIC, AND ALIPHATIC SUBSTITUTION ANALOGUES OF DOCETAXEL

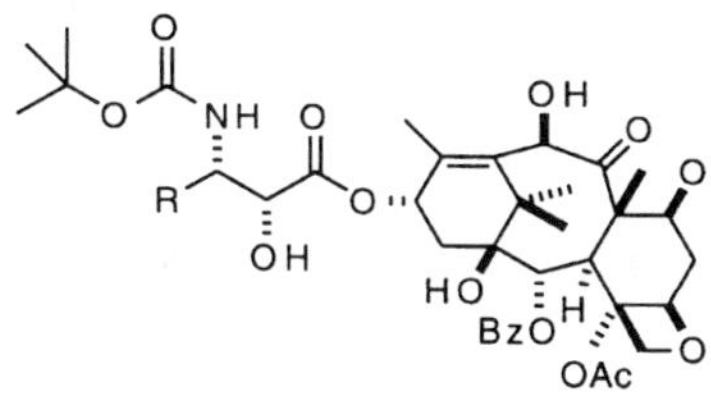

R	EC_{50}^{rel}	R	EC_{50}^{rel}	R	EC_{50}^{rel}
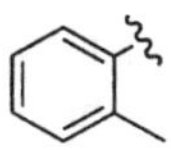 **178**[56]	MTD- 2.2 P388- 20[a]	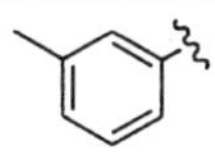 **179**[56]	MTD- 2.8 P388- 18[a]	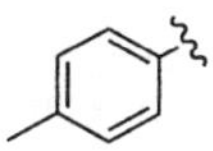 **180**[56]	MTD- 1.3 P388- 1.0[a]
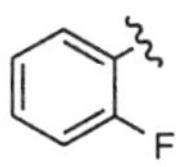 **181**[56]	MTD- 1.1 P388- 1.0[a]	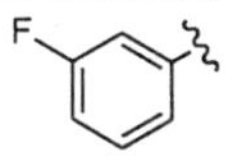 **182**[56]	MTD- 1.0 P388- 2.5[a]	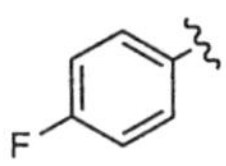 **183**[56,71]	MTD- 0.9 P388- 0.75[a]
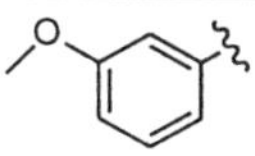 **184**[56]	MTD- 3 P388- 18[a]	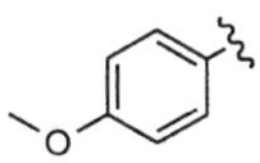 **185**[56]	MTD- 0.7 P388- 1.3[a]	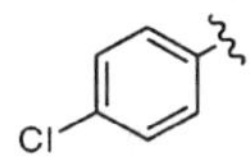 **186**[56]	MTD- 1.0 P388- 1.3[a]
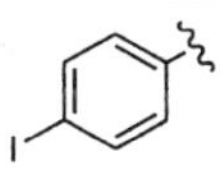 **187**[56]	MTD- 2.2 P388- 7.5[a]	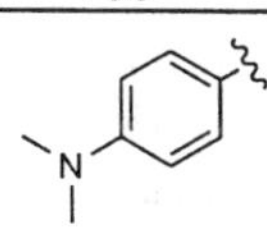 **188**[56,71]	MTD- 3.5 P388- 16[a]	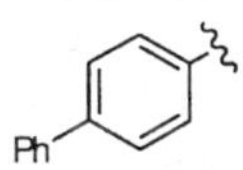 **189**[143]	MTD- 2.0 P388- 16[a]
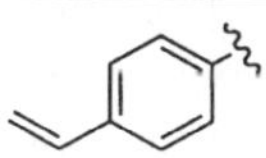 **190**[143]	MTD- 0.7 P388- 1.5[a]	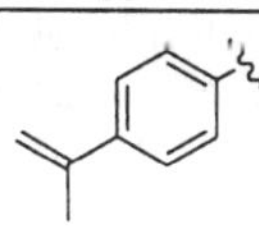 **191**[143]	MTD- 1.8 P388- 8.8[a]	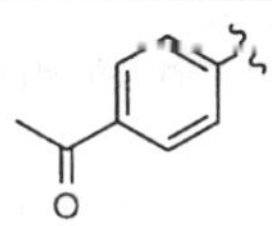 **192**[143]	MTD- 0.05 P388- 4.8[a]
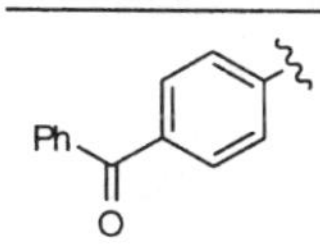 **193**[143]	MTD- 20 P388- >25[a]	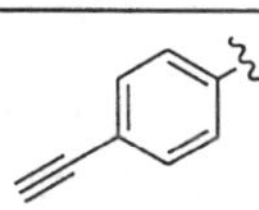 **194**[143]	MTD- 0.6 P388- 2.1[a]	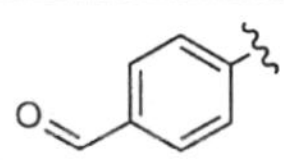 **195**[143]	MTD- 0.75 P388- 3.3[a]
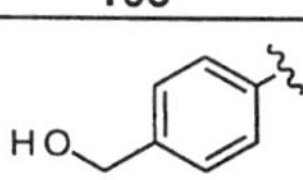 **196**[143]	MTD- 0.4 P388- 1.5[a]	 **197**[144]	MTD- 0.72 P388- 7.9[a]	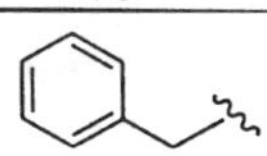 **198**[56]	MTD- 150[a]
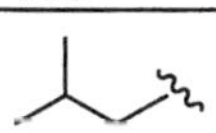 **199**[145]	MTD- 0.78 P388- 1.2[a]	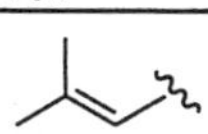 **200**[145]	MTD- 0.64 P388- 1.3[a]	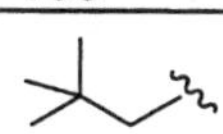 **201**[145]	MTD- 1.5 P388- 13[a]
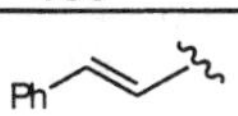 **202**[145]	MTD- 1.5 P388- 28[a]	 **203**[57]	MTA- 0.38 B16- 0.4		

[a] $EC_{50}/EC_{50(docetaxel)}$; docetaxel (**7**, Table 1.2) is twofold more cytotoxic than paclitaxel to P388 cells.

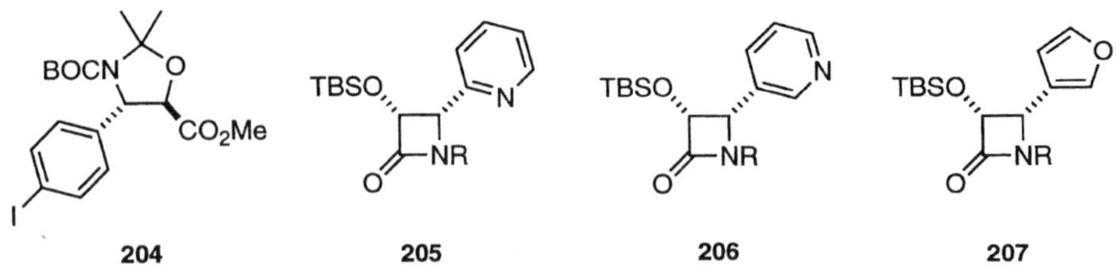

Figure 1.4. Precursors for 3-arylisoserines.

Interestingly, the nonaromatic compound **208** [EC_{50}^{rel} = 0.47 (MTA); EC_{50}^{rel} = 1.6 (B16)], synthesized by platinum-catalyzed hydrogenation of paclitaxel, possesses similar microtubule and cytotoxicity to paclitaxel, indicating none of the aromatic rings is necessary for activity.[142]

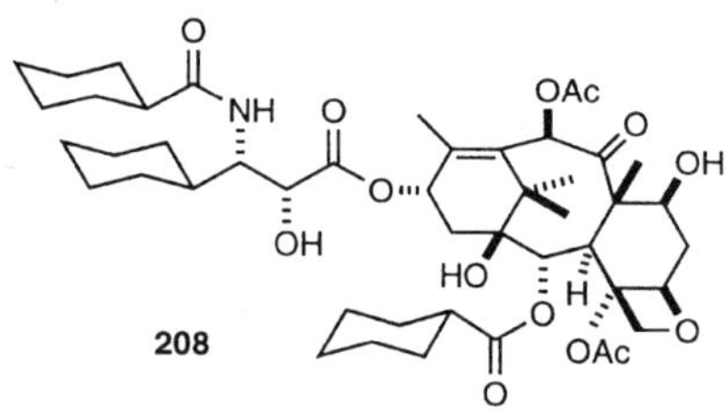

208

Phenylalanine served as the starting material to prepare homologue **198**, which was devoid of MTD activity (the basic synthetic methodology is shown in Scheme 1.6). However, other 3′-alkyl-3′-dephenyldocetaxel analogues **199**–**202**, synthesized via the β-lactam method (eq. 1.9), are potent agents. Isobutyl **199** and isobutenyl **200** are the best of these analogues, with **201** and **202** showing diminished cytotoxicity. Since clinical tumor treatment with paclitaxel and docetaxel suffer from the development of resistance, it is not surprising that **199**–**202** show reduced efficacy against multiple-drug-resistant cell lines (P388/Dox).[33] However the 10-O-acetyl analogues of **199** and **200** were markedly more effective toward MDR cells (MCF7 vs. MCF7-R).[33,145]

The 3′-(*t*-butyl)-3′-dephenylpaclitaxel and docetaxel analogues prepared from *t*-butylvaline (**35**; see Scheme 1.7) are also potent analogues; however, the docetaxel analogues **203** (Table 1.7) exhibits superior cytotoxicity compared to **177** (Table 1.6). In addition to **177** and **203**, investigation of a limited number of N3′ modified analogues (**209**–**217**, Table 1.8), based on the 3-(*t*-butyl)isoserine framework, resulted in SARs similar to the 3-phenylisoserine analogues found in Tables 1.4 and 1.5. Interestingly, it appears that a chain length optimum for N3′ substituents has been exceeded for **211** (compare to **148** and **153**, Table 1.5). The activity of the urea analogues **216** and **217** confirm that this underinvestigated functionality is a worthy N3′ substituent. The bioisosteric replacement of the carbonyl oxygen of **216** with sulfur resulted in thiourea **218** [EC_{50}^{rel} = 3.6 (MTA); EC_{50}^{rel} >41 (B16)] with greatly diminished activity.[57] An increase in water solubility (87–110-fold greater than paclitaxel; 4–5-fold greater than docetaxel) was noted

Table 1.8 C3′ *t*-BUTYL ANALOGUES

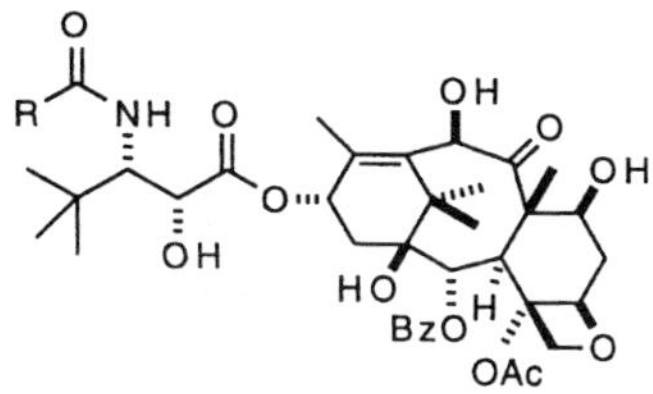

R	EC_{50}^{rel}	R	EC_{50}^{rel}	R	EC_{50}^{rel}
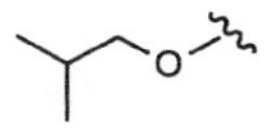 209[57]	MTA- 0.54 B16- 1.5	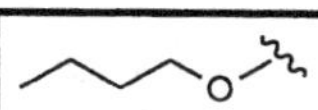 210[57]	MTA- 0.86 B16- 2.1	*n*-$C_6H_{13}O$ 211[57]	MTA- 2.3 B16- 12
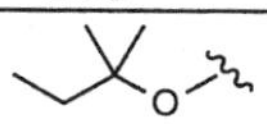 212[57]	MTA- 0.45 B16- 0.58	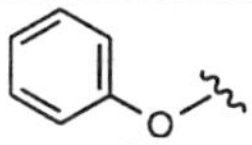 213[57]	MTA- 1.4 B16- >32	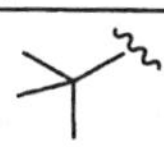 214[57]	MTA- 1.3 B16- 32
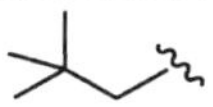 215[57]	MTA- 0.50 B16- 3.6	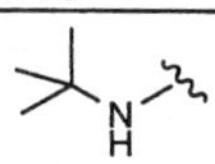 216[57]	MTA- 0.60 B16- 0.90	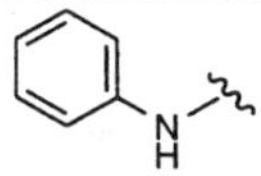 217[57]	MTA- 0.32 B16- 0.70

for **203–210**, **212**, **215**, and **216**; however, the water solubility of paclitaxel and docetaxel is so poor[146] that even these analogues would require addition of a surfactant vehicle for administration.

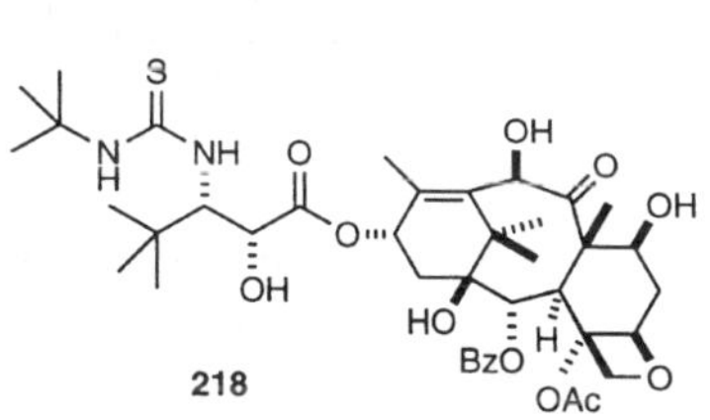

218

The SAR of the diterpene system of paclitaxel and docetaxel has also received considerable attention,[41] but is beyond the scope of this review. However, an extensive set of O13-modified side chains has been presented based on the natural product 13-O-acetyl-9-dihydrobaccatin III.[148,148] 9(*R*)-Dihydropaclitaxel (**219**) and analogues **220–252** (Tables 1.9 and 1.10) were synthesized from 9-dihydrobaccatin III (derived from the natural product) via the β-lactam methodology (see eq. 1.9). The N-modified 3′-phenylisoserines **221–236** complement the SARs established in Tables 1.4 and 1.5. Most notable in this series are the outstanding cytotoxicities of **227–229** and **231** against P388 and B16. The adamantyl carbamate **234** and the urea analogue **235** are similar in the MTD assay and B16 cytotoxicity compared to paclitaxel. Nevertheless, they are more cytotoxic towards P388 cells. Perhaps the

Table 1.9 N3' ANALOGUES OF 9(*R*)-DIHYDROTAXOL

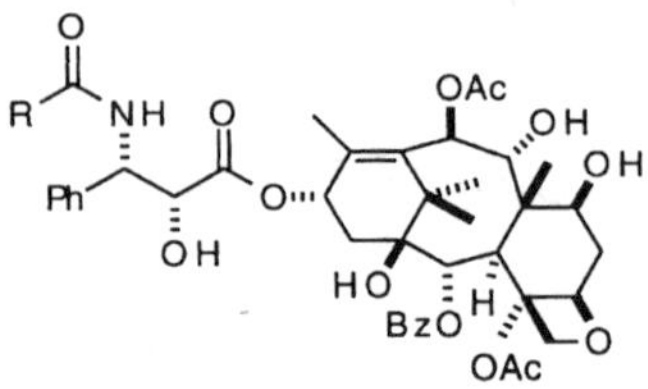

R	EC_{50}^{rel}	R	EC_{50}^{rel}	R	EC_{50}^{rel}
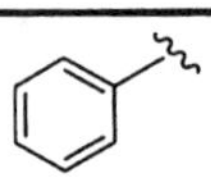 **219**[74,149-151]	MTA- 0.76 P388- 1.2 B16- 6.1	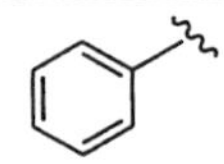 **220**[149]	MTA- 0.83[a] P388- ~14[a] B16- ~8.0	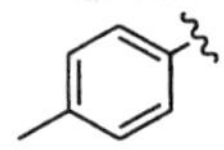 **221**[151]	MTA- 0.75 P388- 1.4 B16- 6.1
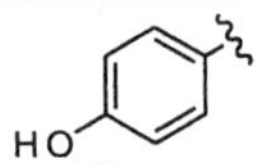 **222**[151]	MTA- 1.1 P388- >2.2 B16- >2.4	Me_3 **223**[151]	MTA- 3.2 P388- 44 B16- 268	 **224**[151]	MTA- 2.4 P388- 1.8 B16- 8.5
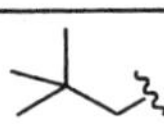 **225**[151]	MTA- 1.0 P388- 1.1 B16- 1.4	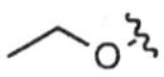 **226**[151]	MTA- 0.82 P388- 0.9 B16- 4.4	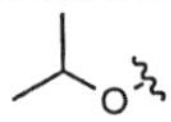 **227**[151]	MTA- 0.75 P388- 0.2 B16- 0.5
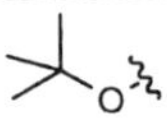 **228**[74,151]	MTA- 0.87 P388- 0.05 B16- 0.1	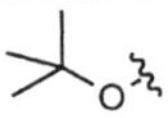 **229**[149]	MTA- 1.3[a] P388- ~0.28[a] B16- ~0.08	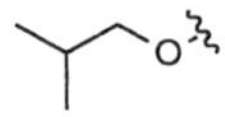 **230**[151]	MTA- 1.0 P388- 0.61 B16- 1.7
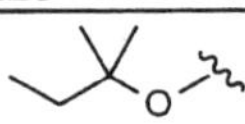 **231**[151]	MTA- 0.73 P388- 0.10 B16- 0.07	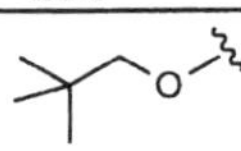 **232**[151]	MTA- 0.82 P388- 0.41 B16- 1.8	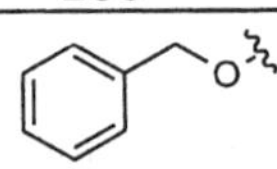 **233**[151]	P388- 1.2 B16- 10.7
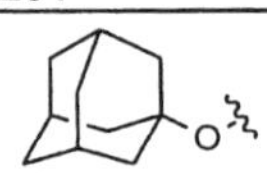 **234**[151]	MTA- 1.5 P388- 0.22 B16- 1.2	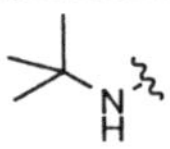 **235**[151]	MTA- 1.1 P388- 0.37 B16- 1.1	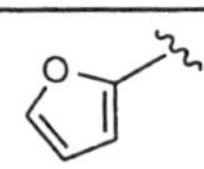 **236**[151]	MTA- 0.73 P388- 0.74 B16- 2.9

[a] 10-Deacetyl analogue.

highly lipophilic character of the adamantyl moiety aids cellular absorption for this compound.

A variety of C3′ modifications have been reported based on 9(*R*)-dihydropaclitaxel (**219**) and the very active 10-O-acetyl-9(*R*)-dihydrodocetaxel (**228**). A convergent strategy utilized the versatile β-lactam **69** (Figure 1.1) for the synthesis of the O13 side-chain equivalents, leading to **237**, **240**–**252** (Table 1.10 and compounds **238** and **239**).[74,149]

Table 1.10 C3′ ANALOGUES OF 10-O-ACETYL-9(*R*)-DIHYDROTAXOTERE

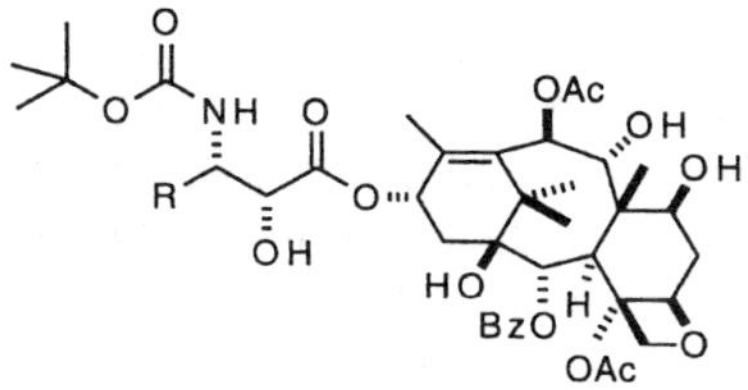

R	EC_{50}^{rel}	R	EC_{50}^{rel}	R	EC_{50}^{rel}
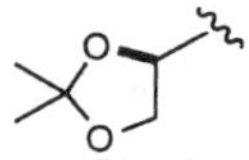 237[74,149]	MTA- 7.9 P388- >40[a] B16- 215[a]	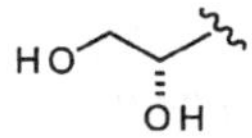 240[74,149]	MTA- 2.7 P388- >40[a] B16- >250[a]	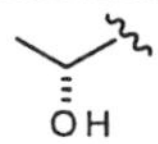 241[74]	MTA- 0.80[b] P388- >40[a,b] B16- >250[a]
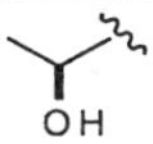 242[74]	MTA- 1.2[c] P388- >40[a,c] B16- >250[a]	Me$_3$ 243[74,149]	MTA- 1.1 P388- 17.2[a] B16- 48[a]	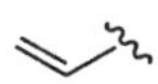 244[74,149]	MTA- 0.92 P388- 6.4[a] B16- 11[a]
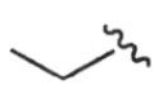 245[74,149]	MTA- 0.61 P388- 4.4[a] B16- 8.0[a]	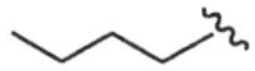 246[74]	MTA- 2.2 P388- 7.2[a] B16- 24[a]	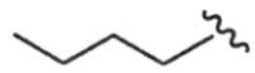 247[74]	MTA- 9.7[d] P388- >40[a,d] B16- >250[a]
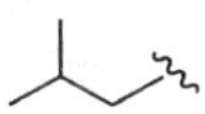 248[74,149]	MTA- 0.95 P388- 0.14[a] B16- 2.3[a]	C_5H_{11} 249[74]	MTA- 2.0 P388- 6.4[a] B16- 35[a]	 250[74,149]	MTA- 0.57 P388- 5.6[a] B16- 14[a]
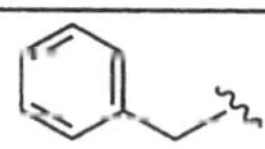 251[74,149]	MTA- >17 P388- >40[a] B16- >250[a]	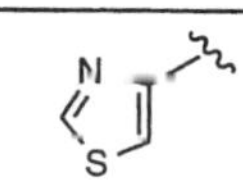 252[74]	MTA- 1.4 P388- 0.56[a] B16- 0.75[a]		

[a] $EC_{50}/EC_{50(analogue\ 228)}$; analogue **228** (Table 1.9) is 20-fold more active than paclitaxel against P388 cells.
[b] *R* or *S*.
[c] *S* or *R*.
[d] 3′*R*.

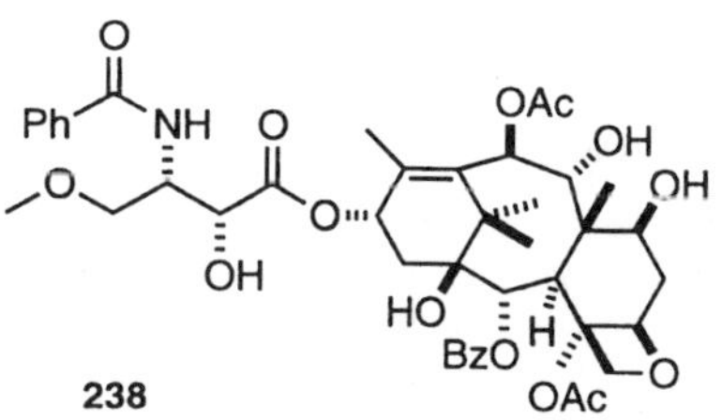

238

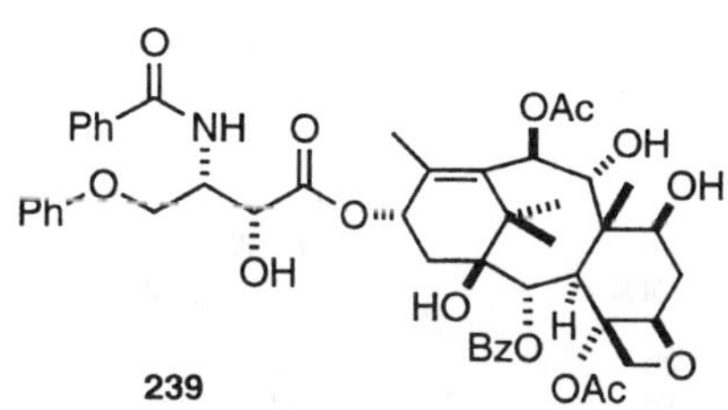

239

The polar C3′ substituents of **237–242** are generally tolerated in the MTA assay. In the 3′-alkyl series, analogues with small (**243–245**) or compact (**248** and **250**) groups are active MTA agents, while larger, more flexible groups (**246**, **249**, and **251**; see also **239**) show reduced activity in this assay. Cytotoxicity against P388 and B16 is reduced for all C3′ alkyl-substituted analogues compared to the parent 10-O-acetyl-9(*R*)-dihydrodocetaxel (**228**), with the exception of **248** and **252**, which are nearly ten- and twofold more cytotoxic toward P388 cells.

1.5 Summary

The potent activity of the anticancer agent paclitaxel, combined with its limited availability from natural sources, provided an opportunity for the chemical community to solve the paclitaxel supply problem. After the discovery by the Potier group that 10-deacetylbaccatin III could be obtained in quantity from regenerable sources, methodologies for the asymmetric preparation of the 3-phenylisoserine side chain of paclitaxel and docetaxel were invented in many laboratories. Novel chemistry was developed to overcome the difficult O13 acylation of baccatin III derivatives.

The β-lactam route to paclitaxel is now used by Bristol-Myers Squibb for the industrial-scale semisynthesis of paclitaxel and Rhone Poulenc Rorer prepares docetaxel via the oxazolidine route. Novel lead compounds, including Oread's 3-*t*-butyldocetaxel and Abbott's 9-dihydrodocetaxel analogues, were discovered. The innovative synthetic methodologies devised allowed detailed structure–activity studies of the N-benzoyl-3-phenylisoserine side chain, useful for the development of future generation anticancer taxanes.

1.6 Abbreviations

Ac	acetyl
Bn	benzyl
BOC	*t*-butoxycarbonyl
Bu	butyl
t-Bu	tertiary butyl
Bz	benzoyl
10-DAB	10-deacetylbaccatin III
DCC	dicyclohexylcarbodiimide
DDQ	2,2-dichloro-5,6-dicyano-1,4-benzoquinone
de	diastereomeric excess
DEAD	diethyl azodicarboxylate
DET	diethyl tartrate
$(DHQ)_2PHAL$	hydroquinine 1,4-phthalazinediyl diether
DMAP	4-dimethylaminopyridine
DPC	di-2-pyridyl carbonate
DQCB	dihydroquinidine 4-chlorobenzoate
ee	enantiomeric excess
EE	1-(1-ethoxy)ethyl

ent	enantiomeric
Et	ethyl
hfc	3-(heptafluoropropylhydroxymethylene)-*d*-camphorato
LAH	lithium aluminum hydride
LDA	lithium diisopropylamide
LiHMDS	lithium hexamethyldisilazide
Me	methyl
MTA	microtubule assembly
MTD	microtubule disassembly
MTI	microtubule initial (slope)
NCS	N-chlorosuccinimide
NMO	4-methylmorpholine N-oxide
Pd/C	palladium on carbon
Ph	phenyl
PMB	*p*-methoxybenzyl
PMP	*p*-methoxyphenyl
4-PPNO	4-phenyl-pyridine N-oxide
i-Pr	isopropyl
PTSA	*p*-toluenesulfonic acid
PPTS	pyridinium *p*-toluenesulphonate
SAR	structure-activity relationship
TBAF	tetrabutylammonium fluoride
TBS	*t*-butyldimethylsilyl
TES	triethylsilyl
Tf	trifluoromethylsulfonyl
TFA	trifluoroacetic acid
TIPS	triisopropylsilyl
TMS	trimethylsilyl
Troc	2,2,2-trichloroethoxycarbonyl
Ts	*p*-toluenesulfonyl
B16	murine melanoma cell line
P388	murine lymphocytic leukemia cell line
J774.2	murine macrophagelike cell line
KB	epidermoid carcinoma cell line
HCT116	human colon carcinoma cell line
L1210	murine lymphoid leukemia cell line
P388/Dox	doxorubicin-resistant leukemia cell line
MCF7	human mammary carcinoma cell line
MCF7-R	doxorubicin-resistant mammary carcinoma cell line

1.7 References and Notes

1. For review: C. N. C. Drey, in *Chemistry and Biochemistry of the Amino Acids*, G. C. Barrett, Ed. (Chapman and Hall, London, 1985), p. 25.

2. For review: C. N. C. Drey, in *Chemistry and Biochemistry of Amino Acids, Peptides and Proteins*, B. Weinstein, Ed. (Marcel Dekker, New York, 1977), vol. 4, p. 242.

3. For review: E. Juaristi, D. Quintana, J. Escalante, *Aldrichim. Acta* **27**, 3 (1994).

4. For review: D. C. Cole, *Tetrahedron* **50**, 9517 (1994).

5. S. Matsunaga, N. Fusetani, *J. Org. Chem.* **60**, 1177 (1995).

6. G. Trimurtulu, I. Ohtani, G. M. L. Patterson, R. E. Moore, T. H. Corbett, F. A. Valeriote, L. Demchik, *J. Am. Chem. Soc.* **116**, 4729 (1994).

7. G. L. Helms, R. E. Moore, W. P. Niemczurca, G. M. L. Patterson, *J. Org. Chem.* **60**, 1177 (1988).

8. B. E. Maryanoff, M. N. Greco, H.-C. Zhang, P. Andrade-Gordon, J. A. Kauffman, K. C. Nicolaou, A. Liu, P. H. Brugs, *J. Am. Chem. Soc.* **117**, 1225 (1995).

9. For review: R. B. Morin, M. Gorman, Eds., *Chemistry and Biology of β-Lactam Antibiotics* (Academic, New York, 1982).

10. For review: G. Lukacs, M. Ohno, Eds., *Recent Progress in the Chemical Synthesis of Antibiotics and Related Natural Products* (Springer, Berlin, 1990).

11. For review: P. C. Cherry, C. E. Newall, in *Chemistry and Biology of β-Lactam Antibiotics*, R. B. Morin, M. Gorman, Eds. (Academic, New York, 1982), vol. 2, p. 361.

12. For review: G. I. Georg, Ed., *Recent Advances in the Chemistry and Biology of β-Lactams and β-Lactam Antibiotics. Bioorg. Med. Chem. Lett.* **3**, pp. 2135–486 (1993).

13. D. Burnett, M. A. Caplen, H. R. Davis, R. E. Burrier, J. W. Clader, *J. Med. Chem.* **37**, 1733 (1994).

14. For review: M. B. Smith, *Methods of Non-α-Amino Acid Synthesis* (M. Dekker, New York, 1995).

15. For review: A. F. Spatola, in *Chemistry and Biochemistry of Amino Acids and Proteins*, B. Weinstein, Ed. (Marcel Dekker, New York, 1983), p. 268.

16. For review: G. I. Georg, T. T. Chen, I. Ojima, D. M. Vyas, Eds., *Taxane Anticancer Agents: Basic Science and Current Status* (American Chemical Society, Washington, D.C., 1995).

17. For review: M. Suffness, Ed., *Taxol® Science and Applications* (CRC, Boca Raton, FL, 1995).

18. For review: M. Suffness, M. Wall, in *Taxol® Science and Applications*, M. Suffness, Ed. (CRC, Boca Raton, FL, 1995), p. 3.

19. M. C. Wani, H. L. Taylor, M. E. Wall, P. Coggon, A. T. McPhail, *J. Am. Chem. Soc.* **93**, 2325 (1971).

20. H. Lataste, V. Sénilh, M. Wright, D. Guénard, P. Potier, *Proc. Natl. Acad. Sci. U.S.A.* **81**, 4090 (1984).

21. S. B. Horwitz, *Trends Pharm. Sci.* **13**, 134 (1992).

22. J. J. Manfredi, S. B. Horwitz, *Pharmacol. Ther.* **25**, 83 (1984).

23. The naturally occurring cytotoxic epothilones were recently identified to possess a paclitaxel-like mechanism of action. D. M. Bollag, P. A. McQueney, J. Zhu, O. Hensens, L. Koupal, J. Liesche, M. Goetz, E. Lazarides, C. M. Woods, *Cancer Res.* **55**, 2325 (1995).

24. Discodermolide is a cytotoxic marine natural product that stabilizes microtubules more potently than paclitaxel. E. ter Haar, R. J. Kowalski, E. Hamel, C. M. Lin, R. E. Longley, S. P. Gunasekera, H. S. Rosenkrantz, B. W. Day, *Biochemistry* **35**, 243 (1996).

25. For review: F. A. Holmes, A. P. Kudelka, J. J. Kavanagh, M. H. Huber, J. A. Ajani, V. Valero, in *Taxane Anticancer Agents: Basic Science and Current Status*, G. I. Georg, T. T. Chen, I. Ojima, D. M. Vyas, Eds. (American Chemical Society, Washington, D.C., 1995), p. 31.

26. M. E. Stearns, M. Wang, *Cancer Res.* **52**, 3776 (1992).

27. D. D. L. Woo, S. Y. P. Miao, J. C. Pelayo, A. S. Woolf, *Nature* **368**, 750 (1994).

28. R. A. Holton, H.-B. Kim, C. Somoza, F. Liang, R. J. Biediger, P. D. Boatman, M. Shindo, C. C. Smith, S. Kim, H. Nadizadeh, Y. Suzuki, C. Tao, P. Vu, S. Tang, P. Zhang, K. K. Murthi, L. N. Gentile, J. H. Liu, *J. Am. Chem. Soc.* **116**, 1599 (1994).

29. R. A. Holton, C. Somoza, H.-B. Kim, F. Liang, R. J. Biediger, P. D. Boatman, M. Shindo, C. C. Smith, S. Kim, H. Nadizadeh, Y. Suzuki, C. Tao, P. Vu, S. Tang, P. Zhang, K. K. Murthi, L. N. Gentile, J. H. Liu, *J. Am. Chem. Soc.* **116**, 1597 (1994).

30. K. C. Nicolaou, Z. Yang, J. J. Liu, H. Ueno, P. G. Nantermet, R. K. Guy, C. F. Claiforne, J. Renaud, E. A. Couladouros, K. Paulvannan, E. J. Sorensen, *Nature* **367**, 630 (1994).

31. J. J. Masters, J. T. Link, L. B. Snyder, W. B. Young, S. J. Danishefsky, *Angew. Chem., Int. Ed. Engl.* **34**, 1723 (1995).

32. J.-N. Denis, A. E. Greene, D. Guénard, F. Guéritte-Voegelein, L. Mangatal, P. Potier, *J. Am. Chem. Soc.* **110**, 5917 (1988).

33. For recent reviews, see Refs. 34–39.

34. R. A. Holton, R. J. Biediger, P. D. Boatman, in *Taxol: Science and Applications*, M. Suffness, Ed. (CRC Press, Boca Raton, FL, 1995), p. 97.

35. K. C. Nicolaou, W.-M. Dai, R. K. Guy, *Angew. Chem., Int. Ed. Engl.* **33**, 15 (1994).

36. G. I. Georg, S. M. Ali, J. Zygmunt, L. R. Jayasinghe, *Exp. Opin. Ther. Pat.* **4**, 109 (1994).

37. M. Hepperle, G. I. Georg, *Drugs Future* **19**, 573 (1994).

38. A. N. Boa, P. R. Jenkins, N. J. Lawrence, *Contemp. Org. Syn.* **1**, 47 (1994).

39. D. G. I Kingston, A. A. Molinero, J. M. Rimoldi, in *Progress in the Chemistry of Organic Natural Products*, W. Herz, G. W. Kirby, R. E. Moore, W. Steglich, C. Tamm, Eds. (Springer, New York, 1993), vol. 61, p. 1.

40. G. I. Georg, G. C. B. Harriman, D. G. Vander Velde, T. C. Boge, Z. S. Cheruvallath, A. Datta, M. Hepperle, H. Park, R. H. Himes, L. Jayasinghe, in *Taxane Anticancer Agents: Basic Science and Current Status*, G. I. Georg, T. T. Chen, I. Ojima, D. M. Vyas, Eds. (American Chemical Society, Washington, D.C., 1995), p. 217.

41. For review: G. I. Georg, T. C. Boge, Z. S. Cheruvallath, J. S. Clowers, G. C. B. Harriman, M. Hepperle, H. Park, in *Taxol® Science and Applications*, M. Suffness, Ed. (CRC, Boca Raton, FL, 1995), p. 317.

42. M. Z. Hoemann, D. G. Vander Velde, J. Aubé, G. I. Georg, L. R. Jayasinghe, *J. Org. Chem.* **60**, 2918 (1995).

43. L. Mangatal, M. T. Adeline, D. Guénard, F. Guéritte-Voegelein, P. Potier, *Tetrahedron* **45**, 4177 (1989).

44. F. Guéritte-Voegelein, D. Guénard, F. Lavelle, M.-T. Le Goff, L. Mangatal, P. Potier, *J. Med. Chem.* **34**, 992 (1991).

45. I. Ringel, S. B. Horwitz, *J. Natl. Cancer Inst.* **83**, 288 (1991).

46. D. G. I. Kingston, *Pharmacol. Ther.* **52**, 1 (1991).

47. J.-N. Denis, A. E. Greene, A. A. Serra, M.-J. Luche, *J. Org. Chem.* **51**, 46 (1986).

48. L. Deng, E. N. Jacobsen, *J. Org. Chem.* **57**, 4320 (1992).

49. J.-N. Denis, A. Correa, A. E. Greene, *J. Org. Chem.* **55**, 1957 (1990).

50. A. Commerçon, D. Bézard, F. Bernard, J. D. Bourzat, *Tetrahedron Lett.* **33**, 5185 (1992).

51. Z.-M. Wang, H. C. Kolb, K. B. Sharpless, *J. Org. Chem.* **59**, 5104 (1994).

52. A. M. P. Koskinen, E. K. Karvinen, J. P. Siirila, *J. Chem. Soc., Chem. Commun.*, 21 (1994).

53. J.-N. Denis, A. Correa, A. E. Greene, *J. Org. Chem.* **56**, 6939 (1991).

54. A. M. Kanazawa, A. Correa, J.-N. Denis, M.-J. Luche, A. E. Greene, *J. Org. Chem.* **58**, 255 (1993).

55. For a similar approach, see: A. Dondoni, D. Perrone, T. Semola, *Synthesis*, 181 (1995).

56. A. Commerçon, J. D. Bourzat, E. Didier, F. Lavelle, in *Taxane Anticancer Agents: Basic Science and Current Status*, G. I. Georg, T. T. Chen, I. Ojima, D. M. Vyas, Eds. (American Chemical Society, Washington, D.C., 1995), p. 233.

57. S. M. Ali, M. Z. Hoemann, J. Aubé, L. A. Mitscher, G. I. Georg, R. McCall, L. R. Jayasinghe, *J. Med. Chem.* **38**, 3821 (1995).

58. F. A. Davis, R. T. Reddy, R. E. Reddy, *J. Org. Chem.* **57**, 6387 (1992).

59. M. E. Bunnage, S. G. Davies, C. J. Goodwin, *J. Chem. Soc., Perkin Trans. 1*, 1375 (1993).

60. J. Escalante, E. Juaristi, *Tetrahedron Lett.* **36**, 4397 (1995).

61. K. Hattori, M. Miyata, H. Yamamoto, *J. Am. Chem. Soc.* **115**, 1151 (1993).

62. A. M. Kanazawa, J.-N. Denis, A. E. Greene, *J. Chem. Soc., Chem. Commun.*, 2591 (1994).

63. A. M. Kanazawa, J.-N. Denis, A. E. Greene, *J. Org. Chem.* **59**, 1238 (1994).

64. C. S. Swindell, M. Tao, *J. Org. Chem.* **58**, 5889 (1993).

65. C. Mukai, I. J. Kim, M. Hanaoka, *Tetrahedron: Asymm.* **3**, 1007 (1992).

66. C. Mukai, I. J. Kim, E. Furu, M. Hanaoka, *Tetrahedron* **37**, 8323 (1993).

67. For review: G. I. Georg, Ed., *The Organic Chemistry of β-Lactams* (VCH Publishers, New York, 1993).

68. For review: G. I. Georg, V. T. Ravikumar, in *The Organic Chemistry of β-Lactams* , G. I. Georg, Ed. (VCH Publishers, New York, 1993), p. 295.

69. V. Farina, S. I. Hauck, D. G. Walker, *Synlett*, 761 (1992).

70. G. I. Georg, E. Akgün, P. M. Mashava, M. Milstead, H. Ping, Z. Wu, D. Vander Velde, F. Takusagawa, *Tetrahedron Lett.* **33**, 2111 (1992).

71. J. D. Bourzat, A. Commerçon, *Tetrahedron Lett.* **34**, 6049 (1993).

72. Y. Hashimoto, A. Kai, K. Saigo, *Tetrahedron Lett.* **36**, 8821 (1995).

73. G. I. Georg, Z. Wu, *Tetrahedron Lett.* **35**, 381 (1994).

74. L. Li, S. A. Thomas, L. L. Klein, C. M. Yeung, C. J. Maring, D. J. Grampovnik, P. A. Lartey, J. J. Plattner, *J. Med. Chem.* **37**, 2655 (1994).

75. C. Palomo, J. M. Aizpurua, J. I. Miranda, A. Mielgo, J. M. Odriozola, *Tetrahedron Lett.* **34**, 6325 (1993).

76. R. A. Holton, J. H. Liu, *Bioorg. Med. Chem. Lett.* **3**, 2475 (1993).

77. S. M. Tenneson, B. Belleau, *Can. J. Chem.* **58**, 1605 (1980).

78. A. K. Bose, M. S. Manhas, J. M. van der Veen, S. S. Bari, D. R. Wagle, V. R. Hedge, L. Krishnan, *Tetrahedron Lett.* **26**, 33 (1985).

79. C. Hubschwerlen, G. Schmid, *Helv. Chim. Acta* **66**, 2206 (1983).

80. A. K. Bose, V. R. Hedge, D. R. Wagle, S. S. Bari, M. S. Manhas, *J. Chem. Soc., Chem. Commun.*, 161 (1986).

81. L. Li, S. A. Thomas, L. Klein, C. M. Yeung, C. J. Maring, D. J. Grampovnik, J. J. Plattner, in *207th ACS National Meeting of the American Chemical Society* (American Chemical Society, Washington, D. C., 1994), Abs. MEDI 103.

82. D. A. Evans, E. B. Sjogren, *Tetrahedron Lett.* **26**, 3783 (1985).

83. D. R. Kronenthal, C. Y. Han, M. K. Taylor, *J. Org. Chem.* **47**, 2765 (1082).

84. For review: D. J. Hart, D.-C. Ha, *Chem. Rev.* **89**, 1447 (1989).

85. For review: M. J. Brown, *Heterocycles* **29**, 2225 (1989).

86. I. Ojima, I. Habus, M. Zhao, G. I. Georg, L. R. Jayasinghe, *J. Org. Chem.* **56**, 1681 (1991).

87. For review: J. K. Whitesell, *Chem. Rev.* **92**, 963 (1992).

88. G. I. Georg, Z. S. Cheruvallath, R. H. Himes, M. R. Mejillano, C. T. Burke, *J. Med. Chem.* **35**, 4230 (1992).

89. G. I. Georg, G. C. B. Harriman, M. Hepperle, R. H. Himes, *Bioorg. Med. Chem. Lett.* **4**, 1381 (1994).

90. G. Cainelli, D. Giacomini, M. Panunzio, G. Martelli, G. Spunta, *Tetrahedron Lett.* **28**, 5369 (1987).

91. T. Uyehara, I. Suzuki, Y. Yamamoto, *Tetrahedron Lett.* **30**, 4275 (1989).

92. I. Ojima, Y. H. Park, C. M. Sun, T. Brigaud, M. Zhao, *Tetrahedron Lett.* **33**, 5737 (1992).

93. G. I. Georg, Z. S. Cheruvallath, G. C. B. Harriman, M. Hepperle, H. Park, *Bioorg. Med. Chem. Lett.* **3**, 2467 (1993).

94. G. Helmchen, R. Wierzchowski, *Angew. Chem.* **96**, 59 (1984).

95. G. I. Georg, T. C. Boge, G. C. B. Harriman, M. Hepperle (unpublished results).

96. H. E. Zimmerman, M. D. Traxler, *J. Am. Chem. Soc.* **79**, 1920 (1957).

97. X. Wang, C. Lee, *Tetrahedron Lett.* **34**, 6241 (1993).

98. R. N. Patel, A. Banerjee, R. Y. Ko, J. M. Howell, W.-S. Li, F. T. Comezoglu, R. A. Partyka, L. Szarka, *Biotechnol. Appl. Biochem.* **20**, 23 (1994).

99. R. Brieva, J. Z. Crich, C. J. Sih, *J. Org. Chem.* **58**, 1068 (1993).

100. D.-M. Gou, Y. C. Liu, C.-S. Chen, *J. Org. Chem.* **58**, 1287 (1993).

101. H. Hönig, P. Seufer-Wasserthal, H. Weber, *Tetrahedron* **46**, 3841 (1990).

102. J. Kearns, M. M. Kayser, *Tetrahedron Lett.* **35**, 2845 (1994).

103. R. N. Patel, A. Banerjee, J. M. Howell, C. G. McNamee, D. Brozozowski, D. Mirfakhrae, V. Nanduri, J. K. Thottathill, L. J. Szarka, *Tetrahedron: Asymm.* **4**, 2069 (1993).

104. J. Kant, W. S. O'Keefe, S.-H. Chen, V. Farina, C. Fairchild, K. Johnston, J. F. Kadow, B. H. Long, D. Vyas, *Tetrahedron Lett.* **35**, 5543 (1994).

105. J.-N. Denis, A. M. Kanazawa, A. E. Greene, *Tetrahedron Lett.* **35**, 105 (1994).

106. E. Didier, E. Fouque, A. Commerçon, *Tetrahedron Lett.* **35**, 3063 (1994).

107. E. Didier, E. Fouque, I. Taillepied, A. Commerçon, *Tetrahedron Lett.* **35**, 2349 (1994).

108. D. G. I. Kingston, A. G. Chaudhary, A. A. L. Gunatilaka, M. L. Middleton, *Tetrahedron Lett.* **35**, 4486 (1994).

109. I. Ojima, I. Habus, M. Zhao, M. Zucco, Y. H. Park, C. M. Sun, T. Brigaud, *Tetrahedron* **48**, 6985 (1992).

110. I. Ojima, C. M. Sun, M. Zucco, Y. H. Park, O. Duclos, S. Kuduk, *Tetrahedron Lett.* **34**, 4149 (1993).

111. For review: W. C. Rose, in *Taxol® Science and Applications*, M. Suffness, Ed. (CRC, Boca Raton, 1995), p. 209.

112. C. S. Swindell, N. E. Krauss, S. B. Horwitz, I. Ringel, *J. Med. Chem.* **34**, 1176 (1991).

113. F. Guéritte-Voegelein, V. Sénilh, B. David, D. Guénard, P. Potier, *Tetrahedron* **42**, 4451 (1986).

114. D. Guénard, F. Guéritte-Voegelein, P. Potier, *Acc. Chem. Res.* **26**, 160 (1993).

115. V. Sénilh, F. Guéritte-Voegelein, D. Guénard, M. Colin, P. Potier, *C. R. Acad. Sc. Paris t. 299, Serie II*, **15**, 1039 (1984).

116. G. I. Georg, T. C. Boge, Z. S. Cheruvallath, G. C. B. Harriman, M. Hepperle, H. Park, R. H. Himes, *Bioorg. Med. Chem. Lett.* **4**, 335 (1994).

117. H. J. Williams, A. I. Scott, R. A. Dieden, C. S. Swindell, L. E. Chirlian, M. M. Francl, J. M. Heerding, N. E. Krauss, *Can. J. Chem.* **72**, 252 (1994).

118. F. Guéritte-Voegelein, D. Guénard, L. Mangatal, P. Potier, J. Guilhem, M. Cesario, C. Pascard, *Acta Crystallogr.* **C46**, 781 (1990).

119. J. Kant, S. Huang, H. Wong, C. Fairchild, D. Vyas, V. Farina, *Bioorg. Med. Chem. Lett.* **3**, 2471 (1993).

120. N. F. Magri, D. G. I. Kingston, *J. Nat. Prod.* **51**, 298 (1988).

121. W. Mellado, N. F. Magri, D. G. I. Kingston, R. Garcia-Arenas, G. A. Orr, S. B. Horwitz, *Biochem.* **124**, 329 (1984).

122. L. R. Jayasinghe, A. Datta, S. M. Ali, J. Zygmunt, D. G. Vander Velde, G. I. Georg, *J. Med. Chem.* **37**, 2981 (1994).

123. J. G. Topliss, *J. Med. Chem.* **15**, 1006 (1972).

124. J. G. Topliss, Y. C. Martin, in *Drug Design*, E. J. Ariens, Ed. (Academic, New York, NY, 1975), vol. V, p. 1.

125. J. G. Topliss, *J. Med. Chem.* **20**, 463 (1977).

126. K.-H. Czaplinski, G. L. Grunewald, *Bioorg. Med. Chem. Lett.* **4**, 2211 (1994).

127. C. S. Swindell, J. M. Heerding, N. E. Krauss, S. B. Horwitz, S. Rao, I. Ringel, *J. Med. Chem.* **37**, 1446 (1994).

128. S. Rao, N. E. Krauss, J. M. Heerding, C. S. Swindell, I. Ringel, G. A. Orr, S. B. Horwitz, *J. Biol. Chem.* **269**, 3132 (1994).

129. D. Dasgupta, H. Park, G. C. B. Harriman, G. I. Georg, R. H. Himes, *J. Med. Chem.* **37**, 2976 (1994).

130. C. Combeau, A. Commerçon, C. Mioskowski, B. Rousseau, F. Aubert, M. Goeldner, *Biochemistry* **33**, 6676 (1994).

131. G. I. Georg, G. C. B. Harriman, M. Hepperle, J. S. Clowers, D. G. Vander Velde, R. H. Himes, *J. Org. Chem.* **61**, 2664 (1996).

132. G. I. Georg, Z. S. Cheruvallath, R. H. Himes, M. R. Mejillano, *Bioorg. Med. Chem. Lett.* **2**, 295 (1992).

133. G. I. Georg, T. C. Boge, Z. S. Cheruvallath, G. C. B. Harriman, M. Hepperle, H. Park, R. H. Himes, *Bioorg. Med. Chem. Lett.* **4**, 1825 (1994).

134. G. I. Georg, Z. S. Cheruvallath, R. H. Himes, M. R. Mejillano, *Bioorg. Med. Chem. Lett.* **2**, 1751 (1992).

135. C. S. Swindell, J. M. Heerding, N. E. Krauss, S. B. Horwitz, I. Ringel, *Bioorg. Med. Chem. Lett.* **4**, 1531 (1994).

136. G. I. Georg, G. C. B. Harriman, H. Park, R. H. Himes, *Bioorg. Med. Chem. Lett.* **4**, 487 (1994).

137. G. Chauviere, D. Guénard, F. Picot, V. Sénilh, P. Potier, *C. R. Acad. Sc. Paris t. 293, Serie II*, **35**, 501 (1981).

138. C. H. O. Huang, D. G. I. Kingston, N. F. Magri, G. Samaranayake, *J. Nat. Prod.* **49**, 665 (1986).

139. W. Ma, G. L. Park, G. A. Gomez, M. H. Nieder, T. L. Adams, J. S. Aynsley, O. P. Sahai, R. J. Smith, R. W. Stahlhut, P. J. Hylands, F. Bitsch, C. Shackleton, *J. Nat. Prod.* **57**, 116 (1994).

140. G. I. Georg, Z. S. Cheruvallath, G. C. B. Harriman, M. Hepperle, H. Park, R. H. Himes, *Bioorg. Med. Chem. Lett.* **4**, 2331 (1994).

141. B. Monsarrat, E. Mariel, S. Cros, M. Garés, D. Guénard, F. Guéritte-Voegelein, M. Wright, *Drug Metab. Dispos.* **18**, 895 (1990).

142. T. C. Boge, R. H. Himes, D. G. Vander Velde, G. I. Georg, *J. Med. Chem.* **37**, 3337 (1994).

143. J. D. Bourzat, F. Lavelle, A. Commerçon, *Bioorg. Med. Chem. Lett.* **5**, 809 (1995).

144. I. Ojima, O. Duclos, M. Zucco, F. Lavelle, *J. Med. Chem.* **37**, 2602 (1994).

145. I. Ojima, Y. H. Park, I. Fenoglio, O. Duclos, C.-M. Sun, S. D. Kuduk, M. Zucco, G. Appendino, P. Pera, J. M. Veith, R. J. Bernacki, M.-C. Bissery, C. Combeau, P. Vrignaud, J. F. Riou, F. Lavelle, in *Taxane Anticancer Agents: Basic Science and Current Status*, G. I. Georg, T. T. Chen, I. Ojima, D. M. Vyas, Eds. (American Chemical Society, Washington, D.C., 1995), p. 264.

146. D. M. Vyas, H. Wong, A. R. Crosswell, A. M. Casazza, J. O. Knipe, S. W. Mamber, T. W. Doyle, *Bioorg. Med. Chem. Lett.* **3**, 1357 (1993).

147. G. P. Gunawardana, U. Premachandran, N. S. Burres, D. N. Whittern, R. Henry, S. Spanton, J. B. McAlpine, *J. Nat. Prod.* **55**, 1686 (1992).

148. L. O. Zamir, M. E. Nedea, S. Belair, F. Sauriol, O. Mamer, E. Jacqmain, F. I. Jean, F. X. Garneau, *Tetrahedron Lett.* **33**, 5173 (1992).

149. L. L. Klein, L. Li, C. M. Yeung, C. J. Maring, S. A. Thomas, D. J. Grampovnik, J. J. Plattner, in *Taxane Anticancer Agents: Basic Science and Current Status*, G. I. Georg, T. T. Chen, I. Ojima, D. M. Vyas, Eds. (American Chemical Society, Washington, D.C., 1995), p. 276.

150. L. L. Klein, *Tetrahedron Lett.* **34**, 2047 (1993).

151. C. J. Maring, D. J. Grampovnik, C. M. Yeung, L. L. Klein, L. Li, S. A. Thomas, J. J. Plattner, *Bioorg. Med. Chem. Lett.* **4**, 1429 (1994).

CHAPTER

2

Biological Activity of β-Amino Acids and β-Lactams

Joaquín Tamariz

2.1 Introduction

The importance of β-amino acids has been focused, particularly in the past several decades, on the β-lactam antibiotics, since naturally occurring penicillin derivatives have been discovered as broad antibiotic activity agents.[1] Over the years, countless numbers of these compounds have been prepared and tested,[2] and a variety of new β lactam ring systems have been introduced such as cephems,[3] cephalosporins,[4] oxacephems,[5] penems,[6] carbapenems,[7] oxapenams,[8] as well as monocyclic[9] and policyclic[10] ring systems. Moreover, β-amino acids have been known to play an important role in both primary and secondary metabolism.[11] In recent years, they proved also to have biological activity as monomeric[12] or as peptidic[13] components, and as their α-hydroxy derivatives.[14] Among the latter, Taxol® (paclitaxel)[15] and Taxotere® (docetaxel),[16] a semisynthetic analogue, are currently considered the most promising anticancer agents.[17]

Since the literature is abundant concerning the biological activity of β-amino acids[18] and previous reviews have appeared covering the bioorganic and medicinal chemistry of these important targets,[19] this review attempts to point out some relevant and recent discoveries about β-amino acids, under their structural feature of β-lactams, peptides, or as more complex structure components. Drug design approaches to face traditional pharmacological challenges, including bacterial β-lactamase stability,[20] methicillin-resistant *Staphylococcus aureus* (MRSA),[21] or directed enzyme prodrug therapy,[22] are discussed. Furthermore, new biologically active β-amino acid skeletons, as well as recent structure–activity relationship (SAR) studies on these antibiotics, are described. Special attention will be paid to very active research on the biological evaluation of taxol analogues and new naturally occurring taxoids.

2.2 β-Amino Acids

2.2.1 β-Amino Acids and Diamino Acids

The glycosyl dipeptide amipurimycin (**1**)[23] presented activity against the causative agent for rice blast disease *Pyricularia oryzae*.[24]

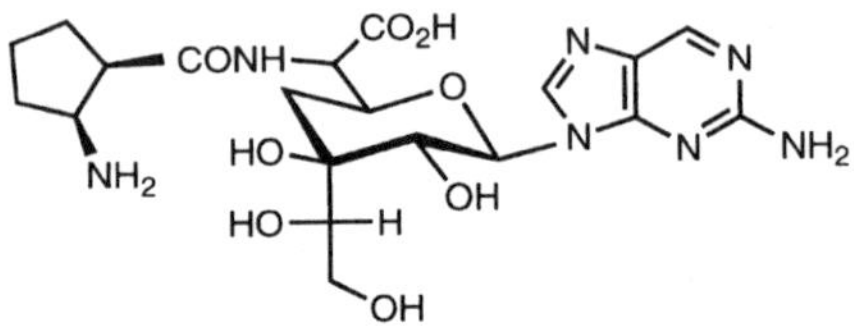

1

Inhibitory activity of long-chain fatty acid oxidation in rat liver mitochondria by emeriamine (**2**) has been observed. SAR was studied with N-linear-long-chain alkane amides. Compounds **3** and **4** exhibited stronger inhibition of fatty acid oxidation than **2**.[25] Hypoglycemic and antiketogenic activities were determined in fasting rats after oral administration of **2**.[25,26]

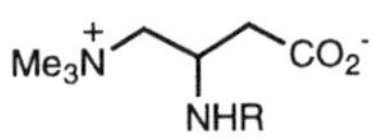

2, R = H
3, R = $CH_3(CH_2)_{12}CO$
4, R = $CH_3(CH_2)_{14}CO$

(2*S*,3*S*)-2,3-Diaminobutanoic acid is an unusual β-amino acid and a common component of linear heptapeptide antibiotics antrimycins (**5**),[27] which are produced by *Streptomyces (St.) xanthocidicus MG-125-CF1*. The synthesis of this α,β-diamino acid has recently been accomplished.[28]

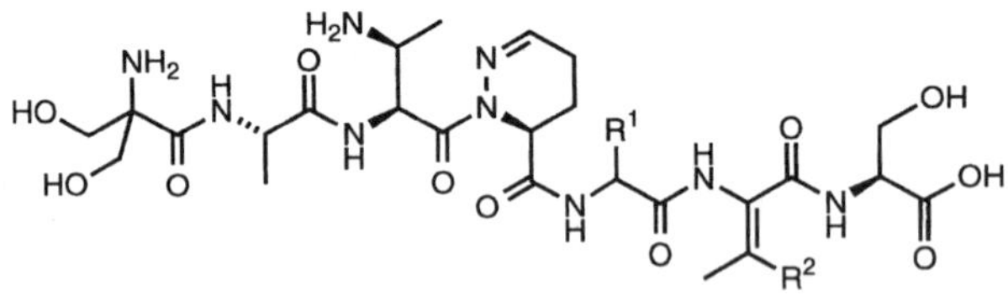

5a, $R^1 = R^2$ = Me
5b, R^1 = Et, R^2 = Me
5c, R^1 = *n*-Pr, R^2 = Me
5d, R^1 = *i*-Bu, R^2 = Me
5e, R^1 = Me, R^2 = Et
5f, $R^1 = R^2$ = Et
5g, R^1 = *n*-Pr, R^2 = Et
5h, R^1 = *i*-Bu, R^2 = Et

2.2.2 β-Amino Acids with HIV-Protease Inhibitor Activity

HIV-protease inhibitors have been identified as promising chemotherapy agents of HIV infections,[29] since an effective inhibition of the viral enzyme gives rise to immature and noninfectious virions.[30] It has been demonstrated that HIV protease is functioning as a C_2 symmetric homodimer.[31] Applying the concept of mimetism of the structure of natural substrate in order to replace the interaction sites of the enzyme, novel peptide-based inhibitors of HIV-1 proteinase[32] have then been designed that present a C_2 symmetry,[33] that is, analogues that resemble the transition state for the natural polyproteic substrate used by the HIV proteinase as the late stage in the replication of HIV. Thus, considering that the human immunodeficiency virus type 1 (HIV-1) encodes an aspartic proteinase for its replication, β-amino acids were incorporated as a central building block in *bis*-symmetric lipophilic side chains (e.g., **6**). These derivatives showed a high inhibition activity of HIV protease.[34]

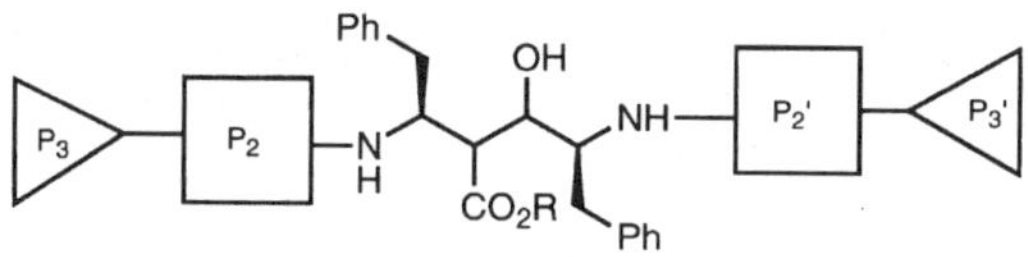

6a, R = H
6b, R = Me

This central building block carrying a peptide modified chain has also been tested for HIV-protease inhibitory activity. However, the diastereomeric compounds **7**, **8**, and **9** only exhibited weak inhibition of HIV-1 protease.[34]

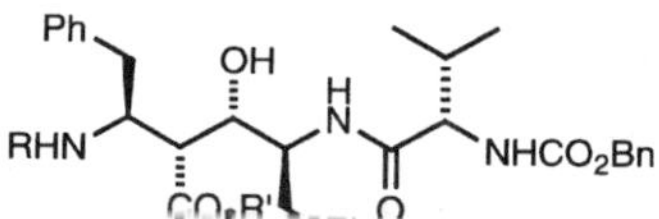

7a, R = Boc, R' = H
8a, R = Boc, R' = Me
9a, R = Cbz-Val, R' = H

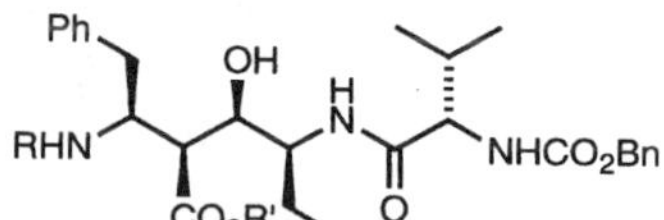

7b, R = Boc, R' = H
8b, R = Boc, R' = Me
9b, R = Cbz-Val, R' = H

In contrast, a series of penicillin-derived C_2 symmetric dimers **10** showed potent enzyme inhibition and activity against the virus in vitro.[35]

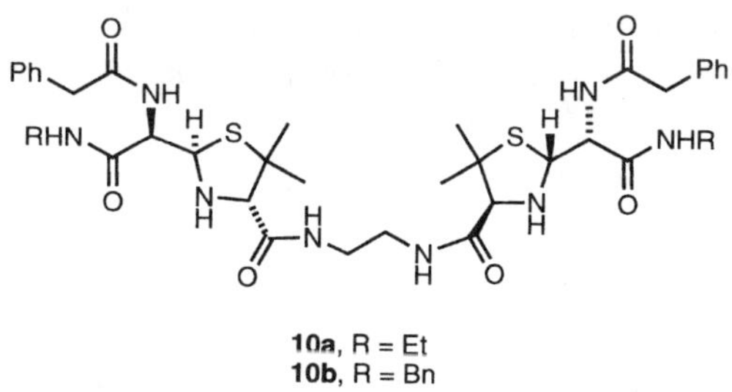

10a, R = Et
10b, R = Bn

With reduced molecular weights, monomers **11a–11d** having potent inhibitor characteristics and antiviral activity were also developed.[36] Computer-assisted molecular modeling of these compounds was based on the X-ray co-crystal structure of **10a** with HIV proteinase.[37] Unfortunately, a poor pharmacokinetic profile of these compounds in dogs was observed.[36b]

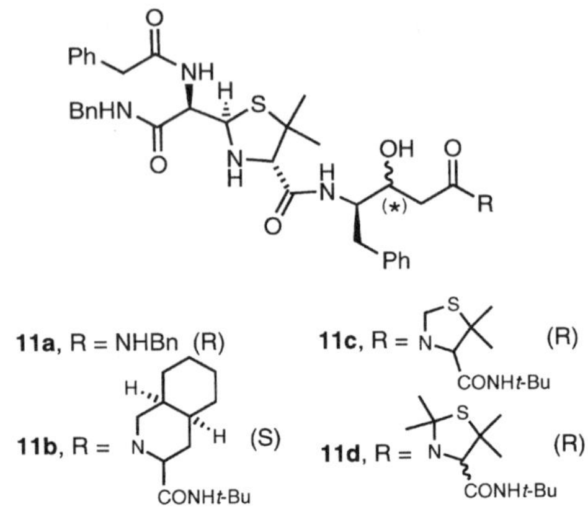

Ro 31-8959/003 (**12**) is another potent and highly selective inhibitor of HIV proteinases in vitro with similar activity against both the HIV-1 and the HIV-2 enzymes.[38]

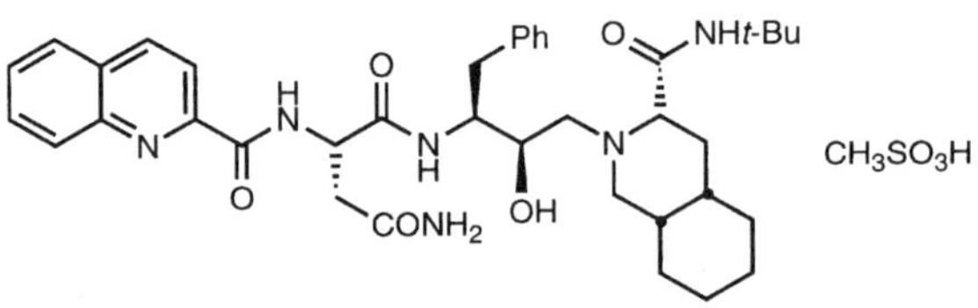

12

2.2.3 β-Amino Acids with Antibiotic Activity

β-Lysine (**13**) is important in both primary and secondary metabolism.[11] (3*S*)-β-Lysine is the first intermediate in the anaerobic catabolism of (2*S*)-α-lysine in *Clostridia*,[39] and it is also widely produced by *Streptomyces* species, and incorporated into a large family of broad spectrum antibiotics,[40] such as streptothricin F (**14**).[12] β-Tyrosine (**15**) is another example of these compounds, which is a component of (±)-jasplakinolide (**16**).[41] The latter is a cyclodepsipeptide isolated from a soft-bodied sponge, *Jaspis* sp., and with potent insecticidal, antifungal, and anthelminthic properties.

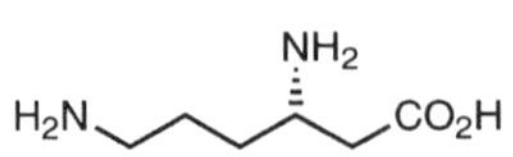

13

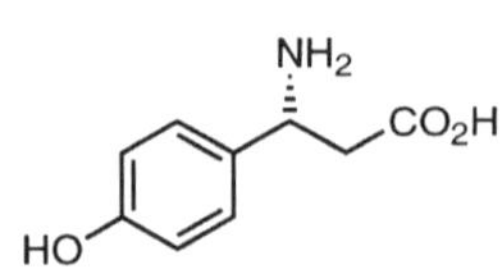

15

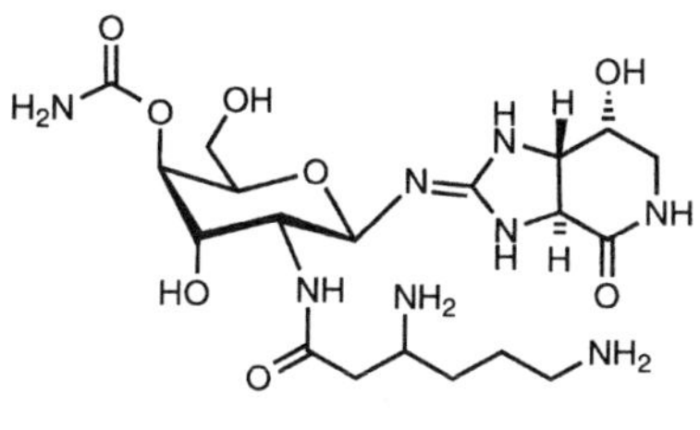

14

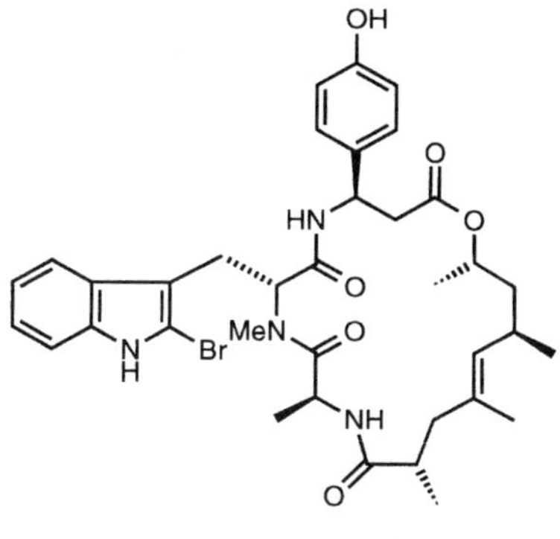

16

Negamycin (**17**) is a structurally unique peptidelike natural product,[42] which exhibits striking activity against Gram-negative bacteria, including *Pseudomonas aeruginosa*.[43]

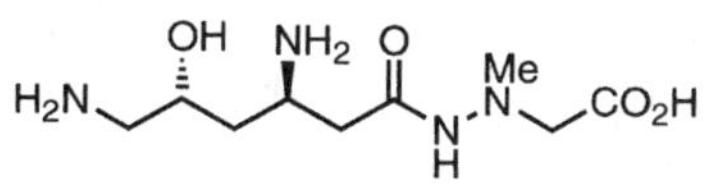

17

2.2.4 β-Amino-α-Hydroxy Acids

2.2.4.1 Peptides

β-Amino-α-hydroxy acids have acquired great interest because of their presence in peptides with enzyme-inhibition activity.[44] Bestatin (**18**) and amastatin (**19**) are two important members of this class of compounds with immunoregulatory,[45] antitumor, and antimicrobial properties.[46] Interestingly, the latter contains a second β-amino acid with the presence of terminal aspartic acid.

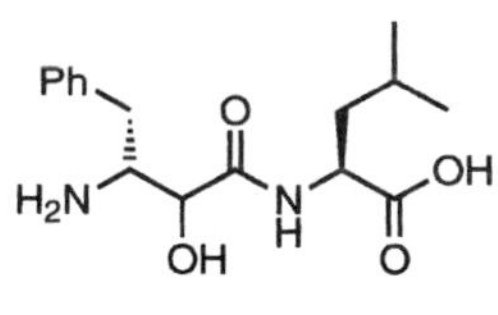

18

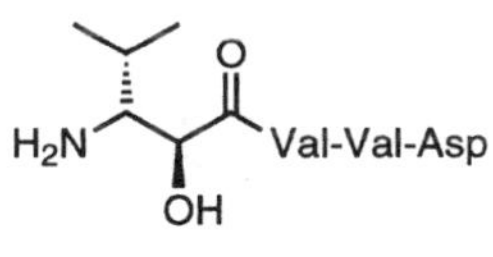

19

Renin is an enzyme that plays a central role in the regulation of blood pressure, since it catalyzes the cleavage of the hormone angiotensinogen during formation of the potent vasoconstrictor angiotensin II.[47] Inhibition of angiotensin-converting enzyme (ACE) is an effective therapy for the majority of hypertensive patients and represents a major advance in the treatment of hypertension and congestive heart failure. β-Amino-α-hydroxy-peptides KR1 1314 (**20a**) and KR1-1230 (**20b**),[48] among other renin inhibitors used as transition-state mimics, have shown a high renin inhibition.

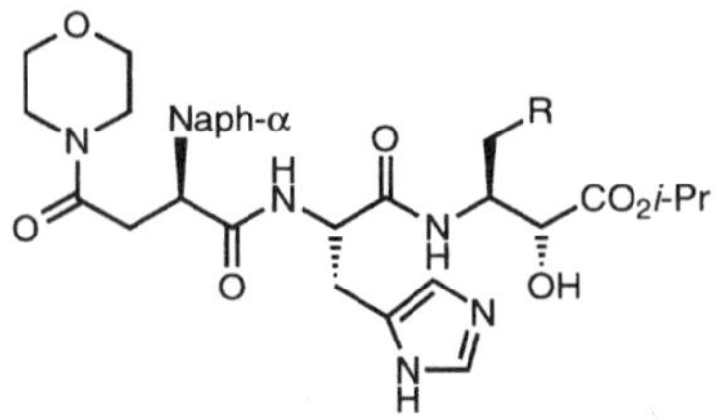

20a, R = cyclohexyl
20b, R = isopropyl

2.2.4.2 Taxoids

The clinical efficacy of Taxol® (paclitaxel) (**21a**), a natural product isolated from the Western yew, *Taxus brevifolia*,[15] in advanced cancer chemotherapy,[49] and its potent analogue Taxotere® (docetaxel) (**21b**),[50] have been studied extensively in recent years. Taxol® has been approved in clinical usage for the treatment of breast and ovarian cancers[51] and is also demonstrating promising antitumor activity against lung, skin, and head and neck cancers,[52] with minimal acute side effects. Their potent anticancer properties have prompted considerable effort to prepare them by total[53] and partial[54] syntheses.

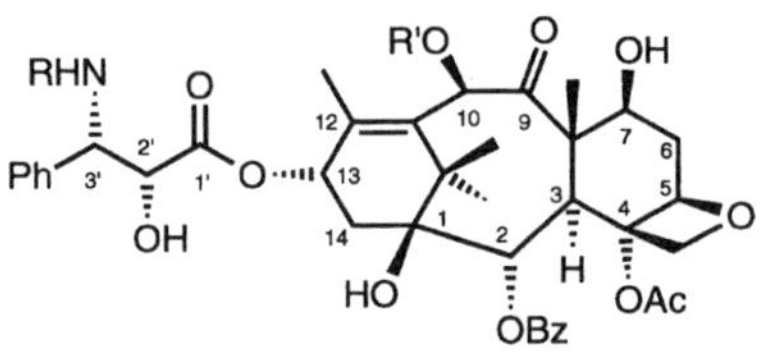

21a, R = COPh, R' = Ac
21b, R = CO_2t-Bu, R' = H

Taxol comprises a diterpene moiety and by a C-13 N-benzoyl phenylisoserine side chain. Biosynthetic studies[55] have shown that baccatin-III (**22**) is the precursor of the diterpenoid skeleton, while the β-amino acid side chain (**23**) is formed from phenylalanine (**24**) via β-phenylalanine (**25**).

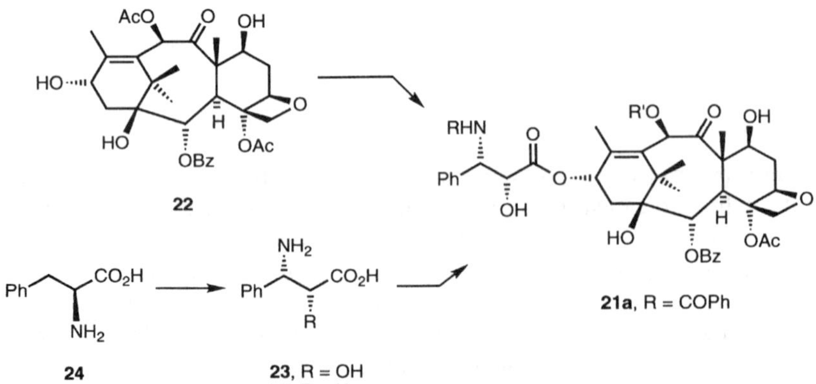

Compounds **21a** and **21b** are, however, inactive against certain tumor types, such as colon and renal cell carcinoma, and they have a number of undesirable side effects in high doses.[56] Therefore, an intensive effort has been undertaken to develop new anticancer analogues of **21a**, through chemical studies for elucidating the SAR of Taxol.[17b,54,57] It has been shown that the C-13 phenylisoserine side chain, as well as the diterpene part, are of importance for biological activity.[15] Studies centered on the side chain have demonstrated that any modification or deletion of the 3′-amino group and the 2′-hydroxyl group can alter cytotoxicity.[58] For instance, cytotoxic activity in vivo was enhanced against the adriamycin-resistant human breast cancer cell line (MCF7-R), human ovarian (A121), human nonsmall lung cell (A549), and human colon (HT-29) cancers in comparison with those of **21a** and **21b**, with 3′-alkyl and 3′-alkenyl analogues of **21b**, especially the 3′-isobutenyl **26a** and 3′-crotyl **26b** derivatives.[59]

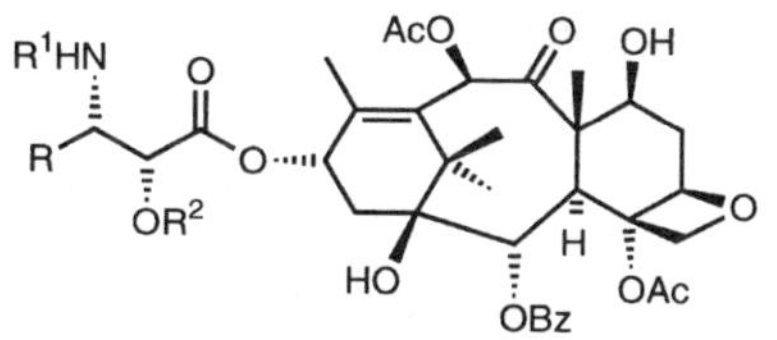

26a, R = CH=C(Me_2), R^1 = CO_2*t*-Bu, R^2 = H
26b, R = (E)-CH=CHMe, R^1 = CO_2*t*-Bu, R^2 = H
26c, R = Ph, R^1 = COPh, R^2 = $COCH_2OPEG$
26d, R = Ph, R^1 = COPh, R^2 = $CO(CH_2)_2CONHPEG$
26e, R = Ph, R^1 = COPh, R^2 = $COCH_2NHCO_2PEG$

Water-soluble prodrugs of **21a** have been the object of recent investigations[60] with successful side-chain functionalization. Thus the 2′-polyethylene glycol (PEG) esters of Taxol **26c–26e** were found to be essentially water soluble, and with short enough half-lives and similar in vitro activity as Taxol (**21a**) to be considered useful prodrugs.[61] Onium salts linked to the C-2′ hydroxy group of **21a** have been evaluated in cytotoxic and microtubule binding assays as well as in vivo studies.[62] Pyridinium salts

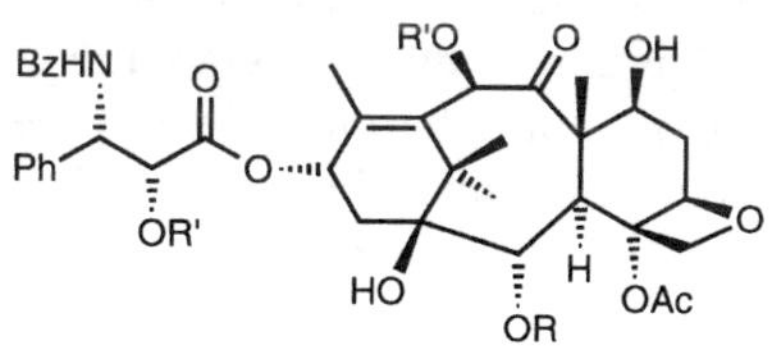

27a, R = COPh; R' = 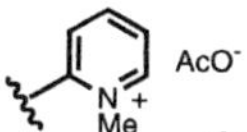

27b, R = CO 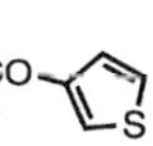

; R' = 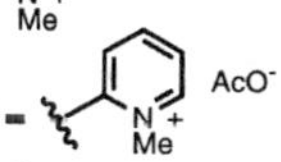

27c, R = COPh; R' = H

27d, R = CO 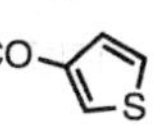

; R' = H

27a and **27b** gave the expected cytotoxicity profile, functioning as prodrugs by liberating their precusor **27c** and **27d**, respectively.

2.3 β-Lactams

2.3.1 Monocyclic β-Lactams

Monobactams or sulfazecins[9b,9c] are structurally characterized by a monocyclic β-lactam or 2-azetidinone ring with an SO_3^- group bonded to the ring nitrogen.[1d,63] The relatively unstrained β-lactam confers to these compounds a considerable bacteria β-lactamase stability. Among the clinical use monobactams, azthreonam (**28a**) and carumonam (**28b**) and derivatives **28c**–**28d** possess such high stability and a specific activity against Gram-negative bacteria.[64]

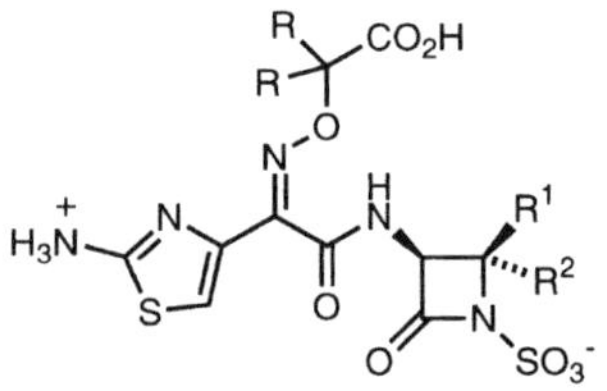

28a, R = Me; R^1 = H; R^2 = Me
28b, R = R^2 = H; R^1 = CH_2OCONH_2
28c, R = R^1 = R^2 = H
28d, R = Me; R^1 = R^2 = H

Nocardicins[65] are also monocyclic β-lactams with very high antibiotic activity, particularly against Gram-negative microorganisms.[66] Nocardicin A (**29a**) has proven to have high and broad-spectrum antimicrobial activity.[66] It appears that the phenolic OH group plays an important role in the biological activity of nocardicins, since the dehydroxy derivative **29b** did not exhibit any biological activity.[67] In vivo oxidation to quinone methide metabolite **29c** has been proposed as a possible mechanism for explaining the biological properties of nocardicins. This has also been suggested by the antimicrobial activity in vitro of masked *p*-quinone methide β-lactam **29d**.

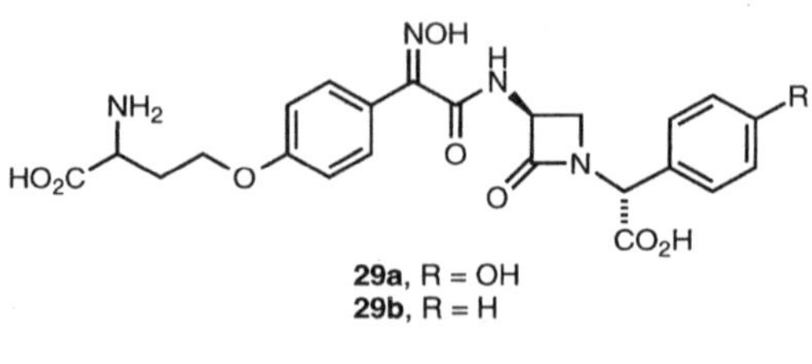

29a, R = OH
29b, R = H

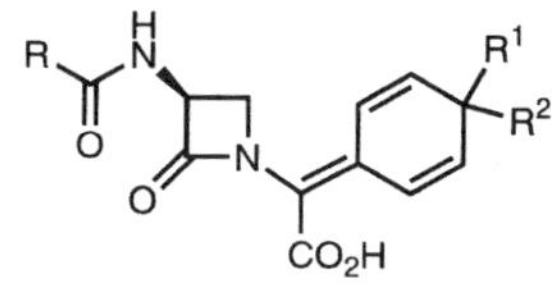

29c, R = $C(NOH)C_6H_4$*p*-$O(CH_2)_2CH(NH_2)CO_2H$; R^1, R^2 = O
29d, R = CH_2Ph; $R^1 = R^2$ = OMe

Levogyric 2-azetidinone **30** has been synthetized[68] among a series of cholesterol intestinal absorption inhibitors, showing a satisfactory inhibition of acyl-CoA: cholesterol acyltransferase (ACAT), the major enzyme associated with cholesterol esterification.[69]

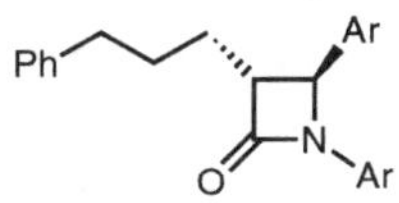

30, Ar = C_6H_4*p*-OMe

2.3.2 Bicyclic β-Lactams

2.3.2.1 Carbapenems

A countless number of penicillin derivatives, **31**, have been prepared and tested in antibiotic assays.[2a,70] Because of the rapid increase of penicillin-resistant strains of bacteria,[71] a variety of new β-lactam ring systems have been discovered and created in order to improve the effectiveness of these antimicrobial agents. Penems, **32**, and carbapenems, **33**, have been among the most widely studied clinically and chemically and they have demonstrated a high biological activity.[2b,72]

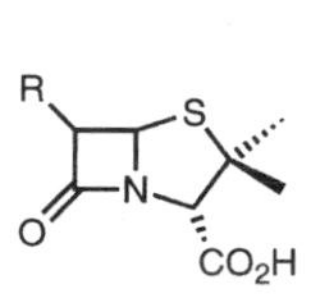

31

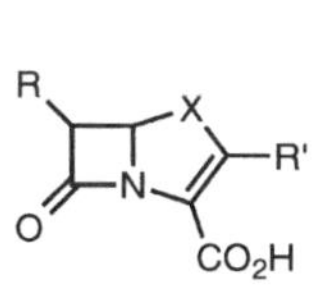

32, X = S
33a, X = CH_2
33b, X = CH-β-(CH_3)

Thienamycin (**34a**) and analogues[73] such as PS-5 (**34b**) and **34c** are very potent antibacterial compounds, among the most representative carbapenems. They show a broad antibiotic activity spectrum against Gram-positive and Gram-negative bacteria and are stable to β-lactamases. Nevertheless, a very limited pH range tolerance and accelerated inactivation encountered in concentrated solutions of the antibiotic restricts the use of **34a** as a practical therapeutic.[74] It has been found that the stability against dehydropeptidase I (DHP-I) can be improved with the introduction of a 1β-methyl group into C-1 of the carbapenem skeleton, as exemplified by meropenem (SM-7338) (**35a**)[75] and biapenem (L-627) (**35b**).[76]

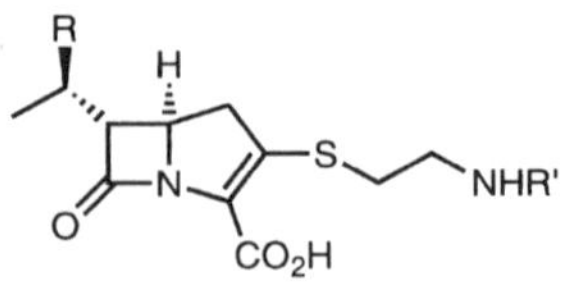

34a, R = OH, R' = H
34b, R = H, R' = Ac
34c, X = OH, R' = CH=NH

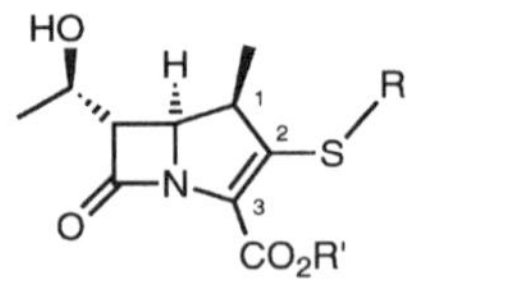

35a, R = 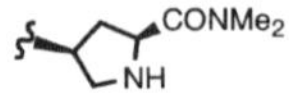

, R' = H

35b, R = 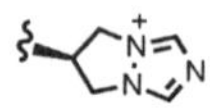

, R' = (-)

β-Lactam antibiotics act by acylating enzymes, which are essential in the synthesis of the bacteria cell wall[77]; such enzymes are named penicillin-binding proteins (PBPs). These enzymes are the DD-transpeptidases,[78] which catalyze the transpeptidation process in the biosynthesis of the peptidoglycan, a cross-linked polymer that encloses the cell.[79] The loss of activity of β-lactams against bacteria that have become resistant may be caused by a low affinity of these enzymes to β-lactams,[80] as is the case of pathogen bacteria methicillin-resistant *Staphylococcus aureus* (MRSA)[21b]; thereby, only a limited number of chemotherapeutic agents are available for treatment of MRSA infection.

Recently, Sunagawa and colleagues[81] have described the preparation of new 1β-methyl carbapenems **36a–36e**, having a thiazol-2′-ylthio side chain on C-2, which exhibit good anti-MRSA activity and maintain stability against DHP-I. The introduction of a phenyl group on C-4′ (**36b**) seemed to be more effective than alkyl substitution. Aryl substituents on C-4′ did not affect the anti-MRSA activity (**36c**). However, all these compounds have a binding affinity to human serum albumin (HSA) too high for therapeutic use. Cationic carbapenems such as **36d** and **36e** decreased HSA bindings to less than 60%, and they displayed improved DHP-I stability and higher activity, not only against MRSA but also against Gram-negative bacteria.

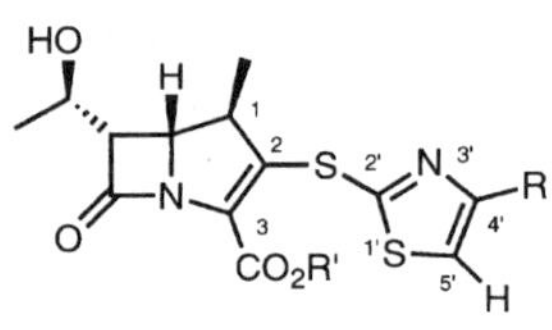

36a, R = R = H
36b, R = Ph, R' = H
36c, R = Ar, R' = H

36d, R = 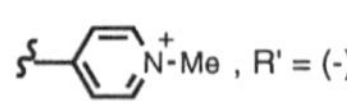

, R' = (-)

36e, R = 

, R' = (-)

2.3.2.2 Cephems, Carbacephems, and 2-Oxaisocephems

Cephalosporins (**37a**) are the naturally occurring compounds of the cephem class of antibiotics and are isolated originally from the fungus *Cephalosporium acremonium*.[82] They have been widely used to treat infections by bacteria that are resistant to penicillins,[83] including cefuzonam (**37b**) and cefmenoxime (**37c**), which are members of the third-generation antibiotics. The carbacephems (**37d**) are the carba-dethio analogues of the cephems and are obtained by synthesis.[84] They have proven to be as effective as the cephems and are associated with an enhanced physiological half-life, as in the case of the clinically used 3-chlorocarbacephem lorabid (**37e**), which possesses a spectrum of biological activity similar to the oral antibiotic cephem analogue cefaclor (**37f**), but is superior in chemical stability. This trend is confirmed in the cephalosporin and carbacephalosporin pairs **37g/37h**, **37i/37j**, and **37k/37l**.[85]

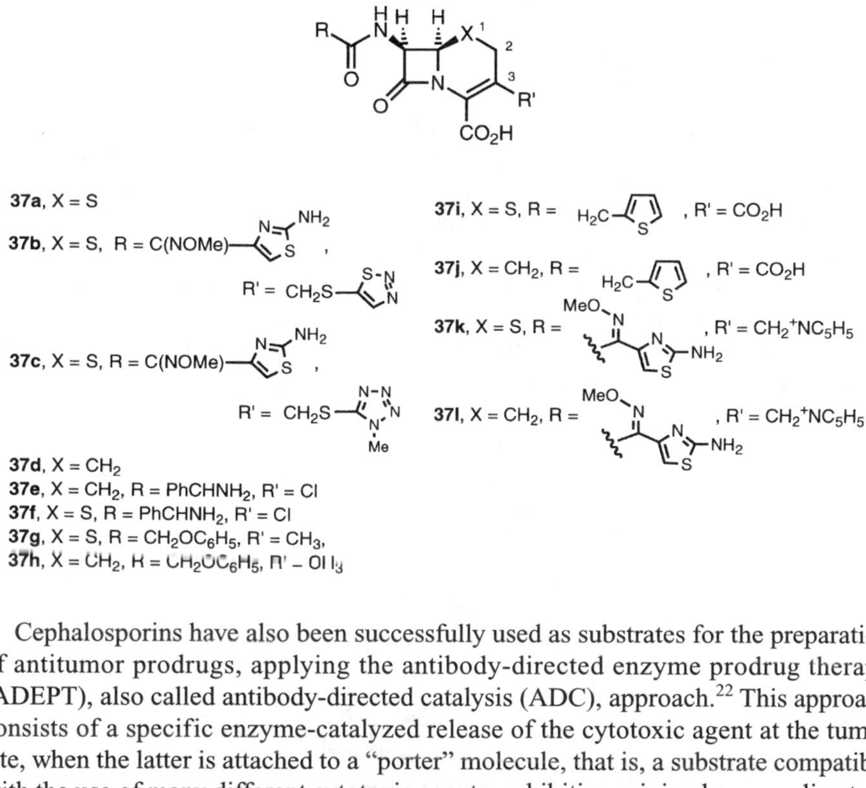

Cephalosporins have also been successfully used as substrates for the preparation of antitumor prodrugs, applying the antibody-directed enzyme prodrug therapy (ADEPT), also called antibody-directed catalysis (ADC), approach.[22] This approach consists of a specific enzyme-catalyzed release of the cytotoxic agent at the tumor site, when the latter is attached to a "porter" molecule, that is, a substrate compatible with the use of many different cytotoxic agents, exhibiting minimal mammalian toxicity and a high affinity for the enzyme. The enzyme of interest is covalently bound to a monoclonal antibody (mAb) that is localized on the targeted tumor cell surface. Cephalosporins have been used as the "porter" substrate, conjugated with the cytotoxic agent, and this has provided a prodrug with significantly reduced cytotoxicity relative to the free drug. The ADC system utilizes β-lactamases (β-Lac) as the releasing enzyme, because they are the bacteria's natural defense mechanism against β-lactam antibiotics. Thus the prodrugs **38a** and **38b** were efficient in releasing the free cytotoxic agents methotrexate (**39**) and 5-fluorouracil (**40**), respectively, when they were treated with β-lactamase, indicating that prodrugs are good substrates for the enzyme. However, the cytotoxicity of prodrugs **38a** and **38b** in vitro was not significantly different from that of free **39** and **40**.

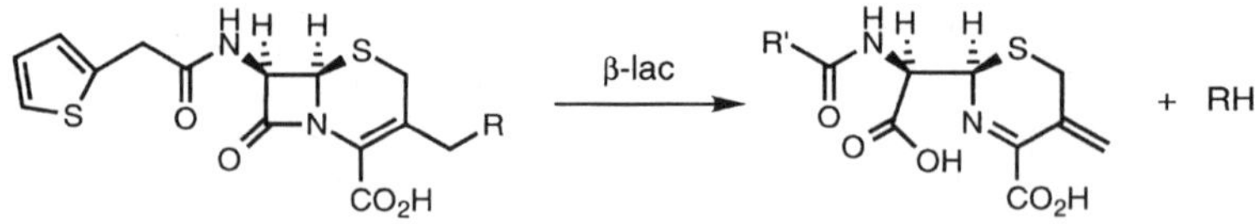

39

40

38a, R =

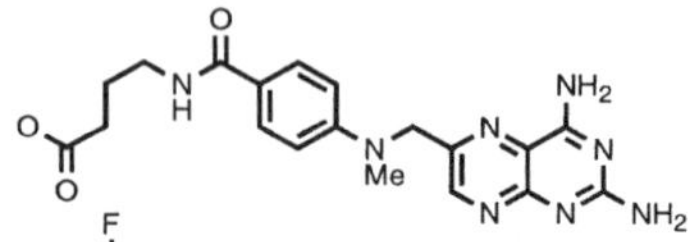

38b, R =

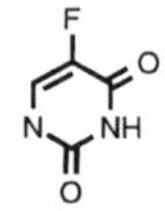

More promising prodrugs for activation by mAb-β-Lac were the doxorubicin (**41**) derivatives **42a**,[86] **42b**,[87] and **42c**–**42g**.[88] The combination of **42a** and **42b** with mAb-β-Lac conjugate resulted in a rapid releasing of **41**. In vivo experiments utilizing the prodrug **42a** demonstrated that this is able to achieve antigen-mediated tumor suppression against LS174T colon and OVCAR-3 human ovarian tumor xenografts. The prodrugs **42b**, **42c**, and **42g** showed similar cytotoxicity to **41**, when they were tested in the presence of *Enterobacter cloacae* P99 β-lactamase (EClbL) and on the human lung adenocarcinoma cell line H2981. These results establish that the side chain has an influence on prodrug cytotoxicity.[88]

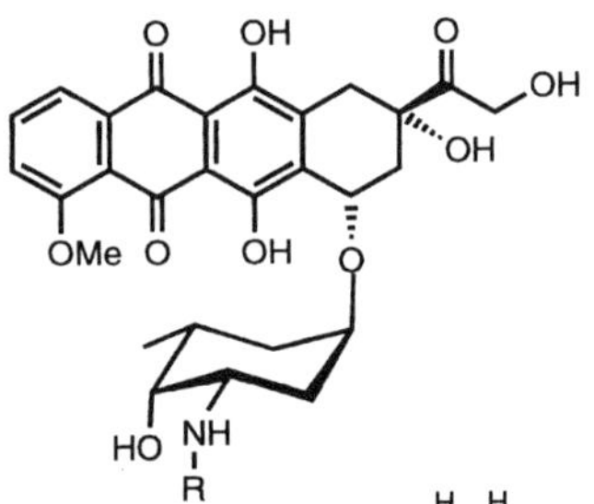

41, R = H

42, R =

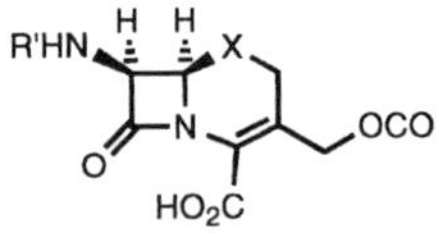

42a, X = S^{+}-O^{-}, R' =

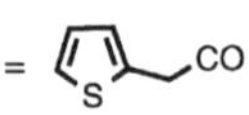

42b, X = S, R' = $PhCH_2CO$

42c, X = S, R' = H

42d, X = S, R' = $HO_2C(CH_2)_3CO$

42e, X = S, R' = $PhCH(SO_3H)CO$

42f, X = S, R' =

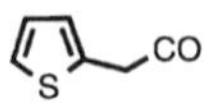

42g, X = S, R' =

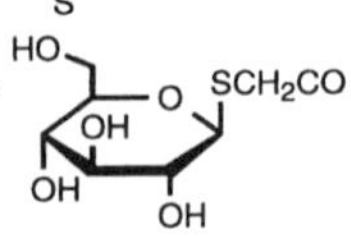

Among other successful cephalosporin-based ADC systems, prodrug **43** was also found to be a good substrate for the *B. cereus* enzyme, and its cytotoxic activity on H2981 human lung adenocarcinoma cell line was similar to that of free drug **44**.[89]

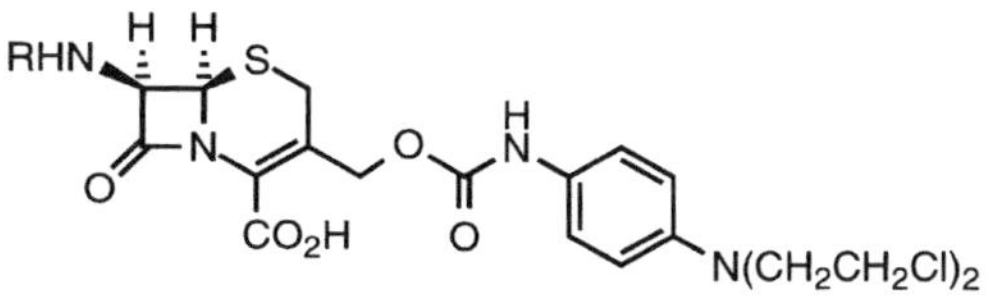

43, R = OC 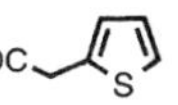

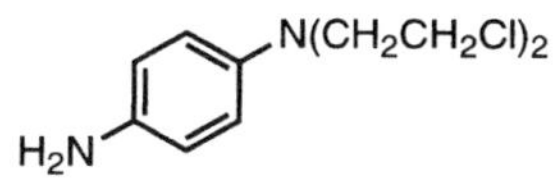

44

Carbacephems have also been evaluated as potential MRSA antibiotics. Compounds **45**, bearing various thiazolethio moieties, exhibit both high potency against MRSA and high affinity to PBP-2′.[90] In spite of the excellent in vitro activities, the high binding affinity to HSA seems to limit **45a** and **45b** in their therapeutic efficacy.

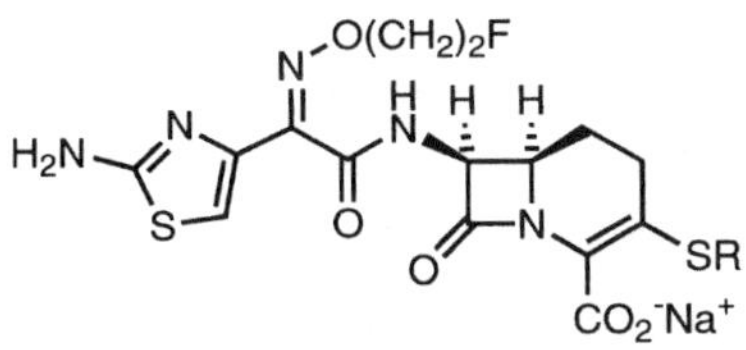

45a, R = 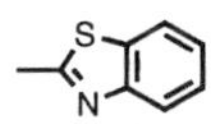

45b, R = 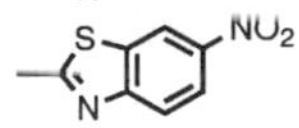

45c, R = 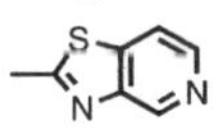

45d, R = 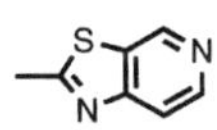

45e, R = 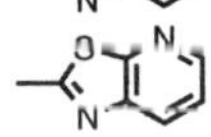

However, compounds **45c** and **45d** did show good MRSA activity with lowered serum binding. In particular, the in vitro microbiological evaluation of **45c** indicated that it is a very potent compound against several strains of MRSA, and exhibited good affinity for the PBP2a enzyme. Moreover, the pharmacokinetic profile of **45c** in monkeys compared well with those obtained for ceftriaxone (**46a**) and cefoperazone (**46b**).

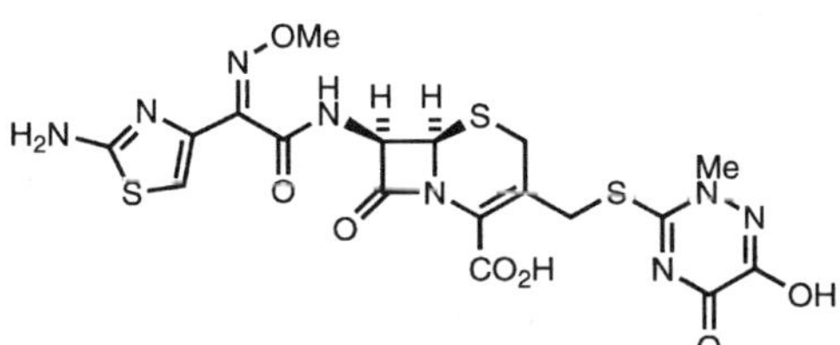

46a

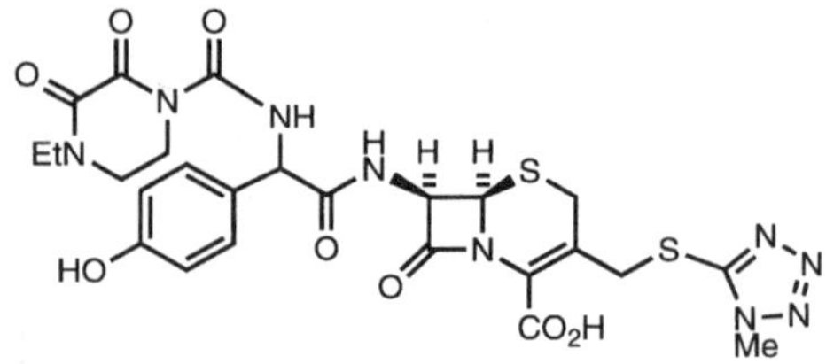

46b

The synthesis and biological evaluation of carbacephems with thienamycin (**34a**)-type characteristics have recently been described by Hanessian.[91] Novel hybrid compound **47a** showed only weak activity against *E. faecium* and against sensitive and resistant strains of *S. aureus*. Analogues **47b** and **47c** were rapidly hydrolyzed in aqueous solution and in pH 7 phosphate buffer; however, **47c** showed good stability against the metallo-β-lactamases CcrA and L1 (*X. maltophilia*). The hydrolysis stability was improved by preparing 3-methyl analogue **47d**, which was, however, devoid of antibacterial activity.

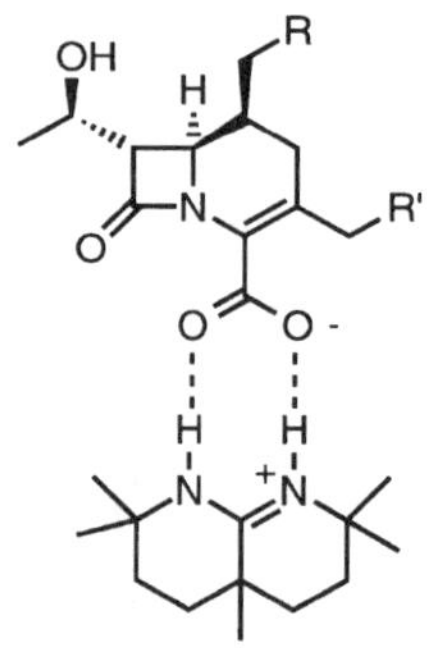

47a, R = OH, R' = OAc
47b, R = H, R' = OAc
47c, R = NHAc, R' = OAc
47d, R = OH, R' = H

Even though β-lactamase inhibitors do not often possess antibiotic activity, they can be helpful as active antibiotics that do not present β-lactamase resistance. These synergistic mixtures are successfully used as favorable clinical therapeutic forms.

The uncommon 7-vinylidene cephems **48** have been described as more potent inhibitors against the β-lactamase derived form *E. cloacae* P99, when compared to clavulanic acid (**49**).[92] Prominent activity of derivative **48b** was observed, which is *ca.*

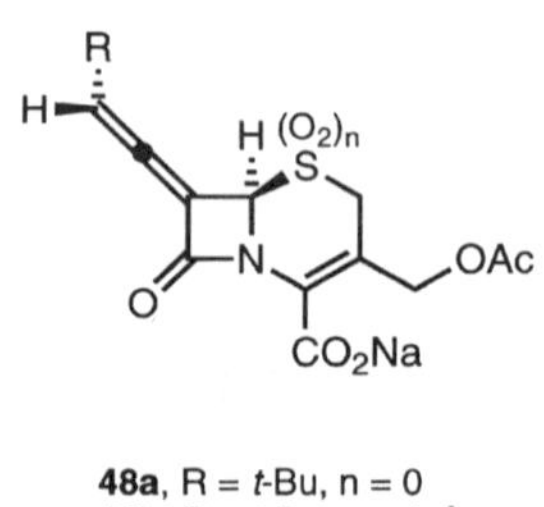

48a, R = *t*-Bu, n = 0
48b, R = *t*-Bu, n = 1
48c, R = H, n = 1

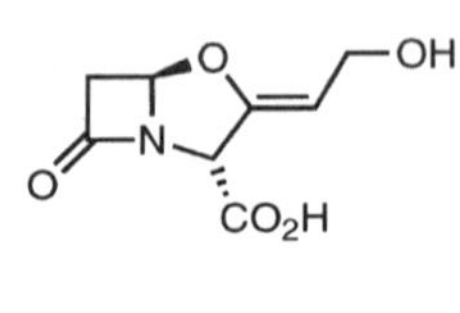

49

350-fold higher than **49** and 5-fold more effective than tazobactam. The role of the allene seems to increase the acidity of the terminal allenyl proton, leading to highly stabilized acyl enzyme enyne toward hydrolysis.

Flomoxef (**50**)[93] is one of the most efficient third-generation antibiotics, which is structurally related to cephems by a similar bicyclic β-lactam core, but with different the heteroatom. The 2-oxaisocephems **51** present the oxa group at the vicinal position of the cyclohexyl ring without losing their antibiotic properties.[94] Very recently, Tsubouchi et al.[95] have prepared a series of optically active 3,7-disubstituted 2-oxaisocephems **52** and have established their SAR. Among the most biologically active derivatives, **52a–52d** showed in vitro and in vivo activities against Gram-positive and Gram-negative bacteria. Compounds **52b–52d** show enhanced antibacterial activity against MRSA and *Enterococcus faecalis* in comparison with cefuzonam (**37b**) and **50**. Moreover, **52b** exhibited more potent activity than cefmenoxime (**37c**) against the systemic infection with *S. aureus* Smith in mice. This suggests an improved effect of the combination of the C-3 and C-7 substituted analogues of these antibiotics with the 2-oxaisocephem moiety.

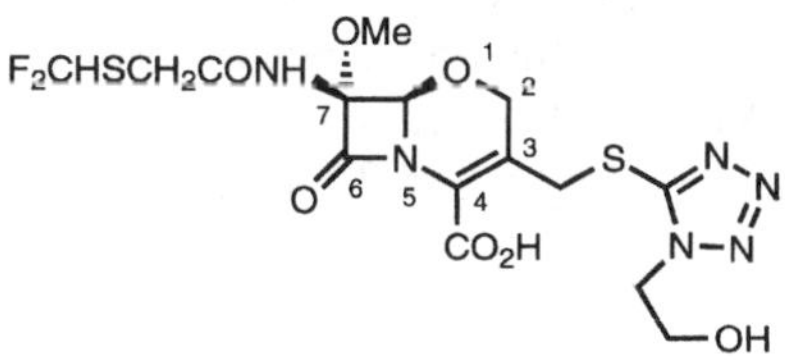

50

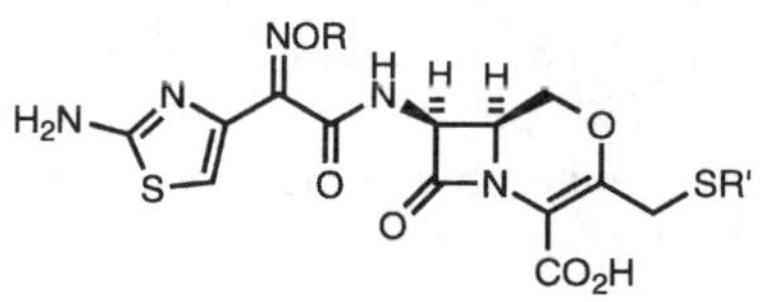

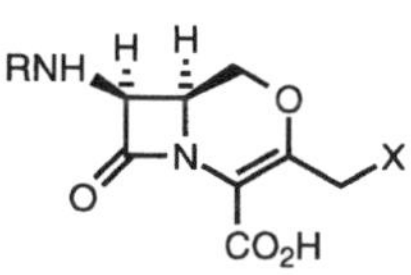

51, X = SR', OR', H, Alkyl, vinyl

52a, R = Me, R' = 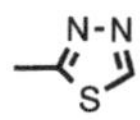

52b, R = cyclopentyl, R' = 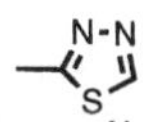

52c, R = cyclopentyl, R' = 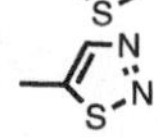

52d, R = cyclopentyl, R' = 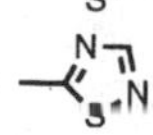

2.3.3 Tricyclic β-lactams

Tricyclic carbapenems, the tribactams **53**, are a new family of synthetic β-lactam antibiotics recently discovered[96] with promising clinical usage. On the other hand, Hanessian and Reddy[97] were able to prepare tricyclic β-lactams incorporating a carbacephem structure. Derivatives **54a** and **54b** were, however, inactive against selected bacterial strains. This may be due to the absence of an appropriate leaving group at C-3, necessary to carry out the acylation process with PBP of the bacteria.

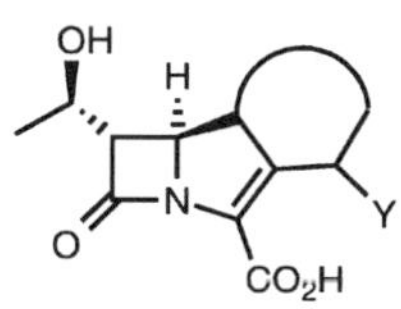

53

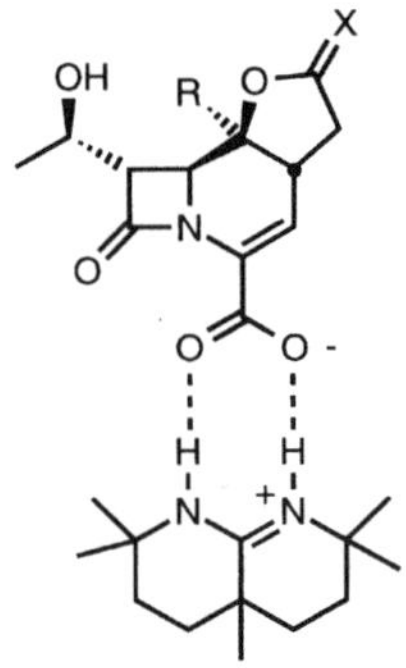

54a, X = H, H, R' = H
54b, X = O, R' = Me

The tricyclic β-lactam antibiotic GV104326 (**55**) is the first member of the tribactam series to be extensively evaluated in vitro. It is a highly potent agent with a broad antibacterial spectrum that encompasses a wide range of Gram-negative, Gram-positive, and anaerobic pathogenic bacteria. It is also resistant to β-lactamases and to renal dehydropeptidases.[98]

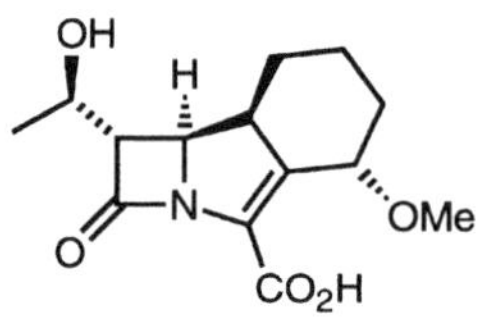

55

Furo[3,4-c]cepham sulfones **56a** and **56b** were readily obtained from keto sulfones **57a** and **57b**.[99] The latter were studied as potent inhibitors of human leukocyte elastase (HLE),[100] a serine endopeptidase believed to be involved in the pathogenesis of pulmonary emphysema and other connective tissue diseases, with moderate hydrolytic stability.[101] The cyclization of **57** to tricyclic cephems **56** improved both stability of the β-lactam towards hydrolytic cleavage at physiological pH (7.4) and HLE-inhibitory activity.[99]

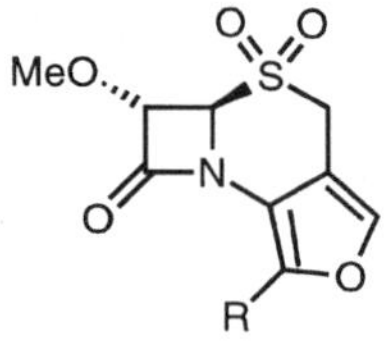

56a, R = Ph
56b, R = *t*-Bu

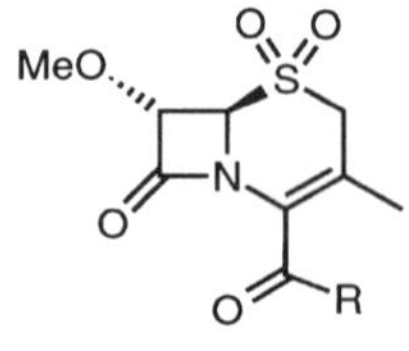

57a, R = Ph
57b, R = *t*-Bu

2.4 References

1. (a) G. I. Georg, Ed., *The Organic Chemistry of β-Lactams* (VCH, New York, 1993).
   (b) N. H. Georgopapadakou, in *Annual Reports in Medicinal Chemistry*, J. A. Bristol, Ed. (Academic Press, San Diego, 1993), Vol. 28, p. 119.
   (c) A. G. Brown, S. M. Roberts, Eds., *Recent Advances in the Chemistry of β-Lactam Antibiotics* (The Royal Society of Chemistry, Burlington House, Cambridge, 1984).
   (d) W. Dürckheimer, J. Blumbach, R. Lattrell, K. H. Scheunemann, *Angew. Chem., Int. Ed. Engl.* **24**, 180 (1985).
   (e) B. G. Spratt, *J. Gen. Microbiol.* **129**, 1247 (1983).
   (f) R. B. Morin, M. Gorman, Eds., *Chemistry and Biology of β-Lactam Antibiotic* (Academic Press, New York, 1982), Vols. 1–3.
   (g) A. G. Brown, *Pure Appl. Chem.* **59**, 475 (1987).

2. (a) P. H. Bentley, R. Ponsford, Eds., *Recent Advances in the Chemistry of Anti-Infective Agents* (The Royal Society of Chemistry, Cambridge, 1993).
   (b) R. N. Guthikonda, B. G. Christensen, *Pure Appl. Chem.* **59**, 455 (1987).

3. M. Narisada, *Pure Appl. Chem.* **59**, 459 (1987).

4. (a) D. C. Hodgkin, E. N. Maslen, *J. Biochem.* **79**, 393 (1961).
   (b) G. G. F. Newton, E. P. Abraham, *Nature* **175**, 548 (1955).
   (c) E. P. Abraham, G. G. F. Newton, *J. Biochem.* **79**, 377 (1961).

5. M. Narisada, T. Yoshida, H. Onoue, M. Ohtani, T. Okada, T. Tsuji, I. Kikkana, N. Haga, H. Satoh, H. Itani, W. Nagata, *J. Med. Chem.* **22**, 757 (1979).

6. (a) I. Ernest, J. Gosteli, R. B. Woodward, *J. Am. Chem. Soc.* **101**, 6301 (1979).
   (b) H. R. Pfaendler, J. Gosteli, R. B. Woodward, *J. Am. Chem. Soc.* **102**, 2039 (1980), and references cited therein.
   (c) G. Frenceschi, M. Alpegiani, C. Battistini, A. Bedeschi, E. Perrone, F. Zarini, *Pure Appl. Chem.* **59**, 467 (1987).

7. M. S. Manhas, J. S. Chib, Y. H. Chiang, A. K. Bose, *Tetrahedron* **25**, 4421 (1969).

8. (a) A. G. Brown, *J. Antimicrob. Chemother.* **7**, 15 (1981).
   (b) P. H. Bentley, G. Brooks, M. L. Gilpin, E. Hunt, *J. Chem. Soc., Chem. Commun.* 905 (1977).
   (c) E. Hunt, P. H. Bentley, G. Brooks, M. L. Gilpin, *J. Chem. Soc., Chem. Commun.* 906 (1977).
   (d) A. G. Brown, D. Butterworth, M. Cole, G. Hanscomb, J. D. Hood, C. Reading, *J. Antibiotics* **29**, 668 (1976).

9. (a) N. S. Isaacs, G. Sunman, C. Reading, *J. Antibiotics* **35**, 589 (1982).
   (b) A. Imada, K. Kitano, K. Kintaka, M. Murio, M. Asai, *Nature* **289**, 590 (1981).
   (c) R. B. Sykes, C. M. Cimarusti, D. P. Bonner, K. Bush, D. M. Floyd, N. H. Georgopapadakou, W. H. Koster, W. C. Liu, W. L. Parker, P. A. Principe, M. L. Rathnum, W. A. Slusarchyk, W. H. Trejo, J. S. Wells, *Nature* **291**, 489 (1981).
   (d) T. Kamiya, M. Hashimoto, O. Nakaguchi, T. Oku, *Tetrahedron* **35**, 323 (1979).

10. (a) J. G. Phillips, D. Chu, S. Spanton, R. Henry, J. J. Plattner, *Tetrahedron Lett.* **33**, 3733 (1992).
    (b) T. Wollmann, U. Gerlach, R. Horlein, N. Krass, R. Lattrell, M. Limbert, A. Markus, in *Recent Advances in the Chemistry of Anti-Infective Agents*; P. H. Bentley, R. Ponsford, Eds. (The Royal Society of Chemistry, Cambridge, 1993), pp. 50–66.

11. (a) O. W. Griffiths, *Ann. Rev. Biochem.* **55**, 855 (1986).
    (b) C. N. C. Drey, *The Chemistry and Biochemistry of β-Amino Acids*, in *Chemistry and Biochemistry of Amino Acids, Peptides and Proteins* (Marcel Dekker, New York, 1972), Vol. 4.

12. T. K. Thiruvengadam, S. J. Gould, D. J. Aberhart, H.-J. Lin, *J. Am. Chem. Soc.* **105**, 5470 (1983).

13. (a) T. M. Zabriskie, J. A. Klocke, C. M. Ireland, A. H. Marcus, T. F. Molinski, D. J. Faulkner, C. Xu, J. C. Clardy, *J. Am. Chem. Soc.* **108**, 3123 (1986).
(b) P. Crews, L. Y. Manes, M. Boehler, *Tetrahedron Lett.* **27**, 2797 (1986).
(c) H. Suda, T. Takita, T. Aoyagi, H. Umezawa, *J. Antibiotics* **29**, 100 (1976).

14. C. W. Jefford, Z.-H. Lu, J. B. Wang, *Pure Appl. Chem.* **66**, 2075 (1994).

15. M. C. Wani, H. L. Taylor, M. E. Wall, P. Coggon, A. T. McPhail, *J. Am. Chem. Soc.* **93**, 2325 (1971).

16. D. Guénard, F. Guéritte-Voegelein, P. Potier, *Acc. Chem. Res.* **26**, 160 (1993).

17. (a) G. I. Georg; T. C. Chen, I. Ojima, D. M. Vyas, Eds., *Taxane Anticancer Agent: Basic Science and Current Status* (American Chemical Society, Washington, D.C., 1995).
(b) M. Suffness, in *Annual Reports in Medicinal Chemistry*; J. A. Bristol, Ed. (Academic Press, San Diego, 1993), Vol. 28, pp. 305–14.
(c) D. G. I. Kingston, *Pharm. Ther.* **52**, 1 (1991).

18. (a) A. Meister, in *Biochemistry of the Amino Acids* (Academic Press, New York, 1965).
(b) P. R. Carnegie, *J. Biochem.* **89**, 459 (1963).

19. J. J. Plattner, in *Annual Reports in Medicinal Chemistry*, J. A. Bristol, Ed. (Academic Press, San Diego, 1994), Vol. 29, pp. 113–22.

20. (a) H. C. Neu, *Science* **257**, 1064 (1992).
(b) D. H.-M. Shih, F, Baker, L. D. Cama, B. G. R. Christensen, *Heterocycles*, **21**, 29 (1984).

21. (a) S. J. Eykyn, *Lancet* **8577**, 100 (1988).
(b) D. L. Saravolatz, D. J. Pohlod, L. M. Arking, *Ann. Intern. Med.* **97**, 325 (1982).

22. L. N. Jungheim, T. A. Sheperd, *Chem. Rev.* **94**, 1553 (1994).

23. S. Harada, T. Kishi, *J. Antibiotics* **30**, 6523 (1977).

24. P. Garner, in *Studies in Natural Products Chemistry*, Vol. 1, Stereoselective Synthesis, Part A; Atta-ur-Rahman, Ed. (Elsevier, Amsterdam, 1988), pp. 397–434.

25. S. Shinagawa, T. Kanamaru, S. Harada, M. Asai, H. Okażaki, *J. Med. Chem.* **30**, 1458 (1987).

26. T. Kanamaru, S. Shinagawa, M. Asai, H. Okazaki, Y. Sugiyama, T. Fujita, H. Iwatsuka, M. Yoneda, *Life Sci.* **37**, 217 (1985).

27. N. Shimada, K. Morimoto, H. Naganawa, T. Takita, M. Hamada, K. Maeda, T. Takeuchi, H. Umezawa, *J. Antibiotics* **34**, 1613 (1981).

28. Y. Nakamura, M. Hirai, K. Tamotsu, Y. Yonezawa, C.-g. Shin, *Bull. Chem. Soc. Jpn.* **68**, 1369 (1995), and references cited therein.

29. J. R. Huff, *J. Med. Chem.* **34**, 2305 (1991).

30. A. H. Kaplan, J. A. Zack, M. Knigge, D. A. Paul, D. J. Kempf, D. W. Norbeck, R. Swanstrom, *J. Virol.* **67**, 4050 (1993).

31. (a) A. Wlodawer, M. Miller, M. Jaskólski, B. K. Sathyanarayana, E. Baldwin, I. T. Weber, L. M. Selk, L. Clawson, J. Schneider, S. B. H . Kent, *Science* **245**, 616 (1989).
(b) M. A. Navia, P. M. D. Fitzgerald, B. M. McKeever, C.-T. Leu, J. C. Heimbach, W. K. Herber, I. S. Sigal, P. L. Darke, J. P. Springer, *Nature* **337**, 615 (1989).

32. (a) G. B. Dreyer, B. W. Metcalf, T. A. Tomaszek, Jr., T. J. Carr, A. C. Chandler III, L. Hyland, S. A. Fakhoury, V. M. Magaard, M. L. Moore, J. E. Strickler, C. Debouck, T. D. Meek, *Proc. Natl. Acad. Sci. USA* **86**, 9752 (1989).
(b) T. J. McQuade, A. G. Tomasselli, L. Liu, V. Karacostas, B. Moss, T. K. Sawyer, R. L. Heinrikson, W. G. Tarpley, *Science* **247**, 454 (1990).

(c) J. P. Vacca, J. P. Guare, S. J. deSolms, W. M. Sanders, E. A. Giuliani, S. D. Young, P. L. Darke, J. Zugay, I. S. Sigal, W. A. Schleif, J. C. Quintero, E. A. Emini, P. S. Anderson, J. R. Huff, *J. Med. Chem.* **34**, 1225 (1991).

33. (a) D. J. Kempf, D. W. Norbeck, L. M. Codacovi, X. C. Wang, W. E. Hohlbrenner, N. E. Wideburg, D. A. Paul, M. F. Knigge, S. Vasavanonda, A. Craig-Kennard, A. Saldivar, W. Rosenbrook, Jr., J. J. Clement, J. J. Plattner, J. Erickson, *J. Med. Chem.* **33**, 2687 (1990).
(b) G. B. Dreyer, J. C. Boehm, B. Chenera, R. L. DesJarlais, A. M. Hassell, T. D. Meek, T. A. Tomaszek, Jr., M. Lewis, *Biochemistry* **32**, 937 (1993).

34. P. Ettmaher, M. Hübner, A. Billich, B. Rosenwirth, H. Gstach, *Bioorg. Med. Chem. Lett.* **4**, 2851 (1994).

35. D. C. Humber, N. Cammack, J. A. V. Coates, K. N. Cobley, D. C. Orr, R. Storer, G. G. Weingarten, M. P. Weir, *J. Med. Chem.* **35**, 3080 (1992).

36. (a) D. S. Holmes, R. C. Bethell, N. Cammack, I. R. Clemens, J. Kitchin, P. McMeekin, C. L. Mo, D. C. Orr, B. Patel, I. L. Paternoster, R. Storer, *J. Med. Chem.* **36**, 3129 (1993).
(b) J. Kitchin, R. C. Bethell, N. Cammack, S. Delon, D. N. Evans, S. Holman, D. S. Holmes, P. McMeekin, C. L. Mo, N. Nieland, D. C. Orr, J. Saunders, B. E. V. Shenoy, I. D. Starkey, R. Storer, *J. Med. Chem.* **37**, 3707 (1994).

37. A. Wonacott, R. Cooke, F. R. Hayes, M. M. Hann, H. Jhoti, P. McMeekin, A. Mistry, P. Murray-Rust, O. M. P. Singh, M. P. Weir, *J. Med. Chem.* **36**, 3113 (1993).

38. J. A. Martin, *Drugs Fut.* **16**, 210 (1991).

39. D. J. Aberhart, S. J. Gould, H.-J. Lin, T. K. Thiruvengadam, B. H. Weiller, *J. Am. Chem. Soc.* **105**, 5461 (1983), and references cited therein.

40. B. W. Bycroft, Ed., *Dictionary of Antibiotics and Related Substances* (Chapman and Hall, London, 1988), p. 665.

41. P. A. Grieco, Y. S. Hon, A. Perez-Medrano, *J. Am. Chem. Soc.* **110**, 1630 (1988), and references cited therein.

42. M. Hamada, T. Takeuchi, S. Kondo, Y. Ikeda, H. Naganawa, K. Maeda, H. Umezawa, *J. Antibiotics* **23**, 170 (1970).

43. T. Korzybski, Z. Kowszyk-Gindifer, W. Kurytowicz, *Antibiotics* (American Society of Microbiology, Washington, D.C., 1978), Vol. 1, pp. 343–46.

44. D. H. Rich, in *Comprehensive Medicinal Chemistry*, P. G. Sammes, Ed. (Pergamon Press, Oxford, 1990), Vol. 2, p. 391.

45. (a) K. Miura, T. Sawa, T. Takeuchi, H. Umezawa, *J. Antibiotics* **39**, 734 (1986).
(b) H. Umezawa, *Small Molecular Immunomodifiers of Microbial Origin. Fundamental and Clinical Studies of Bestatin*, H. Umezawa, Ed. (Pergamon Press, Oxford, 1981).
(c) M. Ishizuka, T. Masuda, N. Kanbayashi, S. Fukasawa, T. Takeuchi, T. Aoyagi, H. Umezawa, *J. Antibiotics* **33**, 642 (1980).

46. H. Blomgren, J. Wasserman, *J. Canc. Lett.* **11**, 303 (1981).

47. W. J. Greenlee, *Med. Res. Rev.* **10**, 173 (1990).

48. K. Iizuka, T. Kamijo, T. Kubota, K. Akahane, H. Umeyama, Y. Kiso, *J. Med. Chem.* **31**, 701 (1988).

49. (a) E. K. Rowinsky, L. A. Cazenave, R. C. Donehower, *J. Natl. Cancer Inst.* **82**, 1247 (1990).
(b) W. P. McGuire, E. K. Rowinsky, N. B. Rosenshein, F. C. Grumbine, D. S. Ettinger, D. K. Armstrong, R. C. Donehower, *Ann. Int. Med.* **111**, 273 (1989).

50. (a) M. Colin, D. Guénard, F. Guéritte-Voegelin, P. Potier, Eur. Patent 253,739. January 20, 1988.

51. (a) F. A. Holmes, R. S. Walters, R. L. Theriault, A. D. Forman, L. K. Newton, M. N. Raber, A. U. Buzdar, D. K. Frye, G. N. Hortobagyi, *J. Natl. Cancer Inst.* **83**, 1797 (1991).
(b) M. Hepperle, G. I. Georg, *Drugs Fut.* **19**, 573 (1994).

52. (a) D. S. Ettinger, *Semin. Oncol., Suppl. 3*, **20**, 46 (1993).
(b) S. S. Legha, S. Ring, N. Papadopoulos, M. Raber, R. S. Benjamin, *Cancer* **65**, 2478 (1990).

53. (a) K. C. Nicolaou, Z. Yang, J. J. Liu, H. Ueno, P. G. Nantermet, R. K. Guy, C. F. Claiborne, J. Renaud, E. A. Couladouros, K. Paulvannan, E. J. Sorensen, *Nature* **367**, 630 (1994).
(b) R. A. Holton, C. Somoza, H.-B. Kim, F. Liang, R. J. Biediger, P. D. Boatman, M. Shindo, C. C. Smith, S. Kim, H. Nadizadeh, Y. Suzuki, C. Tao, P. Vu, S. Tang, P. Zhang, K. K. Murthi, L. N. Gentile, J. H. Liu, *J. Am. Chem. Soc.* **116**, 1597 (1994).
(c) R. A. Holton, H.-B. Kim, C. Somoza, F. Liang, R. J. Biediger, P. D. Boatman, M. Shindo, C. C. Smith, S. Kim, H. Nadizadeh, Y. Suzuki, C. Tao, P. Vu, S. Tang, P. Zhang, K. K. Murthi, L. N. Gentile, J. H. Liu, *J. Am. Chem. Soc.* **116**, 1599 (1994).

54. K. C. Nicolaou, W.-M. Dai, R. K. Guy, *Angew. Chem., Int. Ed. Engl.* **33**, 15 (1994).

55. P. E. Fleming, H. G. Floss, M. Haertel, A. R. Knaggs, A. Lansing, U. Mocek, K. D. Walker, *Pure Appl. Chem.* **66**, 2045 (1994).

56. J. Werweij, M. Clavel, B. Chevalier, *Ann. Oncol.* **5**, 459 (1994).

57. (a) G. I. Georg, S. M. Ali, J. Zygmunt, L. R. Jayasinghe, *Exp. Opin. Ther. Pat.* **4**, 109 (1994).
(b) A. Datta, L. R. Jayasinghe, G. I. Georg, *J. Med. Chem.* **37**, 4258 (1994).

58. F. Guéritte-Voegelein, D. Guénard, F. Lavelle, M.-T. Le Goff, L. Mangatal, P. Potier, *J. Med. Chem.* **34**, 992 (1991).

59. I. Ojima, O. Duclos, S. D. Kuduk, C.-M. Sun, J. C. Slater, F. Lavalle, J. M. Veith, R. J. Bernacki, *Bioorg. Med. Chem. Lett.* **4**, 2631 (1994).

60. (a) D. M. Vyas, H. Wong, A. R. Crosswell, A. M. Casazza, J. O. Knipe, S. W. Mamber, T. W. Doyle, *Bioorg. Med. Chem. Lett.* **3**, 1357 (1993).
(b) K. C. Nicolaou, C. Riemer, M. A. Kerr, D. Rideout, W. Wrasidlo, *Nature* **364**, 464 (1993).
(c) A. E. Mathew, M. R. Mejillano, J. P. Nath, R. H. Himes, V. J. Stela, *J. Med. Chem.* **35**, 145 (1992).

61. R. B. Greenwald, A. Pendri, D. Bolikal, C. W. Gilbert, *Bioorg. Med. Chem. Lett.* **4**, 2465 (1994).

62. K. C. Nicolaou, J. Renaud, P. G. Nantermet, E. A. Couladouros, R. K. Guy, W. Wrasidio, *J. Am. Chem. Soc.* **117**, 2409 (1995).

63. Reviews of monobactams:
(a) C. M. Cimarusti, R. B. Sykes, *Chem. Br.* 302 (1983).
(b) W. H. Koster, C. M. Cimarusti, R. B. Sykes, in *Chemistry and Biology of β-Lactam Antibiotics*, R. B. Morin, M. Gorman, Eds. (Academic Press, New York, 1982), Vol. 3, pp. 339–75.

64. (a) C. M. Cimarusti, R. B. Sykes, *Med. Res. Rev.* **4**, 1 (1984).
(b) R. B. Sykes, D. P. Bonner, K. Bush, N. H. Georgopapadakou, *Antimicrob. Ag. Chemother.* **21**, 85 (1982).
(c) T. Matsuo, T. Sugawara, H. Masuya, Y. Kawano, N. Noguchi, M. Ochiai, *Chem. Pharm. Bull.* **31**, 1874 (1983).
(d) Y. Matsuhisa, K. Umemoto, K. Itakura, Y. Usui, *Chem. Pharm. Bull.* **34**, 3779 (1986).

65. (a) M. Hashimoto, T.-a. Komori, T. Kamiya, *J. Am. Chem. Soc.* **98**, 3023 (1976).
(b) H. Aoki, H.-i. Sakai, M. Kohsaka, T. Konomi, J. Hosoda, Y. Kubochi, E. Iguchi, H. Imanaka, *J. Antibiotics* **29**, 492 (1976).

66. Y. Mine, S. Nonoyama, H. Kojo, S. Fukada, M. Nishida, S. Goto, S. Kuwahara, *J. Antibiotics* **30**, 932 (1977).

67. G. H. Hakimelahi, S.-C. Tsay, J. R. Hwu, *Helv. Chem. Acta* **78,** 411 (1995).

68. D. A. Burnett, *Tetrahedron Lett.* **40**, 7339 (1994).

69. D. A. Burnett, M. A. Caplen, H. R. Davis, Jr., R. E. Burrier, J. W. Clader, *J. Med. Chem.* **59**, 1733 (1994).

70. M. L. Sassiver, A. Lewis, in *Structure–Activity Relationships among the Semisynthetic Antibiotics*, D. Perlman, Ed. (Academic Press, New York, 1977), pp. 87–160.

71. (a) G. J. Chin, J. Marx, Eds., *Resistance to Antibiotics, Science* **264**, 359–93 (1994).
(b) L. E. Bryan, Ed., *Antimicrobial Drug Resistance* (Academic Press, New York, 1984).

72. (a) M. Sunagawa, H. Matsumura, T. Inoue, M. Fukasawa, M. Kato, *J. Antibiotics* **44,** 459 (1991).
(b) R. N. Guthikonda, L. D. Cama, M. Quesada, M. F. Woods, T. N. Salzmann, B. G. Christensen, *J. Med. Chem.* **30**, 871 (1987).
(c) S. M. Schmitt, T. N. Salzmann, D. H. Shih, B. G. Christensen, *J. Antibiotics* **41**, 780 (1988).

73. (a) G. Albers-Schönberg, B. H. Arison, O. D. Hensens, J. Hirshfield, K. Hoogsteen, E. A. Kaczka, R. E. Rhodes, J. S. Kahan, F. M. Kahan, R. W. Ratcliffe, E. Walton, L. F. Ruswinkle, R. B. Morin, B. G. Christensen, *J. Am. Chem. Soc.* **100**, 6491 (1978).
(b) K. Okamura, S. Hirata, Y. Okumura, Y. Fukagawa, Y. Shimauchi, K. Kuono, T. Ishikura, *J. Antibiotics* **31**, 480 (1978).
(c) W. J. Leanza, K. J. Wildonger, T. W. Miller, B. G. Christensen, *J. Med. Chem.* **22**, 1435 (1979).

74. J. S. Kahan, F. M. Kahan, R. Goegelman, S. A. Currie, M. Jackson, E. O. Stapley, T. W. Miller, A. K. Miller, D. Hendlin, S. Mochales, S. Hernández, H. B. Woodruff, J. Birnbaum, *J. Antibiotics* **32**, 1 (1979).

75. M. Sunagawa, H. Matsumura, T. Inoue, M. Fukasawa, M. Kato, *J. Antibiotics* **43**, 519 (1990).

76. Y. Nagao, Y. Nagase, T. Kumagai, H. Matsunaga, T. Abe, O. Shimada, T. Hayashi, Y. Inoue, *J. Org. Chem.* **57**, 4243 (1992).

77. B. G. Spratt, *Phil. Trans. Royal Soc. Lond. (B)* **289**, 273 (1980).

78. H. H. Martin, *J. Gen. Microbiol.* **36**, 441 (1964).

79. G. D. Shockman, J. F. Barrett, *Ann. Rev. Microbiol.* **37**, 501 (1983).

80. K. Ubukata, N. Yamashita, M. Konno, *Antimicrob. Ag. Chemother.* **27**, 851 (1985).

81. M. Sunagawa, H. Yamaga, H. Shinagawa, H. Houchigai, Y. Sumita, *Bioorg. Med. Chem. Lett.* **4**, 2793 (1994).

82. (a) R. P. Elander, H. Aoki, in *Chemistry and Biology of β-Lactam Antibiotics*, R. B. Morin, M. Gorman, Eds. (Academic Press, New York, 1982), Vol. 3, pp. 83–153.
(b) E. P. Abraham, in *β-Lactam Antibiotics: Mode of Action, New Developments and Future Prospects*, M. Salton, G. Shockman, Eds. (Academic Press, New York, 1981), pp. 311–25.

83. (a) R. B. Kammer, in *Chemistry and Biology of β-Lactam Antibiotics*, R. B. Morin, M. Gorman, Eds. (Academic Press, New York, 1982), Vol. 3, pp. 287–301.
(b) H. C. Neu, in *β-Lactam Antibiotics: Mode of Action, New Developments and Future Prospects*, M. Salton, G. Shockman, Eds. (Academic Press, New York, 1981), pp. 545–66.

84. R. D. G. Cooper, in *The Chemistry of Beta-Lactams*, M. I. Page, Ed. (Chapman & Hall, London, 1992), pp. 272–305.

85. L. C. Blaszczak, R. F. Brown, G. K. Cook, W. J. Hornback, R. C. Hoying, J. M. Indelicato, C. L. Jordan, A. S. Katner, M. D. Kinnick, J. H. Mc Donald III, J. M. Morin, Jr., J. E. Munroe, C. E. Pasini, *J. Med. Chem.* **33**, 1656 (1990).

86. L. N. Jungheim, T. A. Shepherd, J. K. Kling, *Heterocycles* **35**, 339 (1993).

87. T. W. Hudyma, K. Bush, K. L. Colson, R. A. Firestone, H. D. King, *Bioorg. Med. Chem. Lett.* **3**, 323 (1993).

88. V. M. Vrudhula, H. P. Svensson, P. D. Senter, *J. Med. Chem.* **38**, 1380 (1995).

89. R. P. Alexander, N. R. Beeley, M. O'Driscoll, F. P. O'Neill, T. A. Millican, A. J. Pratt, F. W. Willenbrock, *Tetrahedron Lett.* **32**, 3269 (1991).

90. R. J. Ternansky, S. E. Draheim, A. J. Pike, F. W. Bell, S. J. West, C. L. Jordan, C. Y. Ernie Wu, D. A. Preston, W. Alborn, Jr., J. S. Kasher, B. L. Hawkins, *J. Med. Chem.* **36**, 1971 (1993).

91. S. Hanessian, M. J. Rozema, *Bioorg. Med. Chem. Lett.* **4**, 2279 (1994).

92. J. D. Buynak, D. Khasnis, B. Bachmann, K. Wu, G. Lamb, *J. Am. Chem. Soc.* **116**, 10955 (1994).

93. (a) T. Tsuji, H. Satoh, M. Narisada, Y. Hamashima, T. Yoshida, *J. Antibiotics* **38**, 466 (1985).
(b) H. C. Neu, N.-X. Chin, *Antimicrob. Ag. Chemother.* **30**, 638 (1986).

94. (a) T. W. Doyle, J. L. Douglas, B. Belleau, T. T. Conway, C. F. Ferrari, D. E. Horning, G. Lim, B.-y. Luh, A. Martel, M. Menard, L. R. Morris, M. Misiele, *Can. J. Chem.* **58**, 2508 (1980).
(b) H. Mastalerz, M. Menard, V. Vinet, J. Desiderio, J. Fung-Tomc, R. Kessler, Y. Tsai, *J. Med. Chem.* **31**, 1190 (1988).
(c) S. M. Tenneson, B. Belleau, *Can. J. Chem.* **58**, 1605 (1980).

95. H. Tsubouchi, H. Ishikawa, *Bioorg. Med. Chem.* **3**, 143 (1995), and references cited therein.

96. (a) B. Tamburini, A. Perboni, T. Rossi, D. Donati, D. Andreotti, G. Gaviraghi, R. Carlesso, C. Bismara (Glaxo S.p.A.) *Eur. Pat. Appl.*, 0416953 A2, 1991, Bulletin 91/11; *Chem. Abst.* **116**, 235337t (1992).
(b) B. Tamburini, A. Perboni, T. Rossi, D. Donati, G. Gaviraghi, G. Tarzia, *Recent Advances in the Chemistry of Anti-Infective Agents*, P. H. Bentley, R. Ponsford, Eds. (The Royal Society of Chemistry, Cambridge, 1992), pp. 21–35.
(c) A. Padova, S. M. Roberts, D. Donati, C. Marchioro, A. Perboni, *J. Chem. Soc., Chem. Commun.* 661 (1995).

97. S. Hanessian, G. B. Reddy, *Bioorg. Med. Chem. Lett.* **4**, 2285 (1994).

98. E. Di Modugno, I. Erbetti, L. Ferrari, G. Galassi, S. H. Hammond, L. Xerri, *Antimicrob. Ag. Chemother.* **38**, 2362 (1994).

99. M. Alpegiani, P. Bissolino, R. Corigli, V. Rizzo, E. Perrone, *Bioorg. Med. Chem. Lett.* **5**, 687 (1995).

100. P. R. Berstein, P. D. Edwards, J. C. Williams, in *Progress in Medicinal Chemistry*, G. P. Ellis, D. K. Luscombe (Elsevier, Amsterdam, 1994), Vol. 31, pp. 59–120.

101. M. Alpegiani, P. Bissolino, R. Corigli, S. Del Nero, E. Perrone, V. Rizzo, N. Sacchi, G. Cassinelli, G. Franceschi, A. Baici, *J. Med. Chem.* **37**, 4003 (1994).

CHAPTER

3

Selective Transformations of N,N-Dibenzyl Protected Asparagine and Aspartic Acid Derivatives

Peter Gmeiner

3.1 Introduction

Among the proteinogenic α-amino acids the difunctional representatives aspartic acid (Asp) and asparagine (Asn) are of special interest, which is due to their β-amino carboxylic acid and β-amino carboxamide moiety, respectively. Thus both compounds can be regarded as natural β-amino acid derivatives as well. In the binding sites of many receptor proteins, enzymes, and (catalytic) antibodies, β-carboxylate or carboxamide functions of Asp and Asn residues play a critical role for the recognition of selective agonists or substrates as well as for signal transduction or catalysis (for example, general acid–base catalysis).[1] Furthermore, asparagine-linked glycoproteins are known as a common feature in many glycoconjugates.[2]

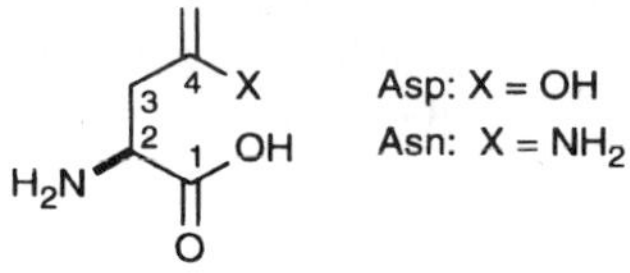

For the chemical synthesis of enantiomerically pure compounds, aspartic acid and asparagine can serve as perfect chiral building blocks for the following reasons:

- Asp and Asn are cheap and commercially available in both absolute configurations.
- The chiral C-4 moiety can be modified at any position.
- These transformations can be done chemo-, regio-, as well as stereoselectively.

This has been demonstrated convincingly in the literature. A few examples are sketched in the following schemes.

McGarvey and co-workers developed practical methodology for a regioselective reduction of the C-1 carboxyl group.[3] Starting from the N-Cbz protected aspartic acid **1**, a cyclic anhydride was formed that could be efficiently reduced by $NaBH_4$ to give the β-amino lactone **2**. Valuable applications of **2** were described resulting in the synthesis of the C21–C37 half of the polyene macrolide amphotericin B[4] (**4**) or L-daunosamine (**3**), an important component of the anthracycline antibiotics daunorubicin and adriamycin.[5]

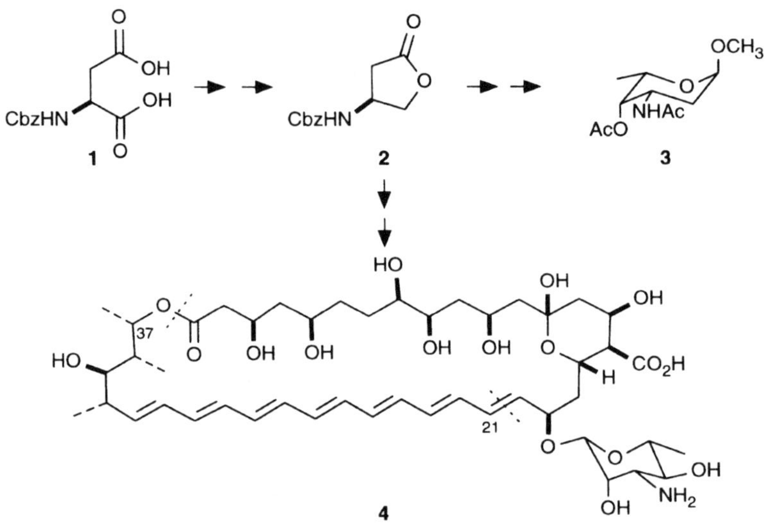

According to Rapoport and co-workers, selective intra- and intermolecular alkylations of N-phenylfluorenyl protected asparagine and aspartic acid derivatives at C-3 gave access to the chiral pyridine carboxylates **5**, **6** and **7**.[6–8] When starting from asparagine, differentiation between the two carboxyl equivalents can be facilitated by dehydration of the carboxamide. When aspartic acid was used as the chiral building block, regioselective esterification of the β-carboxylate or highly regiocontrolled Cu(II)-assisted ester hydrolysis of the α-carboxylate was successful. **5**–**7** served as the key intermediates for the syntheses of (+)-apovincamine (**8**),[6] (+)-vincamine (**9**),[7] and (–)-vindoline (**10**).[8]

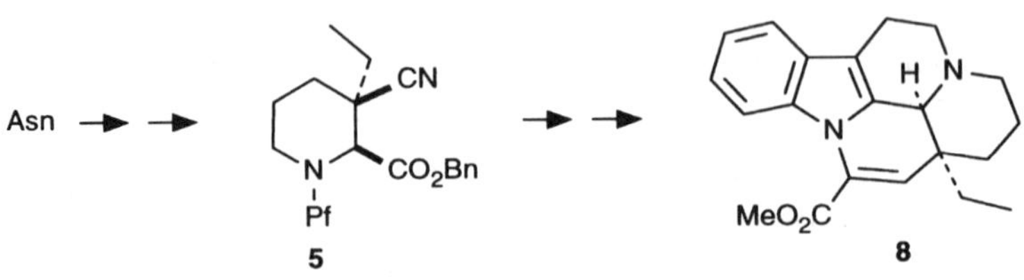

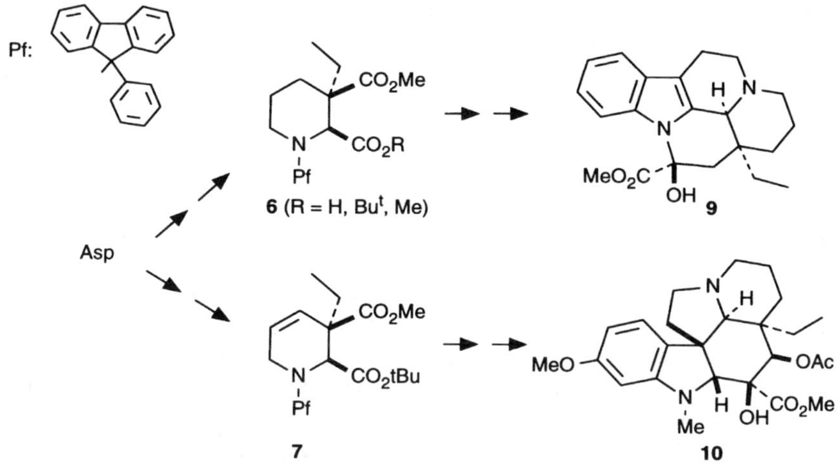

Besides for diastereoselective preparations of enantiomerically pure 2,3-diamino acids,[9] γ-lactam-bridged dipeptides,[10] and 2,3-pyrrolidine dicarboxylates[11] the above-mentioned Rapoport-type C-3 alkylation of aspartate derivatives served as a key step for a total synthesis of (+)-pilocarpine (**13**).[12] This work also includes an example for a stereoselective reaction at the C-2 position, namely, an exchange of the amino group by a bromide substituent (→ **12**). Under classical conditions (HBr, KBr, $NaNO_2$) complete retention was observed, which is due to the anchimeric effect of the α-carboxyl group.

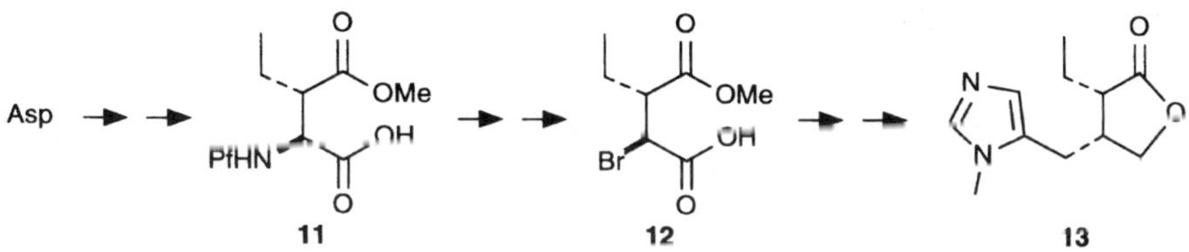

Representatives for selective reactions at C-4 are the synthesis of 2,3-diamino-propanoic acids (DAPA) from N-Boc, N-Cbz, or N-tosyl protected asparagine under degradation of the carboxamide[13] as well as the preparation of L-phosphinothricin (**15**), a natural occurring inhibitor of glutamine synthase, from aspartic acid, when the protected homoserine lactone **14** served as the key intermediate.[14] Among the described preparations of homoserine lactone from Asp or Asn,[15] the method of Uzar, including precipitation of a dicyclohexylammonium salt of the methoxycarbonyl pro-

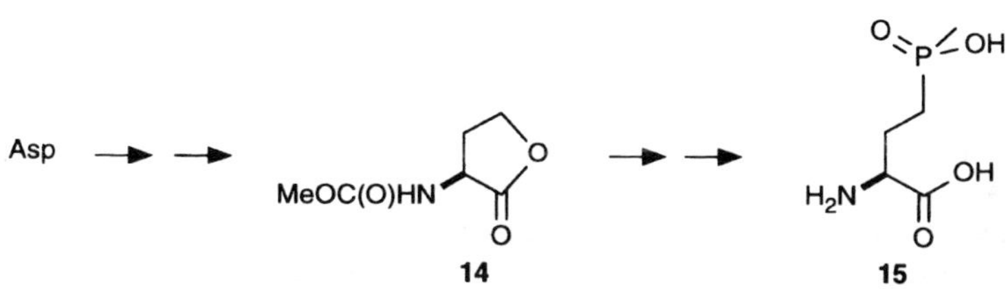

tected β-methyl aspartate followed by a chemoselective reduction with $LiBH_4$, seems to be the most practical.[16]

3.2 Chemoselective Transformations Starting from Asparagine

In connection with structure–activity relationship studies at the dopamine D-2 receptor, an efficient approach to enantiomerically pure 6- and 7-aminotetrahydroindolizines was required. For the construction of these compounds it was intended to use Asn as a flexible chiral building block. Thus protection of the primary amine, reduction of both terminals, and Paal Knorr pyrrole synthesis should lead to the 1,2-amino alcohol **16**, as a cyclization precursor for the 7-aminoindolizines (**17**). On the other hand, selective modification of the α-carboxyl function, followed by hydrolysis and reduction of the C-4 position, was envisioned as a possible route to the 1,3-amino alcohol **19** and, thus, to the 6-amino isomers **20**. The key step of the second route was a chemoselective transformation of the α-carboxylate in order to generate a β-amino acid–type intermediate (**18**). Thus the first goal was to develop a general method for the synthesis of enantiomerically pure β-amino acids.

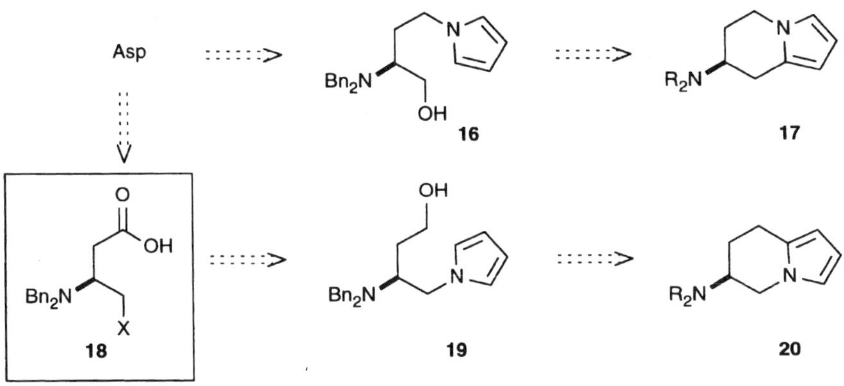

16 17 18 19 20

3.2.1 General Synthesis of β-Amino Acids

For the protection of the amino group N,N-dibenzyl protection was selected.[17] This moiety was expected to be stable towards a variety of reaction conditions. According to results from Reetz et al., the N,N-dibenzylamino function should provide configurational stability at the stereogenic center.[18] Furthermore, a convenient way to deprotect by hydrogenolysis was expected.

The usual procedure for N,N-dibenzyl protection, which includes heating of the amino acid with benzyl bromide in aqueous NaOH or Na_2CO_3 solution,[17] was not successful for Asn. Due to the nucleophilic character of the carboxamine, these reaction conditions gave not the expected N,N-dibenzylasparagine benzyl ester but the imide **21**, together with further reaction products. To overcome this problem, a reductive alkylation process was elaborated. Thus reaction of an aqueous solution of Asn with an excess of benzaldehyde and $NaCNBH_3$ at pH 7 resulted in 83% of N,N-dibenzyl asparagine

22. For a stabilization of the amide function and an effective differentiation between the positions 1 and 4, **22** was transformed into the cyanoalanine derivative **23** by esterification and subsequent dehydration. Chemoselective reduction of the α-carboxylate with $LiAlH_4$ at –78 to –40°C, followed by activation of the resulting alcohol with methanesulfonyl chloride, gave the activated β-homoserine derivative **24** in 60% overall yield (based on Asn). Compound **24** should serve as a central intermediate for a general β-amino acid synthesis. Therefore, the mesyloxy group needed to be exchanged by various types of nucleophiles. For C,C bond formation, organo cuprates were reacted with **24**. It turned out that treatment of **24** with the lower-order Gilman cuprates Me_2CuLi, Bu_2CuLi, and Ph_2CuLi resulted in formation of the displacement products **25** (R = Me, Bu, Ph) in 48–72% yield.[19,20] $Me_2Cu(CN)Li_2$ as an example of a higher-order Lipshutz cuprate did not afford the desired substitution product. Synthesis of the 3-aminobutane nitrile **25** (R = H), as a precursor of *(R)*-3-aminobutanoic acid, was performed by a chemoselective reduction of the β-homoserine derivative **24** under cleavage of the C–O bond.[21] As a suitable hydride source $LiBH_4$ was employed. The last two steps for the transformation of **25** into the free β-amino acids **26** involved acidic hydrolysis of the nitriles by aqueous HCl, followed by hydrogenolytic deprotection of the benzyl groups using $Pd(OH)_2/C$ as a catalyst. The overall yield of the *(R)*-configured β-amino acids **26** from Asn was 23–36%. Following the identical reaction sequence the respective *(S)*-enantiomers can be obtained from *(R)*-asparagine.

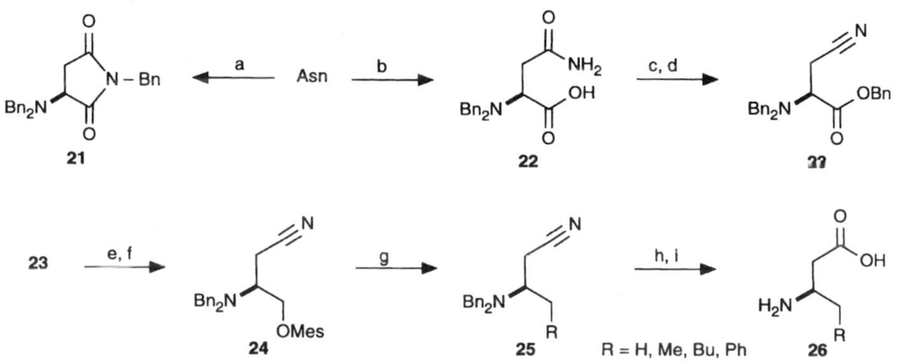

a: BnBr, K_2CO_3/H_2O (26%); b: PhCHO, $NaCNBH_3$, H_2O, pH 7 (83%); c: BnBr, Et_3N, THF (85%); d: TosCl, pyridine, CH_2Cl_2 (92%); e: $LiAlH_4$, THF, –78°C (94%); f: MesCl, Et_3N, THF (97%); g: $LiBH_4$, THF, 50°C (52%), R = H), R_2CuLi, Et_2O, –60 to –20°C (65% for R = Me, 72% for R = Bu, 48% for R = Ph); h: HCl/H_2O (89–91%); i: $Pd(OH)_2/C$, H_2, MeOH (90–95%).

Using 3-N,N-dibenzylaminopentanoic acid as an example the configurational integrity of the synthesis could be established by esterification, subsequent debenzylation, and coupling with *(R)*-1-phenylethyl isocyanate. Examination of ^1H NMR spectra of the respective urea, calibrated by doping experiments [with the addition product derived from *(S)*-1-phenylethyl isocyanate], revealed the synthetic material to be configurationally pure (ee > 99).

3.2.2 Preparation of Tetrahydroindolizines

Application of the above-described β-amino acid synthesis yielded the azide substituted β-amino acid **28** as an important intermediate for the synthesis of enantiomerically pure 6-aminotetrahydroindolizines (**20**).[22] In practice, the activated β-homoserine derivative **24** was reacted with NaN_3, resulting in a displacement of the mesyloxy group. Subsequent acidic hydrolysis afforded the synthetic intermediate **28**. Upon treatment with BH_3xTHF, the azide as well as the carboxylic acid could be reduced to give the diamino alcohol **29**. Finally, Paal–Knorr reaction and triflic anhydride induced 6-exo-tet ring closure gave access to the dibenzylamine substituted bicycle **20** (*R* = benzyl). Modification of the N-substituents is possible by hydrogenolytic deprotection and subsequent reductive alkylation. Employing an analogous reaction sequence, the 7-amino regioisomers **17** could be synthesized from the diamino alcohol **27** via **16**.[23] **27** could be prepared by borane reduction of N,N-dibenzyl asparagine **22**. During the following receptor binding studies and functional experiments, which also included the enantiomers of **17** and **20** [synthesized from *(R)*-Asn], the *(S)*-configured test compound **17** (for *R* = *n*-propyl) turned out to be a strong and selective dopamine D-2 autoreceptor agonist.[24] Thus **17** might be of interest for the development of atypical neuroleptics, a family of drugs that are expected to provide an efficient treatment of schizophrenia.[25]

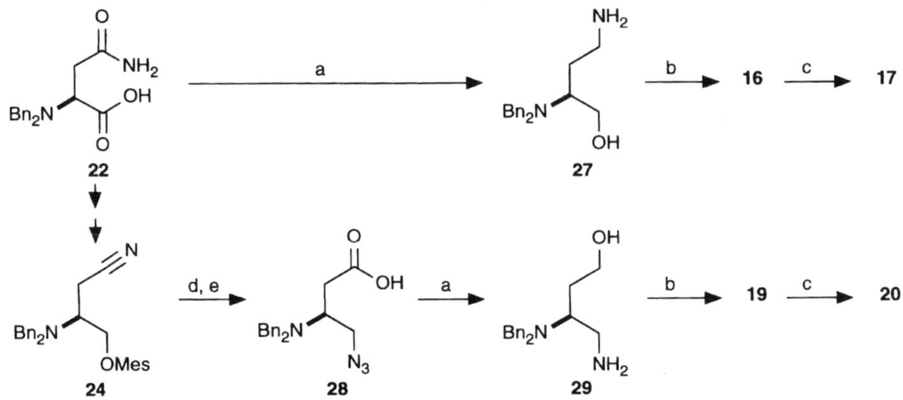

a: BH_3 x THF, THF (85–86%); b: 2,5-dimethoxytetrahydrofuran, HOAc/NaOAc (75–90%); c: Tf_2O, CH_2Cl_2 (65–94%); d: NaN_3, DMF (82%); e: HCl/H_2O (86%).

3.2.3 Synthesis of β,γ-Diaminobutanenitriles and β,γ-Diaminobutyric Acids

β,γ-Diaminobutyric acids are of special interest as aspartic acid analogs in reduced peptide bond isosters and have been successfully applied in peptidomimetics of the growth releasing factor, secretin, cholecystokinin, and tetragastrin.[26] Due to their γ-aminobutyric acid (GABA) substructure, these compounds can potentially interact with the GABA ergic system.[27] On the other hand, β,γ-diaminobutanenitriles have been reported rarely.[28] As far as we know, these family of compounds had not been described in nonracemic form.

The above-described β-homoserine equivalent **24** was expected to serve as a precursor for the envisioned target compounds. Employing amines as nucleophiles in polar solvents, however, did not give the projected displacement reaction. Instead of this, **24** was converted into a 2:1 trans/cis mixture of the aminobutenenitrile **31**.[19] This side reaction obviously proceeds through an aziridinium intermediate (**30**). Subsequent deprotonation and ring opening leads to **31**. To overcome this problem the amino alcohol **32**, as the direct synthetic precursor of **24**, was reacted with phthalimide as a nitrogen equivalent under the very mild *Mitsunobu* conditions[29] when the protected diaminonitrile **33** could be isolated in 67% yield.[30] As an alternative, the diaminobutane nitriles **36** could be synthesized by Swern oxidation[31] of **32** and subsequent reductive amination. Reductive coupling of the intermediate **35** with *(S)*-alanine ethyl ester, followed by carefully performed NMR analysis, revealed the configurational integrity of the synthesis. The moderate yield of the amination products is due to a side reaction of the amino aldehyde **35**, resulting in formation of the aminopentenenitrile **37**. Formally, this elimination includes the removal of a formyl anion, which is very uncommon. Using **33** as an example, the nitrile function was hydrolyzed under acidic conditions. Subsequent hydrazinoloysis yielded the β,γ-diaminobutyric acid **34**.

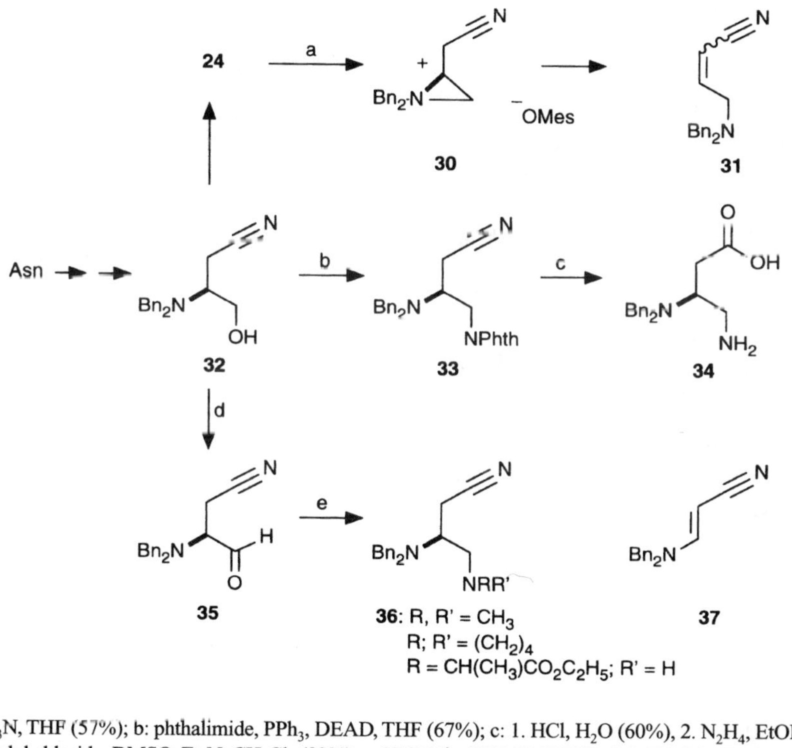

a: Et_3N, THF (57%); b: phthalimide, PPh_3, DEAD, THF (67%); c: 1. HCl, H_2O (60%), 2. N_2H_4, EtOH (95%); d: oxalyl chloride, DMSO, Et_3N, CH_2Cl_2 (82%); e: HNRR′ x HCl, $NaCNBH_3$, MeOH (20–42%).

3.2.4 Synthesis of α-Substituted β-Amino Nitriles and β-Amino Acids

α,β-Disubstituted β-amino acids are components in a number of molecules that display interesting biological properties. For example, *(2S,3R)*-2-methyl-3-aminopentanoic acid could be isolated as a hydrolysis fragment of the marine peptide antitumor agents dolastatin 11 and 12[32] and of the antifungal products majusculamide C and 57-normajusculamide C.[33]

Besides the opportunity to modify position 4 of the above-described β-homoserine equivalents, the CH-acidity of the nitrile α-position was anticipated to make possible the introduction of substituents into position 2. This should allow an individual construction of both the α-substituent and the β-side chain. To facilitate regioselective introduction of substituents by deprotonation and subsequent alkylation, the OH group of the building block **39**, prepared from *(R)*-asparagine (**38**), was protected by silylation. Upon deprotonation of **40** by LDA and subsequent treatment with the electrophiles methyl iodide or benzyl bromide at –78°C, 1:1 mixtures of diastereomers were formed in 85 and 92% yield, respectively.[34]

When changing the solvent to Et_2O or employing $KN(SiMe_3)_2$ as a base, a modest selectivity was observed (threo–erythro 2:1). After O-deprotection by HOAc in aqueous THF and separation by flash chromatography, the threo configured products **42** and the erythro diastereomers **43** were isolated. An efficient and diastereoselective approach to **42** was found when unprotected **39** was treated with an excess of base

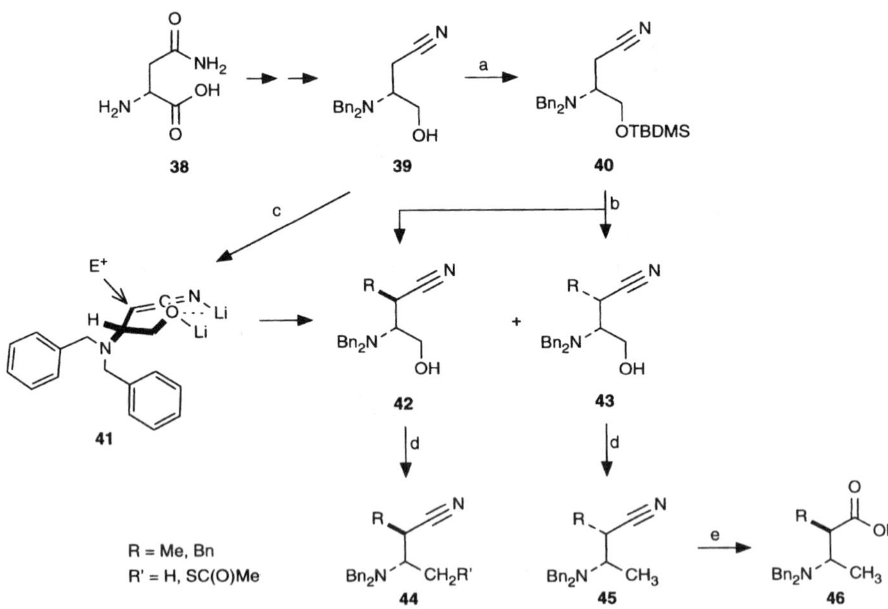

a: TBDMS-Cl, imidazole, DMF (82%); b: 1. LDA, MeI (BnBr), THF; 2. HOAc/THF/H_2O (**42:** 40–42%, **43:** 40–44%); c: LDA, THF (**42:** 39–43%, **43:** 5–6%); d: 1. MesCl, Et_3N, THF (69–96%); 2. $LiBH_4$ (K-thioacetate), THF (DMF) (37–70%); e: HCl, MeOH/H_2O (59–73%).

and, subsequently, reacted with alkyl halides (threo–erythro 7:1). This can be explained by a preferred attack of the *Re* face of the dianion **41** when existing in a chelated structure. As an example of the possible modifications at C-4, the hydroxy substituents of **42** and **43** were displaced by hydrogen. Thus activation of **42** or **43** with MesCl and subsequent $LiBH_4$ reduction gave the α-substituted β-homoalanine derivatives **44** and **45**. Introduction of a thioacetate group resulted in formation of a substituted β-homocystein equivalent. Hydrolysis of the nitrile function was only possible with the erythro isomers **45**, resulting in the isomerically pure β-amino acids **46**. Epimerization at the chiral α-center was not detected. Starting from α-amino aldehydes an alternative synthesis of the β-amino nitriles **44** and **45**, including Barton deoxygenation of the corresponding cyano hydrins and alkylation, is described by Reetz and co-workers.[35]

3.3 Regioselective Functionalization of Aspartic Acid Derivatives

As a complement to methodology based on transformation of N,N-dibenzyl protected asparagine derivatives, it was envisaged to utilize N,N-dibenzyl aspartates as chiral building blocks.

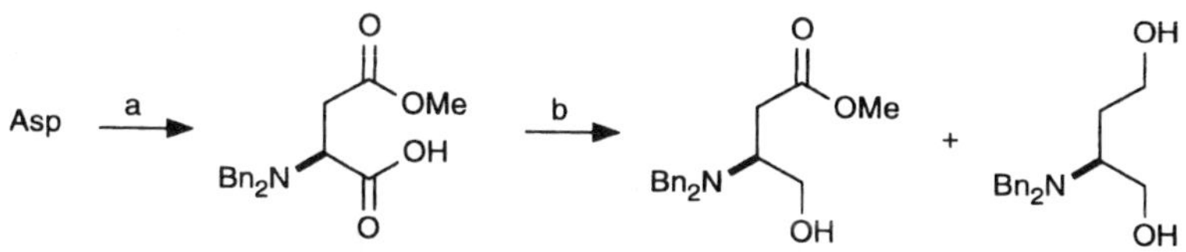

47 **48** **49**

a: 1. HCl, MeOH (91%); 2. PhCHO, $NaCNBH_3$, H_2O, pH 7 (31%); b: BH_3 x THF, THF (**48**: 13%, **49**: 56%).

An initial route that was expected to provide differentiation between the two carboxyl functions involved the protected aspartate β-methyl ester **47**, which was readily available from aspartic acid by selective esterification (HCl, MeOH)[7] and N,N-dibenzylation under reductive conditions (benzaldehyde, $NaCNBH_3$, aqueous pH 7 buffer).[19] Chemoselective borane reduction of **47** should give access to the β-homoserine derivative **48**. However, this reduction required fairly drastic conditions, resulting in only 13% of **48** and 56% of the diol **49** indicating low chemoselectivity. Furthermore, **48** turned out to be unstable towards lactone formation during workup and storage and thus seemed not be suitable as a central intermediate for further studies. On the other hand, it should be easy to optimize the synthesis of **49**. Due to the bulky dibenzylamino group in position 2, an efficient differentiation between the two primary alcohols was anticipated. This should be an interesting playground for regioselective transformations, as sketched in the following scheme. Selective protection of the more accessible HO group in position 4, followed by converting the alcohol in position 1 into a leaving group, should give a useful intermediate (**52**) for a general synthesis of 1,3-amino alcohols (**53**). Performing activation and protection in a reversed sequence followed by nucleophilic displacement and O deprotection should lead to the 1,2-amino alcohols **51** through an intermediate of type **50**.

Besides the opportunity to oxidize the alcohol functions of the products (in order to synthesize amino aldehydes or amino acids), this strategy should be applied for natural product synthesis and the preparation of conformationally restricted dopamine analogs.

Additionally, a subsequent introduction of two different nucleophiles (Nu_1 and Nu_2) at both terminal positions of the bis-electrophilic intermediate (**54**) should be investigated (→**55**). For Nu_1 = CN, this route should lead to β-amino nitriles and thus to β-amino acids (**56**).

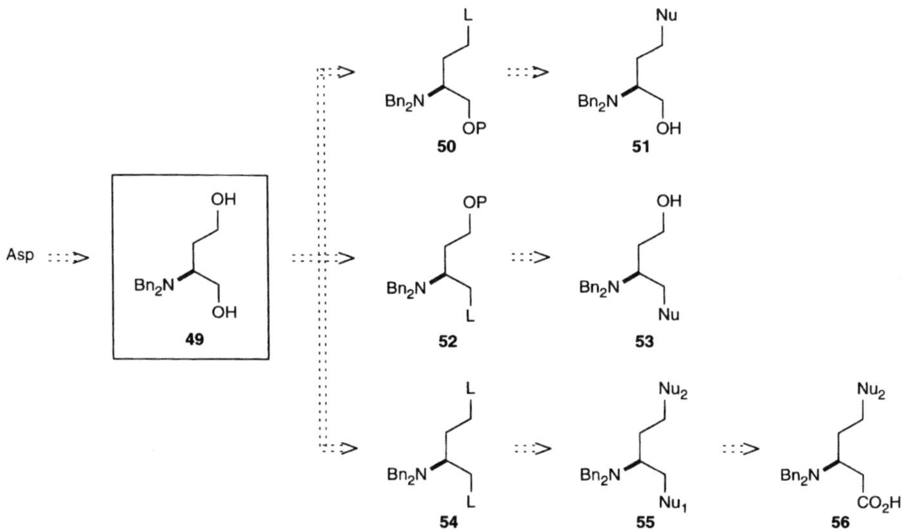

3.3.1 Synthesis of 1,2-Amino Alcohols, Application for the Construction of Amino Aldehydes, Amino Acids, and Tetrahydrobenzindoles

For the conversion of aspartic acid into the diol **49**, an efficient and convenient two-step synthesis was elaborated. Thus Asp was refluxed with benzyl bromide in aqueous K_2CO_3 to give the tetrabenzyl derivative **57**, which could be reduced by $LiAlH_4$ to give **49**.[36,37] In order selectively to activate the position 4, the diol **49** was treated with 1.1 equivalents of methanesulfonyl chloride. Subsequently, the crude reaction product was blocked with *tert*-butyldimethylsilyl chloride (TBDMS-Cl) in the presence of imidazole to afford the central intermediate **58**. For a nucleophilic displacement of the mesyloxy substituent, compound **58** was reacted with classical nucleophiles (CN^-, phthalimide, N_3^-), lower-order Gilman cuprates (Me_2CuLi, Ph_2CuLi) and heterocyclic anions (pyrrolide, pyrazolide, indolide) to give the respective substitution products in 51–99% yield. Finally, O-deprotection was accomplished using acetic acid or sodium hydroxide to afford the 1,2-amino alcohols **51**.[38]

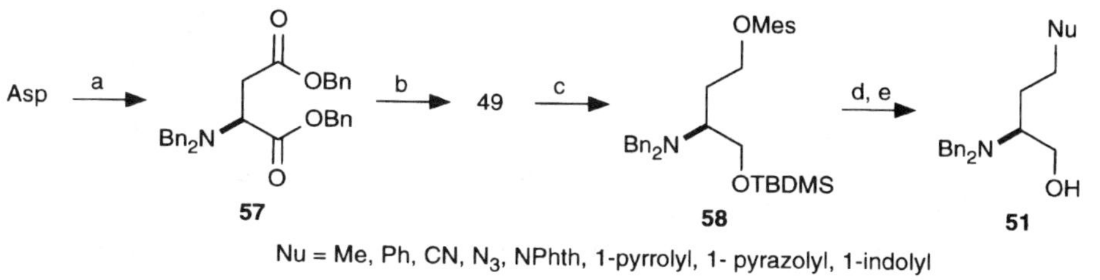

a: BnBr, K_2CO_3/H_2O (77%); b: $LiAlH_4$, THF −78 to −55°C (80%); c: 1. MesCl (1.1 equ), Et_3N, $CHCl_3$; 2. TBDMS-Cl, imidazole, DMF (57%); d: Me_2CuLi (Et_2O) or Ph_2CuLi (Et_2O) or NaCN (DMF) or NaN_3 (DMF) or Phth-K (DMF) or K-pyrrolide (THF) or K-pyrazolide (THF) or K-indolide (THF) (51–99%); e: HOAc/THF/H_2O or NaOH/EtOH/H_2O (78–99%).

Employing the pyrazolyl-substituted amino alcohol **51** as an example, Swern oxidation was performed to give the respective 1,2-amino aldehyde **59** (*X* = H). Subsequent oxidation with sodium chlorite resulted in formation of the amino acid **59** (*X* = OH). Thus the N,N-dibenzyl protected amino alcohols can serve as precursors for amino aldehydes and amino acids with side chains different from those of proteinogenic amino acids.[37]

Furthermore, the 1-indolyl substituted 1,2-amino alcohol **51** was used as a cyclization precursor for the synthesis of the 8-aminopyrido[1,2-a]indole **60**, which revealed selective affinity for the dopamine D-2 autoreceptor, according to receptor binding studies.[38]

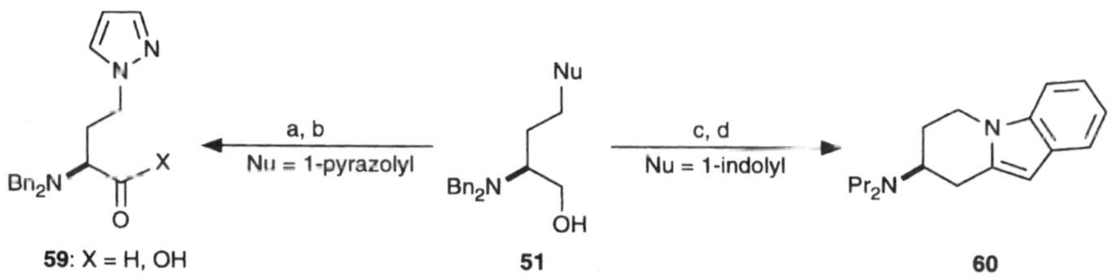

a: Oxalyl chloride, DMSO Et_3N, CH_2Cl_2 (67%); b: Na-chlorite, 2-methyl-2-butene, NaH_2PO_4, CH_3CN/t-BuOH (72%); c: Tf_2O, Et_3N, CH_2Cl_2 (86%); d: 1. H_2, $Pd(OH)_2/C$, MeOH (79%); 2. propionaldehyde, $NaCNBH_3$, MeOH, 74%).

3.3.2 Synthesis of 1,3-Amino Alcohols

Treatment of the diol **49** with TBDMS-Cl resulted in preferred attack at the less hindered position 4, affording **61** in 66% yield besides its regioisomer (2%) and the bis-protected derivative (11%). Activation of **61** with MesCl/Et_3N gave the projected intermediate **62a**. However, the mesylate **62a** could be only detected in pure form (by NMR spectroscopy) immediately after addition of the reagents since, in a following reaction step, rearrangement occurred. After 5 h, the secondary chloride **64b**, which was obviously formed through the aziridinium intermediate **63**, was isolated as a single product in 94% yield. It is assumed that the exclusive formation of the secondary alkyl halide **64b**, instead of a sterically favored ring opening at the less substituted position to give **62b**, is due to thermodynamic control. By analogy, the rearranged methanesulfonic ester **64a** could be prepared from **49** and Mes_2O/Et_3N. In the course of surveying the reactivity of **64a,b** with nucleophiles, it was found that the displacement reactions proceeded again through the aziridinium intermediate **63**, resulting in migration of the dibenzylamine group.[40] Reaction of **64b** with NaCN, phthalimide-K, or NaN_3 resulted in preferred attack at the less crowded position (kinetic control), when the protected 1,3-amino alcohols **65** were formed as the main products, besides the regioisomers **66**. For substitutions with organocuprates the chloride **64b** was not reactive enough. However, the more electrophilic mesylate **64a** gave a smooth reaction with Me_2CuLi or Bu_2CuLi to afford the protected 1,3-amino alcohols **65** (Nu = Me, Bu), exclusively. The reaction of **64a** with the potassium salts of indole as well as its 4- and 5-methoxy derivatives proceeded also under high regiocontrol (ratio of isomers: 11–18:1). On the other hand, employment of nucleophiles with leaving group character (LiCl, LiBr), allowing thermodynamic control, resulted in formation of the protected 1,4-amino alcohols **66** (Nu = Cl, Br). Using phthalimide as an example, it was shown that nucleophiles can also be introduced by reacting **61** under Mitsunobu conditions, affording 57% of **65** besides 23% of the easily separable regioisomer **66** (Nu = NPhth). For a selective cleavage of the TBDMS protecting group, the substitution products **65** and **66** were treated with aqueous NaOH or HOAc to afford the N,N-dibenzylamino alcohols **53** and **67**, respectively.

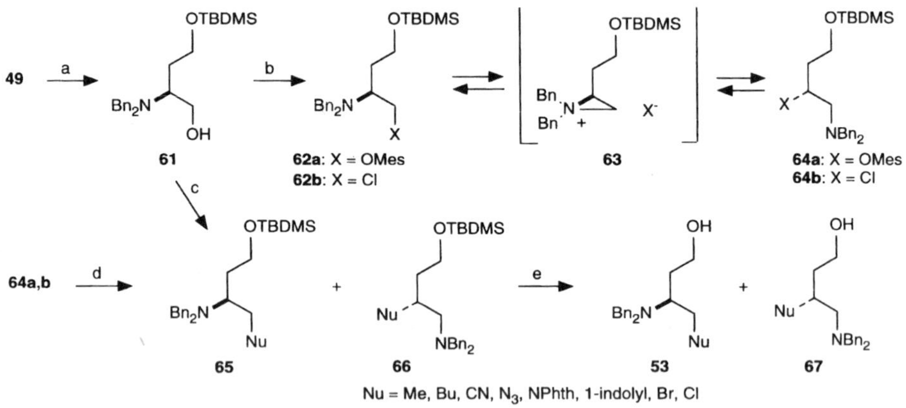

a: TBDMS-Cl, imidazole, DMF (66%); b: MesCl (Mes_2O), Et_3N, CH_2Cl_2 (94%); c: HNPhth, PPh_3, DEAD, THF (**65:** 57%, **66:** 23%); d: Me_2CuLi (Et_2O) or Bu_2CuLi (Et_2O) or NaCN (DMF) or NaN_3 (DMF) or PhthN-K (DMF) or K-indole (THF) or K-4- or 5-methoxyindolide) (**65:** 32–95%, **66:** 0–19%); LiBr or LiCl (CH_2Cl_2) (**66:** 68–72%), e: HOAc/THF/H_2O or NaOH/EtOH/H_2O (60–90%).

Starting from the protected 3,4-diaminoalcohols **65** and **66** (*R* = NPhth), we could demonstrate the optical integrity and the stereospecifity of the synthesis by hydrazinolysis and subsequent derivatization with optically pure *(R)*-1-phenylethylisocyanate. Furthermore, the formation of enantiomeric tetrabenzyl derivatives (opposite optical rotation values) by reductive benzylation indicated that the synthesis of **66** occurred exclusively through the aziridinium intermediate **63** and not by a direct S_N2 reaction of **64b**.

The thus-prepared enantiomerically pure 1,3-amino alcohols as well as the diamino derivatives can serve as valuable intermediates. Starting from the diaminobutanol **53** ($R = NH_2$), a practical synthesis of 8a-epi- and 1,8a-diepislaframine was elaborated.[37]

3.3.3 General Synthesis of β-Amino Nitriles and β-Amino Acids Including β-Proline Derivatives

Starting from the enantiomerically pure diol **49**, a concise and versatile synthesis of optically active β-amino acids including β-proline derivatives was reported when regioselective transformations of the 1,4-bis-electrophile **68** were facilitated by an anchimeric participation of the dibenzylamino group.[41] The method depends on activation of both hydroxyl groups of the building block **49**. Reaction of **49** with 2.5 equivalents of methanesulfonic chloride resulted in formation of the 1,4-bis-electrophile **68**, which could be characterized by immediately performed ^1H NMR spectroscopy. Treatment of crude **68** with 1.1 equivalents of LiCN gave a regioselective functionalization of position 1, affording the activated β-amino nitrile **69** in 74% yield. This shows that an activating neighboring group participation of the dibenzylamino group overrides its protecting steric effect. Modification of the amino ni-

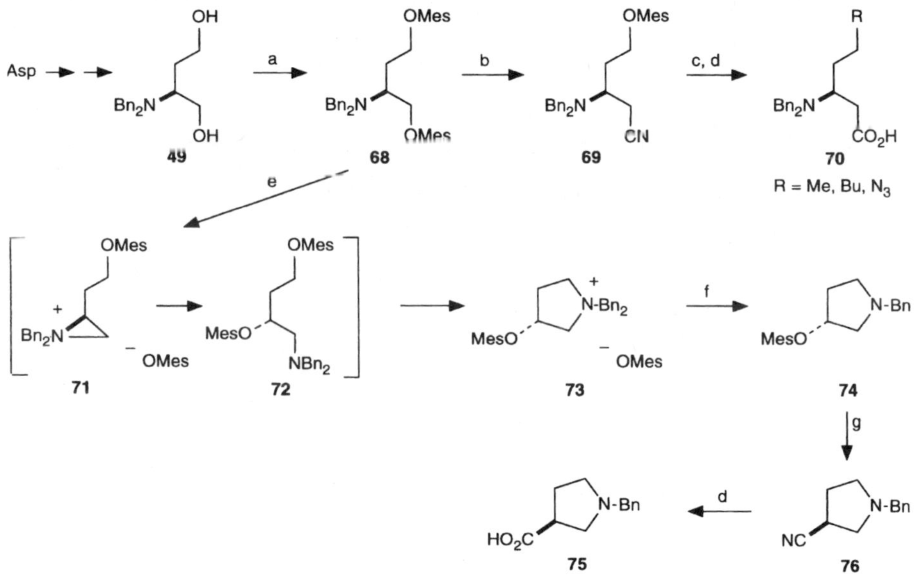

a: MesCl, Et_3N, THF; b: LiCN (1.1 eq), DMF, (74%, based on **49**). c: Me_2CuLi (Et_2O) or Bu_2CuLi (Et_2O) or NaN_3 (DMF) (47–85%); d: HCl, H_2O (30–79%). e: 1. flash chromatography (silica gel, petroleum ether–EtOAc 1:1); 2. 12 h RT (60%, based on **49**). f: H_2, $Pd(OH)_2/C$, H_2, MeOH, (78%). h: LiCN, DMF (61%).

trile side chain could be achieved by treatment of **69** with Me_2CuLi, Bu_2CuLi, or NaN_3 to give the respective displacement products. Subsequent acidic hydrolysis by aqueous HCl resulted in formation of the N-protected β-amino acids **70**. For a projected synthesis of β-proline derivatives, it was envisioned to utilize the previously observed tendency of α-dibenzylamino mesylates to undergo rearrangement via an aziridinium intermediate.[36,37] Thus migration of the dibenzylamino group via **71** was expected to give the 1,4-amino mesylate **72**, which was anticipated to react intramolecularly. In fact, separation of a colorless oil was observed after flash chromatography of the dimesylate **68** and subsequent storage of the product solution for 12 h at room temperature. The product, which crystallized after removal of the solvent, proved to be the pyrrolidinium salt **73**. Subsequent hydrogenolytic mono-debenzylation [$Pd(OH)_2/C$, H_2, 17 min] afforded the tertiary amine **74**. Introduction of the cyanide group gave the β-proline precursor **76**. Finally, synthesis of the β-amino acid **75** was accomplished by acidic hydrolysis. Further investigations of **75**, including esterification, N-deprotection, and subsequent coupling with *(R)*- and *(R,S)*-1-phenylethyl isocyanate, proved the synthetic material to be isomerically pure.

3.4 References

1. For examples, see: R. B. Silverman, *The Organic Chemistry of Drug Design and Drug Action*, Academic Press, 1992; S. Trumpp-Kallmeyer, J. Hoflack, A. Bruinvels, M. Hibert, *J. Med. Chem.* **35**, 3448 (1992); J. K. Chen, S. L. Schreiber, *Angew. Chem.* **107**, 1041 (1995); *Angew. Chem. Int. Ed. Engl.* **34**, 953 (1995); R. A. Lerner, S. J. Bencovic, P. G. Schulz, *Science*, **252**, 659 (1991).
2. For examples, see: H. Paulsen, *Angew. Chem.* **102**, 851 (1990); *Angew. Chem. Int. Ed. Engl.* **29**, 823 (1990); J. Lechner, F. Wieland, *Ann. Rev. Biochem.* **58**, 173 (1989).
3. G. M. McGarvey, J. M. Williams, R. N. Hiner, Y. Matsubara, T. Oh, *Tetrahedron Lett.* **24**, 2733 (1983).
4. G. M. McGarvey, J. M. Williams, R. N. Hiner, Y. Matsubara, *J. Am. Chem. Soc.* **108**, 4943 (1986).
5. J. Jurczak, J. Kozak, A. Golebiowski, *Tetrahedron* **48**, 4231 (1992).
6. B. D. Christie, H. Rapoport, *J. Org. Chem.* **50**, 1239 (1985).
7. P. Gmeiner, P. L. Feldman, M. Chu-Moyer, H. Rapoport, *J. Org. Chem.* **55**, 3068 (1990).
8. (a) P. L. Feldman, H. Rapoport, *J. Org. Chem.* **51**, 3882 (1986). (b) P. L. Feldman, H. Rapoport, *J. Am. Chem. Soc.* **109**, 1603 (1997).
9. P. J. Dunn, R. Häner, H. Rapoport, *J. Org. Chem.* **55**, 5017 (1990).
10. J.-P. Wolf, H. Rapoport, *J. Org. Chem.* **54**, 3164 (1989).
11. J. M. Humphrey, R. J. Bridges, J. A. Hart, A. R. Chamberlin, *J. Org. Chem.* **59**, 2467 (1994).
12. J. M. Dener, L.-H. Zhang, H. Rapoport, *J. Org. Chem.* **58**, 1159 (1993).
13. (a) J. Rudinger, K. Poduska, M. Zaoral, *Collect. Czech. Chem. Commun.* **25**, 2022 (1960). (b) M. Waki, Y. Katajima, N. Izumiya, *Synthesis* **1981**, 266. (c) M. Otsuka, A. Kittaka, T. Iimori, H. Yamashita, S. Kobayashi, M. Ohno, *Chem. Pharm. Bull.* **33**, 509 (1985).
14. M. G. Hoffmann, H.-J. Zeiss, *Tetrahedron Lett.* **33**, 2669 (1992).
15. (a) J. E. Baldwin, M. North, A. Flinn, *Tetrahedron* **44**, 637 (1988). (b) B. Gong, D. G. Lynn, *J. Org. Chem.* **55**, 4763 (1990).

16. H. C. Uzar, *Synthesis* **1991**, 526.

17. L. Velluz, G. Amiard, R. Heymes, *Bull. Soc. Chim. Fr.* **1954**, 1012; M. T. Reetz, M. W. Drewes, A. Schmitz, *Angew. Chem.* **99**, 1186 (1987).

18. For a review, see: M. T. Reetz, *Angew. Chem.* **103**, 1559 (1991).

19. P. Gmeiner, *Arch. Pharm (Weinheim, Ger.)* **324**, 551 (1991).

20. P. Gmeiner, *Tetrahedron Lett.* **31**, 5717 (1990).

21. P. Gmeiner, *Liebigs Ann. Chem.* **1991**, 501.

22. P. Gmeiner, *Heterocycles* **32**, 1499 (1991).

23. P. Gmeiner, H. Lerche, *Heterocycles* **31**, 9 (1990).

24. P. Gmeiner, J. Mierau, G. Höfner, *Arch. Pharm (Weinheim, Ger.)* **325**, 57 (1992).

25. For a review, see: P. H. Anderson, E. B. Nielson, *Drug News & Perspectives* **4**, 150 (1991).

26. S. J. Hocart, W. A. Murphy, D. H. Coy, *J. Med. Chem.* **33**, 1954 (1990); B. M. Haffar, S. J. Hocart, D. H. Coy, S. Mantey, H.-C. V. Chiang, R. T. Jensen, *J. Biol. Chem.* **266**, 316–22 (1991); M. Rodriguez, M.-F. Lignon, M.-C. Galas, P. Fulcrand, C. Mendre, A. Aumelas, J. Laur, J. Martinez, *J. Med. Chem.* **30**, 1366–73 (1987); K. Shiosaki, C. W. Lin, M. R. Leanna, H. E. Morton, T. R. Miller, D. Witte, M. Stashko, A. M. Nadzan, *BioMed. Chem. Lett.* **3**, 855–60 (1993); J. Martinez, J.-P. Bali, M. Rodriguez, B. Castro, R. Magous, J. Laur, *J. Med. Chem.* **28**, 1874–79 (1985).

27. N. Upton, *TiPS* **15**, 456 (1994) and references cited therein.

28. D. S. Jones, A. Srinivasan, S. Kasina, A. R. Fritzberg, D. W. Wilkening, *J. Org. Chem.* **54**, 1940–43 (1989).

29. O. Mitsunobu, *Synthesis* **1981**, 1.

30. P. Gmeiner, E. Hummel, C. Haubmann, *Arch. Pharm. (Weinheim, Ger.)* **328**, 265 (1995).

31. A. J. Mancuso, D. Swern, *Synthesis* **1981**, 165.

32. G. R. Pettit, Y. Kamano, H. Kizu, C. Dufresne, C. L. Herald, R. J. Bontemps, J. M. Schmidt, F. E. Boettner, R. A. Nieman, *Heterocycles* **28**, 553–58 (1989).

33. D. C. Carter, R. E. Moore, J. S. Mynderse, W. P. Niemczura, J. S. Todd, *J. Org. Chem.* **49**, 236–41 (1984); J. S. Mynderse, H. Hunt, *Journal of Natural Products* **51**, 1299–301 (1988).

34. P. Gmeiner, E. Hummel, C. Haubmann, *Liebigs Ann. Chem.* **1995**, 1987–1992.

35. M. T. Reetz, F. Kayser, K. Harms, *Tetrahedron Lett.* **35**, 8769–72 (1994).

36. P. Gmeiner, A. Kärtner, D. Junge, *Tetrahedron Lett.* **34**, 4325 (1993).

37. P. Gmeiner, D. Junge, A. Kärtner, *J. Org. Chem.* **59**, 6766 (1994).

38. P. Gmeiner, A. Kärtner, *Synthesis* **1995**, 83.

39. P. Gmeiner, J. Kärtner, J. Mierau, *Arch. Pharm.–Pharm. Med. Chem.* **328**, 626 (1995).

40. For related migration reactions, see: A. Shanzer, L. Somekh, *J. Am Chem. Soc.* **104**, 5836 (1982); H. Setoi, H. Takeno, M. Hashimoto, *Heterocycles* **24**, 1261 (1986).

41. P. Gmeiner, F. Orecher, C. Thomas, K. Weber, *Tetrahedron Lett.* **36**, 381 (1995).

CHAPTER

4

Synthesis of Enantiomerically Pure β-Amino Acids Starting from L- and D-Aspartic Acids

Charles W. Jefford and James McNulty

4.1 Introduction

L-Aspartic acid (**1**) is an ideal precursor to many enantiomerically pure β-amino acids (**2**), since it already possesses the butanoic backbone and only requires that the carboxyl group at C1 be differentiated from the other at C4 and transformed into the desired 4-alkyl or aryl substituent (Scheme 4.1). The assembly of *syn* and *anti* α-substituted β-amino acids (**3**) is more demanding, requiring an additional step, diastereoselective functionalization of the C3 position. Obviously, a repetition of these processes with D-aspartic acid (**4**) would provide the enantiomeric β-amino acids **5** and

Scheme 4.1

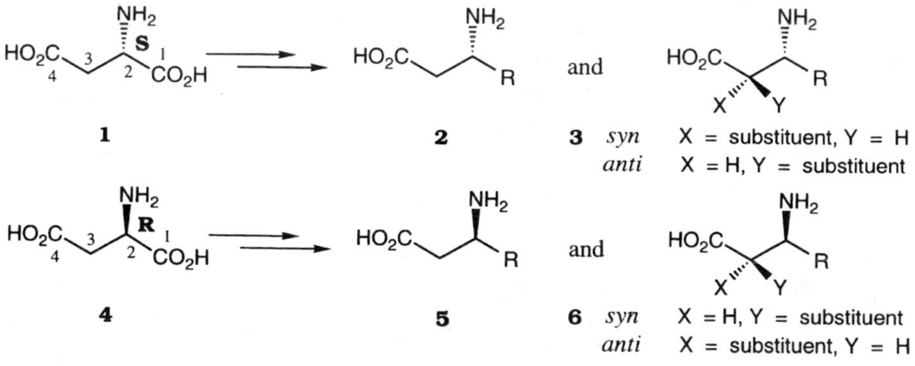

1 **2** **3** *syn* X = substituent, Y = H; *anti* X = H, Y = substituent

4 **5** **6** *syn* X = H, Y = substituent; *anti* X = substituent, Y = H

the α-substituted diastereomers **6**. It will be seen that most syntheses entail the intermediacy of β-lactams or γ-lactones; however, some are surprisingly effective although they only make use of simple acyclic intermediates.

4.2 Synthesis of β-Amino Acids. Acyclic Route

The problem of differentiating the carboxyl groups is conveniently solved by the selective formation of the half-ester. Mere dissolution of aspartic acid in rigorously dried alcohol containing dry hydrogen chloride for a few minutes and addition of diethyl ether was enough to deposit the 4-mono-ester hydrochloride.[1] Consequently, thanks to this method and selective enzymic hydrolysis of the diesters, various mono-esters are now commercially available.

(a) A typical approach starts with 4-*t*-butyl L-aspartate (**7**) in which the free carboxyl acid function is ready to be transformed into the required substituent.[2] Protection as the N-benzyloxycarbonyl derivative **8** and treatment with ethyl chloroformate afforded the mixed anhydride, which on reduction with sodium borohydride gave the alcohol **9** (Scheme 4.2). The tosylate **10** and iodide **11** were next prepared by standard methods, and by addition to lithium dialkylcuprates, the well-known Gilman reagents, alkylation was achieved. Tetrahydrofuran (THF) was the best solvent for reaction of **11**. The tosylate **10** reacted best in ethyl ether containing a little THF to solubilize the organocuprate. Primary alkyl groups were the easiest to introduce, and the alkylated product **12** was obtained in high yield regardless of the nature of the leaving group. However, nucleophilic displacement by secondary alkyl cuprates was only successful with the iodide **11**. The tosylate **10** invariably led to poor yields or elimination products. The reaction of **11** with lithium di-(8-methyldecyl)cuprate is illustrative. It gave **13** in 86% yield necessarily as a diastereomeric mixture since the dialkylcuprate itself was racemic. Acid hydrolysis to the acid **14** followed by deprotection by catalytic hydrogenolysis was straightforward and furnished the β-amino acid **15**, a component of the antifungal peptide, iturin A.[3] Hydrolysis and deprotection of the other esters **12** (where *R* = Bu or i-Pr) similarly led to the corresponding (3*R*)-3-aminooctanoic and heptanoic acids **16** in six steps from **8** and overall yields of 35–48%.

The optical purity of the amino esters was checked by ^{1}H-NMR analysis after deprotection and derivatization with (1*S*)-1-phenylethyl isocyanate. Enantiomeric excesses better than 98% were observed.

(b) A method similar to that preceding exploited the sterically differentiated 1,4-diol **18**, obtained from L-aspartic acid (**1**) by treating it with excess benzyl bromide and reducing the resulting tetrabenzyl derivative **17** with lithium aluminum hydride (Scheme 4.3).[4] The diol was converted to the *bis*-methanesulfonate **19**. Its reaction with lithium cyanide in dimethylformamide (DMF) was regioselective, giving exclusively the 1-cyano derivative **20** from **17**. Despite apparent hindrance at C1, the electronic assistance of the nearby dibenzylamino group has clearly predominated. Next, nucleophilic displacement on the remaining primary methanesulfonate with lithium dibutylcuprate or sodium azide provided the corresponding substitution products **21**

Scheme 4.2

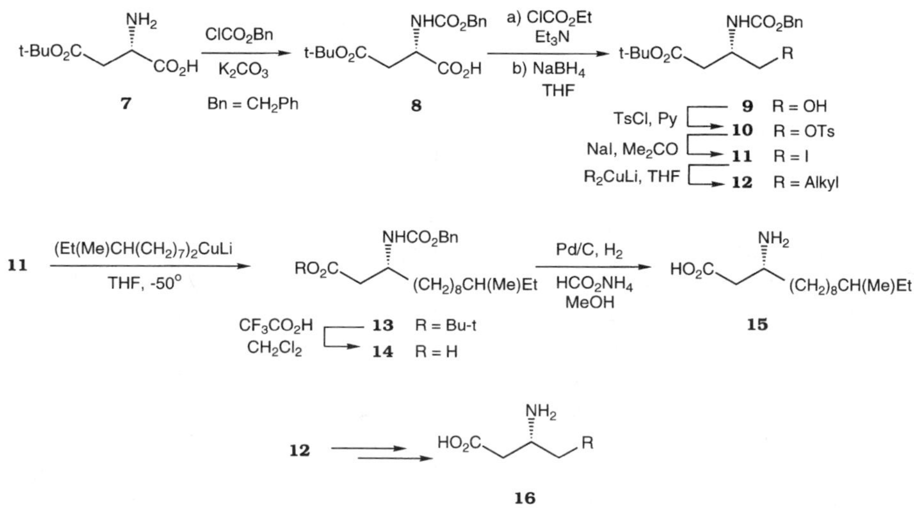

and **22**. Subsequent acid hydrolysis liberated the N-protected β-aminopentanoic acids **23** and **24**.

The propensity to neighboring group participation was turned to advantage for the preparation of β-proline. The 1,4-*bis*-methanesulfonate **19** was rearranged to the pyrrolidinium salt **26** presumably through the intervening 1,3-isomer **25** simply by exposure to silica gel for several hours (Scheme 4.4). The free base **27** was generated by catalytic hydrogenolysis over palladium on charcoal. Substitution by cyanide ion with inversion of configuration gave **28** which on acid hydrolysis delivered (3S)-*N*-benzyl-β-proline (**29**).

Scheme 4.3

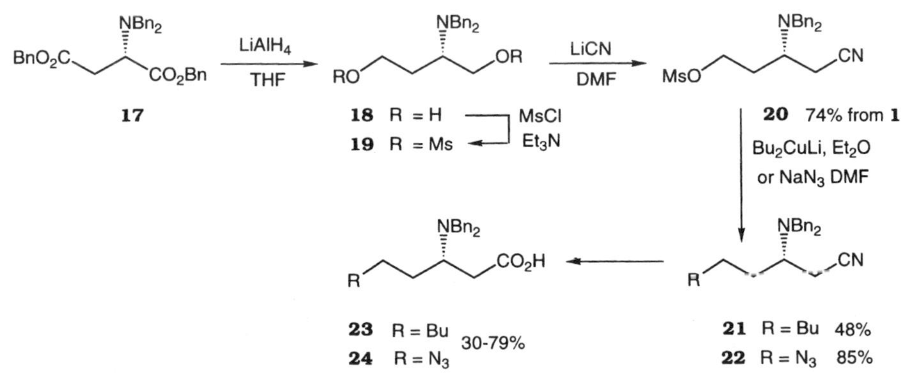

Scheme 4.4

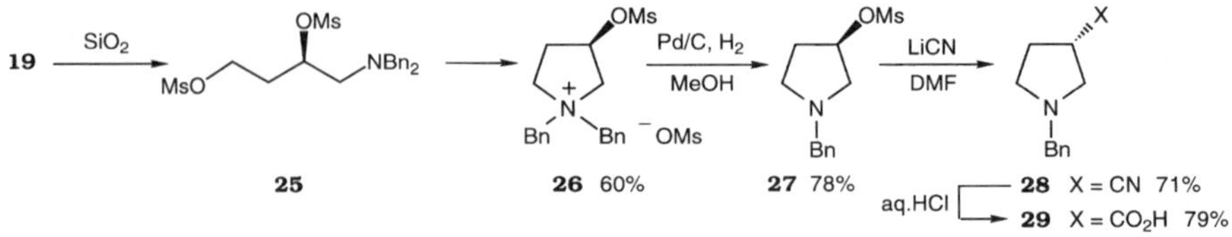

The configuration of **29** was corroborated by converting (3*R*)-3-hydroxyproline (**30**) to an authentic sample of **27** (Scheme 4.5). It was found to be spectroscopically identical to **27** derived from **1** by the above-mentioned route. Moreover, both samples had the same optical rotation. As for the N-benzyl-β-proline **29**, its optical purity was verified by esterification, deprotection, reaction with (1*R*,*S*)-and (1*R*)-1-phenylethyl isocyanate and inspection of their NMR spectra. In general, the yields of the above reactions were moderate to good. The key intermediate **20** was obtained in 74% from **1**. Nevertheless it was to be expected that reaction of mesylates with Gilman reagents (e.g., **20**) would be middling (48%). Hydrolysis of the nitriles **20**, **21**, and **28** gave variable yields of the acids (30–79%).

(c) An example of the homologation of the free carboxyl group of a half-ester is provided by 4-benzyl N-benzyloxycarbonyl-L-aspartate (**31**).[5] Formal reduction to the aldehyde **32** was easy, occurring in 59% yield (Scheme 4.6). Unfortunately, Wittig olefination with *t*-butyl triphenylphosphoranylacetate consistently gave a 1:1 mixture of

Scheme 4.5

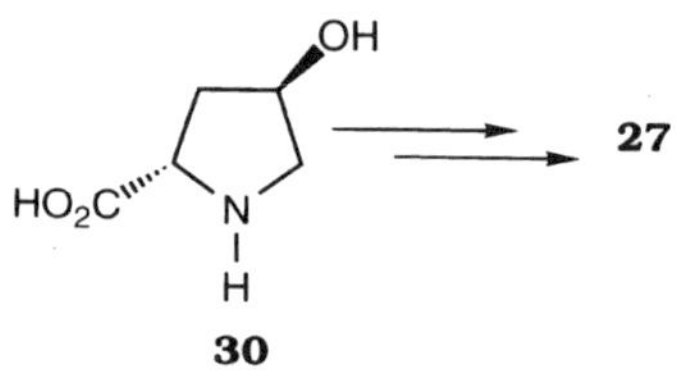

Scheme 4.6

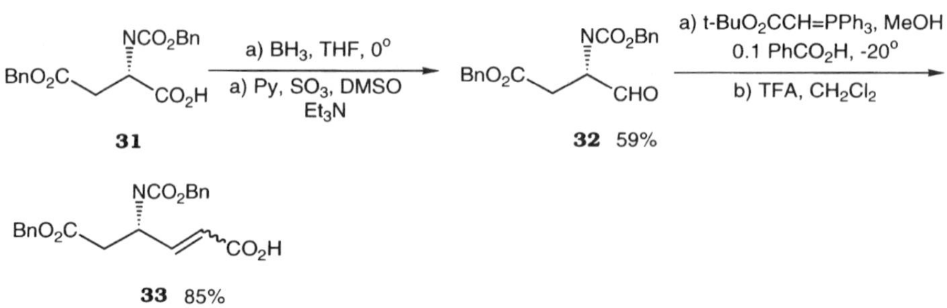

the *E*- and *Z*-isomers (**33**), albeit in high yield. The *Z* isomer could be made to crystallize out of the reaction mixture in about 37% yield; so the procedure is of practical value. The products were not saponified or deprotected to the free acids, but used for other synthetic ends.

4.3 Synthesis of α-alkyl-β-amino acids. Acyclic Route

The alkylation of aspartic esters would appear to be a perfectly straightforward way of gaining access to the corresponding acids.

(a) The double deprotonation of di-*t*-butyl N-formyl-L-aspartate (**34**) was cleanly accomplished in THF at –78° with lithium diethylamide (but *not* with lithium di-*iso*-propylamide) to give the *bis*-lithium salt **35** (Scheme 4.7).[6] Active alkyl halides, such as methyl and ethyl iodide, allyl and benzyl bromide, on reaction with **35** gave the 3- and 2-monoalkylated products **36** and **37** in a ratio of about 3.5:1 and a yield of 60–70%. Chromatographic separation revealed that the 3-alkylated esters **36** were pure diastereomers. Therefore, alkylation at C3 was completely directed by the adjacent chiral center. Even **37** was optically active. By the use of chiral shift reagents, ^{1}H-NMR spectroscopy indicated an enantiomeric excess of 60%. The configuration of the methyl derivative **36** was confirmed by hydrolysis to the known (2*S*,3*R*)-3-methylaspartic acid (**38**). Thus all the other alkylated esters **36** were assumed to have the *erythro* or *anti* configuration. These alkylations, although not really providing a synthetic method as such, are of mechanistic significance.

(b) A reasonable supposition is that other aspartic esters would also undergo alkylation in the same *anti* sense and that it could be limited to the C3 position by a suit-

Scheme 4.7

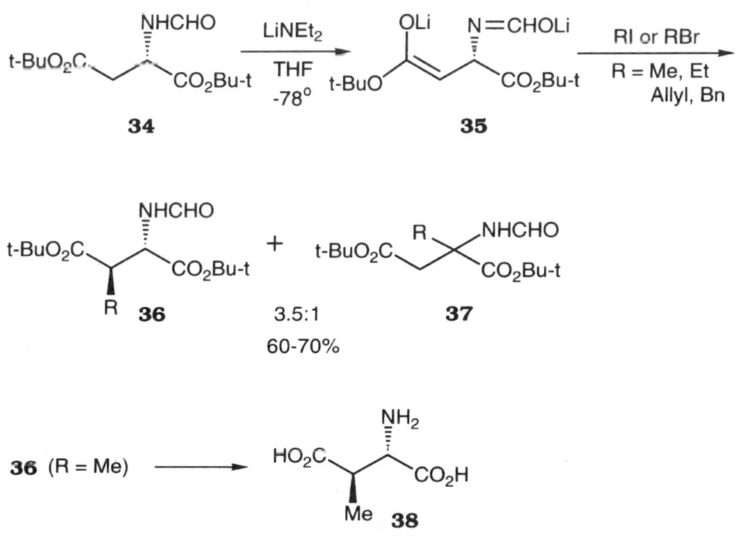

able choice of substituents and N-protection. Just such an example is the *t*-butyl methyl ester **41** (Scheme 4.8). L-Aspartic acid (**1**) was regioselectively esterified to 4-methyl-L-aspartate (**39**), protected as the N-benzyloxycarbonyl derivative (**40**), and treated with *iso*-butene in sulfuric acid, thereby giving the diester **41**.[7] Alkylation of **41** was performed through the anion generated either by lithium di-*iso*-propylamide (LDA) or lithium hexamethyldisilazide (LiHMDS). Benzyl and allyl bromides were effective electrophiles, whereas primary iodides failed to react. Methyl iodide resulted in both C- and N-alkylation. The 3-benzyl derivatives **42** and **43** and 3-allyl derivatives **44** and **45** were obtained in moderate yields as mixtures of *erythro* and *threo* isomers in ratios of 5:1 and 3:1, respectively. This is a surprising result and shows that the enolate ion manifests little diastereofaciality. No configurations were assigned; however, in keeping with the above precedent, the major diastereomer may be assumed to have the *erythro* or *anti* configuration.

The choice of base turned out to be critical for controlled alkylation. Deprotection of **42** and reaction of the resulting amine with Mosher's acid chloride [(–)-(*R*)-3,3,3-trifluoromethyl-2-methoxy-2-phenylpropanoyl chloride] gave an amide, the NMR spectrum of which was compared with that of the analogous amide obtained by the same sequence from racemic aspartic acid. When LDA was used for alkylation, the enantiomeric excess was estimated at 85%. Thus at some stage, epimerization had occurred at the C2 position, without leading to any doubly alkylated product. In contrast, LiHMDS behaved impeccably, revealing no trace of racemization. Finally, hydrolysis of **42** with lithium hydroxide in aqueous methanol liberated the 2-benzyl-3-N-(benzyloxycarbonyl)amino acid **46** of the presumed 2*R*,3*S* configuration. Despite the formation of diastereomeric mixtures, which are at least separable, the

Scheme 4.8

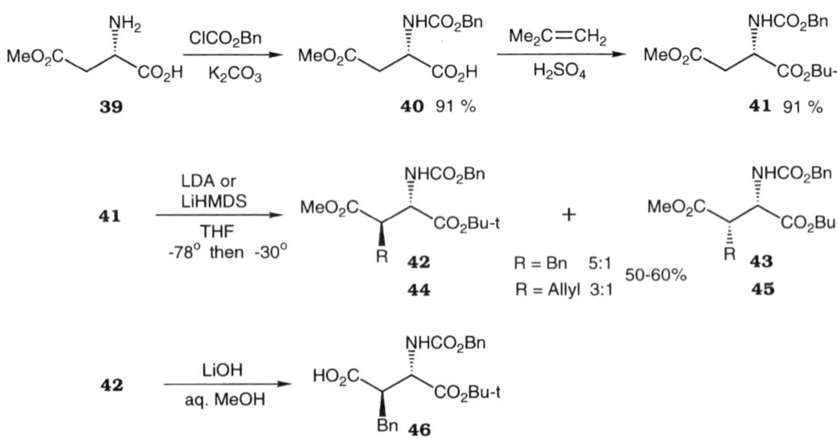

conservation of the original configuration of the aspartic ester is noteworthy, provided, of course, that LiHMDS be used as base.

4.4 Synthesis of α,β-Diamino Acids. Acyclic Route

The previous two sets of results appear to demonstrate that acyclic diastereoselection favors the *anti* configuration. Further experiments show that the nature of the ester at the C1 position and the size of the N-protecting group exercise a subtle effect on the stereochemical outcome of alkylation.

(a) The synthesis of (2*S*,3*R*) and (2*S*,3*S*)-2,3-diamino-4-phenylbutanoic acids (**47** and **48**) is a telling illustration. Such α,β-diamino acids are fairly rare in nature, particular examples being the *syn* and *anti* butanoic acids (**49** and **50**), both of which form part of the peptide antibiotic, aspertocin (Scheme 4.9).[8]

The starting material was the methyl half-ester **39**, which by standard procedures was converted into the diester **51** bearing the large N-(9-phenylfluoren-9-yl)amino group (Scheme 4.10).[9] Thus there is no chance of α-alkylation. Indeed, submission of **51** to KHMDS and benzyl bromide or iodide proceeded efficiently. The *syn* and *anti* diastereomers **52** and **53** were formed in ratios of 2:1 and 7:1, respectively. Alkaline hydrolysis of the product mixture to the acids **54** and Curtius degradation with diphenyl phosphorazidate afforded the epimeric cyclic ureas **55** and **56**. Chromatographic separation and successive exposure to trifluroacetic acid (TFA) and 2N hydrochloric acid provided the *syn* and *anti* acids **47** and **48**.

Although the better electrophile, benzyl iodide, certainly brings out greater diastereoselectivity, the size of the N-protecting group helps, too. Alkylation of the highly sterically encumbered derivative **57** with benzyl bromide and iodide was extremely selective, giving the *syn* and *anti* diastereomers **58** and **59** in high yield and ratios of 12:1 and 25:1, respectively (Scheme 4.11). Catalytic hydrogenolysis of **58** in the presence of di-*t*-butyl dicarbonate [$(BOC)_2O$] yielded the BOC-protected acid **60**. Finally, as before, Curtius degradation, cyclization to the urea **61**, and hydrolysis all took place efficiently and delivered the *anti* acid **48**.

A further experiment revealed that hindrance on the amino group must be abetted by the correct alkoxy substituent at C1. The 1-methyl ester **62** on alkylation with benzyl iodide underwent a dramatic reversal of behavior to favor the *anti* diastereomer **63** over the *syn* **64** by 1.5:1 (Scheme 4.12).

Scheme 4.9

47	R = Bn	**48**
49	R = Me	**50**

Scheme 4.10

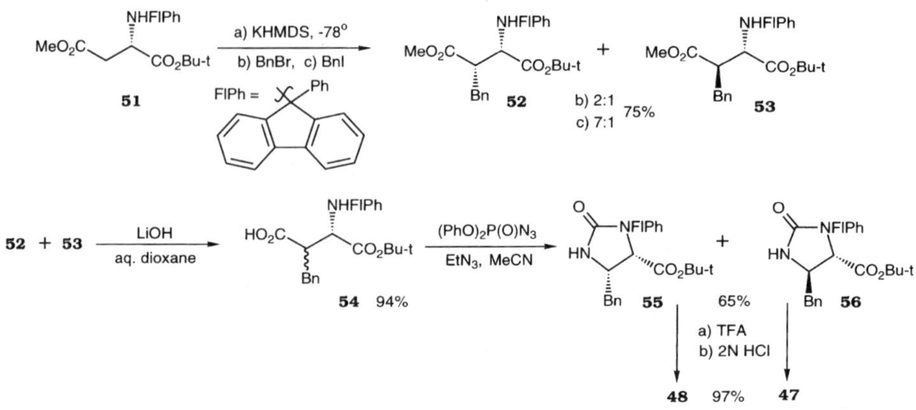

Scheme 4.11

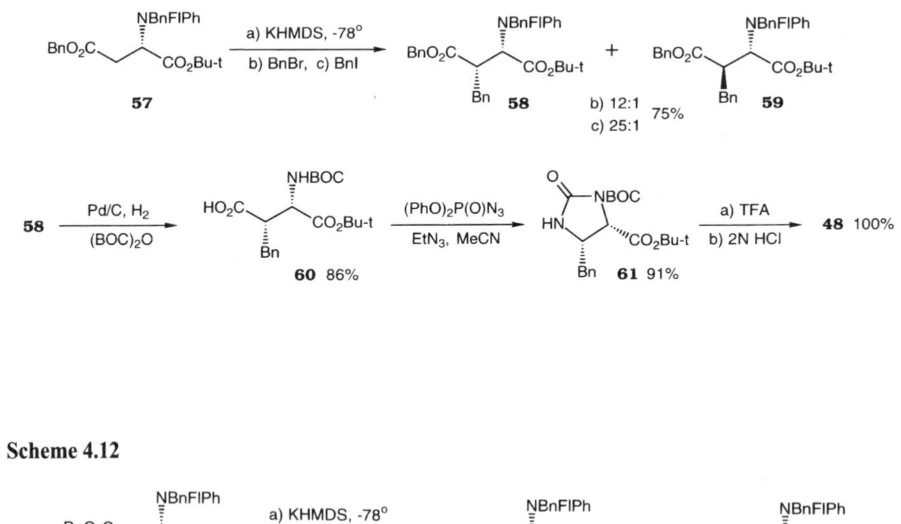

Scheme 4.12

The steric course is noteworthy; the *syn* benzylated aspartates **52** and **58** are transformed into the *anti* acid **48**. The results also show that great encumbrance around the C1 and C2 positions imposes *syn* diastereoselectivity.

4.5 Synthesis of α-Alkyl-β-Amino Acids. Acyclic versus Cyclic Enolates

(a) Diastereoselectivity should also improve when alkylation is effected within or external to five-membered ring derivatives of aspartic acid where much greater steric control might be expected. In addition, both *syn* or *anti* diastereoselection should be possible. Aspartic acid (**1**) was protected as its N-benzyloxycarbonyl derivative, which was closed to its anhydride **65** by exposure to acetic anhydride and acetic acid (Scheme 4.13). Regioselective reduction with sodium borohydride in THF furnished the γ-lactone **66**. Alkylation of the dianion derived from **66** by the action of LDA with the usual active halides, here typified by methyl iodide, gave mainly the *trans* monomethylated product **67**, but accompanied by the *cis* isomer **68** in ratios of about 7:1 in high yield.[10] The N-benzoyl derivative **69**, derived from **66**, behaved similarly, but exhibited higher diastereoselectivity giving the *trans* and *cis* products **70** and **71** in a ratio of 10:1. By further varying the conditions of alkylation, it was concluded that the product ratios were essentially insensitive to the nature of the electrophile and the base.

The opposite diastereoselectivity was observed for the 2-phenyloxazoline derivatives **72** and **73**, which were prepared from **66** and **69** by standard methods (Scheme 4.14). This time, the action of LDA generated the chelated monoanion, which on quenching with methyl iodide gave the *anti* monomethylated derivative as the major and the *syn* derivative as the minor product in high yields. Ester **72** gave **74** and **75** in a ratio of 2:1, whereas amide **73** was more sterically discriminating, producing **76** and **77** in a ratio of 5:1. The complementarity of this reaction with the previous one was

Scheme 4.13

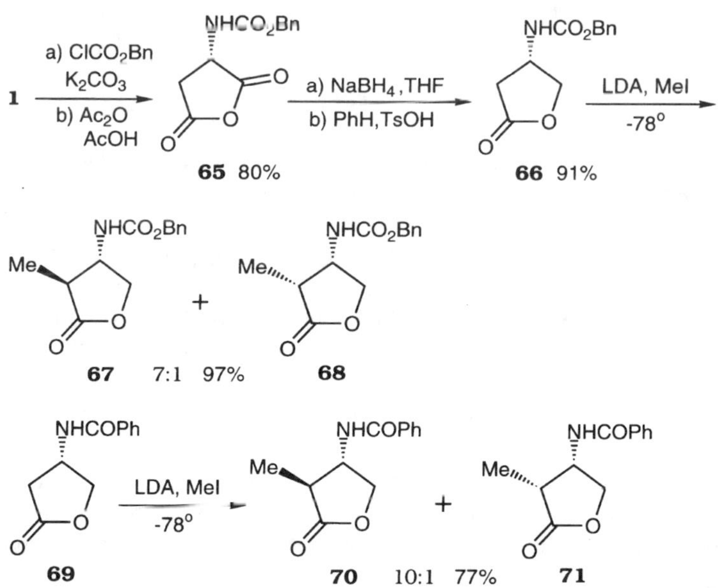

Scheme 4.14

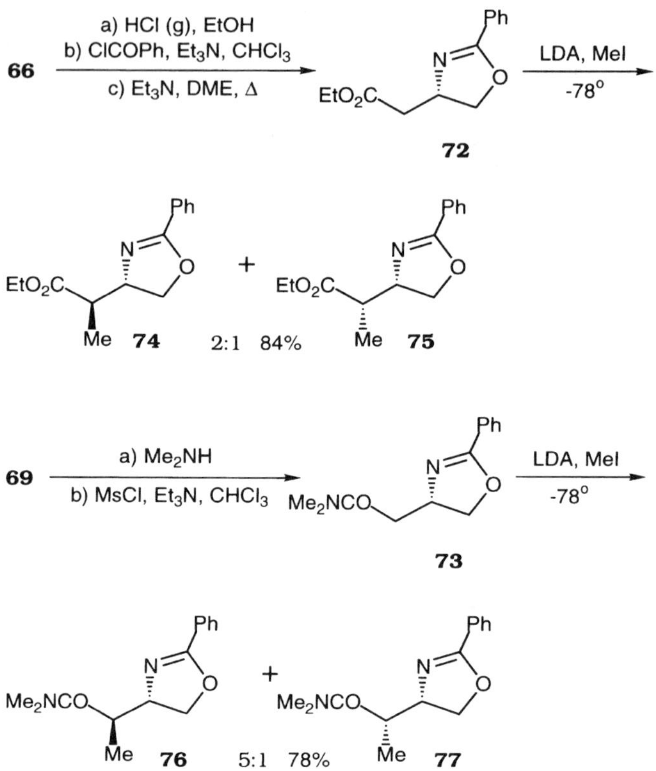

demonstrated by the conversion of **70**, the major product from the lactone **69**, to **77**, the minor product obtained from the oxazoline **73**. This correlation also substantiated the configurational attributions.[10,11]

The diastereomeric excesses for all these methylations, regardless of the kind of control, by hindrance or chelation, were not exceptional and ranged from 34 to 82%. In mitigation, the yields were uniformly high and the mixtures were readily separable by column chromatography. Although not performed, the products should be easily convertible into pure, diastereomeric α-methyl-β-amino acids.

By way of conclusion, it may be noted that stereoselection for the lactones springs from within the ring and would lead mainly to a *syn* product in the open-chain form, whereas the oxazolines provide it by a ring that is external to the reaction site, favoring an *anti* product. In either case, the cyclic intermediates do not appear to be superior to the acyclic ones. Further experiments, however, will prove otherwise.

4.6 Synthesis of α-Alkyl-β-Amino Acids. β-Lactam Route

β-Lactams and β-amino acids are complementary to each other. The first can be cyclized to the second and the reverse process achieved by hydrolysis. However, the β-lactam has the stereoelectronic advantage of being a small ring, which controls very effectively events such as substitution by electrophiles. These two features taken together provide a basis for the synthesis of α-substituted β-amino acids possessing the *anti* configuration.

(a) The cyclization of L-aspartic acid, although simple in principle, requires careful preparation. First, dibenzyl L-aspartate (**78**) was silylated with trimethylsilyl chloride (Scheme 4.15). Treatment of the N-trimethylsilyl derivative **79** with an equivalent of *t*-butylmagnesium chloride brought about condensation to the chiral azetidinone ester **80**.[12] Removal of the trimethylsilyl protecting group to give **81** and its replacement by *t*-butyldimethylsilyl followed by debenzylation of the better protected derivative **82** under catalytic conditions gave the carboxylic acid (**83**), setting the stage for alkylation.[13] By deprotonation with LDA, the dianion derived from **83** on treatment with methyl iodide afforded exclusively the *trans*-monomethyl isomer **84** in an isolated yield of 37%. Acid hydrolysis cleaved the ring to give the known (2*S*,3*R*)-3-methylaspartic acid (**38**).[14] As expected, alkylation of **83** with ethyl iodide and allyl and benzyl bromides was equally successful. A shortcut to **83** was found by directly converting **78** to its *t*-butyldimethylsilyl derivative **85** and effecting cyclization to the azetidinone ester **82** followed by catalytic hydrogenolysis.[15]

Scheme 4.15

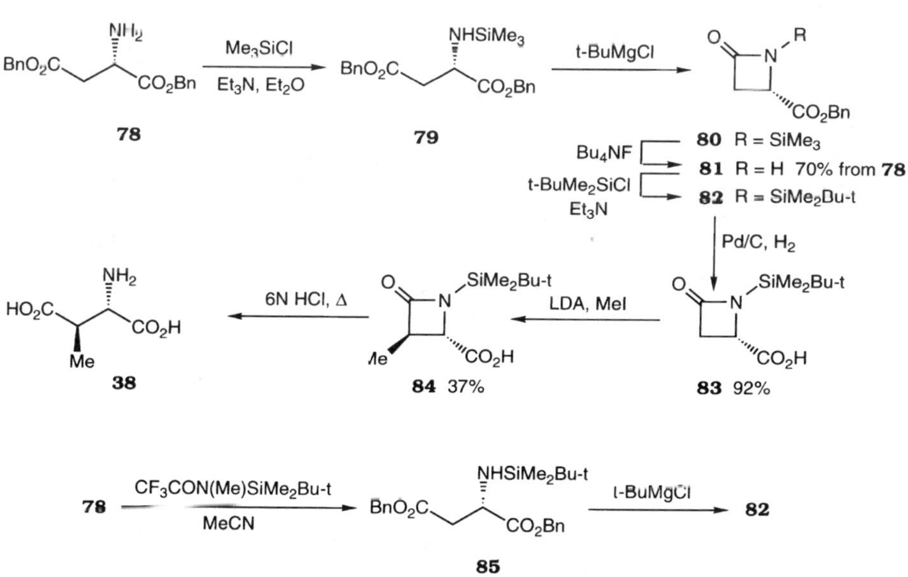

Scheme 4.16

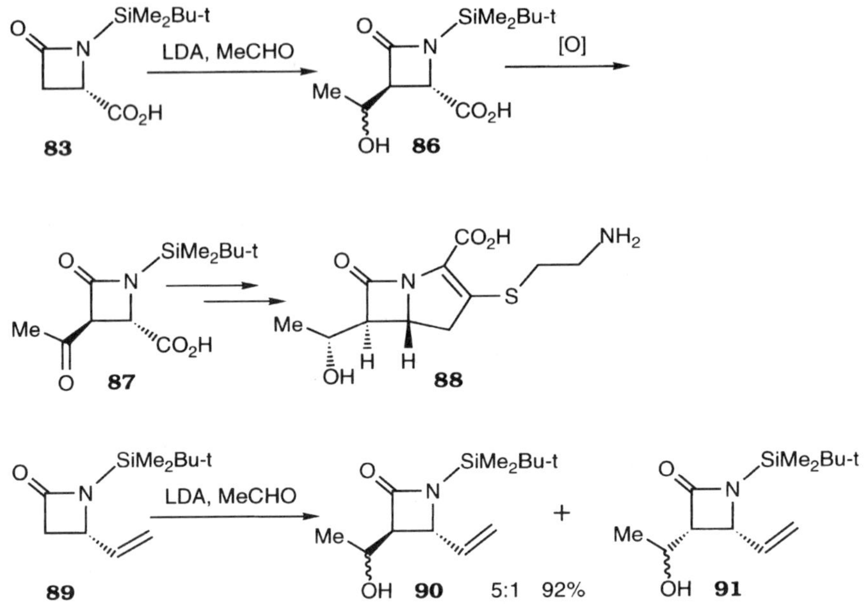

It should be mentioned that the groundwork for the above methodology had been laid earlier by studies directed to the synthesis of the bicyclic antibiotic thienamycin (**88**) (Scheme 4.16). In one approach, the hydroxyethyl side chain was introduced with complete stereocontrol by quenching the dianion derived from **83** with excess acetaldehyde.[13] The epimeric hydroxy acids **86** on oxidation gave the *trans* keto acid **87** as the sole product. The carboxyl function of **87** was then subsequently elaborated to provide a shortened version of a chiral synthesis of **88**.

There are some signs that the substituent at C4 on a β-lactam may not always overwhelmingly favor *trans* attack by the electrophile on the corresponding enolate anion. Its electronic or steric nature may play a role. The 4-vinyl derivative **89** serves as an illustration.[16] Hydroxyethylation under the previously used conditions produced the *trans* and *cis* isomers **90** and **91** in a ratio of 5:1 and in 92% yield. Nonetheless, the carboxyl substituent exerts greater diastereoselectivity. The utility of 2-azetidinone-4-carboxylic acid (**83**) as a chiral template is incontestable. Apart from its use as a precursor to bicyclic β-lactams, **83** acts as a cyclic form of L-aspartic acid, thereby enabling *anti* 3-substituted derivatives to be prepared with high diastereomeric purity.

4.7 Synthesis of β-Amino Acids. γ-Lactone Route

The previously described conversion of N-protected aspartic anhydride to the γ-lactone requires amplification. It was recognized long ago that unsymmetrical cyclic anhydrides undergo reduction with sodium borohydride on the carbonyl group adjacent to the more substituted center.[17] Relevant examples are the reduction of the protected malic and aspartic anhydrides **92** and **65** to the γ-lactones or 3-substituted 4-butanolides **93** and **66** in preparatively useful yields (Scheme 4.17).[11,18] Even an exceptionally voluminous substituent, as in the case of N-triphenylmethylaspartic anhydride, does little to deter hydride attack on the apparently more hindered α-carbonyl group.[19] Although the reason for this remarkable regioselectivity has never been satisfactorily explained, the result has unquestioned, synthetic potential since it provides a means of differentiating between the carboxyl groups of the parent acids. In principle, hydrolysis of **93** and **66** and appropriate deprotection should not disturb the chiral center and would be predicted to liberate the 3,4-dihydroxy and 3-amino-4-hydroxybutanoic acids **94** and **95**, respectively. In fact, reduction and acid hydrolysis of N-benzyloxycarbonyl-L-aspartic anhydride (**65**) confirmed this assumption by giving solely (3*S*)-3-(benzyloxycarbonyl)amino-4-hydroxybutanoic acid (**96**).[20]

In order to translate the foregoing idea into a general method for preparing β-amino acids, the lactone ring needs to be opened in a controlled fashion to give an acyclic intermediate that is amenable to further chemical modification. The reagent of choice is trimethylsilyl iodide.[21] In dichloromethane as solvent containing an alcohol, it behaves as an acid surrogate and opens lactones (e.g., **97**) to iodo-silyl esters, which are instantly trans-esterified by the alcohol present to the more stable alkyl esters **98**.

(a) The combination of the aforementioned reactions was successfully realized by taking L-aspartic acid (**1**) as starting material.[22] The amino function was protected with the p-toluenesulfonyl group on account of its resilience towards the reagents employed in the subsequent steps (Scheme 4.18). Submission of N-tosylaspartic acid (**99**) to excess acetic anhydride for 16 hours at 20°C gave optically pure anhydride **100** in 93% yield. Heating to accelerate closure was scrupulously avoided, as it caused racemization. Reduction with sodium borohydride in THF, followed by solution in benzene containing a drop of p-toluenesulfonic acid (TsOH), was entirely re-

Scheme 4.17

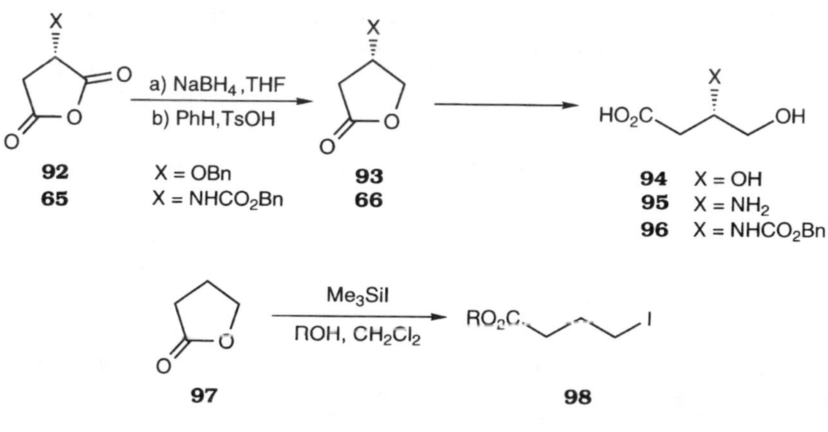

Scheme 4.18

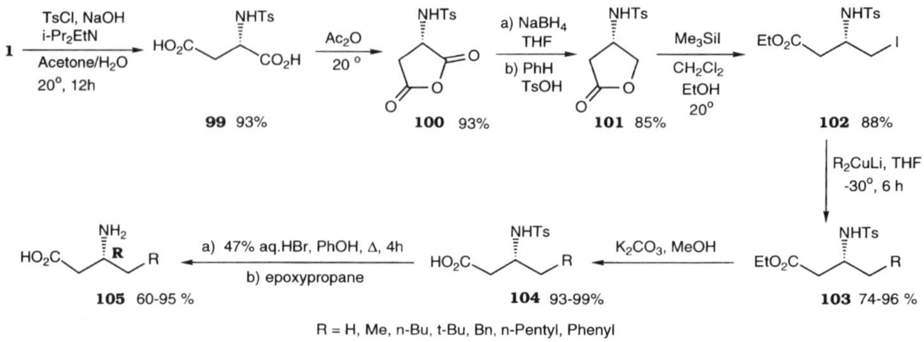

gioselective and gave (3*S*)-3-(N-tosylamino)-4-butanolide (**101**). Next, treatment of **101** with trimethylsilyl iodide and ethanol in dichloromethane cleaved the ring, giving the key chiral intermediate, the iodo-homoserine ester **102** in high yield. Nucleophilic displacement on **102** was effected with various lithium organocuprates. Methyl, butyl, pentyl, benzyl, and phenyl groups were all introduced in high yield (74–96%). The bulky *t*-butyl group was also substituted, but in lower yield. Best results were ensured by adding a twofold excess of reagent at –30°C and subsequently allowing the mixture to warm to room temperature. Reductive de-iodination of **102** with tributyltin hydride gave the parent butanoic ester **103** (*R* = H). Alkaline hydrolysis of the 4-substituted butanoic esters **103** proceeded normally to the corresponding series of (3*R*)-3-N-tosylamino acids **104**. Finally, reductive removal of the tosyl group with aqueous hydrobromic acid and phenol delivered the β-amino acids **105** in overall yields of 27–52% from aspartic acid.

The successive exposure to alkaline and acid conditions might appear to be drastic, but had no effect on the initially implanted chirality. By using D-aspartic acid, (3*S*)-3-aminopentanoic acid was prepared. Conversion of the latter and its 3*R* enantiomer to their methyl-N-camphanoyl derivatives and comparison of their ^{1}H-NMR spectra showed that each was diastereomerically pure.

4.8 Synthesis of α-Alkylated β-Amino Acids. γ-Lactone Route

N-Protected 3-amino-4-butanolides have been used with varying degrees of success for the synthesis of α-alkylated-β-amino acids, which occur as components of certain biologically active, naturally occurring products.

(a) An indirect approach was undertaken for the preparation of (2*S*,3*S*,8*S*,9*S*)-3-amino-9-methoxy-2,6,8-trimethyl-10-phenyldeca-4,6-dienoic acid (**106**), the crucial part of the toxic cyclic peptide, nodularin (**107**) (Scheme 4.19).[23]

Scheme 4.19

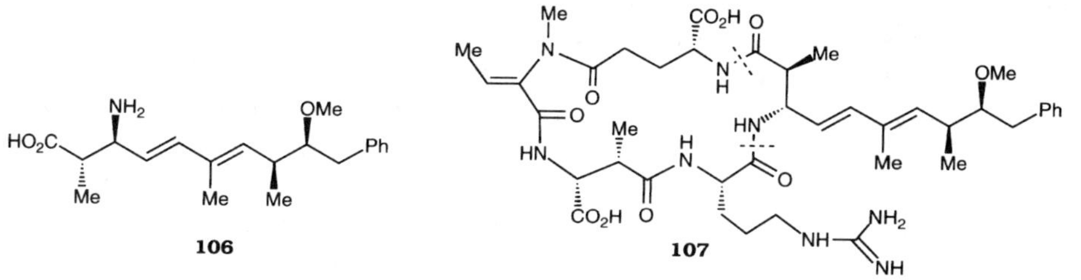

(3*R*)-3-(N-benzyloxycarbonylamino)-4-butanolide (**108**), obtained from D-aspartic acid (**4**), was methylated to give a mixture of the 2*R*,3*R* and 2*S*,3*R* diastereomers **109** and **110** in a 4:1 ratio and yield of 68% (Scheme 4.20). Adjustment of the unfavorable ratio was achieved by heating the mixture in benzene and triethylamine (9:1) for 2 days to favor the desired *cis* isomer **110**. Saponification followed by esterification furnished the *anti* 3-(N-benzyloxycarbonyl-amino)-4-hydroxy-2-methylbutanoic ester **111**. Oxidation to the aldehyde **112** and Wittig olefination with the independently prepared, triphenylphosphonium bromide **113** gave the intermediate diene having the *E*,*E* configuration, which on saponification and deprotection attained the synthetic goal, the decadienoic acid **106**.

(b) The preceding results indicate that the N-benzyloxycarbonyl protecting group does not confer high diastereoselection on methylation of the lactone entity. In the ex-

Scheme 4.20

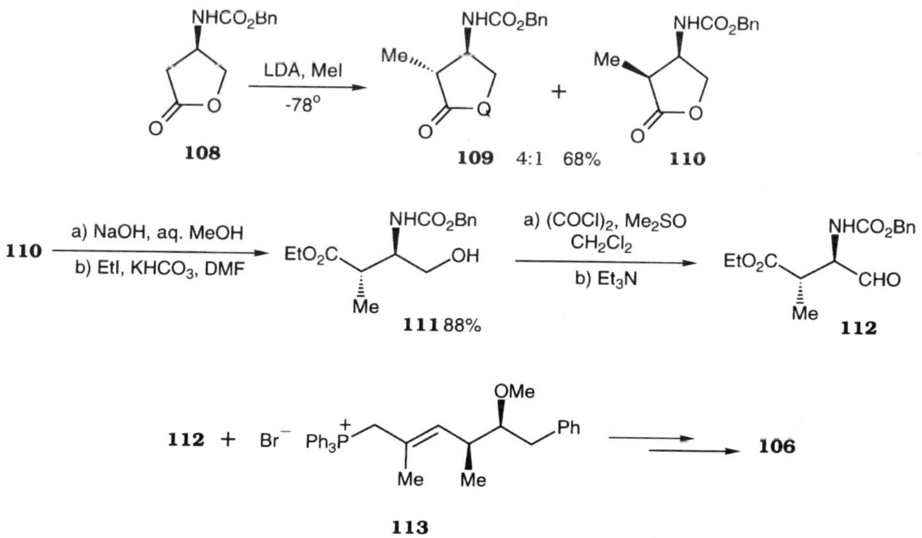

ample cited, it does not really matter since the final product had the wrong, or directly inaccessible, *anti* configuration, which necessitated the additional equilibration step.

Earlier studies on the alkylation of the 3*S* enantiomer **66** revealed similar trends.[24] The dianion of **66**, generated by LDA in THF at −78°, on treatment with methyl and ethyl iodides gave the *trans* (**114**, **116**) and *cis* isomers (**115**, **117**) in a ratio of 4:1 and good yield (Scheme 4.21). Bulkier electrophiles, such as *i*-propyl iodide and acetone, favored the *trans* (**118**, **120**) over the *cis* isomers **119**, **121**) in ratios of better than 19:1. However, yield was sacrificed in the case of **118**. Aldolization was highly efficient, giving essentially **120** with barely a trace of the *cis* isomer **121**. The superior yield and stereoselectivity were attributed to the greater thermodynamic stability of the *trans* isomer. If true, this last result casts doubt on the equilibration of **109** to **110**.

Analogous experiments with 3-amino-γ-lactones equipped with other N-protecting groups, such as *t*-butyloxycarbonyl (BOC) and benzoyl, showed that diastereoselectivity towards alkylation and aldolization could be enhanced, but not completely. In all cases,[11,25] the *trans* isomer was favored in ratios between 9:1 and 11:1.

(c) Despite the less-than-perfect *trans* diastereofaciality displayed by the various N-protected 3-amino-γ-lactones, they have the singular advantage of offering a neat solution to the challenging problem of preparing α-alkyl-β-amino acids of the *syn* configuration. A pertinent illustration is presented by the potent antineoplastic, cyclic peptide, dolastatin 11 (**122**), which was isolated from the sea hare, *Dolabella auricularia*.[26] The molecule is composed of an array of the common α-amino acids together with an odd one, (2*S*,3*R*)-3-amino-2-methylpentanoic acid (**123**) (Scheme 4.22).

Scheme 4.21

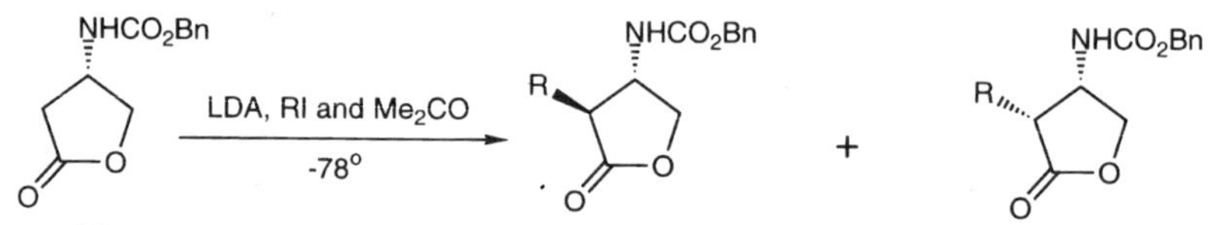

66

R = Me	**114**	4:1	70%	**115**
R = Et	**116**	4:1	50%	**117**
R = i-Pr	**118**	19:1	15%	**119**
R = Me2(HO)C	**120**	19:1	84%	**121**

Scheme 4.22

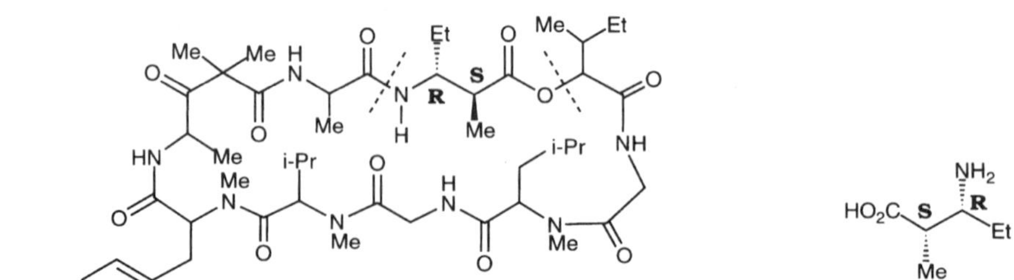

122 **123**

In view of the medicinal importance of **122**, a simple synthesis of **123** is desirable so that it may be utilized as an intermediate. The deployment of (3*S*)-3-(N-tosylamino)-4-butanolide (**101**) provides the required chirality and the means for the practical synthesis of **123** (Scheme 4.23). Successive treatment of **101** with LDA and methyl iodide at –78°C with subsequent warming to –60°C furnished (2*S*,3*S*)-2-methyl-3-(N-tosylamino)-4-butanolide (**124**) in 70% yield.[27] No trace of the corresponding *cis* isomer was observed. In other words, the bulk of the amino substituent completely controlled the direction of attack by the electrophile. A small amount of the *gem*-dimethyl derivative **125** was also formed, but easily removed by chromatography.

Opening of **124** with trimethylsilyl iodide and ethanol gave the iodo-ester **126**, which was directly treated with lithium dimethylcuprate. Ethyl (2*S*,3*R*)-2-methyl-3-(N-tosylamino)-pentanoate (**127**) was obtained in 79% for the two steps. Alkaline hydrolysis afforded the tosylamino acid **128**, which by reductive deprotection with aqueous hydrobromic acid and phenol delivered the free β-amino acid **123** in 84% yield. The hydrochloride of **123** was prepared and shown to have spectroscopic properties identical to those of natural and previously prepared samples.[28,29] The optical purity of **128** was verified by its reduction to the primary alcohol and conversion of the latter to Mosher's ester. Inspection of the ^{1}H and ^{13}C-NMR spectra revealed a single diastereomer.

This procedure gave **123** in an overall yield of 34% from L-aspartic acid and compares favorably with other methods.[29,30]

The choice of the tosyl group for protection was originally dictated by its stability towards iodo-esterification; however, it brought a bonus, high diastereoselectivity. It

Scheme 4.23

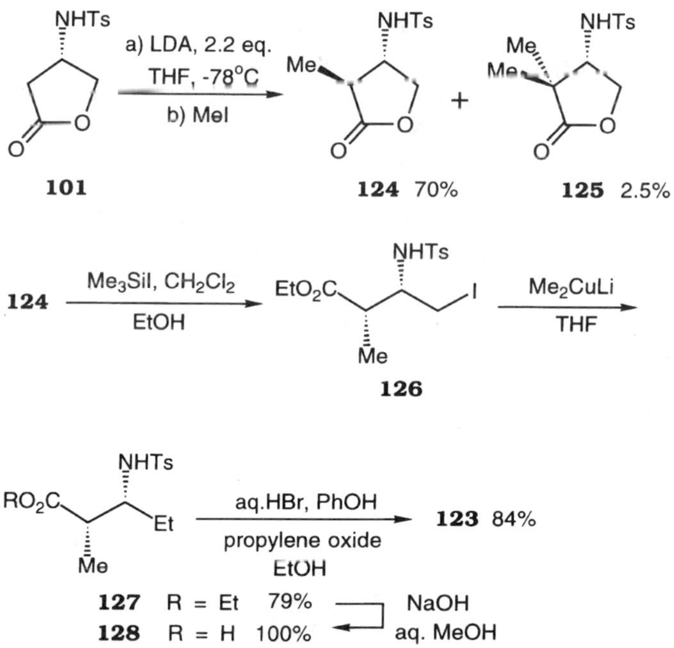

is superior to the N-benzyloxycarbonyl and like groups, which elicit *cis*-alkylation to a minor extent, perhaps through the operation of a complex-induced proximity effect.[31] Evidently, the sulfonyl function does not satisfy the criteria for the presumed complexation.

4.9 Synthesis of α-Hydroxy-β-Amino Acids. γ-Lactone Route

The inherent, high diastereoselectivity of 3-(N-tosylamino)-4-butanolides can be turned to advantage for other electrophilic reactions. An obvious application is the synthesis of α-hydroxy-β-amino acids. These acids are often found as the terminal components of a variety of natural products endowed with valuable pharmaceutical properties, nearly all of which possess the *syn* configuration. Clearly, such entities are ideal candidates for synthesis by the γ-lactone route.

(a) Hydroxylation of the 3*S* enantiomer **101** was best accomplished by successive treatment with NaHMDS in THF at –78°C and racemic 2-phenylsulfonyl-3-phenyloxaziridine (Scheme 4.24).[32] A single product, the *trans*-hydroxy-N-tosylaminolactone **129**, was formed in 64% yield. Other conventional hydroxylating agents, such as *m*-chlorobenzoic acid, lead tetracetate, and *bis*-(trimethylsilyl)peroxide, either led to no reaction or complex mixtures. The action of (+)-(camphorylsulfonyl)oxaziridine on **101** and its 3*R* enantiomer **130** was identical. Only the *trans* hydroxy-lactones **129** and **131** were formed in 55 and 68% yield, respectively. Therefore, the matching, or mismatching for that matter, of reaction partners was of no mechanistic importance.

Opening of **129** was accomplished in the usual way with ethanolic trimethylsilyl iodide and gave the pivotal, chiral intermediate, the hydroxyiodo-homoserine ester **132** in 88% yield. Efficient nucleophilic substitution was carried out with a wide range of Gilman reagents in THF at 0°C. The introduction of methyl, ethyl, benzyl, and even phenyl was trouble-free, being characterized by yields of 71–90%. Branched alkyl groups, for example, *t*-butyl and cyclohexyl, encountered some resistance as evidenced by the yields of around 42%. The protected esters **133** so obtained were then saponified as before to the α-hydroxy N-tosylamino acids **134**. Finally, reductive deprotection with sodium naphthalide in dimethoxyethane (DME) was sufficiently mild to preserve the hydroxyl substituent from elimination and liberated the *syn*-configured acids, namely the (2*S*,3*R*)-3-amino-2-hydroxy-4-substituted butanoic acids **135**. Scrutiny of the ^{1}H-NMR spectra of the different products showed that single diastereomers were obtained. The various acids were obtained in yields of 10–25% from L-aspartic acid.

The acid **135** (where *R* = phenyl), is the N-terminal of the dipeptide bestatin (**136**), a well-known immunoregulator and inhibitor of aminopeptidase B (Scheme 4.25).[33] The present synthesis of the phenyl-substituted acid **135** compares well in economy of means and chiral transmission with other methods.[34] It may be added that the general method is practical and requires no special handling.

(b) It is obvious that intermediate **132** can be used to make many other α-hydroxy-β-amino acids of the *syn* configuration. Furthermore, their enantiomeric counterparts

Scheme 4.24

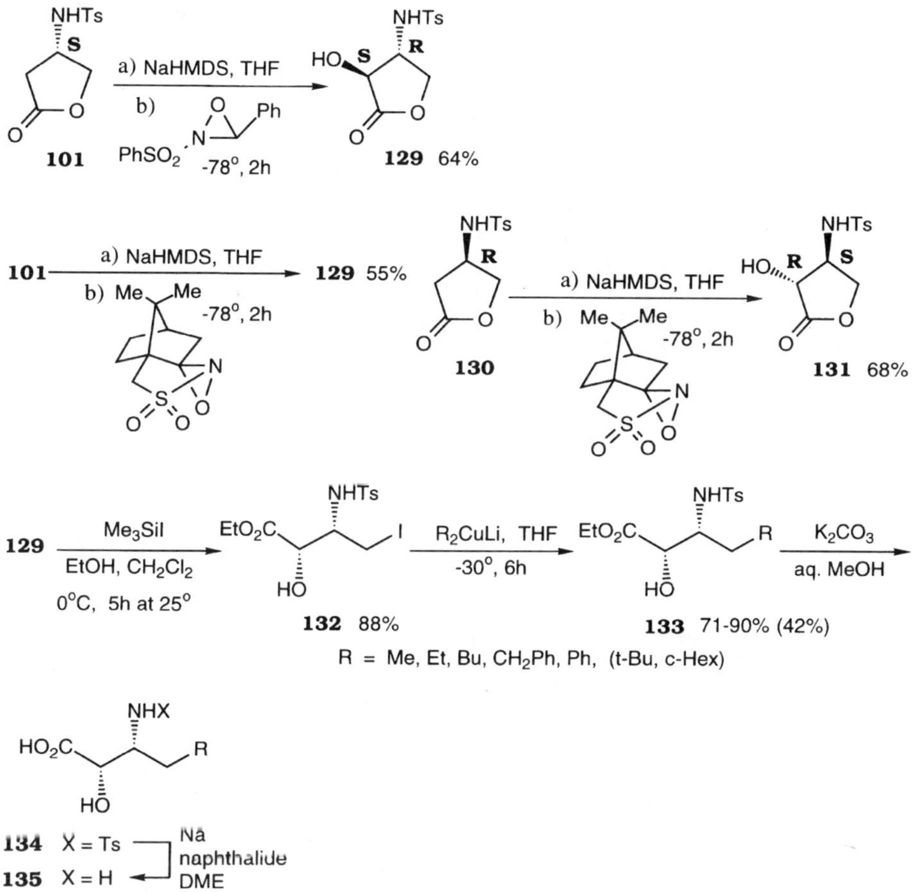

will be procurable by performing the same synthetic sequence on the enantiomeric equivalent **137**, which is available from the hydroxy-lactone **131**, in turn prepared from D-aspartic acid (**4**). Since both intermediates are chiral templates, they will generate products having opposite, but unimpeachable, configurations. Hence, a standard is at hand for checking doubtful structures (Scheme 4.26).

Scheme 4.25

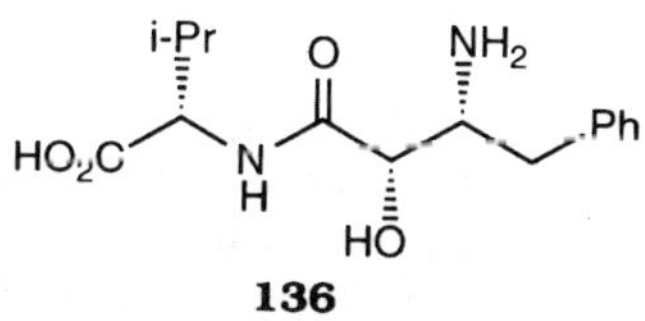

136

Scheme 4.26

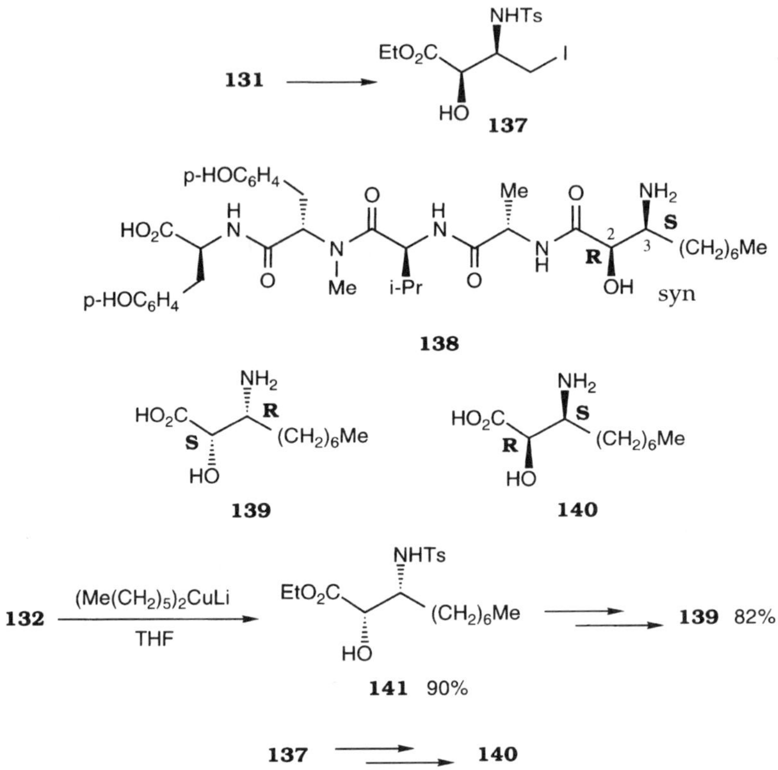

A particular instance is the configuration of the N-terminal acid of microginin (**138**). The latter is a linear pentapeptide, isolated from the freshwater, blue-green alga, *Microcystis aeruginosa*.[35] It is of interest because of its ability to inhibit angiotensin-converting enzyme. Originally, the configuration of the hydroxyl group was determined, incorrectly as it turned out, as *R*. That of the amino group was unassigned, but was supposed to be *S* on the basis of the generally observed *syn* configuration of such acids, as exemplified by the C13 side chain of taxol.[36] The ambiguity was resolved by synthesizing the (2*S*,3*R*) and (2*R*,3*S*)-3-amino-2-hydroxydecanoic acids (**139** and **140**). Treatment of **132** with lithium dihexylcuprate in the customary manner gave the ethyl decanoate **141** in 90% yield.[37] Alkaline hydrolysis and deprotection with sodium naphthalide in DME gave **139** in 82% yield. The same procedure applied to **137** furnished **140** in comparable yield. Both acids had similar melting points (214–215°C) and were obtained in an overall yield of 41% and eight steps from aspartic acid. Their optical rotations were complementary, namely –5.5 and +6.0 (in 0.1 N HCl/MeOH, 2:1) for **139** and **140**, respectively. The same acids have also been made by the asymmetry-induced, formal amino-hydroxylation of *E*-decenoic ester and exhibit slightly different physical properties.[38] The correct N-terminus of microginin is represented by **139**, the opposite of that originally proposed.

4.10 Conclusion

The foregoing examples amply demonstrate that aspartic acids display surprising versatility for the synthesis of a wide range of enantiomerically pure β-amino acids and their α-alkyl, amino, and hydroxy derivatives. 4-Substituted-3-aminobutanoic acids can be conveniently prepared either from N-protected half-esters of aspartic acid or the 3-(N-tosylamino)-4-butanolides. The synthesis of 4-substituted 2,3-diaminobutanoic acids from sterically hindered aspartic esters points the way to other applications. The synthesis of 2-alkyl-3-aminobutanoic acids of the *anti* configuration is best achieved through alkylation of N-protected 2-azetidinone-4-carboxylic acid. On the other hand, if the *syn* diastereomers are required, then 3-(N-tosylamino)-4-butanolide is the template of choice. Similarly, the analogous synthesis of *syn* 3-amino-2-hydroxy-butanoic acids may be accomplished by electrophilic hydroxylation of the aforementioned γ-lactone templates. Although there appear to be no reports so far, it is obvious that electrophilic hydroxylation of β-lactam templates will provide access to *anti* 3-amino-2-hydroxy acids.

4.11 References

1. D. Coleman, *J. Chem. Soc.* 2294 (1951).
2. A. El Marini, M. L. Roumestant, P. Viallefont, D. Razafindramboa, M. Bonato, M. Follet, *Synthesis* 1104 (1992).
3. F. Peypoux, M. Guinand, G. Michel, L. Delcambe, B. C. Das, P. Vareune, E. Lederer, *Tetrahedron* **29**, 3455 (1973).
4. P. Gmeiner, F. Orecher, C. Thomas, K. Weber, *Tetrahedron Lett.* **36**, 381 (1995).
5. R. A. Ward, G. Procter, *Tetrahedron Lett.* **33**, 3359 (1992).
6. D. Seebach, D. Wasmuth, *Angew. Chem. Int. Ed. Engl.* **20**, 971 (1981).
7. J. E. Baldwin, M. G. Moloney, M. North, *Tetrahedron* **45**, 6309 (1989).
8. E. Atherton, J. Meierhofer, *Hoppe-Seyler's Physiol. Chem.* **354**, 689 (1973).
9. P. J. Dunn, R. Häner, H. Rapoport, *J. Org. Chem.* **55**, 5017 (1990).
10. G. J. McGarvey, R. N. Hiner, Y. Matsubara, T. Oh, *Tetrahedron Lett.* **24**, 2733 (1983).
11. G. J. McGarvey, J. M. Williams, R. N. Hiner, Y. Matsubara, T. Oh, *J. Am. Chem. Soc.* **108**, 4943 (1986).
12. T. N. Saltzmann, R. W. Ratcliffe, B. G. Christensen, F. A. Bouffard, *J. Am. Chem. Soc.* **102**, 6161 (1980).
13. P. J. Reider, E. J. J. Grabowski, *Tetrahedron Lett.* **23**, 2293 (1982).
14. S. Hanessian, K. Sumi, B. Vanasse, *Synlett* 33 (1992).
15. J. E. Baldwin, R. M. Adlington, D. W. Gollins, C. J. Scholfield, *Tetrahedron* **46**, 4733 (1990).

16. F. A. Bouffard, B. G. Christensen, *J. Org. Chem.* **46**, 2208 (1981).

17. D. M. Bailey, R. E. Johnson, *J. Org. Chem.* **35** 3574 (1970).

18. D. H. R. Barton, M. Bénéchie, F. Khuong-Huu, P. Potier, V. Reyna-Pinedo, *Tetrahedron Lett.* **23**, 651 (1982).

19. C. Athanassopoulos, C. Tzavara, D. Papaioannou, G. Sindona, L. L. S. Maia, *Tetrahedron* **51**, 2679 (1995).

20. Y. Takahashi, H. Yamashita, S. Kobayashi, M. Ohno, *Chem. Pharm. Bull.* **34**, 2732 (1986).

21. H. R. Kricheldorf, *Angew. Chem. Int. Ed. Engl.* **18**, 689 (1979); G. A. Olah, S. C. Narang, B. G. B. Gupta, R. Malhotra, *J. Org. Chem.* **44**, 1247 (1979); M. Kolb, J. Barth, *Synth. Commun.* **11**, 763 (1981).

22. C. W. Jefford, J. B. Wang, *Tetrahedron Lett.* **34**, 1111 (1993).

23. K. L. Rinehart, K.-I. Harada, M. Namikoshi, C. Chen, C. A. Harvis, M. H. G. Munro, J. W. Blunt, P. E. Mulligan, V. R. Beasley, A. M. Dahlem, *J. Am. Chem. Soc.* **110**, 8557 (1988); M. Namikoshi, K. L. Rinehart, A. M. Dahlem, V. R. Beasley, W. Carmichael, *Tetrahedron Lett.* **30**, 4349 (1989).

24. Y. Takahashi, S. Hasegawa, T. Izawa, S. Kobayashi, M. Ohno, *Chem. Pharm. Bull.* **34**, 3020 (1986).

25. H. Yoda, Y. Nakagami, K. Takabe, *Tetrahedron: Asymmetry* **5**, 169 (1994).

26. G. R. Pettit, Y. Kamano, H. Kizu, C. Dufresne, C. L. Herald, R. J. Bontems, J. M. Schmidt, F. E. Boettner, R. A. Nieman, *Heterocycles* **28**, 553 (1989).

27. C. W. Jefford, J. McNulty, *Helv. Chim. Acta* **77**, 2142 (1994).

28. D. C. Carter, R. E. Moore, J. S. Mynderse, W. P. Niemczura, J. S. Todd, *J. Org. Chem.* **49**, 236 (1984).

29. R. B. Bates, S. Gangwar, *Tetrahedron: Asymmetry* **4**, 69 (1993).

30. S. G. Davies, O. Ichihara, I. A. S. Walters, *J. Chem. Soc. Perkin Trans. I* 1141 (1994).

31. J. E. Resek, P. Beak, *J. Am. Chem. Soc.* **116**, 405 (1994); P. Beak, W. K. Lee, *J. Org. Chem.* **58**, 1109 (1993).

32. C. W. Jefford, J. B. Wang, Z. H. Lu, *Tetrahedron Lett.* **34**, 7557 (1993).

33. H. Suda, T. Takita, T. Aoyagi, H. Umezawa, *Antibiotics* **26**, 100 (1976).

34. B. H. Norman, M. L. Morris, *Tetrahedron Lett.* **33**, 6803 (1992); Y. Kobayashi, Y. Takemoto, Y. Ito, S. Terashima, *Tetrahedron Lett.* **31**, 3030 (1990); C. Palomo, A. Arrieta, F. P. Cossio, J. M. Aizpurua, A. Mielgo, N. Aurrekoetxea, *Tetrahedron Lett.* **31**, 6429 (1990); R. Herranz, J. Castro-Pichel, S. Vinuesa, M. T. García-López, *J. Org. Chem.* **55**, 2232 (1990); W. H. Pearson, J. V. Hines, *J. Org. Chem.* **54**, 4235 (1989); S. Kobayashi, T. Osobe, M. Ohno, *Tetrahedron Lett.* **25**, 5079 (1984).

35. T. Okino, H. Matsuda, M. Murakami, K. Yamaguchi, *Tetrahedron Lett.* **34**, 501 (1993).

36. D. Guénard, F. Guéritte-Voegelein, P. Potier, *Acc. Chem. Res.* **26**, 160 (1993).

37. C. W. Jefford, Z.-H. Lu, J. B. Wang, *Pure Appl. Chem.* **66**, 2075 (1994); C. W. Jefford, Z. H. Lu, J. B. Wang, J. McNulty, *Helv. Chim. Acta*, **79**, 1203 (1996).

38. F. Matsuura, Y. Hamada, T. Shioiri, *Tetrahedron* **50**, 11303 (1994); M. E. Bunnage, A. J. Burke, S. G. Davies, C. J. Goodwin, *Tetrahedron: Asymmetry* **6**, 165 (1995).

CHAPTER

5

Preparation of Enantiopure β-Amino Acids from α-Amino Acids Using the Arndt–Eistert Homologation

Jennifer L. Matthews, Christine Braun, Christel Guibourdenche, Mark Overhand, and Dieter Seebach

5.1 Introduction

The large array of methodologies available for the synthesis of α-amino acids makes them ideal candidates for starting materials in the synthesis of β-amino acids. Furthermore, the methodology introduced[1] almost 70 years ago by Arndt, Eistert, and Partale provides a route for this conversion (Scheme 5.1). The adaptation[2] of this reaction to allow the homologation of phthaloyl-protected α-amino acids by Balenović and co-workers and the use of carbamate-protected α-amino acids as starting materials[3] has served to widen the application of this methodology.

The following section contains a brief introduction to the Arndt–Eistert reaction as applied to the homologation of amino acid derivatives. This is followed by a survey of some of the most recent applications in the field.

Scheme 5.1

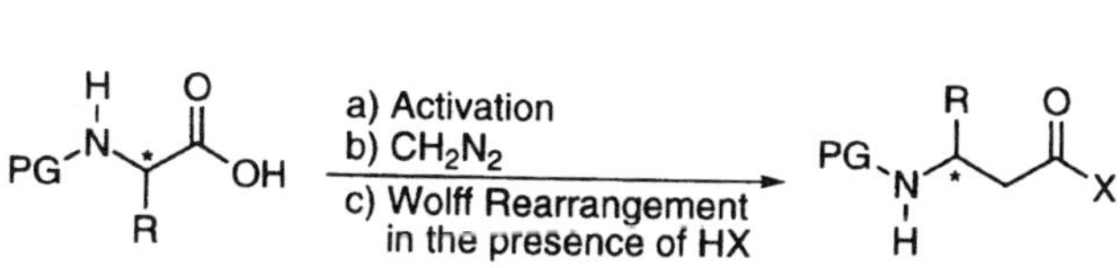

5.2 The Arndt–Eistert Reaction in Amino Acid Synthesis

5.2.1 Introduction

The formation of β-amino acids from α-amino acids using the Arndt–Eistert reaction has seen many applications in natural product syntheses and other areas of organic synthesis. This section will concentrate on recent applications of the Arndt–Eistert reaction in amino acid and peptide synthesis that demonstrate the versatility of the process.

5.2.2 The Arndt–Eistert Reaction—An Overview

The Arndt–Eistert reaction involves the formation of a diazoketone and subsequent loss of nitrogen and Wolff rearrangement under silver ion catalysis, photolysis, or, less commonly, thermolysis. In the synthesis of β-amino acids, the simplest examples of this process are those where a readily available naturally occurring α-amino acid is first converted to the corresponding diazoketone and subsequent rearrangement yields a single β-amino acid derivative. One of the many examples[4,5] of this is shown in Scheme 5.2. This homologation and subsequent trapping of the reactive intermediate allows the introduction of a complementary protecting group at the C-terminus and hence facilitates subsequent peptide formation procedures. The use of a wide range of amino acids is possible and the application of this reaction to N-methylated amino acids has been demonstrated[6,7]; see also Section 5.3.

In most cases, readily prepared ethereal diazomethane is reacted with an activated carboxylic acid derivative to form the diazoketone. Early work concentrated on the use[2] of acid chlorides as the activated derivative. However, many protecting groups do not allow the preparation of amino acid chlorides either because of the acid lability or because oxazolones are formed,[8,9] and for these reasons phthaloyl protection has been used. With the advent of mixed anhydride activation,[8] the use of the more convenient urethane protected amino acids became possible. Activation with dicyclohexylcarbodiimide (DCC) is also possible,[8] but the separation of the urea may cause problems,

Scheme 5.2

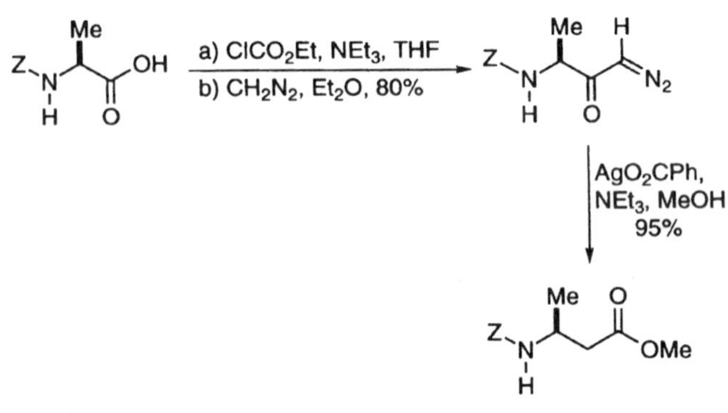

and procedures involving the use of active esters[8,10] or other coupling agents[4] are less well established.

Although mixed anhydride activation is suitable in most cases, the less reactive α-amino acids must be converted to their acid chloride derivatives to enable diazoketone formation, and this therefore necessitates the use of acid stable protecting groups. The acid lability of the *t*-butoxycarbonyl (Boc) protecting group means that formation and handling of the acid chlorides can only be achieved[11] at low temperatures (–30°C), and in the case of benzyloxycarbonyl (Z) protected α-amino acids, the acid chlorides tend to form N-carboxylic anhydrides (Leuch anhydrides),[12] which do not react with diazomethane. As urethane protecting groups are generally preferred in peptide synthesis, mild methods for the preparation of the acid chlorides of the less reactive urethane-protected amino acids have been developed, with the use[13] of 1-chloro-N,N,2-trimethylpropenylamine at 0°C being one example (see Section 5.6). Carpino and co-workers introduced the fluorenylmethoxycarbonyl (Fmoc) protected α-amino acid chlorides, which can be generated[14] using thionyl chloride in dichloromethane at room temperature and, as can be seen in Section 5.6, these react readily with diazomethane.

Racemization has been observed during the Arndt–Eistert homologation of phenylglycine.[4] In this case, the activated phenylglycine derivative can easily form a dihydrooxazole (azlactone) (**1**), in which formation of an aromatic system is favored and cannot be avoided by the use of special activation methods.[15] It is therefore believed that racemization occurs during activation rather than during the course of the rearrangement (Scheme 5.3).

Scheme 5.3

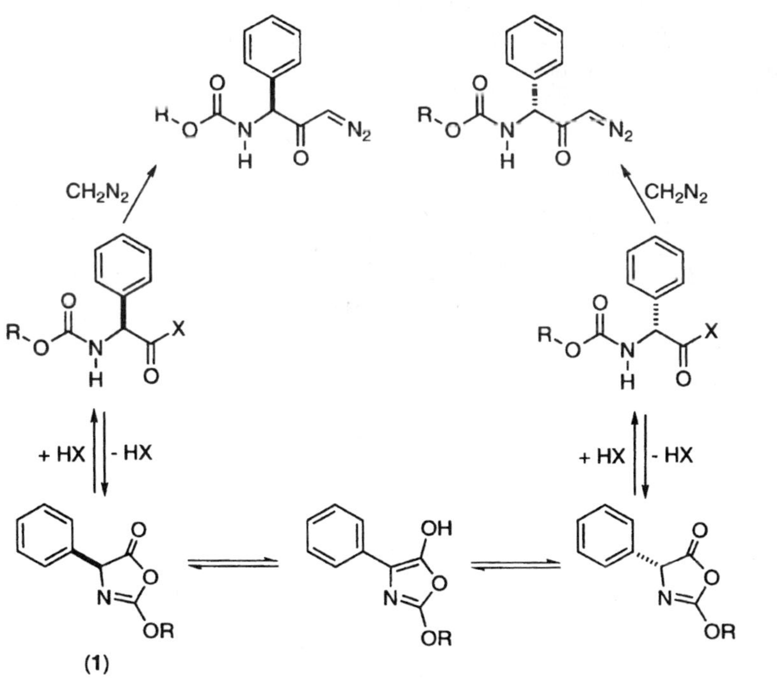

Trimethylsilyldiazomethane[16] has also been used in the Arndt–Eistert reaction, and its use[17] for the homologation of proline (Scheme 5.4) provides an example of the thermal Wolff rearrangement, which sees far less use than either silver ion catalysis or photolysis. The silylated intermediate (**2**) cannot be isolated, but immediate rearrangement to the β-amino acid (**3**) is possible in a "one-pot" procedure. The diazoketone (**4**) can also be isolated by the simple addition of water to the reaction mixture.

The mechanism of the Wolff rearrangement has been discussed by other authors.[5,18] Although there have been many attempts to use other metals, catalysts based on silver have, in general, been found to be the best. Silver oxide catalysis, which involves a heterogeneous reaction at high temperature, has been largely replaced by procedures, first introduced[19] by Newman and co-workers, involving the use of silver benzoate or some silver alkanoates in combination with triethylamine. This results in a homogeneous reaction that proceeds at lower temperatures. It is assumed that, after initial deprotonation of the diazo carbon, a diazoketone-$Ag(NEt_3)_n^+$ complex is formed.[20] Subsequent electron transfer to the silver cation is thought to be followed by a radical chain process.

This mechanism may explain why diazoketones derived from higher diazoalkanes do not rearrange under these conditions (see Section 5.6). Equally, highly substituted diazoketones (see Section 5.6) also seem to be too hindered to rearrange in the presence of this sterically demanding $Ag(NEt_3)_n^+$ complex. Sulfur compounds[21] must also be rearranged under photolytic conditions as the presence of sulfur appears to poison the silver catalyst. In these cases, photolysis becomes the method of choice, and its application will be discussed in greater detail in Sections 5.4 and 5.6. The mechanism of the photolytic rearrangement remains unconfirmed, but low-temperature spectroscopy and matrix isolation experiments have suggested that a ketocarbene is formed, which may be stabilized by the subsequent formation of an oxirene derivative, before rearrangement to a ketene occurs.[5,18]

Scheme 5.4

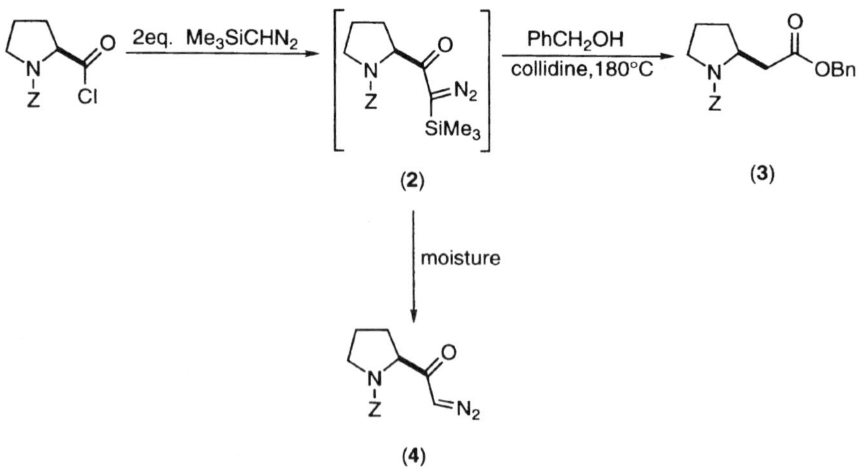

Scheme 5.5

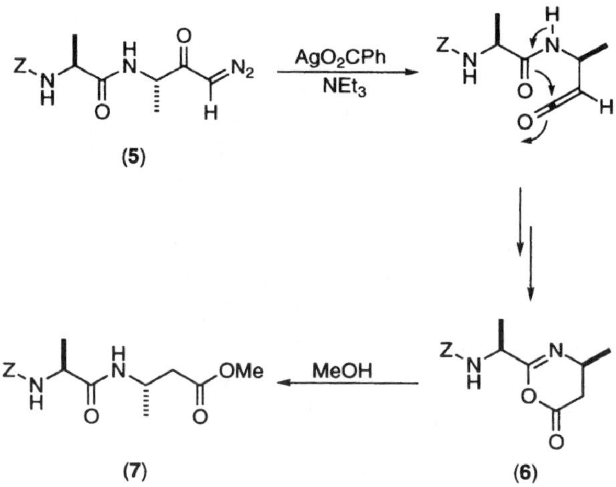

It has been shown that when the Wolff rearrangement of a diazoketone derived from a dipeptide is carried out in the presence of Ag^{+}/triethylamine in inert solvents and without any nucleophile present, a relatively stable cyclic intermediate (**6**) was formed.[22] It was kept in solution for several hours, and subsequent addition of a nucleophile led to the homologated derivative (**7**) (Scheme 5.5).

5.2.3 Some Applications of the Arndt–Eistert Reaction in Peptide Chemistry

The incorporation of one β-amino acid in a peptide containing α-amino acids has seen many applications. The homologation of a terminal α-amino acid residue can be used to introduce one β amino acid residue. McKervey and co-workers have shown[23] that β-amino-α-keto esters (**8**) can be synthesized from dipeptides of α-amino acids using the Arndt–Eistert reaction with a silver-catalyzed Wolff rearrangement and a subsequent diazo transfer reaction (Scheme 5.6). The application of the Arndt–Eistert

Scheme 5.6

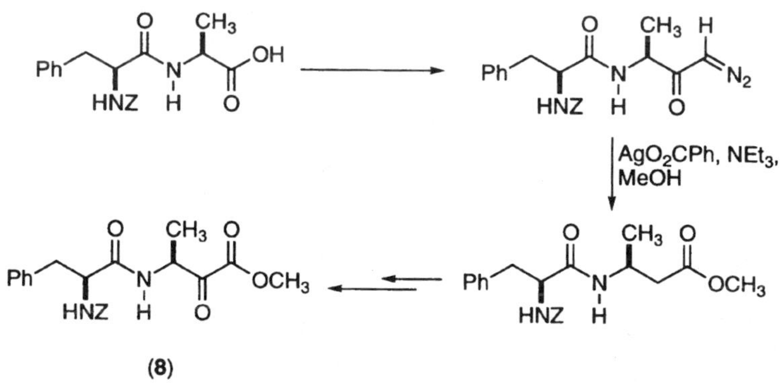

reaction in the synthesis of peptides is an area of considerable current interest and will be discussed further in Section 5.4.

High yielding methods for the synthesis of enantiomerically pure key intermediates used in the synthesis of many classes of naturally occurring compounds are continually being investigated. The Arndt–Eistert reaction has also seen use in this field. One example of this is the synthesis,[24] by Knight and co-workers, of the Geissman–Waiss lactone (**10**), which is a useful precursor to pyrrolizidine alkaloids. This lactone is formed by the homologation and subsequent cyclization of (+)-*cis*-(2*R*,3*S*)-3-hydroxyproline (**9**) (Scheme 5.7).

The synthesis of β-lactam derivatives has also employed the Arndt–Eistert reaction in many cases. The chain extension of oxoazetidine carboxylic acids has been applied[25] in the synthesis of key intermediates for the formation of thienamycin and its analogues. These syntheses also demonstrate the use of the photochemically initiated Wolff rearrangement. Interestingly, Shiozaki and co-workers found that the photolytic

Scheme 5.7

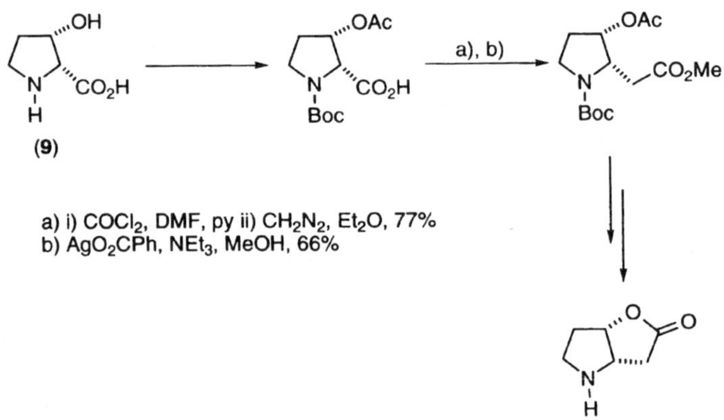

Scheme 5.8

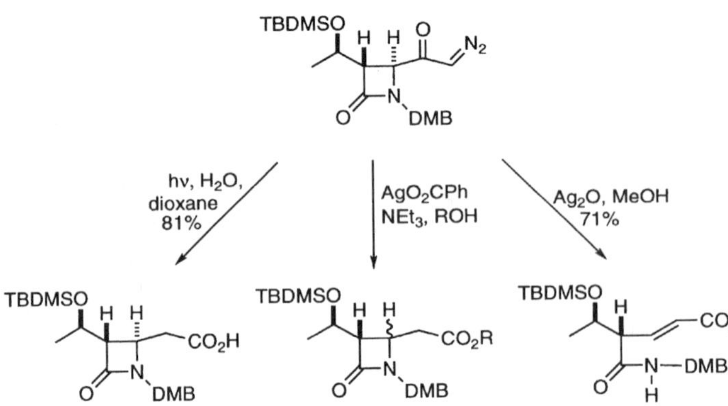

Wolff rearrangement proceeded with complete retention of configuration at the C4 position, whereas the use of silver benzoate as catalyst gave a mixture of *cis* and *trans* isomers and silver oxide in methanol gave a single product, which resulted from cleavage of the β-lactam ring (Scheme 5.8).

The Arndt–Eistert reaction has also featured in many complete natural product syntheses. The stereoselective synthesis[26] of (+)-negamycin, an unusual peptide antibiotic, by Schmidt and co-workers (Scheme 5.9), demonstrates the application of the standard Arndt–Eistert protocol. In this typical procedure, the α-amino acid is converted to the diazoketone in the usual way and subsequent reaction of this compound with silver benzoate in triethylamine and methanol yields the homologated amino acid as its methyl ester in 74% yield.

Scheme 5.9

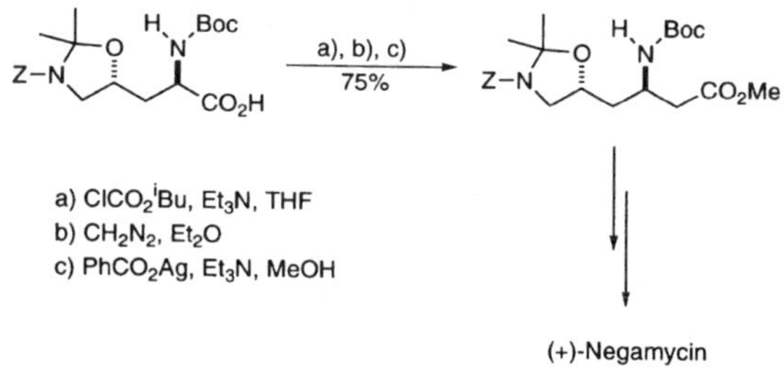

The homologation of the tyrosine derivative (**11**) in the synthesis[27] of (+)-jasplakinolide by Grieco and co-workers (Scheme 5.10) elegantly demonstrates the use of *tert*-butanol as the nucleophile in the Wolff rearrangement. The formation of such sterically hindered esters by this procedure is seen relatively infrequently. However, as can be seen below, the Wolff rearrangement proceeds in the presence of silver benzoate with an acceptable yield of 61%.[28]

Scheme 5.10

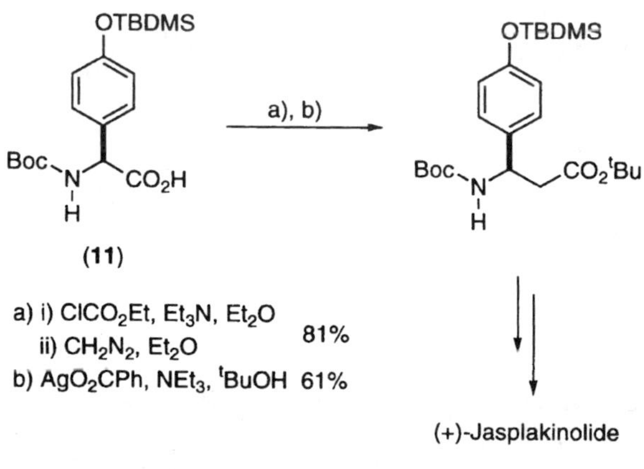

Jefford and co-workers also employed the Arndt–Eistert reaction in the synthesis[29] of (+)-monomorine, a trail pheromone of the Pharaoh's ant (Scheme 5.11). The homologation of the pyrrole alanine derivative (**12**) via the diazoketone (**13**) used silver acetate as the catalyst in aqueous THF to give the β-amino acid (**14**) in 75% yield.[28] Here, the advantage of silver acetate over the more commonly used silver benzoate is made clear, as the acetic acid formed during the necessary acidic workup is simply removed and overcomes the problems associated with the often complex removal of traces of benzoic acid from the amino acid. Silver trifluoroacetate has been used in similar circumstances, and an example of this is discussed in Section 5.3.

Scheme 5.11

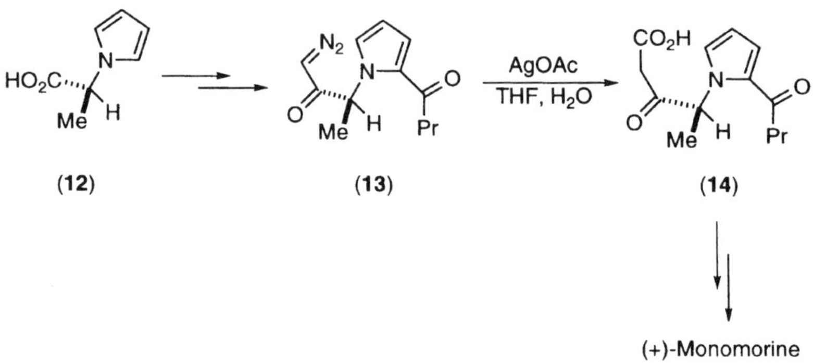

In summary, the formation of β-amino acids by the homologation of α-amino acids using the Arndt–Eistert methodology has seen many applications in a variety of areas of organic synthesis, and the following sections aim to expand on some of this work in more detail.

5.3 The Synthesis of Amide Analogues of Poly(hydroxybutanoate)

The Arndt–Eistert homologation of diazoketones derived from alanine has been used to construct β-aminobutanoic acid derivatives. Boc-protected NMe-β-amino esters can be obtained[7] by this method in yields ranging from 65 to 90% (Scheme 5.12). A

Scheme 5.12

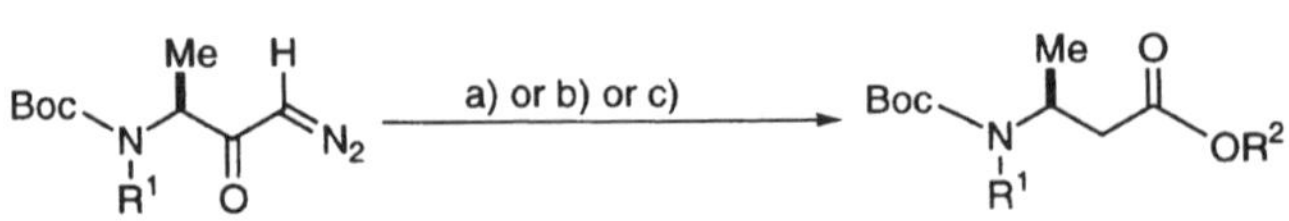

R^1 = H, Me

a) for R_2 = H : AgO_2CCF_3, NEt_3, H_2O, THF
b) for R_2 = Me : AgO_2CPh, NEt_3, MeOH
c) for R_2 = CH_2Ph : AgO_2CPh, NEt_3, $PhCH_2OH$, THF

homogeneous solution of a catalytic amount of silver trifluoroacetate in triethylamine has been used[30] to obtain Boc-protected β-aminobutanoic acid (R^1=R^2=H, Scheme 5.12) from the corresponding diazoketone in 89% yield. Here, the use of silver trifluoroacetate gives the same advantages as those discussed for silver acetate in the pervious section.

These β-amino acid monomers have been used to prepare the linear and cyclic amide analogues of oligo(hydroxybutanoate) by repetitive fragment coupling and subsequent cyclization (Scheme 5.13).

Scheme 5.13

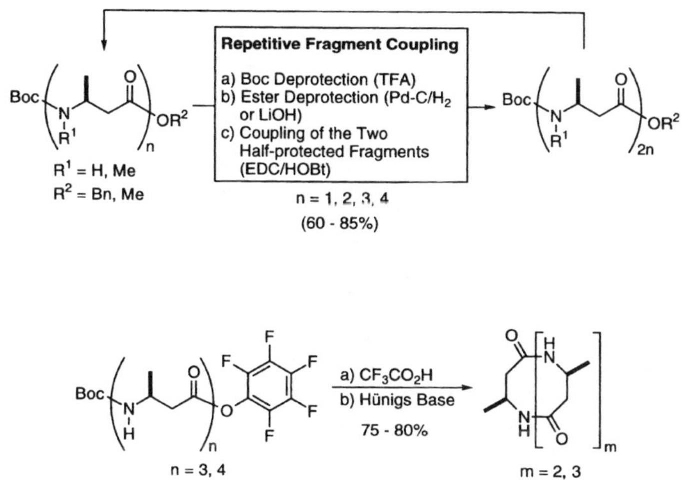

5.4 Arndt–Eistert Homologation with Concomitant Peptide Bond Formation

An intriguing use of the Arndt–Eistert homologation is the trapping of the reactive intermediate, which arises from the Wolff rearrangement of a diazoketone, with the amino functionality of an amino acid derivative. By applying this strategy and commencing with an α-amino acid derived diazoketone, a β-amino acid is produced[31] with concomitant peptide coupling (Scheme 5.14). Using photochemical conditions, diazoketones derived from either diazoethane [(**15**): R^1 = Me] or diazopropane [(**15**): R^1 = Et] have been coupled to the amino group of racemic amino esters.[32] The interesting feature of the application of these diazoketones, which are discussed in greater detail in Section 5.6, is that a homologated dipeptide is produced that possesses an alkyl functionality at the α-position of the β-amino acid residue.

Peptides containing one specific β-amino acid in their sequence have been obtained from the reaction[22] of a peptide diazoketone and an oligopeptide with an amine substituent in the presence of $PhCO_2Ag/NEt_3$ (Scheme 5.15). This reaction only occurs when the free amine of an amino acid derivative is used as the nucleophile. A diazoketone derived from an α-amino acid does not react with the hydrochloride or

Scheme 5.14

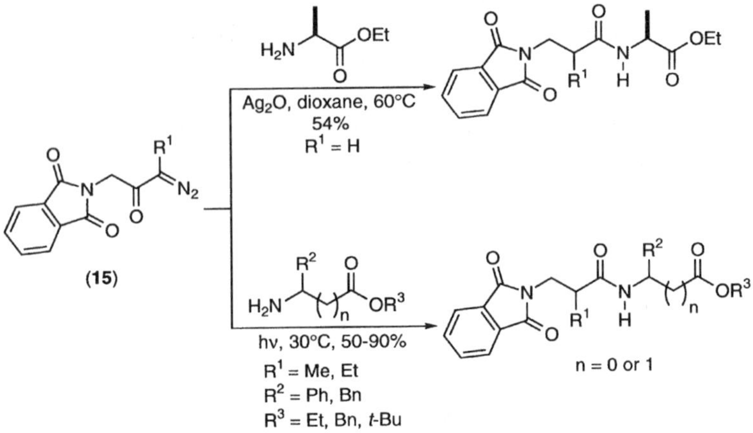

trifluoroacetate salt of an amino acid derivative when a catalytic amount of $PhCO_2Ag$ in triethylamine is added. More than one equivalent (normally approximately 2.5 equivalents) of the free amine must be used in order to obtain a useful yield of the homologated peptide.

A base labile diazoketone was used in the homologation reaction[33] with simultaneous peptide formation by irradiation with 290 nm light (Scheme 5.16). Under these photochemical conditions, the hydrochloride salt of an amino ester can be applied to produce the desired dipeptide, in the presence of triethylamine. Furthermore, a slight excess of the hydrochloride salt of the amino ester (1.5 equivalents) produces the dipeptide in useful yield.

A direct comparison of the reaction of an amino acid derivative with a nonprotonated amino group catalyzed by silver benzoate in triethylamine and the reaction of a

Scheme 5.15

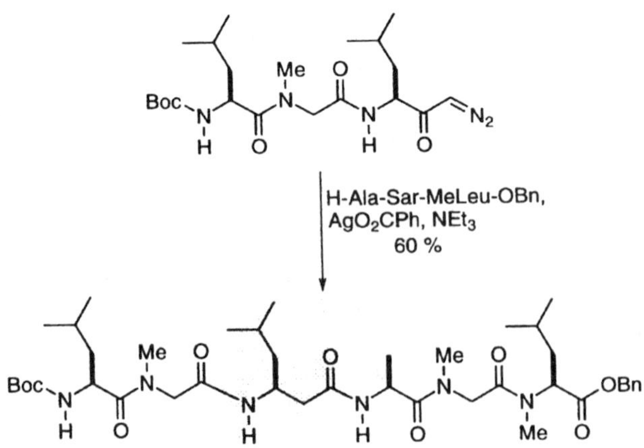

Scheme 5.16

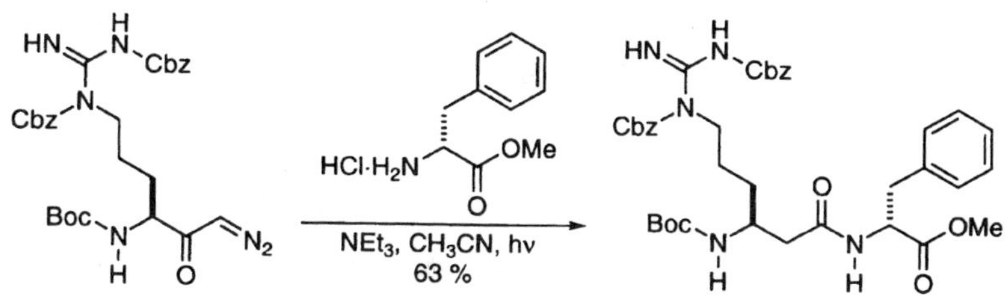

salt of an amino acid derivative under photochemical conditions is outlined in Scheme 5.17. Reaction of the diazoketone (**16**) derived from Boc-Ala-OH with the nonprotonated amino group of a β-amino ester (2.7 equivalents) gave the desired dipeptide (**17**) containing two β-amino acids in 84% yield.[34] Irradiation of the same diazoketone in the presence of the trifluoroacetate salt (1.5 equivalents) and triethylamine gave the dipeptide (**17**) in 70% yield.

The overall yield of a reaction and the number of equivalents of the amino acid derivative necessary to produce the desired peptide in a useful yield are both important factors in determining which of the two methods to choose for a homologation reaction with simultaneous peptide coupling. The dipeptide (**17**) can be further elongated at the N-terminus, by a second photochemical reaction after removal of the Boc protecting group (Scheme 5.18). The desired tripeptide (**18**), which contains β-amino acids exclusively, was obtained in 85% yield, by using 1.5 equivalents of the deprotected dipeptide. The crystal structure of this tripeptide (**18**) shows remarkable structural similarities with part of the crystal structures of oligolides derived from

Scheme 5.17

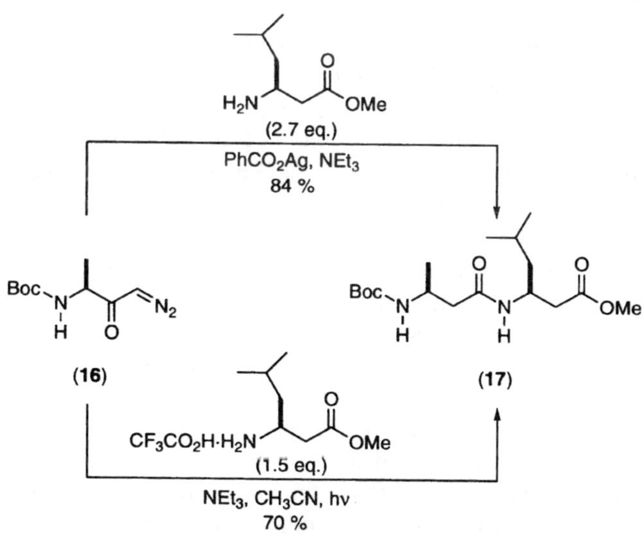

Scheme 5.18

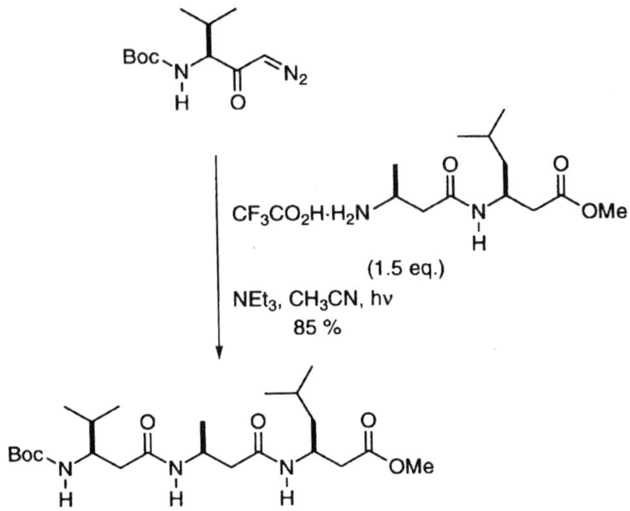

(18)

hydroxybutanoic acid.[34] This tripeptide has been used to construct a hexapeptide that consists of only β-amino acids.

5.5 Competitive Acylation Reactions after the Wolff Rearrangement

5.5.1 Introduction

It has been shown[27] that ketenes generated by the Wolff rearrangement of amino acid derivatives can be trapped by *tert*-butanol to give *tert*-butyl esters when the alcohol is employed as a solvent. This led to the suggestion that it may be possible to employ other compounds that contain highly hindered hydroxy groups but that cannot possibly be employed as the solvent. Carbohydrates containing one unprotected alcohol functionality were selected to examine the feasibility of this suggestion. The bis-acetonide derivatives of α-D-galactopyranose, α-D-mannofuranose, and α-D-glucofuranose, where the hydroxy group becomes increasingly hindered, were chosen. The products resulting from rearrangement of the diazoketone (**19**), which is derived from Z-Ala-OH, in the presence of a carbohydrate derivative were formed with the carbohydrates containing both the least (**20**) and the most hindered (**21**) hydroxy group (Scheme 5.19). When the diazoketone derived from Z-Ala-Ala-OH (**5**) was used, products were only found in the reactions of the two less hindered carbohydrates (**22, 23**) (Scheme 5.19). With the bis-acetonide of α-D-glucofuranose, no product could be detected even after lengthy reaction times.

The obviously slower rate of reaction with these highly hindered nucleophiles led to the suggestion that selectivity between unprotected nucleophiles in multifunctional

Scheme 5.19

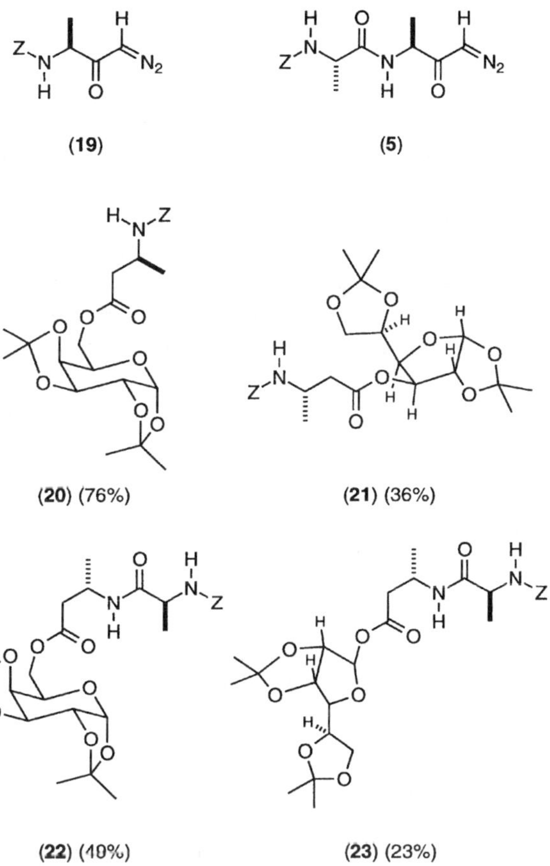

compounds (e.g., an aminoalcohol) could be expected. The following section will discuss examples of competition reactions that result in the formation of a β-amino acid derivative and involve competition between two nucleophilic centers within one molecule.

5.5.2 Competitive Trapping of Products from the Wolff Rearrangement[35]

2-Aminoethanol (**25**) is one of the simplest substrates that can be envisaged for the examination of competition between nucleophiles in the trapping of products from the Wolff rearrangement. The Wolff rearrangement of the diazoketone (**24**), which is derived from Boc-Val-OH, in the presence of 2-aminoethanol and under the standard conditions[4] of silver ion catalysis, yielded the amide product (**26**) with the alternative ester product not being observed (Scheme 5.20). However, replacing the primary amine by a secondary amine [i.e., the use of 2-(methylamino)ethanol] results in a

Scheme 5.20

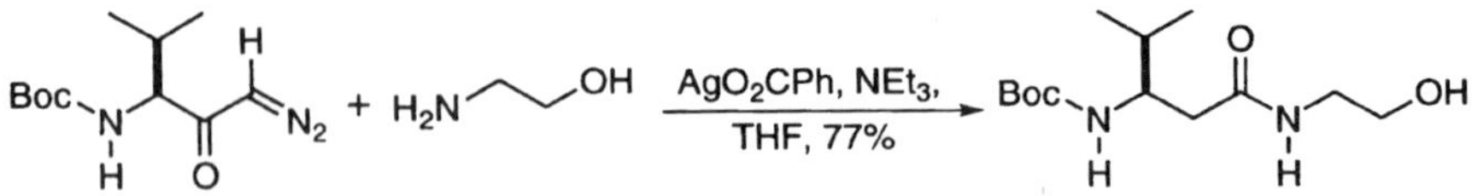

(24) (25) (26)

much lower degree of selectivity, and in the competition between a secondary amine and a secondary alcohol (the use of pseudoephedrine) virtually no selectivity is observed. It is therefore clear that not only the nucleophilicity but also the steric hindrance around the nucleophilic center plays a role in determining the product ratio. The influence of steric hindrance is also demonstrated when the diol 1-methylbutan-1,3-diol (**28**) is used. When the diazoketone (**27**) derived from Z-Val-OH is subjected to the conditions of the Wolff rearrangement in the presence of this diol, the ester product (**29**) resulting from reaction of the primary OH group is formed with a selectivity of 19:1 (Scheme 5.21).

Scheme 5.21

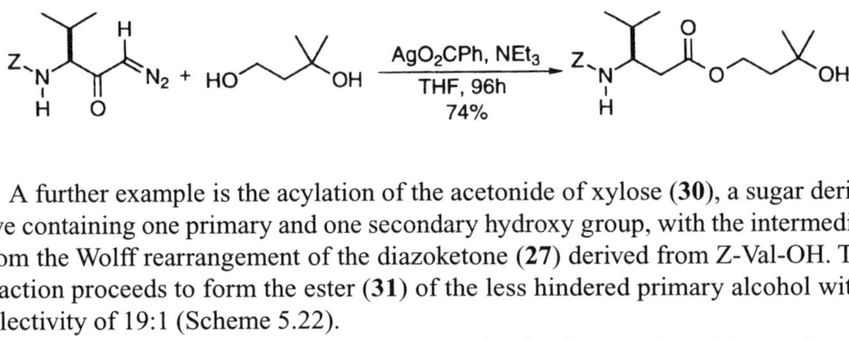

A further example is the acylation of the acetonide of xylose (**30**), a sugar derivative containing one primary and one secondary hydroxy group, with the intermediate from the Wolff rearrangement of the diazoketone (**27**) derived from Z-Val-OH. This reaction proceeds to form the ester (**31**) of the less hindered primary alcohol with a selectivity of 19:1 (Scheme 5.22).

The application of nucleosides as nucleophiles for the trapping of the product derived from the Wolff rearrangement would allow the formation of their acylated derivatives, which have many potential applications.[36] When thymidine (**32**) is used, the nature of the side chain of the diazoketone has a considerable effect on the selectivity. O-Acylation at the 5′ position (**33**) is favored over reaction at the 3′ position with

Scheme 5.22

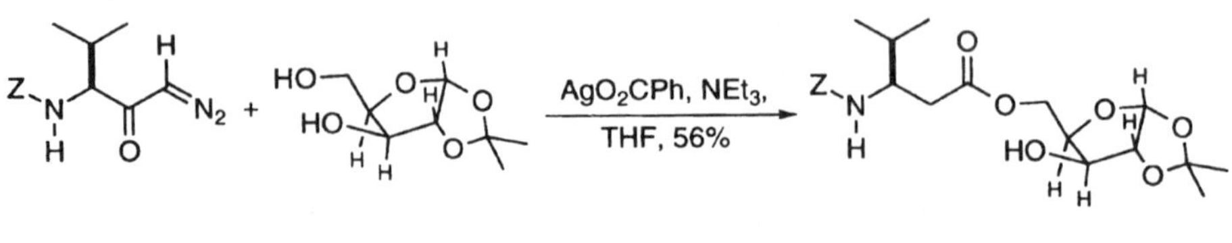

(27) (30) (31)

Scheme 5.23

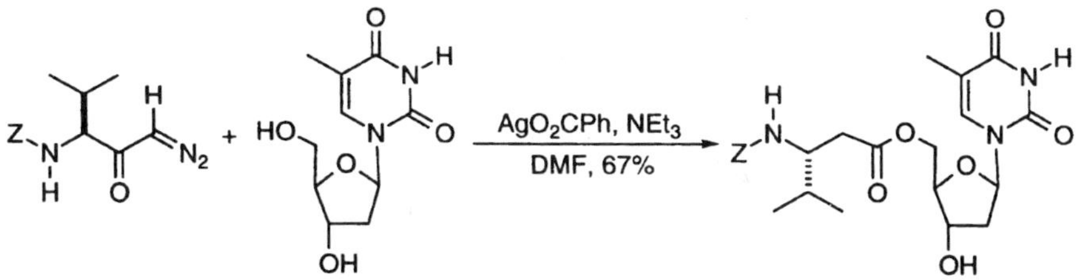

derivatives of Z-Ala-OH and Z-Val-OH. However, the selectivity is 6.5:1 in the former and 14:1 in the latter case (Scheme 5.23). There is no indication for products that might have resulted from reaction at the base NH group.

When a similar reaction sequence is carried out with the diazoketone derived from Z-Ala-Ala-OH (**5**), quite different results are obtained. The remarkably stable but sterically hindered dihydrooxazinone (**6**) (see Scheme 5.5), which is known[22] to be formed when dipeptides are subjected to the conditions of the Wolff rearrangement without a nucleophile being present, can be preformed and then reacted with the nucleophile. When thymidine is added to this acylating agent, the product (**34**) from O-acylation at the 5′ position is favored with a selectivity of 32:1 in 57% yield (Scheme 5.24), whereas if the nucleophile is present at the start of the reaction, the product (**34**) from O-acylation at the 5′ position is formed in 58% yield with a selectivity of 4.5:1. This result can be interpreted as a consequence of the higher selectivity of the dihydrooxazinone (**6**) as an acylating agent as compared to the ketene.

Scheme 5.24

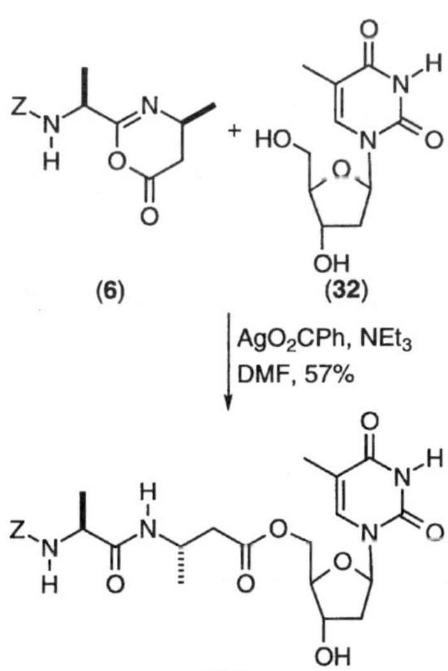

Scheme 5.25

(19) (35) (36)

The reaction of two equivalents of the acetonide of adenosine (**35**) with the product of the Wolff rearrangement of the diazoketone derived from Z-Ala-OH yields the product (**36**) derived only from O-acylation with no product of N-acylation being found (Scheme 5.25). However, the yield is only 48%, and it is therefore impossible to rule out formation of the N-acylated product, which is subsequently not recovered.

Further work in this area suggests that reactions of completely unprotected nucleosides will also display distinct selectivity, and it can be envisaged that these novel combinations between nucleosides and β-amino acid derivatives could form the basis for a combinatorial library.[37]

5.6 β-Amino Acids with Differing Substitution Patterns

In the simple case of homologation of naturally occurring α-amino acids, β-monosubstituted β-amino acids are obtained. However, it can be envisaged that it is possible that this methodology can be extended to allow the synthesis of a whole range of β-amino acids with a variety of substitution patterns.

5.6.1 α-Substituted β-Amino Acids

Diazoketones derived from α-amino acids and diazoalkanes other than diazomethane have been rearranged[32,38] to give α-substituted β-amino acids (Scheme 5.26). This can only be achieved using photolysis or in the presence of Ag_2O at higher temperatures (90–100°C) and not by Ag^+ catalysis in homogenous solution. Previous work in this

Scheme 5.26

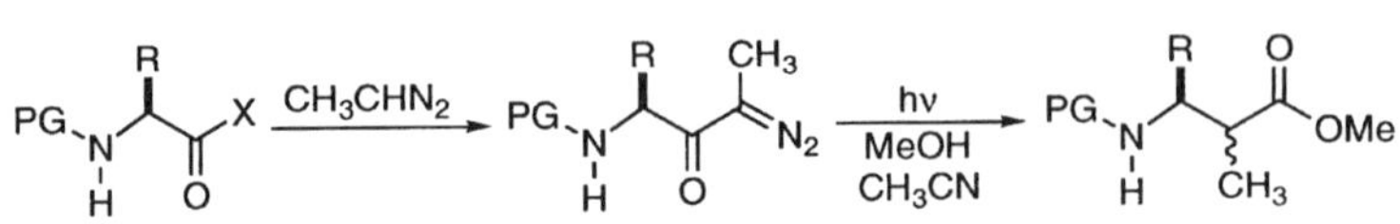

field indicates little about the results of the homologation reaction, but recent work[39] suggests that the yields are at best moderate and the diastereoselectivities are low.

The problems associated with the use of diazoalkanes other than diazomethane mean that the alkylation of simple β-substituted β-amino acid esters is of special interest. The Li_2-dianions of these esters can be diastereoselectively alkylated in the α-position, as Seebach and Estermann have shown.[4,40] Further work has also considered the case of β-homo-Phe (**37**), where double alkylation with high diastereoselectivity has been shown to be possible (Scheme 5.27).

Scheme 5.27

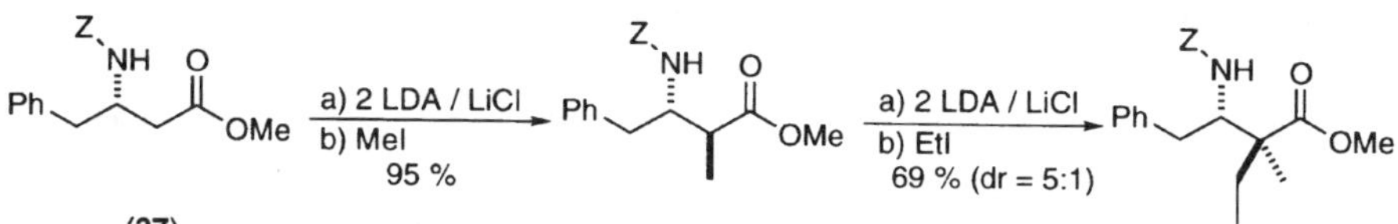

(37)

Other functional groups have been introduced at the α-position by amination[4] of the Li_2-derivative with diethyl azodicarboxylate (DEAD) (**38**) or by the reaction[41] with *p*-methoxybenzyl disulfide (**39**) (Scheme 5.28).

Scheme 5.28

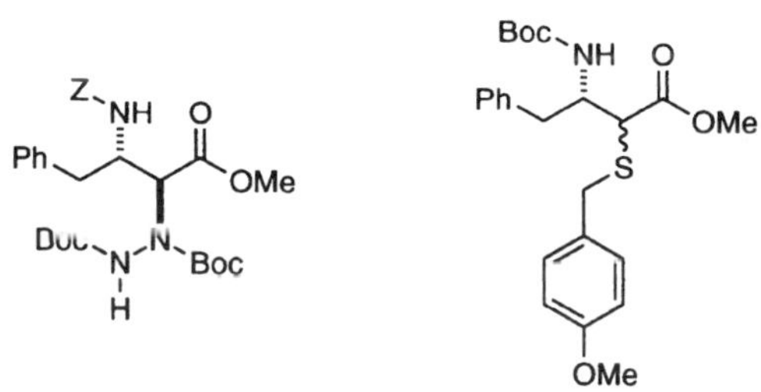

(38) (39)

5.6.2 β,β-Disubstituted β-Amino Acids

α,α-Disubstituted α-amino acids are easily accessible[42] and appear to be a good starting point for the formation of β,β-disubstituted β-amino acids by Arndt–Eistert homologation. However, both the preparation of the diazoketones and their subsequent homologation cause severe problems. This is particularly so in the case of highly hindered α,α-disubstituted α-amino acids.

Hindered carboxylic acids with a tertiary carbon in the α-position have been transformed[43] into diazoketones using acid chlorides as the active species in most cases. The problems arising from the preparation of N-protected α-amino acid chlorides have already been alluded to in Section 5.2. With the exception of 2-methylproline (**40**) (Scheme 5.29), no reaction is observed between the mixed anhydrides of I- or

Scheme 5.29

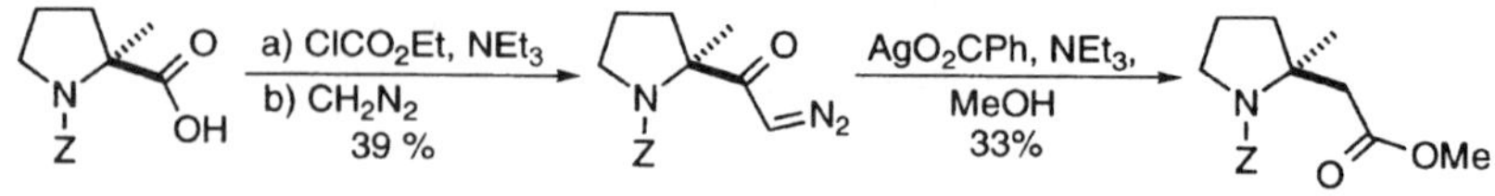

(40)

Boc-protected α,α-disubstituted α-amino acids or the acid chlorides derived from phthaloyl-protected α,α-disubstituted α-amino acids and diazomethane.[44]

Blair and Mander have reported[45] the formation of an oxazolone upon treatment of the trifluoroacetamide protected α-branched amino acid chloride (**41**) with diazomethane. They proposed an equilibrium between the open-chain acid chloride and the cyclic oxazolone hydrochloride (**42**) (Scheme 5.30) to explain the outcome of this reaction. The excess diazomethane functions as an acid scavenger, which traps the HCl. In a complex cascade of reactions, four compounds are formed, of which the diazoketone (**43**) is the main product (60%). This diazoketone was not rearranged but used for a cyclization. Despite the observed oxazolone formation, the acid chloride seems to be the only suitably activated species for the reaction with diazomethane.

The conversion of carboxylic acids into their acid chlorides[46] using α-chloroenamine (**44**), a method that has been developed by Ghosez, has been used for the preparation of Z-protected α-amino acid chlorides.[13] The acid chlorides cannot be

Scheme 5.30

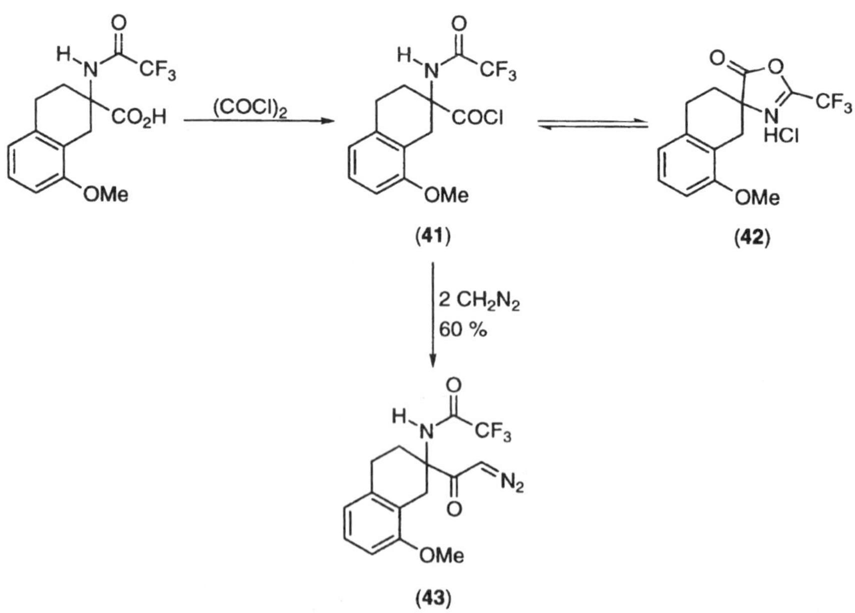

(43)

Scheme 5.31

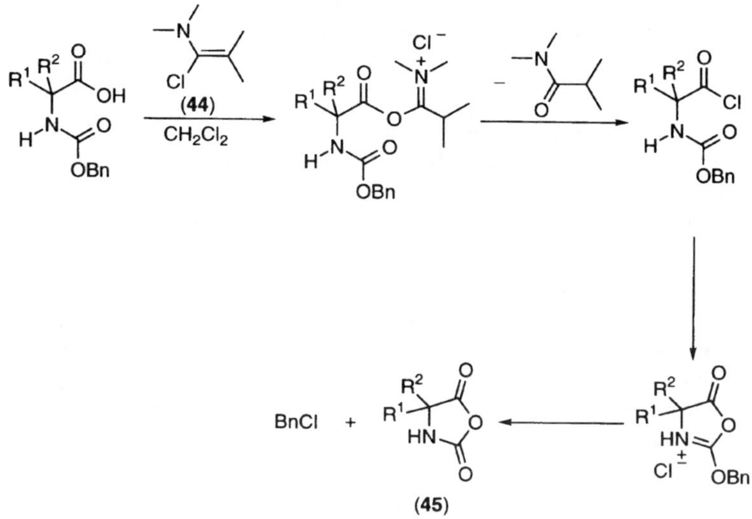

isolated because of their instability. When the α-chloroenamine (**44**) is used, the byproduct (N,N-dimethyl isobutyramide) does not interfere in the subsequent reaction. However, in the case of α,α-disubstituted Z-protected amino acids, much longer reaction times are required (45 min instead of 15 min at 0°C) and under these condi-

Scheme 5.32

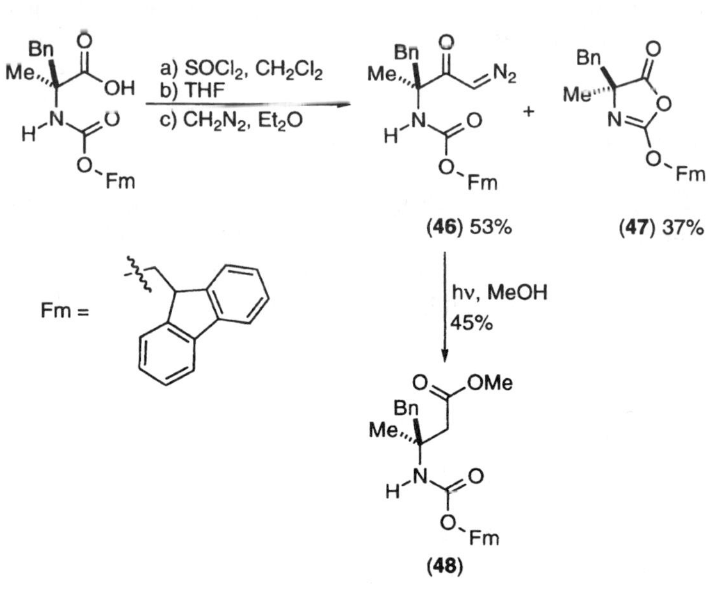

tions, the acid chlorides form the N-carboxylic anhydrides [(**45**), Leuch anhydrides], which do not react with diazomethane[47] (Scheme 5.31).

It is known that Fmoc-protected α-amino acid chlorides are fairly stable and may be prepared by the reaction of Fmoc-protected α-amino acids with thionyl chloride in dichloromethane at room temperature.[14] The treatment of the acid chloride derived from an Fmoc-protected α,α-disubstituted α-amino acid with diazomethane in ether at 0°C affords a mixture of a diazoketone and an oxazolone. This mixture can easily be separated by chromatography, and the oxazolones can be recycled by hydrolysis back to the Fmoc α-amino acids. These highly substituted diazoketones do not rearrange in the presence of Ag^+/triethylamine in methanol, but irradiation in methanol does result in the products of rearrangement (Scheme 5.32).

In the one case shown above,[47] the diazoketone (**46**) is formed in 53% yield with the oxazolone (**47**) in 37% yield. The ratio of diazoketone to oxazolone is independent of the reaction temperature. The benzyl methyl substituted diazoketone (**46**) undergoes the Wolff rearrangement upon irradiation to give the β,β-disubstituted β-amino acid (**48**) in a yield of 45%.

The preliminary results in this area suggest that this method might become a useful route to β-branched β-amino acids with a variety of substitution patterns.

5.7 Summary

The application of the Arndt-Eistert methodology has become a useful and versatile method for the synthesis of β-amino acids of differing substitution patterns, and these have seen many applications. The use of this methodology for the formation of a β-amino acid with concomitant chain extension of a peptide is of particular note. We believe that the applications of this method will continue to grow and result in a wide range of uses being found throughout organic synthesis.

5.8 Acknowledgments

We thank the Royal Society, Great Britain (fellowship to J.L.M.), the Deutsche Forschungsgemeinshaft (fellowship to C.B.), the Eidgenössische Stipendienkommission für ausländische Studierende (fellowship to M.O.), the Schweizerischer Nationalfonds zur Förderung der wissenschaftlichen Forschung, and Sandoz Pharma and Agro Divisions (Basel).

5.9 References

1. F. Arndt, B. Eistert, W. Partale, *Ber. Dtsch. Chem. Ges.* **60**, 1364 (1927).
2. K. Balenović, D. Cerar, Z. Fuks, *J. Chem. Soc.*, 3316 (1952).
3. D. S. Tarbell, J. A. Price, *J. Org. Chem.* **22**, 245 (1957).

4. J. Podlech, D. Seebach, *Liebigs Ann.*, 1217 (1995).
5. T. Ye, M. A. McKervey, *Chem. Rev.* **94**, 1091 (1994).
6. E. Freund, Diplomarbeit, ETH Zürich, 1995.
7. J. L. Matthews, M. Overhand, D. Seebach, manuscript in preparation.
8. B. Penke, J. Czombos, L. Balaspiri, J. Petres, K. Kovacs, *Helv. Chim. Acta* **53**, 1057 (1970).
9. K. Plucińska, B. Liberek, *Tetrahedron* **43**, 3509 (1987).
10. G. R. Pettit, P. S. Nelson, *Can. J. Chem.* **64**, 2097 (1986).
11. H. O. Kim, R. K. Olsen, O.-S. Choi, *J. Org. Chem.* **52**, 4531 (1987).
12. P. Stelzel, in *Methoden der Organischen Chemie*, E. Wünsch, Ed. (Thieme Verlag, Stuttgart, 1974), Volume 15/2, p. 187.
13. U. Schmidt, M. Kroner, U. Beutler, *Synthesis*, 475 (1988).
14. L. A. Carpino, B. J. Cohen, K. E. Stephens, Jr., S. Y. Sadat-Aalaee, J.-H. Tien, D. C. Langridge, *J. Org. Chem.* **51**, 3732 (1986).
15. E. Wünsch, in *Methoden der Organischen Chemie*, E. Wünsch, Ed. (Thieme Verlag, Stuttgart, 1974), Volume 15/1, p. 36.
16. T. Shioiri, T. Aoyama, in *Advances in the Use of Synthons in Organic Chemistry*, A. Dondoni, Ed. (JAI Press Ltd., London, 1993), Volume 1, p. 51.
17. T. Aoyama, T. Shioiri, *Tetrahedron Lett.* **21**, 4461 (1980); T. Aoyama, T. Shioiri, *Chem. Pharm. Bull.* **29**, 3249 (1981).
18. H. Meier, K.-P. Zeller, *Angew. Chem. Int. Ed. Eng.* **14**, 32 (1975).
19. M. S. Newman, P. F. Beal, *J. Am. Chem. Soc.* **72**, 5163 (1950).
20. Y. Yukawa, Y. Tsuno, T. Ibata, *Bull. Chem. Soc. Jpn.* **40**, 2613 (1967); Y. Yukawa, Y. Tsuno, T. Ibata, *Bull. Chem. Soc. Jpn.* **40**, 2618 (1967).
21. B. Paul, W. Korytnyk, *J. Med. Chem.* **19**, 1002 (1976).
22. J. Podlech, D. Seebach, *Angew. Chem. Int. Ed. Engl.* **34**, 471 (1995).
23. P. Darkins, N. McCarthy, M. A. McKervey, K. O'Donnell, T. Ye, B. Walker, *Tetrahedron: Asymmetry* **5**, 195 (1994).
24. J. Cooper, P. T. Gallagher, D. W. Knight, *J. Chem. Soc., Perkin Trans. 1*, 1313 (1993).
25. J. Fetter, K. Lempert, T. Gizur, J. Nyitrai, M. Kajtár-Peredy, G. Simig, G. Hornyák, G. Doleschall, *J. Chem. Soc., Perkin Trans. 1*, 221 (1986); M. Shiozaki, N. Ishida, T. Hiraoka, H. Maruyama, *Tetrahedron* **40**, 1795 (1984).
26. U. Schmidt, F. Stäbler, A. Lieberknecht, *Synthesis*, 482 (1992).
27. P. A. Grieco, Y. S. Hon, A. Perez-Medrano, *J. Am. Chem. Soc.* **110**, 1630 (1988).
28. Although there are obvious similarities to the homologation of phenylglycine, both authors assume no racemization during the homologation step in their syntheses, and the enantiopurity of the final products supports this assumption. However, they did not probe this question directly.
29. C. W. Jefford, Q. Tang, A. Zaslona, *J. Am. Chem. Soc.* **113**, 3513 (1991).
30. M. Overhand, J. L. Matthews, D. Seebach, manuscript in preparation.
31. D. Fles, A. Markovac-Prpić, *Croat. Chem. Acta* **28**, 73 (1956).
32. W. Jugelt, P. Falck, *J. Prakt. Chem.* **38**, 88 (1968).

33. H. M. M. Bastiaans, A. E. Alewijnse, J. L. van der Baan, H. C. J. Ottenheijm, *Tetrahedron Lett.* **35**, 7659 (1994).

34. D. Seebach, M. Overhand, F. N. M. Kühnle, B. Martinoni, L. Oberer, U. Hommel, H. Widmer, *Helv. Chim. Acta*, **79**, 913 (1996).

35. C. Guibourdenche, J. Podlech, D. Seebach, *Liebigs Ann.*, 1121 (1996).

36. S. Chladek, in *Chemistry of Nucleosides and Nucleotides*, L. B. Townsend Ed. (Plenum Press, New York, 1994), Volume 3, p. 107.

37. R. A. Houghten, *Gene* **137**, 7 (1993).

38. K. Balenović, I. Jambresić, *Chem. & Ind.*, 1673 (1955); K. Balenović, I. Jambresić, *Croat. Chem. Acta* **29**, 87 (1957).

39. C. Guibourdenche, unpublished results, ETH Zürich, 1995.

40. D. Seebach, H. Estermann, *Tetrahedron Lett.* **28**, 3103 (1987); D. Seebach, H. Estermann, *Helv. Chim. Acta* **71**, 1824 (1988).

41. E. M. Gordon, J. D. Godfrey, N. G. Delaney, M. M. Asaad, D. Von Langen, D. W. Cushman, *J. Med. Chem.* **31**, 2199 (1988).

42. A. Studer, D. Seebach, *Liebigs Ann. Chem.*, 217 (1995), and references cited therein.

43. P. Ceccherelli, M. Curini, M. C. Marcotullio, O. Rosati, *Tetrahedron* **47**, 7403 (1991); B. Saha, G. Bhattacharjee, U. R. Ghatak, *Tetrahedron Lett.* **27**, 3913 (1986).

44. A. Greppi, unpublished results, ETH Zürich, 1995.

45. I. A. Blair, L. N. Mander, *Aust. J. Chem.* **32**, 1055 (1979).

46. B. Haveaux, A. Dekoker, M. Rens, A. R. Sidani, J. Toye, L. Ghosez, *Org. Synth.* **59**, 26; A. Devos, J. Remion, A.-M. Frisque-Hesbain, A. Colens, L. Ghosez, *J. Chem. Soc., Chem. Commun.*, 1180 (1979).

47. C. Braun, unpublished results, ETH Zürich, 1995.

CHAPTER

6

Synthesis of Nonracemic β-Amino Acids via Chiral Sulfinimines

Franklin A. Davis and Rajarathnam E. Reddy

6.1 Introduction

The development of new and improved methods for the asymmetric synthesis of β-amino acids is of considerable current interest. These compounds exhibit biological activity[1–5] and are key components of antibiotics,[6] peptides,[7–10] and other bioactive materials.[11–13] For example L-phenylalanine is present in the astins A, B, and C (**1a–c**), an antitumor cyclic pentapeptide,[14,15] and in andrimid **2**, a peptide antibiotic.[16] N-Benzoyl-(2*R*,3*S*)-3-phenylisoserine (**3**), the C-13 side chain of the antitumor agent taxol,[17,18] and bestatin (**4**), a potent aminopeptidase B inhibitor, are examples of important α-hydroxy β-amino acids.[19,20] Cyclization of β-amino acids is a useful route to the β-lactam class of antibiotics,[21,22] and substitution for α-amino acids is impor-

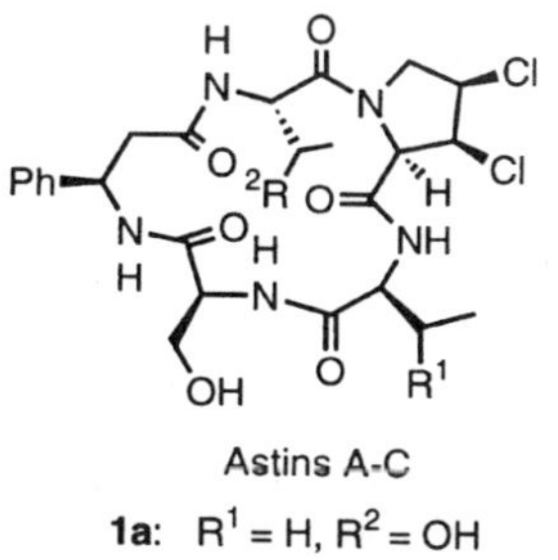

Astins A-C

1a: $R^1 = H, R^2 = OH$

b: $R^1 = OH, R^2 = H$

c: $R^1 = R^2 = H$

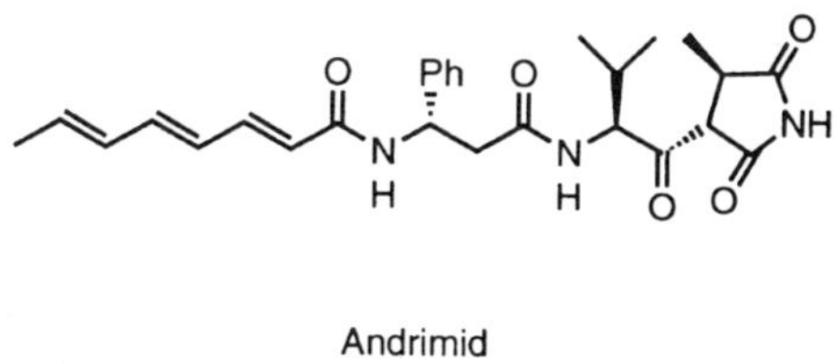

Andrimid

2

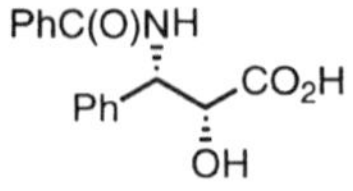

Taxol side chain

3

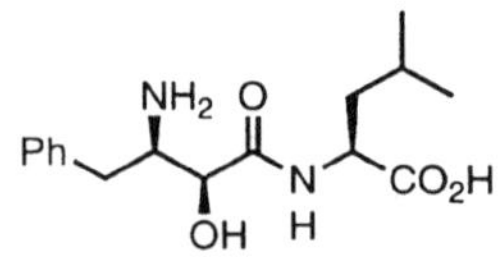

Bestatin

4

tant in the preparation of peptide analogs with increased activity and enzymatic stability.[23] In view of the importance and utility of β-amino acids, a number of strategies for their synthesis[24,25] and asymmetric synthesis have been reported.[26–28]

One important route to β-amino acids involves the condensation of ester enolates with aldimines (Scheme 6.1) and has been viewed by Hart and Ha[21] and by Kleinman.[27] The asymmetric synthesis of β-lactams from chiral enolates and imines, also discussed in these reviews, proceeds with varying degrees of success. The diastereoselectivity is not only sensitive to the chiral auxiliary, but to the geometries of the enolate and the imine as well. Most examples employ α-aryl aldimines (e.g., R^2 = Ar) because they are relatively stable, and lacking enolizable protons they cannot form azaenolates, which lead to undesirable side reactions. Good yields for the addition of boron, zinc, and aluminum enolates to aliphatic imines have, however, been reported, but the diastereoselectivity is variable.[21,27] Removal of the auxiliary (R^1) to give the primary amine derivative is often problematic.

Enantiopure sulfinimines **5** (thiooxime S-oxides) are useful chiral ammonia imine building blocks because nucleophilic addition across the C-N double bond and cleavage of the resultant sulfinamide **6** furnishes the primary amine **7**, which contains a new stereogenic center (Scheme 6.2). In contrast to N-alkyl and N-aryl imines,

Scheme 6.1

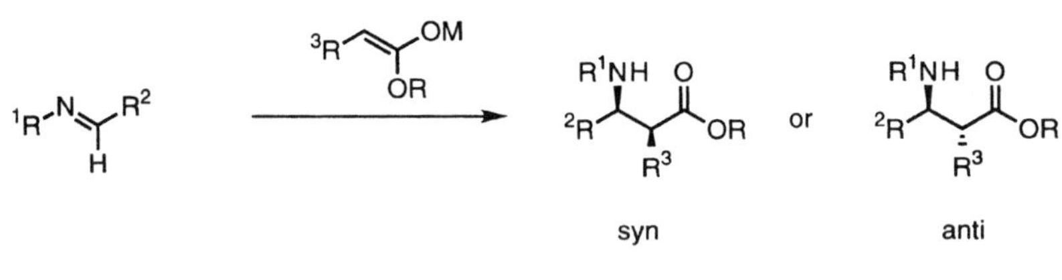

Scheme 6.2

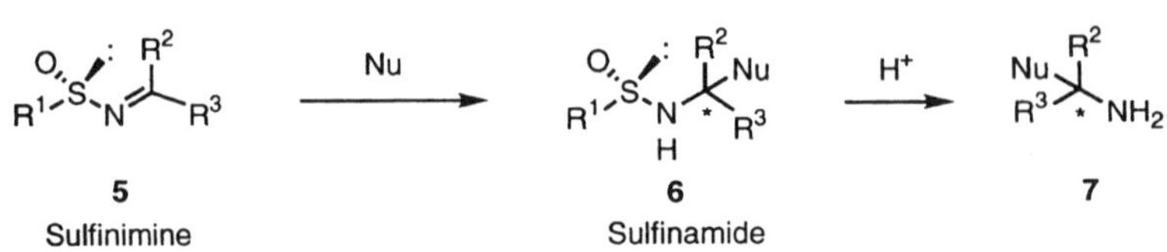

sulfinimines are stable and generally exist in one isomeric form. Since the powerful stereodirecting N-sulfinyl auxiliary activates the C-N double bond toward addition, azaenolate formation with aliphatic examples is generally not a problem. Importantly, the N-sulfinyl auxiliary is easily removed to afford the primary amine derivative without epimerization (Scheme 6.2). Highly diastereoselective asymmetric syntheses of amines,[29–32] α-amino acids,[33,34] N-sulfinyl aziridines,[35,36] and N-sulfinyl cis aziridine-2-carboxylic acids[37,38] have been realized using **5**. The purpose of this chapter is to review recent progress in the asymmetric synthesis of β-amino acids and α-substituted β-amino acids using enantiomerically pure sulfinimine **5**.

6.2 Synthesis of Chiral Sulfinimines

The synthesis of racemic sulfinimines **5** was first reported by Davis and co-workers in 1974 by the oxidation of sulfenimines **9** (Scheme 6.3) with *m*-chloroperbenzoic acid (*m*-CPBA).[39–42] The "silver-assisted" methodology involving the reaction of aromatic disulfides, aldehydes or ketones, silver nitrate, and ammonia was used in the preparation of **9**.[43,44] Other methods are also available for the preparation of sulfenimines.[45,46]

The first synthesis of enantiomerically pure sulfinimines **5** was developed by Cinquini and co-workers[29] and extended by Hua et al. (Scheme 6.4).[30] This procedure involves the reaction of a metalketimine with (1*R*,2*S*,5*R*)-(−)-menthyl-(*S*)-*p*-toluenesulfinate (**10**). Yields are moderate to good. A similar route, using an enantiopure cyclic sulfinamide, has been described by Willis et al.[47] This latter procedure has the advantage of recycling the chiral auxiliary. However, these "Andersen-type" procedures are limited to the synthesis of alkyl, aryl ketimine derived sulfinimines **5** (*R* ≠ H) because the metalketimine is prepared by treatment of an aromatic nitrile with a lithium or Grignard reagent. Thus aliphatic and aromatic aldimine sulfinimines **5** (R^2 = H) cannot be prepared by these routes.

Scheme 6.3

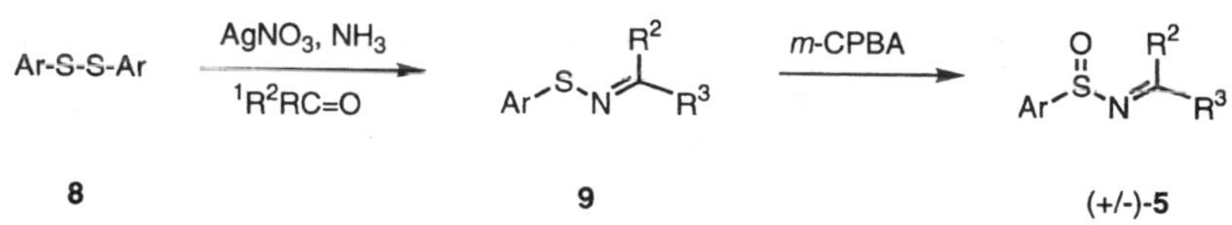

Scheme 6.4

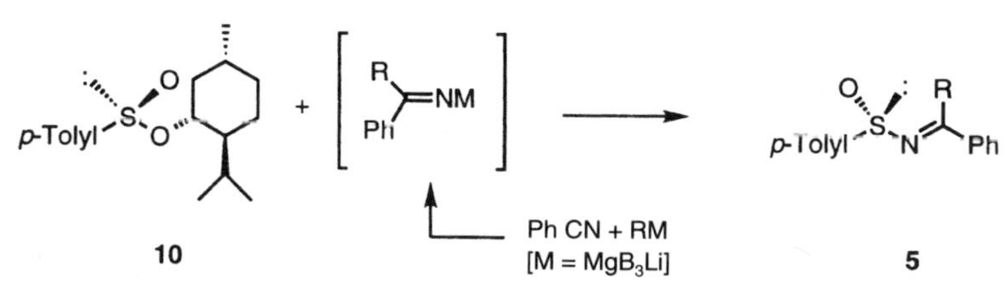

Scheme 6.5

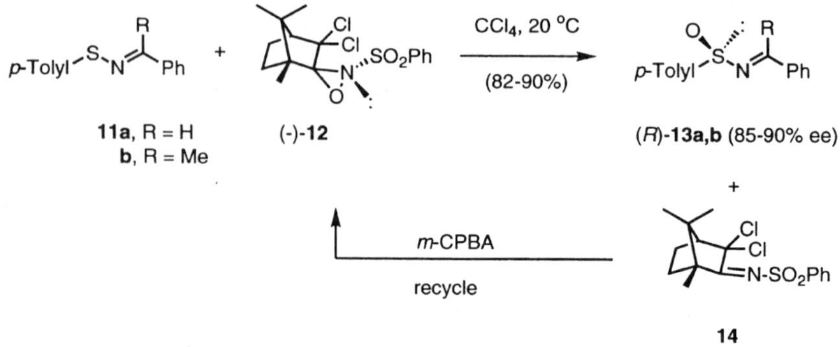

Asymmetric oxidation of ketone and aldehyde derived sulfenimines **11** affords enantiomerically enriched sulfinimines **13** (Scheme 6.5).[48] In this procedure Davis and co-workers carried out the asymmetric oxidation of **11** with (+)- or (–)-N-(phenylsulfonyl)(3,3-dichlorocamphoryl)oxaziridine (**12**), an asymmetric oxidizing reagent developed for the asymmetric oxidation of sulfide to sulfoxides[49] and selenides to selenoxides.[50] Sulfinimines **13a** and **13b** were isolated in 82–90% yield and 85–90% ee.[48] Crystallization from *n*-hexane upgrades the enantiomeric purity of these compounds to better than 97%. Although the sulfonimine byproduct **14** can be isolated in >90% yield and recycled, a limitation of this procedure is that if the sulfinimine is noncrystalline, which is true of most aliphatic examples, the ees cannot be upgraded. Oxidation of chiral nonracemic sulfenimines having a camphor-based mercapto chiral auxiliary with *m*-CPBA affords sulfinimine diastereomers in 44–99% de.[32]

The Davis group has also introduced general methodology for the synthesis of enantiomerically pure aromatic and aliphatic aldimine sulfinimines **17** (Scheme 6.6).[51] (1*R*,2*S*,5*R*)-(–)-menthyl-(*S*)-*p*-toluenesulfinate (**10**), the Andersen reagent, is treated with lithium bis(trimethylsilyl)amide (LiHMDS) at –78°C followed by the in situ treatment of the intermediate N-bis(trimethylsilyl)sulfinamide **15** with an aldehyde and CsF. While this "one pot" Petersen-type olefination, reaction affords good to excellent yields of **17** (70–90%) with aromatic aldehydes, it fails for aliphatic aldehydes (Table 6.1). Apparently the basic lithium menthoxide **16** byproduct causes aldol-type reac-

Scheme 6.6

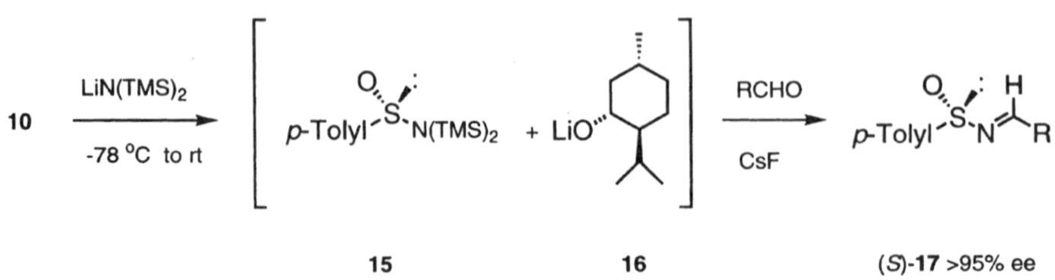

Table 6.1 ASYMMETRIC SYNTHESIS OF SULFINIMINES **17** FROM ALDEHYDES

Entry	Aldehyde	Procedure	Sulfinimine **17** % Yield[a]	Ref.
1	PhCHO	one-step[b]	82	51
2	*p*-MePhCHO	one-step	90	51
3	*p*-MeSPhCHO	one-step	80	38
4	3-PyridylCHO	one-step	70	54
5	*E*-PhCH=CHCHO	one-step	85	51
6	*E*-MeCH=CHCHO	one-step	0	51
7		two-step[c]	57	51
8	*n*-PrCHO	one-step	<10	51
9		two-step	30	

[a] Isolated yields.
[b] One pot procedure.
[c] Isolation of crude **15**.

tions with aliphatic aldehydes because isolation of crude **15** and subsequent treatment with the aldehyde and CsF affords the aliphatic examples in 52–63% (Table 6.1).

6.3 Synthesis of β-Amino Acids

Hua et al. reported the diastereoselective addition of allylmagnesium bromide to sulfinimines **13b** and **13c** (*R* = Me, *n*-Bu) to give sulfinamides **18** in 82–98% de and in 92–96% yield.[30] Following separation of the diastereoisomers, the sulfinamides **18** were hydrolyzed with trifluoroacetic acid (TFA), and the allylic amines were acetylated to the N-acetates **19**. The acetates were transformed to the free β-amino acids **20b** and **20c** by ozonolysis, oxidation with $AgNO_3$, and deacetylation with HCl, as shown in Scheme 6.7.

Scheme 6.7

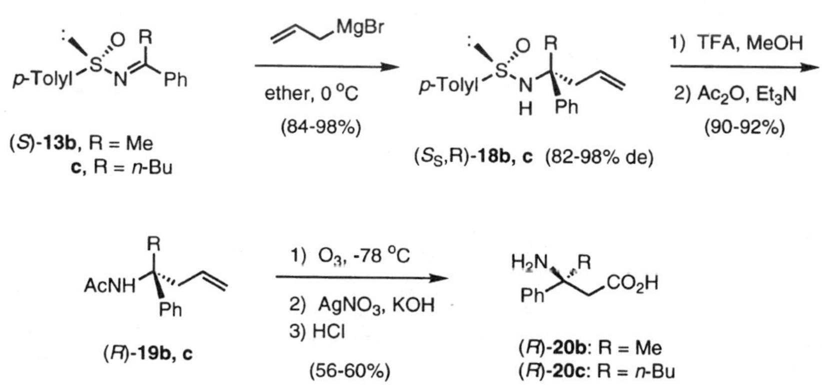

A simpler route to β-amino acids than that depicted in Scheme 6.7 is the highly diastereoselective 1,2-addition of the metal enolates of methyl acetate to sulfinimines (*R*)-**13a-b** (*R* = H, Me) (Scheme 6.8).[48,52,53] Thus addition of the lithium enolate of methyl acetate to (*R*)-**13a** (*R* = H) gave sulfinamide (R_s,*S*)-**21a** in 82% yield and 86% de. Crystallization afforded the diastereomerically pure product in 74% isolated yield.[48] With the sodium enolate in THF the de of **21a** improved to 92% and in ether to 98%.[52] Addition of the lithium enolate of methyl acetate to the acetophenone sulfinimine (R_s*S*)-**21b** resulted in a single diastereomer isolated in 84% yield.[48] Importantly, this result demonstrates that enolization of the sulfinimine does not compete with enolate addition to the C-N double bond. The excellent selectivity observed in these additions was interpreted in terms of a Zimmerman–Traxler chairlike six-membered transition favoring approach of the enolate from the *Si* face of the C-N π bond in (*R*)-**13**.

Hydrolysis of the enantiomerically pure (R_s,*S*)-sulfinamides **21** (*R* = H, Me, Ar = *p*-tolyl) was accomplished with TFA in methanol to give (*S*)-β-phenylalanine methyl ester (**22a**) in 73% and (*S*)-methyl 3-amino-3-methyl-3-phenyl propionate (**22b**) in 85% yield (Scheme 6.8).[48] The byproduct (*R*)-methyl-*p*-tolylsulfinate (**23**) resulting from N-sulfinyl auxiliary was isolated in almost quantitative yield and 64% ee.[54]

The sulfinimine-β-amino acid protocol has recently been employed in the asymmetric synthesis of (*S*)-ethyl-β-amino pyridinepropionate (24), a key subunit of (*S*)-ethyl-β-[[4-(aminoiminomethyl)phenyl]amino-1,4-dioxobutyl]amino]-3-pyridinepropanoate (**25**) (Scheme 6.9).[55] Compound **25** is a peptidomimetic for the Arg-Gly-Asp-Phe sequence of fibrinogen and is an orally active antiplatelet agent that may be useful in preventing myocardial infarction and unstable angina.[55] However, the diastereoselectivity for addition of the lithium or sodium enolates of ethyl acetate to (*R*)-(−)-**26** was modest, for example, 47 to 64% (Scheme 6.10) and contrasts sharply with similar additions to (*R*)-**13** (de >98%).[52] The modest diastereoselectivities observed for sulfinimine **26** were attributed to coordination of the metal enolate with the

Scheme 6.8

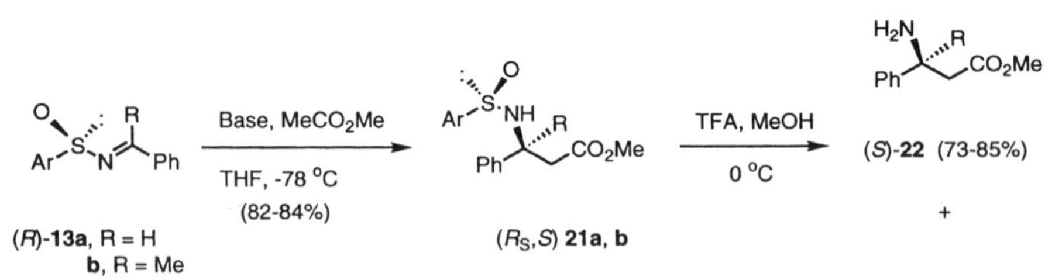

Base	R =	%de
LDA	H	80
NaHMDS	H	92
NaHMDS/ether	H	>98
LDA	Me	>97

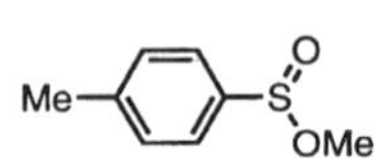

23 (99%)
64% ee

Scheme 6.9

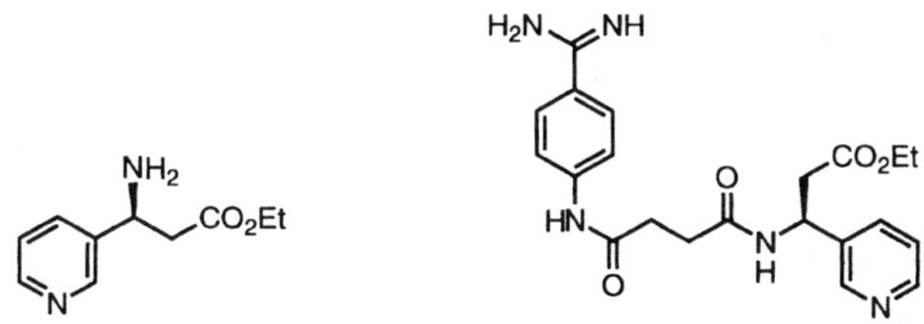

(*S*)-**24** **25**

Scheme 6.10

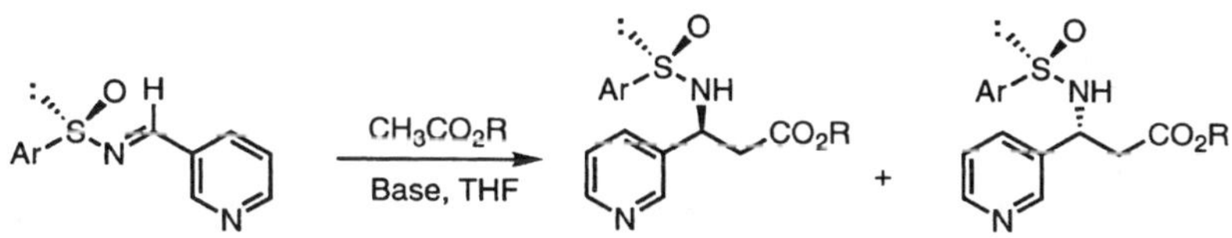

(*R*)-(-)-**26** (R_S,*S*)-**27** (R_S,*R*)-**27**

Ar = *p*-Tolyl

Base	R	% de	% Yield
NaHMDS	Me	47	80
NaHMDS	Et	64	72
LDA	Et	66	56
LiHMDS	Et	52	59

pyridyl nitrogen, which disrupts the transition for enolate addition. All attempts to separate the liquid sulfinamide diastereoisomers of **27** chromatographically were unsuccessful.

Interestingly, addition of the lithium enolate of ethyl acetate to (*R*)-(–)-N-(3-pyridinidine)-2-methoxynaphthalenesulfinimide (**28**) afforded sulfinamide **29** in 86% yield and in 80% de (Scheme 6.11).[54] The major diastereomer (R_s,*S*)-**29** was easily separated chromatography in 76% isolated yield, and hydrolysis with TFA/EtOH afforded the enantiomerically pure (*S*)-ethyl β-amino-3-pyridinepropanoate (**24**) in 74% yield.[54] Sulfinimine **28** was prepared in 70% yield using (+)-(*R*)-menthyl-2-methoxy-1-naphthalenesulfinate (**10**) [*p*-tolyl = 2-methoxynaphthyl], an auxiliary recently introduced by Payne and co-workers,[56] and 3-pyridinecarboxaldehyde as previously described (Scheme 6.6).

The preparations of β-amino acids from chiral nonracemic sulfinimines illustrated in Schemes 6.7–6.11 are examples of "chiral auxiliary" based asymmetric syntheses. Such syntheses have the advantage of producing diastereoisomers that can often be

Scheme 6.11

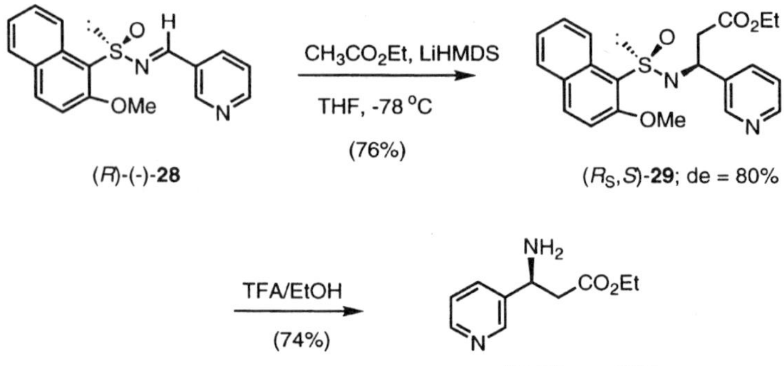

(*R*)-(-)-**28** (*R*S,*S*)-**29**; de = 80% (*S*)-**24**; ee >97%

separated, affording the enantiomerically pure products after hydrolysis. A disadvantage is that the auxiliary needs to be introduced and eventually removed. However, if the auxiliary is recyclable or a precursor regenerated, these limitations are, to a large extent, mitigated. It was found that hydrolysis of (S_sR)-sulfinamide **21a** with TFA in the presence of 1.2 equivalents of (1*R*,2*S*,5*R*)-(−)-menthol afforded the desired optically pure (*R*)-β-phenylalanine (**22a**) in 86% yield (>97% ee).[54] The byproduct, menthyl-*p*-toluenesulfinate (**10**), the sulfinimine precursor (Schemes 6.4 and 6.6), was isolated in 74% yield with a diastereomeric purity of about 10% (Scheme 6.12). The enantiomerically pure auxiliary **10** is obtained by crystallization from acetone/HCl using the Solladie–Posner protocol, which equilibrates the sulfinate ester diastereoisomers with displacement toward the less soluble one.[57,58]

Scheme 6.12

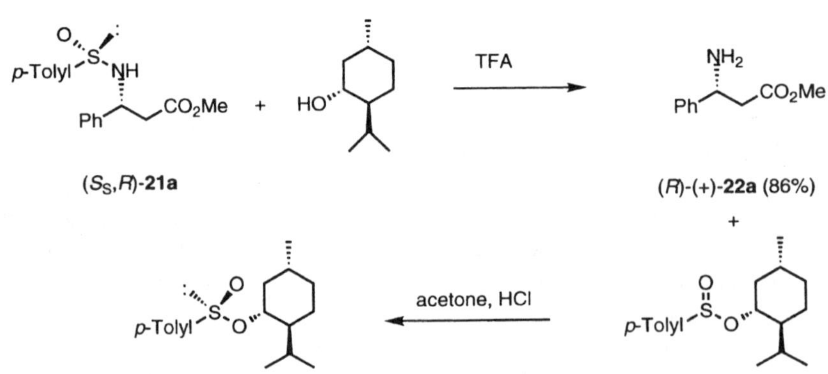

(*S*S,*R*)-**21a** (*R*)-(+)-**22a** (86%) + **10** (74%) (-)-(*S*)-**10** (55-63%)

6.4 Synthesis of α-Substituted β-Amino Acids

Interest in the asymmetric synthesis of (–)-N-benzoyl-(2*R*,3*S*)-3-phenylisoserine (**3**), the C-13 side chain of the remarkable antitumor agent Taxol, has been stimulated by two factors.[59] First, the side chain is essential for its biological activity, and second is the semisynthetic synthesis of Taxol from **3** and 10-deacetyl baccatin III, which is available from a renewable source. (S_s,3*R*)-(+)-Methyl N-(*p*-tolylsulfinyl)-3-amino-3-phenylpropanoate (**21a**) is an advanced precursor of the Taxol side chain **3**, requiring only N-benzoylation and stereospecific induction of the hydroxyl group. One-pot hydrolysis–benzoylation of (S_s,3*R*)-(+)-**21a** was accomplished by treatment with TFA/MeOH, followed by reaction with benzoyl chloride, triethylamine, and a catalytic amount of DMAP to afford the (*R*)-benzoate ester **30** in 76% overall yield (Scheme 6.13).[48] Next the 2-hydroxyl group was installed by hydroxylation of the enolate using the enolate oxidation protocol.[60] The dienolate of **30** was generated using 2.2 eq of LDA followed by hydroxylation with (+)-(camphorylsulfonyl)oxaziridine (**31**) in the presence of LiCl, affording an 86:14 ratio of (–)-(2*R*,3*S*)-**3** and (+)-(2*S*,3*S*)-**3** in 58% yield. Yields were markedly lower in the absence of LiCl (e.g., <15%).

The facial selectivity for enolate hydroxylations by N-sulfonyloxaziridines nearly always occurs at the least hindered face of the enolate.[60] Thus the fact that the major product, the Taxol side chain (2*R*,3*S*)-**3**, appears to involve approach of the oxaziridine from the more hindered face of the enolate of **30** was unexpected. To explain this result it was suggested that the dienolate of **30** forms an eight-membered cyclic chelate whose conformation is such that the oxaziridine approaches the enolate from the sterically least hindered direction, affording (2*R*3*S*)-**3**.

The ability of fluorine to function as a hydrogen bond acceptor, together with the similarity of typical C-F and C-O bond lengths, suggests that replacement of hydroxy

Scheme 6.13

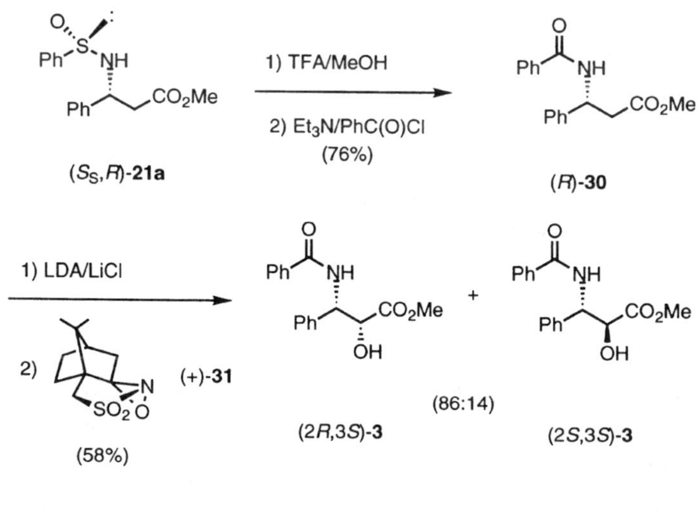

Scheme 6.14

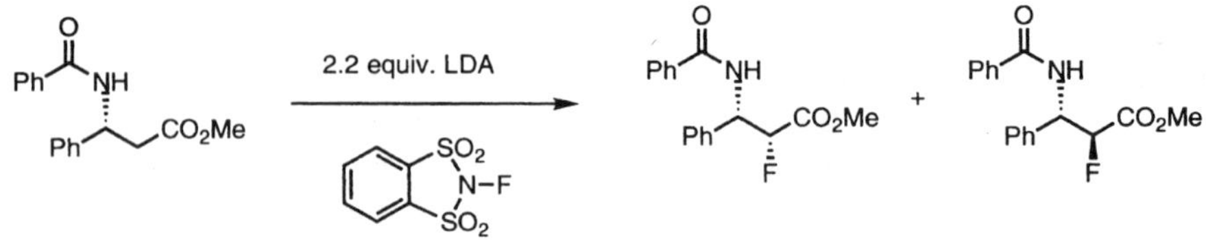

by fluorine in bioactive compounds could result in useful analogs as well as a probe of hydrogen bonding.[61] Accordingly, syn-2*R*,3*S*)- and anti-(2*S*,3*S*)-methyl-2-fluoro-3-(N-benzoylamino)-3-phenylpropanoate (**32**), fluorine analogs of the Taxol C-13 side chain, were prepared from (*R*)-**30**.[62] N-Fluoro-*o*-benzenedisulfonimide (**31**),[63,64] a source of electrophilic fluorine, was employed in the synthesis of **32** in a strategy reminiscent of that used to prepare **3** from **30** (Scheme 6.14).[62] Fluorination of the dianion enolate of **30** with **31** afforded a 35:65 syn–anti mixture of the α-fluoro esters **32** in 82% yield. The yield (95%) and syn-to-anti ratio (44:56) improved in the presence of LiCl. The poor selectivity observed in these electrophilic fluorinations was shown to be due to a base-catalyzed epimerization of these isomers caused by the enhanced acidity of the α-fluoro proton. 2′-Fluoro Taxol has been prepared by Kant and co-workers and found to be about 100 times less active than Taxol itself.[65]

6.5 Summary

Addition of enolates to enantiopure sulfinimines **5**, chiral ammonia imine building blocks, offer easy access to β-amino acids in high diastereomer purity. The N-sulfinyl auxiliary is easily removed with acid and can be transformed (recycled) into the Andersen reagent, menthyl-*p*-toluenesulfinate (**10**), the sulfinimine precursor.

6.6 References

1. S. Shinagawa, T. Kanamaru, M. Asai, H. Okazaki, *J. Med. Chem.* **30**, 1458 (1987).
2. A. Bongini, G. Cardillo, M. Orena, G. Porzi, S. Sandri, *Tetrahedron* **43**, 4377 (1987).
3. N. Kawabata, Y. Inamoto, K. Sakane, T. Iwamoto, S. Hashimoto, *J. Antibiot.* **45**, 513 (1992).
4. T. Oki, M. Hirano, K. Tomatu, K.I.K. Humata, H. Kamei, *J. Antibiot.* **42**, 513 (1989).
5. G. Gasiraghi, L. Colombo, G. Rassu, P. Sapnu, *J. Org. Chem.* **56**, 6523 (1991).
6. For leading references see: D. Kiers, D. Moffat, R. Tomanek, *J. C. S. Perkin Trans. 1*, 1041 (1991).
7. G. C. Barrett, Ed., *Chemistry and Biochemistry of the Amino Acids* (Chapman and Hall, London, 1985).
8. S. M. Hecht, *Acc. Chem. Res.* **19**, 383 (1986).

9. H. Morita, S. Nagashima, K. Takeya, H. Itokava, *Chem. Pharm. Bull.* **41**, 992 (1993).

10. O. W. Griffith, *Ann. Rev. Biochem.* **55**, 855 (1986).

11. N. Mayachim, M. Shibasaki, *J. Org. Chem.* **55**, 1975 (1990).

12. S. Kim, P. H. Lee, T. A. Lee, *J. Chem. Soc., Chem. Commun.* 1242 (1988).

13. D. Tanner, P. Somfai, *Tetrahedron* **44**, 613 (1988).

14. S. Kosemura, T. Ogava, K. Totsuka, *Tetrahedron Lett.* **34**, 1291 (1993).

15. H. Morita, S. Nagashima, K. Takeya, H. Itokawa, *Chem. Pharma. Bull.* **41**, 992 (1993).

16. A. Fredenhagen, S. Y. Tamura, P. T. M. Kenny, H. Komura, Y. Naya, K. Nakanishi, K. Nishiyama, M. Sugiura, H. Kita, *J. Am. Chem. Soc.* **109**, 4409 (1987).

17. J.-N. Denis, A. Correa, A. E. Green, *J. Org. Chem.* **56**, 6939 (1991).

18. I. Ojima, I. Habus, M. Shao, *J. Org. Chem.* **56**, 1681 (1991).

19. W. H. Pearson, J. V. Hines, *J. Org. Chem.* **54**, 4235 (1989).

20. B. H. Norman, M. L. Morris, *Tetrahedron Lett.* **33**, 6803 (1992).

21. D. J. Hart, D. C. Ha, *Chem. Rev.* **89**, 1447 (1989).

22. I. Ojima, I. Habus, *Tetrahedron Lett.* **30**, 4289 (1990).

23. A. F. Spatola, in *Chemistry and Biochemistry of Aminoacids, Peptides and Proteins*, B. Weinstein, Ed. (Marcel Dekker, New York, 1983), Vol. 7, pp. 331–33 and references cited therein.

24. K. Izuka, T. Kamijo, H. Harada, K. Akahana, T. Kubota, H. Umeyama, Y. Kiso, *J. Chem. Soc., Chem. Commun.* 1678 (1989).

25. I. T. Millar, H. D. Springall, *Sidgwic's Organic Chemistry of Nitrogen* (Clarendon, Oxford, 1966), p. 207.

26. E. Juaristi, D. Quintana, J. Escalante, *Aldrichimica Acta* **27**, 3 (1994).

27. E. F. Kleinman. *Comprehensive Organic Synthesis* (Pergamon Press, Oxford, 1991), p. 355.

28. D. C. Cole, *Tetrahedron* **50**, 9517 (1994).

29. R. Annunziata, M. Cinquini, F. Cozzi, *J. C. S. Perkin Trans. 1*, 339 (1982).

30. D. H. Hua, S. W. Miao, J. S. Chen, S. Iguchi, *J. Org. Chem.* **56**, 4 (1991).

31. D. R. J. Hose, T. Raynham, M. Willis, *Tetrahedron: Asymmetry* **4**, 2159 (1992).

32. T.-K. Yang, R.-Y Chen, S.-S. Lee, W.-S Peng, Y.-Z. Jiang, A. Q. Mi, T.-T. Jong, *J. Org. Chem.* **59**, 914 (1994).

33. F. A. Davis, R. E. Reddy, P. S. Portonovo, *Tetrahedron Lett.* **35**, 1351 (1994).

34. D. H. Hua, N. Lagneau, H. Wang, J. Chen, *Tetrahedron: Asymmetry* **6**, 349 (1995).

35. J. L. Garcia Ruano, I. Frenandez, C. Hamdouchi, *Tetrahedron Lett.* **36**, 295 (1995).

36. F. A. Davis, P. Zhou, C.-H. Liang, R. E. Reddy, *Tetrahedron: Asymmetry*, **6**, 1511 (1995).

37. F. A. Davis, P. Zhou, G. V. Reddy, *J. Org. Chem.* **59**, 3243 (1994).

38. F. A. Davis, P. Zhou, *Tetrahedron Lett.* **35**, 7525 (1994).

39. F. A. Davis, A. J. Friedman, E. W. Kluger, *J. Am. Chem. Soc.* **96**, 5000 (1974).

40. F. A. Davis, A. J. Friedman, *J. Org. Chem.* **41**, 898 (1976).

41. F. A. Davis, E. W. Kluger, *J. Am. Chem. Soc.* **98**, 302 (1976).

42. F. A. Davis, A. J. Friedman, Nadir, *J. Am Chem. Soc.* **100**, 2844 (1978).

43. F. A. Davis, W. A. R. Slegeir, S. Evans, A. Schwartz, D. L. Goff, R. Palmer, *J. Org. Chem.* **38**, 2809 (1973).

44. F. A. Davis, J. M. Kaminski, E. W. Kluger, S. Howard, *J. Am. Chem. Soc.* **97**, 7085 (1975).

45. P. K. Claus, in *The Chemistry of Sulphenic Acids and their Derivatives*, S. Patal, Ed. (John Wiley & Sons, New York, 1990), Chap. 17, p. 723.

46. L. Craine, M. Raban, *Chem. Rev.* **89**, 689 (1989).

47. D. R. J. Hose, T. Raynham, M. Willis, *Tetrahedron Lett.* **33**, 1153 (1993).

48. F. A. Davis, R. T. Reddy, R. E. Reddy, *J. Org. Chem.* **57**, 6337 (1992).

49. F. A. Davis, R. Thimma Reddy, W. Han, P. J. Carroll, *J. Am. Chem. Soc.* **114**, 1428 (1992).

50. F. A. Davis and R. Thimma Reddy, *J. Org. Chem.* **47**, 2599 (1992).

51. F. A. Davis, R. E. Reddy, J. M. Szewczyk, P. S. Portonovo, *Tetrahedron Lett.* **34**, 6229 (1993).

52. F. A. Davis, R. E. Reddy, J. M. Szewczyk, *J. Org. Chem.*, **60**, 7037 (1995).

53. J. Jiang, K. K. Schumacher, M. M. Joullie, F. A. Davis, R. E. Reddy, *Tetrahedron Lett.* **35**, 2121 (1994).

54. F. A. Davis, J. M. Szewczyk, R. E. Reddy, *J. Org. Chem.* **61**, 2222 (1996).

55. J. G. Rico, J. R. Lindmark, T. E. Rogers, P. R. Bovy, *J. Org. Chem.* **58**, 7948 (1993).

56. S. G. Pyne, A. R. Hajipour, K. Prabakaran, *Tetrahedron Lett.* **35**, 465 (1994).

57. C. Mioskowski, G. Solladie, *Tetrahedron* **36**, 227 (1980).

58. M. Hulce, J. P. Mallamo, L. L. Frye, T. P. Kogan, G. H. Posner, *Org. Synth.* **64**, 196 (1984).

59. *Taxane Anticancer Agents*, G. I. Georg, T. T. Chen, I. Ojima, D. M. Vyas, Eds., ACS Symposium Series 583 (American Chemical Society, Washington, D.C. 1995).

60. F. A. Davis, B.-C. Chen, *Chem. Rev.* **92**, 919 (1992).

61. J. T. Welch, S. Eswarakrishnan, *Fluorine in Bioorganic Chemistry* (John Wiley & Sons, New York, 1991).

62. F. A. Davis, R. E. Reddy, *Tetrahedron: Asymmetry* **5**, 955 (1994).

63. F. A. Davis, W. Han, *Tetrahedron Lett.* **22**, 1631 (1991).

64. F. A. Davis, W. Han, C. K. Murphy, *J. Org. Chem.* **58**, 4890 (1995).

65. J. Kant, S. Huang, H. Wong, C. Fairchild, D. Vyas, V. Farina, *BioMed. Chem. Lett.* **3**, 2471 (1993).

CHAPTER

7

Asymmetric Addition of Amines to α,β-Unsaturated Esters and Nitriles in the Enantioselective Synthesis of β-Amino Acids

Eusebio Juaristi and Oscar García-Barradas

7.1 Introduction

Conceptually, one of the most attractive methods for the preparation of enantiopure β-amino acids consists of the Michael addition of amines to α,β-unsaturated esters or nitriles. Indeed, a variety of approaches may be employed to induce the stereoselective formation of the Michael products[1]: (1) Incorporation of chiral adjuvants in the olefinic framework renders its faces diastereotopic and thus of unequal reactivity; (2) the use of chiral amines in the reaction leads to diastereoselective addition; and (3) asymmetric catalysis may offer an efficient route for the large-scale preparation of the chiral β-amino ester (or nitrile) product.

In 1965, Terentev and collaborators described the first application of the above concepts when a chiral amine was added to crotonic acid[2]; nevertheless, the enantiomeric ratios (e.r.) obtained were rather low (Scheme 7.1a). Also disappointing were the 1977 studies of Furukawa et al.[3] on the analogous conjugated nitriles (Scheme 7.1b).

In this context, it is important to mention the pioneering studies of Achiwa and Soga,[4] who carried out the catalytic asymmetric hydrogenation of β-(acety-

Scheme 7.1

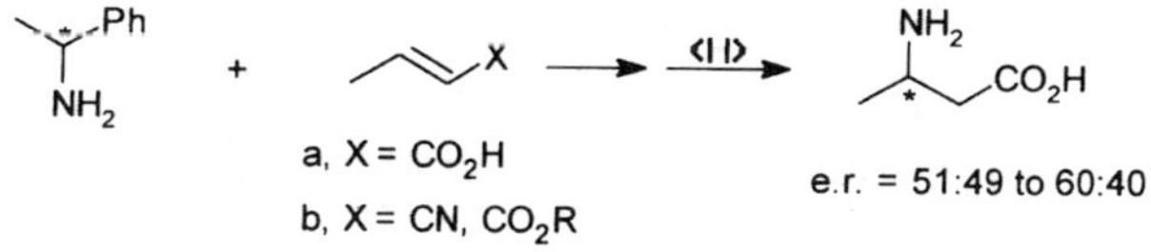

lamino)acrylic acid derivatives to give the optically active β-amino esters in low enantiomeric purities (Scheme 7.2).

Scheme 7.2

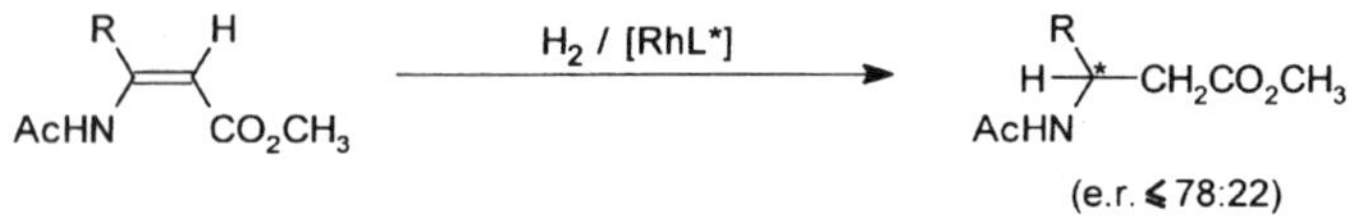

Dramatic improvements in the efficiency of this asymmetric hydrogenation approach were achieved by Noyori and co-workers,[5] who discovered that 2,2′-bis(diphenylphosphino)-1,1′-binaphthyl-ruthenium(II) complexes serve as excellent catalysts for enantioselective hydrogenation of β-substituted (*E*)-β-(acylamino) acrylic acids (Scheme 7.3).

Scheme 7.3

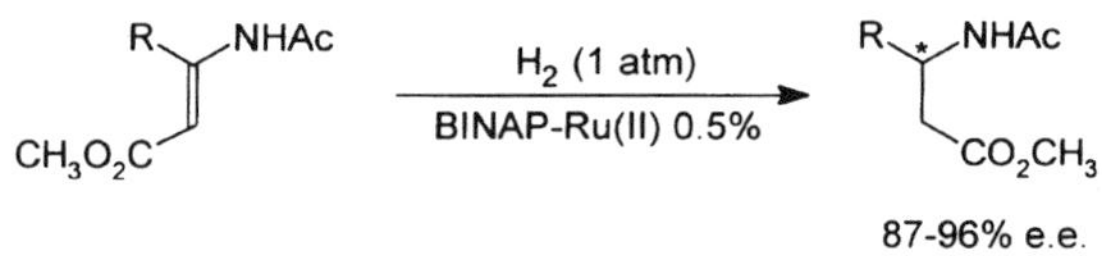

7.2 Addition of Amines to Chiral α,β-Unsaturated Esters

The early work of Furukawa, Stec, and co-workers described above[3] motivated the relevant studies of d'Angelo and Maddaluno,[6] who combined novel chiral auxiliaries in the conjugated ester with the use of high reaction pressure to achieve high stereochemical control as well as excellent yields of amine addition (Scheme 7.4).

In particular, crotonates derived from 8-phenylmentol[7] (*R* = Ph in Scheme 7.4) reacted with benzylamine to give the expected Michael products with 80% diastereoselectivity. Further increase in the induction level was observed by using analogues of 8-phenylmenthol in which the phenyl ring is substituted in the *para*-position by bulky groups (*R* = *p*-phenoxyphenyl in Scheme 7.4; ds = 88%), and essentially complete diastereoselectivity was achieved in the case of 8-(β-naphthyl)menthyl crotonate (*R* = β-naphthyl in Scheme 7.4; ds > 99%).[6]

Scheme 7.4

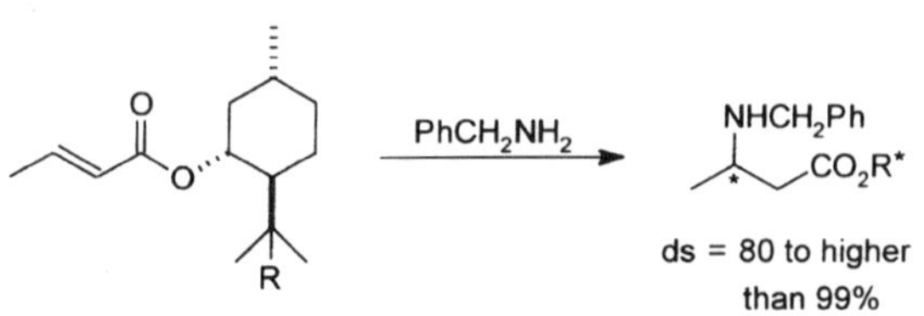

The authors[6] suggest a "π-stacking" model,[8] in which the aryl group of the chiral adjuvant hinders one face of the olefin (Figure 7.1), to explain the stereochemical outcome of the reaction. This interpretation was supported by the shielding effect provoked by the aryl group on the ^{1}H-NMR chemical shifts of the vinylic protons.

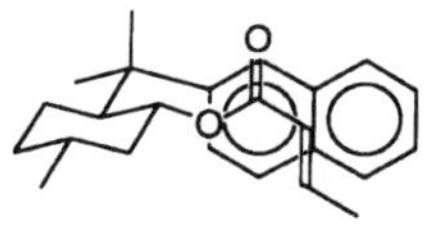

Figure 7.1

In a conceptually related system, Perlmutter and Tabone[9] observed modest to excellent diastereoselectivities in the conjugate addition of benzylamine to chiral 2-hydroxyalkylpropenoates. The best results were obtained with *t*-butyldimethylsilylether **1** (Scheme 7.5).

Scheme 7.5

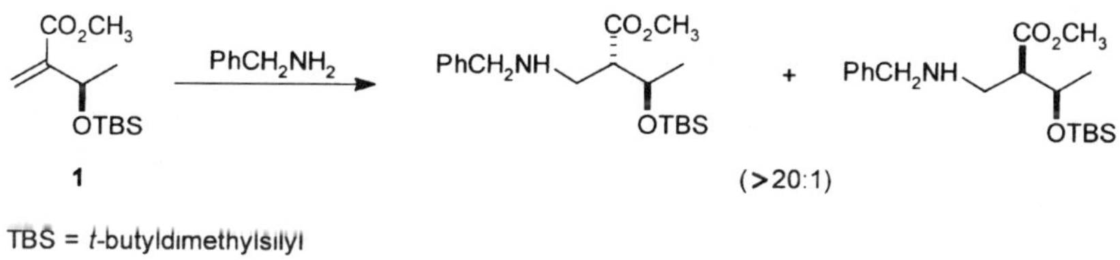

In this context, the addition of lithium enolates to chiral aminomethylacrylate **2** proceeded with excellent diastereoselectivity (up to 99% ds). Thus acrylate (*S,S*)-**2** (or its enantiomer) constitutes a useful synthon for the synthesis of β-amino acid derivatives (Scheme 7.6).[10]

Scheme 7.6

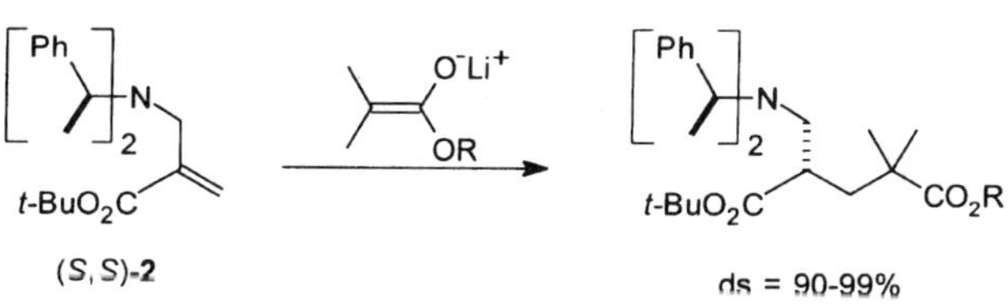

In another elegant application, Feringa et al.[11] have described the synthesis of enantiopure β-amino-γ-butyrolactones via asymmetric conjugate addition of amines

Scheme 7.7

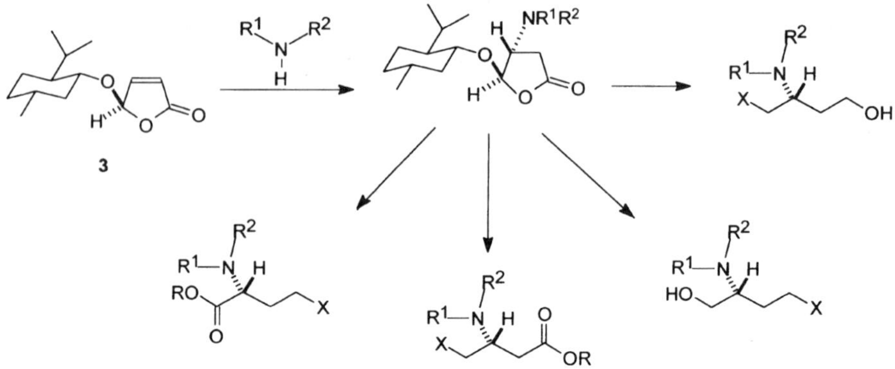

to 5-menthyloxy-2(5H)-furanone (**3**). This procedure provides access to a variety of enantiopure derivatives, including β-amino acids and esters (Scheme 7.7).

7.3 Diastereoselective Michael Additions of Synthetic Equivalents of "Chiral NH_3"

Diastereoselective 1,4-addition of chiral lithium amides to α,β-unsaturated esters, followed by removal of the chiral substituent(s) at nitrogen, provides a convenient protocol for enantioselective preparation of β-amino acids or esters (Scheme 7.8).

Scheme 7.8

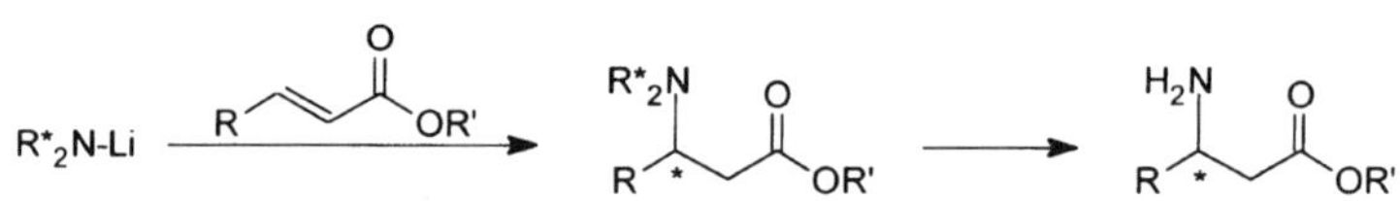

Scheme 7.9

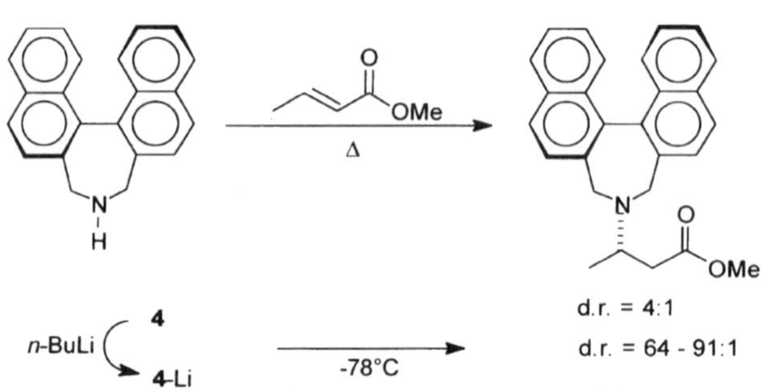

In 1986, Hawkins and Fu[12] described the reaction of C_2 symmetric dihydroazepine **4** and its lithium amide **4**-Li with methyl crotonate. Michael addition of the latter took place with greater diastereoselectivity than the free amine: 64–94:1 compared with 4:1 diastereoisomer ratios, respectively (Scheme 7.9).

The experimental conditions employed in the above procedure worked poorly with methyl decenoate. In order to improve the regio- and stereoselectivity, the solvent and the ester alkoxy group were varied (Scheme 7.10).[13]

Scheme 7.10

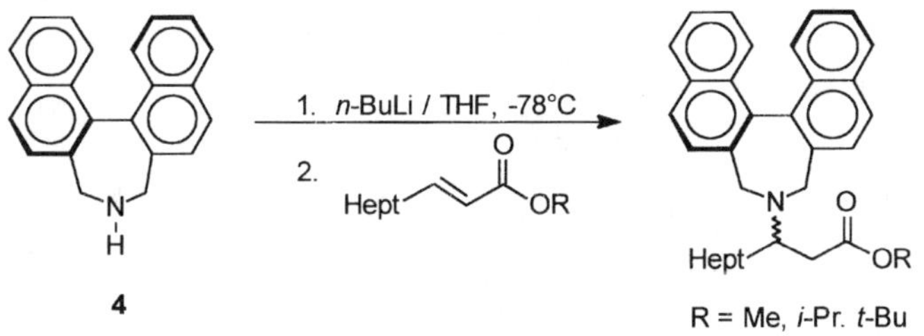

R	Solvent	Equiv. **4**-Li	dr (yield)
CH_3	THF	1.0	20:1 (25)
CH_3	THF	4.0	20:1 (44)
CH_3	THF/HMPA (4:1)	1.0	3:1 (18)
i-Pr	THF/12-crown-4	4.0	17:1 (65)
t-Bu	THF/12-crown-4	4.0	37:1 (88)
t-Bu	DME	1.5	53:1 (80)

Increasing the size from methyl to isopropyl to *tert*-butyl improved the yield and stereoselectivity, and hydrogenolysis of the resulting Michael adducts gave the desired β-amino esters, high diastereoisomeric ratios of the former affording, of course, high enantiomeric purities in the latter. Finally, for a series of different substrate esters, a given enantiomer of the chiral lithium amide consistently approached the same face of the olefin.[13]

More recently, *erythro*- and *threo*-α-substituted-β-amino esters were prepared by addition of (*S*)-**4**-Li to *tert*-butyl crotonate followed by addition of methyl iodide to the intermediate enolate.[14] Michael adduct **5** was produced as a 65:2:2:1

Scheme 7.11

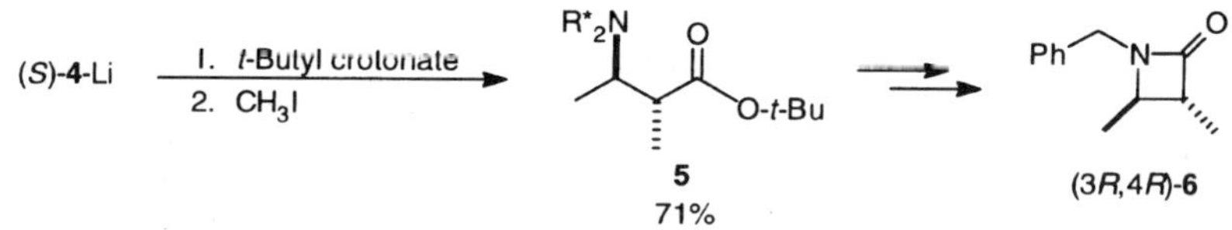

mixture of the four possible diastereomeric products. The absolute configuration of the predominant product was established by conversion to the known β-lactam **6** (Scheme 7.11).

Michael additions of (*S*)-1-phenylethylamine to methyl crotonate affords a 3:2 mixture of the (3*R*,1′*S*) and (3*S*,1′*S*) diastereoisomers **7**, which can be separated by preparative HPLC[15] or by flash chromatography.[16] Hydrogenolytic removal of the phenethyl group with concomitant ester hydrolysis gives enantiomerically pure 3-aminobutanoic acids, (*R*)- and (*S*)-**8** (Scheme 7.12).

Scheme 7.12

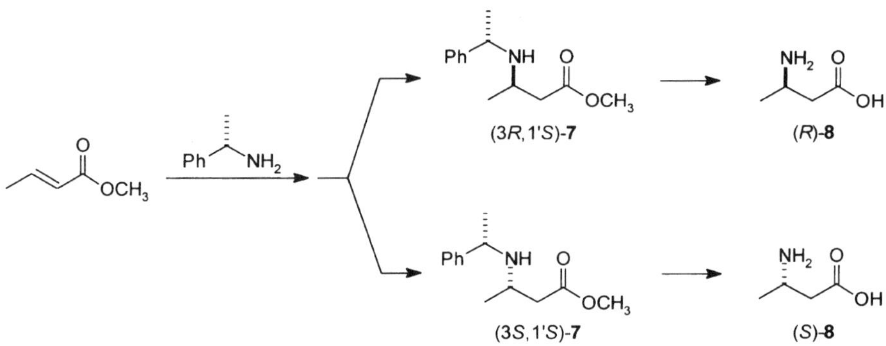

α-Alkylated-β-aminobutanoic acid derivatives were obtained by Seebach and Estermann[17] via double deprotonation of N-acyl-β-aminobutanoates (Scheme 7.13). The selectivity with which these reactions takes place is usually very high, affording mainly the product that originates from addition of the electrophile (E^+) to the enolate with relative *like* topicity (*lk*-1,2-induction[18]) (Figure 7.2).

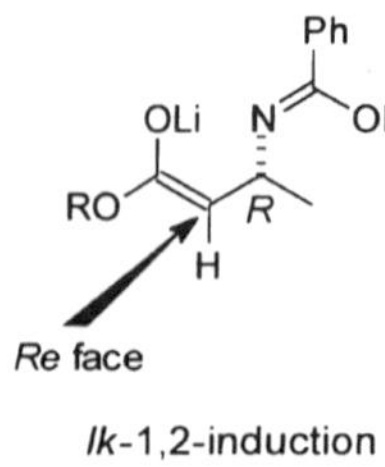

Figure 7.2

Scheme 7.13

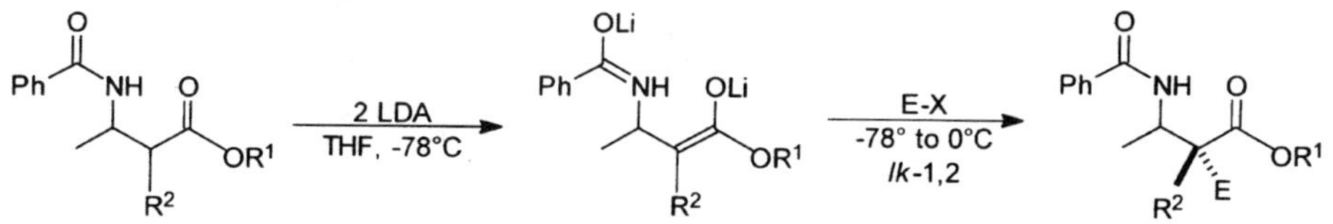

R^1	R^2	E-X	Yield (%)	dr	mp (°C)
C_2H_5	H	CH_3I	73	4:1	81–82
C_2H_5	H	C_2H_5I	38	13:1	55–56
C_2H_5	H	C_6H_5CHO	44	3:1	109–110
CH_3	H	CH_3I	45–50	4:1	77–78
CH_3	H	C_2H_5I	73	16:1	58–59
CH_3	H	$CH_2{=}CHCH_2Br$	70–90	31:1	66–67
CH_3	H	$C_6H_5CH_2Br$	70–90	36:1	120–121
CH_3	H	C_6H_5CHO	60–70	5:1	122–123
CH_3	C_2H_5	$C_6H_5CH_2Br$	80	99:1	104–105

The synthetic usefulness of secondary amines **9**, **10**, and **11** as "chiral ammonia" equivalents in Michael additions has been most masterly demonstrated by Davies and co-workers[19,20] (Scheme 7.14)

Scheme 7.14

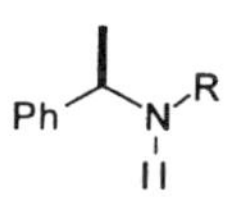

9, R = CH_2Ph
10, R = (3,4)-$CH_2C_6H_3(OCH_3)_2$
11, R = (*R*)-$Ph(CH_3)OH$

For example, Michael addition of the lithium amide derived from (*R*)-N-(α-phenylethyl)benzylamine (**9**) to benzyl (*E*)-crotonate was highly stereoselective (dr = 97.5:2.5), giving, after debenzylation with Pearlman's catalyst, enantiomerically pure (*R*)-β-aminobutanoic acid (Scheme 7.15a).[19] Similarly, addition to methyl (*E*)-(*p*-benzyloxy)cinnamate was completely stereoselective, leading to enantiomerically pure (*S*)-β-tyrosine (Scheme 7.15b).[19]

In a further elegant application, the antifungal antibiotic (–)-(1*R*,2*S*)-2-aminocyclopentane-1-carboxylic acid (cispentacin, **12**) was prepared via the highly stereose-

Scheme 7.15

a) 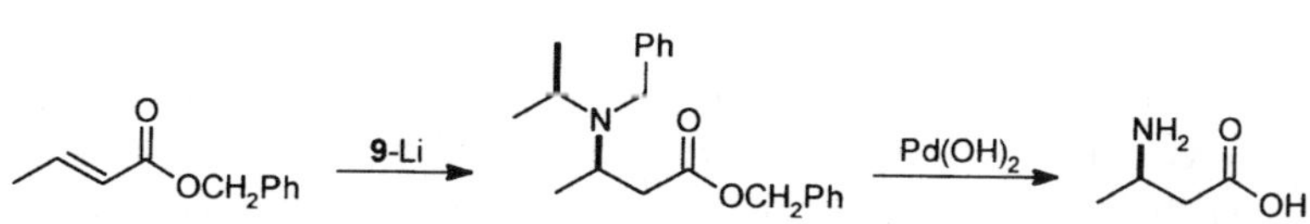

b)

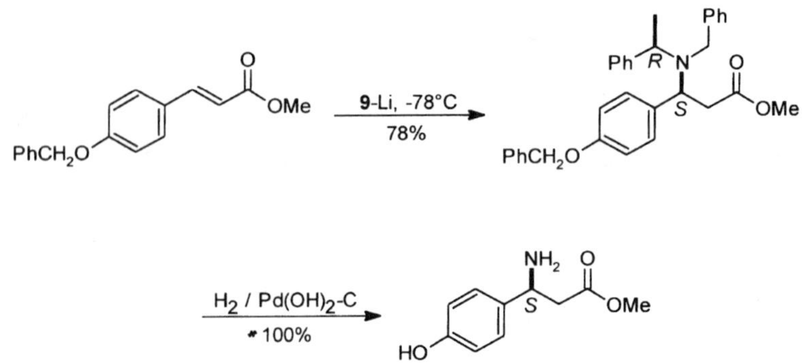

lective Michael addition of (*S*)-**9**-Li to *tert*-butyl 1-cyclopentene-1-carboxylate[21,22] (Scheme 7.16).

Scheme 7.16

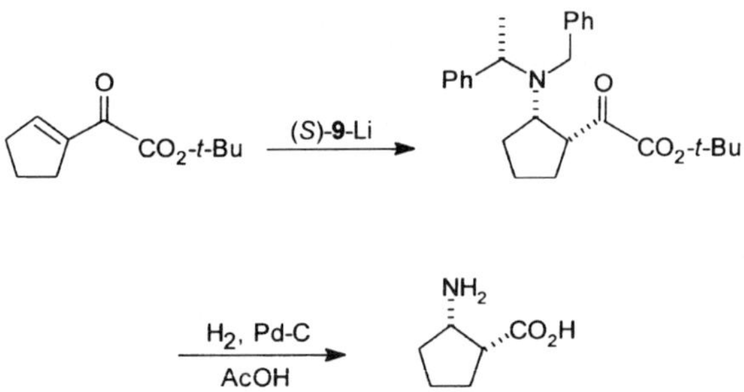

12, cispentacin

Another important achievement is the enantioselective preparation of various derivatives of 3-phenylisoserine, the side chain in Taxol, the potent antitumour/antileukemic agent (see Chapters 1 and 2). Conjugate addition of enantiopure lithium amide (*R*)-**9**-Li to *tert*-butyl cinnamate (reaction proceeds with 95% diastereoselectivity[23]) was followed by hydroxylation of the intermediate enolate with (+)-(camphorsulfonyl)oxaziridine **13** to give amino ester **14** in 86% yield and 96% diastereoselectivity. A single diastereoisomer of **14** was obtained after column chromatography. Debenzylation, transesterification, and benzoylation led then to the *anti* diastereoisomer **15**, which was treated under Mitsunobu conditions to give, after hydrolysis and base-catalyzed O to N benzoyl transfer, the *syn* diastereoisomer (2*S*,3*R*)-N-benzoyl-3-phenylisoserine methyl ester **16** (Scheme 7.17).[24]

The availability of both enantiomers of **9** allows for the convenient preparation of all four stereoisomers of the side chain in Taxol. Furthermore, the synthetic approach described in Scheme 7.17 was successfully applied in the enantioselective preparation of (2*S*,3*R*)-3-amino-2-hydroxydecanoic acid **17**, the α-hydroxy-β-amino acid com-

Scheme 7.17

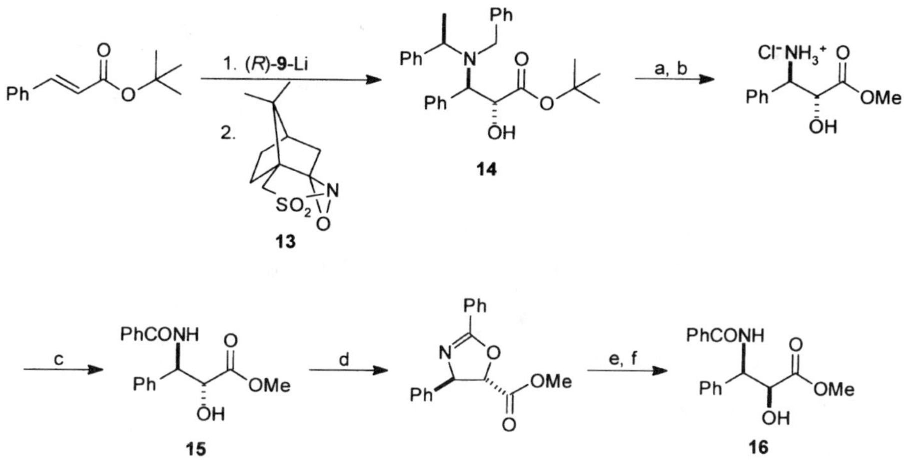

a: H_2 / Pd-C, AcOH b: HCl (g), MeOH c: PhCOCl, Et_3N
d: diethyl azodicarboxylate, Ph_3P e: HCl, MeOH f: $NaHCO_3$ (aq)

ponent of microginin,[25] and of allophenylnorstatine **18**, a novel amino acid of therapeutic interest.[26]

17 **18**

While Davies and co-workers[27] have presented evidence that Michael addition of enantiopure magnesium amide (*R*)-**19**-MgBr occurs with excellent stereoselectivity, Sewald et al.[28] have recently described the development of efficient methodology for the a symmetric synthesis of α-deuterated β-amino acids by means of highly stereoselective conjugate addition of N-[(*R*)-1-phenylethyl]trimethylsilylamidocuprate (*R*)-**20**-CuI (Scheme 7.18).

(*R*)-**19**-MgBr (*R*)-**20**-CuI

Scheme 7.18

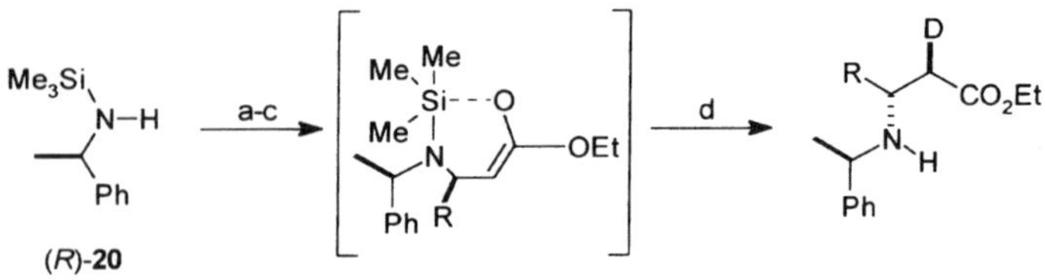

(*R*)-**20**

a: *n*-BuLi, Et_2O, -20°C b: CuI, -78°C c: $(EtO)_3P$, (*E*)-RCH=$CHCO_2Et$, -78°C d: D_2O, RT, then H_3O^+

7.4 Acknowledgment

We are grateful to Dr. Jennifer Matthews and Professor Dieter Seebach for a careful revision of this manuscript.

7.5 References

1. See, for example: (a) E. Juaristi, *Introduction to Stereochemistry and Conformational Analysis* (Wiley, New York, 1991). (b) M. Nógrádi, *Stereoselective Synthesis: A Practical Approach*, 2nd ed. (VCH Publishers, New York, 1994). (c) E. L. Eliel, S. H. Wilen, L. N. Mander, *Stereochemistry of Organic Compounds* (Wiley, New York, 1994).
2. A. P. Terentev, R. A. Gracheva, T. F. Dendeko, *Dokl. Akad. Nauk SSSR* **163**, 674 (1965).
3. (a) M. Furukawa, T. Okawara, Y. Terawaki, *Chem. Pharm. Bull.* **25**, 1319 (1977). See also: (b) R. Kinas, K. Pankiewicz, W. J. Stec, P. B. Farmer, A. B. Foster, M. Jarman, *J. Org. Chem.* **42**, 1650 (1977).
4. (a) K. Achiwa, T. Soga, *Tetrahedron Lett.* 1119 (1978). See also: (b) M. Furukawa, T. Okawara, Y. Noguchi, Y. Terawaki, *Chem. Pharm. Bull.* **27**, 2223 (1979). (c) S. Jolidon, T. Meul, *Eur. Pat.* Nr. 0144980 A2; *Chem. Abstr.* **105**, 43325p (1987).
5. W. D. Lubell, M. Kitamura, R. Noyori, *Tetrahedron: Asymmetry* **2**, 543 (1991).
6. (a) J. d'Angelo, J. Maddaluno, *J. Am. Chem. Soc.* **108**, 8112 (1986). See also: (b) D. Potin, F. Dumas, J. d'Angelo, *J. Am. Chem. Soc.* **112**, 3483 (1990).
7. Originally developed by Corey and Ensley: E. J. Corey, H. E. Ensley, *J. Am Chem. Soc.* **97**, 6908 (1975).
8. Cf. W. Oppolzer, C. Robbiani, K. Bättig, *Helv. Chim. Acta* **63**, 2015 (1980).
9. P. Perlmutter, M. Tabone, *Tetrahedron Lett.* **29**, 949 (1988).
10. I. T. Barnish, M. Corless, P. J. Dunn, D. Ellis, P. W. Finn, J. D. Hardstone, K. James, *Tetrahedron Lett.* **34**, 1323 (1993).
11. B. de Lange, F. van Bolhuis, B. L. Feringa, *Tetrahedron* **45**, 6799 (1989).
12. J. M. Hawkins, G. C. Fu, *J. Org. Chem.* **51**, 2820 (1986).

13. J. M. Hawkins, T. A. Lewis, *J. Org. Chem.* **57**, 2114 (1992).

14. J. M. Hawkins, T. A. Lewis, *J. Org. Chem.* **59**, 649 (1994).

15. H. Estermann, D. Seebach, *Helv. Chim. Acta* **71**, 1824 (1988).

16. E. Juaristi, J. Escalante, B. Lamatsch, D. Seebach, *J. Org. Chem.* **57**, 2396 (1992).

17. D. Seebach, H. Estermann, *Tetrahedron Lett.* **28**, 3103 (1987).

18. (a) D. Seebach, V. Prelog, *Angew. Chem., Int. Ed. Engl.* **21**, 654 (1982). (b) See also: Ref. 1a, pp. 52–54.

19. S. G. Davies, O. Ichihara, *Tetrahedron: Asymmetry* **2**, 183 (1991).

20. For additional synthetic equivalents of "chiral NH_3," see: (a) P. J. Cox, N. S. Simpkins, *Tetrahedron: Asymmetry* **2**, 1 (1991). (b) E. Juaristi, P. Murer, D. Seebach, *Synthesis* 1243 (1993).

21. S. G. Davies, O. Ichihara, I. A. S. Walters, *Synlett* 461 (1993).

22. See also: S. G. Davies, O. Ichihara, I. Lenoir, I. A. S. Walters, *J. Chem. Soc., Perkin Trans. 1*, 1411 (1994).

23. (a) S. G. Davies, N. M. Garrido, O. Ichihara, I. A. S. Walters, *J. Chem. Soc., Chem. Commun.* 1153 (1993).

24. (a) M. E. Bunnage, S. G. Davies, C. J. Goodwin, *J. Chem. Soc., Perkin Trans. 1*, 1375 (1993). (b) See also: J. Escalante, E. Juaristi, *Tetrahedron Lett.* **36**, 4397 (1995).

25. M. E. Bunnage, A. J. Burke, S. G. Davies, C. J. Goodwin, *Tetrahedron: Asymmetry* **5**, 203 (1994).

26. M. E. Bunnage, S. G. Davies, C. J. Goodwin, O. Ichihara, *Tetrahedron* **50**, 3975 (1994).

27. M. E. Bunnage, S. G. Davies, C. J. Goodwin, I. A. S. Walters, *Tetrahedron: Asymmetry* **5**, 35 (1994).

28. N. Sewald, K. D. Hiller, B. Helmreich, *Liebigs Ann.* 925 (1995).

CHAPTER

8

Asymmetric Synthesis of β-Amino Acids through $TiCl_4$ Mediated Addition to Imines

Cesare Gennari and Anna Vulpetti

8.1 Introduction

β-Amino acids, although less abundant than their α-counterparts, are components of natural peptides,[1a] and show interesting biological properties. They also display a wide spectrum of synthetic applications, including the preparation of modified peptides[1b] and β-lactam antibiotics.[1c] Numerous methods for the synthesis of β-amino acids exist, and have been recently reviewed.[1d–g] One of the most useful methods for the preparation of β-amino acids and esters, leading to β-lactam derivatives, involves the reaction of imines with enolates. In this chapter we review a particular aspect of this chemistry, the stereoselective addition of (chiral) enolates to (chiral) imines mediated by titanium tetrachloride.

8.2 Results and Discussion

The first synthesis of β-amino esters via a $TiCl_4$-mediated reaction of enol silanes with imines was reported by Ojima and co-workers in 1977.[2] The authors showed that O-alkyl-O-(trimethylsilyl)ketene acetals react with imines in the presence of $TiCl_4$ to afford β-lactams or β-aminoesters, depending on the nature of the imine: N-alkylimines afforded β-lactams, while N-arylimines gave β-amino esters. The $TiCl_4$-mediated addition of enol silanes to carbonyls and imines has been extensively reviewed, under both a mechanistic and a synthetic point of view.[3]

The $TiCl_4$-mediated addition of dimethylketene methyl trimethylsilyl acetal (**1**) to chiral imines (**2**) and (**3**) was reported by Ojima and Inaba to proceed with high di-

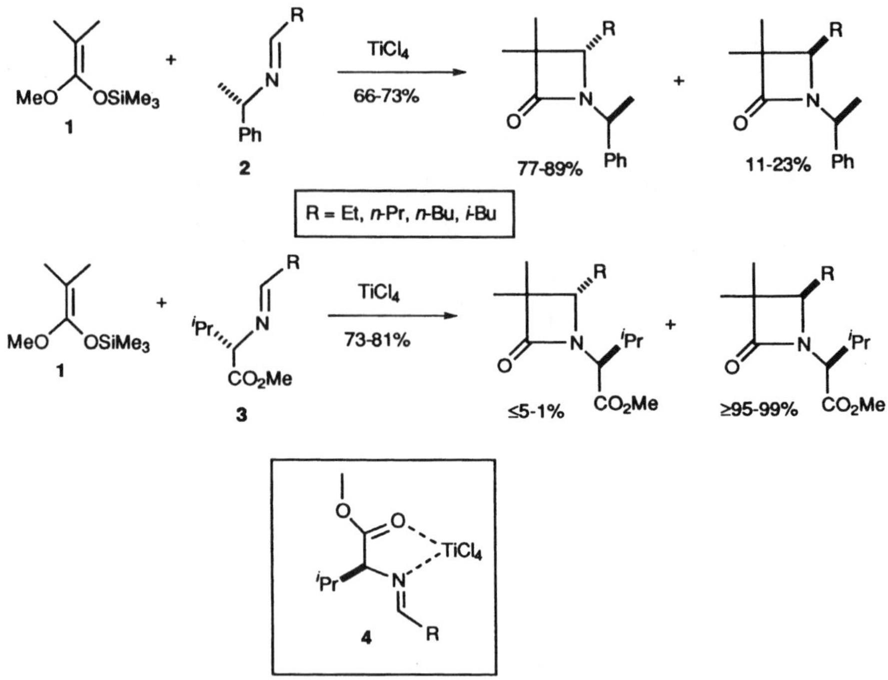

Figure 8.1. $TiCl_4$-mediated addition of dimethylketene methyl trimethylsilyl acetal to chiral imines.[4]

astereoselectivity, leading directly to β-lactams (Figure 8.1).[4] In the case of imines (**2**), the diastereomeric excesses were in the range of 54–78%, and the (*S*) configuration was induced at the C-4 position of the resulting β-lactams. The best diastereomeric ratios (up to > 99:1) were obtained using imines (**3**), derived from (*S*)-valine methyl ester, which form the chelated complex (**4**) with $TiCl_4$.

Chiral N-methylephedrine-derived silyl ketene acetals (**5**,**6**) were reported by Gennari and co-workers to react with imines in the presence of 2 mol. equiv. $TiCl_4$ to give β-amino esters with moderate to high diastereoselectivity (Figure 8.2).[5,6] With benzylideneaniline (**7**) the reaction is *anti* selective, while with imino esters (**8**,**9**) that chelate $TiCl_4$ to give complexes such as (**10**), the reaction is *syn* selective, in agreement with the general principle that "electrophiles able to form a chelated complex with the Lewis acid control (usually invert) the simple stereoselection of the reaction."[3] β-Amino esters were transformed into optically active *trans* and *cis* β-lactams.[5,6]

Titanium enolates of α-halo and α-alkoxyacetates were reported by Fujisawa and co-workers to react with a chiral imine possessing a 1,3-dioxolane ring derived from (2*S*,3*S*)-1,4-dimethoxy-2,3-butanediol as a chiral auxiliary to give *trans* β-lactams with excellent diastereoselectivity.[7,8]

Evans and co-workers described the diastereoselective reaction of chiral titanium enolates with a variety of electrophiles, including N-chloromethylbenzamide, which allows efficient amidoalkylation (Figure 8.3).[9]

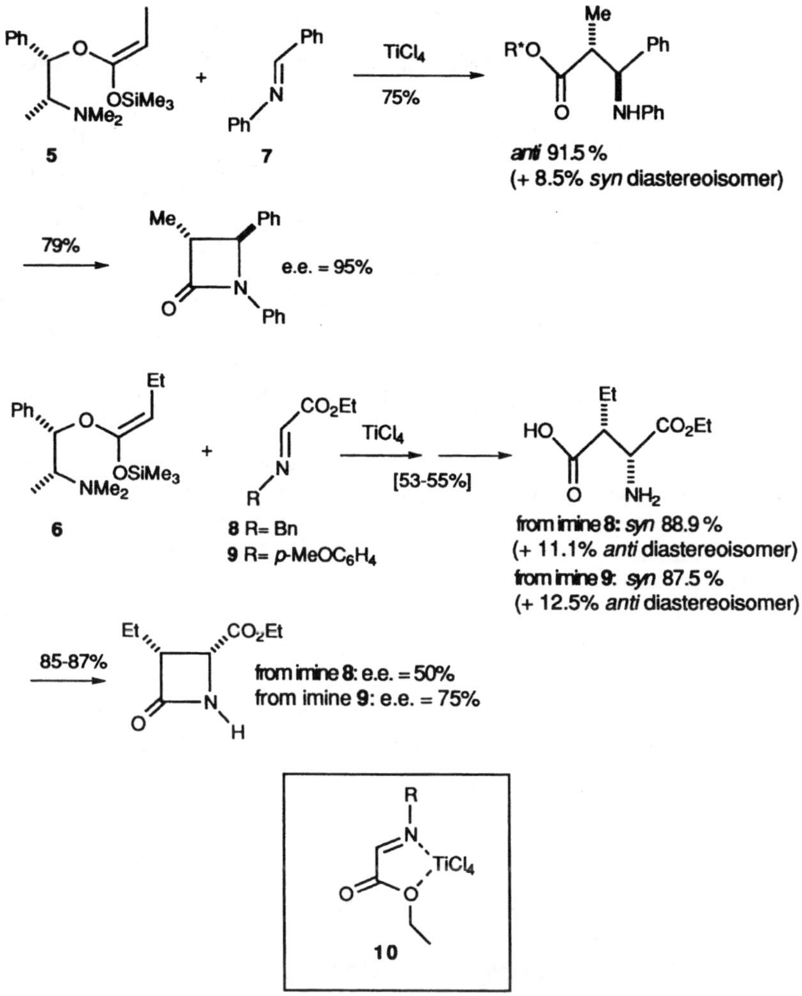

Figure 8.2. $TiCl_4$-mediated addition of chiral silyl ketene acetals to imines.[5,6]

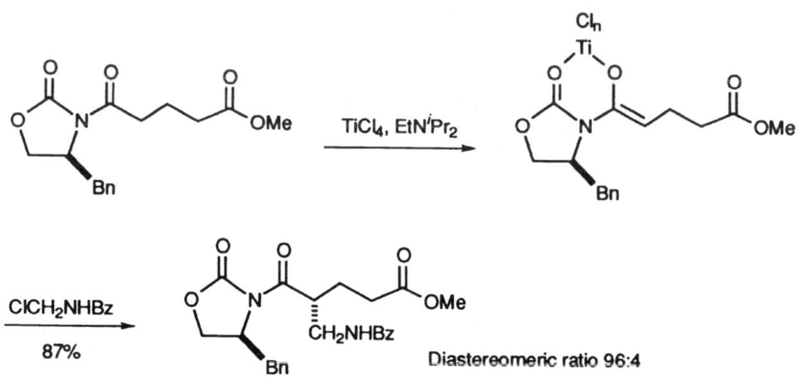

Figure 8.3. $TiCl_4$-mediated addition of a chiral enolate to N-chloromethylbenzamide.[9]

Cozzi, Cinquini, and co-workers reported that treatment of 2-pyridyl thioesters with triethylamine and $TiCl_4$ affords titanium enolates that add to imines to give β-lactams in fair to excellent yields with moderate to good 3,4-*trans:cis* stereoselectivity.[10] The reaction was applied to chiral imines, derived from enantiomerically pure alkoxy aldehydes, and to a chiral enolate, derived from 3-hydroxybutyrate, to give chiral β-lactams with high diastereofacial selectivities (Figures 8.4–8.6).[11] Advanced precursors of (+)-PS-5, (+)-PS-6, thienamycin, 1β-methylthienamycin, and bestatin were prepared to illustrate the potential of the method.[11,12] Different factors concur in determining the observed stereochemical results. Among these, enolate and imine structures seem the most important. The condensation of the titanium enolates of C-2 substituted 2-pyridylthioesters with imines affords β-lactams in *trans:cis* ratios that depend on the structure of both the C-imine residue and the thioester C-2 substituent (Figure 8.5). The *trans:cis* ratios increase on increasing the bulkiness of the thioester C-2 alkyl substituent (*cf.* R = Et and R = *i*-Pr; Figure 8.5), while a strong *cis* preference is observed with C-2 alkoxy substituents (*cf.* R = alkyl and R = OBn, OAc; Figure 8.5). The stereochemical results were rationalized based on NMR studies of the enolization process.[13]

The reaction of 2-pyridyl thioesters with chiral imines derived from chiral amines was also studied. The chiral amines were selected among benzylic amines, β-aminoalcohols, β-heterosubstituted α-aminoesters.[14] Inexpensive and available in both enantiomeric forms, α-methylbenzylamine was identified as the chiral auxiliary

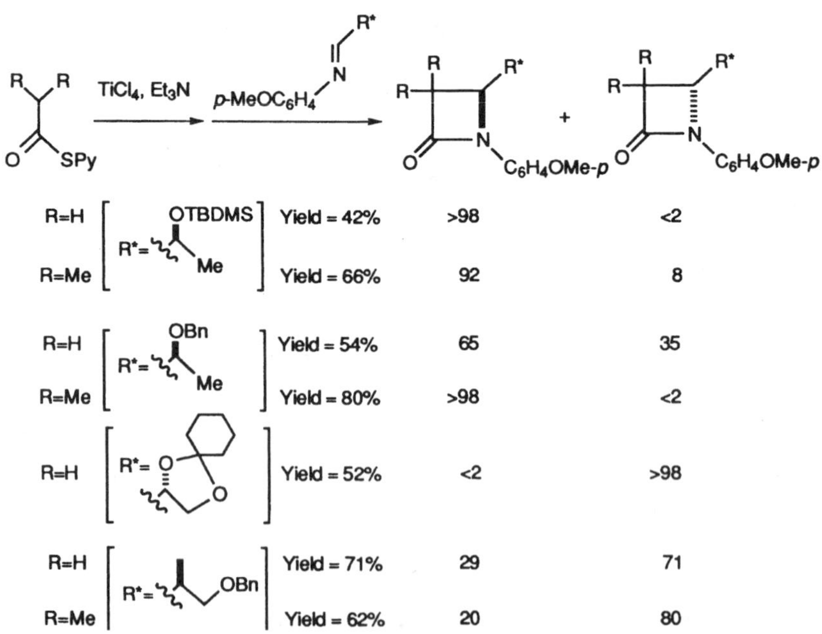

Figure 8.4. $TiCl_4$-mediated addition of 2-pyridyl thioester enolates to chiral imines.[10–14]

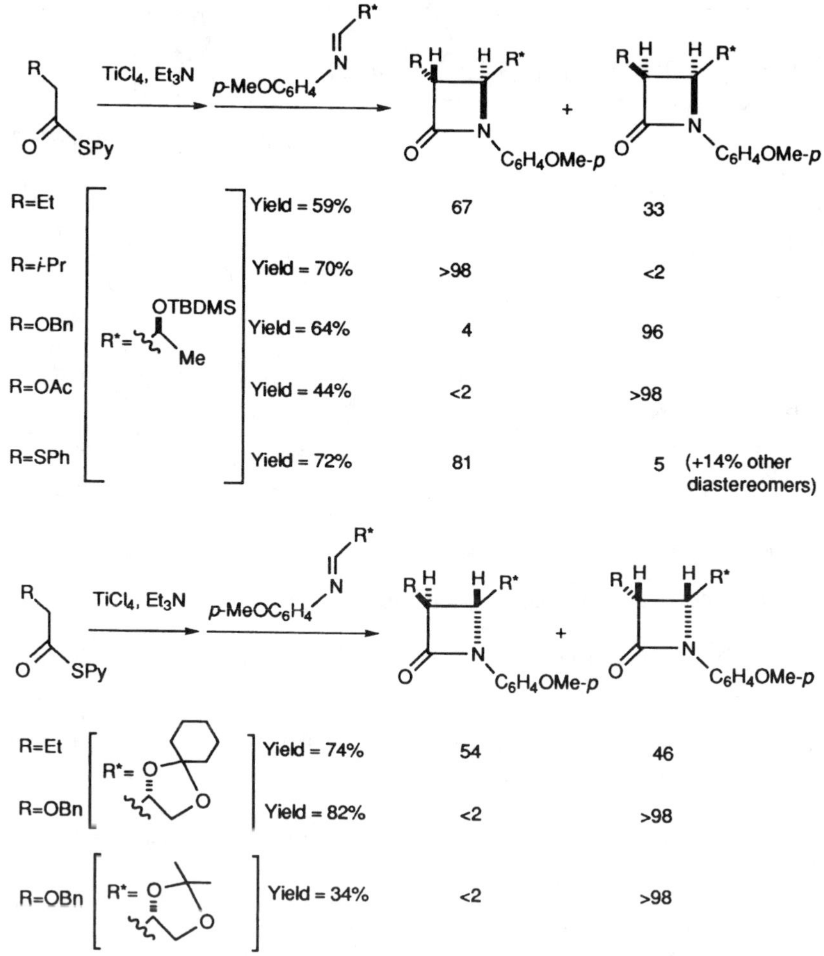

Figure 8.5. $TiCl_4$-mediated addition of 2-pyridyl thioester enolates to chiral imines.[10–14]

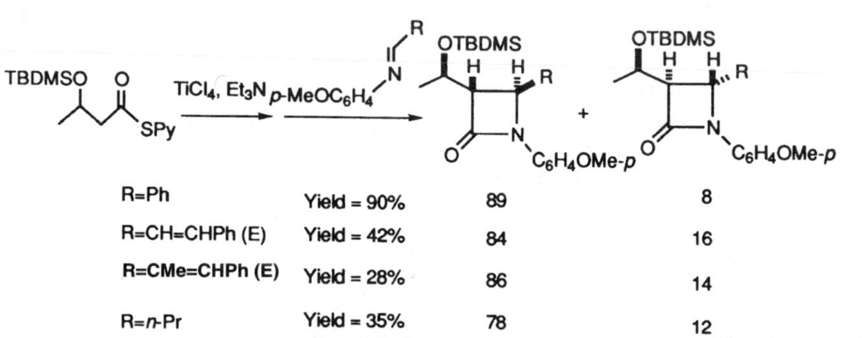

Figure 8.6. $TiCl_4$-mediated addition of chiral 2-pyridyl thioester enolates to imines.[10–14]

(Continued)

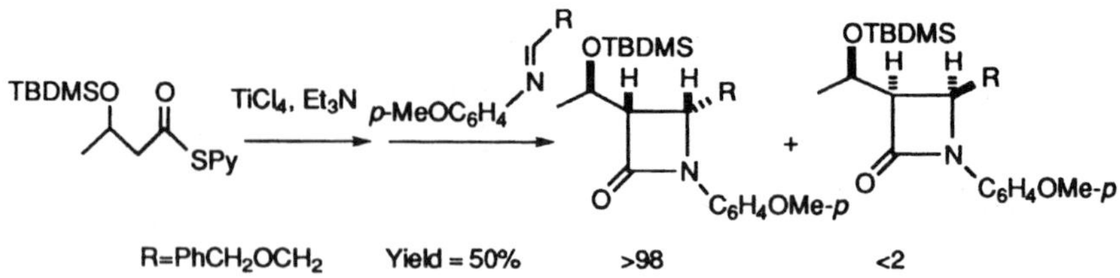

Figure 8.6. (*Continued*)

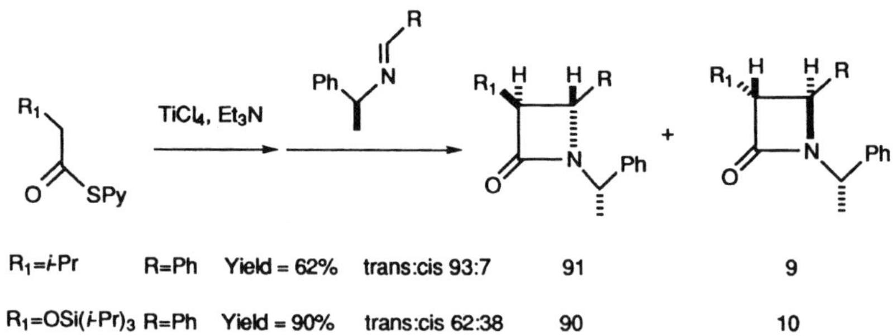

Figure 8.7. $TiCl_4$-mediated addition of 2-pyridyl thioester enolates to chiral imines.[10–14]

that generally combines good reactivity and satisfactory levels of stereocontrol (Figure 8.7).[14]

8.3 References

1. (a) C. N. C. Drey, in *Chemistry and Biochemistry of Amino Acids*, G. C. Barrett, Ed. (Chapman and Hall, London, 1985), p. 25. (b) W. P. Frankmolle, G. Knubel, R. E. Moore, G. M. L. Patterson, *J. Antibiot.* **45**, 1458 (1992). (c) *The Organic Chemistry of β-Lactams*, G. I. Georg, Ed. (VCH, New York, 1993). (d) E. Juaristi, D. Quintana, J. Escalante, *Aldrichimica Acta* **27**, 3 (1994). (e) D. C. Cole, *Tetrahedron* **50**, 9517 (1994). (f) D. J. Hart, D.-C. Ha, *Chem. Rev.* **89**, 1447 (1989). (g) E. F. Kleinman, in *Comprehensive Organic Synthesis*, B. M. Trost and I. Fleming, Eds. (Pergamon Press, Oxford, 1991), Vol. 2, p. 893.
2. I. Ojima, S.-I. Inaba, K. Yoshida, *Tetrahedron Lett.* **18**, 3643 (1977).
3. (a) C. Gennari, in *Comprehensive Organic Synthesis*, B. M. Trost and I. Fleming, Eds. (Pergamon Press, Oxford, 1991), Vol. 2, p. 629. (b) C. Gennari, in *Selectivities in Lewis Acid Promoted Reactions*, D. Schinzer, Ed. (Kluwer Academic Publishers, Dordrecht, 1989), p. 53.
4. I. Ojima, S.-I. Inaba, *Tetrahedron Lett.* **21**, 2077 (1980); *Tetrahedron Lett.* **21**, 2081 (1980).
5. C. Gennari, I. Venturini, G. Gislon, G. Schimperna, *Tetrahedron Lett.* **28**, 227 (1987).
6. C. Genneri, G. Schimperna, I. Venturini, *Tetrahedron* **44**, 4221 (1988).
7. T. Fujisawa, R. Hayakawa, M. Shimizu, *Tetrahedron Lett.* **33**, 7903 (1992).
8. M. Shimizu, T. Ishida, T. Fujisawa, *Chem. Lett.* 1403 (1994).

9. D. A. Evans, F. Urpi, T. C. Somers, J. S. Clark, M. T. Bilodeau, *J. Am. Chem. Soc.* **112**, 8215 (1990).

10. M. Cinquini, F. Cozzi, P. G. Cozzi, E. Consolandi, *Tetrahedron* **47**, 8767 (1991).

11. (a) R. Annunziata, M. Cinquini, F. Cozzi, P. G. Cozzi, *Tetrahedron Lett.* **33**, 1113 (1992); (b) *J. Org. Chem.* **57**, 4155 (1992).

12. R. Annunziata, M. Benaglia, M. Cinquini, F. Cozzi, F. Ponzini, *J. Org. Chem.* **58**, 4746 (1993).

13. R. Annunziata, M. Benaglia, M. Cinquini, F. Cozzi, F. Ponzini, L. Raimondi, *Tetrahedron* **50**, 2939 (1994).

14. R. Annunziata, M. Benaglia, M. Cinquini, F. Cozzi, L. Raimondi, *Tetrahedron* **50**, 9471 (1994).

CHAPTER

9

Highly Stereoselective Synthesis of β-Amino Esters via Double Stereodifferentiation

Kazuaki Ishihara, Kouji Hattori, and Hisashi Yamamoto

9.1 Introduction

Imines are important starting materials for nitrogen-containing compounds, especially significant compounds such as alkaloids, amino acids, β-lactams, amino sugars, and terpenes. These compounds should be prepared in optically active form since it is usually observed that only one of the two enantiomers shows biological activity. However, there are many difficult problems in the stereoselective reaction of imines. Being neutral species, imines have low reactivity; therefore, active methylene compounds with pK_a values greater than 15~16 must be preactivated as an enolate or enolisilane derivative for the condensation reaction. Moreover, abstraction of enolizable α-proton of imine and self-condensation of enolizable imine sometimes occur in the course of a reaction.[1] Consequently, despite the enormous progress made in the asymmetric reactions of carbonyl compounds affording chiral alcohols, development of the corresponding studies of imines to chiral amines remains a challenging and exciting field of research.[2]

In recent years, the development of chiral Lewis acids has been one of the most important approaches to the production of optically active compounds. Their use is no

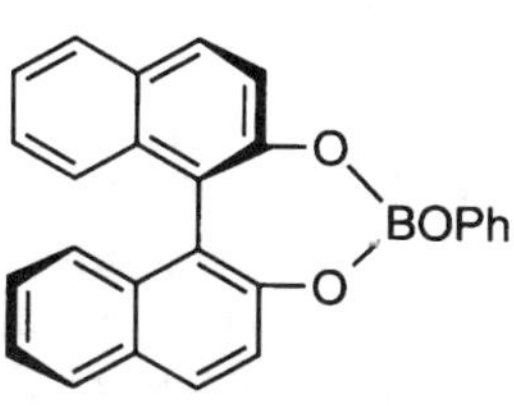

(*R*)-**1**

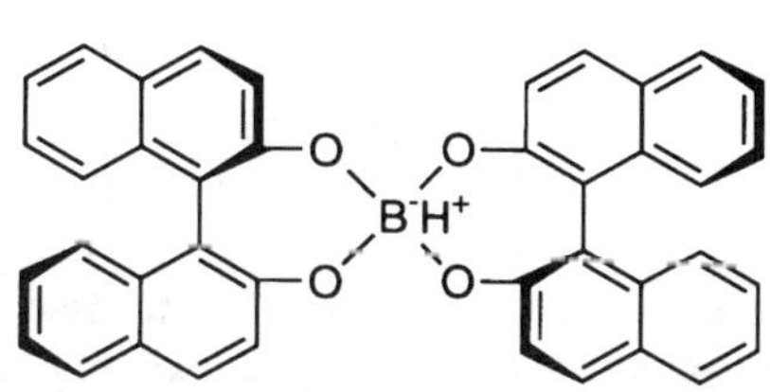

(*R*)-**2**

longer restricted only to reduction and oxidation since it is now possible to perform stereoselectively carbon–carbon bond formation. We developed effective chiral Lewis acids **1** and **2** for asymmetric reactions of imine and a new methodology for stereocontrolled synthesis of biological active compounds.

9.2 Asymmetric Aza Diels–Alder Reaction

9.2.1 Enantioselective Diels–Alder Reaction of Imines Mediated by Chiral Lewis Acid 1[3]

The asymmetric aza Diels–Alder reaction of imines is promoted by an in situ generated boron complex of type **1**. The method is successful with both aromatic and aliphatic aldimines and affords products of high enantiomeric purity.

The chiral boron complex (*R*)-**1** is prepared in situ simply by mixing a 1:1 molar ratio of optically active binaphthol and triphenyl borate in dichloromethane at ambient temperature for 1 h (Eq. 9.1). The aza Diels–Alder reaction of aldimine **3** with Danishefsky diene **4** was promoted by this reagent in the presence of molecular sieves 4A at –78°C for several hours. After the usual workup, purification by column chromatography furnished the cyclo-adduct **5** in 75% yield with 82% ee (eq. 9.2). Absolute configuration was ascertained to be *R* after converting to the known (*R*)-2-phenylpiperidine. As the source of boron reagent, triaryl borate has high ability in both reactivity and enantioselectivity, and the use of a structurally more bulky aryloxy reagent slightly improved the optical yield: B(O-2-tolyl)$_3$→76% yield, 84% ee; B(O-3,5-xylyl)$_3$→75% yield, 86% ee. The choice of solvent is crucial for high optical yields, and the reaction is more efficient in methylene chloride than in other solvents. Generally, a benzyl-type substituted imine is preferable to other substitutions.

$$B(OPh)_3 + (R)\text{-Binaphthol} \xrightarrow[\text{rt}]{CH_2Cl_2,\ 4A\ MS} (R)\text{-}\mathbf{1} \qquad \text{(eq. 9.1)}$$

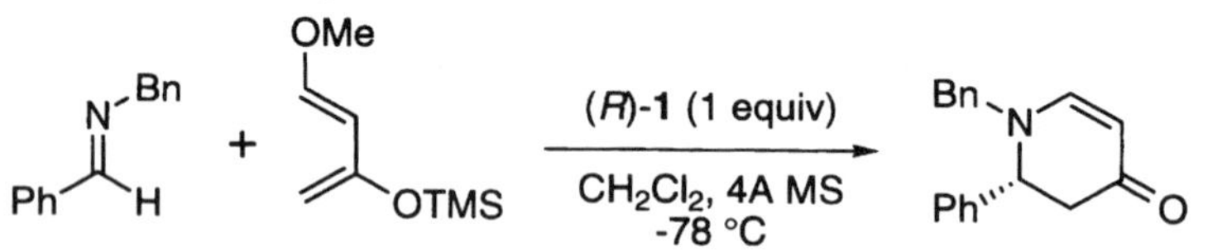

3 **4** **5**

75% yield, 82% ee

(eq. 9.2)

The present method is applicable to a variety of aldimines and Danishefsky-type dienes to provide the Diels–Alder adduct in high enantiomeric purity (Scheme 9.1).

The new chiral boron reagent disclosed herein offers the following advantages: (1) Both chiral binaphthol and phenyl borate are commercially available, and the catalyst can be generated without any complex technology. (2) No difficult isolation procedure is required, and recovery of binaphthol is quantitative. (3) Either the *R* or *S* form

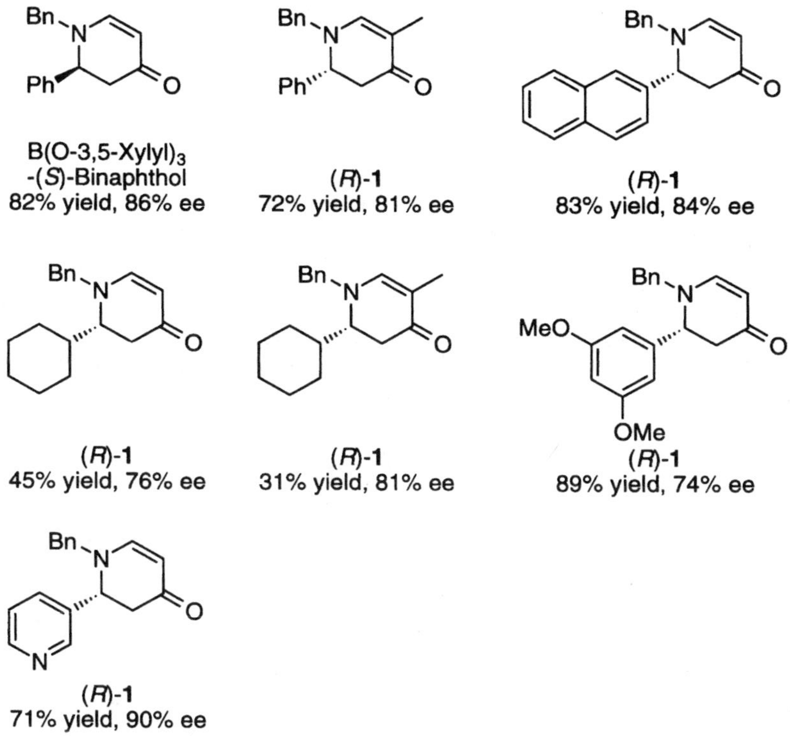

Scheme 9.1

of the product may be synthesized since the required reagents are readily accessible in both enantiomeric forms.

The practical advantage of this method is demonstrated by the efficient synthesis of anabasine, piperidine alkaloid. A mixture of 3-pyridyl aldimine and diene **4** was exposed to the chiral boron catalyst (*S*)-**1** derived from (*S*)-binaphthol to obtain dehydropiperidinone in 68% yield and with 90% ee. After simple trituration with ether, the optically pure product was transformed into (–)-anabasine by known methods (Scheme 9.2).

9.2.2 Double Asymmetric Induction of Diels–Alder Reaction of Imines Mediated by Chiral Lewis Acid 1[3]

There have been a number of investigations into reactions exhibiting diastereofacial selectivity with imines containing a chiral auxiliary. In a preliminary work, Larsen and Grieco[4] demonstrated an aza Diels–Alder reaction of chiral iminium salts using α-methylbenzylamine as an auxiliary, and the resulting stereoselectivity was moderate (ratio of 4:1). Recently, several inactivated imines having chiral auxiliaries have been developed in aza Diels–Alder reactions with a Danishefsky diene. For example, Waldmann[5] utilized amino acids for the chiral auxiliary and Kunz[6] and Midland[7] effec-

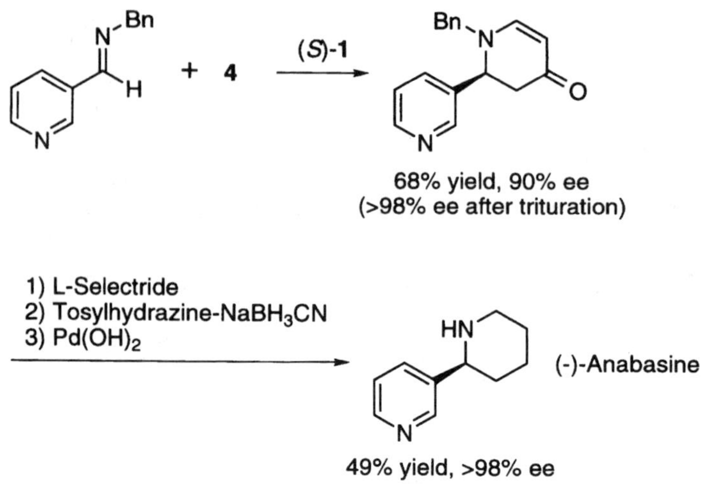

Scheme 9.2

tively utilized other auxiliaries including sugar or α-aldehyde for this type of reaction. These auxiliaries have thus been demonstrated to be quite useful for the diastereoselective condensation; however, there have been certain practical difficulties in preparing and/or removing them. The possibility of a simple α-methylbenzylamine auxiliary is appealing since this functional group is easy to prepare and to remove in an aza Diels–Alder reaction. As indicated in Eq. (9.3), we have found that excellent diastereoselectivity is realized with an achiral reagent derived from triphenyl borate and biphenol. The use of too structurally bulky an aryloxy reagent reduces asymmetric induction.

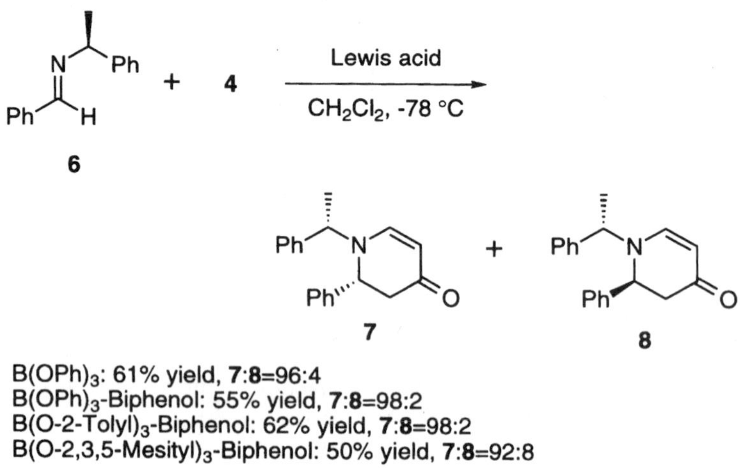

(eq. 9.3)

As a logical extension of this work, we have succeeded in the double asymmetric induction of chiral imine using α-methylbenzylamine as a chiral auxiliary in the aza Diels–Alder reaction mediated by the chiral boron complex **1**. The effectiveness of this method is apparent from the results summarized in Table 9.1. Under optimum

Table 9.1 DOUBLE ASYMMETRIC INDUCTION OF AZA DIELS–ALDER REACTION MEDIATED BY BORON REAGENTS

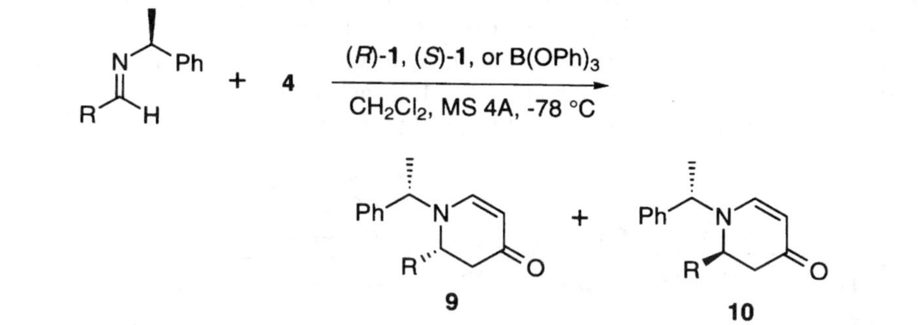

Imine: *R*	Lewis Acid (1 equiv.)	Yield (%)	**9**:**10**
Phenyl	(*R*)-**1**	61	99:1
	$B(OPh)_3$	57	96:4
	(*S*)-**1**	30	93:7
3-Pyridyl	(*R*)-**1**	63	99:1
	$B(OPh)_3$	53	89:11
	(*S*)-**1**	35	86:14
c-Hexyl	(*R*)-**1**	31	99:1
	$B(OPh)_3$	40	90:10
	(*S*)-**1**	20	89:11
Propyl	(*R*)-**1**	49	95:5
	$B(OPh)_3$	59	91:9
	(*S*)-**1**	31	91:9

conditions with a better matching pair, almost complete diastereoselectivitics were obtained for a variety of aldimines. Furthermore, the reactions were faster with the matching pair than with the mismatching pair. The phenomenon was actually confirmed that the chiral catalyst discriminated to some extent between the racemic imines of (+)-**6** and (–)-**6** to accomplish the effective kinetic resolution (Eq. 9.4).

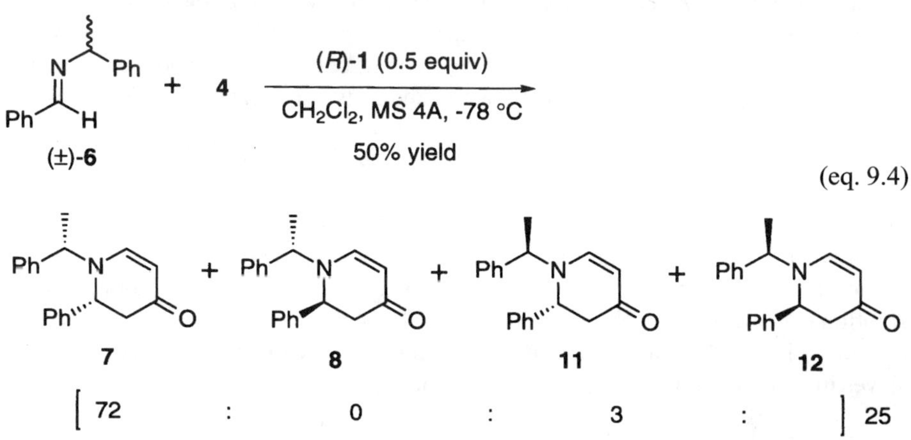

(eq. 9.4)

The practical advantage of this method is illustrated by the synthesis of piperidine alkaloid (+)-coniine **15**. A mixture of proply aldimine **13** and diene **4** was exposed at –78°C with the chiral boron mediator (*R*)-**1** to obtain dihydropyridinone **14** as a pure diastereomer in 41% yield. Reduction of **14** with L-selectride at –78°C afforded quantitatively the conjugate reduction product. After thioketalization, treatment with Raney nickel resulted in simultaneous desulfurization and the removal of benzyl group to give (+)-coniine **15** (54% yield from **14**) (Eq. 9.5).

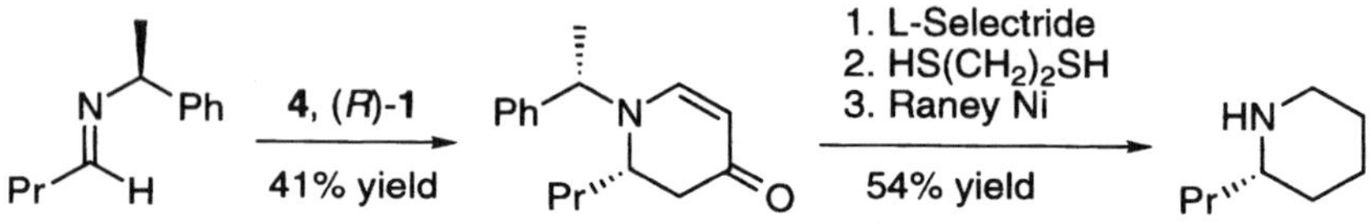

13 **14** **15**

(eq. 9.5)

Under optimum conditions with the chiral boron reagent **1**, almost complete diastereoselectivities were again obtained for a variety of aldimines. The success of the present aza Diels–Alder reaction is particularly due to the structural flexibility of the binaphthol ligand. The absolute configuration of the adducts agrees with predictions based on Figure 9.1. The nucleophile would approach the *Re* face of the (*S*)-**6**·(*R*)-**1** complex. In a 2D-NOESY study of the (*S*)-**6**·(*R*)-**1** complex, a NOE of proton H^c was observed with H^a and CH_3^b; thus the boron reagent would effectively block the *si* face of the imine. In fact, the imine proton of H^c was strongly shielded by the aromatic ring of the boron reagent. Thus the NMR spectrum of the imine (*S*)-**6** indicated that this proton, originally appearing at δ 8.44, gives rise to a signal δ 8.04 when exposed to (*R*)-**1**. The preferred *E*-geometry of the imine (*S*)-**6** could clearly be demonstrated by ^{1}H-NMR analysis of the imine (*S*)-**6** and the complex (*S*)-**6**·(*R*)-**1**.

9.2.3 Asymmetric Diels–Alder Reaction of Imines Mediated by Brønsted Acid Assisted Chiral Lewis Acid 2[8]

During our investigation of the structure of the chiral boron catalyst **1**, we obtained crystals suitable for X-ray analysis by addition of (*S*)-**6** from a dichloromethane–hexane bilayer at ambient temperature. Figure 9.2 displays the structure of the complex. Interestingly, X-ray diffraction analysis indicated that the chiral boron complex and imine are joined by phenol: Three components are linked by two hydrogen bonds. The structure of the boron complex revealed **2** rather than **1**. *The X-ray analysis of the boron complex 2 demonstrates that it exists as a Brønsted acid assisted chiral Lewis acid (BLA).*[43]

The disclosed structure of **2** led us to change the preparation procedure for the chiral boron complex. The new chiral Lewis acid **2** was conveniently prepared in situ simply by mixing a 1:2 molar ratio of triphenyl borate and optically pure binaphthol in dichloromethane at ambient temperature for 1 h (method A). Also, a phenol-free solution of **2** was prepared by the reaction of $B(OMe)_3$ (1 equiv.) with (*R*)-binaphthol (2 equiv.) in dichloromethane at reflux with removal of methanol (4A molecular sieves in a Soxhlet thimble) for 2~3 h (method B).

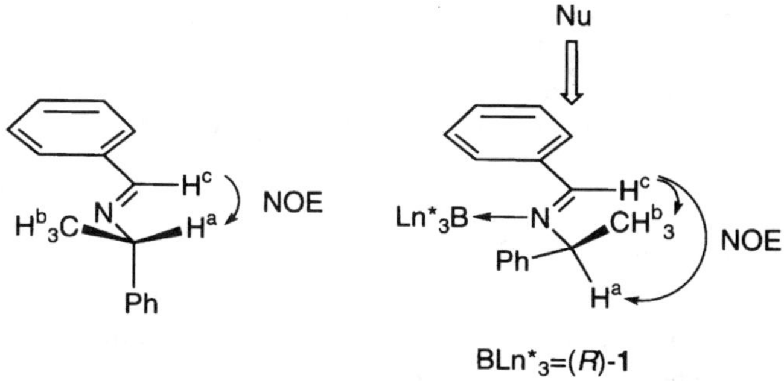

Figure 9.1

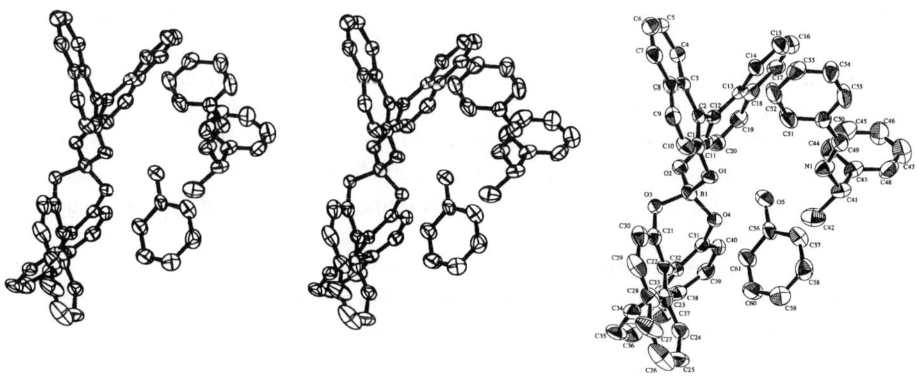

Figure 9.2. ORTEP diagram of (*S*) **6**·(*S*) **2**·PhOH·CH_2Cl_2 (CH_2Cl_2 is excluded from the figure).

The aza Diels–Alder reaction of **3** with **4** in the presence of (*R*)-**2** (1 equiv.) gave a 78% yield of **5** with an ee of 86% (eq. 9.6). Furthermore, when **6** was employed as chiral imine in the reaction with **4** in the presence of (*R*)-**2** (1 equiv.), a 64% yield of

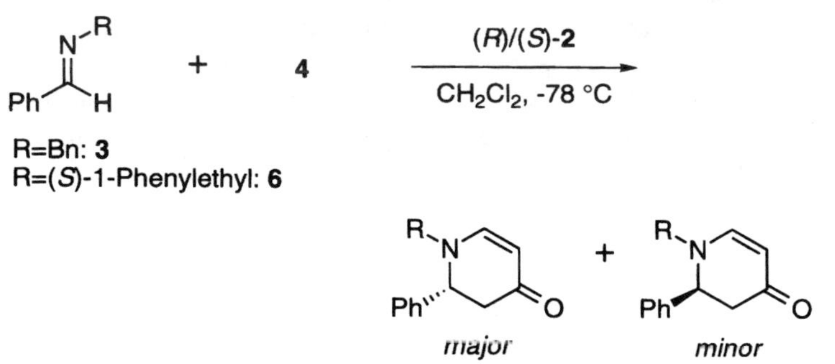

(eq. 9.6)

3, (*R*)-**2**, method B: 78% yield, 86% ee (*R*)
6, (*R*)-**2**, method B or C: 64% yield, ≥99% de (*R*)
6, (*S*)-**2**, method B or C: 49% yield, 83-84% de (*R*)

the aza Diels–Alder adduct **7** with a de of 99% was obtained after a reaction time of 12 h at –78°C. These results clearly indicate double stereodifferentiation of the aza Diels–Alder reaction. The cycloaddition using yellow crystals of (*R*)-**1**·(*S*)-**3**·PhOH resulted in extraordinary (>99.5:0.5) diastereoselectivity (method C).

9.3 Asymmetric Aldol-Type Reaction[8,9]

9.3.1 Double Asymmetric Induction of Aldol-Type Reaction of Imines Mediated by Chiral Lewis Acid 1 or 2[8,9a]

Largely stimulated by the synthesis of β-lactam antibiotics, there have been a great number of investigations into stereochemical aspects of imine condensations, especially those exhibiting diastereofacial selectivity with an imine containing a chiral auxiliary.[10–17] Initial efforts used α-methylbenzylamine as the auxiliary, but the resulting enantiomeric excesses were moderate (33–78% de).[11,12] Other auxiliaries are more efficient for diastereoselective condensation reactions; however, there are certain practical difficulties in removing or preparing them.[13–17] We developed highly stereoselective aldol-type reaction of chiral imines derived from α-methylbenzylamine and aldehydes with ketene silyl acetals in the presence of chiral Lewis acid **1**[9] or BLA **2**.[8]

The reagent, derived from equimolar amounts of (*R*)- or (*S*)-binaphthol and triphenylborate (formed in 1 h at room temperature in CH_2Cl_2), promoted smooth condensation of the imine **6** and the ketene acetal **16** derived from *tert*-butyl acetate at –78°C during 8 h to afford the β-amino ester **17** in good yield after the usual workup. The binaphthol was efficiently recovered for reuse. The reaction with (*R*)-**1** produced the (*R*)-adduct **17** in 92% de, whereas the reaction with (*S*)-**1** gave the adduct in 74% de. In a similar way, aliphatic imine **13** can be converted to the β-amino ester **18** with 94% de by using (*R*)-**1** (Table 9.2).

The use of **2** in the double stereodifferentiation of aldol-type reaction of chiral imines met with striking success. Reaction of (*S*)-**6** with trimethylsilyl ketene acetal **16** in the presence of (*R*)-**2** (1 equiv., method A) at –78°C for 12 h provided the corresponding aldol-type adduct (63% yield) with 94% de. The use of (*R*)-**2** prepared by method B gave similar results. Phenol included in the reaction mixture did not influence the reactivity or the diastereoselectivity. The aldol-type reaction using yellow crystals of (*R*)-**1**·(*S*)-**6**·PhOH proceeded with unprecedented (>99.5:0.5) diastereoselectivity (method C). In general, **2** is a more efficient chiral Lewis acid promoter than **1**. Some examples are listed in Table 9.2.

The enantioselective synthesis of α-methyl-β-amino ester equivalents is the most challenging task in organic synthesis, and the aldol reaction is a prospective tool for this purpose. Our present method can be applied to the reaction of imine with ketene silyl acetal derived from propionic ester. The boron reagent (*R*)- or (*S*)-**1** similarly promoted smooth condensation of the imine **6** and (*E*)-ketene silyl acetal **19** derived from *tert*-butyl propionate at –78°C to afford the α-methyl-β-amino ester **20** in good yield with complete *anti* selectivity (*anti*:*syn*=>99:1). It is noteworthy that the matching and

Table 9.2 DOUBLE ASYMMETRIC INDUCTION OF ALDOL-TYPE REACTION MEDIATED BY BORON REAGENTS

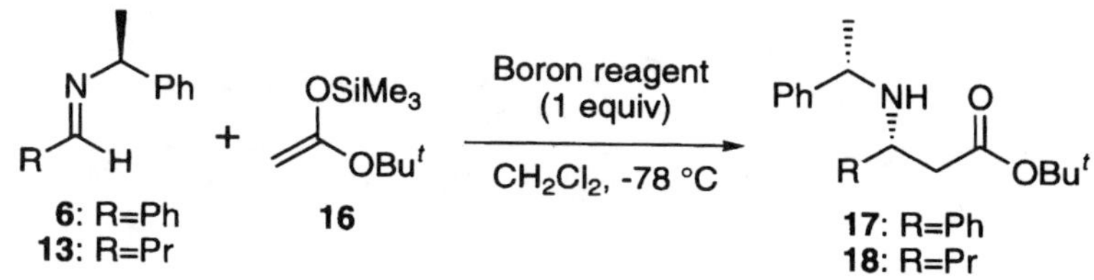

Imine	Boron Reagent[a]		Yield (%)	de (%)
6	(*R*)-**1**		59	92
	B(OPh)$_3$		67	78
	(*S*)-**1**		56	74
	(*R*)-**2**	(Method A)	63	94
		(Method B)	63	95
		(Method C)	65	>99
	(*S*)-**2**	(Method A)	18	66
		(Method B)	54	74
		(Method C)	64	62
13	(*R*)-**1**		56	94
	B(OPh)$_3$		64	66
	(*S*)-**1**		49	86

[a] See Section 9.2.3 for Methods A, B, and C.

mismatching pairs in double stereodifferentiation depend on the structure of the ketene silyl acetals. In the reaction with the ketene silyl acetal **19**, (*S*)-**1** matches chiral Lewis acid to produce a better selective adduct in 94% de, while the use of (*R*)-**1** gave the adduct in 78% de. The corresponding (*Z*)-ketene silyl acetal had no reactivity toward (*S*)-**6** under these reaction conditions (Table 9.3).

Table 9.3 DOUBLE ASYMMETRIC INDUCTION OF ALDOL-TYPE REACTION MEDIATED BY BORON REAGENTS

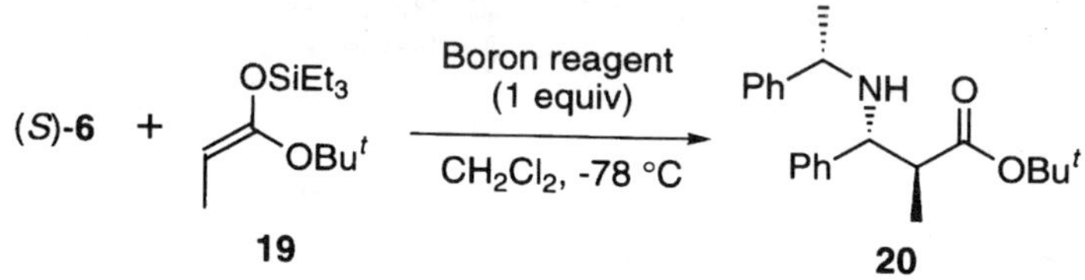

Boron reagent	Yield (%)	*Anti/syn*	Anti de (%)
(*R*)-**1**	65	100/0	78
(*S*)-**1**	35	100/0	94

Table 9.4 DOUBLE ASYMMETRIC INDUCTION OF ALDOL-TYPE REACTION MEDIATED BY BORON REAGENTS

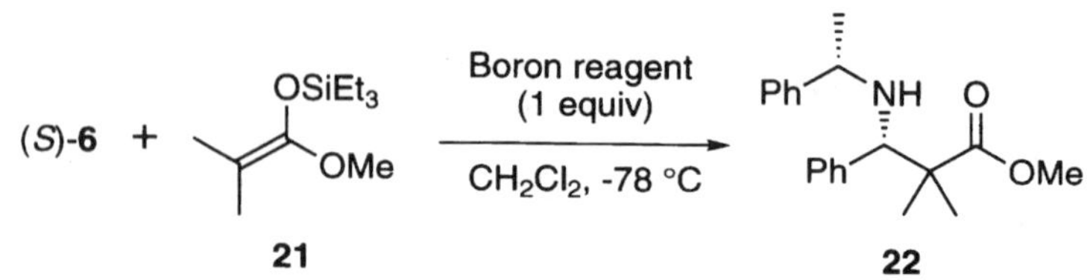

Boron reagent	Yield (%)	de (%)
(*R*)-**1**	60	76
$B(OPh)_3$	78	72
(*S*)-**1**	71	94

We further examined the reaction of (*S*)-**6** with dimethylketene triethylsilyl acetal **21** derived from methyl *iso*-butylate. (*R*)-**1** is a matching boron reagent in this reaction, and the β-amino ester **22** was produced in 94% de (Table 9.4).

Transformation to β-lactam compounds from the β-amino esters is easily performed. **17** and **20** were converted to the β-lactams **23** and **24** using Ohno's method[18]: treatment with trifluoroacetic acid, followed by Mukaiyama's reagent to β-lactam compounds, respectively, in 56% and 82% yield (Eq. 9.7). Methylation of **23** with LDA followed by MeI produced the same *trans* β-lactam **24**. On the other hand, exposure of **22** to vinyl magnesium bromide in THF at 0°C furnished the β-lactam **25** in 64% yield (Eq. 9.8).

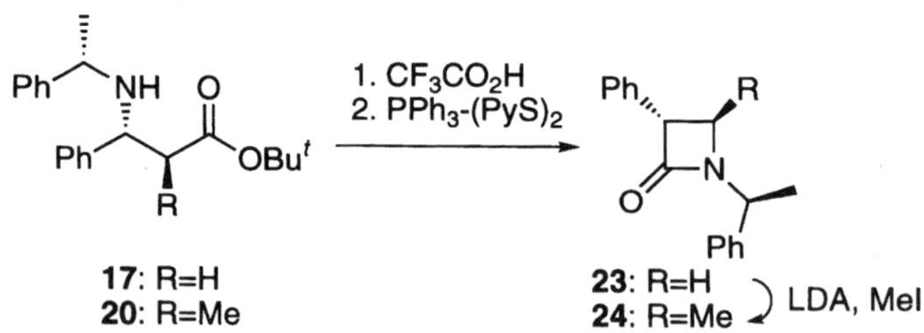

(eq. 9.7)

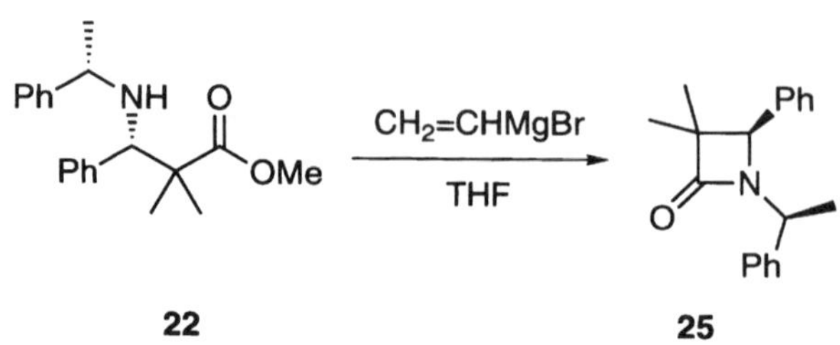

(eq. 9.8)

9.3.2 Stereoselective Synthesis of Carbapenem Antibiotics[9]

As an extension of highly diastereoselective aldol-type reactions of chiral imines and ketene silyl acetals in the presence of a chiral boron complex **1** we developed, we applied this method to the novel stereoselective synthesis of (+)-PS-5 and optically pure 3-(1′-hydroxyethyl)-2-azetidinone, useful as a synthetic intermediate of (+)-thienamycin (Scheme 9.3).

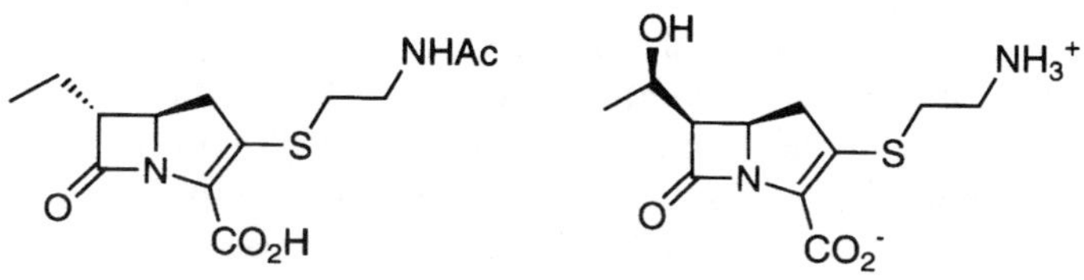

(+)-PS-5 (+)-Thienamycin

Scheme 9.3

9.3.2.1 Synthesis of (+)-PS-5[9a]

We chose acetylenic imine **26** prepared from trimethylsilylethynylaldehyde and (*S*)-α-methylbenzylamine, which transformed the formation of C(4) alkyl substituents of the β-lactam nucleus in (+)-PS-5. The chiral boron reagent **1** promoted smooth condensation of the imine **26** and the ketene silyl acetal **27** at –78°C during 8 h to afford the β-amino ester **28** in good yield. The reaction with (*R*)-**1** as the Lewis acid catalyst produced the aldol adduct **28** with extremely high *anti* selectivity (anti:syn=40:1) and diasetereoselectivity (97% de *anti*); **28** was converted to the β-lactam **29** using Ohno's method.[18] Shibasaki and co-workers synthesized **29** from the imine **26** with a boron enolate and succeeded in transforming it to (+)-PS-5 with moderate stereoselectivity.[19]

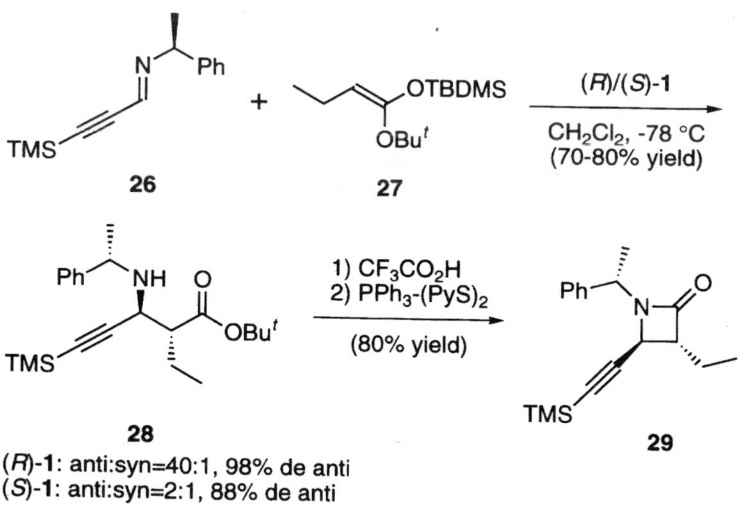

(eq. 9.9)

The absolute configuration of the adducts agrees with predictions based on the following mechanism. In the reaction of imine **6** and several silyl ketene acetals or dienes, the nucleophile approaches the *Re* face of the *E*-imine (see Section 9.2.3). However, the acetylenic imine consists of a mixture of *E* and *Z* isomers, and the ratio is influenced by Lewis acids. The *E*/*Z* ratio of imines is shown in Table 9.5. The preferred *E*-geometry of the imine **6** was clearly demonstrated by ^{1}H-NMR analysis of the complex of the imine and chiral Lewis acid **1** in CD_2Cl_2 at –60°C. (*S*)-Imine **26**, under the influence of the Lewis acid **1**, was gradually rearranged to *Z*, which could be the actual intermediate in the synthesis of **29**.

We also studied the boron reagent-induced shift (BIS) NMR analysis of the imines. Figure 9.3 indicates the plots of BIS magnitude (Δδ) for the imine proton and the Δδ for the imine carbon against the concentration of $B(OPh)_3$. In the Δδ of both proton and carbon, the values of the (*E*)-**6** and (*Z*)-**26** at a given concentration of the boron reagent increase, while those of the *E*-imine **26** do not change. Thus the *Z*-isomer is the actual intermediate in the reaction of **26** and the nucleophile would approach the *Si* face of the *Z*-imine. That would be consistent with the stereochemical results.

9.3.2.2 Synthesis of Thienamycin[9b,c]

This method was also applied to the novel stereoselective synthesis of optically pure 3-(1′-hydroxyethyl)-2-azetidinone, useful as a synthetic intermediate of (+)-thienamycin. Our approach is shown in eq. 9.10. The question with which we were concerned was whether simple diastereoselection (*syn–anti*) and diastereoface selectivity of the products are related to the structural features of the chiral ketene silyl acetal derived from 3-hydroxybutyrate. Acetylenic imines (*S*)- and (*R*)-**26** were prepared from chiral α-methylbenzylamine and trimethylsilylethylnyl aldehyde. The *E*- and *Z*-ketene acetals were prepared selectively in the following manner. The *Z*-ketene acetal **30** (*E*:*Z*=4:96) was prepared by treatment of the lithium dianion from (3*R*)-hydroxy-

Table 9.5 *E*/*Z* GEOMETRY RATIO OF IMINES[a]

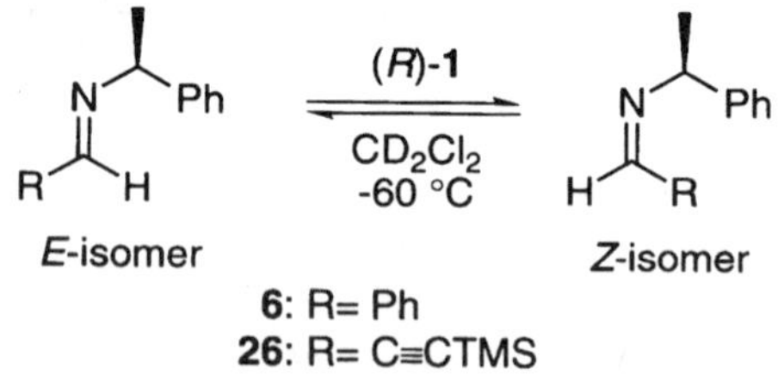

(*R*)-**1** (equiv)	0	0.3	0.7	1.0	2.0
E/*Z* ratio of **6**	2/1	1.5/1	0.7/1	0.4/1	0/1
E/*Z* ratio of **26**	1/0	1/0	1/0	1/0	1/0

[a] The ratio was determined by ^{1}H-NMR analysis.

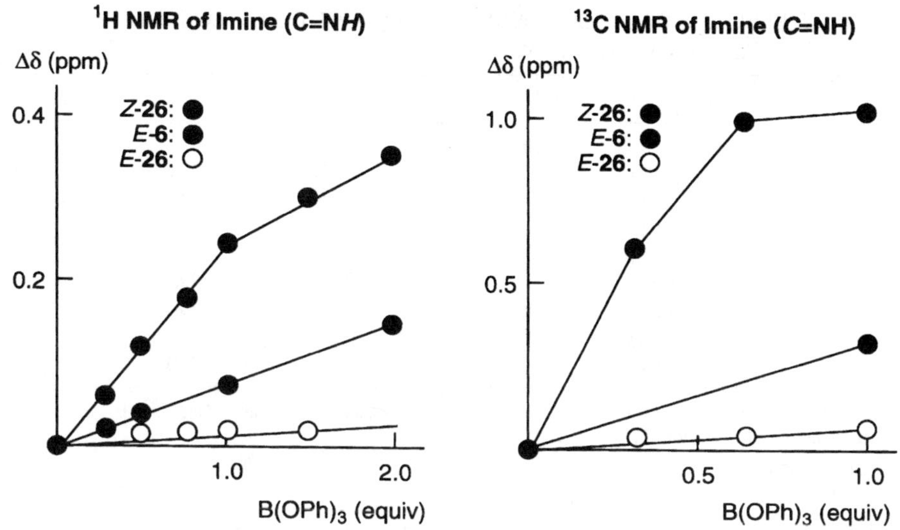

Figure 9.3. $B(OPh)_3$-induced shift NMR analysis of imines.

butyrate with triethylsilyl chloride; the internal quench procedure was used in preparation of the corresponding *E*-ketene silyl acetal **31**.[21] Treatment of mixture of (3*R*)-triethylsilyloxy butyrate and trimethylsilyl chloride with LDA gave the *E*-ketene acetal **31** (*E*:*Z*=>99:1).[20] The reaction of the imines with the ketene silyl acetals was promoted smoothly in the presence of equimolar boron reagents at –78°C for 8 h to afford the β-amino esters (eq. 9.10). It should be noted that *E*-imine **26**, under the influence of the boron reagents, was gradually rearranged to *Z*, which could be the real intermediate in the reaction. The remarkable feature of the reaction is shown in Table 9.6. With triphenyl borate as the Lewis acid, the *syn*-adduct **32** (*syn*:*anti*=92:8, diastereoface ratio=98:2) was stereoselectively produced. In sharp contrast, the use of

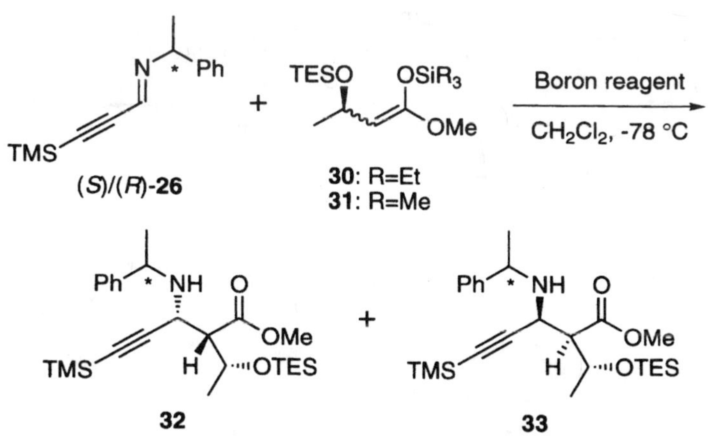

(eq. 9.10)

Table 9.6 STEREOSELECTIVE CONDENSATION OF A CHIRAL SILYL KETENE ACETAL AND A CHIRAL IMINE[a]

Entry	Imine	Boron reagent	Ketene silyl acetal	Yield (%)	*Syn/anti*	**32**:**33**
1	(*S*)-**26**	(*R*)-**1**	**30** (*E*:*Z*=4:94)	53	>99:1	33:67
2		$B(OPh)_3$		75	92:8	98:2
3		(*S*)-**1**		83	>99:1	2:98
4[b]		(*R*)-**1**	**31** (*E*:*Z*=99:1)	36	>99:1	73:27
5[b]		$B(OPh)_3$		51	90:10	>99:1
6[b]		(*S*)-**1**		60	>99:1	50:50
7	(*R*)-**26**	(*R*)-**1**	**30** (*E*:*Z*=4:94)	70	>99:1	>99:1
8		$B(OPh)_3$		55	90:10	>99:1
9		(*S*)-**1**		72	>99:1	>99:1

[a] Unless otherwise specified, reactions were carried out in dichloromethane using 1 equiv. of the boron reagent and 2 equiv. of ketene silyl at –78°C for 8 h.

[b] 1.5 equiv. of **31** was used.

(*S*)-**1** produced the *syn*-adduct **33** with almost complete stereoselectivity (*syn*:*anti*=>99:1, diastereoface=2:98), whereas the use of (*R*)-**1** decreased the selectivity. The effect of ketene silyl acetal geometry on the stereochemical course is apparent in entries 4–6. The reactions with *E*-ketene silyl acetal **31** revealed similar selectivity in simple diastereoselection and lower selectivity in diastereoface selectivity compared with *Z*-ketene acetal **30**. The mechanistic reason for this is not clear. However, it should be noted that the reaction is closely related to the assigned geometry of the intermediary ketene silyl acetals, and the reaction thus proceeds through an acyclic transition state.[22] (*R*)-**26** and **30** would be expected to have the same trend in diastereoface selectivity. Indeed, the reaction of (*R*)-**1** produced the *syn*-adduct **32** with any of the boron reagents (entries 7–9).

Transformation to the key intermediate for the synthesis of thienamysin is shown in Eqs. 9.11 and 9.12. Phenylmagnesium bromide or vinylmagnesium bromide was found to be an effective reagent for cyclization of the β-amino esters. Thus exposure

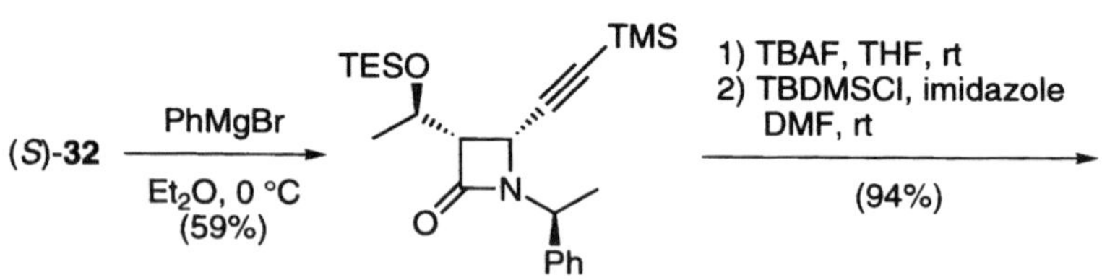

34

(eq. 9.11)

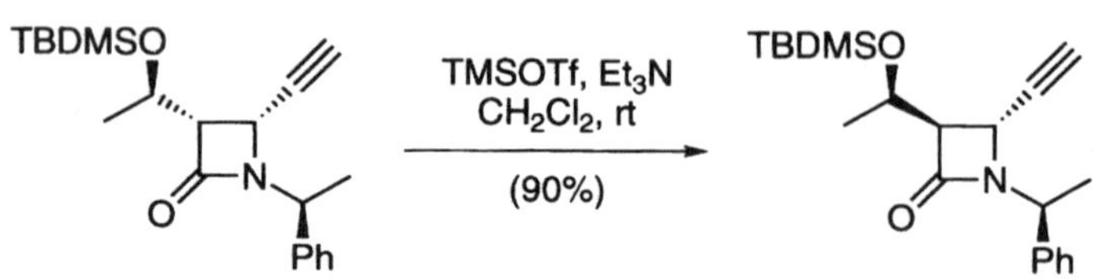

35 **36**

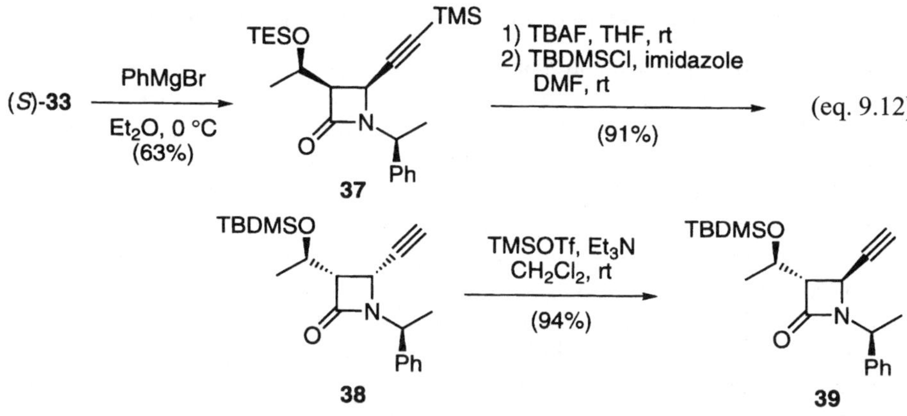

(eq. 9.12)

of **32** and **33** to phenylmagnesium bromide in ether at 0°C for 2 h furnished *cis*-β-lactams **34** and **37**, respectively, in 59 and 63% yields. The protective group of **34** and **37** was transformed from a triethylsilyl ether to a *tert*-butyldiemethylsilyl ether, and *cis* β-lactams **35** and **38** were then converted to *trans*-β-lactams **36** and **39** by treatment with trimethylsilyl triflate in dichloromethane.[23] All possible β-lactam ring isomers **35**, **36**, **38**, and **39** were therefore selectively obtained in optically pure form.

Subsequently, deprotection of benzyl group **39** under Birch reduction condition gave **40** in 94% yield. *Trans*-β-lactam **40** was then converted to **41** by partial reduction in 84% yield. *Trans*-β-lactam **40** could also be transformed into 4-acetoxy-3-[(*R*)-1-(*tert*-butyldimethylsiloxy)ethyl]-2-azetizinone (4-ABA) according to the reported procedure.[24] β-Lactam **41** was previously transformed into **42**,[25] and 4-ABA and **42** were established as synthetic intermediates leading to thienamycin (Eq. 9.13).[26]

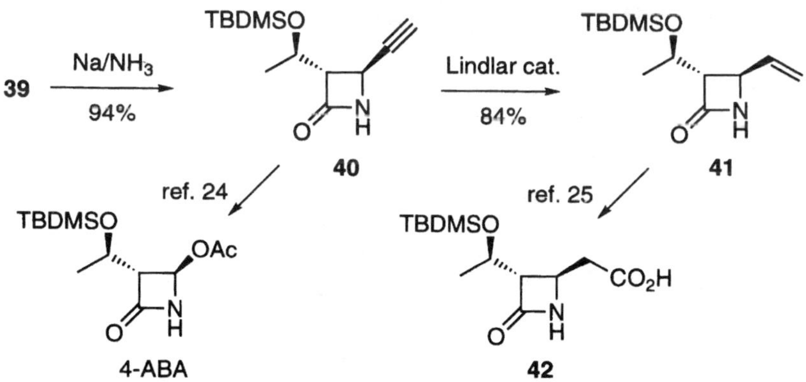

(eq. 9.13)

In conclusion, the present method provides a novel 3-hydroxybutyrate route for the construction of optically active 3-hydroxyethyl-2-azetidinones, and can be applied to the efficient asymmetric synthesis of thienamycin and related structures.

9.3.3 Synthesis of Both Diastereomers of α-Hydroxy-β-Amino Esters[27]

Some medicinally important compounds involve optically active α-hydroxy-β-amino ester units as their key components: Taxol, for example, a complex diterpene, is currently considered the most exciting substrate in cancer chemotherapy.[28] Taxol possesses high cytotoxicity and strong antitumor activity against cancers that have not responded well to treatment by existing antitumor drugs. It should be noted that the C-13 side chain, N-benzoyl-(2*R*,3*S*)-3-phenylisoserine moiety, is essential for this potent antitumor activity.[29] Accordingly, investigation of the structure–activity relationship of the Taxol side-chain analogs with some modification appears quite promising to find more effective pharmacological properties. Another example, renin, an aspartic protease, generates angiotensin I from angiotensinogen, and numerous inhibitory peptides of human renin have been studied for their potential as agents of antihypertensive therapy.[30] KRI-1314, which is one of the renin inhibitors developed for its promise of antihypertensive activity with oral efficacy, bears isopropyl (2*R*,3*S*)-3-amino-4-cyclohexyl-2-hydroxybutyrate, cyclohexylnorstatine, at its active site.[31]

Optically active α-hydroxy-β-amino ester, therefore, is essential to promote the activity, making development of a general method for this unit extremely important to further research in the therapeutic area. Unfortunately, however, although a number of methods have been developed, only a few have proven useful.[32,33] Our present method is a novel technique that is flexible and practical, both for the preparation of diastere-

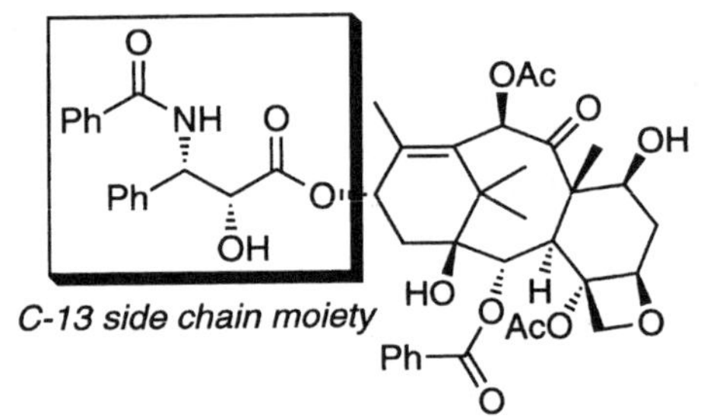

Taxol

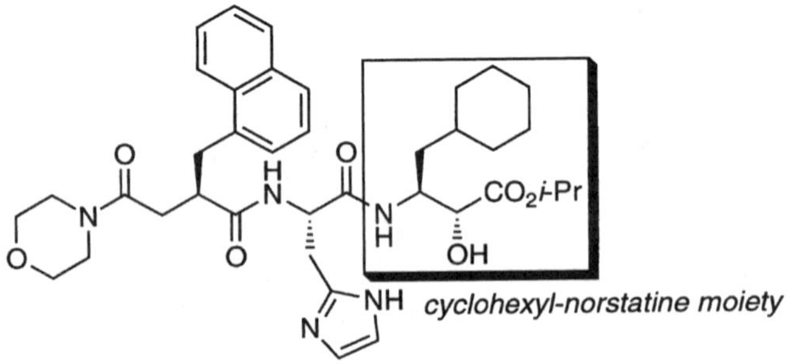

KRI 1314

Scheme 9.4

omers of optically active α-hydroxy-β-amino esters and for application in the stereoselective synthesis of the optically pure Taxol side chain and the norstatine family.

We set about constructing the α-hydroxy-β-amino ester units by aldol reaction of chiral imines with (*E*)- and (*Z*)-ketene silyl acetals derived from α-siloxy esters with *syn*/*anti* selectivity and diastereofacial selectivity of the products. First, several Lewis acid mediators were utilized in the aldol-type reaction of the chiral imine **6** with (*E*)-2-*tert*-butyldimethylsiloxy-1-methoxy-1-trimethylsiloxy-ethylene **43** and (*Z*)-1-methoxy-1,2-di(triethylsiloxy)ethylene **46**.[21] After screening several Lewis acids, it was found that the *anti* aldol adduct was obtained in the case of the $B(OPh)_3$, mediated reaction with *E*-ketene acetal **43** in moderate selectivity, while the *syn* aldol adduct was preferentially produced in the reaction with *Z*-ketene acetal **46** by employing every Lewis acid shown in Table 9.7. However, the use of these typical Lewis

Table 9.7 DIASTEREOSELECTIVE ALDOL-TYPE REACTION OF (*S*)-**6** MEDIATED BY LEWIS ACIDS[a]

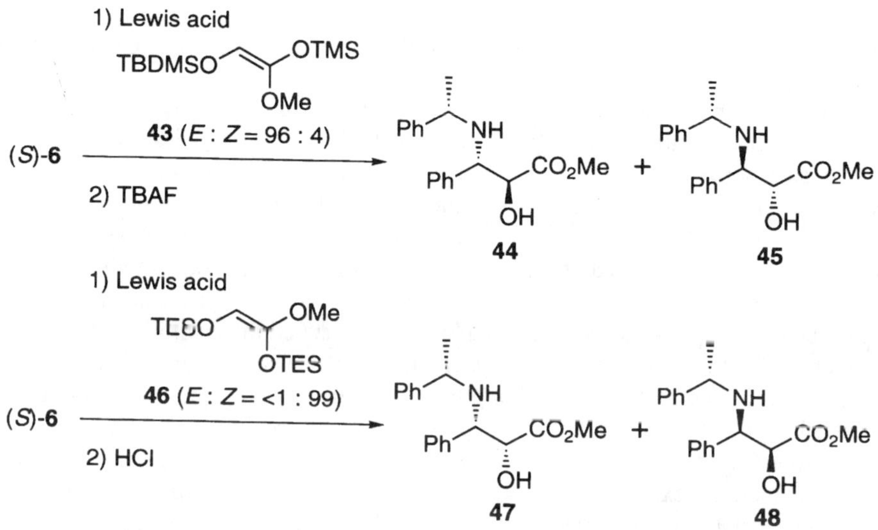

Entry	Lewis acid	Ketene silyl acetal	Yield (%)	*Anti* (**44**:**45**)/*Syn* (**47**:**48**)
1	$TiCl_4$	**43**	79	56 (89:11)/44 (89:11)
2		**46**	88	21 (84:16)/79 (91:9)
3	$BF_3{\cdot}Et_2O$	**43**	80	69 (94:6)/31 (89:11)
4		**46**	91	11/89 (92:8)
5	$ZnCl_2$	**43**	58	67 (92:8)/33 (80:20)
6		**46**	40	15/85 (95:5)
7	TMSOTf	**43**	52	55 (94:6)/45 (90:10)
8		**46**	47	3/97 (95:5)
9	$B(OPh)_3$	**43**	75	80 (93:7)/20
10		**46**	89	13/87 (93:7)

[a] Aldol-type reaction was carried out in dichloromethane using 1 equiv. of Lewis acid and 1.2 equiv. of ketene silyl acetal at −78°C for 8 h.

Table 9.8 DIASTEREOSELECTIVE ALDOL-TYPE REACTION OF (*S*)-6-MEDIATED BY (*R*)/(*S*)-1[a]

Entry	Lewis acid	Ketene silyl acetal	Yield (%)	*Anti* (**44**:**45**)/*Syn* (**47**:**48**)
1	(*R*)-**1**	**43** (*E*:*Z*=96:4	95	83 (95:5)/18
2	(*S*)-**1**		81	76 (90:10/24
3	(*R*)-**1**	**49** (*E*:*Z*=90:10)[b]	91	98 (96:4)/2
4	(*S*)-**1**		96	93 (95:5)/7
5	(*R*)-**1**	**46** (*E*:*Z*=<1:99)	94	6/94 (97:3)
6	(*S*)-**1**		91	1/99 (99:1)

[a] Aldol-type reaction of (*S*)-**6** was carried out in dichloromethane using 1 equiv. of (*R*)/(*S*)-**1** and 1.2 equiv. of ketene silyl acetals at –78°C for 8 h.

[b] **49**: 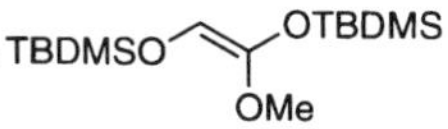

acids poses problems in selectivity or reactivity for the preparation of α-hydroxy-β-amino esters.

Next the "double stereodifferentiation technology" using chiral Lewis acid **1** applied to the novel construction of the optically active α-hydroxy-β-amino ester units. The results are summarized in Table 9.8. It is noteworthy that the chiral boron reagents product almost optically pure α-hydroxy-β-amino esters and that the stereoselectivity of the products completely depends on the geometry of the ketene silyl acetal. The reaction of the (*E*)-ketene acetal **49** mediated (*R*)-**1** produced the *anti* aldol adduct **44** with high stereoselectivity (*anti*:*syn*=98:2, diastereofacial ratio=96:4). In sharp contrast, the reaction of the (*Z*)-ketone acetal **46** mediated by (*S*)-**1** produced the enantiomerically pure *syn* adduct **47** (*syn*:*anti*=99:1, diastereofacial ratio=99:1). The use of (*R*)-**1** as a Lewis acid was generally effective in obtaining the *anti* adduct, while using (*S*)-**1** effectively realized the *syn* adduct. A kind of silyl group acting to protect hydroxy ester influenced the selectivity of the products. The mechanistic reason for this is not clear; however, the diastereoselectivity exhibited here is closely related to the assigned geometry of the ketene silyl acetals and the expected structure of the acyclic transition state.[22] Thus our methodology provides the first practical and efficient route for the selective preparation of both diastereomers of the α-hydroxy-β-amino esters. Furthermore, the *syn* adduct **47** was easily transformed into the Taxol side chain. **47** was subjected to hydrogenolysis over a palladium catalyst, followed by a Schotten–Baumann reaction to give the desired N=benzoyl-(2*R*,3*S*)-phenylisoserine methyl ester **50**, $[\alpha]_D = -48°$ (*c* 0.80, methanol) [lit.[34] $[\alpha]_D = -48°$ (*c* 1.0, methanol)], in 68% yield (eq. 9.14).

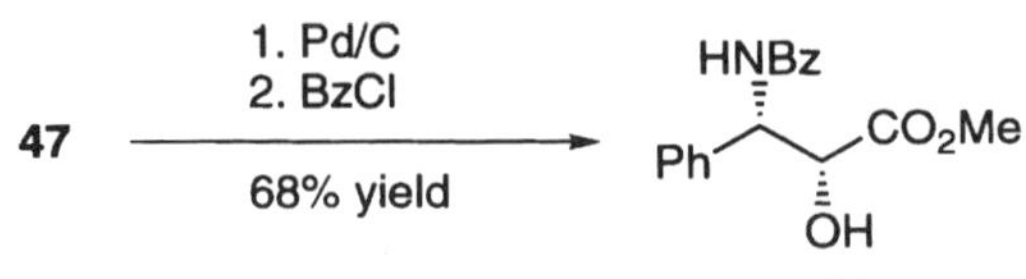

(eq. 9.14)

We also investigated the reaction of aliphatic imines under the same conditions used for the norstatine in Table 9.9. The imine **51** (*R*=*c*-C_6H_{11}) reacted with *E*-ketene silyl acetal **49** to furnish the corresponding aldol adducts in good yields; however, *anti*/*syn* selectivity was low with any of the boron reagents. When **51** (*R*=*c*-C_6H_{11}) was treated with *Z*-ketene silyl acetal **46** in the presence of (*S*)-**1**, the *syn* adduct **54** could be obtained in high yield with almost complete diastereoselectivity (*syn*:*anti*=99:1, diastereofacial ratio=99:1). Similarly, **51** (*R*=*i*-Pr) smoothly reacted under the same conditions to give the optically pure *syn* adduct **54**. The *syn* adducts **54** (*R*=*c*-C_6H_{11}) was deprotected by the benzyl group by hydrogenolysis over the palladium catalyst, followed by hydrolysis with 6*N* HCl to afford the hydrochloride salt of cyclohexylnorstatine, $[\alpha]_D=-14.6°$ (*c* 0.30, 1 *N* HCl) [lit.[35] $[\alpha]_D=-13.6°$ (1 *N* HCl)], in 88% yield. Thus the desired norstatine family could be easily obtained with high selectivity.

In conclusion, the present method offers a means for the construction of optically active α-hydroxy-β-amino ester units. The flexibility of chiral Lewis acid **1** and the novel procedure make this a very practical approach for the asymmetric synthesis of the Taxol side chain and the norstatine family.

Table 9.9 DIASTEREOSELECTIVE ALDOL-TYPE REACTION OF (*S*)-**51** MEDIATED BY (*R*)/(*S*)-**1**

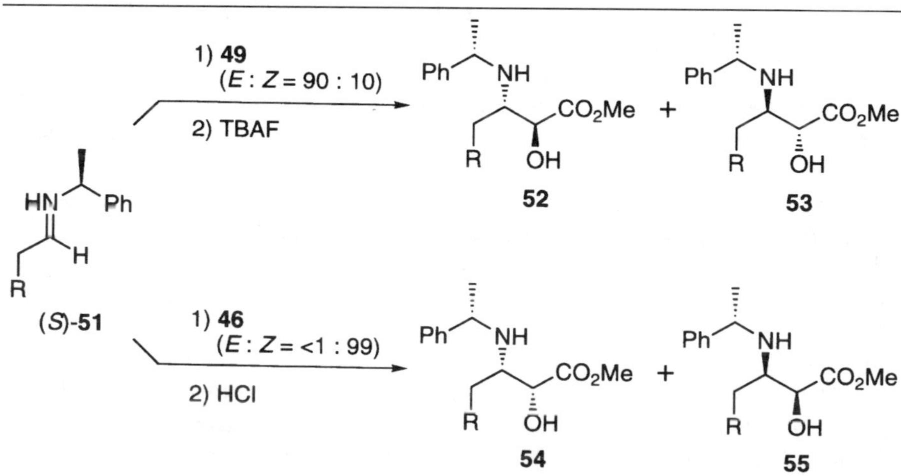

Entry	(*S*)-**51**	KSA[a]	Lewis acid	Yield (%)	*Anti* (**52**:**53**)/*syn* (**54**:**55**)
1	*R*=*c*-C_6H_{11}	**49**	(*R*)-**1**	85	62 (88:12)/38 (86:14)
2			(*S*)-**1**	73	65 (92:8)/35 (91:9)
3		**46**	(*R*)-**1**	85	7/93 (91:9)
4			(*S*)-**1**	87	1/99 (99:1)
5	*R*=*i*-Pr	**49**	(*R*)-**1**	81	54 (96:4)/46 (93:7)
6			(*S*) **1**	90	52 (94:6)/48 (97:3)
7		**46**	(*R*)-**1**	99	4/96 (91:9)
8			(*S*)-**1**	98	1/99 (99:1)

[a] Ketene silyl acetal.

9.3.4 Enantioselective Aldol-Type Reaction of Imines Mediated by Brønsted Acid Assisted Chiral Lewis Acid (BLA)

On the basis of the successful experimental results of double stereodifferentiation of aldol-type reaction of chiral imines, we developed the first method for the enantioselective synthesis of chiral β-amino acid esters from *achiral* imines and ketene silyl acetals using chiral BLA **2**.[8] As a first step, we studied the N-substituent-dependent enantioselectivity of the BLA **2** (1 equiv.; Method B; see Section 9.2.3) promoted aldol-type reaction of imines (1 equiv.) derived from benzaldehyde and several amines with **16** (2 equiv.) in dichloromethane at –78°C. The results of these experiments are summarized in Table 9.10. The enantioselectivity of the aldol-type reaction was dramatically increased by substituting sterically bulky N-substituents. The condensation of the imine **56** derived from benzhydrylamine occurred with high enantioselectivity (90% ee). The best result (96% ee) was achieved by using a mixture of 1:1 volumetric ratio of toluene and dichloromethane as solvents.

The enantioselective aldol-type reaction of a variety of N-benzhydryl imines with **16** by (*R*)-**2** under optimum conditions is summarized in Table 9.11. Excellent enantioselectivity (95% ee or better) was achieved in the reactions of aromatic aldehyde-derived imines. Unfortunately, enantioselectivity was low in the cases of the condensation of aliphatic aldehyde-derived imines (R=C_3H_7: 59% yield, 56% ee). The removal of the N-benzhydryl protecting group from β-aryl β-amino acid esters was easily advanced by catalytic hydrogenation (10% Pd/C, H_2, MeOH). In general, benzylamino compounds are debenzylated by hydrogenolysis over palladium on carbon in the order of N-substituent, Ph_2CH >> $PhCH_2$ > PhMeCH > $PhCHCH_2CO_2$*t*-Bu.[36]

Thus we have defined a new method for the synthesis of β-aryl β-amino acids in enantiomerically pure form, which are important fragments of spermine alkaloids.[37] For instance, (*S*)-β-phenyl-β-alanine derivatives (entry 1) are optically active units of (*S*)-dihydroperiphylline,[38] (*S*)-celacinnine,[39] and (*S*)-verbascenine.[40]

Finally, we studied the solution conformations of the uncomplexed (*S*)-**6** and the complex (*R*)-**2**·(*S*)-**6** using ^{1}H-NMR analysis and difference NOE measurements to understand the stereochemical outcome of the above reactions. Irradiation of H^a in the complex in CD_2Cl_2 at –60°C resulted in a weak NOE (5%) to CH_3^c and no NOE to H^b (Figure 9.4).[41] At that time, the imine proton H^a was strongly deshielded by the phenyl ring of N-substituent and (*R*)-**2**, while H^b and H^c were weakly shielded by the naphthyl ring of (*R*)-**2**: Chemical shifts of H^a, H^b, and H^c for the uncomplexed (*S*)-**6** were δ 8.39, 4.50, and 1.52, respectively, while those for the uncomplexed (*S*)-**6** were δ 12.65, 3.50, and 1.22. These results indicate the phenyl group of N-substituent is close of H^a.[41] Thus the absolute configuration of the adducts can be understood in terms of a rational model **57** involving an intramolecular hydrogen binding interac-

Table 9.10 N-SUBSTITUENT EFFECT (R^1) ON THE ALDOL-TYPE REACTION OF IMINES [(*E*)-PhCH=NR^1] WITH **16** BY (*R*)-**2**

N-substituent (R^1)	Ph	Bn	1-naphthylmethyl	Ph_2CH	(*S*)-PhMeCH
ee or de (%)	<5	29 (*R*)	46	90 (*R*)	95 (*R*)

Table 9.11 ENANTIOSELECTIVE ALDOL-TYPE REACTION OF N-BENZHYDRYLIMINES WITH **16** BY (*R*)-**2**

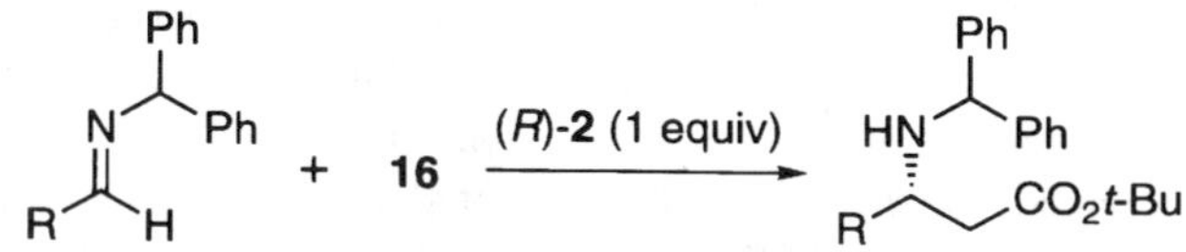

Entry	*R*	Yield (%)	ee (%)
1	C_6H_5	58	96 (*R*)
2	*p*-MeC_6H_4	35	97
3	*p*-ClC_6H_4	45	98
4	*p*-$AcOC_6H_4$	52	98
5	2,4-$Cl_2C_6H_3$	49	95
6	2-Naphthyl	43	96

tion via Brønsted acid (Figure 9.4). This hydrogen binding interaction would cause Lewis acidity of boron and π-basicity of naphthoxy moiety to increase, and the transition-state assembly would be stabilized.[43] In the complex **57**, the 2-hydroxynaphthyl group fixed by the intramolecular hydrogen bond would effectively block the *Si* face of the (*E*)-imine **6**[42] complexed with (*R*)-**2**, and the nucleophile would approach the *Re* face. In contrast, although ^{1}H-NMR assignment of the complex (*R*)-**1·56** was not clear, absolute enantioselectivity in the aldol-type reaction of N-benzhydryl imine can be understood by the similar mechanistic model.

A new concept for the design of enantioselective Lewis acid has been supported by experiments that demonstrate that it is a very practical and promising methodology for enantioselective synthesis of optically pure amino compounds. We believe that the experimental results outlined above will stimulate further exciting advances for the BLA system and offer essential information on the direction of future design of catalytic asymmetric reactions of imines by chiral Lewis acid.

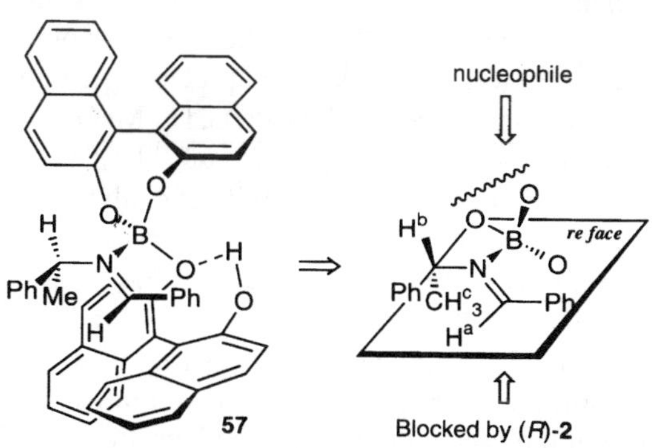

Figure 9.4. The matched pair complex (*R*)-**2**·(*S*)-**6**.

9.3.4.1 A Concise Synthesis of (+)-(S)-Dihydroperiphylline[44]

(+)-(*S*)-Dihydroperiphylline (**58**) is one of six alkaloids isolated from the leaves of *Periptergia marginata*.[45] All these alkaloids contain a 13-membered ring system derived from dicinnamoylspermidine. They are of special interest because of the notable biological activity shown by members of the polyamine family.[46] The total synthesis of (±)-**58** was first accomplished by Wasserman et al. in 1981.[47,48] More recently, an enantioselective route of (+)-(*S*)-**58** from N-Boc-(*S*)-β-phenyl-β-alanine prepared from diethyl L-tartrate in 15 steps was reported by Kibayashi et al.[38,49]

We previously developed boron-templated cyclization of triamino esters as an efficient method to induce long open chains with α- and ω-functional groups to condense internally into large rings, and applied this to total synthesis of (±)-celacinnine, which was the isomeric lactam of **58**.[50] On the basis of this methodology, a short, highly efficient and enantioselective synthesis of (+)-(*S*)-**58** was attained as described here.

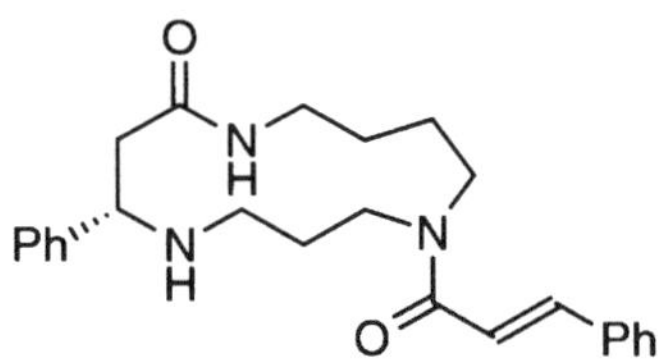

58

The present strategy employed for the synthesis of (+)-(*S*)-**58** is outlined in Schemes 9.5 and 9.6. The retrosynthetically derived plan of synthesis led to (*S*)-β-phenyl-β-alanine methyl ester (**60**) and the triamino ester **65** as key intermediates. Precursor **59** of **60** was accessible in 96% ee from N-benzylidenebenzhydrylamine (**56**) via highly enantioselective aldol-type reaction with ketene trimethylsilyl acetal **16** using the chiral BLA **2**. A triamino acid derivative such as **65** was envisaged as an ideal progenitor of the macrocyclic spermidine skeleton **67**, which could then be elaborated via boron-templated cyclization and selective acylation to **58**.

(*S*)-Methyl 3-(benzhydrylamino)-3-phenylpropionate (**59**) was prepared in 56% yield and 96% ee by the enantioselective aldol-type reaction of **56** with **16** in the presence of the (*S*)-BLA **2** and subsequent transesterification. (*S*)-Binaphthol was recovered in >95% yield (Scheme 9.5). The removal of N-benzhydryl protecting group of **59** was selectively accomplished by catalytic hydrogenation (10% Pd/C, H_2, MeOH) without debenzylation of 3-phenylpropionate moiety.[36] Optically pure (*S*)-**60**,[51] $[\alpha]_D^{26}=-18.2°$ (*c* 1.46, $CHCl_3$), was obtained by resolution of 96% ee **60** (1 equiv. of oxalic acid in methanol, one recrystallization from only methanol at 25°C and treatment of the salt with 1*M* NaOH) in good yield. The optical purity of **60** was shown to be >99% ee, as judged by comparison with the optical rotation value reported in the literature[51b,c] and HPLC analysis.

Next the diamino unit **63** was prepared from 3-amino-1-propanol (**61**) and 4-bromobutyronitrile, as outlined in Scheme 9.6. N-(2-naphthyl)methylation of **61** occurred in 97% conversion by sequential imine condensation with 2-naphthylaldehyde and reduction with sodium borohydride, and then the cyanopropylation of **62** with 4-bromobutyronitrile gave **63** in 98% yield.

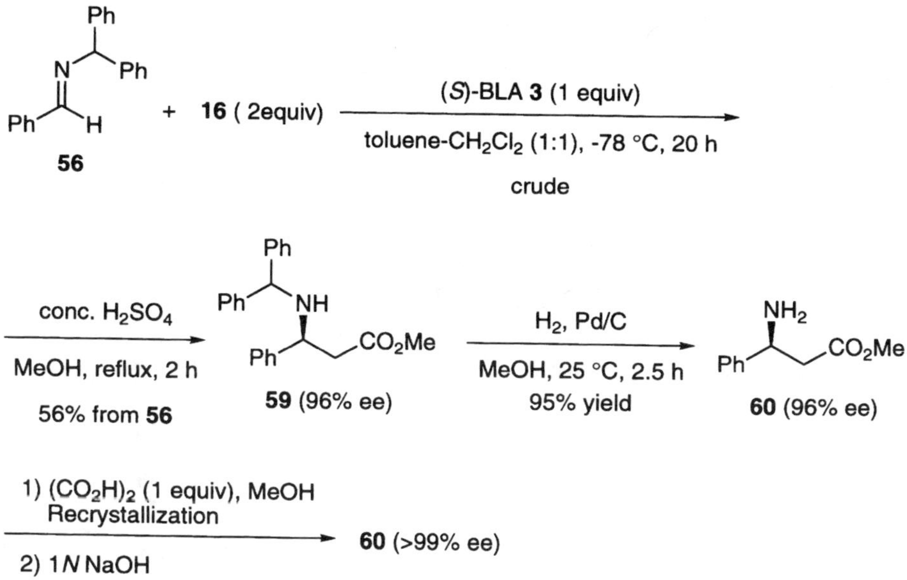

Scheme 9.5. Preparation of (*S*)-β-phenyl-β-alanine methyl ester.

The coupling of the key components **60** and **63** proceeded smoothly in the presence of triflic anhydride and excess N,N-diisopropylethylamine to form the triamino unit **64** in 92% yield. Hydrogenolysis of N-(2-naphthyl)methyl group of **64** over 10% activated palladium on carbon and subsequent hydrogenation of cyano group over platinum oxide[52] in the presence of concentrated hydrochloric acid gave the triaminoester **65** in 74% yield from **64**. However, debenzylation of **64**, which was protected by the benzyl group in place of the (2-naphthyl)methyl group, was very slow over palladium catalysts [Pd/C, $PdCl_2$, and $Pd(OH)_2/C$] under hydrogen atmosphere, and the undesired pyrrolidine derivative was gradually produced by hydrogenolysis and successive annulation between the debenzylated amino group and the cyano group.

The cyclization of the triamino ester **65** was effected with tris(dimethylamino)borane in xylene under reflux to furnish the desired diaminolactam **67** in 74% yield. We believe that the possible intermediate **66** generated by transamination of triaminoborane could be reasonably expected to undergo a facile, sterically driven cyclization to **67** as well as the cyclization step in the synthesis of cellacinnine.[50] It is noted that, when the *tert*-butyl ester was used in place of the methyl ester **65**, the present cyclization proceeded and obtained **67** in 25% yield even at reflux over 2 days.

The selective acylation at N-7 of **67** was accomplished with cinnamoyl chloride in the presence of 4-(dimethylamino)pyridine at low temperature to give **58** in almost quantitative yield. The optical rotation determined for **58** ($[\alpha]_D^{26}$=+3.6° (c=1.19, $CHCl_3$)) and the NMR spectra were in good agreement with the reported value for synthetic (+)-(*S*)-dihydroperiphylline.[47,49] Finally, (+)-(*S*)-dihydroperiphylline was synthesized by an efficient six-step sequence using (*S*)-β-phenyl-β-alanine methyl ester as a chiral unit. Tris(dimethylamine)borane proved to be highly useful for the metal-templated cyclizations of not only ethyl 3-phenyl-4,9,13-triazatridecanoate, the precursor of celacinnine, but also methyl 3-phenyl-4,8,13-triazatridecanoate.

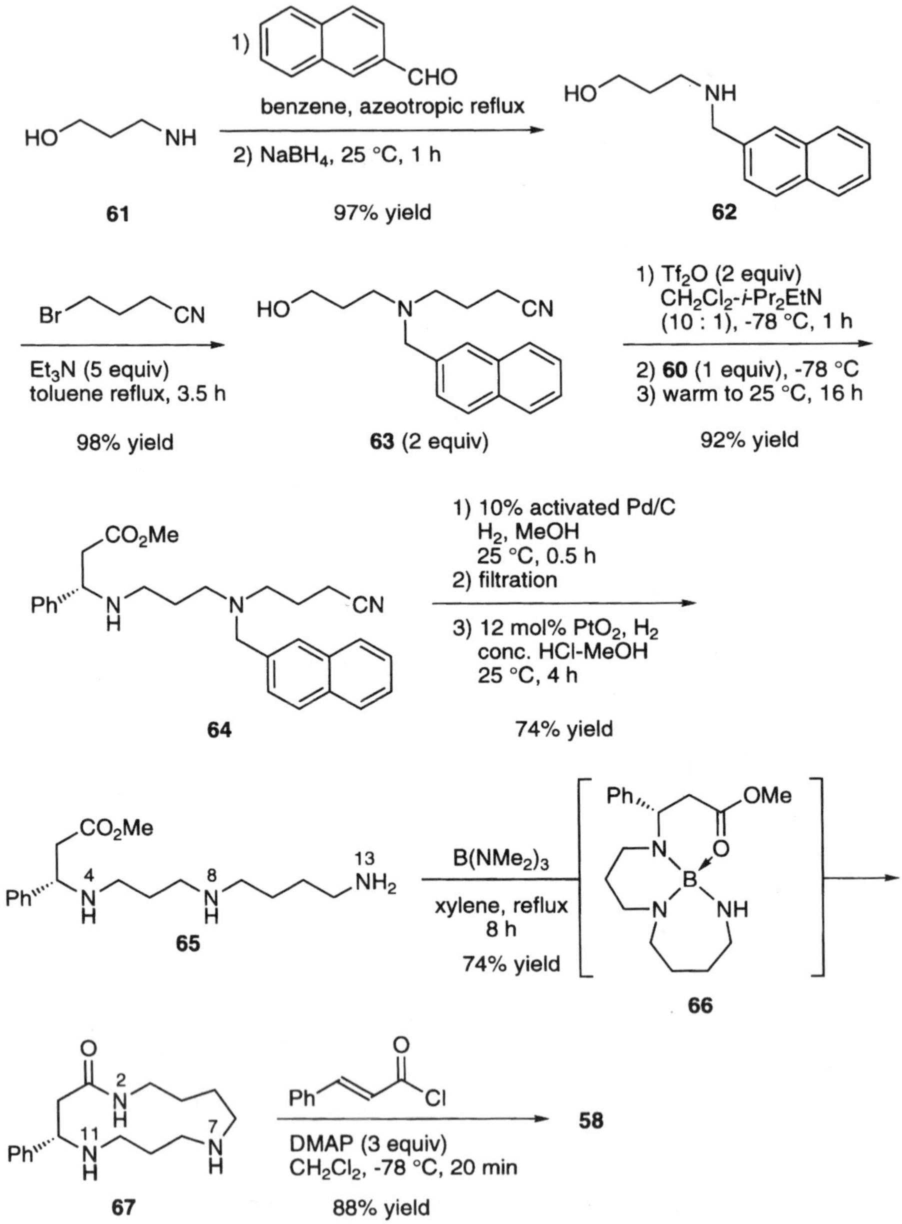

Scheme 9.6 Enantioselective synthesis of (+)-(*S*)-dihydroperiphylline

9.4 References

1. General review articles on the chemistry of imines: (a) H. Schiff, *Ann. Chem.* **131**, 118 (1864). *Ann. Chem. (Suppl.)* **3**, 343, (1864–65). (b) M. M. Sprung, *Chem. Rev.* **26**, 297 (1940). (c) W. R. Layer, *Chem. Rev.* **43**, 489 (1963).
2. For more recent work: (a) S. Itsuno, Y. Sakurai, K. Shimizu, K. Ito, *J. Chem. Soc., Perkin Trans. 1* 1548 (1989). (b) B. T. Cho, Y. S. Chun, *J. Chem. Soc.., Perkin Trans. 1* 3200 (1990). (c) D. E. Gibbs, D. Barnes, *Tetrahedron Lett.* **31**, 5555 (1990). (d) Y. N. C. Chan, J. A. Osborn, *J. Am. Chem. Soc.* **112**, 9400 (1990). (e) F. Spindler, B. Puginm, H.-U. Blaser, *J. Am. Chem. Soc.* **114**, 7562 (1992). (f) M. Nakagawa, T. Kawate, T. Kakikawa, Yamada, T. Matsui, T. Hino, *Tetrahedron* **49**, 1739 (1993).
3. (a) K. Hattori, H. Yamamoto, *J. Org. Chem.* **47**, 3264 (1992). (b) K. Hattori, H. Yamamoto, *Tetrahedron* **49**, 1749 (1993).
4. S. D. Larsen, P. A. Grieco, *J. Am. Chem. Soc.* **107**, 1768 (1985).
5. (a) H. Waldmann, M. Braun, M. Dräger, *Angew. Chem. Int. Ed. Engl.* **29**, 1468 (1990). (b) H. Waldmann, M. Braun, M. Weymann, M. Gewehr, *Synlett* **1991**, 881. (c) H. Waldmann, M. Braun, M. Dräger, *Tetrahedron: Asymmetry* **2**, 1231 (1991). (d) H. Waldmann, M. Braun, *J. Org. Chem.* **57**, 4444 (1992).
6. W. Pfrengle, H. Kunz, *J. Org. Chem.* **54**, 4261 (1989).
7. (a) M. M. Midland, J. I. McLoughlin, *Tetrahedron Lett.* **29**, 4653 (1988). (b) M. M. Midland, R. W. Koops, *J. Org. Chem.* **57**, 1158 (1992).
8. K. Ishihara, M. Miyata, K. Hattori, H. Yamamoto, T. Tada, *J. Am. Chem. Soc.* **116**, 10520 (1994).
9. (a) K. Hattori, M. Miyata, H. Yamamoti, *J. Am. Chem. Soc.* **115**, 1151 (1993). (b) K. Hatoori, H. Yamamoto, *Synlett* **1993**, 239. (c) K. Hattori, H. Yamamoto, *BioMed. Chem. Lett.* **3**, 2337 (1993).
10. Review: (a) E. F. Kleinnman, in B. M. Trost, I. Fleming, C. H. Heathcock, Eds. *Comprehensive Organic Synthesis*, Vol. 2, Chapter 4.1, Pergamon Press, 1991. (b) D. J. Hart, D.-C. Ha, *Chem. Rev.* **89**, 1447 (1989).
11. N. Furukawa, T. Okawara, H. Noguchi, Y. Tetrakawa *Heterocycles* **6**, 1323 (1977).
12. I. Ojima, S.-I. Inaba, *Tetrahedron Lett.* 2077 (1980). For a similar condensation with a thioester tin enolate, see T. Yamada, H. Suzuki, T. Mukaiyama, *Chem. Lett.* **1987**, 293.
13. I. Ojima, S.-I. Inaba, *Tetrahedron Lett.* **32**, 2081 (1980).
14. More recently, Ojima reported a lithium enolate-imine condensation with optically active 2-phenylcyclohexanol as a chiral control unit. See I. Ojima, I. Habus, M. Zhao, *J. Org. Chem.* **56**, 1681 (1991) and references cited therein.
15. H. Waldmann, M. Braun, M. Draeger *Angew. Chem. Int. Ed. Engl.* **29**, 1468 (1990). H. Waldmann, Braun, M. *J. Org. Chem.* **57**, 444 (1992).
16. W. Pfrengle, H. Kunz, *J. Org. Chem.* **57**, 444 (1992).
17. M. M. Midland, R. W. Koops, *J. Org. Chem.* **57**, 1158 (1992).
18. S. Kobayashi, T. Iimori, T. Izawa, M. Ohno, *J. Am. Chem. Soc.* **103**, 2406 (1981).
19. M. Shibasaki, Y. Ishida, G. Iwasaki, T. Iimori, *J. Org. Chem.* **52**, 3489 (1987).
20. *E,Z*-Geometry of imines was determined by NOE experiments.
21. *E/Z* ratio and geometry were determined by ^{1}H-NMR and NOESY spectra. For preparation method, see K. Hattori, H. Yamamoto *J. Org. Chem.* **58**, 5301 (1993).
22. E. W. Colvin, D. McGarry, M. J. Nugent *Tetrahedron* **44**, 4157 (1988).

23. T. Chiba, T. Nakai *Tetrahedron Lett.* **26**, 4647 (1985).

24. T. Chiba, T. Nakai *Chem. Lett.* **1985**, 651.

25. Y. Ito, Y. Kobayashi, T. Kuwabata, M. Takase, S. Tetrashima *Tetrahedron* **45**, 5767 (1989).

26. G. I. Georg, J. Kant, H. S. Gill, *J. Am. Chem. Soc.* **109**, 1129 (1987). G. I. Georg, *Tetrahedron Lett.* **25**, 3779 (1984). D. C. Ha, D. J. Hart, T. K. Yang, *J. Am. Chem. Soc.* **106**, 4819 (1984). D. G. Melillo, I. Shinkai, T. Liu, K. Ryan, M. Sletzinger, *Tetrahedron Lett.* **21**, 2783. D. G. Melillo, I. Shinkai, T. Liu, K. Ryan, M. Sletzinger, *ibid.* **22**, 913 (1981). I. Shinkai, T. Liu, R. A. Reamer, M. Sletzinger, *ibid.* **23**, 4899 (1982). M. Shiozaki, N. Ishida, T. Hiraoka, H. Yanagisawa, *ibid.* **22**, 5205 (1981). N. Ikota, O. Yoshino, K. Koga, *Chem. Pharm. Bull.* **30**, 1929 (1982).

27. K. Hattori, H. Yamamoto, *Tetrahedron* **50**, 2785 (1994).

28. S. Borman, *Chem. Eng. News* **69**(35), 11 (1991). R. W. Miller *J. Nat. Prod.* **43**, 425 (1980). M. C. Wani, H. L. Taylor, M. E. Wall, P. Coggon, A. T. McPhail, *J. Am. Chem. Soc.* **93**, 2325 (1971).

29. L. Mangatal, M. T. Adeline, D. Guénard, F. Guéritte-Voegelein *Tetrahedron* **45**, 4177 (1989). C. S. Swindell, N. E. Krauss *J. Med. Chem.* **34**, 1176 (1991). A. E. Mathew, M. R. Mejillano, J. P. Nath, R. H. Himes, V. J. Stella *J. Med. Chem.* **35**, 145 (1992).

30. H. Koike *Gendai Kagaku* **55** (1989). D. H. Rich *J. Med. Chem.* **28**, 263 (1985).

31. K. Iizuka, T. Kamijo, H. Harada, K. Akahane, T. Kubota, H. Umeyama, T. Ishida, Y. Kiso *J. Med. Chem.* **33**, 2707 (1990).

32. For the recent synthesis of the side chain of Taxol: (a) I. Ojima, I. Habus, M. Zhao, M. Zucco, Y. H. Park, C. M. Sun, T. Brigaud, *Tetrahedron* **34**, 6985 (1992). (b) L. Deng, E. N. Jacobsen, *J. Org. Chem.* **57**, 4320 (1992). (c) G. I. Georg, P. M. Mashava, E. Akgün, M. W. Milstead, *Tetrahedron Lett.* **32**, 3151 (1991).

33. For the recent synthesis of norstatine analog: (a) I. Ojima, Y. H. Park, C. M. Sun, T. Brigaud, M. Zhao, *Tetrahedron Lett.* **39**, 3737 (1992). (b) Y. Kobayashi, Y. Takemoto, T. Kamijo, H. Harada, Y. Ito, S. Terashima, *Tetrahedron* **48**, 1853 (1992). c) K. Iizuka, T. Kamijo, H. Harada, K. Akahane, T. Kubota, H. Umeyama, T. Ishida, Y. Kiso, *J. Med. Chem.* **33**, 2707 (1990).

34. J.-N. Denis, A. Correa, A. E. Greene, *J. Org. Chem.* **55**, 1957 (1990).

35. Y. Kobayashi, T. Matsumoto, Y. Takemoto, K. Nakatani, Y. Ito, T. Kamijo, H. Harada, S. Terashima, *Chem. Pharm. Bull.* **39**, 2550 (1991).

36. The order of the reactivity in the hydrogenolysis was determined by control experiments using N-benzhydrylheptylamine, N-benzylheptylamine, N-(α-phenylethyl)heptylamine, and *tert*-butyl 3-(α-phenylethyl)amino-3-phenylpropionate as starting substrates.

37. For review, see: H. H. Wasserman, J. S. Wu, *Heterocycles* **17**, 581 (1982).

38. T. Kaseda, T. Kikuchi, C. Kibayashi, *Tetrahedron Lett.* **30**, 4539 (1989) and references therein.

39. C. Seguineau, P. Richomme, J. Pusset, *Helv. Chim. Acta.* **75**, 2283 (1992) and references therein.

40. W. J. Fiedler, M. Hesse, *Helv. Chim. Acta.* **76**, 1511 (1993) and references therein.

41. In NOE experiments of the uncomplexed (*S*)-**6** in CD_2Cl_2 at –78°C, irradiation of H^a resulted in a strong NOE (25%) to H^b and a weak NOE (4%) to CH_3^c. These results indicate H^b is close to H^a.

42. The preferred (*E*) geometry of **56** could clearly be demonstrated by ^{1}H-NMR analysis and NOE experiments of the complex.

43. K. Ishihara, H. Yamamoto, *J. Am. Chem. Soc.* **116**, 1562 (1994).

44. K. Ishihara, Y. Kuroki, H. Yamamoto, *Synlett* **1995**, 41.

45. A. Hocquemiller, A. Cave, H.-P. Husson, *Tetrahedron* **33**, 645 (1977).

46. (a) *Advances in Polyamine Research*, Vols. 1 and 2, R. A. Campbell, D. R. Morris, D. Bartos, G. D. Daves, F. Bartos, Eds., Raven Press, New York, 1978. (b) M. M. Badawi, K. Bernauer, P. Van den Broek, G. Groger, A. Guggisberg, S. Johne, L. Kompis, F. Schneider, H.-J. Veith, M. Hesse, H. Schmid, *Pure Appl. Chem.* **33**, 81 (1973).

47. H. H. Wasserman, H. Matsuyama, *J. Am. Chem. Soc.* **103**, 461 (1981).

48. For other synthetic studies see: (a) R. Hocquemiller, A. Cave, H.-P. Husson, *Tetrahedron* **33**, 653 (1977). (b) L. Crombie, R. C. F. Jones, D. Haigh, *Tetrahedron Lett.* **27**, 5151 (1986).

49. $[\alpha]_D^{23}$=+3.5° (*c* 0.4, $CHCl_3$) for (+)-(*S*)-**1**. The experimental results by Kibayashi et al. indicate that the originally reported optical rotation value (–21°)[45] for natural dihydroperiphylline is erroneous and should be corrected to +3.5°.[38]

50. H. Yamamoto, K. Maruoka, *J. Am. Chem. Soc.* **103**, 6133 (1981).

51. (a) M. Pais, R. Sarfati, F.-X. Jarreau, *Bull. Soc. Chim. Fr.* **1973**, 331. (b) F. A. Davis, F. R. T. Reddy, R. E. Reddy, *J. Org. Chem.* **57**, 6387 (1992). $[\alpha]_D^{20}$=–20° (*c* 1.80, $CHCl_3$) for (*S*)-**60** (>95% ee). (c) J. Jiang, K. K. Schumacher, M. M. Joullie, F. A. Davis, R. E. Reddy, *Tetrahedron Lett.* **35**, 2121 (1994). $[\alpha]_D^{20}$=+22.3° (*c* 1.99, $CHCl_3$) for (*R*)-**60** (>98% ee). $[\alpha]_D^{20}$=+22.3° (*c* 1.99, $CHCl_3$) for (*R*) =**60** (>98% ee).

52. J. A. Secrist III, M. W. Louge, *J. Org. Chem.* **37**, 335 (1972).

CHAPTER

10

Diastereo- and Enantioselective Synthesis of β-Amino Acids via SAMP Hydrazones and Hetero Michael Addition Using TMS-SAMP as a Chiral Equivalent of Ammonia

Dieter Enders, Wolfgang Bettray, Jürgen Schankat, and Jürgen Wiedemann

10.1 Introduction

β-Amino acids **A**, although of less importance than the parent α-amino acids, are crucial structural features of numerous biologically active natural products[1] as well as building blocks for β-lactam antibiotics.[2] This is reflected by the ever-increasing research activity in the field of their stereoselective synthesis.[3] Besides numerous syntheses employing C-C bond formations, we have developed C-C and C-N bond connections according to the retrosynthetic analysis shown in Figure 10.1.

One general access developed in our group is based on the 1,2-addition of organometallic reagents to the C-N double bond of SAMP [(*S*)-(−)-1-amino-2-(methoxymethyl)pyrrolidine] hydrazones generating the amino function. Two different approaches turned out to be a practical method for the asymmetric synthesis of β-amino acids. First, the 1,2-addition of organocerium reagents to 3,3-ethylenedioxyalkylaldehyde SAMP hydrazones **E** (n=1) leads to β-amino acetals, which are converted to β-amino acids by oxidative transformation of the acetal moiety into the carboxylic acid group (retrosynthetic pathway **A** ⇒ **B**). Second, the 1,2-addition of allylic organometallics to aldehyde SAMP hydrazones **F** gives rise to homoallylic amines, which are converted to β-amino acids by oxidative cleavage of the double bond (retrosynthetic pathway **A** ⇒ **C**). Following the third retrosynthetic pathway **A** ⇒ **D**, the aza-analogous Michael addition of enantiopure nitrogen nucleophiles **G** as chiral equivalents of ammonia to enoates opened a highly stereoselective access to a wide range of β-amino acids.

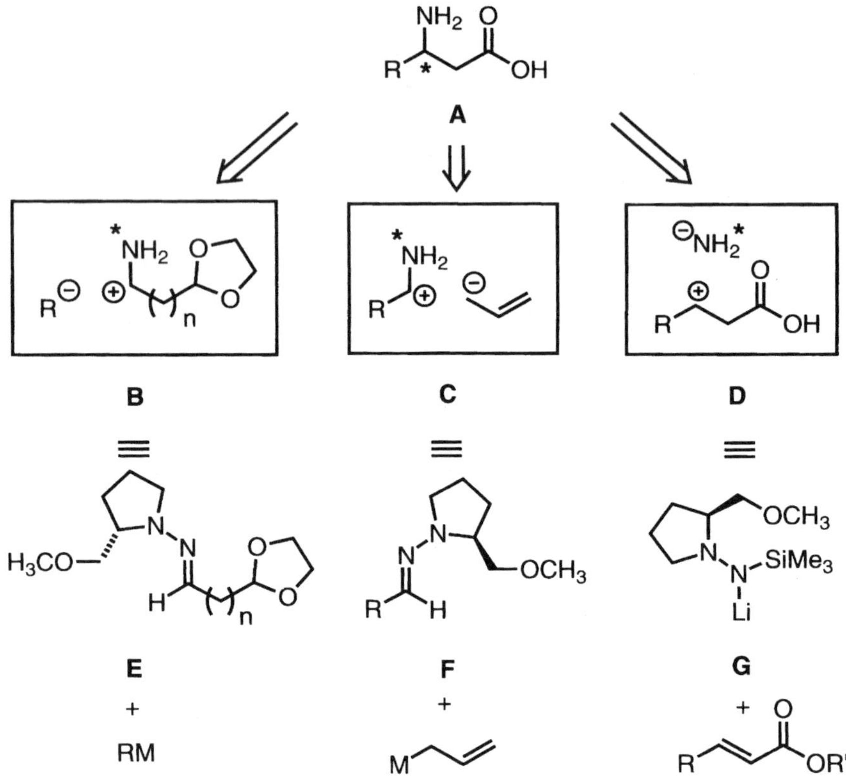

Figure 10.1. Retrosynthetic pathways to β-amino acids.

10.2 Enantioselective Synthesis of Protected Homoallylic Amines and β-Amino Acetals as Precursors of β-Amino Acids via 1,2-Addition of Organometallic Reagents to SAMP/RAMP Hydrazones

The 1,2-addition of organometallic reagents to the C-N double bond of imines, hydrazones, and other imine equivalents is an important method for the generation of the amino functionality from carbonyl groups.[4] Consequently β-amino acids have been synthesized by nucleophilic 1,2-additon of organometallic reagents including enolates to chiral imine equivalents. The asymmetric synthesis of β-amino acids by the addition of enolates to prochiral imines is an especially well known and established method in organic synthesis.[3,5] A new strategy for the synthesis of β-amino acids is based on our SAMP/RAMP (*R*)-(+)-amino-z-(methoxymethyl) pyrrolidinehydrazone method for the synthesis of primary amines.[6] The addition of organometallic reagents to 3,3-ethylenedioxypropanal SAMP hydrazone **1** with subsequent cleavage of the auxiliary affords the protected β-amino acetals **2**.[7] Oxidative cleavage of the acetal moiety with ozone and deprotection leads to the desired β-amino acids **3** (Figure 10.2).[8]

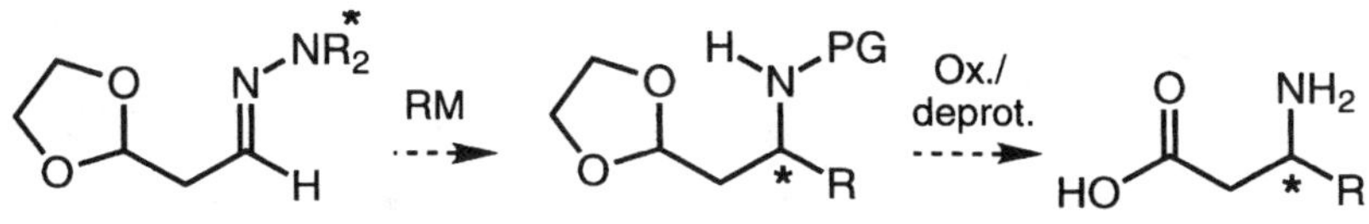

1 2 3

Figure 10.2. 1,2-Addition of organometallic reagents to chiral hydrazones **1** leading to protected β-amino acetals **2** as precursors for β-amino acids **3**.

An alternative synthetic strategy published by Hua et al.[9] and Kunz et al.[10] is the addition of allyl-type polar organometallics to the C-N double bond of chiral sulfinamides and galactopyranosylimines **4**. Subsequent cleavage of the auxiliary affords protected homoallylic amines **5**,[11] in which the C-C double bond can be oxidatively cleaved to the carboxyl group to give β-amino acids **6** (Figure 10.3).[12] Following this strategy using allyl cerium or allyl Grignard reagents in the 1,2-addition to SAMP hydrazones, we have developed a flexible synthesis of β-amino acids. The combination of both strategies makes it possible to extend this method to β-amino diacids **9** from chiral hydrazones **7** (Figure 10.4).

10.2.1 1,2-Addition of Organocerium Reagents to 3,3-Ethylenedioxypropanal SAMP Hydrazone[8]

The reaction of the aldehyde **10**[13] with SAMP or RAMP[6,14] gave the hydrazone **(*S*)-11** or **(*R*)-11** in good yield (90%). Subsequent 1,2-additions of in situ generated organocerium reagent of high nucleophilicity and low basicity to the C-N double bond of hydrazone **11** at –100°C occurred with good yield (49–95%) and with high diastereoselectivity (de = 75–99%) (Figure 10.5, Tables 10.1, and 10.2) Organocerium reagents can be prepared according to Imamoto et al.[15] from a carefully dried suspension of cerium(III)chloride in THF with organolithium reagents at –78°C or Grignard reagents at 0°C. Upon warming to room temperature overnight, the primary addition products were trapped with methyl chloroformate (MOCCl, Method A, *R* = alkyl) or with saturated aqueous $NaHCO_3$ solution (Method B, *R* = aryl), leading to the protected hydrazines **12a–e** and free hydrazines **12f–h**. Purification with diastereomeric enrichment was found possible by flash column chromatography. The carbamate protected hydrazines **12a–e** were cleaved by lithium in liquid ammonia[16] (4 equiv. Li-wire, 1 h) at –33°C, affording the β-amino acetals

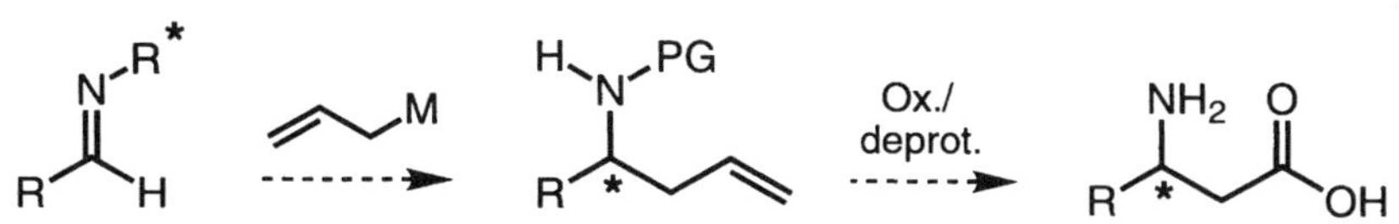

4 5 6

Figure 10.3. Nucleophilic addition of allyl-type polar organometallics to imines **4** leading to protected homoallylic amines **5** as precursors of β-amino acids.

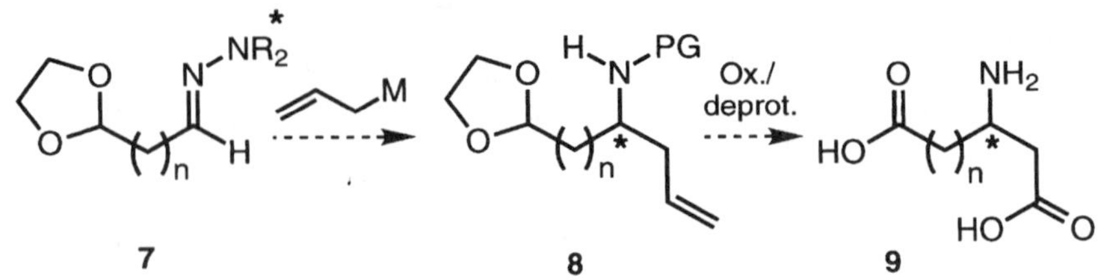

Figure 10.4. Enantioselective synthesis of β-amino diacids **9**.

13a–e in good yields (56–95%) and without racemization (ee = 92–96%). The ee values of **13a–e** were determined by GC analysis [CSP, heptakis-(2,3,6-tri-O-methyl)-β-cyclodextrin in OV 1701] as the acetyl protected derivatives of **13a–e** and by ^{1}H-NMR-LIS experiments. As the carbamate protected hydrazines **12f–h** were susceptible to cleavage of the benzylic C-N bond, the free hydrazines **12f–h** were

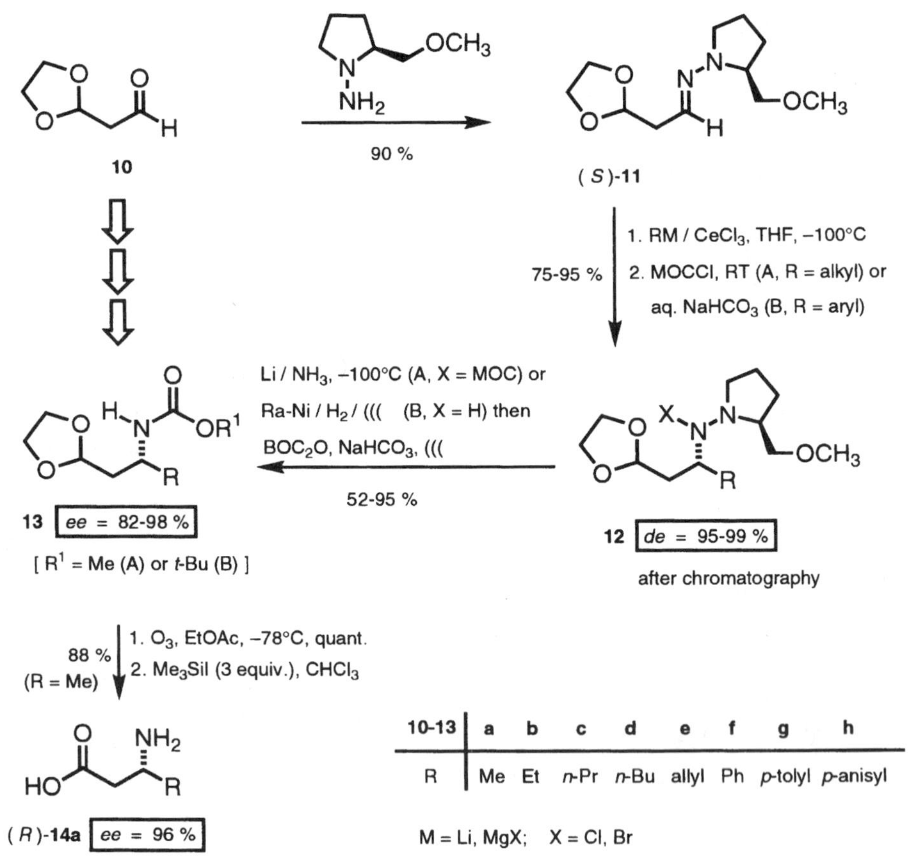

10-13	a	b	c	d	e	f	g	h
R	Me	Et	n-Pr	n-Bu	allyl	Ph	p-tolyl	p-anisyl

M = Li, MgX; X = Cl, Br

Figure 10.5. Enantioselective synthesis of β-amino acids via 1,2-addition to 3,3-ethylendioxypropanal-SAMP hydrazone **(*S*)-11**.

Table 10.1 SYNTHESIS OF β-HYDRAZINO ACETALS **12a–h** BY 1,2-ADDITION OF ORGANO-CERIUM REAGENTS TO 3,3-ETHYLENEDIOXYPROPANAL SAMP/RAMP HYDRAZONES 11

Product	*R*	*M*/$CeCl_3$	Yield **11**→**12**[a] (%)	de[b] (%)	$[\alpha]_D^{RT}$ (*c*, $CHCl_3$)[c]	Config.
12a	Me	Li	75	>99	–68.4 (0.95)	(*S*,*R*)
12a[d]	Me	Li	95	97	+70.6 (1.26)	(*R*,*S*)
12b	Et	MgBr	60 (88)	95 (79)	–51.5 (0.99)	(*S*,*R*)
12c	*n*-Pr	Li	54 (92)	94 (75)	–65.0 (0.82)	(*S*,*R*)
12d	*n*-Bu	Li	78 (88)	97 (92)	–58.6 (1.10)	(*S*,*R*)
12e	allyl	MgCl	99	>98 (95)	–55.5 (0.99)	(*S*,*R*)
12f	Ph	MgBr	97	>99	–94.8 (0.97)	(*S*,*S*)
12g	*p*-tolyl	MgBr	49 (67)	99 (88)	–73.4 (0.92)	(*S*,*S*)
12h	*p*-anisyl	MgBr	55 (67)	>99 (90)	–62.4 (0.98)	(*S*,*S*)

[a] Yield of both diastereomers in parentheses.

[b] Determined by ^{1}H- and ^{13}C-NMR spectroscopy or by GC analysis; in parentheses are de values before chromatography.

[c] **12a–e** as MOC-protected hydrazines, **12f–h** as free hydrazines.

[d] RAMP was used as chiral auxiliary.

cleaved with Raney–Nickel[17] (W-2, 1 spatula/mmol) catalyzed hydrogenation at atmospheric pressure and ultrasound conditions according to the procedure of Alexakis et al.[18] The free β-amino acetals were directly protected with $BOC_2O/NaHCO_3$ under ultrasound conditions[19] (20 min) in good yields (52–70%). Unfortunately, a remarkable racemization occurred (up to 17%), as observed by NMR of the MTPA-amides.[20] Attempts to cleave the N-N bond by treatment with samarium diiodide[21] also proved unsuccessful.[22] The utility of this novel class of β-amino acetals as precursors for β-

Table 10.2 SYNTHESIS OF PROTECTED β-AMINO ACETALS **13a–h** BY N-N BOND CLEAVAGE VIA $LiNH_3$ (**13a–e**) OR RANEY NICKEL (**13f–h**)

Product	*R*	Yield **13**→**14** (%)	$[\alpha]_D^{RT}$ (*c*, $CHCl_3$)	ee[a] (%)	Config.[b]
13a	Me	95	–14.7 (1.02)	98	(*R*)
13a[c]	Me	67	+14.2 (1.10)	97	(*S*)
13b	Et	92	–8.0 (0.97)	95	(*R*)
13c	*n*-Pr	88	–7.7 (1.11)	92	(*R*)
13d	*n*-Bu	70	–7.2 (1.14)	97	(*R*)
13e	allyl	56	–22.6 (1.33)	95	(*R*)
13f	Ph	70	–54.0 (0.99)	82[d]	(*S*)
13g	*p*-tolyl	65	–43.8 (0.97)	82[d]	(*S*)
13h	*p*-anisyl	52	–34.3 (0.88)	82[d]	(*S*)

[a] Determined by GC-CSP, heptakis (2,3,6-tri-O-methyl)-β-cyclodextrin in OV 1701 or ^{1}H-NMR-LIS, configuration in parentheses.

[b] Major enantiomer.

[c] RAMP was used as chiral auxiliary.

[d] Determined by MTPA-amides of unprotected β-amino acetals **13f–h**.

amino acids was demonstrated in the two-step synthesis of (*R*)-3-aminobutyric acid **(*R*)-14a** starting from **(*R*)-13a**. The oxidative cleavage of the acetal group with ozone in ethyl acetate at –78°C according to the procedure of Deslongchamps et al.[23] afforded quantitatively the corresponding pure β-hydroxyethyl ester, which was deprotected by three equivalents of $Me_3SiI/CHCl_3$[24] in good yield (88%).

The ee value (96%) was determined by GC and NMR analysis of the MTPA-amide[20] of the methyl ester of **14a**, whereas the absolute configuration was proved by correlation of the optical rotation of **14a** with literature data.[25,26] The method is also applicable for the synthesis of higher homologues of amino acetals and amino acids.[22]

10.2.2 1,2-Addition of Allyl-Type Polar Organometallics to Alkyl Aldehyde SAMP Hydrazones and 3,3-Ethylenedioxyalkylaldehyde SAMP Hydrazones[12]

The SAMP hydrazones **(*S*)-16a–h** and **(*R*)-16e** were simply synthesized from aldehydes **15**[13,22,27] and SAMP or RAMP.[14] A great variety of allyl-type polar organometallics have been used for the 1,2-addition to imines and their equivalents to synthesize homoallylic amines.[11] In our case we have tested two different reagents: First we used allyl cerium reagents at –100°C (method A), or alternatively we carried out the 1,2-addition with the more practicable allyl Grignard reagents in toluene at –78°C (method B). The allyl Grignard reagents were prepared from allyl magnesium chloride in THF (1 mL/mmol) or allyl magnesium bromide in ether (1 mL/mmol) and dissolved in toluene (30–50 mL/mmol) according to Alexakis et al.[28] The preliminary addition products were trapped with methyl chloroformate to obtain the carbamate protected hydrazines **17a–h** (Figure 10.6, Tables 10.3 and 10.4). In both cases the addition occurred with good yields (63–99%) and with high diastereoselectivities (89–99%). It is conceivable that the formation of the acetal substituted carbamates **18e–h** may be responsible for the optimized results via precoordination of the organocerium reagent to the oxygen of the acetal group. The use of the allyl cerium reagent instead of the allyl Grignard reagent gave a slight improvement in both the diastereomeric excesses (3%, 9%, **17d,f**) and yields (7%, 0%, **17d,f**). Grignard reagents, not dissolved in toluene, reacted poorly.[22,29] It was possible to obtain the pure diastereomers of **17e–h** but not of **17a–d** by flash column chromatography. The cleavage of the N-N bond of the protected hydrazines **17a–h** to homoallylic amines **18a–h** was carried out without racemization (ee = 90–98%) in good yields (54–89%) according to the procedure described in Section 10.2.1. Hydrogenation of the C-C double bond and thus formation of alkylated products occurred occasionally on a small scale (<5%).

Reportedly the oxidation of the allylic side chain of protected homoallylic amines has been achieved by $KMnO_4/NaIO_4$ in water[10] or ozone in CH_2Cl_2 and subsequent oxidation with $AgNO_3$-KOH.[9] We have tested the procedure of Marshall et al.[30] using ozone in basic MeOh/CH_2Cl_2 for the oxidation of the olefin moiety to the methyl ester. The selective ozonolysis conditions used for the oxidation of **18b,c,e** afforded the carbamate protected β-amino methyl ester **19b,c** and the carbamate protected β-amino methyl ester acetal **19e** in 66%, 74%, 56% yield, respectively (Figure 10.7). The intact acetal moiety in **19e** could be transformed into the hydroxy ester group of

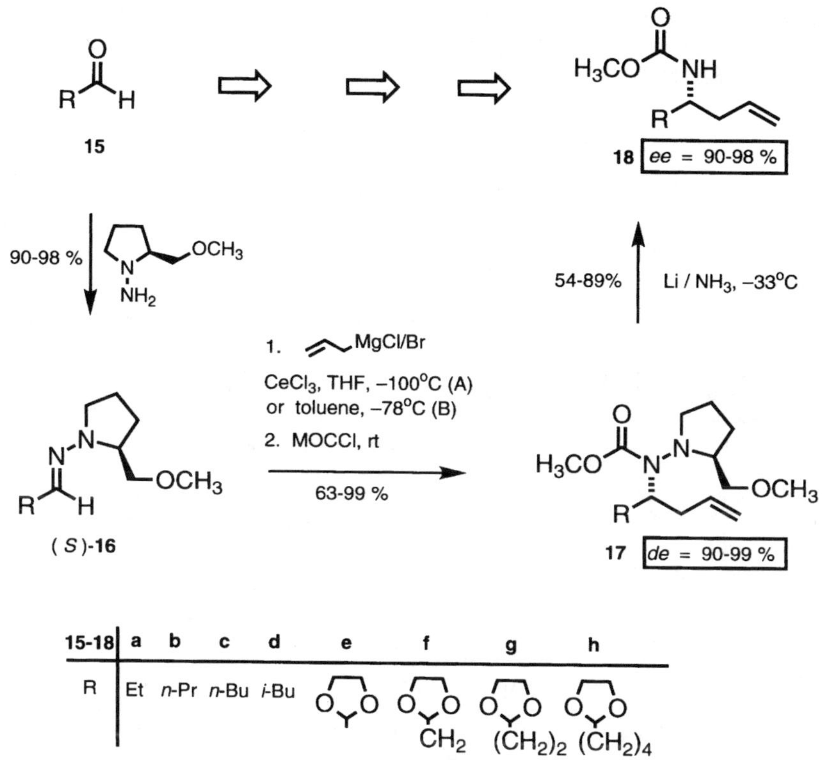

Figure 10.6. Enantioselective synthesis of protected homoallylic amines **18**.

Table 10.3 SYNTHESIS OF CARBAMATE PROTECTED HYDRAZINES **17a-h** VIA 1,2-ADDITION OF ALLYLIC ORGANOMETALLICS TO ALDEHYDE SAMP/RAMP HYDRAZONES **16**

Product	*R*	Metal	Yield **16→17** (%)	de[a] (%)	$[\alpha]_D^{RT}$ (*c*, $CHCl_3$)
17a	Et	$MgCl/CeCl_3$	83	90	−59.0 (0.95)
17b	*n*-Pr	$MgCl/CeCl_3$	81	91	−56.5 (0.96)
17c	*n*-Bu	$MgCl/CeCl_3$	77	92	−51.4 (2.41)
17d	*i*-Bu	$MgCl/CeCl_3$	78	92	−60.8 (1.01)
17d	*i*-Bu	MgBr	71	90 (89)	
17e	dioxolan	MgCl	75	97 (92)	−58.0 (0.90)
17e[b]	dioxolan	MgCl	92	98 (91)	+57.4 (0.91)
17f	dioxolan-CH_2	$MgCl/CeCl_3$	99	99	−55.5 (0.52)
17f	dioxolan-CH_2	MgCl	99	98 (90)	
17g	dioxolan-$(CH_2)_2$	$MgCl/CeCl_3$	84	98 (98)	−39.4 (0.72)
17h	dioxolan-$(CH_2)_4$	$MgCl/CeCl_3$	63	98 (95)	−47.6 (1.24)

[a] Determined by ^1H- and ^{13}C-NMR spectroscopy or by GC analysis; in parentheses are de values prior to chromatography.
[b] RAMP was used as chiral auxiliary.

Table 10.4 SYNTHESIS OF PROTECTED HOMOALLYLIC AMINES **18a–d** AND PROTECTED HOMOALLYLIC AMINO ACETALS **18e–h** BY N-N-BOND CLEAVAGE WITH $LiNH_3$ OF THE PROTECTED HYDRAZINES **17**

Product	*R*	Yield **17→18** (%)	$[\alpha]_D^{RT}$ (*c*, $CHCl_3$)	ee (%)	Config.
18a	Et	58	−54.1 (0.86)	90	(*S*)
18b	*n*-Pr	67	−46.9 (1.11)	91[a]	(*S*)
18c	*n*-Bu	61	−22.1 (0.71)	93[a]	(*S*)
18d	*i*-Bu	61	−58.0 (0.92)	92	(*S*)
18e	dioxolan	60	−5.9 (0.99)	97[b]	(*R*)
18e[c]	dioxolan	55	+6.8 (1.07)	97[d]	(*S*)
18f	dioxolan-CH_2	61	−22.6 (1.33)	97[d]	(*R*)
18g	dioxolan-$(CH_2)_2$	70	−4.7 (1.08)	98[d]	(*S*)
18h	*p*dioxolan-$(CH_2)_4$	89	−5.4 (0.98)	97[d]	(*S*)

[a] ee Values were determined on MTPA-amides of the β-amino methyl esters. The configuration was proven by correlation of the optical rotation of the corresponding β-amino acids with literature data.

[b] Determined on the MTPA-amide of the aspartic acid dimethyl diester; the (*R*) configuration was proven by correlation of the optical rotation of aspartic acid **[(*R*)-18e]** obtained with literature data.

[c] RAMP was used as chiral auxiliary.

[d] Determined on the MTPA-amide of the unprotected homoallylic amino acetal.

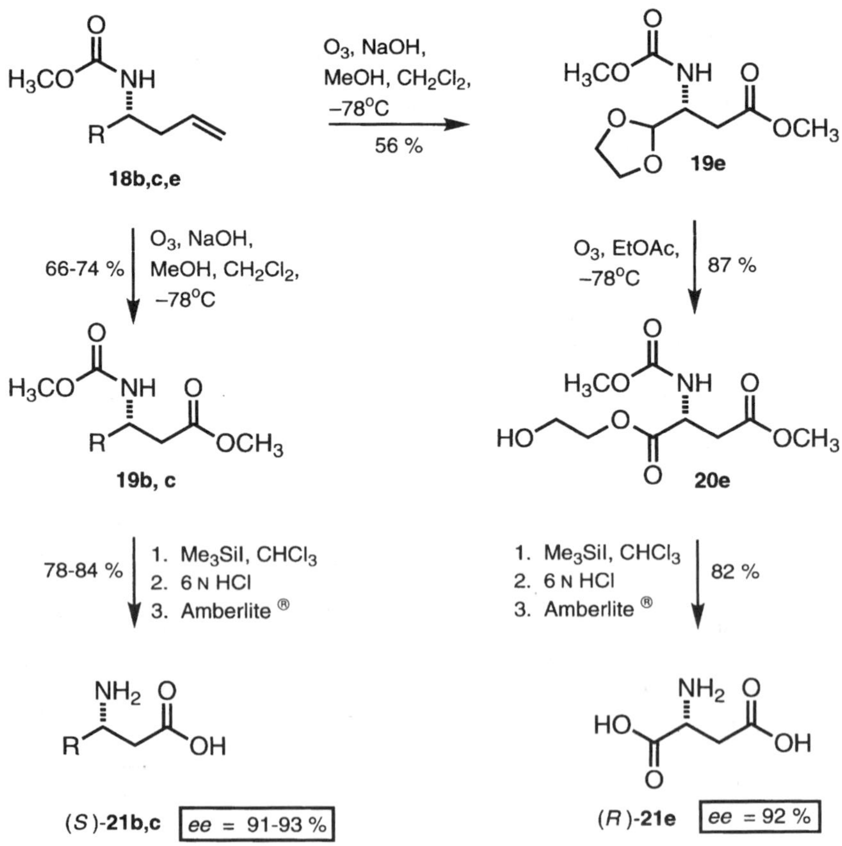

Figure 10.7. Functionalization of **18** to β-amino acids **21b,c** and β-amino diacids **21e**.

20e with ozone in ethyl acetate according to the procedure presented in Section 10.2.1. Removal of the N-(methoxy-carbonyl)amino protecting group using 1.5 or 3 equivalents of Me_3SiI[24] in $CHCl_3$ with subsequent hydrolysis of the ester group (aqueous 6*N* HCl) afforded the target (*S*)-(+)-β-amino acids **21b,c** in good yields (78%, 84%) and without racemization (ee = 91%, 93%). Upon deprotection, (*R*)-(–)-aspartic acid **21e** was obtained in 82% yield with only slight racemization (ee = 92%).

The ee values of the β-amino acids **(*S*)-21b,c** and aspartic acid **(*R*)-21e** were determined by GC analysis and ^1H- and ^{19}F-NMR spectroscopy of the MTPA-amides[20] of the methyl ester and of the dimethyl diester. The absolute configuration was proven by correlation of the optical rotation of the β-amino acids[26] **(*S*)-21b,c** and aspartic acid[31] **(*R*)-21e** with literature data.

10.3 Enantioselective Synthesis of β-Amino Acids Using TMS-SAMP as a Chiral Ammonia Equivalent for the Aza-Analogous Michael Addition[32]

As indicated in Section 10.1, a second pathway to generate β-amino acids via formation of a new C-N bond is the aza-analogous 1,4-addition of nitrogen nucleophiles to various Michael acceptors.[33–35] To achieve this, achiral N-nucleophiles can be added to chiral Michael acceptors.[33] Alternatively, chiral amine functionality donors, especially N-silylated amines,[35] have proven to be versatile soft nucleophiles for that purpose.[35]

10.3.1 Enantioselective Synthesis of β-Amino Acids by Nucleophilic Michael Addition of Lithiated TMS-SAMP to Enoates[32a]

Following the strategy of nucleophilic addition of nitrogen nucleophiles as chiral ammonia equivalents depicted in Figure 10.1 (**A** ⇒ **D**) we have developed a highly enantioselective method for the synthesis of β-amino acids by hetero-Michael addition of the new (*S*)-2-methoxymethyl-1-trimethylsilylaminopyrrolidine (TMS-SAMP)[36] or the enantiomeric TMS-RAMP to (*E*)-enoates.[26]

The new hydrazine auxiliary can be prepared from SAMP and RAMP,[14] respectively, via deprotonation with *n*-butyllithium in diethyl ether at 0°C with subsequent silylation with chlorotrimethylsilane. TMS-SAMP (TMS-RAMP) was obtained by distillation in 95% yield as a colorless liquid, which can be stored at –25°C under argon. The desired chiral nucleophile **G** can be formed via deprotonation of TMS-SAMP with *n*-butyllithium at –78°C in tetrahydrofuran (30 min) (Figure 10.8).

In the synthesis of β-amino acids (Figure 10.9),[32a] 1.3 equivalents of the N-nucleophile **G** (prepared as described above) were treated with α,β-unsaturated esters **22** at –78°C, affording the N-silylated β-hydrazinoesters **(*S,S*)-23** through Michael addition. Using simple methyl esters as Michael acceptors, 1,2-addition occurs as a side reaction leading to the generation of the corresponding α,β-unsaturated acid hydrazides in 10% yield. The same products could be prepared exclusively if the unsilylated SAMP is used. Employing the sterically more demanding *tert*-butyl esters ($R' = t$-Bu), the problem of 1,2-addition can be avoided. In addition, better yields are obtained (see **24b,i**; Table 10.5). In principle it is possible to obtain the N-silylated β-hydrazinoester by dis-

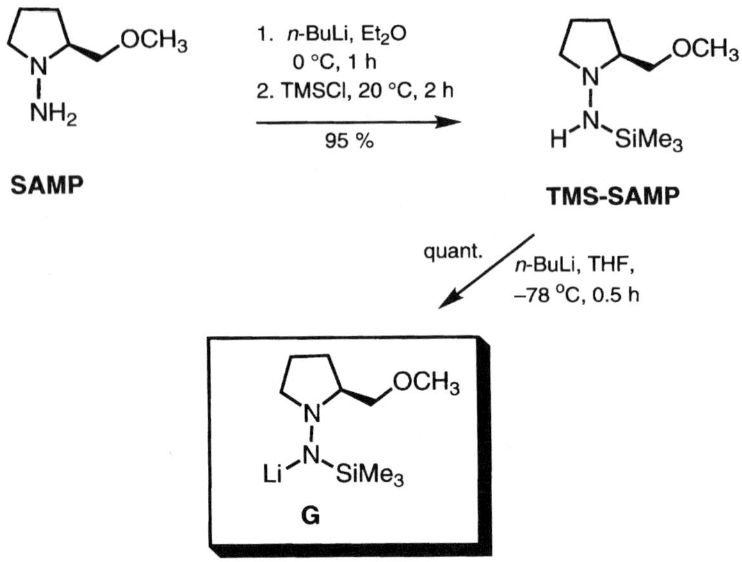

Figure 10.8. Synthesis of the hetero-Michael donor TMS-SAMP.

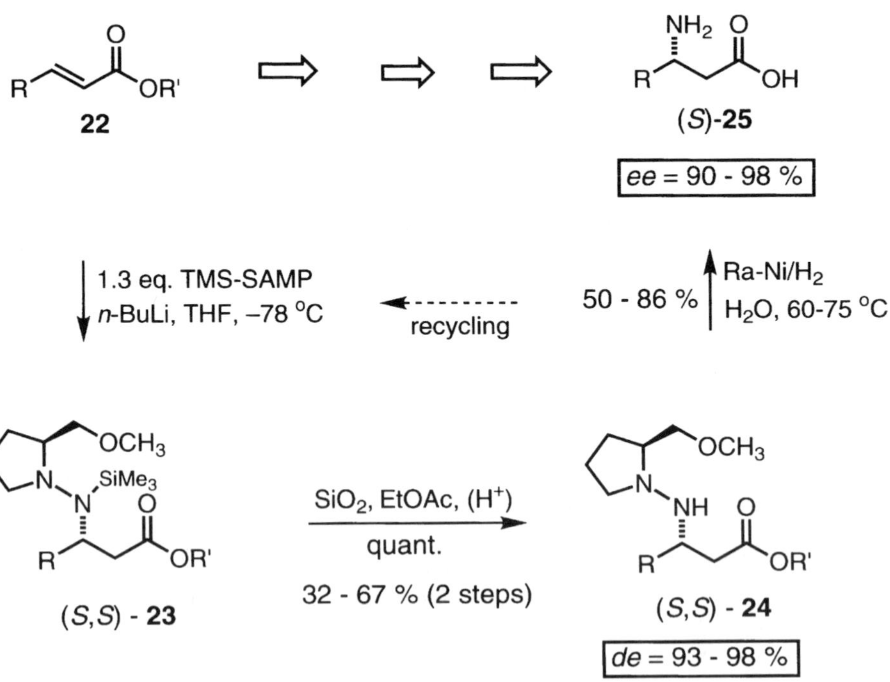

Figure 10.9. Enantioselective synthesis of β-amino acids **25** by aza analogous Michael addition of lithiated TMS-SAMP to enoates.

Table 10.5 PREPARATION OF β-HYDRAZINOESTER 24 VIA AZA ANALOGOUS MICHAEL ADDITION REACTION

Product	*R*	*R′*	Yield (%)[a]	de (%)[b]	$[\alpha]_D^{22}$ (*c*, $CHCl_3$)	Config.
24a	CH_3	CH_3	48	95	–97.6 (1.04)	(*S,S*)
24a[c]	CH_3	CH_3	47	95	+96.4 (0.87)	(*R,R*)
24b	C_2H_5	CH_3	37	95	–100.0 (1.02)	(*S,S*)
24b	C_2H_5	*t*-C_4H_9	49	98	–93.2 (1.09)	(*S,S*)
24c	*n*-C_3H_7	CH_3	40	95	–105.0 (1.00)	(*S,S*)
24d	*i*-C_3H_7	*t*-C_4H_9	32	98	–69.4 (1.08)	(*S,R*)
24e	*n*-C_4H_9	CH_3	39	≥94	–108.8 (0.98)	(*S,S*)
24f	*i*-C_4H_9	CH_3	35	≥93	–113.8 (1.02)	(*S,S*)
24g	*n*-C_5H_{11}	CH_3	39	≥94	–102.0 (1.05)	(*S,S*)
24h	*n*-C_7H_{15}	CH_3	36	95	–87.9 (1.78)	(*S,S*)
24i	*n*-$C_{11}H_{23}$	CH_3	35	95	–67.0 (1.10)	(*S,S*)
24i	*n*-$C_{11}H_{23}$	*t*-C_4H_9	67	97	–98.5 (1.21)	(*S,S*)
24j	C_6H_5	*t*-C_4H_9	68	97	–98.0 (1.20)	(*S,R*)

[a] Total yield over two steps.

[b] Determined by ^{1}H- and ^{13}C-NMR spectroscopy.

[c] TMS-RAMP was used as chiral auxiliary.

tillation, as demonstrated for **(*S,S*)-23a** [30% yield; de = 95%]. Nevertheless, it is recommended directly to desilylate the crude products by stirring with a suspension of silica gel in ethyl acetate (4 to 72 h depending on the *R* group) to prevent lowering of the yield. In order to reduce the desilylation time to 3 h, the procedure should be modified by the addition of a catalytic amount of hydrochloric acid.

After elution of the silica gel with methanol and subsequent chromatographic purification of the crude products, the β-hydrazinoesters **(*S,S*)-24** are obtained as pale yellow oils in 32–67% yield and with high diastereomeric excesses [de = 93–98% (determined by ^{1}H- and ^{13}C-NMR spectroscopy, Table 10.5)]. A conceivable diastereomeric enrichment by column chromatography can be ruled out due to the facile detection of both diastereoisomers by TLC (phosphomolybdic acid without heating). Furthermore, comparison of the de values of **(*S,S*)-23a,b** (determination by gas chromatography) and **(*S,S*)-24a,b** suggests that the acidic desilylation proceeds without epimerization.

Finally, the hydrazinoesters **24** were converted to the corresponding β-amino acids **(*S*)-25** by hydrogenolytic N-N bond cleavage with concomitant saponification of the ester function in an autoclave using freshly prepared Raney nickel (W9, 1 g Ni-Al alloy per mmol of the β-hydrazinoester **24** suspended in 15 mL H_2O, treated with solid NaOH).[17] These are obtained after chromatographic purification in 50–86% yield and with enantiomeric excesses of ee ≥ 90–98% (Table 10.6). Attempts to convert the β-hydrazinoester **24j** to the corresponding β-amino acid revealed partial reduction of the phenyl group. Therefore, pure **25j** could not be prepared using this method.

In order to determine the enantiomeric excesses by ^{1}H- and ^{19}F-NMR spectroscopy as well as GC analysis, the β-amino acids **25** were transformed to the corresponding methyl esters[37] and directly converted with (*S*)-(+)-3,3,3,-trifluoro-2-methoxy-2-phenyl-

Table 10.6 PREPARATION OF β-AMINO ACIDS 25 VIA N-N BOND CLEAVAGE OF THE HYDRAZINOESTERS 24

Product	*R*	Yield (%)	ee (%)[a]	$[\alpha]_D^{22}$ (c, H_2O)[b]	MP (°C)[b,c]	Config.
25a	CH_3	86	93	+36.4 (1.04)	209	(*S*)
25a[d]	CH_3	79	91	−35.7 (0.75)	205	(*R*)
25b	C_2H_5	82	91[e]	+39.5 (0.97)	180	(*S*)
25c	n-C_3H_7	85	92	+35.5 (1.01)	188	(*S*)
25d	i-C_3H_7	60	98	+40.3 (1.02)	182	(*R*)
25e	n-C_4H_9	80	≥90[f]	+33.3 (0.91)	202	(*S*)
25f	i-C_4H_9	85	≥90[f]	+31.2 (0.76)	207	(*S*)
25g	n-C_5H_{11}	76	≥90[f]	+31.1 (1.11)	201	(*S*)
25h	n-C_7H_{15}	50	94	+26.6 (0.93)	199	(*S*)
25i	n-$C_{11}H_{23}$	76	93[e]	+28.4 (0.90)	194	(*S*)

[a] The ee values of the β-amino acid methyl esters were determined as their corresponding Mosher amides prior to recrystallization.[32a]

[b] After one recrystallization from MeOH/EtOH [(S)-**25a**], MeOH/H_2O [(***R***)-**25a**], EtOH [(***S***)-**25b,c,f–i**, (***R***)-**25d**], MeOH/Et_2O [(***S***)-**25e**].

[c] With decomposition.

[d] TMS-RAMP was used as chiral auxiliary.

[e] Starting from the methyl esters (R' = Me).

[f] These results are based on the corresponding de values assuming a divergence of 2–4%.

propanoyl chloride into the corresponding Mosher amides.[20] Comparison of the determined de and ee values showed a slight divergence of 2–4%, which is in line with our experiences concerning N-N bond cleavage of similar hydrazines with Raney nickel.[16a]

The determination of the absolute configuration of the synthetic β-aminobutyric acid **(*S*)-25a** was carried out by comparison of the optical rotation with literature data (lit.: $[\alpha]_D^{22}$ = +38.8 (0.48, H_2O) for (*S*)-aminobutyric acid.[38] The β-amino acids **25b,c** and **25e–i** were also assigned the (*S*)-configuration on the consistent assumption that all substrates react along the same reaction pathway. The β-amino acid **25d** was assigned the (*R*)-configuration because of the formal inversion of the substituent priorities in accordance with the CIP rules. These assumptions could be corroborated by X-ray structure analysis carried out on the corresponding tandem Michael addition products (see Section 10.3.2).

The synthesis of the enantiomers **(*R*)-25**, as exemplified by the preparation of **(*R*)-25a**, is feasible in an analogous fashion by the use of TMS-RAMP as the N-Michael donor (Tables 10.5 and 10.6).[32a]

10.3.2 Diastereo- and Enantioselective Synthesis of α-Substituted β-Amino Acid Esters by Tandem Michael Addition/α-Alkylation[32b]

Following the retrosynthetic pathway via aza analogous Michael addition with respect to the high enantioselective synthesis of various β-amino acids, it is obvious to extend this approach towards α-substituted β-amino acids based on a tandem hetero Michael addition/α-esterenolate alkylation protocol.

As is depicted in Figure 10.10, the addition of lithiated TMS-SAMP as chiral N-nucleophile to enoates **22** followed the same protocol as in the synthesis of unsubstituted β-amino acids (Section 10.3.1). After 15 h the reaction mixture was treated with hexamethylphosphoric triamide (HMPA, 3 equiv.) and the appropriate halide R^2X was added. Along with the experience gathered concerning the synthesis of unsubstituted β-amino acids, the crude tandem products **26** were directly desilylated by stirring over a suspension of silica gel/ethyl acetate and a trace of concentrated hydrochloric acid. Correspondingly the α-substituted β-hydrazinoesters **27** were obtained in 26–68% overall yield after purification by chromatography (Figure 10.10, Table 10.7).

The diastereomeric excesses were determined by ^{13}C-NMR spectroscopy and ranged from de = 63 to 96% for aliphatic groups R^1, whereas with an aromatic R^1 the asymmetric inductions were virtually complete (de ≥ 96%). Because in simple 1,4-additions of lithiated TMS-SAMP to enoates the generation of the new stereogenic center through C-N bond formation revealed excellent inductions (ee ≥ 93%),[32a] the de values measured were those for the α-epimers of **27**. Assuming a uniform relative topicity for all Michael additions described here, the relative (*anti*) and absolute configurations [(*S,S*) or (*S,R*) in the case of R^1 = aryl] of the two new stereocenters are based on a X-ray structure analysis of **(*S,S,R*)-271** (Figure 10.11). This result is furthermore in agreement with an assignment of the configuration regarding the β-cen-

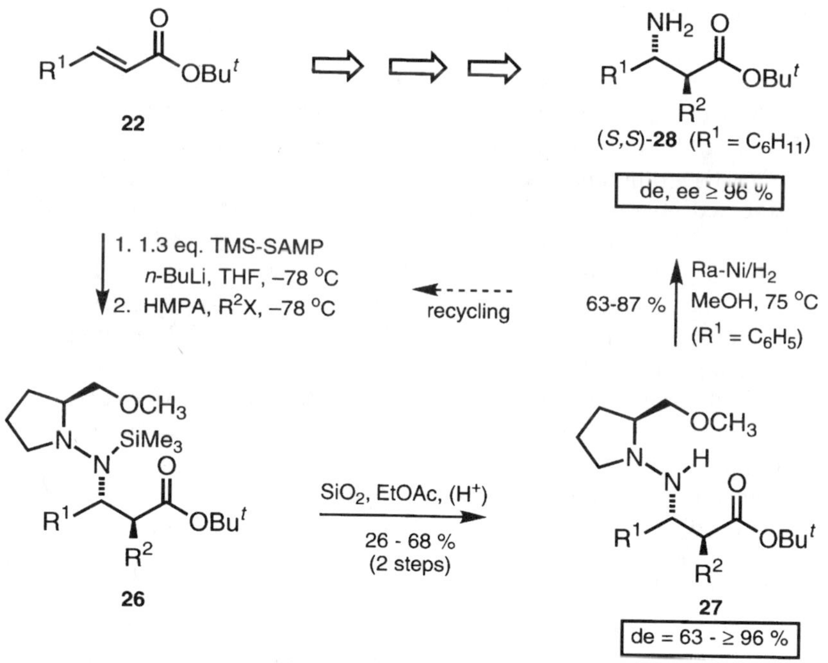

Figure 10.10. Diastereo- and enantioselective synthesis of α-substituted β-amino acid esters by aza analogous Michael addition/α-alkylation of lithiated TMS-SAMP to enoates.

Table 10.7 α-SUBSTITUTED β-HYDRAZINOESTERS 27

Product	R^1	R^2[a]	Yield (%)	$\alpha]_D^{RT}$ (*c*, $CHCl_3$)	de[b] (%)	Config.
27a	C_2H_5	CH_3	56	–52.0 (1.29)	77	(*S,S,S*)
27b	C_2H_5	C_2H_5	48	–60.9 (1.08)	66	(*S,S,S*)
27c	C_2H_5	*n*-C_3H_7	53	–78.4 (0.74)	68	(*S,S,S*)
27d	C_sH_5	*n*-C_4H_9	33	–69.9 (p.90)	65	(*S,S,S*)
27e	C_2H_5	$C_6H_5CH_2$	26	–71.2 (1.20)	76	(*S,S,S*)
27f	C_2H_5	$(CH_2)_3OH$[c]	51	–41.4 (0.80)	>96	(*S,S,S*)
27g	*n*-$C_{11}H_{23}$	*n*-C_3H_7	41	–68.0 (1.01)	63	(*S,S,S*)
27h	C_6H_5	C_2H_5	34	–35.2 (1.41)	>96	(*S,S,R*)
27i	C_6H_5	*n*-C_3H_7	68	–63.4 (0.95)	>96	(*S,S,R*)
27j	C_6H_5	*n*-C_4H_9	64	–58.8 (0.67)	>96	(*S,S,R*)
27k	C_6H_5	CH_2=$CHCH_2$	52	–48.3 (1.13)	>96	(*S,S,R*)
27k[d]	C_6H_5	CH_2=$CHCH_2$	8	–57.7 (1.16)	>96	(*S,R,S*)
27l	C_6H_5	$C_6H_5CH_2$	67	–64.7 (1.00	>96	(*S,S,R*)
27l[e]	C_6H_5	$C_6H_5CH_2$	48	+68.6 (0.90)	>96	(*R,R,S*)
27m	*p*-$H_3COC_6H_4$	CH_3	53	–64.8 (1.23)	>96	(*S,S,R*)

[a] Except for benzyl bromide and allyl bromide (*X* = Br), the corresponding iodides (*X* = I) were used as electrophiles.

[b] Determined by ^{13}C-NMR spectroscopy.

[c] Electrophile: $I(CH_2)_3OSi(CH_3)_3$.

[d] *syn*-Diastereomer via secondary deprotonation with LDA.

[e] TMS-RAMP was used instead of TMS-SAMP as nitrogen nucleophile.

ter by polarimetry.[32a] Needless to mention that the enantiomeric auxiliary TMS-RAMP leads to the optical antipodes, as was demonstrated in the case of **(*S,S,R*)-** and **(*R,R,S*)-27l** (Table 10.7). In order to avoid the toxic HMPA, N,N′-dimethyltetrahydro-2(1H)-pyrimidinone (DMPU) was tried as cosolvent, but this led to either lower yields or a decrease of the diastereoselectivity. As Y. Yamamoto et al.[35] demonstrated, the use of the enolate geometry control leads to the generation of either the *syn-* or *anti*-configurated β-amino acid esters, respectively. Following this strategy, we were able to demonstrate, although in lower chemical yield (8%), that after secondary deprotonation of the β-hydrazinoester **24j** (see Section 10.3.1) with LDA and sequent trapping with allyl bromide followed by desilylation the *syn*-β-hydrazinoester **27k** could be prepared with high diastereo- and enantiomeric excess. In conclusion, this example demonstrates that in principle both diastereomers of the β-hydrazinoester **27** are accessible.

Finally, reductive N-N bond cleavage of the hydrazino esters **27** applying modified conditions (Raney nickel in methanol at 80°C and a hydrogen pressure of 8.5 bar) afforded the α-substituted β-amino acid esters (*S,S*)-28h–j,l in good yields. In order to avoid the known epimerization of the stereogenic centers α to a carboxyl group, the freshly prepared Raney nickel was previously washed with aqueous formic acid (5%). By this way the prepared amino acid esters were formed with high diastereo- and enantiomeric excesses (de, ee ≥ 96%) (Table 10.8). Due to the reductive reaction conditions, the phenyl group R^1 was reduced to the cyclohexane moiety, although the benzyl group of **27l** was not reduced under these conditions.

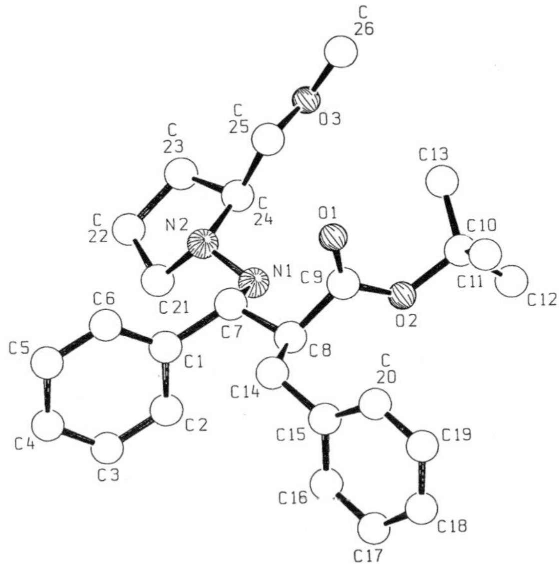

Figure 10.11. Molecular structure of one of two independent molecules **(*S*,S,*R*)-271** in the solid state (Schakal[46] plot).

10.3.3 Diastereo- and Enantioselective Synthesis of *trans*-2-Aminocycloalkanoic Acids[32c]

As described in the previous section, we have been able to develop a highly diastereo- and enantioselective synthesis of α-alkyl-β-amino acid esters of *anti*-configuration via alkylation of the intermediate ester enolates[32b] derived from the addition of the chiral nitrogen nucleophile **G** to enoates **22**. The logical extension of this method is the intramolecular variant, leading to diastereo- and enantiomeric pure *trans*-configured 2-aminocycloalkanoic acids of type **29** based on an intramolecular Michael initiated ring closure (MIRC reaction).[39] Amino acids of this type show an extraordinary variety of interesting properties, both as derivatives and in the free form.[40] Starting from ω-halo substituted α,β-unsaturated esters **31**, the intermediate ester enolates **30**

Table 10.8 α-SUBSTITUTED β-AMINO ACID ESTERS 28 PREPARED BY REDUCTIVE N-N BOND CLEAVAGE

Product	R^1	R^2	Yield (%)	$[\alpha]_D^{RT}$ (*c*, $CHCl_3$)	de[a] (%)	ee (%)	Config.
28h	C_6H_{11}	C_2H_5	87	+7.6 (0.87)	>96	>96	(*S*,*S*)
28i	C_6H_{11}	*n*-C_3H_7	63	+2.4 (0.88)	>96	>96	(*S*,*S*)
28j	C_6H_{11}	*n*-C_4H_9	83	−1.7 (0.91)	92[b]	>96	(*S*,*S*)
28l	C_6H_{11}	$C_6H_5CH_2$	63	+4.6 (0.99)	>96	>96	(*S*,*S*)

[a] Determined by ^{13}C-NMR spectroscopy.

[b] After chromatography (silica gel; ethanol).

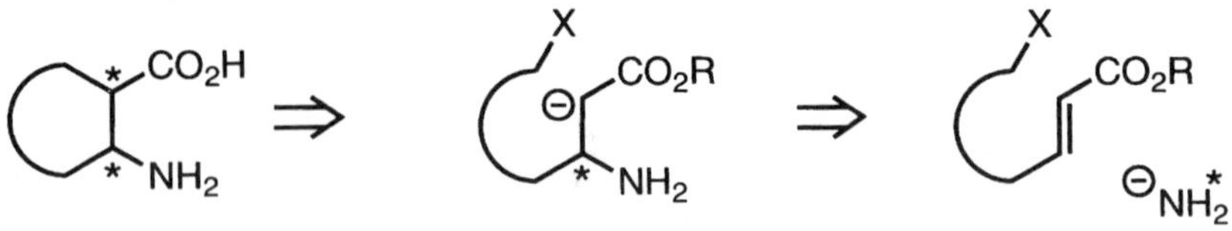

Figure 10.12. Retrosynthetic pathway to *trans*-2-aminocycloalkanoic acids via aza analogous MIRC-reaction.

were prepared via conjugate addition of the chiral ammonia equivalent TMS-SAMP with subsequent ring closure by intramolecular alkylation (Figure 10.12).[32c]

As shown in Figure 10.13, the enoate **32** was added dropwise to a solution of lithiated TMS-SAMP in THF and the reaction mixture was stirred at –78°C for 15 h ac-

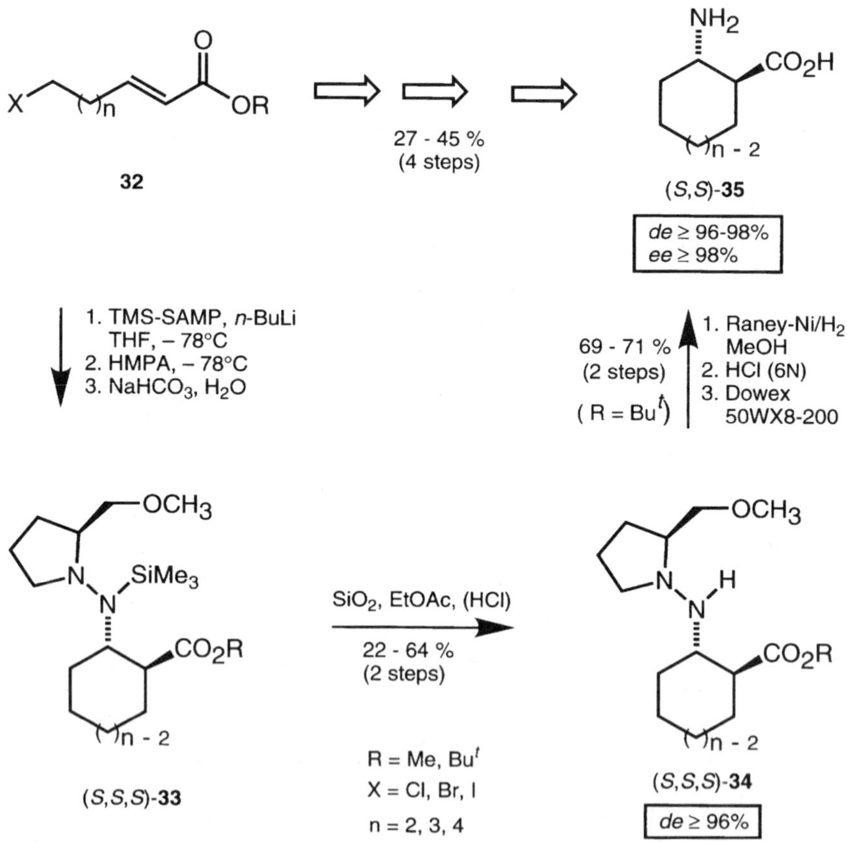

Figure 10.13. Diastereo- and enantioselective synthesis of *trans*-2-amino-cycloalkane-1-carboxylic acids **35** via intramolecular tandem Michael addition/α-alkylation (MIRC) reaction.

cording to the general reaction procedure (Section 10.3.1). Depending on the desired ring size, cyclization of the intermediate ester enolate must be promoted by addition of HMPA to activate the ester enolate towards intramolecular cyclization (Table 10.9). Whereas cyclization to the seven-membered ring system could only be achieved if the iodide was used and HMPA was added, the five- and six-membered rings were obtained using the bromides without the addition of cosolvent. Use of the iodides did not prove advantageous. In the case of the cyclopentane ring **34b** (n=2), an increase of yield occurred in the absence of cosolvent, although a slight decrease in yield of the six-membered ring system **34d** (n=3) must be accepted under these conditions. This observation is in contrast to the intermolecular alkylation described in Section 10.3.2, where all electrophiles except methyl iodide needed the HMPA activated ester enolates. The diastereomeric excess was unaffected by the cosolvent or a change of the solvent system to dimethoxyethane (DME). The crude trimethylsilyl protected hydrazines **33** were directly disilylated by stirring over a suspension of silica gel, ethyl acetate, and trace amounts of hydrochloric acid. The desired β-hydrazinoesters **34a–e** were prepared under the optimized conditions in 22–64% overall yield in two steps from **32** (Figure 10.13, Table 10.9).

No β-hydrazinoester **34c** was obtained via the initial route[32a] in the 1,4-addition of lithiated TMS-SAMP to methyl 1-cyclohexenyl carboxylate. In this example the corresponding hydrazide was formed via 1,2-addition of the nucleophile to the carboxyl group.

The diastereomeric excesses of the β-hydrazinoesters **34** were determined to be greater than 96% by ^{13}C-NMR spectroscopy. Reductive N-N bond cleavage of **34** by neutral Raney nickel in methanol at 70–75°C under a hydrogen pressure of 9 bar afforded the β-aminoesters, which were immediately hydrolyzed with hydrochloric acid. Ion-exchange chromatography of the resulting hydrochloride salts led to the cor-

Table 10.9 SYNTHESIS OF THE β-HYDRAZINOESTERS 34

Product	R	n	X	Solvent	Yield (%)	de[a] (%)	$[\alpha]_D^{RT}$ (c, $CHCl_3$)	Config.
34a	Me	2	I	THF/HMPA	22	≥96	−33.5 (0.87)	(*S,S,S*)
34b	*t*-Bu	2	Br	THF/HMPA	45	≥96	−75.9 (0.73)	(*S,S,S*)
34b	*t*-Bu	2	Br	THF	64	≥96		
34b	*t*-Bu	2	Cl	THF	4	≥96		
34c	Me	3	I	THF/HMPA	47	≥96	−40.4 (0.98)	(*S,S,S*)
34d	*t*-Bu	3	Br	THF/HMPA	54	≥96	−80.0 (0.95)	(*S,S,S*)
34d	*t*-Bu	3	Br	THF	37	≥96		
34d	*t*-Bu	3	Br	DME	14	≥96		
34e	*t*-Bu	4	I	THF/HMPA	38	≥96	−84.1 (1.15)	(*S,S,S*)
34f[b]	*t*-Bu	4	Br	THF/HMPA	20	≥96		
34g[b]	*t*-Bu	4	I	THF	16	≥96	−57.6 (1.00)	(*S,S,S*)

[a] Determined by ^{13}C-NMR spectroscopy.

[b] Under these reaction conditions no cyclization to the seven-membered ring system occurred, and the open-chain β-hydrazinoesters **34f** and **34g** were isolated.

responding analytically pure cyclic β-aminoacids **35** in 69–71% yield from **34** with excellent stereoselectivities (de ≥ 96–98%, ee ≥ 98%).

The de values of the β-amino acids **35** were determined by ^{1}H- and ^{13}C-NMR spectroscopy, whereas the enantiomeric excesses were obtained by GC analysis of the appropriate Mosher amides of the corresponding methyl esters.[37,20] The relative configuration of the two new stereocenters was determined to be *trans* by ^{1}H-NMR spectroscopy and NOE experiments, while the absolute configurations were established as (1*S*,2*S*) by comparison of the optical rotations with literature data.[41,42] These results are in total agreement with investigations showing that if the electrophile reacts with the initial Michael addition enolate, it is introduced *anti* to the nitrogen nucleophile TMS-SAMP, thus furnishing the (1*S*) configuration. Application of the antipode TMS-RAMP should analogously lead to the opposite enantiomer with the (1*R*,2*R*) configuration.[32]

10.3.4 Diastereo- and Enantioselective Synthesis of β-Hydrazino-β′-Hydroxy Esters[43]

A logical extension of our method for the synthesis of α-substituted β-amino acids is the use of aldehydes as electrophiles in a tandem Michael aldol reaction (Figure 10.14).[43] The products prepared by this method are highly interesting precursors for the important class of carbapenem antibiotics.[44] Following the retrosynthetic pathway **38** ⇒ **36** it should be possible to generate three new stereogenic centers in a "one-pot" synthesis.

In an analogous fashion to the tandem Michael additions described in Sections 10.3.2 and 10.3.3, TMS-SAMP was deprotonated with *n*-butyllithium at –78°C for 1 h (Figure 10.15). The enoate acceptor **22** was then added dropwise, and the reaction mixture was stirred for 18 h at –78°C. In order to achieve the highly diastereoselective addition to the aldehyde, two variants were tested.

In method A the intermediate ester enolate was activated by the addition of HMPA. The reaction mixture was stirred at –30°C for 3 h and then cooled to –78°C. The aldehyde was added dropwise to the reaction mixture, which was then stirred at this temperature for 18 h. Upon desilylation of the crude products **39** with silica gel, ethyl

Table 10.10 SYNTHESIS OF THE β-AMINO ACIDS 35

Product	*R*	*n*	Yield (%)	de (%)	ee[a] (%)	$[\alpha]_D^{RT}$ (*c*, H_2O)	Config.
35b	*t*-Bu	2	70	97[b]	≥98	+81.2 (0.50)[c]	(*S*,*S*)
35d	*t*-Bu	3	69	≥98[b]	≥98	+67.4 (0.46)[d]	(*S*,*S*)
35e	*t*-Bu	4	71	96[e]	≥98	+29.9 (0.52)	(*S*,*S*)

[a] Determined by GC analysis of the appropriate Mosher amides of the methyl esters.[13]

[b] Determined by ^{1}H-NMR spectroscopy.

[c] Lit.[41]; $[\alpha]_D^{20}$ +51.8 (*c* = 1.0, H_2O) (ee = 80%).

[d] Lit.[42]: $[\alpha]_D^{17}$ + 66.5 (*c* = 2.0, H_2O).

[e] Determined by ^{13}C-NMR spectroscopy.

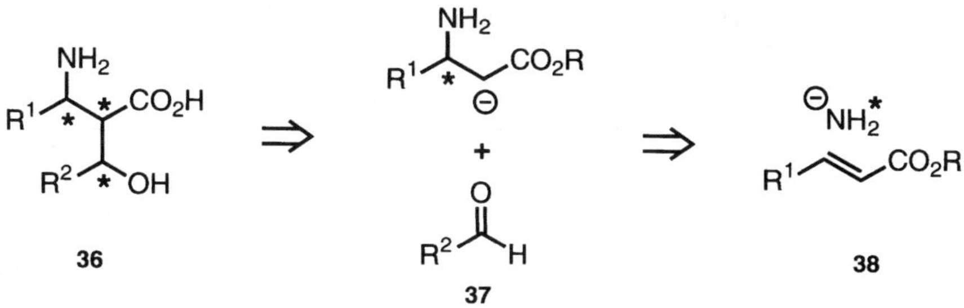

Figure 10.14. Retrosynthetic analysis of β-amino-β′-hydroxy acids (aza analogous tandem-Michael-aldol-reaction).

acetate, and trace amounts of hydrochloric acid, the products **40** were obtained after chromatography as white-yellow solids in 19–67% yield from **22**. The diastereomeric excesses ranged from 20 to greater than 96% (Table 10.11).

Chiral titanium complexes have proven to be excellent reagents for the highly enantioselective addition of nucleophiles to carbonyl compounds.[45] This is explained

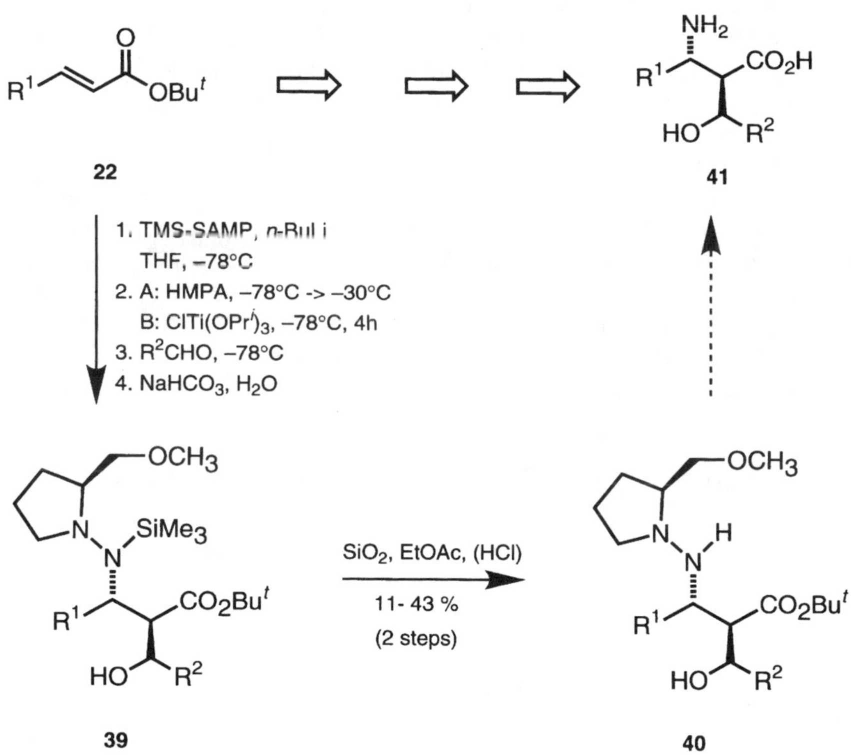

Figure 10.15. Diastereo- and enantioselective synthesis of β-hydrazino-β′-hydroxy esters **40** via aza analogous tandem Michael aldol addition.

Table 10.11 SYNTHESIS OF 40 VIA METHOD A

Product	R^1	R^2	Yield (%)	de[a,b] (%)	Config.
40a	C_6H_5	CH_3	28	20 (≥ 96)	(*S,R,S,R*)
40b	C_6H_5	$(CH_3)_2CH$	63	≥ 96	(*R,R,S,S*)
40c	C_6H_5	$H_3CCH{=}CH$	23	≥96	(*R,R,S,S*)
40d	C_6H_5	FC_6H_4	53	90 (≥ 96)	(*S,S,R,R*)
40e	C_6H_5	$H_3COC_6H_4$	19	92 (≥ 96)	(*S,S,R,R*)
40e[c]	C_6H_5	COC_6H_4	67	87 (≥ 96)	(*R,R,S,S*)
40f	C_6H_5	C_6H_5	46	84 (≥ 96)	(*S,S,R,R*)
40g	$H_3COC_6H_4$	FC_6H_4	35	61 (≥ 96)	(*S,S,R,R*)

[a] The diastereomeric excesses were determined by ^{13}C-NMR-spectroscopy.
[b] In parentheses is the de value after chromatography.
[c] Lithiated TMS-RAMP as nitrogen nucleophile leading to the opposite enantiomer.

by a precomplexation of the titanium with both the carbonyl compound and the nucleophile, leading to a greater diastereofacial differentiation. For this reason, in a second synthetic strategy, the intermediate ester enolate was not activated by the addition of HMPA but treated with chlorotriisopropoxytitanium chloride at –78°C for 4 h to generate the chiral titanium enolate complex. After addition of the aldehyde and stirring for 18 h, the silyl protected β-hydrazino-β′-hydroxy-esters **39** were directly desilylated using the common procedure. The desired products **40** were obtained in 11–43% yield from **22**.

By comparison with method A a distinct increase of the diastereoselectivities could be achieved where, after enrichement via column chromatography, all diastereoselectivities were greater than 96%. One explanation would be an intermolecular chelation of the nucleophile to one diastereotypic face of the carbonyl group. The diastereomeric excesses of the β-hydrazino-β′-hydroxy esters **40** were determined by ^{13}C-NMR spectroscopy. Out of eight possible stereoisomers only two were observed, demonstrating the high diastereoselectivity of this reaction. By comparison with the results obtained in the tandem Michael addition (Section 10.3.2) and the high stere-

Table 10.12 SYNTHESIS OF 40 USING THE TRANSMETALIATION PROTOCOL (METHOD B)

Product	R^1	R^2	Yield (%)	de[a,b] (%)	Config.
40c	C_6H_5	$H_3CCH{=}CH$	23	≥ 96	(*R,R,S,S*)
40e	C_6H_5	COC_6H_4	24	≥ 96	(*S,S,R,R*)
40f	C_6H_5	C_6H_5	43	≥96	(*S,S,R,R*)
40h	$H_3COC_6H_4$	C_6H_5	31	94 (≥ 96)	(*S,S,R,R*)
40i	$H_3CCH{=}CH$	C_6H_5	11	68 (≥ 96)	(*S,S,R,R*)

[a] The diastereomeric excesses were determined by ^{13}C-NMR-spectroscopy.
[b] In parentheses after enrichment via chromatography.

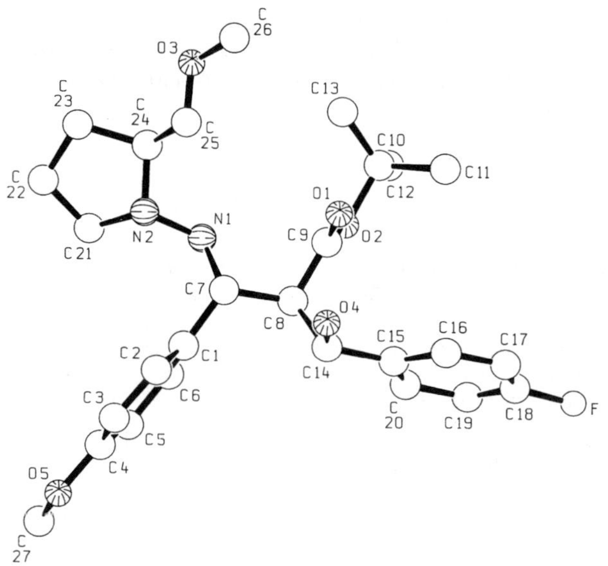

Figure 10.16. Molecular structure of **(*S*,*S*,*R*)-40g** in the solid state (Schakal[46] plot).[47].

oselectivity of these reactions, the observed diastereomers are presumably epimers of the β′-center. The absolute configuration of the products was determined by ^{1}H-NMR spectroscopy and confirmed by X-ray analysis carried out on compound **40g** (Figure 10.16, Table 10.12).

In both variants the absolute configuration of the products was determined as (1′*S*,2*R*,2′′*S*,3*R*) in the cases of additions of aromatic aldehydes, meaning that the nucleophile attacks the carbonyl compound from the *re*-face (entries **d–i**). In the case of 1,2-additions of the nucleophiles to aliphatic aldehydes, the stereochemical progress of the reaction is different. The absolute configuration was determined to be (1′*S*,2*R*,2′′*S*,3*R*), too, showing that the nucleophile (in entries **a–c**) approaches from the *si*-face (inversion of priorities according to the CIP rules). The cieavage of the N-N bond to remove the auxiliary leading to the free β-amino-β′-hydroxyesters **41** is still under investigation within our laboratories.

10.4 Conclusion

In summary, the addition of organometallic reagents to the C-N double bond of SAMP/RAMP hydrazones has proven to be a versatile method in the synthesis of both novel homoallylic amines and β-amino acids. By variation of the metal counterion it is possible to generate a great variety of protected amino acetals of high diastereo- and enantiomeric purity. The extension of this method to the synthesis of β-amino acids once again demonstrates the utility of the SAMP/RAMP hydrazone method in the asymmetric synthesis of interesting and widely applicable compounds.

Alternatively, we have demonstrated the great potential of our new versatile auxiliary TMS-SAMP (TMS-RAMP) in the synthesis of various previously unknown β-amino acids in high optical purity via the aza-analogou8s Michael addition and related tandem reactions.

10.5 References

1. Review: C. N. C. Drey in *Chemistry and Biochemistry of Amino Acids*, G. C. Barret, Ed. (Chapman and Hall, London, New York, 1985), p. 25. Some leading examples: (a) W. H. Pearson, J. V. Hines, *J. Org. Chem.* **54**, 4235 (1989); (b) S.-K. Kang, D.-H. Lee, *Synlett* 175 (1991); (c) M. E. Bunnage, S. G. Davies, C. J. Goodwin, *Synlett* 731 (1993); (d) K. Hattori, H. Yamamoto, *Tetrahedron* **50**, 2785 (1994); (e) M. E. Bunnage, A. J. Burke, S. G. Davies, C. J. Goodwin, *Tetrahedron: Asymmetry* **5**, 203 (1994); (f) M. E. Bunnage, S. G. Davies, C. J. Goodwin, O. Ichihara, *Tetrahedron* **50**, 3975 (1994).

2. Reviews: (a) D. J. Hart, D.-C. Ha, *Chem. Rev.* **89**, 1447 (1989); (b) F. H. van der Steen, G. van Koten, *Tetrahedron* **47**, 7503 (1991). Some leading examples: (a) S. G. Davies, I. Dordor-Hedgecock, K. H. Sutton, J. C. Walker, R. H. Jones, K. Prout, *Tetrahedron* **42**, 5123 (1986); (b) P. A. Jacobi, W. Zheng, *Tetrahedron Lett.* **34**, 2581, 2585 (1993); (c) N. Asao, N. Tsukada, Y. Yamamoto, *J. Chem. Soc., Chem. Commun.* 1660 (1993).

3. Reviews: (a) D. C. Cole, *Tetrahedron* **50**, 9517 (1994); (b) E. Juaristi, D. Quintana, J. Escalante, *Aldrichchimica Acta* **27**, 3 (1994).

4. (a) K. Harada in *The Chemistry of the Nitrogen Double Bond*, S. Patai, Ed. (Wiley-Interscience, Chichester, 1989), pp. 567–677; (b) R. A. Volkmann, in *Comprehensive Organic Synthesis*, B. M. Trost, I. Fleming, Eds. (Pergamon Press, Oxford, 1990), Vol. 1, p. 355; (c) E. Kleinmann, R. A. Volkmann, in *Comprehensive Organic Synthesis*, B. M. Trost, I. Fleming, Eds. (Pergamon Press, New York, 1990), Vol. 2, p. 975.

5. For some leading references see: (a) C. Gennari, I. Venturini, G. Gislon, G. Schimperna, *Tetrahedron Lett.* **28**, 227 (1987); (b) H. Kunz, D. Schanzenbach, *Angew. Chem.* **101**, 1042 (1989); *Angew. Chem. Int. Ed. Engl.* **28**, 1068 (1989); (c) E. J. Corey, C. P. Decicco, R. C. Newbold, *Tetrahedron Lett.* **32**, 5287 (1991).

6. (a) D. Enders; H. Schubert, *Angew. Chem.* **96**, 368 (1984); *Angew. Chem. Int. Ed. Engl.* **23**, 365 (1984); (b) D. Enders, H. Schubert, C. Nübling, *Angew. Chem.* **98**, 1118 (1986); *Angew. Chem. Int. Ed. Engl.* **25**, 1109 (1986); (c) D. Enders, D. Bartzen, *Liebigs Ann. Chem.* 569 (1991); (d) D. Enders, J. Tiebes, *Liebigs Ann. Chem.* 173 (1993).

7. (a) D. Enders, R. Funk, M. Klatt, G. Raabe, E. R. Hovestreydt, *Angew. Chem.* **105**, 418 (1993); *Angew. Chem. Int. Ed. Engl.* **32**, 418 (1993); (b) S. E. Denmark, O. Nicaise, *Synlett* 359 (1993).

8. D. Enders, M. Klatt, R. Funk, *Synlett* 226 (1993).

9. D. H. Hua, S. W. Miao, J. S. Chen, S. Iguchi, *J. Org. Chem.* **56**, 4 (1991).

10. (a) S. Laschat, H. Kunz, *Synlett* 51 (1990); (b) S. Laschat, H. Kunz, *J. Org. Chem.* **56**, 5883 (1991).

11. (a) Y. Yamamoto, *Acc. Chem. Res.* **20**, 243 (1987) and references cited therein; (b) M.-J. Wu, L. P. Pridgen, *Synlett* 636 (1990); (c) H. Tanaka, K. Inoue, U. Pokorski, M. Taniguchi, S. Torii, *Tetrahedron Lett.* **31**, 3023 (1990); (d) Z. Y. Chang, R. M. Coates, *J. Org. Chem.* **55**, 3475 (1990); (e) J. Ukaji, K. Kume, T. Watai, T. Fujisawa, *Chem. Lett.* 173 (1991); (f) P. G. Wuts, Y.-W. Jung, *J. Org. Chem.* **56**, 365 (1991); (g) A. Bocum, C. Boga, D. Savoia, A. Umani-Ronchi, *Tetrahedron Lett.* **32**, 1367 (1991); (h) G. Cainelli, D. Giacomini, E. Mezzina, M. Panunzio, P. Zarantonello, *Tetrahedron Lett.* **32**, 2967 (1991); (i) M. K. Mokhallalati, M.-J. Wu, L. N. Pridgen, *Tetrahedron Lett.* **34**, 47 (1993); (j) S.-I. Murahashi, J. Sun, T. Tsuda, *Tetrahedron Lett.* **34**, 2645 (1993); (k) M. Giammaruco, M. Taddei, P.

Ulivi, *Tetrahedron Lett.* **34**, 3635 (1993); (l) A. Bocoum, D. Savoia, A. Umani-Ronchi, *J. Chem. Soc., Chem. Commun.* 1542 (1993).

12. D. Enders, J. Schankat, M. Klatt, *Synlett* 795 (1994).

13. P. Herdewijn, P. J. Claes, H. Vanderhaeghe, *Nouv. J. Chim.* **7**, 691 (1987).

14. (a) D. Enders, in *Current Trends in Organic Synthesis*, H. Nozaki, Ed. (Pergamon Press, Oxford, 1983), p. 151; (b) D. Enders, in *Selectivity—A Goal for Synthetic Efficiency*, W. Bartmann and B. M. Trost, Eds. (Verlag Chemie, Weinheim, 1984), p. 65.

15. (a) T. Imamoto, T. Kusumoto, Y. Tawarayama, Y. Sugiura, T. Mita, Y. Hatanaka, M. Yokoyama, *J. Org. Chem.* **49**, 3904 (1984); (b) T. Imamoti, N. Takiyama, N. Nakamura, *Tetrahedron Lett.* **26**, 4763 (1985); (c) T. Imamoti, in *Comprehensive Organic Synthesis*, B. M. Trost, I. Fleming, Eds. (Pergamon Press, Oxford, 1990), Vol. 1, p. 231.

16. (a) C. Nübling, Dissertation, Technical University of Aachen (1987); (b) S. E. Denmark, O. Nicaise, J. P. Edwards, *J. Org. Chem.* **55**, 6219 (1990).

17. *Organikum*, 16th ed. (VEB Deutsche Verlag der Wissenschaft, Berlin, 1986).

18. A. Alexakis, N. Lensen, P. Mangeney, *Synlett* 625 (1991).

19. J. Einhorn, C. Einhorn, J. L. Luche, *Synlett* 37 (1991).

20. J. A. Dale, H. S. Mosher, *J. Am. Chem. Soc.* **95**, 512 (1973).

21. M. J. Burk, J. E. Feaster, *J. Am. Chem. Soc.* **114**, 6299 (1992).

22. M. Klatt, Dissertation, Technical University of Aachen (1994).

23. P. Deslongchamps, P. Atlani, D. Frehel, A. Malaval, C. Moreau, *Can. J. Chem.* **52**, 3651 (1974).

24. M. E. Jung, M. A. Lyster, *J. Chem. Soc. Chem. Commun.* 315 (1978).

25. K. Balenovic, D. Cerar, Z. Fuks, *J. Chem. Soc.* 3316 (1952).

26. H. Wahl, Dissertation, Technical University of Aachen (1990).

27. (a) F. Lucchesini, *Tetrahedron* **18**, 9951 (1992); (b) S. L. Schreiber, R. E. Claus, *Tetrahedron Lett.* **23**, 3867 (1982).

28. A. Alexakis, N. Lensen, J.-P. Tranchier, P. Mangeney, *J. Org. Chem.* **57**, 4563 (1992).

29. J. Schankat, Dissertation, Technical University of Aachen (1994).

30. J. A. Marshall, A. W. Garofalo, *J. Org. Chem.* **58**, 3675 (1993).

31. C. G. Baker, H. A. Sober, *J. Am. Chem. Soc.* **75**, 4058 (1953).

32. (a) D. Enders, H. Wahl, W. Bettray, *Angew. Chem.* **107**, 527 (1995); *Angew. Chem. Int. Ed. Engl.* **34**, 455 (1995) and references cited therein; (b) D. Enders, W. Bettray, G. Raabe, J. Runsink, *Synthesis* 1322 (1994) and references cited therein; (c) D. Enders, J. Wiedemann, W. Bettray, *Synlett* 369 (1995) and references cited therein.

33. (a) S. G. Davies, J. Dupont, R. J. C. Easton, *Tetrahedron: Asymmetry* **1**, 279 (1990); (b) D. Potin, F. Dumas, J. d'Angelo, *J. Am. Chem. Soc.* **112**, 3483 (1990); (c) R. Amoroso, G. Cardillo, P. Sabatino, C. Tomasini, A. Trerè, *J. Org. Chem.* **58**, 5615 (1993); (d) S. Matsubara, M. Yoshioka, K. Utimoto, *Chem. Lett.* 827 (1994); (e) G. Cardillo, A. de Simone, L. Gentilucci, P. Sabatino, C. Tomasini, *Tetrahedron Lett.* **35**, 5051 (1994); (f) V. A. Soloshonok, A. G. Kirilenko, N. A. Fokina, V. P. Kukhar, S. V. Galushko, V. K. Svedas, G. Resnati, *Tetrahedron: Asymmetry* **5**, 1119, 1125 (1994).

34. (a) H. Estermann, D. Seebach, *Helv. Chim. Acta* **71**, 1824 (1988); (b) K. Rudolf, J. M. Hawkins, R. J. Loncharich, K. N. Houk, *J. Org. Chem.* **53**, 3879 (1988); (c) L. Wessjohann, G. McGaffin, A. de Miejere, *Synthesis* 359 (1989); (d) S. G. Davies, O. Ichihara, *Tetrahedron: Asymmetry* **2**, 183 (1991);

(e) J. M. Hawkins, T. A. Lewis, *J. Org. Chem.* **57**, 2114 (1992); (f) E. Juaristi, J. Escalante, B. Lamatsch, D. Seebach, *J. Org. Chem.* **57**, 2396 (1992); (g) M. Es-Sayed, C. Gratkowski, N. Krass, A. I. Meyers, A. De Meijere, *Tetrahedron Lett.* **34**, 289 (1993); (h) S. G. Davies, O. Ichihara, I. A. S. Walters, *Synlett* 461 (1993); (i) S. G. Davies, N. M. Garrido, O. Ichihara, I. A. S. Walters, *J. Chem. Soc., Chem. Commun.* 1153 (1993); (j) J. G. Rico, R. J. Lindmark, T. E. Rogers, P. R. Bovy, *J. Org. Chem.* **58**, 7948 (1993); (k) S. G. Davies, O. Ichihara, I. Lenoir, I. A. S. Walters, *J. Chem. Soc., Perkin Trans. 1*, 1411 (1994); (l) M. E. Bunnage, S. G. Davies, C. J. Goodwin, I. A. S. Walters, *Tetrahedron: Asymmetry* **5**, 35 (1994); (m) S. G. Davies, O. Ichihara, I. A. S. Walters, *Synlett* 117 (1994); (n) S. G. Davies, O. Ichihara, I. A. S. Walters, *J. Chem. Soc., Perkin Trans. 1*, 1141 (1994).

35. (a) N. Asao, T. Uyehara, Y. Yamamoto, *Tetrahedron* **44**, 4173 (1988); (b) T. Uyehara, N. Asao, Y. Yamamoto, *J. Chem. Soc., Chem. Commun.* 753 (1989); (c) N. Asao, T. Uyehara, Y. Yamamoto, *Tetrahedron* **46**, 4563 (1990); (d) Y. Yamamoto, N. Asao, T. Uyehara, *J. Am. Chem. Soc.* **114**, 5427 (1992); (e) T. Uyehara, N. Shida, Y. Yamamoto, *J. Org. Chem.* **57**, 3139, 5049 (1992); (f) I. Suzuki, H. Kin, Y. Yamamoto, *J. Am. Chem. Soc.* **115**, 10139 (1993); (g) N. Shida; C. Kabuto, T. Niwa, T. Ebata, Y. Yamamoto, *J. Org. Chem.* **59**, 4068 (1994).

36. (a) A. S. Amaya, Dissertation, University of Bonn (1985); (b) D. Enders, *Acros Organics Acta* **1**, 37 (1995).

37. L-F. Tietze, T. Eicher, *Reaktionen und Synthesen im Organisch-Chemischen Praktikum* (G. Thieme-Verlag, Stuttgart, 1981), p. 115.

38. K. Balenovic, C. Cesar, Z. Fuks, *J. Chem. Soc.* 3316 (1952).

39. (a) R. D. Little, J. R. Dawson, *Tetrahedron Lett.* **21**, 2609 (1980); (b) T. Uyehara, N. Shida, Y. Yamamoto, *J. Org. Chem.* **57**, 3139 (1992).

40. Examples: (a) G. Bernáth, L. Gera, G. Göndös, K. Kovács, I. Jánvary, L. Sebestyén, Z. Ecsery, J. Hermann, *Ger. Offen.* **2**, 624, 290; *Chem. Abstr.* **87**, 102009 (1977); (b) Z. Huang, A. Probstl, J. R. Spencer, T. Yamazaki, M. Goodman, *Int. J. Pept. Protein Res.* **42**, 352 (1993); (c) G. Gondoes, L. Gera, G. Wittmann, L. Balaspiri, K. Kovacs, *Act Chim. Hung.* **124**, 187 (1987); (d) R. Hitzemann, H. H . Loh, *Brain Res.* **144**, 63 (1978); (e) J. Xie, J. M. Soleihac, N. Renwart, J. Peyroux, B. P. Roques, M. C. Fournie-Zaluski, *Int. J. Pept. Protein Res.* **34**, 246 (1989).

41. T. Yamazaki, Y-F. Zhu, A. Probstl, R. K. Chadha, M. Goodman, *J. Org. Chem.* **56**, 6644 (1991).

42. H. Nohira, E. Ehara, A. Miyashita, *Bull. Chem. Soc. Jpn.* **43**, 2230 (1970).

43. W. Bettray, Dissertation, Technical University of Aachen (1993).

44. (a) C. Palomo, in *Recent Progress in the Chemical Synthesis of Antibiotics*, G. Lukacs, M. Ohno, Eds. (Springer Verlag, Berlin, 1990), p. 565; (b) U. Gräfe, *Biochemie der Antibiotika* (Spektrum Akademischer Verlag, Heidelberg (1992).

45. R. O. Duthaler, A. Hafner, *Chem. Rev.* **89**, 459 (1989).

46. E. Keller, *Chem. Unserer Zeit* **20**, 178 (1986).

47. D. Enders, W. Bettray, G. Raabe, unpublished results

CHAPTER

11

Enantioselective Synthesis of β-Amino Acids via 6-Substituted Perhydropyrimidin-4-Ones

Giuliana Cardillo and Claudia Tomasini

Perhydropyrimidin-4-ones are interesting heterocyclic compounds that represent protected forms of β-amino acids and are chiral precursors for the asymmetric synthesis of α-substituted β-amino acids. The presence of the cyclic amidal moiety allows the smooth acid hydrolysis of these compounds, affording directly the corresponding β-amino acids.

The preparation of various β-amino acids[1] is an interesting topic that now has drawn a great deal of attention, because unusual amino acids, such as α-substituted α-amino acids, β-amino acids, and γ-amino acids, allow one to create novel types of peptides that are catalysts or carriers of biologically active residues with pharmacologically interesting properties.[2] To design new peptide ligands, conformational constraints have been introduced, and this approach has provided an important rationalization for peptide ligand development. The incorporation of unusual amino acids[3] results in conformational restriction and increased rigidity, leading to enhanced resistance towards protease enzymes and to the favoring of a particular secondary structure.

With the aim of obtaining α-substituted β-amino acids in enantiomerically pure form, the functionalization at C_5 of the perhydropyrimidin-4-ones through of the corresponding enolates has been designed (Scheme 11.1). The presence in the heterocyclic ring of one or more stereocenters favors the high diastereoselection.

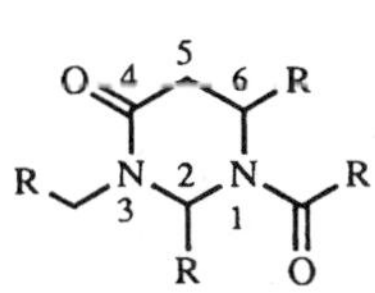

Scheme 11.1

For the preparation of 5,6-disubstituted perhydropyrimidin-4-ones in enantiomerically pure form, it is crucial to start from (*R*)- or (*S*)-6-substituted heterocyclic compounds. Furthermore, the synthesis of perhydropyrimidin-4-ones containing more stereocenters in the ring has been developed. Complete stereoselectivity was observed when a bulky substituent is present at C_2. Moreover, when the (*S*)-phenylethyl moiety was introduced in the heterocycle at $C_{1'}$, interesting results have been obtained both for the resolution of 6-substituted perhydropyrimidin-4-ones and for the transformation of 6-substituted to 5,6-disubstituted heterocycles, so that a new class of useful heterocyclic compounds containing stereocenters at C_6 and at $C_{1'}$ has been developed.

In this chapter the preparation and transformation of a variety of enantiomerically pure 6-alkyl perhydropyrimidin-4-ones will be taken into account. The (*S*)-phenylethyl moiety is always the substituent at N_3, and N_1 is always protected by an easily removable group. Then various functionalizations at C_5 will be described, with the subsequent transformation of the adducts into the corresponding α-substituted β-amino acids (Scheme 11.2).

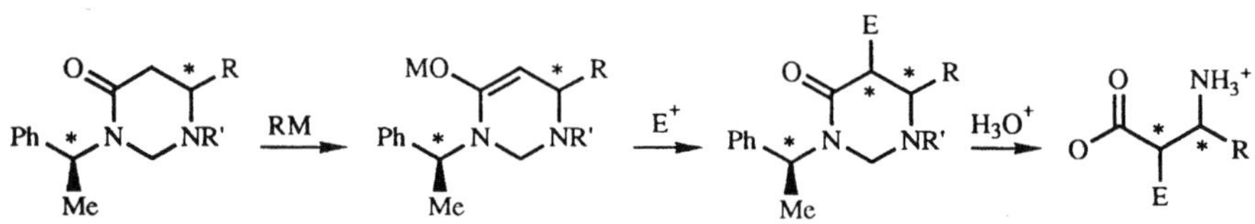

Scheme 11.2

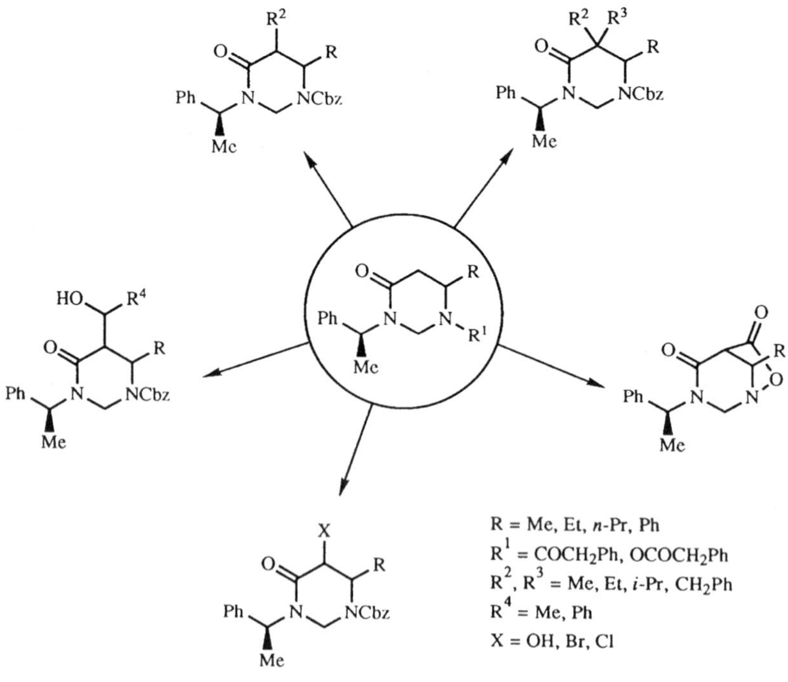

Scheme 11.3

Furthermore, the results obtained by alkylation and hydroxylation of perhydropyrimidin-4-ones with a single chiral center at C_6 will be reported.

In Scheme 11.3 a scheme of the functionalizations at C_5 of (*S*)-3-(1′-phenylethyl)-6-substituted perhydropyrimidin-4-ones is shown. These heterocyclic adducts are the starting materials for the preparation of new complex β-amino acids.

11.1 Synthesis and Separation of 6-Substituted Perphydropyrimidin-4-Ones

11.1.1 Mercury-Mediated Cyclization of α,β-Unsaturated Chiral Amidals

The cyclofunctionalization reaction of double bonds promoted by an electrophile is a valuable tool to create enantiomerically pure molecules containing heteroatoms on stereogenic centers.

Recently many synthetic procedures describing the electrophile-promoted cyclofunctionalization of unsaturated substrates with an internal nucleophile has been reported.[4] In this field new methods have been developed for the synthesis of polyfunctionalized compounds through asymmetric induction from a stereogenic center in the chain to the prostereogenic double bond with the use of suitable derivatives.

The use of cyclic systems to achieve asymmetric induction in acyclic compounds has become a standard useful strategy. In this field methods have been developed for the synthesis of polyfunctionalized compounds through the halocyclofunctionalization of double bonds, as reported in Scheme 11.4. The addition of an electrophile on a suitably prepared derivative promotes a cyclization process that results in the synthesis of an heterocyclic compound under high regio- and stereocontrol.

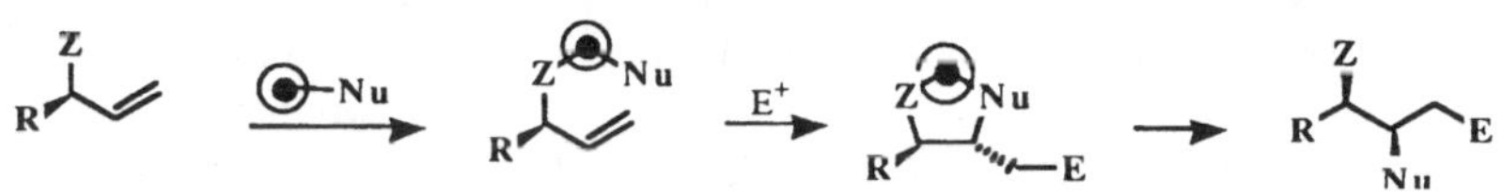

Scheme 11.4

The heterocycle may be considered a temporary protective group whose hydrolysis affords a polyfunctionalized compound with a transfer of chirality from the preexisting stereocenter to the newly formed one.

The synthetic usefulness of electrophile cyclofunctionalization of double bonds is due to the high degree of stereoselectivity obtained when substituents are present in the tether. In fact, substituent interactions play an important role during the formation of the heterocycle due to 1,2- and 1,3-conformational constraints in the transition state. On the other hand, if the stereogenic center is in the molecule carrying the double bond but is out of the tether, much lower diastereoselection should be expected.

As a result of the wide spectrum of application of amino acids, attention has been paid to the development of new methodologies for the preparation of both natural and unnatural amino acids in optically pure form. For this purpose the cyclization of chiral amidals of α,β- and β,γ-unsaturated carboxylic acids promoted by mercury trifluoroacetate has been utilized.[5] In this case the role of a chiral center out of the tether is to allow the resolution of the diastereomeric mixture. This kind of approach was realized, utilizing the commercially available (*S*)-1-phenylethylamine as a chiral source to synthesize enantiomerically pure compounds. This strategy affords diastereomeric mixtures of heterocycles that can be easily separated by flash chromatography.

This method, with the separation of diastereomeric mixtures of chiral imidazolidin-4-ones and perhydropyrimidin-4-ones, was successfully used to obtain a variety of pure amino acids in the *R* or in the *S* configuration in few steps under mild conditions and in good yield (Scheme 11.5).

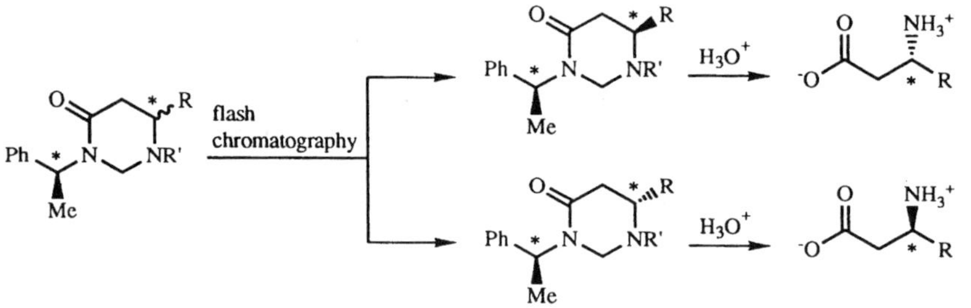

Scheme 11.5

By means of the mercury cyclization of the unsaturated amidals, obtained from the reaction of 1,3,5-tri-(*S*)-phenylethyl-hexahydrotriazine and α,β-unsaturated acyl chlorides, diastereomeric mixtures of imidazolidin-4-ones and perhydropyrimidin-4-one have been synthesized and easily separated by flash chromatography.[6] The subsequent hydrolysis under acid conditions of the separated heterocycles affords, respectively, D or L α- and β-amino acids. The regiochemistry of the cyclization has been studied, depending on the substituents of the double bond.

The amidals **3a–e** have been obtained through simple steps, starting from the chiral 1,3,5-tri-(*S*)-1-phenylethylhexahydrotriazine **1**.[7] It is known that 1,3,5-hexahydrotriazines are highly reactive compounds that afford, by treatment with acyl chlorides, the corresponding N-alkyl-N-chloromethylamides in quantitative yields.[8] Thus **1** was obtained simply by treating (*S*)-1-phenylethylamine either with 40% aqueous solution of formaldehyde or with solid paraformaldehyde in dichloromethane (Scheme 11.6). This compound can be utilized without further purification; nevertheless the crystallization from petroleum ether afforded a white solid.

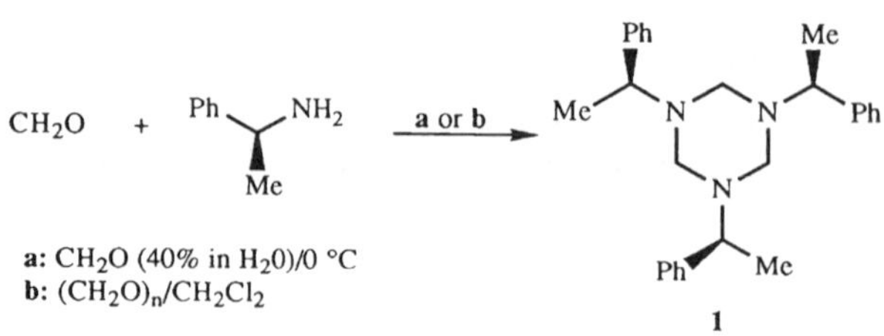

Scheme 11.6

The hexahydrotriazine **1** reacted smoothly with 2,3-unsaturated acyl chlorides in dry dichloromethane at 0°C under argon to give the corresponding N-(*S*)-1-phenylethyl-N-chloromethylamides **2a–e**. The displacement of the chloride group was performed directly by adding the mixture to a saturated solution of ammonia in dry dichloromethane. The amidals **3a–e** were purified with chromatography on neutral alumina or silica gel and obtained as colorless oils in yields ranging from 60 to 80% from the triazine **1** (Scheme 11.7)

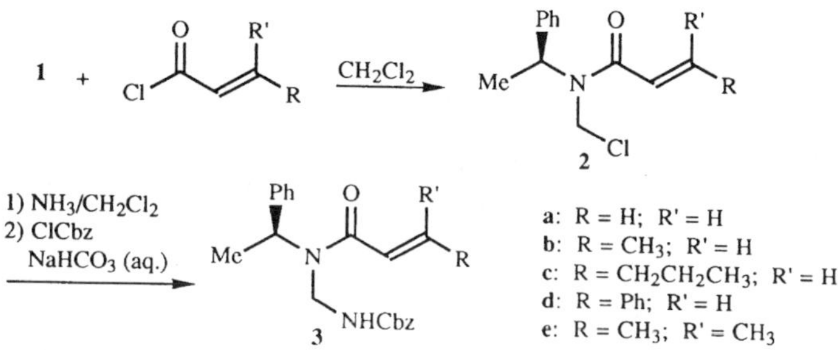

Scheme 11.7

The amidals **3a–c** were cyclized in dry dichloromethane utilizing $Hg(TFA)_2$ as the electrophile.[5] The reaction was complete in 20 min at room temperature, then the solvent was evaporated and replaced with dry acetonitrile. The cleavage of the C-Hg bond was performed by adding 1.1 equivalents of $NaBH_4$ at 0°C. The mercury was removed by filtration, and the analysis of the crude mixture by ^{1}H NMR and capillary column gas chromatography showed[9] the signals of two diastereomers in 1:1 ratio. The diastereomeric mixture, separated on silica gel, afforded the heterocyclic compounds in 80% total yield. The enantiomeric purity and the absolute configuration of each compound were established by ^{1}H-NMR analysis (Scheme 11.8). With these substrates, the cyclization proceeds with the formation of imidazolidin-4-ones, as proved by the IR absorption at 1690 cm^{-1}, characteristic of five-membered rings of this class of compounds.

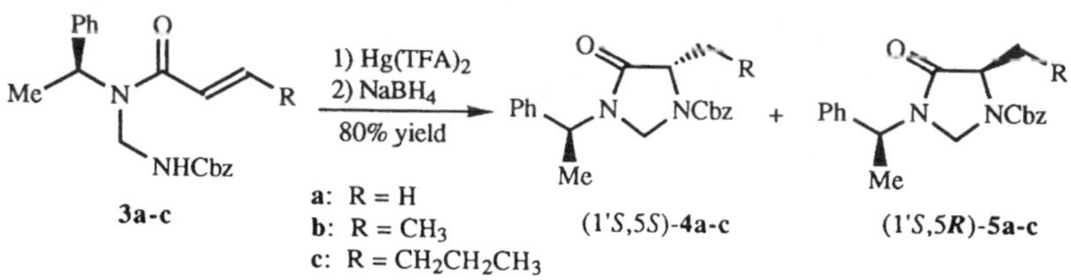

Scheme 11.8

The correct attribution of the stereochemistry of the imidazolidin-4-ones is confirmed by their hydrolysis, which was conducted under acid conditions to avoid racemization and represents an easy step to the synthesis of α-amino acids. Thus the imidazolidin-4-ones, dissolved in methanol and 11*M* HCl were refluxed 24 h to furnish in quantitative yield a mixture of the corresponding α-amino acid hydrochlorides and (*S*)-1-phylethylamine hydrochloride.

The (*S*)-1-phenylethylamine was separated during the workup by treatment with aqueous sodium carbonate followed by extraction with ethyl acetate. On the other hand, the purification of the amino acid from sodium chloride was performed on a column of BIORAD AG 50W-X2 resin using 0.015*M* NH_4OH as eluant (Scheme 11.9). The values of $[\alpha]_D$ are in perfect agreement with those reported for commercial samples.

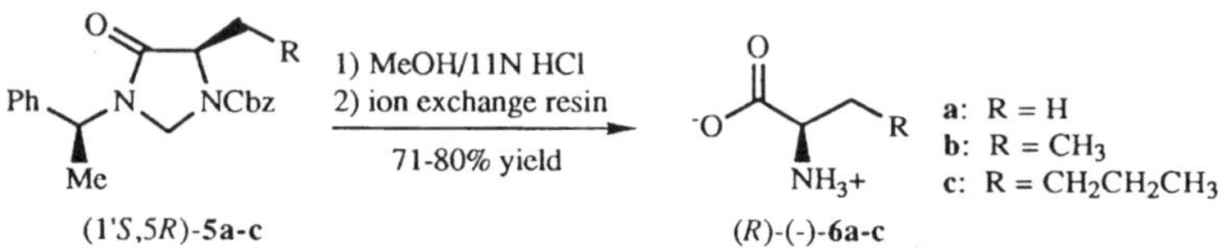

Scheme 11.9

When electron factors are overwhelming, as in the amidals **3d** and **3e**, the cyclization proceeds with the formation of 6-membered rings. The cyclization of **3d** with $Hg(TFA)_2$ in dichloromethane was complete in 2 h. After reduction with $NaBH_4$, a mixture of 6-phenyl perhydropyrimidin-4-ones **7d** and **8d** was obtained in 80% yield and 2:1 diastereomeric ratio, as shown by GC-MS and ^{1}H-NMR analysis. The IR absorption at 1650 cm^{-1} confirmed the formation of six-membered rings (Scheme 11.10).

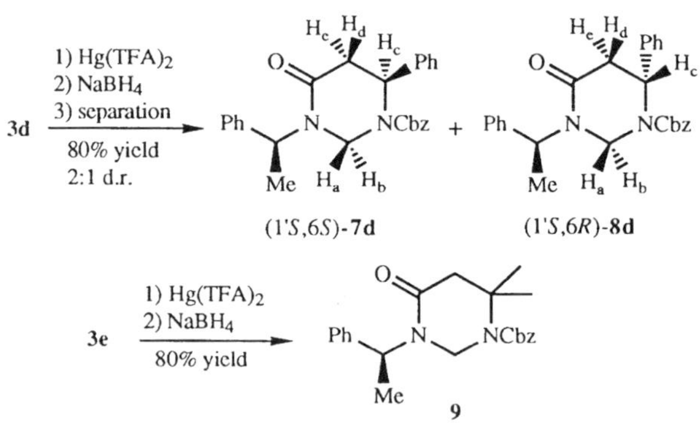

Scheme 11.10

Similar results have been obtained starting from **3e**. In fact, after cyclization with $Hg(TFA)_2$ and reduction with $NaBH_4$, the perhydropyrimidin-4-one **9** was obtained pure in 80% overall yield.

Utilizing the same strategy,[10] a new synthesis of the 6-methyl perhydropyrimidin-4-ones **7a** and **8a**, precursors of (*S*)- and (*R*)-3-aminobutanoic acids, has been envisaged starting from the amidal obtained from the reaction of hexahydrotriazine **1** and 3-butenoyl chloride. Thus by treatment of **1** with 3-butenoyl chloride, the corresponding chloromethyl adduct is obtained in quantitative yield and transformed in situ into the corresponding amidal by reaction with ammonia. The amidal is then protected with benzyloxycarbonyl chloride to give **10** in 85% overall yield from **1**. The adduct **10** is practically pure and can be utilized for further transformations. On the other hand, the purification on silica gel chromatography occurs with partial decomposition (Scheme 11.11).

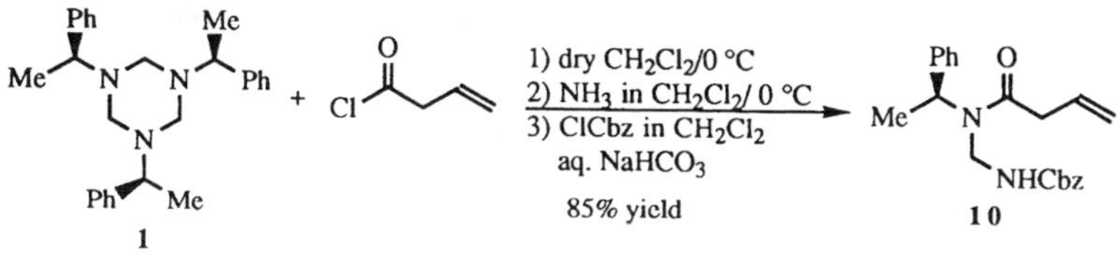

Scheme 11.11

The addition of $Hg(TFA)_2$ to a solution of **10** in dry CH_2Cl_2 at 0°C promotes the cyclization. Under these conditions the reaction monitored with TLC chromatography shows the presence of starting **10** even after prolonged reaction times. On the contrary, the cyclization in acetonitrile with 1.1 equivalents of dry Na_2CO_3 is complete in 2 h.[9b]

As the reductive removal of mercury performed on compound **11** affords a mixture of **7a** and **8a** in low yield, **11** has been reacted with NaBr in methanol to afford **12** in quantitative yield. By treatment of **12** with $LiBH_4$ in THF at –78°C and after the usual workup, a mixture of 6-methyl perhydropyrimidin-4-ones **7a** and **8a** is obtained in 70% yield and 63:37 diastereomeric ratio, as shown by the ^{1}H-NMR spectra and by GC-MS (gas chromatography–mass spectrometer) analysis. The separation of the 6-methyl perhydropyrimidin-4-ones by flash chromatography on silica gel affords pure **7a** ($[\alpha]_D$ –71.6, c 0.7, $CHCl_3$) and **8a** ($[\alpha]_D$ 34.2, c 1.4, $CHCl_3$) in the reported diastereomeric ratio (Scheme 11.12). The IR absorption at 1650 cm^{-1} is diagnostic for the expected six-membered ring formation.

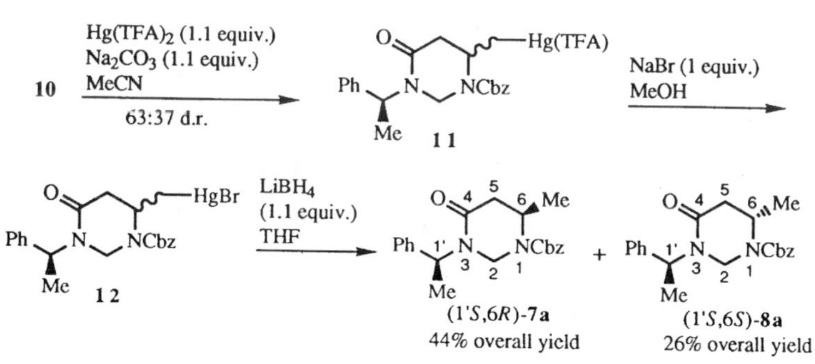

Scheme 11.12

Unfortunately the application of this procedure to the preparation of novel 6-substituted perhydropyrimidin-4-ones failed, because disubstituted β,γ-unsaturated amidals like **10** were unreactive towards mercury cyclization.

11.1.2 Addition of Hydroxylamine to α,β-Unsaturated Acids

In exploring suitable ways to obtain the heterocyclic starting materials on a multigram scale,[11] we chose the preparation of racemic β-amino acids by addition of 2 equivalents of free hydroxylamine to alk-2-enoic acids in ethanol at reflux, following a well-known procedure described by Steiger on cinnamic acid.[12] The hydroxylamine behaves both as nucleophile and reducing agent for the 3-hydroxylamino intermediate. Although the reaction occurs with low yield, the availability of the reagents makes this old reaction still attractive (Scheme 11.13).

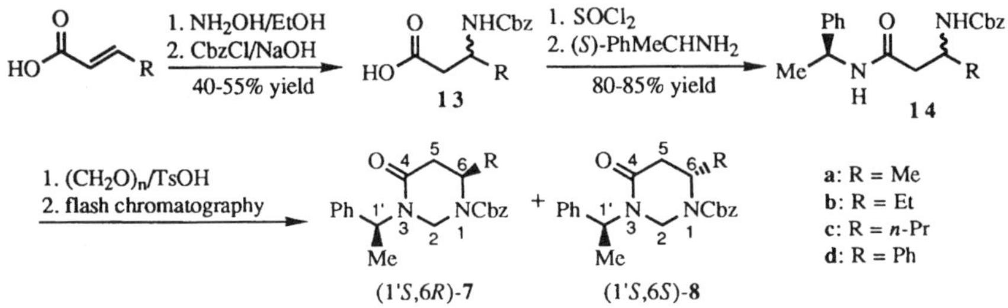

Scheme 11.13

Starting from pent-2-enoic acid, hex-2-enoic acid, and cinnamic acid, following this method, β-amino acids were obtained and protected at the nitrogen with benzyl chloroformate. The products **13b**, **13c**, and **13d** were transformed into the corresponding (*S*)-phenylethylamido derivatives **14b**, **14c**, and **14d**. All these steps occur in a short time with good yield and under mild conditions. The corresponding perhydropyrimidin-4-ones **7** and **8** were easily obtained by treatment with paraformaldehyde and *p*-toluensulfonic acid in benzene at reflux and fully separated by flash chromatography. The synthesis of both (1′*S*,6*R*)- and (1′*S*,6*S*)-6-methyl perhydropyrimidin-4-ones **7a** and **8a** was accomplished following the same method, but starting from commercial (±)-3-butanoic acid.[13]

Thus with this procedure a variety of enantiomerically pure 6-substituted perhydropyrimidin-4-ones can be prepared.

11.1.3 Addition of Hydroxylamine to α,β-Unsaturated Amides

A new synthesis of the enantiomerically pure 3-(1′-phenylethyl)-6-alkyl perhydropyrimidin-4-ones **17** and **18** was achieved in good yield when the free hydroxylamine was reacted with (*S*)-N-(1-phenylethyl)-2-alkenamides **15** in ethanol and water at reflux (Scheme 11.14).[14] Reaction was complete in 3 h, and the addition products were recovered simply by removing the solvent. In contrast with the results obtained with α,β-unsaturated acids and as observed for esters and lactones, the addition of hydroxylamine to the α,β-unsaturated amides **15** affords the corresponding 3-hydroxylamino derivatives without reduction of the N-O bond.

The subsequent reaction with NaOH and benzyl chloroformate in water–acetone afforded exclusively the O-protected derivatives **16** in 70% yield. The IR absorption of **16a** and **16b** at 1758 and 1756 cm^{-1}, respectively, indicates the formation of a carbonate group. The reaction of **16** with paraformaldehyde under acid catalysis affords quantitatively the perhydropyrimidin-4-ones **17** and **18**, which can be easily separated by flash chromatography.

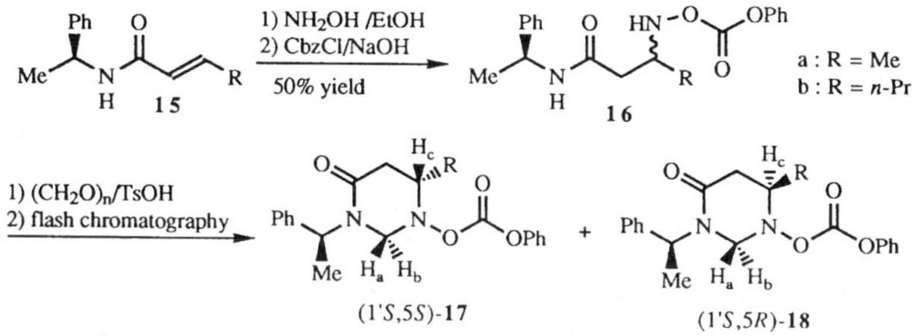

Scheme 11.14

These new compounds are precursors of β-amino and show a particular conformation and an interesting reactivity, which will be further discussed.

11.1.4 Conformational Analysis of 6-Substituted Perhydropyrimidin-4-Ones

The structure of pyrimidin-4-ones **7a** and **8a** has been established by detailed ^{1}H-NMR studies. Two factors are peculiar for the determination of the configuration and of the conformation that these mobile molecules assume. As already observed[15] for heterocycles containing a (*S*)-phenylethylamide moiety, the chiral group on $C_{1'}$ shows a strong tendency to assume a rigid conformation with the hydrogen eclipsing the carbonyl group. ^{1}H-NMR analysis shows that the (*S*) phenylethyl group assumes a preferential conformation, with the hydrogen eclipsing the adjacent carbonyl groups.

This observation has been confirmed by MM2P[16] calculations for perhydrooxazin-2-ones and oxazolidin-2-ones containing the (*S*)-1-phenylethylamine moiety.[17] In this conformation the hydrogen in front of the phenyl group is strongly shielded. Moreover, the carbonyl of the protecting carbobenzoxy group deshields preferentially the hydrogen lying in the same plane, thus H_a or H_b, depending on the conformation of the perhydropyrimidin-4-one (Schemes 11.15 and 11.16).

The presence of the carbobenzoxy group is also responsible for the formation of rotamers, which can be observed recording the ^{1}H-NMR spectra at –30°C.[18] While the ^{1}H-NMR spectra of **7a** and **8a** show a bad resolution at room temperature, due to the coalescence point of the molecule, they show a perfect resolution, if recorded at 50°C.

The ^{1}H-NMR spectrum of the less polar isomer **7a** shows that the hydrogens H_d and H_e, resonating at δ 2.32 and 2.71, respectively, have a large geminal coupling constant ($J_{H_dH_e}$ = 15.7 Hz) and two coupling constants ($J_{H_dH_c}$ = 8.2 Hz; J_{H_e,H_c} – 6.4 Hz) that are in line with the methyl group in the equatorial position. Moreover the doublet at δ 4.14 (*J* = 13.2 Hz) is very shielded: This result suggests that it experiences the presence of the phenyl group, and therefore is attributed to H_a. Furthermore H_b (δ 4.91, *J* = 13.2

Hz) is deshielded by the carbonyl group lying in the same plane. These results agree with a **7a** configuration in the preferential conformation reported in Scheme 11.15.

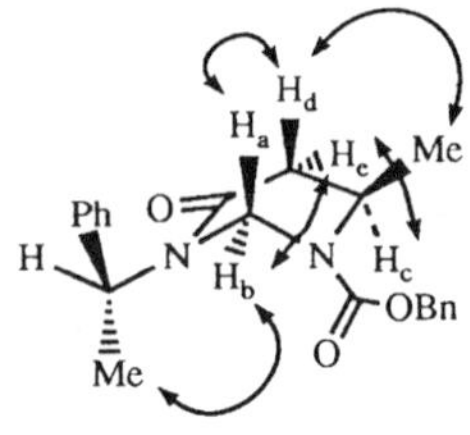

7a

Scheme 11.15

The results of NOEDIF experiments performed on **7a** confirm the stereochemical assignment. Thus the irradiation of the methyl at $C_{1'}$ (δ 1.47) produces an enhancement of the doublet at 4.91, which is therefore assigned to H_b. On the other hand, the irradiation of the dd at δ 2.32 attributed to H_d produced an enhancement of the H_a signal (δ 4.14) and an enhancement of the methyl at C_6 (δ 1.20). Irradiation of H_e (δ 2.71) produces an enhancement of the H_c signal at δ 4.20 and a weak enhancement of H_b (δ 4.91). The irradiation of H_b and of the methyl at C_6 confirm these results.

In a similar way the structure of **8a** has been established by means of ^{1}H-NMR analysis. In fact the coupling constants of H_d-H_c and H_e-H_c ($J_{H_d H_e}$ = 15.7 Hz, $J_{H_d H_c}$ = 6.3 Hz, J_{H_e,H_c} = 6.6 Hz) account for a rather flat conformation with the methyl in the equatorial position. The structure of **8a** is therefore forced in the conformation shown in Scheme 11.16. The chemical shifts of H_a (δ 4.74) and H_b (δ 4,54) agree with this assignment: In this conformation H_a is strongly deshielded by the carbonyl of the carbobenzoxy protecting group, and it does not suffer the shielding effect of the vicinal phenyl group. These effects account for the H_a chemical shift at low field.

The NOEDIF experiments performed on **8a** confirm the attribution and show that the irradiation on H_e causes an enhancement of the methyl on C_6 (δ 1.47) and of the doublet resonating at δ 4.54, which can be attributed to H_b. The irradiation of the methyl on C_6 causes the enhancement of H_e, and the irradiation of H_d causes an enhancement of H_c and a weak enhancement (~1%) of the doublet resonating at δ 4.74 that can be attributed to H_a. The NOEDIF experiment performed on H_a and H_b confirm the above-reported results. Finally the irradiation of the methyl on $C_{1'}$ causes a very weak enhancement of H_a and H_b. These results allow to attribute the conformation of **8a**, as reported in Scheme 11.16.

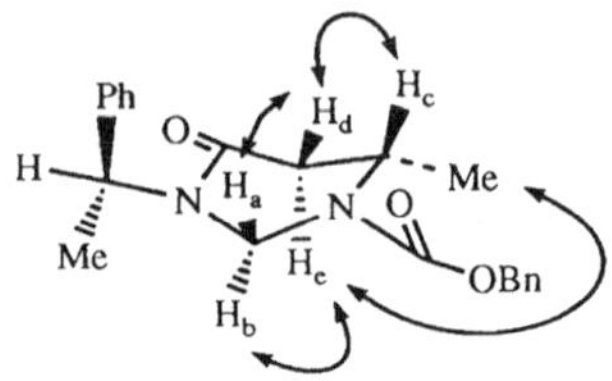

8a

Scheme 11.16

In a similar way, the absolute configuration of pyrimidin-4-ones **7d** and **8d** can be attributed. The structural assignment of **7d** and **8d** is made on the basis of the nonequivalence of H_a and H_b, assuming that the chemical shift of the hydrogen resonating upfield strongly depends on the shielding of the phenyl ring of the 1′*S* stereogenic center.

The phenyl group on 1′*S* shields H_a, which always resonates upfield in **7d** and **8d** (**7d**: δ_{H_a} 4.31, δ_{H_b} 5.05; **8d** δ_{H_a} 4.58, δ_{H_b} 4.95), moreover, due to the additional shielding effect of the substituent at C_6, $\Delta\delta_{H_a,H_b}$ in **7d** is larger than in **8d**. The result of NOEDIF experiment performed on **7d** confirmed the stereochemical assignment. In fact the irradiation of H_d (δ 2.71) caused the enhancement of H_a (δ 4.31), showing the *cis* relationship between H_d and H_a (Scheme 11.17).

Furthermore the coupling constants of H_c–H_d and H_c–H_e in **7d** (J_{H_c,H_d} = 11 Hz, J_{H_c,H_e} = 6 Hz) show that the phenyl substituent occupies the equatorial position. The same trend is observed for **8d** [J_{H_c,H_d} = 5.5 Hz, J_{H_c,H_e} = 10 Hz).

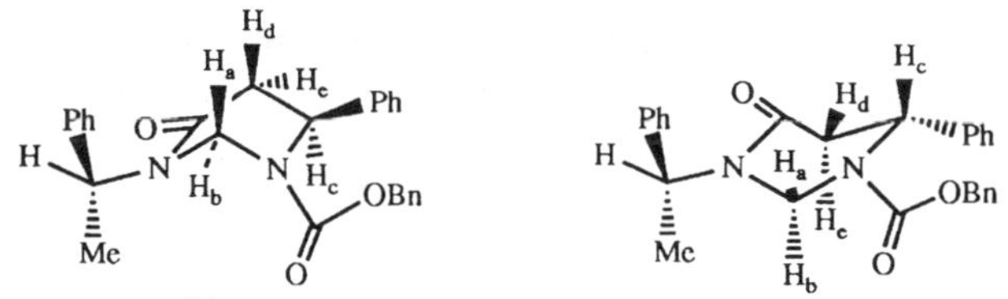

7d **8d**

Scheme 11.17

The absolute configuration of C_6 of **17b** and **18b** could be established utilizing the same method, although the absence of a carbonyl substituent at N_1 makes these heterocycles more flexible. NOEDIF experiments performed on samples of **17b** and **18b** confirmed the stereochemical assignment. Thus irradiation of the methyl at $C_{1'}$ in **17b** (δ 1.41) produces an enhancement of the doublet at δ 4.39, assigned to H_b, while irradiation of H_c (δ 3.20) produces an enhancement of the doublet at δ 3.99, assigned to H_a. Moreover the NOEDIF experiments performed on **18b** confirmed the *R* configuration assigned to C_6. Irradiation of the *d* at δ 1.51, assigned to the methyl on $C_{1'}$, causes an enhancement of H_b (δ 4.41) and vice versa. Finally, on irradiation of H_b, an enhancement of H_c (δ 3.31) is observed and vice versa (Scheme 11.18).

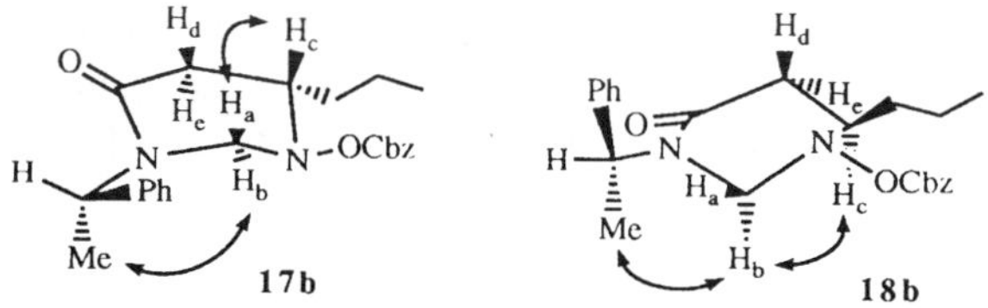

17b **18b**

Scheme 11.18

11.1.5 Hydrolysis of Perhydropyrimidin-4-Ones

The correct attribution of the stereochemistry of the perhydropyrimidin-4-ones is confirmed by the hydrolysis of the compounds, following the same procedure previously described for imidazolidin-4-ones. Thus the hydrolysis was conducted under acid conditions to avoid racemization and represents an easy step to the synthesis of β-amino acids. The perhydropyrimidin-4-ones, dissolved in methanol and 11*M* HCl, were refluxed 24 h to furnish in quantitative yield a mixture of the corresponding β-amino acid hydrochlorides and (*S*)-1-phenylethylamine hydrochloride.

The (*S*)-1-phenylethylamine was separated during the workup, by treatment with aqueous sodium carbonate followed by extraction with ethyl acetate. The purification of the amino acid from sodium chloride was performed on a column of BIORAD AG 50W-X2 resin using 1.5 *M* NH_4OH as eluant.

Therefore, the hydrolysis under acid conditions of the 2:1 mixture of **7d** and **8d** afforded a mixture of (*R*)- and (*S*)-phenylalanine **19d** and **20d**. In fact after elution from the column resin with 1.5 *M* NH_4OH, **15** was obtained in 70% yield and 33% ee ($[\alpha]_D$ −2.3°; *c* 0.7, H_2O; lit.[19] $[\alpha]_D$ −6.9°; *c* 0.8, H_2O), confirming that the more abundant heterocycle has (1′*S*,6*S*) configuration. In the same manner from the 6,6-dimethyl perhydropyrimidin-4-one **9**, the 3-amino-3-methylbutanoic acid **1ge** was obtained in 75% yield (Scheme 11.19).

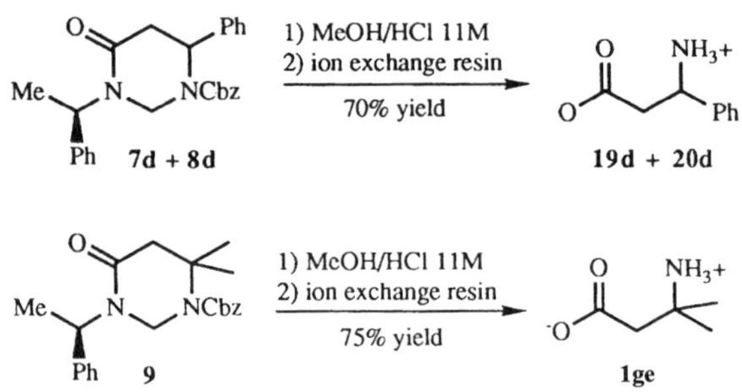

Scheme 11.19

Following the same procedure, the absolute configuration of perhydropyrimidin-4-ones **7a** and **8a** is confirmed by conversion into the corresponding 3-aminobutanoic acids. In fact (*R*)-(−)-**19a** and (*S*)-(+)-**20a** are obtained by acid hydrolysis of **7a** and **8a**, respectively, in 6*M* HCl at reflux. The β-amino acids are purified by treatment with aqueous sodium carbonate followed by extraction with ethyl acetate to separate the (*S*)-phenylethylamine. Acidification of the aqueous layer and elution of the mixture from a cation-exchange resin with 1.5*M* NH_4OH affords pure **19a** and **20a**, which show $[\alpha]_D$ and mp values in agreement with those reported in the literature (Scheme 11.20).[20]

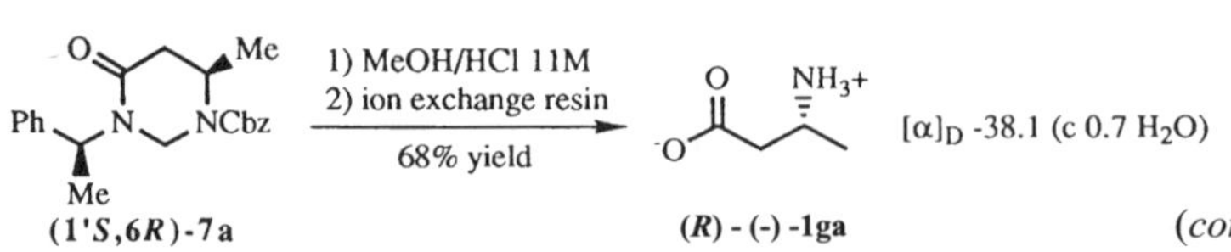

(*continued*)

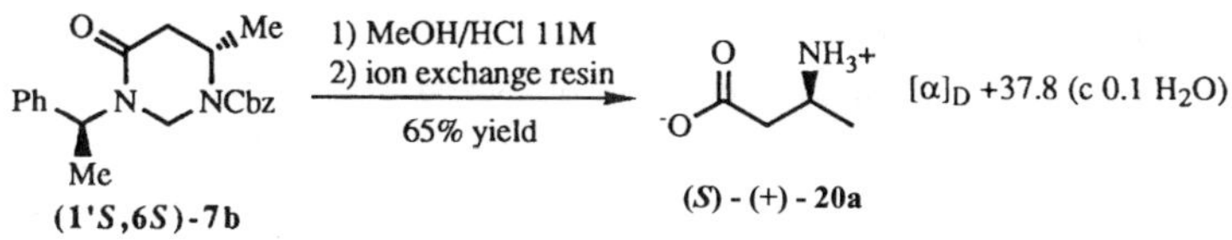

Scheme 11.20

11.2 Synthesis of 5,6-Dialkyl and 5,5,6-Trialkyl-Perhydropyrimidin-4-Ones

Recently increasing interest in the synthesis of α-alkyl β-amino acids has been observed, because these structures are important components of biologically active peptides. Furthermore unnatural α-alkyl β-amino acids are precursors for the preparation of pharmaceutical compounds and are important building blocks for the de novo design of biologically active new polypeptides. In an effort to elaborate a general approach to the asymmetric synthesis of α-alkyl β-amino acids, the preparation of 5,6-dialkyl perhydropyrimidin-4-ones as precursors of α-alkyl β-amino acids has been developed.

11.2.1 Alkylation of 6-Methyl Perhydropyrimidin-4-Ones

The alkylation of (1′*S*,6*R*)- and (1′*S*,6*S*)-3-(1′-phenylethyl)-6-methyl perhydropyrimidin-4-ones **7a** and **8a** produced 5-substituted 6-methyl pyrimidin-4-ones[21] as useful intermediates to α-substituted β-amino acids with high 1,2-*trans* selectivity.[22]

The alkylation of (1′*S*,6*R*)-**7a** was performed by treatment of the corresponding lithium enolate,[23] generated with lithium hexamethyldisilazide (LiHMDS) in dry THF, with ethyl iodide as depicted in Scheme 11.21. The results are reported in Table 11.1.

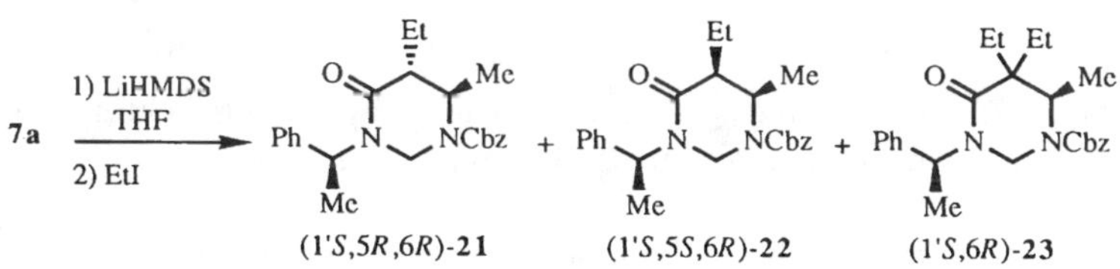

Scheme 11.21

The addition of ethyl iodide was made under a variety of conditions, and the diastereomeric ratios were determined by means of gas chromatography–mass spectrometry (GC-MS) analysis. The 5,6-*trans* and 5,6-*cis* relationships of products **21** and **22** have been determined by means of ^{1}H-NMR analysis and NOEDIF experiments.

As shown in Table 11.1, both the *trans*/*cis* ratio and the amount of dialkyl derivative **23** increase if the reaction temperature decreases; moreover, when more than 1

Table 11.1 DIASTEREOMERIC PRODUCT RATIOS AND CHEMICAL YIELDS FOR THE ALKYLATION REACTION OF 6-METHYL PERHYDROPYRIMIDIN-4-ONE **7a**

Entry	Temp. (°C)	Time (h)	Equiv. LiHMDS	Unreacted **7a** (%)	Yield (%) **21** + **22**	Dialkylated **23** (%)	*trans/cis* ratio[a]
1	0	2	0.9	15.8	84.2	—	87:13
2	0	2	1.1	10.3	78.3	11.4	96:4
3.1	–20	1	1	45.4	53.2	1.4	91:9
3.2	–20	4	1	7.6	89.3	3.1	91:9
4	–78-rt	14	1	4.5	85.8	9.7	94:6

[a] Determined by GC-MS analysis.

equivalent is present (entry 2), the diastereomeric ratio and the amount of **23** increase. Both results suggest that the dialkylation occurs faster on the *cis* diastereoisomer **22** than on the *trans* **21**. In fact when the reaction is performed at –78°C and slowly warmed up at room temperature (entry 4), a good *trans/cis* ratio is observed, although in the presence of an appreciable amount of dialkylated compound **23**. Under these conditions the reaction is complete after 14 h.

In a similar way the lithium enolate of the perhydropyrimidin-4-one (1′*S*,6*S*)-**8a** was generated with LiHMDS in dry THF. The alkylation was performed by adding different alkylating agents at different temperatures (Scheme 11.22). Results of the diastereoselection dependence from the reaction temperature, the employed amount of LiHMDS and the nature of the alkylating agent are reported in Table 11.2.

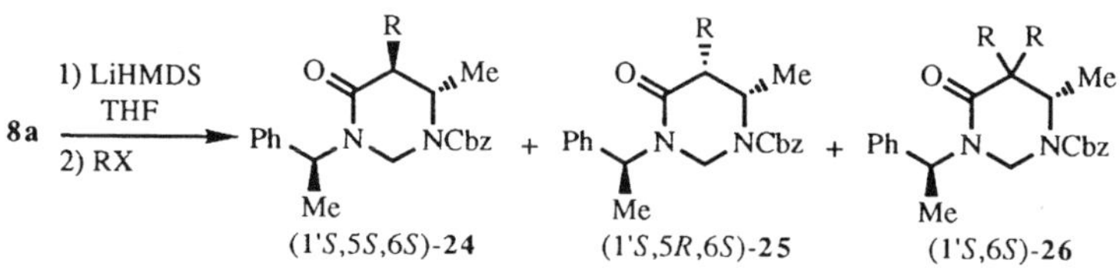

Scheme 11.22

Products **24**, **25**, and **26** have been easily separated by silica gel chromatography, and their structures have been assigned by means of ^{1}H-NMR spectroscopy and NOEDIF experiments.

It is worth noting that the diastereomeric ratio increases with the use of a bulkier substituent (entries 1, 4, 5). Furthermore, in the presence of more than 1 equivalent of LiHMDS, both the *trans/cis* ratio and the amount of dialkylated compound **26** increase, confirming that the dialkylation occurs faster on the *cis* derivative than on the *trans* one (entries 1 and 2). Moreover when the alkylation is carried out at –78°C and the reaction is slowly warmed up, a good *trans/cis* ratio is obtained, both after 4 h (entry 3.1) and after 14 h (entry 3.2), despite the presence of a considerable amount of **26**.

On the basis of the results reported in Tables 11.1 and 11.2, it appears that these substrates are alkylated with high *trans* diastereoselectivity and in high yields. The

Table 11.2 DIASTEREOMERIC PRODUCT RATIOS AND CHEMICAL YIELDS FOR THE ALKYLATION REACTION OF 6-METHYL PERHYDROPYRIMIDIN-4-ONE **8a**

Entry	Temp. (°C)	Time (h)	Equiv. LiHMDS	Alkylating agent	Yield (%) **24** + **25**	Dialkylated **24** (%)	*trans/cis* ratio[a]
1	0	21	0.8	EtI	76.8	—	85:15
2	0	2	1.1	EtI	92.1	7.9	94:6
3.1	−78-rt	4	1	EtI	87.2	6.2	94:6
3.2	−78-rt	14	1	EtI	88.5	10.0	97:3
4	0	50	1	*i*-BuI	56.1	—	87:13
5	0	5	1	BnBr	89.8	—	93:7

[a] Determined by GC-MS analysis.

perhydropyrimidin-4-ones **7a** and **8a** contain two chiral centers at C_6 and at $C_{1'}$. On the other hand, a totally stereoselective reaction was observed in the addition of the lithium enolates of 1-benzoyl-2(*S*)-*tert*-butyl-3,6(*R*)- and 1-benzoyl-2(*R*)*tert*-butyl-3,6(*S*)-dimethyl perhydropyrimidinones to methyl iodide and benzyl bromide[24] (Scheme 11.23). The (2)-*tert*-butyl derivatives were easily prepared starting from (*R*)- and (*S*)-β-amino butanoic acid and pivalaldehyde. Also these compounds contain two chiral centers at C_6 and at C_2.

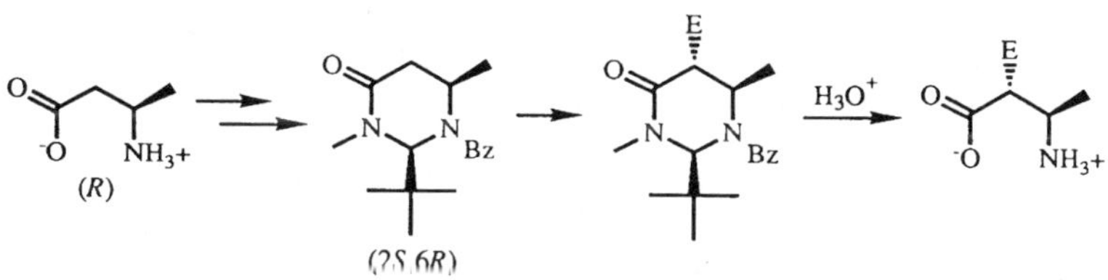

Scheme 11.23

Of interest is the behavior of the 1-benzoyl-3,6(*S*)-dimethyl perhydropyrimidin-4-one **29** when it is submitted to the alkylation under the above-discussed conditions.[25] Indeed the asymmetric induction in this compound is due exclusively to the stereogenic center at C_6 (Scheme 11.25).

The synthesis of 1-benzoyl-3,6(*S*)-dimethyl perhydropyrimidin-4-one (*S*)-**29** was performed by conversion through the ester (*S*)-**27**[24,26] to the corresponding amide followed by Schiff base formation with paraformaldehyde. Imine (*S*)-**28** was used for the cyclization step without purification: Treatment with benzoyl chloride/DMAP gave excellent yields of the desired perhydropyrimidin-4-one (*S*)-**29** (Scheme 11.24).

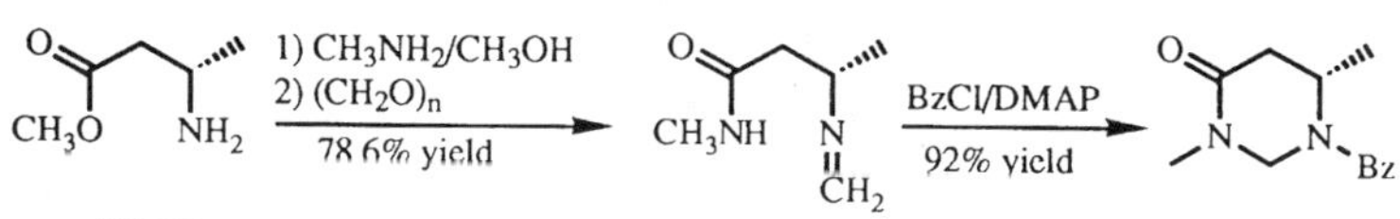

(*S*)-**27** (*S*)-**28** (*S*)-**29**

Scheme 11.24

The alkylation products **30a** and **30b** were formed by treatment of the lithium enolate of (*S*)-**29**, generated with lithium diisopropylamide (LDA) in THF, with methyl iodide and benzyl bromide at –75°C (Scheme 11.25).

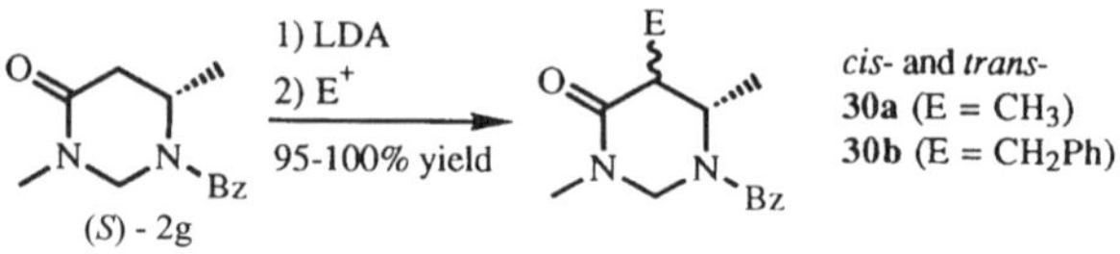

Scheme 11.25

^{13}C-NMR analysis of the crude products indicated the formation of two products in a 4:1 ratio. That the major product corresponds to the isomer of the *trans* relative configuration [(5*S*,6*S*)-**30a** or (5*S*,6*S*)-**30b**, depending on the starting material] was ascertained by chemical correlation with the known α,β-disubstituted β-amino acids obtained by hydrolysis.

The lower diastereoselectivity observed in the alkylation of the lithium enolate of (*S*)-**29**, relative to that observed with the lithium enolates of (1′*S*,6*R*)-**7a**, (1′*S*,6*S*)-**8a**, 1-benzoyl-2(*S*)-*tert*-butyl-3,6(*R*)- and 1-benzoyl-2(*R*)-*tert*-butyl-3,6(*S*)-dimethyl perhydropyrimidinones suggests the relative ability of the *tert*-butyl, methyl, and phenethyl groups as stereodirecting entities when located, respectively, on the stereogenic centers C_2, C_6, and $C_{1'}$ of the 1,3-perhydropyrimidin-4-one system. This sort of information is quite relevant in the design of chiral substrates for the development of asymmetric synthesis.

11.2.2 Methylation of 6-Alkyl Perhydropyrimidin-4-Ones

α-Methyl β-amino acids are becoming of great interest, because they are components of a number of molecules that display interesting biological properties. For instance, (2*S*,3*R*)-2-methyl-3-aminopentanoic acid has been isolated as a hydrolysis fragment of several biologically interesting natural products: the marine peptide antifungal agents Majuscolamide C[27] and 57-Normajascolamide C, and the antitumor agents Dolastatins 11 and 12.[28] Thus we have studied the methylation reaction at C_5 of perhydropyrimidin-4-ones in some detail.

Perhydropyrimidin-4-one (1′*S*,6*R*)-**7** was submitted to the methylation reaction[11] (Scheme 11.26) under the different reaction conditions reported in Table 11.3.

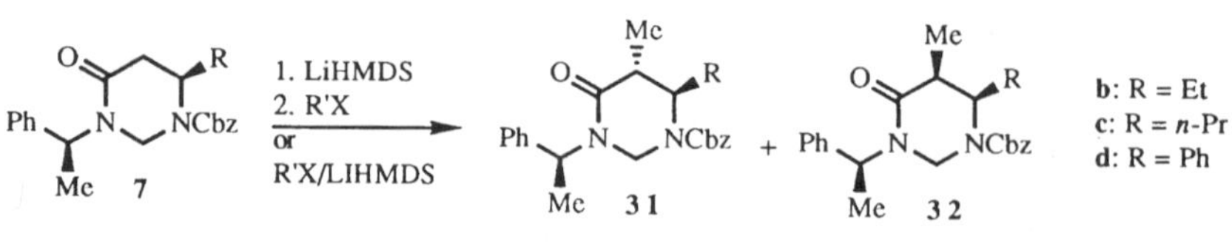

Scheme 11.26

For **7b**, while with Method A (entry 1) the diastereomeric ratio is more satisfactory than with Method B (entry 2), the reaction yield proved to be poorer. Perhydropyrimidin-4-one **7c** (entry 3) has the same behavior as **7b** (entry 2), while **7d** gives rise to better di-

Table 11.3 DIASTEREOMERIC PRODUCT RATIOS AND CHEMICAL YIELDS FOR THE METHYLATION REACTION OF PERHYDROPYRIMIDIN-4-ONES **7b**, **7c**, AND **7d**

Entry	*R*	*R′*	Temp. (°C)[a]	Time (h)[a]	Alkylating agent (equiv.)	Yield (%)	*trans*/*cis* ratio **31**/**32**
1	Et[b]	Me	−78 - rt	20	MeI (6)	60	84:16
2	Et[c]	Me	−20	2	Me_2SO_4 (2)	80	78:22
3	*n*-Pr[c]	Me	−20	2	Me_2SO_4 (2)	81	78:22
4	Ph[c]	Me	−20	2	Me_2SO_4 (2)	50	92:8

[a] Time and temperature refer to the alkylation.

[b] The formation of the enolate was performed with 1 equiv. of LiHMDS at 0°C for 1 h, then the alkylating agent was introduced (Method A).

[c] The base (LiHMDS) was introduced in the presence of the alkylating agent (Method B).

astereoselection but a worse yield (entry 4). A variety of reaction conditions (temperature, time, equivalents of alkylating agents) were explored for this alkylation, but none led to greater selectivity and/or reaction yield than those reported in Table 11.3 (Scheme 11.27).

Furthermore an interesting result to obtain complete diastereoselection towards the formation of 5,6-*trans* compounds is the observation that, by reacting a *trans*/*cis* diastereomeric mixture with LiOH in methanol at reflux for 2 h, the 5,6-*cis* dialkyl perhydropyrimidin-4-ones gave epimerization, and only 5,6-*trans* disubstituted products were obtained.

Scheme 11.27

The behavior of perhydropyrimidin-4-ones (1′*S*,6*S*)-**8** is quite similar to that of compounds **7**. In Scheme 11.28 the general structure of compounds **8**, **33**, and **34** are reported, and Table 11.4 shows the results of the alkylation of **8b**, **8c**, and **8d**.

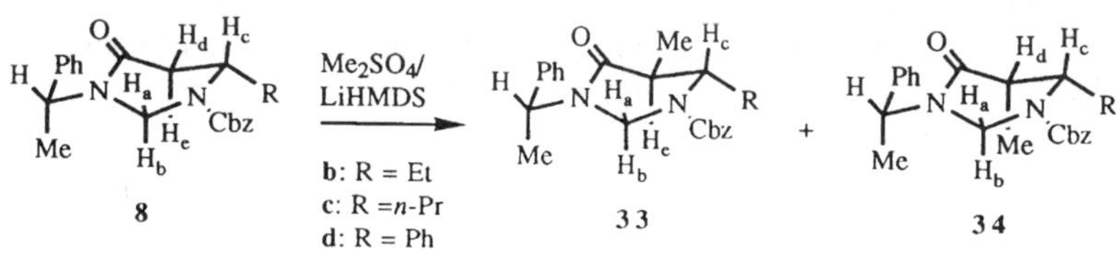

Scheme 11.28

In all the reported examples satisfactory yields and diastereoselectivities have been observed. Thus this kind of reaction is useful to prepare a variety of α-methyl β-amino acids.

Table 11.4 DIASTEREOMERIC PRODUCT RATIOS AND CHEMICAL YIELDS FOR THE METHYLATION REACTION OF PERHYDROPYRIMIDIN-4-ONES **8b, 8c,** AND **8d**

Entry	*R*	*R′*	Time (h)	Alkylating agent (equiv.)[a]	Yield (%)	*trans*/*cis* ratio **33**/**34**	Dialkylated (%)
1	Et	Me	2	Me_2SO_4 (2)	66	80:20	3
2	Et	Me	3	Me_2SO_4 (2)	80	82:18	7
3	*n*-Pr	Me	2	Me_2SO_4 (2)	86	77:23	/
4	Ph	Me	2	Me_2SO_4 (2)	50	92:8	/

[a] The base (LiHMDS) was always introduced in the presence of the alkylating agent at –20°C in dry THF.

11.2.3 Alkylation of 6-Alkyl Perhydropyrimidin-4-Ones

In order to test how the bulkiness of the alkylating agent affects the diasteromeric ratio, the pyrimidin-4-ones **7b**, **7c**, **7d**, **8b**, **8c**, and **8d** were submitted to alkylation with ethyl iodide and/or benzyl bromide[11] (Scheme 11.29). The data resulting from this investigation are reported in Table 11.5.

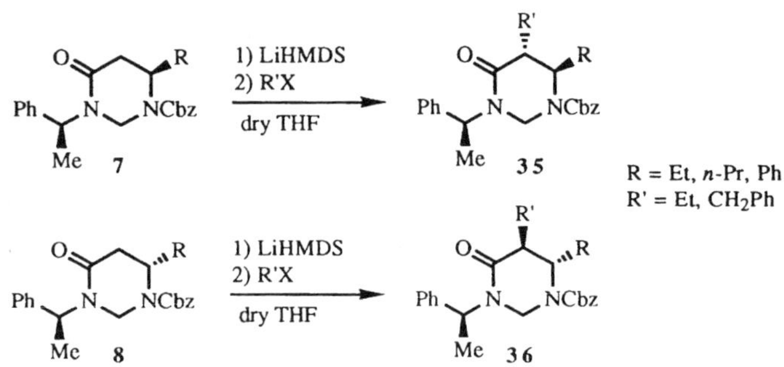

Scheme 11.29

The reaction is highly diastereoselective in all the reported cases, and only entry 1 shows the presence of traces of 5,6-*cis* disubstituted product. All the reactions were performed by adding 1 equivalent of LiHMDS at 0°C for 1 h, then the alkylating agent.

As expected, if both the substituent at C_6 and the alkylating agent are larger than a methyl group, very high diastereoselection is achieved. The excellent yields and diastereoselectivities makes these substrates useful intermediates for the synthesis of α-alkyl β-amino acids.

11.2.4 Conformational Analysis of the 6-Alkyl Perhydropyrimidin-4-Ones

Establishing the structural and stereochemical assignment of mono- and disubstituted perhydropyrimidin-4-ones required an opportune conformational analysis of the products.[11] As above described, the semirigid heterocyclic perhydropyrimidin-4-ones can assume two different half-boat conformations, **U** and **D**, depending on the ring

Table 11.5 DIASTEREOMERIC PRODUCT RATIOS AND CHEMICAL YIELDS FOR THE ALKLYLATION OF 7 AND 8 WITH ETHYL IODIDE AND BENZYL BROMIDE

Entry	Starting material	Product	*R*	*R′*	Alkylating agent (1.5 equiv.)	Yield (%)[a]	*trans/cis* ratio
1	**7b**	**35a**	Et	Et	EtI	78	96:4
2	**7c**	**35b**	*n*-Pr	Et	EtI	90	>99:1[b]
3	**7c**	**35c**	*n*-Pr	CH_2Ph	$PhCH_2Br$	85	>99:1[b]
4	**7d**[c]	**35d**	Ph	Et	EtI	96	>99:1[b]
5	**8b**	**36a**	Et	CH_2Ph	$PhCH_2Br$	95	>99:1[b]
6	**8c**	**36b**	*n*-Pr	Et	EtI	92	>99:1[b]
7	**8c**	**36c**	*n*-Pr	CH_2Ph	$PhCH_2Br$	88	>99:1[b]
8	**8d**[c]	**36d**	Ph	Et	EtI	80	>99:1[b]

[a] No traces of dialkyl derivative were detected.

[b] No traces of *cis* derivative were detected.

[c] Compounds **7d** and **8d** were submitted to alkylation as a 1:1 mixture.

substituents. This is a useful technique to assign the absolute configuration of C_6 of perhydropyrimidin-4-ones by means of ^{1}H-NMR spectroscopy and NOEDIF analysis. Thus we observed that H_a is always more shielded than H_b when the half-boat as **U** conformation, while H_b is always more shielded than H_a in the **D** conformation. Moreover the window for H_a and H_b is larger for **U** than for **D**. As a consequence of these observations, the relative configuration of C_6 can be attributed (Scheme 11.30).

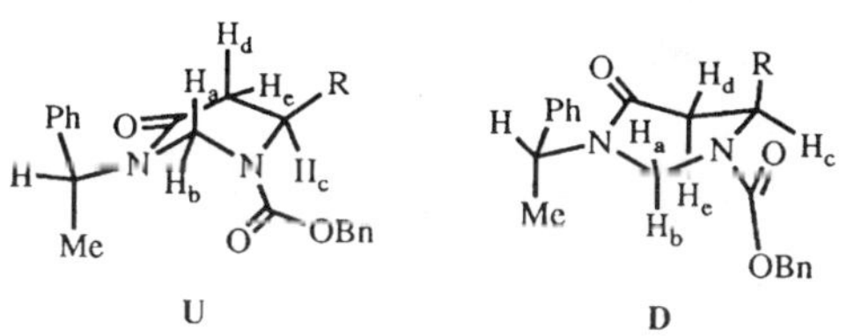

Scheme 11.30

In order to determine the relative configuration of C_5 after the alkylation reaction, three methods can be utilized. First of all, if both isomers are obtained, the coupling constants $J_{H_cH_d}$ and $J_{H_cH_e}$ can be compared. In fact the *cis* coupling constant is always smaller than the *trans* coupling constant, which is usually greater than 6 Hz. Moreover the chemical shift of H_d and/or H_e is always diagnostic. In fact in every perhydropyrimidin-4-one, owing to the effect of the vicinal side chain, the hydrogen at C_5 *cis* to the substituent at C_6 is always more shielded than the *trans* hydrogen, regardless of the nature of the substituent at C_6 and the conformation of the heterocycle.

Finally by means of NOEDIF experiments the configuration can usually be established. For instance, for products **31b** and **32b**, NOE experiments furnished the following results (Scheme 11.31): on irradiation of Me at C_5 a large NOE enhancement (5.4%) of H_c occurred, indicating the *cis* relation between Me and H_c for **31b**.

Likewise on irradiation of H_d in **31b**, a small enhancement (2.0%) was observed for H_c, while a large NOE enhancement (5.4%) of H_c was observed on irradiation of H_e for **32b**.

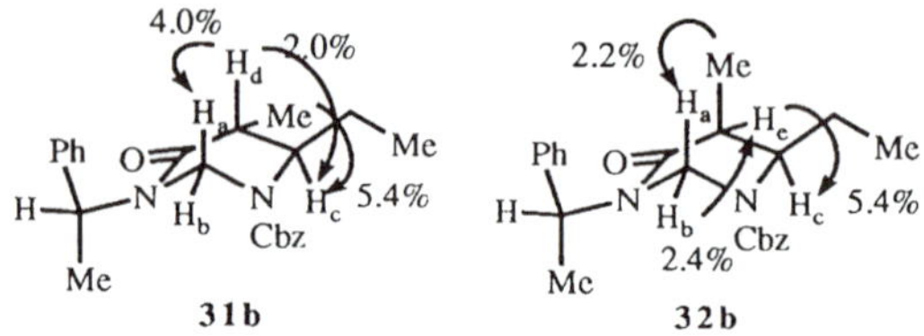

31b **32b**

Scheme 11.31

11.2.5 Hydrolysis of Perhydropyrimidin-4-One Adducts To Give the α-Substituted β-Amino Acids

The highly diastereoselective addition of the lithium enolates of pyrimidin-4-ones **7**, **8**, and **29** to several electrophiles was described. The hydrolysis of the resulting adducts proceeds with 6*N* HCl to afford α-substituted β-amino acids in good yield.

In particular the final step of the overall conversion of (*S*)-3-aminobutanoic acid to (2*S*,3*S*)- and (2*R*,3*S*)-2-alkyl-3-aminobutanoic acids, the hydrolysis of *cis* and *trans* **30a** and **30b**, was achieved by heating the acid to 115–120°C with HCl in a sealed tube[25] (Scheme 11.32). The free amino acids were purified by chromatography on a ion-exchange column; the yields were ca. 84% of the dimethylated compounds **37a** and **37b** and ca. 70% of the benzyl derivatives **37c** and **37d** (Scheme 11.33).

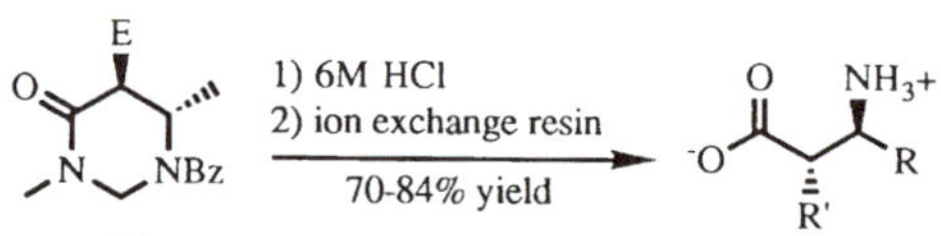

30 **37**

Scheme 11.32

Both the specific rotations and the melting points of the free amino acids were sensitive to conditions of measurement, as evident from discrepancies between measured and previously reported values.[24] Therefore, the nonhygroscopic N-benzoyl methyl esters **38a–d** were prepared for characterization and comparison with the literature values (Scheme 11.33).

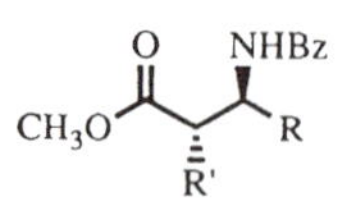

38

a: (2*S*,3*S*), R = CH_3 $[\alpha]_D$ -37.6, $CHCl_3$
b: (2*R*,3*S*), R = CH_3 $[\alpha]_D$ -42.9, $CHCl_3$
c: (2*S*,3*S*), R = CH_2Ph $[\alpha]_D$ -78.0, $CHCl_3$
d: (2*R*,3*S*), R = CH_2Ph $[\alpha]_D$ -12.2, $CHCl_3$

Scheme 11.33

At this point, the specific rotations, the melting points, and the ^{1}H- and ^{13}C-NMR spectra of **38a** and **38c** were found to be essentially identical with those already present in the literature.[24] This chemical correlation permits the assignment of the relative configuration of the main product during alkylation as *trans*.

The hydrolysis of the compounds **33b** and **36c** was performed under acid conditions[11]; thus **33b** and **36c** were suspended in 6*M* HCl and refluxed for 24 h, delivering a mixture of the corresponding β-amino acid hydrochlorides **39a** and **39b** and (*S*)-phenylethylamine hydrochloride (Scheme 11.34).

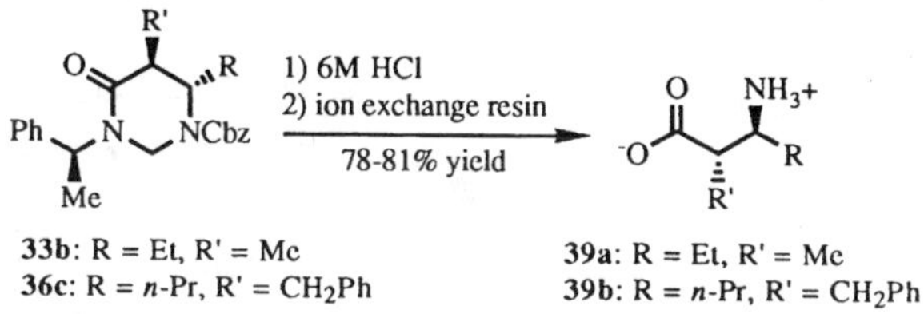

Scheme 11.34

The (*S*)-1-phenylethylamine was separated during the workup, and the purification of the amino acid from sodium chloride was performed on a column of cation-exchange resin using 1.5 *M* NH_4OH as eluant. The α-substituted β-amino acids were obtained pure in high yield, and the $[\alpha]_D$ and the melting point of compound **39a** coincided with reported values.[29]

Moreover, to show the versatility of this synthetic method, the compounds **39a** and **39b** were transformed into the corresponding β-lactams **40a** and **40b**. The β-amino acids were submitted to cyclization with 2-chloro-1-methylpyridinium iodide in the presence of triethyl amine, following the procedure described by Mukayama.[30] The corresponding 3,4-*trans* disubstituted β-lactams **40** were obtained pure in high yield after silica gel chromatography (Scheme 11.35).

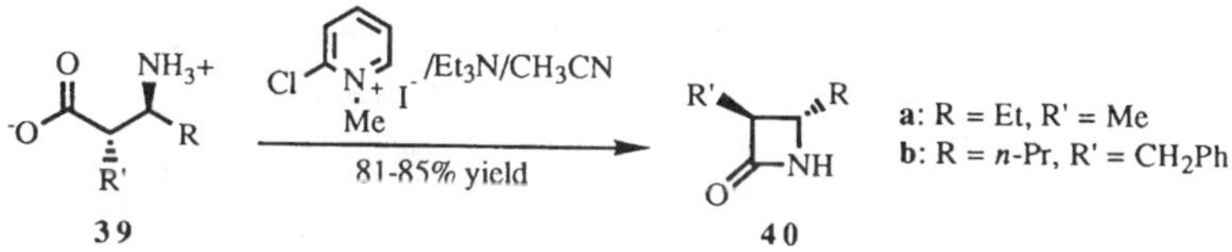

Scheme 11.35

Analysis of the crude reaction mixtures by ^{1}H- and ^{13}C-NMR spectroscopy and MS-GC chromatography showed the presence of a single stereoisomer. The 3,4-*trans* relationship of the substituents was confirmed by means of ^{1}H-NMR analysis; in fact the coupling constants J_{H_3,H_4} of 2.0 Hz for **40a** and 1.9 Hz for **40b** are typical for the 3,4-*trans* disubstituted azetidin-2-ones.

11.2.6 Dialkylation

Previously[21] we hypothesized that the dialkylation occurs faster on the 5,6-*cis* disubstituted than on the 5,6-*trans* disubstituted pyrimidin-4-one, strongly affecting the diastereomeric ratio. In order to demonstrate this assumption and to investigate the possibility of synthesizing enantiomerically pure α,α-disubstituted β-amino acids further, products **33b** and **34b** were easily isolated and submitted separately to alkylation with ethyl iodide (Scheme 11.36, Table 11.6).

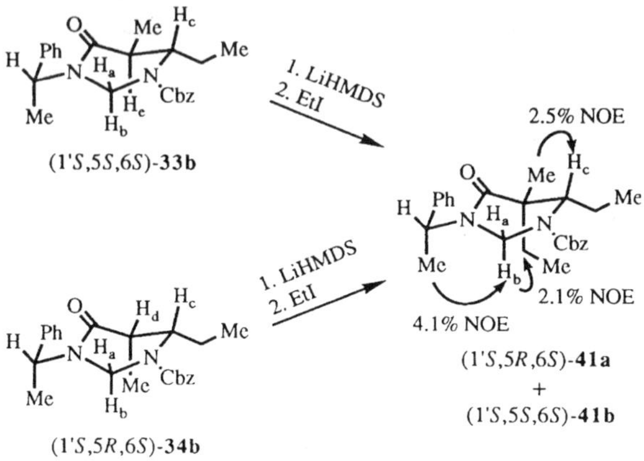

Scheme 11.36

The results reported in Table 11.6 confirm our hypothesis. In fact while the alkylation of the lithium enolate of 5,6-*trans* **33b** with ethyl iodide affords the dialkyl derivatives **41a** and **41b** in 5% overall yield (entry 1), the alkylation of 5,6-*cis* **34b** produces a mixture of **41a** and **41b** in quantitative yield (entry 4). This outcome establishes that the relative configuration of C_5 and C_6 of the heterocycle strongly influences the dialkylation rate, and thus the diastereomeric ratio of the monoalkylation.

Moreover, **41a** was obtained as the major product, whether starting from **33b** or **34b**. The absolute configuration of the major isomer **41a** was established by means of NOEDIF experiments and shows that the preferred isomer is obtained from the attack of the ethyl group from the same side of the ethyl at C_6 (Scheme 11.36).

This behavior may be explained by considering the possible attack directions of the electrophile, as shown in Scheme 11.37. Recently the nonvertical attack has been

Table 11.6 DIASTEREOMERIC PRODUCT RATIOS AND CHEMICAL YIELDS FOR THE DIALKYLATION REACTION OF PERHYDROPYRIMIDIN-4-ONES 33b AND 34b

Entry	Starting material	Temp.[a] (°C)	Time[a] (h)	Diastereomeric ratio **41a/41b**	Yield (%)
1	**33b**	0	1	96:4	5
2	**33b**	rt	1	91:9	35
3	**33b**	rt	72	90:10	47
4	**34b**	0	1	91:9	99

[a] Time and temperature refer to the alkylation. The formation of the enolate was always performed with 1 equiv. of LiHMDS at 0°C for 1 h in dry THF.

shown to be preferred in the enolate–electrophile reaction.[31] Thus for enolate **A**, derived from **8b**, owing to the *S* configuration of C_6, the preferred nonvertical attack direction is from the *Si*-face, leading to a 5,6-*trans* disubstituted product. On the other hand, the methyl group at C_5 and the pseudoaxial hydrogen at C_6 disfavor any attack from the *Si*-face in the enolate **B**, derived from **33b** or **34b**, and lead to the 5,6-*cis* derivative by means of vertical attack from the *Re*-face.

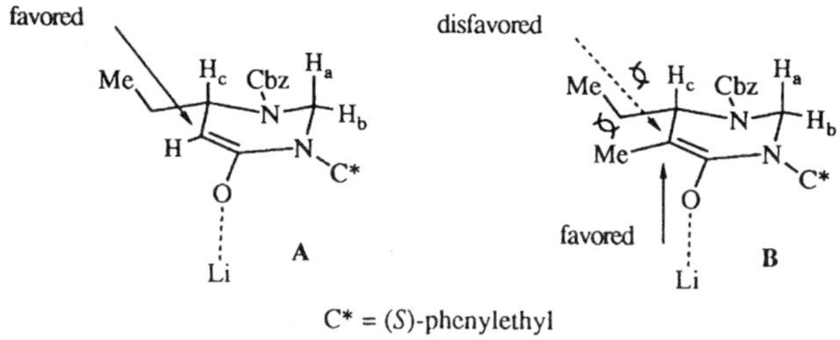

Scheme 11.37

This new class of compounds could receive attention in the insertion of α-dialkylated β-amino acids in polypeptide chains, due to the bulky hindrance at carbon α, which could cause a strong constraint in the polypeptide chain.

11.3 Synthesis of 5-Hydroxymethyl-6-Alkyl-Perhydropyrimidin-4-One Derivatives

The introduction at C_5 of an alkyl substituent that bears a hydroxyl group at the $C_{1'}$ position is a convenient way to synthesize perhydropyrimidin-4-one derivatives as precursors of β-hydroxy-β′-amino acids, which can be subsequently cyclized to useful β-lactam derivatives. The typical way to obtain these adducts is the aldol condensation of metal-enolates with aldehydes. This reaction has been studied in detail; moreover an unusual functionalization at C_5 with a formaldehyde equivalent has been accomplished by means of base-catalyzed cyclization of perhydropyrimidin-4-ones **17** and **18**.

11.3.1 Aldol Condensation of 6-Alkyl-Perhydropyrimidin-4-Ones

The aldol condensation is a very efficient method for the formation of new carbon–carbon bonds and several laboratories developed methods to carry out the reaction under high stereocontrol.[32] It is well known that the *syn*/*anti* ratio of the products is a function of the enolate counterion, the solvent, and the temperature reaction, and strongly depends on the aromatic or aliphatic nature of the aldehyde. The use of specific metal enolates, prepared by exchange reaction of the corresponding lithium enolates with a salt of the selected metal,[33] allows the development of better conditions in order to obtain the desired stereoisomer.

The reaction of benzaldehyde with the lithium enolate of compound **7a**,[13] generated under kinetic conditions by reactions with LiHMDS at 0°C in dry THF, afforded aldol products with satisfactory yield (Scheme 11.38). Among the four possible stereoisomers only the products **42a** and **42b** were obtained as a single spot in a 63:37 ratio, as indicated by the ^{1}H- and ^{13}C-NMR data. The two adducts were separated and fully characterized through the corresponding benzoates **43a** and **43b**.

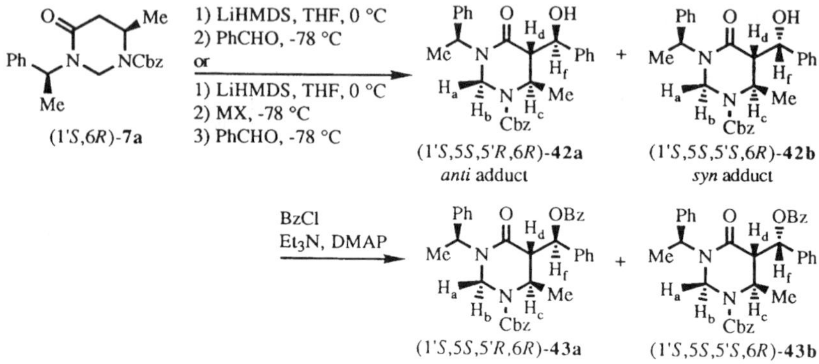

Scheme 11.38

Since enolates of chiral perhydropyrimidin-4-ones have not been previously explored under aldol condensation conditions, in an effort to rationalize the selectivity of this reaction, other common metal cations were examined. Table 11.7 shows the results that we obtained for the aldol reaction of various enolates of **7a** with benzaldehyde. It shows that the zinc-mediated aldol reaction (entry 1) improves the stereoselection, producing an excess of the *anti* adduct. Also if the lithium enolate of **7a** is exchanged with $ClTi(Oi\text{-}Pr)_3$ (entry 2), an enrichment of the *anti* adduct is also observed (*anti*/*syn* ratio 81:19).[34]

By exchanging the lithium enolate of **7a** with a 1*M* solution of $ClAlMe_2$ in hexanes (entry 4), **42a** and **42b** have been obtained in 50:50 *anti*/*syn* ratio, while a

Table 11.7 DIASTEREOMERIC PRODUCT RATIO AND CHEMICAL YIELDS FOR ALDOL REACTIONS OF (1′*S*,6*R*)-6-METHYL PERHYDROPYRIMIDIN-4-ONE **7a**

Entry	Electrophile	*M*	Solvent	*anti*/*syn* ratio[a]	Total yield[b]
1	PhCHO	ZnCl	ether/THF	85:15	75
2	PhCHO	$Ti(i\text{-}Pr)_3$	THF	81:19	82
3	PhCHO	Li	THF	67:33	90
4	PhCHO	$AlMe_2$	hexane/THF	50:50	43
5	PhCHO	$B(OMe)_2$	CH_2Cl_2/THF	20:80	80
6	PhCHO	$B(Oi\text{-}Pr)_2$	CH_2Cl_2/THF	20:80	80
7	PhCHO	$B(OBu)_2$	CH_2Cl_2/THF	20:80	80
8	MeCHO	Li	THF	25:75	90

[a] The *anti*/*syn* ratios were determined on crude reaction mixtures by means of ^{1}H-NMR and ^{13}C-NMR.

[b] All reported yields were based on isolated and purified products.

20:80 *anti/syn* ratio was observed using boron enolates (entries 5, 6, and 7). The chloroborates were obtained simply by mixing a 1*M* solution of BCl_3 in CH_2Cl_2 with $B(OR)_3$ in 1:2 molar ratio.[35] This result proves that the simple diastereoselection is regulated by the nature of the metal; in fact the *anti/syn* ratio improves going from the zinc enolate to the titanium, to the lithium, to the aluminum, and to the boron enolate.

The diastereomeric ratios of **42a** and **42b** have been determined by means of the ^1H- and ^{13}C-NMR spectra of the crude reaction mixtures, and all the hydrogens have been attributed through decoupling experiments. In particular the ^1H-NMR signals of H_d of **42a** and **42b** are easily distinguishable as double doublets at δ 2.57 and 2.67 ppm, respectively.

As mentioned above, only two compounds have been obtained from all the reactions of enolates of **7a** and benzaldehyde. Both the compounds **42a** and **42b** show a *trans* relationship between the hydrogens H_c and H_d, due to the facial diastereoselection induced by the heterocyclic methyl substituent on C_6. The *trans* relationship was confirmed by means of NOEDIF experiments performed on **42a** (Scheme 11.39).

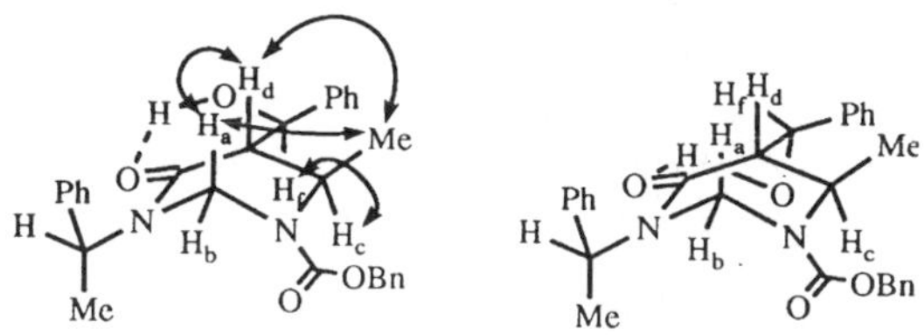

(1'*S*,5*S*,5'*R*,6*R*)-**42a** (1'*S*,5*S*,5'*S*,6*R*)-**42b**

Scheme 11.39

Owing to the carbobenzoxy group, the perhydropyrimidin-4-ones **42a** and **42b** show the presence of two stable rotamers, already observed for this class of compounds. Unfortunately 20°C is the coalescence temperature, and at 50°C the spectrum is nicely resolved, but the hydroxyl group hydrogen was not identified as a doublet. So by recording the spectra at –40°C, we obtained a good resolution and the presence of two rotamers. The hydroxyl group signals appear as doublets at δ 4.79 ppm (J = 3.9 Hz) and at δ 3.61 ppm (J = 3.7 Hz), respectively, and do not change upon varying the concentration of the substrate. Thus at –40°C in both aldol derivatives **42a** and **42b** a hydrogen bond between the carbonyl of the heterocycle and the hydroxyl group is present and forces the molecule in a preferential conformation that allows us to attribute the relative configuration at the newly introduced stereogenic center $C_{6'}$. In fact the ^1H-NMR spectrum of **42a** is characterized by the value of the chemical shift of H_d at δ 2.57 ppm in $CDCl_3$ with J_{H_d,H_f} = 7.4 Hz indicative of an *anti* stereochemistry, while in the spectrum of **42b** H_d resonates at δ 2.67 ppm with J_{H_d,H_f} = 3.7 Hz, typical of a *syn* stereochemistry.[32a]

Furthermore the low values of J_{H_d,H_c} for **42a** (5.8 Hz) and for **42b** (5.5 Hz) show that the heterocyclic rings are rather flat, due to the newly introduced bulky substituent at C_5.

The data are in agreement with a conformation of **42a** and **42b**, similar to that of the starting material **7a** with a *trans* relationship between H_d and H_c as well as an *anti* and a *syn* stereochemistry between H_d and H_f, respectively.

In order to establish the influence of the aldehyde on the simple diastereoselection, the lithium enolate of **7a** was reacted with acetaldehyde in THF (Scheme 11.40). The reaction proceeds in high yield and with inversion of diastereoselectivity compared with the aldols of benzaldehyde, giving **44a** and **44b** in 25:75 *anti*/*syn* ratio (Table 11.7, entry 8).

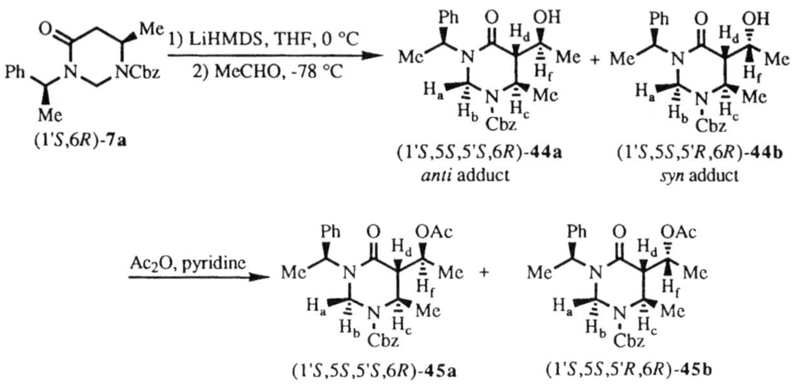

Scheme 11.40

The stereochemical attribution is based on the values of the coupling constants of H_c, H_d, and H_f. In fact in the minor compound **44a** the $J_{H_dH_f}$ = 7.3 Hz is in agreement with an *anti* relationship between H_d and H_f, while for **44b** the value of J_{HdHf} = 5.9 Hz is appropriate for a *syn* stereochemistry.[32a] Moreover by NOEDIF experiments both acetates **45a** and **45b** show an enhancement of the Me on C_6 and of H_a, upon irradiation of H_d. This implies a *trans* relationship between H_d and H_c, although a low value of $J_{H_dH_c}$ (4.6 Hz) was recorded for **44a** and **44b** (Scheme 11.41).

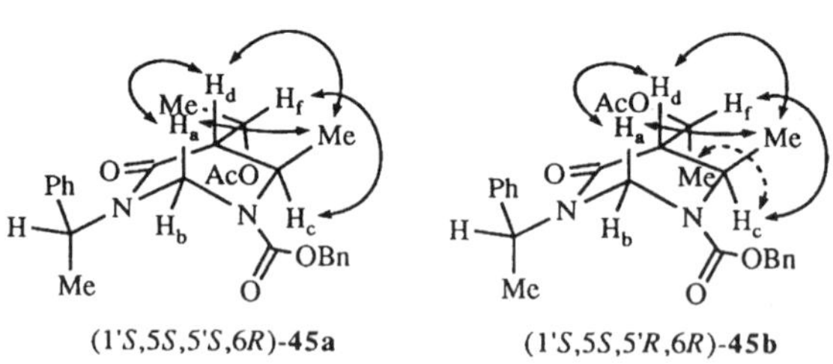

Scheme 11.41

These results confirm the capability of the lithium enolate to afford an *anti* selectivity for aromatic and a *syn* selectivity for aliphatic aldehydes, already reported in the literature.[33b,c,d]

In order to establish the reactivity of perhydropyrimidin-4-one **8a** towards aldol condensation, the lithium and zinc enolates of **8a** have been reacted with benzaldehyde (Scheme 11.42) and acetaldehyde (Scheme 11.43), and the results are reported in Table 11.8.

The metal enolates of **8a** react with benzaldehyde, affording the aldols **46a** and **46b** in good yield, and these are easily separable by flash chromatography. The compounds were characterized by means of ^{1}H- and ^{13}C-NMR spectroscopy.

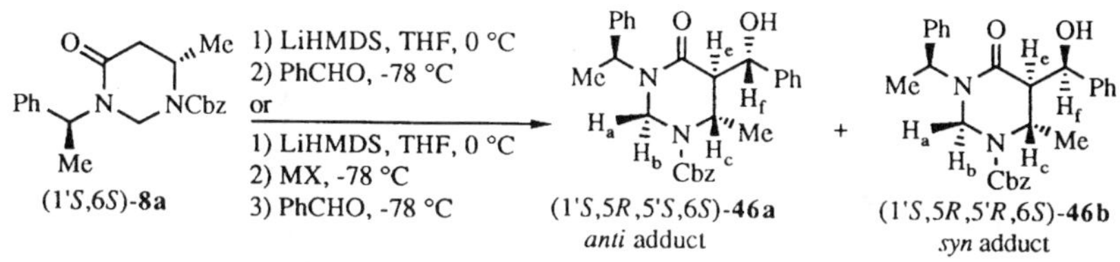

Scheme 11.42

The aldols **47a** and **47b** were obtained in good yield by reaction of the enolates of **8a** with acetaldehyde, and the mixtures of **47a** and **47b** were transformed into the corresponding acetates **48a** and **48b** in 90% yield and easily separated by silica gel chromatography. The structural assignment is made on the basis of the ^{1}H-NMR spectra of **47a** and **47b**.

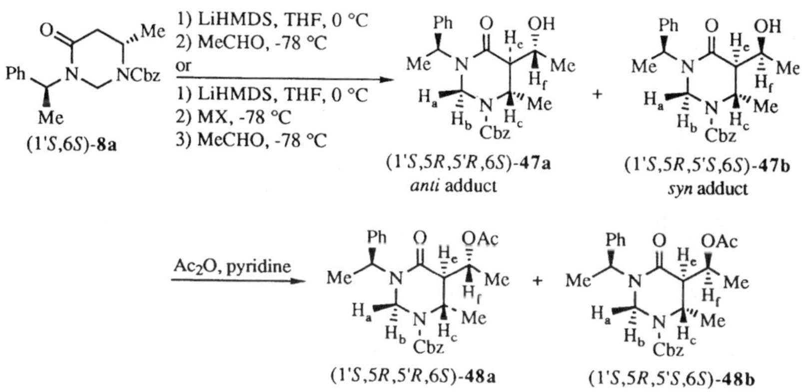

Scheme 11.43

Table 11.8 DIASTEREOMERIC PRODUCT RATIO AND CHEMICAL YIELDS FOR ALDOL REACTIONS OF (1′S,6S)-6-METHYL PERHYDROPYRIMIDIN-4-ONE **8a**

Entry	Electrophile	*M*	Solvent	*anti/syn* ratio[a]	Total yield[b]
1	PhCHO	Li	THF	66:34	82
2	PhCHO	ZnCl	ether/THF	90.10	76
3	MeCHO	Li	THF	25:75	85
4	MeCHO	ZnCl	ether/THF	67:33	82

[a] The *anti/syn* ratios were determined on crude reaction mixtures by means of ^{1}H-NMR and ^{13}C-NMR.

[b] All reported yields were based on isolated and purified products.

The NOEDIF experiments performed on **46a** show the enhancement of the Me on C_6 and H_b upon irradiation of H_e and vice versa, which implies a *trans* relationship between the substituents on C_5 and C_6, as previously observed (Scheme 11.44).

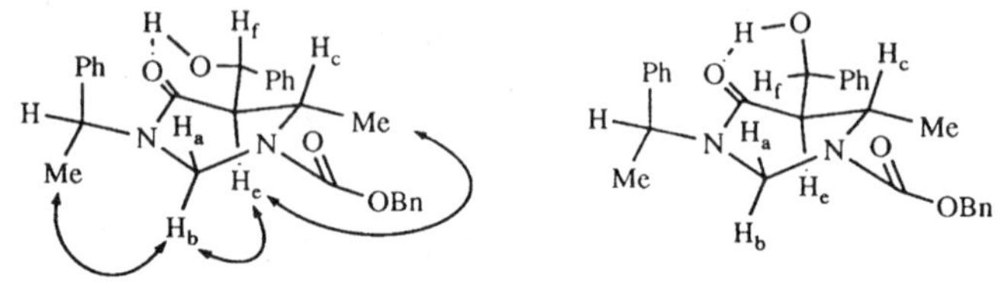

(1'*S*,5*R*,5'*S*,6*S*)-**46a** (1'*S*,5*R*,5'*R*,6*S*)-**46b**

Scheme 11.44

Furthermore, the ^{1}H-NMR spectra of the compounds **46a** and **46b** recorded in $CDCl_3$ at –40°C show a preferential conformation with a hydrogen bond between the hydroxyl group and the carbonyl of the heterocycle. This conformation was confirmed by the hydroxyl hydrogen signal of the rotamers, which appears as a doublet at δ 4.41 and 4.35 (J = 6.0 and 6.3 Hz) and δ 3.93 and 4.17 (J = 5.7 and 6.7 Hz), respectively, and does not change upon varying the concentration of the sample (Table 11.9). The values of the coupling constant J_{H_e,H_f} of 6.9 and 3.2–3.6 Hz accounts for the *anti* and for the *syn* stereochemistry, respectively.[32a]

Although the origin of the different stereoselections among zinc, lithium, titanium, aluminum, and boron enolates is not completely understood, a possible rationalization of the results obtained for pyrimidin-4-ones with various aldehydes takes into account the interactions between the enolate and the electrophile in the transition state, as a function of the metal nature.

The *anti*/*syn* ratio strongly depends on the length of the metal–oxygen bond and increases with its increase. For instance, for metals with longer metal–oxygen distances like zinc, titanium, and lithium (1.7–2.0 Å),[32a,33e] the formation of the *anti* adduct is favored, while for the shorter oxygen–boron bond (1.3–1.4 Å),[32a,33e] the preferential formation of the *syn* adduct is observed.

Table 11.9 SIGNIFICANT ^{1}H-NMR DATA OF ALDOLS 42 AND 46 IN $CDCl_3$ AT 300 MHz AT –40°C

Product	δH_d	J_{H_d,H_f}	δH_e	J_{H_e,H_f}	δ_{H_f}	δ_{OH}	J_{OH,H_f}
42a	2.57	7.4	—	—	4.81	4.79	3.9
					4.87[a]		
42b	2.67	3.7	—	—	5.46	3.61	3.7
						4.35	6.3
46a	—	—	2.60	6.9	4.74	4.41[a]	6.0[a]
			2.71	3.6		3.93	5.7
46b	—	—	2.75[a]	3.2[a]	5.37	4.17[a]	6.7[a]

[a] Owing to the presence of two conformers, two chemical shift values have been reported.

Furthermore entries 1 and 8 of Table 11.7 show a reversal of diastereoselectivity if acetaldehyde or benzaldehyde is used as electrophile. This result has already been observed[33b,c,d] and can be ascribed not only to the differences of steric hindrance between the methyl and the phenyl group, but also to stereoelectronic factors.

Thus the facial diastereoselectivity is controlled by the substituent on the perhydropyrimidin-4-one, while the simple diastereoselectivity is under the control of the metal, due to the length of the metal–oxygen bond, and strongly depends on the nature of the aldehyde.

11.3.2 Synthesis of α-Hydroxymethyl β-Amino Acids

As a consequence of the metallation of perhydropyrimidin-4-ones **17** and **18**, a simple route to a diastereoselective synthesis of bicycles **49** and **51** has been developed.[14] These compounds exhibit peculiar chemical and spectroscopic behavior and can be easily transformed through few easy steps into enantiomerically pure α-hydroxymethyl β-amino acids.

The synthesis of the bicycles can be accomplished readily by reacting compounds **17** or **18** with 1 equivalent of LiHMDS in THF at –78°C. The cyclization reaction is complete in 30 min, and the products are recovered in high yield after the usual workup (Scheme 11.45).

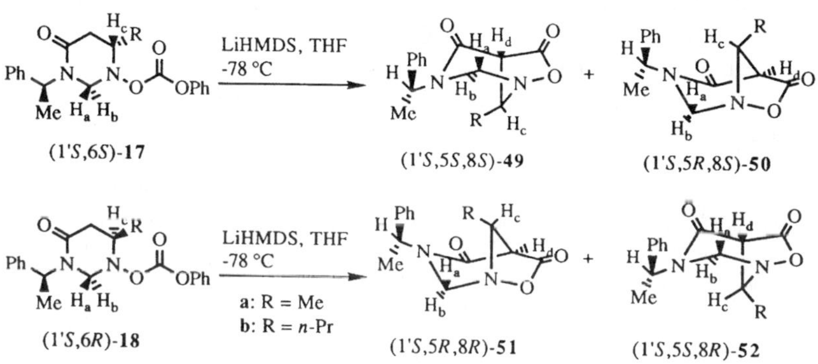

Scheme 11.45

The reaction shows good diastereoselectivity when **17a** and **18a** are reacted under the reported conditions, affording preferentially the *trans* isomers **49a** and **51a** (Table 11.10). At –78°C a small amount of the *cis* derivatives **50a** and **52a** is observed and

Table 11.10 DIASTEREOMERIC RATIOS FOR THE CYCLIZATION OF 17 AND 18 AT –78°C

R	**49/50** ratio[a]	**51/52** ratio[a]
Me	86/14	87/13
n-Pr	>99/1	>99/1

[a] Determined by ^{1}H-NMR analysis

Table 11.11 ^{1}H-NMR DATA OF BICYCLES 49–52

Product	δ_{H_a}	δ_{H_b}	δ_{H_c}	δ_{H_d}	δ_{CH_3}	J_{H_c,H_d}	J_{H_a,H_d}	J_{H_b,H_d}
(1′*S*,5*S*,6*S*)-**49a**	4.06	3.96[a]	2.35	2.87	1.22	5.4	—	1.5
(1′*S*,5*R*,6*S*)-**50a**	3.85[a]	4.47	1.60	2.58	1.18	1.9	1.9	—
(1′*S*,5*R*,6*R*)-**51a**	3.64[a]	4.49	2.21	2.82	0.72	5.5	1.6	—
(1′*S*,5*R*,6*R*)-**52a**	4.06	4.15[a]	1.83	2.58	1.25	2.4	—	1.8
(1′*S*,5*S*,6*S*)-**49b**	4.08	3.99[a]	2.26	2.91	—	6.1	—	1.8
(1′*S*,5*R*,6*R*)**51b**	3.70[a]	4.49	2.09	2.83	—	6.1	2.0	—

[a] These signals show a long-range coupling with H_d.

increases if the reaction takes place at a higher temperature (i.e., –20°C). On the other hand, with larger substituents on **17b** and **18b**, like the *n*-propyl group, no *cis* derivative is recovered even if the reaction is carried out at –20°C.

The configuration of the newly formed asymmetric center at C_5 was attributed by means of the ^{1}H-NMR spectra of the product. The coupling constants J_{H_c,H_d} of **49**, **50**, **51**, and **52** are diagnostic and account for the required stereochemistry (Table 11.11). In fact products **49** and **51** have J_{H_c,H_d} larger than products **50** and **52** (5.4–6.1 Hz versus 1.9–2.4 Hz). Furthermore, owing to the rigid conformation of the molecule, the hydrogens abutting the phenyl of the (*S*)-phenylethyl moiety are strongly shielded (i.e., H_c in **50** and CH_3 in **51**; see Table 11.11). A 1,5 long-range coupling constant (1.5–2.0 Hz) was also recorded between H_d and H_a (products **50** and **51**) or H_d and H_b (products **49** and **52**) depending on the molecular structure.

The NOEDIF experiments performed on **49a**, **50a**, and **51b** are in agreement with the assigned structures (Scheme 11.46).

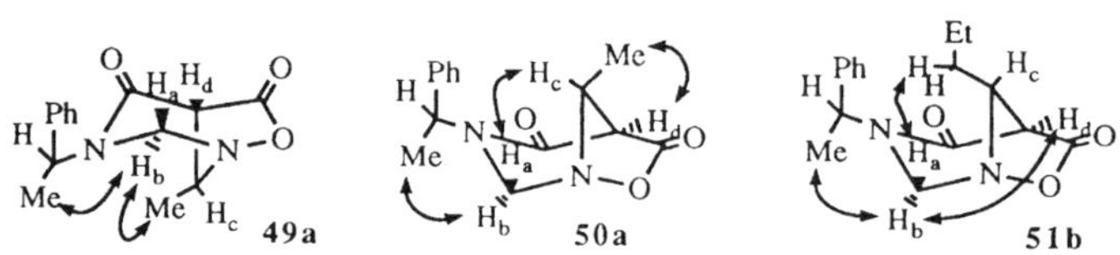

Scheme 11.46

In order to synthesize the optically active α-hydroxymethyl β-amino acid derivative **54**, the bicycle (1′*S*,5*R*,6*R*)-**51b** was easily transformed in high yield by treatment with dry methanol in the presence of Amberlyst A-26 at reflux for 3 h to the ester **53**, which was selectively reduced into the corresponding primary alcohol **54** with $LiAlH_4$ at 0°C in THF (Scheme 11.47). Compound **54** was fully characterized and

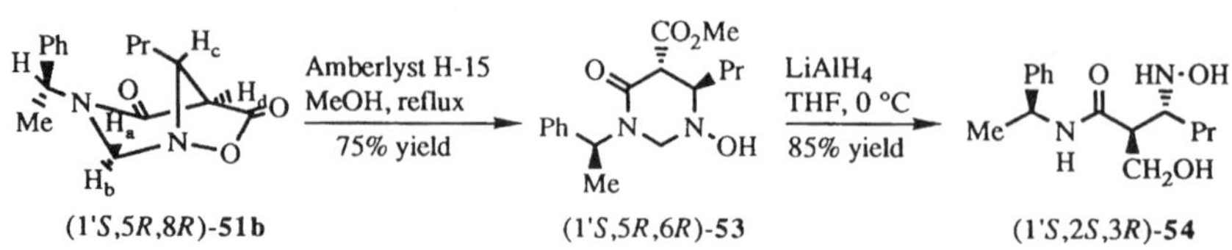

Scheme 11.47

can be easily transformed into the corresponding amino acids by known synthetic methods: reduction of the N–O bond[36] and hydrolysis of the amide group.

11.4 Introduction of a Heteroatom at C_5 of Perhydropyrimidin-4-Ones

The metal-enolates of enantiomerically pure 6-alkyl perhydropyrimidin-4-ones **7**, **8** and 1-benzoyl-3-methyl-6(*S*)-phenyl-perhydropyrimidin-4-one have been functionalized at C_5 by reaction with reagents able to donate a positive hydroxyl group or a positive halogen. With this procedure 5-hydroxy and 5-halogen perhydropyrimidin-4-ones have been obtained, as useful precursors of enantiomerically pure 2-hydroxy-3-amino acids or 2,3-diamino acids.

A variety of biologically interesting compounds containing α-hydroxy β-amino acids are promising anticancer, anti-AIDS, or antihypertensive agents. One of the most famous molecule containing an α-hydroxy β-amino acid is Taxol,[37] which is constituted by a polyoxygenated diterpene and (2*R*,3*S*)-phenylisoserine. It is a strong antineoplastic agent, and physicians are currently studying its effect on nearly every known neoplasm.

11.4.1 Hydroxylation of Perhydropyrimidin-4-Ones: Synthesis of α-Hydroxy β-Amino Acids

Recently many reagents that are able to donate a positive hydroxyl group have been prepared. They belong to two families: the oxaziridine derivatives[38] and the peroxomolybdenum derivatives.[39] The C_5-hydroxylation has been performed with several reagents, in order to compare the diastereoselectivities and yields of the reactions and to find the hydroxylating agent of choice for this class of compounds. Indeed 5-hydroxy 6-alkyl perhydropyrimidin-4-ones furnish the corresponding α-hydroxy β-amino acids after a simple acid hydrolysis. Thus the enolates of (1′*S*,6*R*)-1′-phenylethyl-6-alkyl perhydropyrimidin-4-ones **7a** and **7c** were treated with MoOPH, (–)- and (+)-camphorsulfonyloxaziridine[40] (Scheme 11.48). The results are reported in Table 11.12.

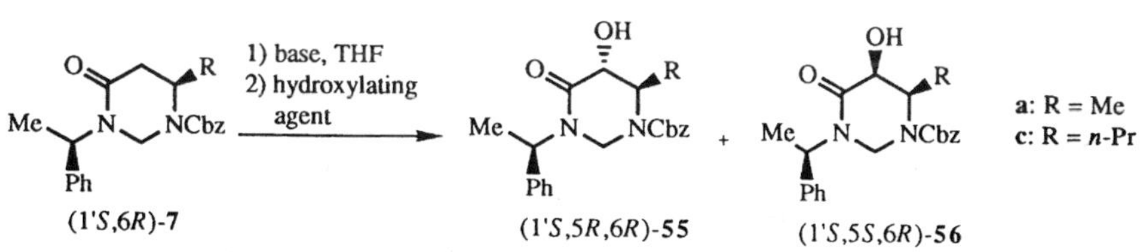

Scheme 11.48

As Table 11.12 shows, different results have been obtained by changing the hydroxylating agent. Indeed when MoOPH was utilized as hydroxylating agent (entry 1), an excess of the 5,6-*trans* derivatives **55** was obtained, confirming the high 5,6-*trans* selectivity previously achieved in the alkylation and in the aldol condensation

Table 11.12 DIASTEREOMERIC PRODUCT RATIOS AND CHEMICAL YIELD FOR THE HYDROXYLATION REACTION OF PERHYDROPYRIMIDIN-4-ONES **7a** AND **7c**

Entry	SM	*R*	Base	Hydroxylating agent (equiv.)	Temp. (°C)	Time (h)	Yield (%)	DR[a] **55/56**
1	**7a**	Me	LiHMDS	MoOPH (2)	–78-r.t.	16	61	80:20
2	**7c**	*n*-Pr	LiHMDS	MoOPH (2)	–78-r.t.	16	40	99:1
3	**7a**	Me	LiHMDS	(–)-camphor. (1.2)	–78-r.t.	16	66	78:22
4	**7c**	*n*-Pr	LiHMDS	(–)-camphor. (1.2)	–78-r.t.	16	23	>99:1
5	**7c**	*n*-Pr	LiHMDS	(–)-camphor. (1.2)	–20-r.t.	16	29	>99:1
6	**7a**	Me	LiHMDS	(+)-camphor. (1.2)	–78-r.t.	16	78	66:34
7	**7c**	*n*-Pr	LiHMDS	(+)-camphor. (1.2)	–78-r.t.	16	40	>99:1
8	**7c**	*n*-Pr	LiHMDS	(+)-camphor. (1.2)	–20-r.t.	16	72	>99:1

[a] The diastereomeric ratios were determined by means of HPLC analysis.

of these substrates. Furthermore (–)- and (+)-camphorsulfonyloxaziridine were employed as hydroxylating agents. Both enantiomers afforded complete diastereoselection in the hydroxylation of **7c**, but (+)-camphorsulfonyloxaziridine furnished better yield.

The configuration of the newly formed asymmetric center was determined by means of ^{1}H-NMR analysis. Indeed the 5,6-*trans* derivatives present a H_5-H_6 coupling constant (J_{H_5,H_6} = 9–11 Hz) always larger than the 5,6-*cis* derivatives (J_{H_5,H_6} = 6–7 Hz) and typical of a 1,2-*trans* relationship of the hydrogens in six-membered rings.

Also (1′*S*,6*S*)-1-phenylethyl-6-alkyl perhydropyrimidin-4-ones **8a** and **8c** were hydroxylated at C_5 under the same conditions (Scheme 11.49). The results are shown in Table 11.13.

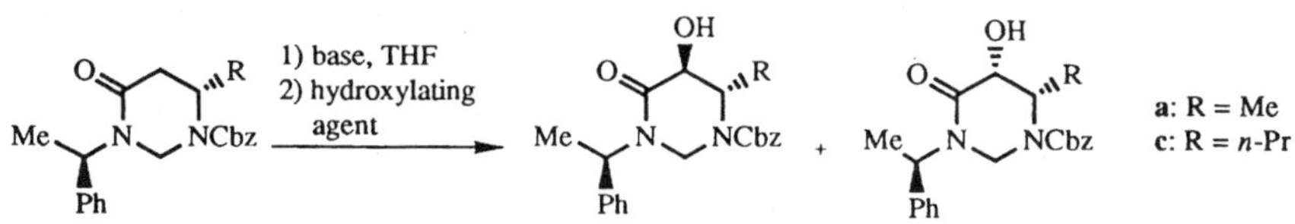

(1′*S*,6*S*)-**8** (1′*S*,5*S*,6*S*)-**57** (1′*S*,5*R*,6*S*)-**58**

Scheme 11.49

A global correlation of these results with what we obtained in Table 11.12 shows that the hydroxylation of (1′*S*,6*S*)-perhydropyrimidin-4-ones **8** proceeds with different diastereoselection from (1′*S*,6*R*)-perhydropyrimidin-4-ones **7**. This is quite a surprising result, because the two diastereoisomers behaved in the same way in the previously analyzed reactions. This dissimilarity can probably be attributed to the steric hindrance of the hydroxylating agents, which are bulkier than the alkylating agents or benzaldehyde. Thus these reagents interact both with the substituent at C_6 of the heterocycle, which can have *R* or *S* configuration, and with the phenylethyl group at $C_{1'}$, which always has the *S* configuration.

Table 11.13 DIASTEREOMERIC PRODUCT RATIOS AND CHEMICAL YIELD FOR THE HYDROXYLATION REACTION OF PERHYDROPYRIMIDIN-4-ONES **8a** AND **8c**

Entry	SM	*R*	Base	Hydroxylating agent (equiv.)	Temp. (°C)	Time (h)	Yield (%)	DR[a] **57/58**
1	**8a**	Me	LiHMDS	MoOPH (2)	–78-r.t.	16	63	72:28
2	**8c**	*n*-Pr	LiHMDS	MoOPH (2)	–78-r.t.	16	44	67:33
3	**8a**	Me	LiHMDS	(–)-camphor. (1.2)	–78-r.t.	16	75	29:71
4	**8c**	*n*-Pr	LiHMDS	(–)-camphor. (1.2)	–78-r.t.	16	6	82:18
5	**8c**	*n*-Pr	LiHMDS	(–)-camphor. (1.2)	–20-r.t.	16	35	99:1
6	**8a**	Me	LiHMDS	(+)-camphor. (1.2)	–78-r.t.	16	78	45:55
7	**8c**	*n*-Pr	LiHMDS	(+)-camphor. (1.2)	–78-r.t.	16	31	72:28
8	**8c**	*n*-Pr	LiHMDS	(+)-camphor. (1.2)	–20-r.t.	16	60	83:17

[a] The diastereomeric ratios were determined by means of HPLC analysis.

The hydroxylation of **8a** and **8c** furnishes 5-hydroxy-6-alkyl-perhydropyrimidin-4-ones with a lower diastereoselection (Table 11.13). When MoOPH is utilized (entries 1 and 2), the 5,6-*trans* derivative is preferentially obtained and the diastereoselection decreases with the increase of the alkyl group at C_6.

(–)- And (+)-camphorsulfonyloxaziridine furnish different results, thus showing a strong matching and mismatching effect.[41] Indeed (–)-camphorsulfonyloxaziridine affords preferentially the 5,6-*cis* derivative with **8a** and reacts with low yield with **8c**, while (+)-camphorsulfonyloxaziridine reacts with no diastereoselection with **8a** and furnishes preferentially the 5,6-*trans* adduct **57c** with **8c**.

Moreover the 2-hydroxy-3-amino acids L-isothreonine **59** and D-alloisothreonine **60** can be easily obtained by acid hydrolysis of the corresponding 5-hydroxy perhydropyrimidin-4-ones, utilizing the same method previously described for 6-alkyl and 5,6-dialkyl perhydropyrimidin-4-ones (Scheme 11.50). Thus, by reaction of pyrimidin-4-ones (1′*S*,5*S*,6*R*)-**56a** and (1′*S*,5*R*,6*R*)-**55a** with 6*N* HCl at reflux for 30 h, complete hydrolysis of the heterocyclic rings was achieved with no racemization (2*S*,3*R*)-L-isothreonine **59** and (2*R*,3*R*)-D-alloisothreonine **60** were obtained pure after chromatography on a cation-exchange resin, as confirmed by ^{1}H-NMR analysis and by the comparison of the $[\alpha]_D$ values with the data reported in the literature.[42]

Finally a complete inversion of the absolute configuration at C_5 can be obtained by means of the Mitsunobu reaction.[43] Indeed (1′*S*,5*S*,6*S*)-5-hydroxy-6-methyl-perhydropyrimidin-4-one **57a** was treated with triphenylphosphine, benzoic acid, and DEAD in dry THF for 6 h at room temperature (Scheme 11.51). The benzoate deriv-

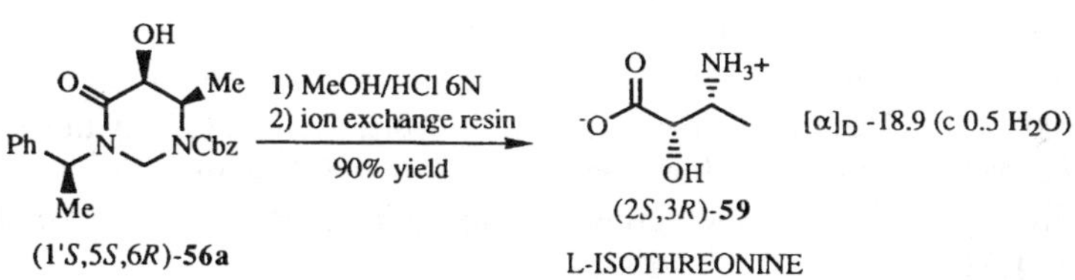

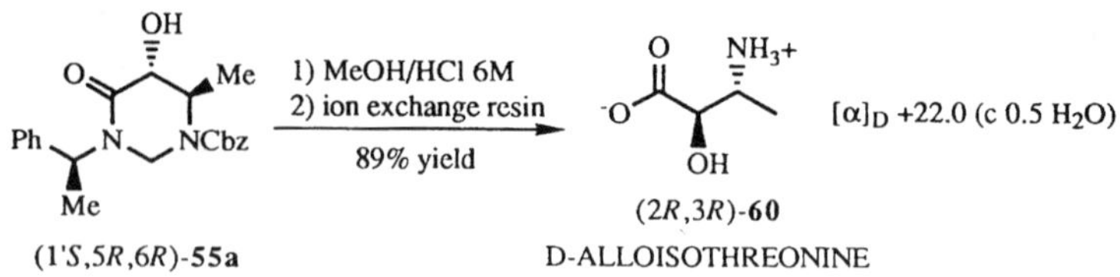

Scheme 11.50

ative of the C_5-epimer **61** was readily obtained with high yield and total inversion of configuration.

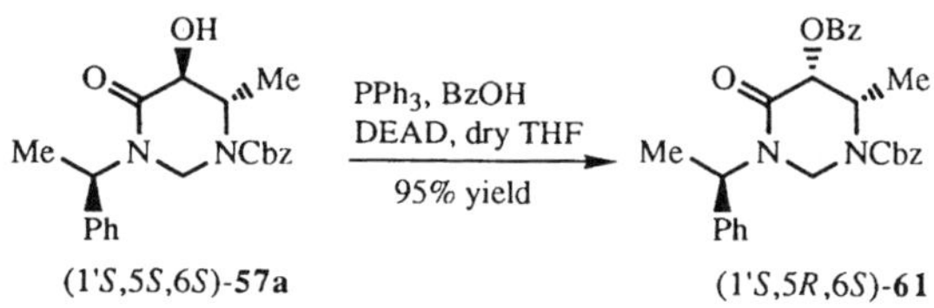

Scheme 11.51

These results demonstrate that the hydroxylation at C_5 of (1′S)-phenylethyl-6-alkyl-perhydropyrimidin-4-ones affords the desired hydroxy derivative with high yield and good diastereoselection, if the reaction conditions are carefully studied for each substrate.

Following the same pathway, N-benzoyl (2*R*,3*R*)-3-phenylisoserine **65** has been obtained by hydroxylation of 1-benzoyl-3-methyl-6(*S*)-phenylperhydropyrimidin-4-one **63**,[44] which has been prepared via the highly diastereoselective addition of the lithium amide of (*R*)-benzyl(α-methylbenzyl)amine[45] to *tert*-butyl cinnamate, followed by hydrogenolysis and esterification. The methyl ester (*S*)-**62** was transformed into the perhydropyrimidin-4-one (*S*)-**63**, following the method previously described for the preparation of (*S*)-**29** (Scheme 11.52).

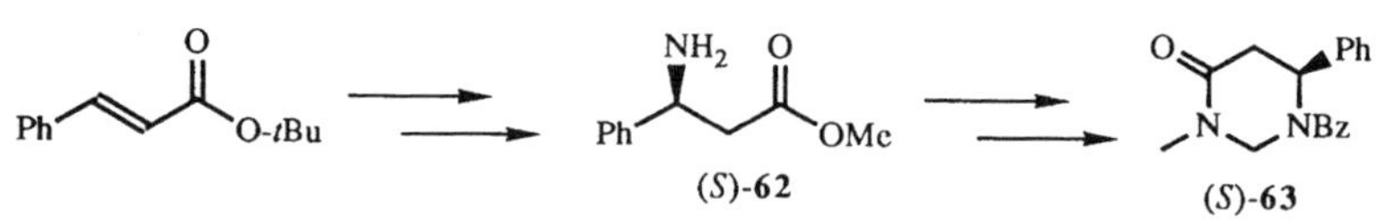

Scheme 11.52

The enolization of (*S*)-**63** using lithium diisopropylamide (LDA)[24] was followed by hydroxylation at –78°C with (+)-camphorsulfonyloxaziridine[46] to afford the expected *trans* product (5*R*,6*R*)-**64** (Scheme 11.53). The high diastereoselectivity observed in the hydroxylation of (*S*)-**63** suggests a remarkable stereodirecting ability of the phenyl group at C_6 in the perhydropyrimidin-4-one system. Finally the hydrolysis of the heterocycle was achieved by heating to 100–105°C with 6*N* HCl in a sealed tube. The hydrochloride was then converted to the known[47] N-benzoylated methyl ester (2*R*,3*R*)-**65**.

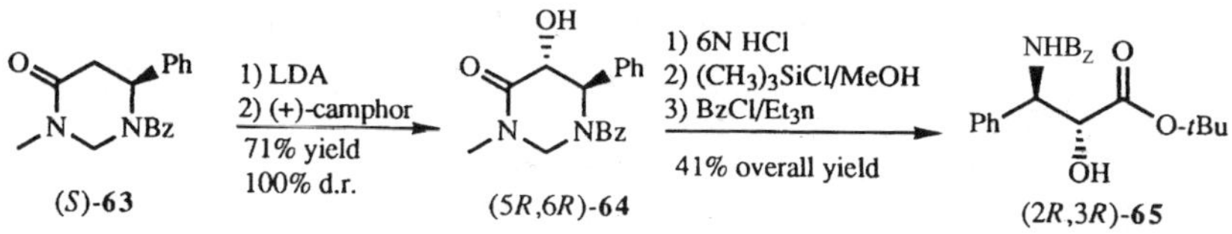

Scheme 11.53

11.4.2 Halogenation of Perhydropyrimidin-4-Ones

6-Methyl-perhydropyrimidin-4-ones **7a** and **8a** have been halogenated at C_5 by reaction of the corresponding lithium-enolates with benzensulfonyl bromide to obtain 5-bromo-6-methyl-perhydropyrimidin-4-ones and with *p*-toluensulfonyl chloride to obtain 5-chloro-6-methyl-perhydropyrimidin-4-ones[48] (Scheme 11.54). The halogen at C_5 can subsequently be transformed into a hydroxyl group or an amino group to obtain hydroxy amino acids or diamino acids.

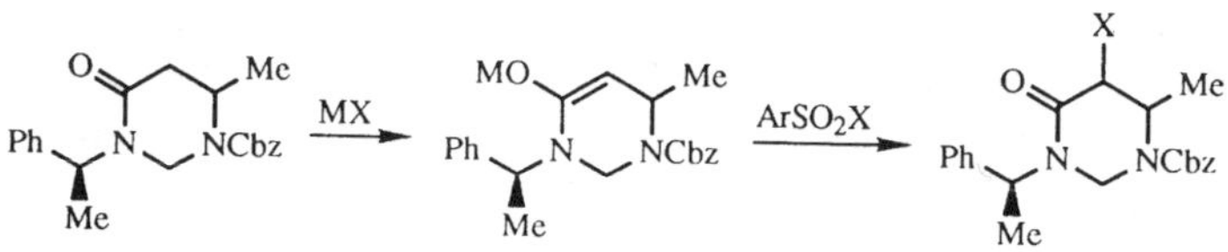

Scheme 11.54

The halogenation of (1′*S*,6*R*)-3-(1′-phenylethyl)-6-methyl-perhydropyrimidin-4-one **7a** was performed by treatment of the corresponding lithium enolate, generated with lithium hexamethyldisilazide (LiHMDS) in dry THF, with benzenesulfonyl bromide and *p*-toluenesulfonyl chloride for 1 h at –30°C, as depicted in Scheme 11.55. The results are reported in Table 11.14.

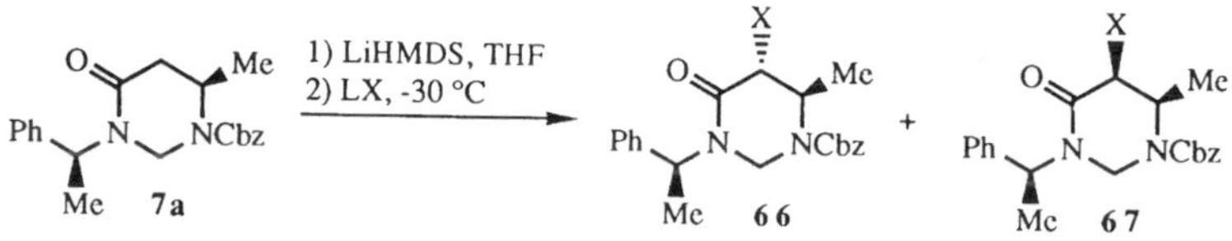

Scheme 11.55

Table 11.14 DIASTEREOMERIC PRODUCT RATIOS AND CHEMICAL YIELD FOR THE HALOGENATION REACTION OF 6-METHYL PERHYDROPYRIMIDIN-4-ONE 7a

Entry	*X*	Reagent	Yield (%)	DR **66/67**[a]
1	Br	$PhSO_2Br$	98	38:62
2	Cl	TsCl	74	10:90

[a] The diastereomeric ratios were determined by means of HPLC analysis.

On the other hand, the halogenation of the lithium enolate of (1′*S*,6*S*)-3-(1′-phenylethyl)-6-methyl-perhydropyrimidin-4-one **8a** with benzensulfonyl bromide and *p*-toluensulfonyl chloride for 1 h at –30°C furnished the corresponding 5-bromo- and 5-chloro-6-methyl-perhydropyrimidin-4-ones **68** and **69**, as summarized in Scheme 11.56. The results are reported in Table 11.15.

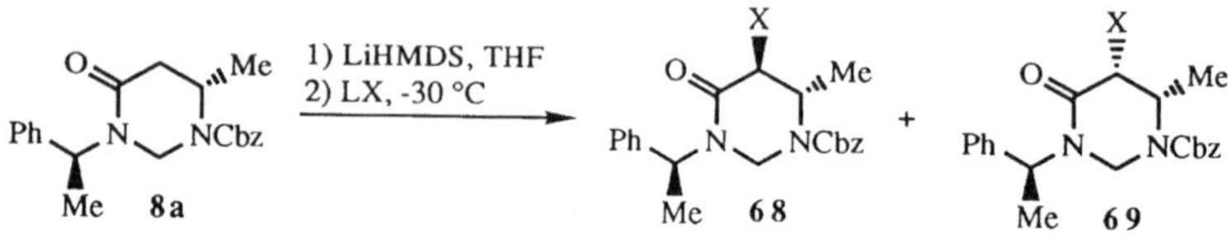

Scheme 11.56

The bromination reaction of **7a** and **8a** proceeds with high yield and modest diastereoselection with an excess of the 5,6-*trans* derivative in both cases.

In contrast, the reaction of **7a** and **8a** with *p*-toluensulfonyl chloride furnishes the 5-chloro-6-methyl-perhydropyrimidin-4-ones with different yields and diastereomeric ratios. Indeed the chlorination of **7a** affords the adducts with good yield and increased diastereoselectivity, while **8a** yields the 5-chloro derivatives with lower diastereoselectivity (58:42 dr) and excellent yield.

Table 11.15 DIASTEREOMERIC PRODUCT RATIOS AND CHEMICAL YIELD FOR THE HALOGENATION REACTION OF 6-METHYL PERHYDROPYRIMIDIN-4-ONE 8a

Entry	*X*	Reagent	Yield (%)	DR **68/69**[a]
1	Br	$PhSO_2Br$	98	36:64
2	Cl	TsCl	82	28:72

[a] The diastereomeric ratios were determined by means of HPLC analysis.

11.5 References

1. For recent reviews on β-amino acids, see: (a) D. C. Cole, *Tetrahedron* **50**, 9517 (1994); (b) E. Juaristi, D. Quintana, J. Escalante, *Aldrichimica Acta* **27**, 3 (1994).

2. (a) V. J. Hruby, *Life Science* **32**, 189 (1982); (b) A. F. Spatola, in *Chemistry and Biochemistry of Amino Acids, Peptides and Proteins*, B. Weinstein, Ed. (Marcel Dekker, N.Y., 1983) Vol. VII, p. 267; (c) R. A. Wiley, D. H. Rich, *Medicinal Research Review* **13**, 328 (1993).

3. J. Brunner, *Chem. Soc. Review* 183 (1993).

4. (a) P. A. Bartlett, *Tetrahedron* **36**, 2 (1980); (b) P. A. Bartlett, in *Asymmetric Synthesis*, Morrison J. D., Ed. (Academic Press, New York, 1984), Vol. 3; (c) G. Cardillo, M. Orena, *Tetrahedron* **46**, 3321 (1990).

5. For the use of amidals in cyclofunctionalization reactions, see: K. E. Harding, T. H. Marman, D. H. Nam, *Tetrahedron* **44**, 5605 (1988) and references therein.

6. R. Amoroso, G. Cardillo, C. Tomasini, P. Tortoreto, *J. Org. Chem.* **57**, 1082–87 (1992).

7. H. E. Zaugg, *Synthesis* 85 and 181 (1984).

8. S. Gronowitz, Z. Lidert, *Synthesis* 810 (1979).

9. For enantioselective cyclizations starting from chiral amidals, see: (a) K. E. Harding, D. R. Hollingsworth, J. Reibenspies, *Tetrahedron Lett.* **30**, 4775 (1989); (b) J. M. Takacs, M. A. Helle, L. Yang, *Tetrahedron Lett.* **30**, 1977 (1989); (c) R. Amoroso, G. Cardillo, C. Tomasini, *Tetrahedron Lett.* **32**, 1971 (1991).

10. Reprinted from R. Amoroso, G. Cardillo, C. Tomasini, *Heterocycles* **34**, 349–55, Copyright 1992, with kind permission from The Japan Institute of Heterocyclic Chemistry, 1-1-7-804, Motoakasaka, Minato-Ku, Tokio 107, Japan.

11. I. Braschi, G. Cardillo, C. Tomasini, R. Venezia, *J. Org. Chem.* **59**, 7292–98 (1994).

12. R. E. Steiger, *Organic Syntheses, Coll. Vol. 3* 664 (1955).

13. Reprinted from *Tetrahedron: Asymmetry* **4**, R. Amoroso, G. Cardillo, G. Mobbili, C. Tomasini, Metal-Assisted Aldol Condensation of Chiral 6-Methyl Perihydropyrimidin-4-ones, 2241–54, Copyright 1993, with kind permission from Elsevier Science Ltd, The Boulevard, Langford Lane, Kidlington OX5 1GB, UK.

14. Reprinted from *Tetrahedron* **50**, I. Braschi, G. Cardillo, C. Tomasini, Synthesis and Stereochemical Behaviour of 8-Alkyl-3-(1′-phenylethyl)-7-oxa-1,3-diazabicyclo-[3.2.1]-octane-4,6-diones Directed Towards the Synthesis of α-Substituted β-Amino Acids, 2955–64, Copyright 1994, with kind permission from Elsevier Science Ltd, The Boulevard, Langford Lane, Kidlington OX5 1GB, UK.

15. (a) G. Cardillo, M. Orena, S. Sandri, C. Tomasini, *Tetrahedron* **43**, 2505 (1987); (b) A. Bongini, G. Cardillo, M. Orena, G. Porzi, S. Sandri, *Tetrahedron* **43**, 4377 (1987).

16. N. L. Allinger, Y. H. Yuh, *Program MM2P (QCPE)*, updated by D. C. Rohrer (1984).

17. A. Bongini, G. Cardillo, M. Orena, G. Porzi, S. Sandri, *Chemistry Lett.* 87 (1988).

18. H. Hiemstra, H. P. Fortgens, S. Stegenga, W. N. Speckamp, *Tetrahedron Lett.* **26**, 3151 (1985).

19. E. Graf, H. Boeddeker, *Liebigs Ann. Chem.* **613**, 111 (1958).

20. K. Balenovic, D. Cerar, Z. Fuks, *J. Chem. Soc.* 3316 (1952).

21. Reprinted from *Tetrahedron Lett.* **33**, R. Amoroso, G. Cardillo, C. Tomasini, Highly Diastereoselective Alkylation of 6-Methyl Perihydropyrimidini-4-Ones Directed Towards the Synthesis of α-Substituted β-Amino Acids, 2725–28, Copyright 1992, with kind permission from Elsevier Science Ltd, The Boulevard, Langford Lane, Kidlington OX5 1GB, UK.

22. For 1,2 asymmetric induction in acyclic systems in the synthesis of α-substituted β-amino acids, see: (a) De. Seebach, H. Estermann, *Tetrahedron Lett.* **28**, 3103 (1987); (b) H. Estermann, D. Seebach, *Helv. Chim. Act* **71**, 1824 (1988); (c) G. J. McGarvey, J. M. Williams, R. N. Hiner, Y. Matsubara, T. Oh, *J. Am. Chem. Soc.* **108**, 4943 (1986).

23. For the alkylation of chiral 1,3-dioxanones in the synthesis of α-substituted β-hydroxy acids, see: (a) D. Seebach, J. Zimmermann, *Helv. Chim. Acta* **69**, 1147 (1986); (b) J. Zimmermann, D. Seebach, *Helv. Chim. Acta* **70**, 1104 (1987); (c) J. P. Genet, S. Juge, S. Mallart, *Tetrahedron Lett.* **29**, 6765 (1988).

24. E. Juaristi, J. Escalante, B. Lamatsch, D. Seebach, *J. Org. Chem.* **57**, 2396 (1992).

25. E. Juaristi, J. Escalante, *J. Org. Chem.* **58**, 2282–85 (1993).

26. H. Estermann, D. Seebach, *Helv. Chim. Acta* **71**, 1824 (1988).

27. D. C. Carter, E. R. Moore, J. S. Mynderse, W. P. Niemczura, J. S. Todd, *J. Org. Chem.* **49**, 236 (1984).

28. G. R. Pettit, Y. Kamano, H. Kizu, C. Dufresne, C. L. Herald, R. J. Bontems, J. M. Schmidt, F. E. Boettner, R. A. Nieman, *Heterocycles* **28**, 553 (1989).

29. R. B. Bates, S. Gangwar, *Tetrahedron: Asymmetry* **4**, 69 (1993) and references therein.

30. H. Huang, N. Iwasawa, T. Mukayama, *Chemistry Letters* 1465 (1984).

31. D. A. Evans, in *Asymmetric Synthesis*, J. D. Morrison, Ed. (Academic Press, New York, 1984), Vol. 3, p. 25, and references therein.

32. (a) C. Heathcock, in *Asymmetric Synthesis*; J. D. Morrison, Ed. (Academic Press, New York, 1984), Vol. 3, p. 111; (b) S. Masamune, W. Choy, J. S. Petersen, L. R. Sita, *Angew. Chem., Int. Ed. Engl.* **24**, 1 (1985).

33. (a) H. O. House, D. S. Crumrine, A. Y. Teranishi, H. D. Olmstead, *J. Am. Chem. Soc.* **95**, 3310 (1973); (b) A. Abdel-Magid, I. Lantos, L. N. Pridgen, *Tetrahedron Lett.* **25**, 3373 (1984); (c) A. Abdel-Magid, L. N. Pridgen, D. S. Eggleston, I. Lantos, *J. Am. Chem. Soc.* **108**, 4595 (1986) and references therein; (d) L. Pridgen, A. Abdel-Magid, I. Lantos, *Tetrahedron Lett.* **30**, 5539 (1989); (e) M. Nerz-Stormes, E. R. Thornton, *J. Org. Chem.* **56**, 2489 (1991) and references therein.

34. M. T. Reetz, R. Steinbach, K. Kesseler, *Angew. Chem. Suppl.* 1899 (1982).

35. W. Gerrard, M. F. Lappert, *J. Chem. Soc.* 3084 (1955).

36. E. Santaniello, A. Manzocchi, *Synthesis* 698 (1977).

37. K. C. Nicolau, W.-M. Dai, R. K. Guy, *Angew. Chem. Int. Ed. Engl.* **33**, 15 (1994) and references therein.

38. For recent review see: F. A. Davis, B.-C. Chen, *Chem. Rev.* **92**, 919 (1992).

39. E. Vedejs, D. A. Engler, J. E. Telschow, *J. Org. Chem.* **43**, 188 (1978).

40. G. Cardillo, A. Tolomelli, C. Tomasini, *Tetrahedron* **51**, 11831 (1995).

41. S. Masamune, W. Choy, J. S. Petersen, L. R. Sita, *Angew. Chem. Int. Ed. Engl.* **24**, 1 (1985).

42. J.-P. Wolf, H. Pfander, *Helv. Chem. Acta* **70**, 116 (1987).

43. O. Mitsunobu, *Synthesis* 1 (1981).

44. J. Escalante, E. Juaristi, *Tetrahedron Lett.* **36**, 4397 (1987).

45. (a) M. E. Bunnage, A. N. Chernega, S. G. Davies, C. J. Goodwin, *J. Chem. Soc., Perkin Trans. I* 2373 (1994); (b) for the preparation of enantiopure alkyl (α-methylbenzyl)amines, see: E. Juaristi, P. Murer, D. Seebach, *Synthesis* 1243 (1993).

46. F. A. Davis, R. T. Reddy, R. E. Reddy, *J. Org. Chem.* **56**, 6387 (1992).

47. Z.-M. Wang, H. C. Kolb, K. B. Sharpless, *J. Org. Chem.* **59**, 5104 (1994).

48. G. Cardillo, A. Tolomelli, C. Tomasini, unpublished results.

CHAPTER

12

Asymmetric Synthesis of β-Aryl- and β-Alkyl-β-Amino Acids via Enantionmerically Pure Dihydropyrimidinones

Joseph P. Konopelski

12.1 Introduction

The marriage between total synthesis and synthetic methodology development has been a long and fruitful one. The ability to construct complex molecular structures is defined by the availability of appropriate reaction schemes and reagents. The limits of these synthetic tools are tested and refined to the utmost in a total synthesis venture. This is particularly true with the synthesis of natural products, where the target is defined by Nature, and thus there is no room for compromise.

12.2 Asymmetric Synthesis of β-Aryl-β-Amino Acids

The cytotoxic marine natural product jasplakinolide (**1**[1], Figure 12.1) is now in preclinical studies at the National Cancer Institute by virtue of both interesting activity and an unusual mechanism of action.[2,3] We became aware of this compound through our colleague, Professor Phil Crews, who headed one of the groups responsible for determining the structure of this unusual mixed amino acid/polypropionate compound.[4,5] When the details of the structure became fully known, we were faced with the prospect of a total synthesis in which only the alanine residue could be directly employed as received from the vendor. Most intriguing was the β-tyrosine unit. At the time we started this work in 1988, we were unaware of any asymmetric synthetic methodology that would be sufficient for the synthesis of this β-aryl-β-amino acid in enantiomerically pure form. On the other hand, a general solution to the synthesis of β-amino acids was highly desirable, as it would have impact on such diverse medicinal fields as β-lactam production[6] and anticancer drug development, as seen in the isoserine side chain of Taxol **2** and Taxotére **3**[7] (Figure 12.1). Therefore, the development of such methodology became a parallel goal to the total synthesis venture.

Consider the basic structure of an enantiomerically pure β-amino acid (Figure 12.2). As is the case with any carbon bearing four different substituents, any one of the bonds to the β-carbon of generalized structure **4** (labeled with an asterisk) in theory can be the last one formed and therefore will be the strategic event in the determination of absolute configuration. Figure 12.2 gives a number of approaches employed over the past few years to achieve this goal with regard to **4**. All these routes are more fully discussed in other chapters of this book. Our own approach, which is highlighted, was to form the carbon–carbon bond between the β-carbon and the *R* group in the asymmetric event.

This methodology was designed for synthetic versatility. In drug development it is often necessary to prepare a series of derivatives with slight structural variations for optimization studies. In a set of β-amino acids the alkyl or aryl group bonded at the β-carbon is one key structural variable. One approach to the desired synthetic versatility would be to develop a "template" that would contain all the elements of the de-

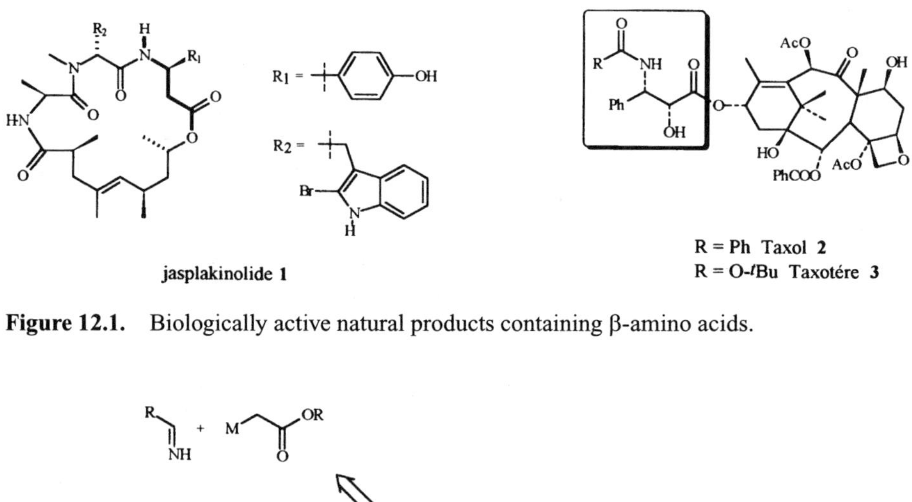

Figure 12.1. Biologically active natural products containing β-amino acids.

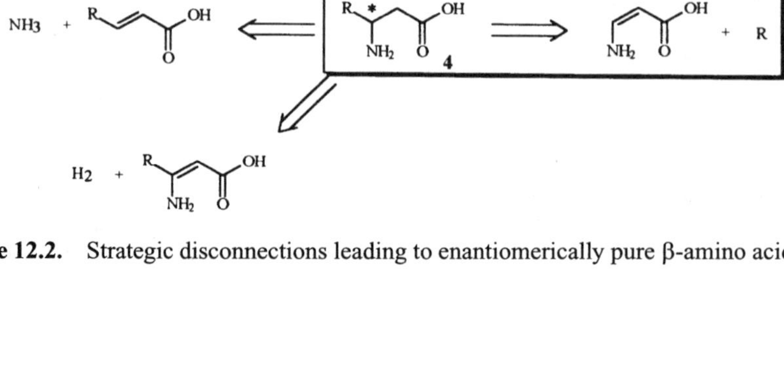

Figure 12.2. Strategic disconnections leading to enantiomerically pure β-amino acids.

sired final product save the variable one. Several attributes were desirable in such a template. For example, it must be easily prepared in high yield and be available in both enantiomeric forms. It should be a stable compound, so that it could be prepared in large quantities and stored until needed. Finally, it should possess structural elements that would allow the new C-C bond to be formed with both a predictable and a high degree of absolute configuration. Analysis of these attributes led to the consideration of structures **5** and **6**. Both structures pay tribute to the work of Seebach and, particularly for compound **6**, the concept of "self-reproduction of stereogenic centers,"[8] articulated in the Zurich laboratories.

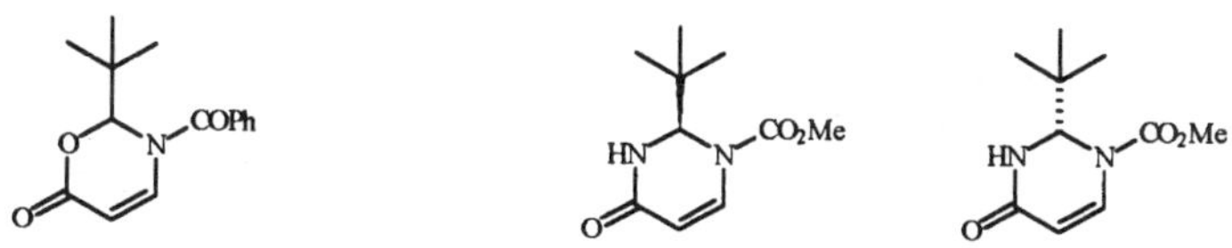

5 **(S)-6** **(R)-6**

Oxazenone **5** was a disappointment. In our hands the synthetic yield of this compound, and other members of this class of compounds, proved to be very low (≤20%). In addition, key derivatives in the synthetic scheme appeared to be hydrolytically unstable. However, some limited success in conjugate addition reactions to **5**, as shown below, prompted the continued exploration of this general approach.

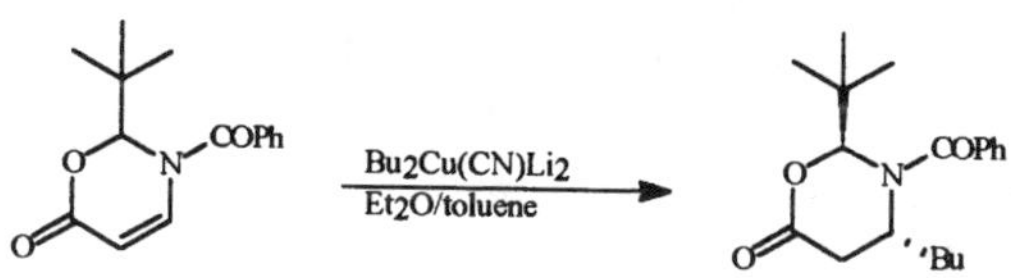

5

Some years ago, Hardy and Samworth[9] demonstrated that the acid salt of asparagine **7** condensed with acetone to form the corresponding pyrimidone. If the same reaction were to occur with aldehydes instead of acetone, and the requisite protection and oxidative decarboxylation steps could be controlled, this sequence would lead to type **6** molecules. In the event (Scheme 12.1), treatment of the potassium salt of (*S*)-asparagine (**7**) with pivalaldehyde (**8**) in methanol for 6–8 h at reflux produced a white powder with spectral characteristics consistent with desired cyclized product **9** as a single isomer. Analysis of the ^{1}H-NMR along with molecular mechanics calculations indicated a conformation in which both the acid and *tert*-butyl groups are equatorial substituents. Protection of the amine nitrogen of **9** with methyl chloroformate ($NaHCO_3$, H_2O) yielded the corresponding carbamate as an amorphous solid that resisted all attempts at crystallization. A modified procedure employed in these laboratories currently[10] yields **10** in a one-pot reaction in water from asparagine. Thus the salt of asparagine is prepared in aqueous solution with KOH and is stirred at room temperature with pivalaldehyde until the solution is clear. Addition of $NaHCO_3$ and methyl chloroformate followed by precipitation with acid and collection of the solid affords **10** in 79% yield. It is interesting to note that this experimentally facile reac-

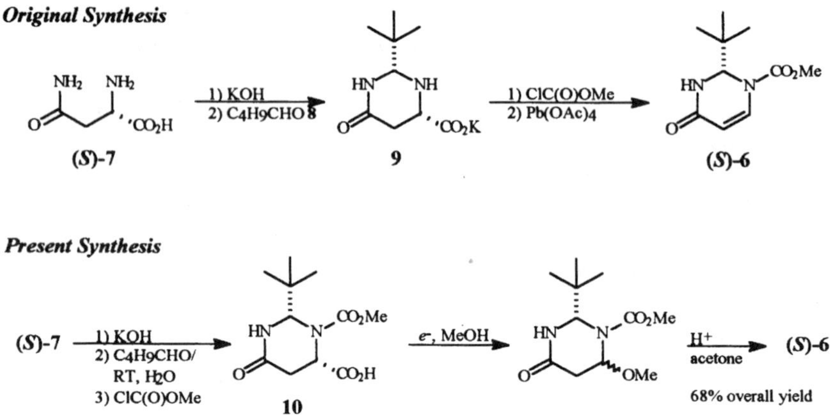

Scheme 12.1

tion sequence is best carried out in water, a solvent that usually is excluded from ketalization reactions.

The initial transformation of **10** to **6** was accomplished with lead(IV) acetate.[11,12] The modified sequence is a two-step method that is much cleaner. Simultaneously, the cost and hazards of using and disposing of lead are precluded. Anodic oxidative decarboxylation employing graphite electrodes is a simple, high-yielding process to the 6-methoxy material.[13] While there is no control of stereochemistry in the decarboxylation, the subsequent resin-promoted elimination to **6** makes this inconsequential. The two enantiomers of **6** are easily prepared from the corresponding enantiomer of asparagine, which is readily available and inexpensive in both the D- and L-series. The material is highly crystalline and robust, with a large specific rotation that allows for easy determination of enantiomeric purity. The entire sequence occurs without chromatography in 68% yield from the amino acid.

Recently the synthesis of 2-aryl pyrimidinones related to **10** have been the subject of study in our laboratory. The synthesis is very similar except that the purely aqueous conditions that were so advantageous with pivalaldehyde do not work for crystalline aromatic aldehydes with poor water solubility. In these cases the aldehyde is added as a DME solution to the asparagine salt in water. After reaction with $NaHCO_3/ClCO_2Me$, the desired heterocycle is isolated by acid precipitation.[14]

A single crystal x-ray structure determination of **6** reveals that the ring is in a sofa-like conformation, with C2 above the nominal plane of the other atoms. The most striking structural element is the axial *tert*-butyl group at C2. Examination of space-filling models of **6** very clearly indicate that this *tert*-butyl group exerts a powerful steric presence over the face of the six-membered ring, effectively blocking all reagents from *syn* approach. Thus another of the criteria for a successful template molecule has been met; namely, both the stereochemistry of reagent attack to compound **6** and the resultant absolute stereochemistry at the newly formed chiral center(s) are completely predictable.

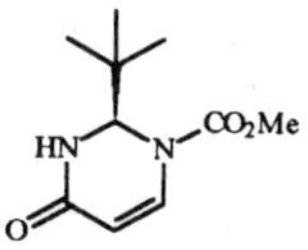

(*R*)-6

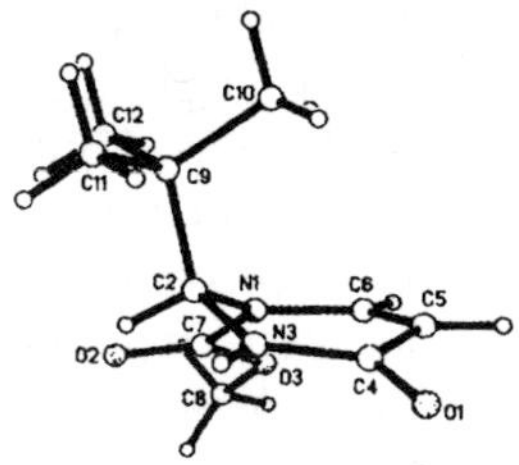

The immediate synthetic goal of a β-aryl-β-amino acid synthesis required the formation of a C-C bond between C6 of the heterocyclic system and a suitably substituted aryl group. Initially, a study of the copper-mediated conjugate addition of alkyls and aryls to derivatives of **6** was undertaken. Substantial efforts in this area bore little in the way of positive results. A one-pot method for the derivatization of **6** was developed that involved: (1) one equiv. of *n*-BuLi followed by reaction with an appropriate electrophile; and (2) subsequent treatment with two equiv. of *n*-BuLi (to cleave the carbomethoxy group) followed by a second electrophile. In this way, a number of base-stable derivatives of **6** could be prepared with ease, as exemplified below. However, with the exception of the formation of **11**, these endeavors were unsuccessful.

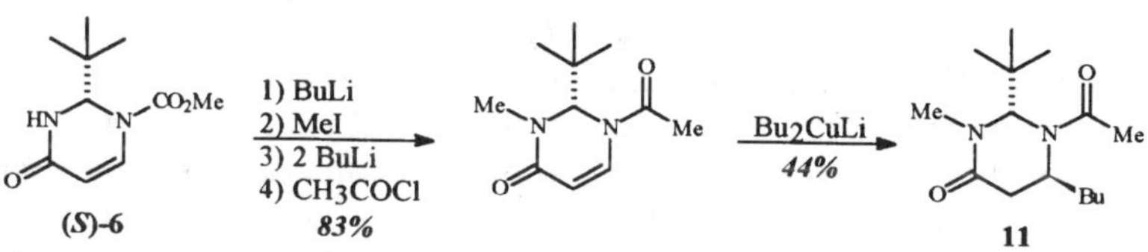

Attention was given next to other organometallic mediated C-C coupling protocols. Perhaps most notable in this regard are organopalladium methods. However, these notions were entertained with some misgivings. While the Heck[15] reaction is firmly established as a premier method for aryl–vinyl coupling in acyclic systems, the accepted mechanism of this reaction would appear to make it unsuitable for use in a cyclic system such as **6**. As depicted in Scheme 12.2, this mechanism involves an arylpalladium complex, formed by the oxidative addition of an aryl iodide to a palladium(0) complex. *Syn* addition to the substrate alkene, followed by rotation around

Heck Reaction - vinyl substitution

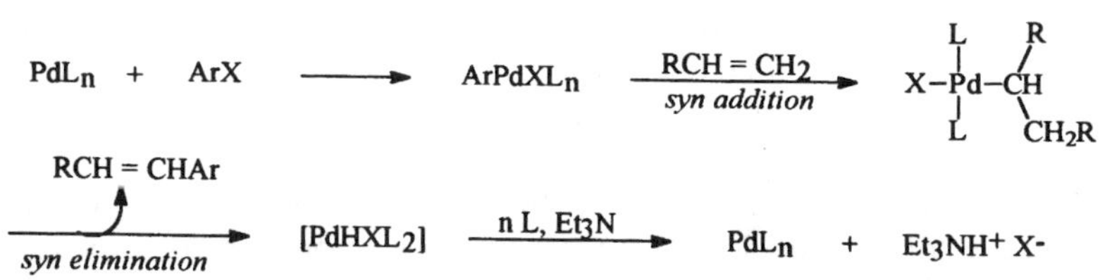

Scheme 12.2

the new single bond, allows for the *syn* elimination of a hydridopalladium species producing the vinyl substitution product. The hydridopalladium compound subsequently reacts with the amine base to afford palladium(0) and trialkylammonium halide. In the present case, however, there would be no hydrogen in a *syn* orientation to the palladium, since rotation is precluded by the cyclic nature of the substrate.

Cyclic systems had been employed as substrates for the Heck reaction on a limited basis, although in most cases the conventional mechanism can be accommodated.[16] Most intriguing for the present purpose were the results of Stokker[17] on the palladium-catalyzed conjugate arylation of 5,6-dihydro-2H-pyran-2-ones. In this instance, the author presents evidence to support reduction of the C-Pd bond by triethylamine to yield the formal conjugate addition reaction rather than the substitution reaction shown in Scheme 12.2. In the event, **6** is treated with 4-iodoanisole (1.1 equiv.) and triethylamine (1.5 equiv.) in dimethylformamide (DMF) containing a catalytic amount of palladium acetate and tri-o-tolylphosphine at 100°C for 24 h to give **12** as a crystalline product in 78% isolated yield. Use of 4-iodophenol, 4-((*tert*-butyldimethylsilyl)oxy)iodobenzene, and iodobenzene results in analogous products being formed in 50, 76, and 60% yield, respectively. Biaryls, presumably formed by oxidative coupling of the aryl iodide substrate, are also obtained. The use of acetonitrile as solvent affords lower yields, as does the use of lower temperature. Increases in the amount of amine base relative to the substrate and attempts to perform the reaction without DMF, using triethylamine as the solvent, results in a multitude of uncharacterized products with little or no formation of **12**. Tributylamine can be substituted for triethylamine with no apparent change in the reaction performance. Similarly, tetrakis(triphenylphosphine) palladium(0), palladium acetate/tri-*o*-tolylphosphine, and palladium acetate/triphenylphosphine are approximately equivalent in efficacy. A recent modification[10] employs diethylamine in place of triethylamine. While the reaction is sluggish (48 h reaction time versus 24 h with NEt_3), negligible amounts of biaryls are formed.

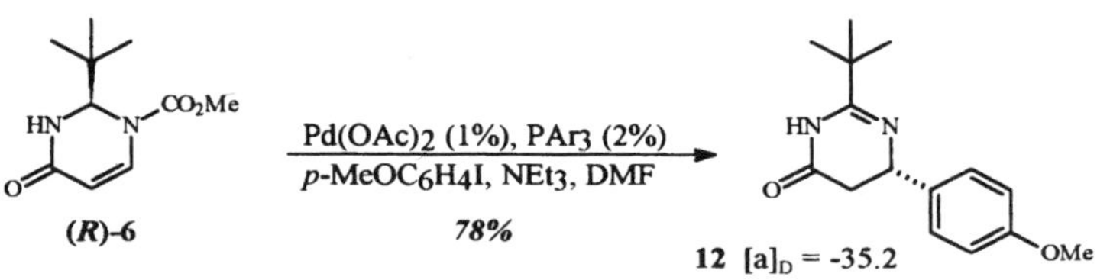

The basic nature of the heterocyclic product makes isolation facile.[18] The crude reaction mixture is partitioned between an aqueous HCl solution at pH 2 and an organic solvent. Multiple extractions remove the biaryl product. Adjustment of the pH of the aqueous layer to 12 and extraction with organic solvent affords **12** sufficiently pure for subsequent transformations. Treatment of **12** with $NaBH_4/H_3O^+$ followed by hydrolysis with 3*N* HCl affords the corresponding derivative of β-tyrosine hydrochloride in 85% yield.

Heterocycle **12** was subjected to the protocol described above followed by reaction with acetic formic anhydride[19] to yield the known formamide of (*S*)-β-tyrosine-O-methyl ether.[20] Comparison of the reported rotation for 91% ee material with the compound derived from **12** indicated that the chirality transfer in the palladium-catalyzed reaction was complete. Additional evidence for the enantiomeric purity of this β-tyrosine derivative comes from its successful use in the total synthesis of jasplakinolide.[1]

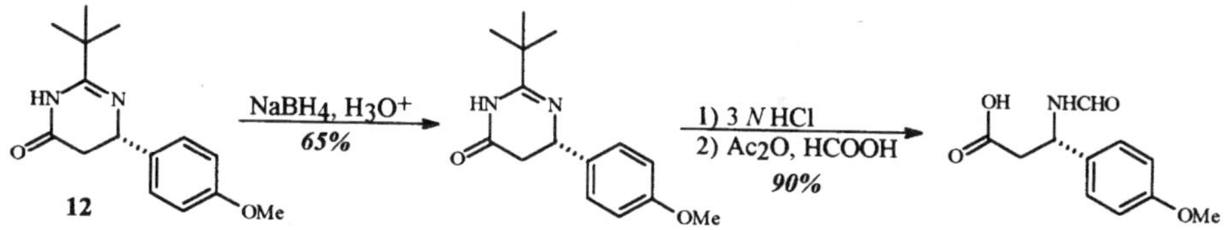

12.3 Mechanistic Considerations

A number of experiments were performed to elucidate the mechanism by which this formal conjugate addition occurs to the C5-C6 bond of heterocycle **6**. These experiments include deuterium labeling studies and, taken together with established chemistry, suggest the following mechanism (Figure 12.3). Given the high fidelity of chirality transfer, an initial assumption is that the formation and addition of the aryl-palladium(II) complex to **6** occurs in the normal Heck fashion, affording intermediate **13**. Since there is no *syn* C-H bond adjacent to the C5-Pd bond, the normal *syn* elimination mechanism cannot be in force. Triethylamine enters the coordination sphere of palladium to afford **A** and **D** from **13**. The presence of deuterium in the reaction mixture has no effect on this event. Insertion into the C_α-H bond of triethylamine by palladium produces hydridopalladium/imminium species **E**.[21] This reaction is less competitive when NEt_3-d_{15} is employed, leaving **A** as the predominant reactive intermediate when deuterated base is employed. Either iodide or tertiary amine[22] acts as nucleophile toward the methyl group of the carbamate, precipitating an electron reorganization with oxidation of the N1-C2 bond and formation of hydride or deuteride ion. We have independently verified the iodide route as shown below.

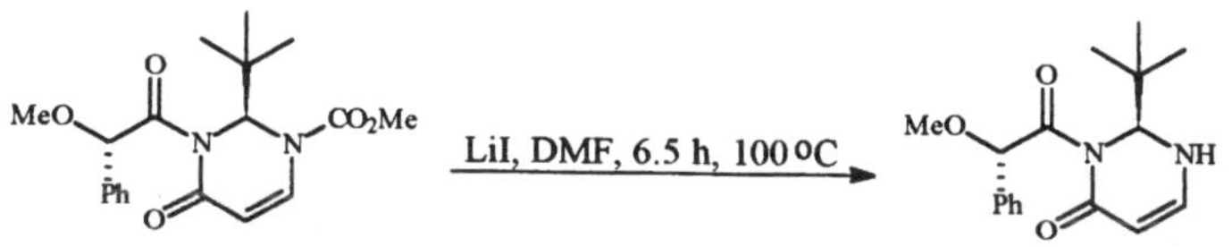

The palladium present on the molecule in **A** and **E** offers a ready acceptor for the hydride from C2. Without such an acceptor, decomposition is competitive. As the rehybridization of C2 proceeds, the hydrogen (or deuterium) at C2 moves from an equatorial position (see X-ray of **6**) to an axial position *cis* to the palladium atom. Thus transfer can be efficiently made only as a result of the release of the $A^{(1,2)}$ strain between the *tert*-butyl and the carbomethoxy groups.

The exact coordination sphere of palladium in this sequence is unknown. For example, it is possible that the hydride acceptor in **A** is a coordinatively unsaturated palladium species that arises from ligand dissociation rather than the direct displacement reaction depicted in Scheme 12.3. Alternatively, the intermediacy of an oxa-π-allyl palladium species cannot be dismissed.

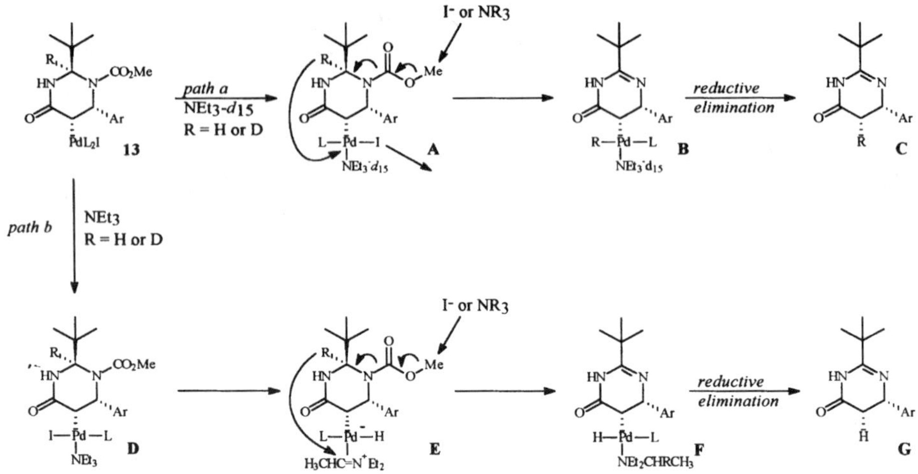

Figure 12.3. Proposed mechanism for the formation of aryl conjugate addition product. See text for discussion. [Reprinted from K. S. Chu et al., *J. Am. Chem. Soc.* **114**, 1800 (1992) with kind permission from the American Chemical Society.]

In the case of **A**, hydridopalladium species **B** is formed directly, followed by reductive elimination to **C**. The *R* group originally at C2 is efficiently and stereospecifically transferred to C5. In the case of **E**, an hydridopalladium species is already in place, and the best acceptor is the imminium carbon.[23] Hydride transfer from C2 to this electron-deficient carbon regenerates the requisite hydridopalladium species (**F**) capable of reductive elimination to **G**. The two pathways appear to be balanced on the strength of the C-H versus C-D bond with regard to the formation of **E**. To our knowledge, this is the first example of a transannular hydride transfer in a palladium-catalyzed reaction.

12.4 Synthesis of β-Alkyl-β-Amino Acids

What about β-*alkyl*-β-amino acids? Could the present template strategy be extended to this class of compounds? The answer is yes,[24] although the current procedure is somewhat less efficient than the corresponding arylation sequence.

Compound **14** is derived from **6** by protection of the nitrogen atoms of the heterocycle with base-stable groups. As opposed to the one-pot procedure described for the early organocopper studies, **14** is best prepared by a more conventional two-pot approach. First, the amide is deprotonated with *n*-butyllithium, followed by methylation with methyl iodide. This reaction requires at least 3 h at room temperature. Alternatively, potassium hydride can be used at low temperature for this methylation.[14] The carbomethoxy group in **6** is then cleaved with sodium hydroxide. The resulting amine is protected with the MOM group with an overall 70% yield from **6**. Excess chloromethyl methyl ether is avoided in this step since it causes a decrease in yield.

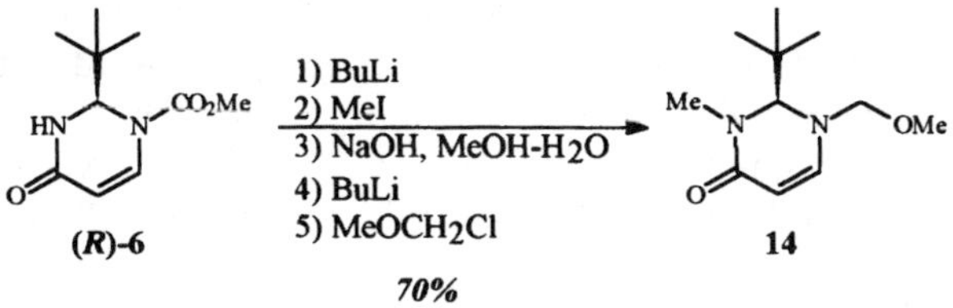

Compound **14** can be directly reacted with *tert*-butyllithium at –78°C in a manner similar to the β-aminoacrylic acid derivatives reported by Schmidt and co-workers.[25] The vinyl anion so formed is treated with electrophiles such as alkyl halides and aldehydes to give C6-substituted heterocycles. Results of these reactions are summarized in Table 12.1. Methylation is quite successful. Other alkyl halides afford yields that, while not as high, are still useful. Arylation with phenyl iodide affords modest amounts of product.[26] Treatment of **15d** with *tert*-BuLi/MeI afforded **15f**.[27,28]

Aldehydes are excellent electrophiles in this reaction. The product alcohols (as a mixture of diastereomers that can be separated on column chromatography) can be converted to their corresponding acetates by treatment with acetic anhydride. The acetate group can then be reduced with sodium borohydride in high yield to afford the C6-alkylated heterocycles. Thus, whereas direct alkylation of the anion of **14** with isobutyl bromide is unsuccessful, the formation of **15b** and subsequent reduction of the derived acetate affords **15h** (Scheme 12.2) in 76% yield from **14** in enantiomerically pure form (vide infra). Likewise, the overall yield of **15e** is improved by this three-step protocol.

Table 12.1 REACTIONS OF THE VINYL ANION OF 14 WITH SELECTED ALDEHYDES AND ALKYL HALIDES

Entry	Electrophile	Yield of **15**
a	CH_3CHO	81%
b	$(CH_3)_2CHCHO$	85%
c	PhCHO	84%
d	CH_3I	95%
e	$PhCH_2Br$	55% (78%[a])
f	CH_3CH_2I	55%
g	PhI	27%
h	$(CH_3)_2CHCH_2I$	0% (76%[b])

[a] Substrate prepared from **15c**; see adjacent sketch.

[b] Substrate prepared from **15b**; see adjacent sketch.

Scheme 12.3

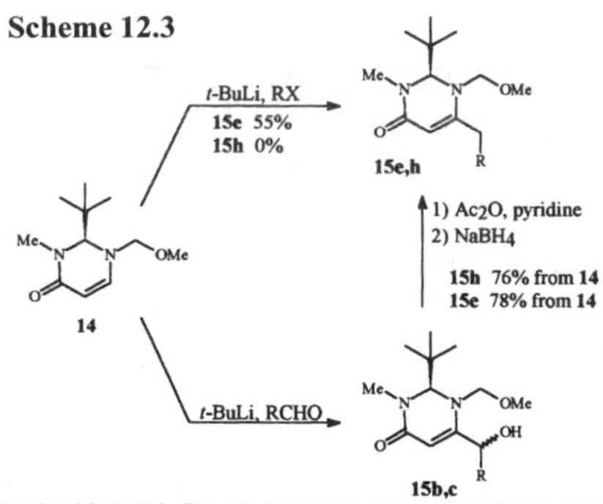

Although there are a few reports of reductions of similar substrates,[29] hydrogenation of β-amino conjugated systems has not been straightforward. Attempted hydrogenation of **15d** using $Pd(OH)_2$, Pd-C, PtO_2, Raney Ni, and Rh on alumina as catalyst and with hydrogen pressure up to 50 psi gave no reaction. However, compounds **15d** and **15e** were cleanly reduced in good yield with sodium cyanoborohydride[30] to the corresponding saturated heterocycles **16**. The *cis* relationship between the *tert*-butyl group at C2 and the new chiral center at C6 was established by NMR (NOESY). High diastereomeric selectively (>95%) was consistently observed. The temperature of this reduction is important; at higher temperatures (≥0°C) the carbonyl group is also reduced. Unexpectedly, in addition to the double bond, the MOM group is also reduced.

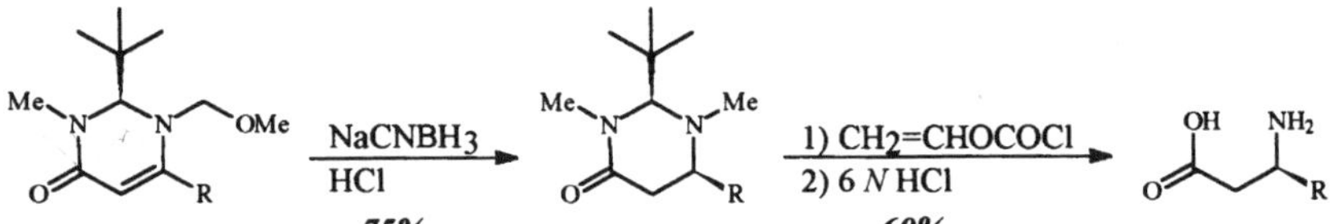

15d,e **16d,e** **17d,e**

Demethylation of the amine nitrogen and hydrolysis were combined in the last step to give the desired β-amino acids **17**.[31] The heterocycle is first treated with vinyl chloroformate in refluxing 1,2-dichloroethane. The resulting carbamate is then refluxed with ethanolic HCl to achieve dealkylation, and after removal of the solvent the reaction mixture is hydrolyzed with 6*N* HCl to liberate the β-amino acids.

In conclusion, the above route to β-alkyl-β-amino acids, coupled with the palladium-catalyzed approach to β-aryl-β-amino acids, allows for the first time a unified and general set of protocols towards obtaining this important class of compounds in enantiomerically pure form. Additionally, compound **6** has been shown to function as a very effective chiral auxiliary for the production of enantiomerically pure α-substituted carboxylic acids.[32] Studies into the use of **6** both as an auxiliary for aldol reactions and as a reagent for the synthesis of highly substituted α-amino acids are currently underway.

12.5 Acknowledgments

It is a pleasure to acknowledge the wonderful co-workers whose names grace the publications in which this work has appeared. Funding from the American Cancer Society and the National Institutes of Health was instrumental in bringing this work to fruition. Finally, the American Chemical Society and Elsevier Science Ltd. are thanked for their kind permission to reprint some of the figures and tables herein that originally appeared in Refs. 12 and 24, respectively.

12.6 References

1. K. S. Chu, G. R. Negrete, J. P. Konopelski, *J. Org. Chem.* **56**, 5196 (1991).
2. M. R. Bubb, A. M. J. Senderowicz, E. A. Sausville, K. L. K. Duncan, E. D. Korn, *J. Biol. Chem.* **269**, 14869 (1994).

3. A. M. J. Senderowicz, G. Kaur, E. Sainz, C. Laing, W. D. Inman, J. Rodríguez, P. Crews, L. Malspeis, M. R. Grever, E. A. Sausville, K. L. K. Duncan, *J. Natl. Cancer Inst.* **87**, 46 (1995).

4. P. Crews, L. V. Manes, M. Boehler, *Tetrahedron Lett.* **27**, 2797 (1986).

5. T. M. Zabriskie, J. A. Klocke, C. M. Ireland, A. H. Marcus, T. F. Molinski, D. J. Faulkner, C. Xu, J. C. Clardy, *J. Am. Chem. Soc.* **108**, 3123 (1986).

6. For example, see L. S. Liebeskind, M. E. Welker, R. W. Fengl, *J. Am. Chem. Soc.* **108**, 6328 (1986).

7. K. C. Nicolaou, W. M. Dai, R. K. Guy, *Angew. Chem. Inst. Ed. Engl.* **33**, 15 (1994).

8. D. Seebach, M. Boes, R. Naef, W. B. Schweizer, *J. Am. Chem. Soc.* **105**, 5390 (1983).

9. P. M. Hardy, D. J. Samworth, *J. Chem. Soc., Perkin Trans. 1*, 1954 (1977).

10. F. J. Lakner, K. S. Chu, G. N. Negrete, J. P. Konopelski, *Org. Synth.* **73**, 201 (1995).

11. J. P. Konopelski, K. S. Chu, G. R. Negrete, *J. Org. Chem.* **56**, 1355 (1991).

12. K. S. Chu, G. R. Negrete, J. P. Konopelski, F. J. Lakner, N.-T. Woo, M. M. Olmstead, *J. Am. Chem. Soc.* **114**, 1800 (1992).

13. D. Seebach, R. Charczuk, C. Gerber, P. Renaud, H. Berner, H. Schneider, *Helv. Chim. Acta* **72**, 401 (1989).

14. L. K. Filonova, T. A. Ritsema, unpublished results.

15. R. F. Heck. *Palladium Reagents in Organic Synthesis* (Academic Press, New York, 1985).

16. K. Nilsson, A. Hallberg, *J. Org. Chem.* **55**, 2464 (1990).

17. G. E. Stokker, *Tetrahedron Lett.* **28**, 3179 (1987).

18. I. R. Politzer, A. I. Meyers, *Org. Synth.* **51**, 24 (1971).

19. L. I. Krimen, *Org. Synth.* **50**, 1 (1970).

20. W. Keller-Schierlein, A. Klaus, H. U. Naegeli, G. Wolf, H. Zähner, *Experientia* **31**, 1001 (1975).

21. S.-I. Murahashi, T. Watanabe, *J. Am. Chem. Soc.* **101**, 7429 (1979).

22. J. McMurry, *Org. React.* **24**, 1878 (1976).

23. (a) S.-I. Murahashi, T. Hirano, T. Yano, *J. Am. Chem. Soc.* **100**, 350 (1978). (b) S.-I. Murahashi, T. Yano, *J. Am. Chem. Soc.* **102**, 2456 (1980).

24. K. S. Chu, J. P. Konopelski, *Tetrahedron* **49**, 9183 (1993).

25. R. R. Schmidt, R. Hirsenkorn, *Tetrahedron* **39**, 2043 (1983).

26. This reaction may involve a benzyne intermediate. See H. O. House. *Modern Synthetic Reactions* (Benjamin, Menlo Park, 1972), pp. 537–40.

27. The analog of **14** that bears an N1 methyl also undergoes this C6 alkylation reaction, whereas the same structure with an N1-Cbz group affords a multitude of products under identical conditions.

28. D. Seebach, U. Misslitz, P. Uhlmann, *Angew. Chem. Int. Ed. Engl.* **28**, 472 (1989).

29. E. Juaristi, D. Quintana, *Tetrahedron: Asymmetry* **3**, 723 (1992).

30. M. Furukawai, T. Okawara, Y. Noguchi, Y. Terawaki, *Chem. Pharm. Bull.* **27**, 2223 (1979).

31. C. H. Mitch, D. M. Zimmerman, J. D. Snoddy, J. K. Reel, D. E. Cantrell, *J. Org. Chem.* **56**, 1660 (1991).

32. G. R. Negrete, J. P. Konopelski, *Tetrahedron: Asymmetry* **2**, 105 (1991).

CHAPTER

13

Enantioselective Synthesis of α-Substituted and α,β-Disubstituted β-Amino Acids via Chiral Derivatives of 3-Aminopropionic Acid

Eusebio Juaristi and Dieter Seebach

13.1 Introduction

Over a decade ago, Seebach and co-workers[1–3] carried out transformations in which α-hydroxy and α-amino acids were α-alkylated without racemization. For example, the starting α-amino acid was first converted to one of two possible diastereoisomeric cyclic aminals with pivalaldehyde such that a *temporary* auxiliary stereogenic center was created. An enolate was then generated by deprotonation with base: *the original stereogenic center is destroyed* (tetrahedral to trigonal center conversion). In the third step, the reaction of the enolate with an electrophile, the bulky substituent on the N,N-acetal center induces the *stereoselective regeneration* of the original center of chirality. Finally, the temporary stereogenic aminal center is removed, affording the new α-branched α-amino acid (Scheme 13.1). Because the above protocol does not em-

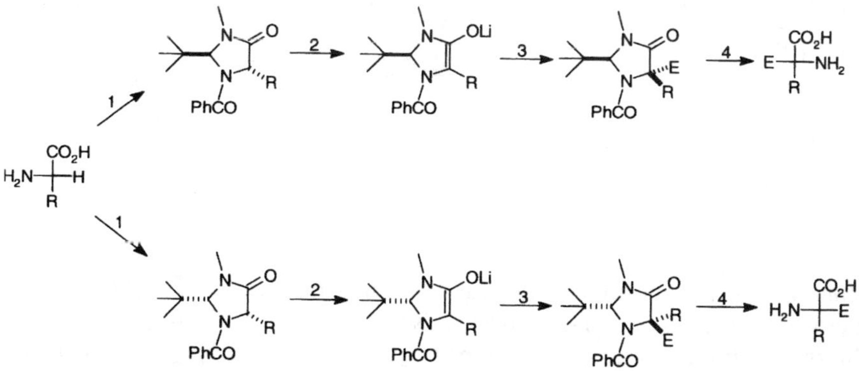

Scheme 13.1

ploy a chiral auxiliary reagent, the synthetic transformation is said to take place with "self-regeneration of stereogenic centers."[4]

Furthermore, removal of the C(5)-substituent on the initial 1,3-imidazolidin-4-one afforded enantiopure aminal **1** (Scheme 13.2), formally derived from glycine. The flexibility of the method was thus substantially increased because not only could α-amino acids be branched without racemization (Scheme 13.1), but now the preparation of monosubstituted (natural or unnatural) α-amino acids also became feasible[5,6] (Scheme 13.2).

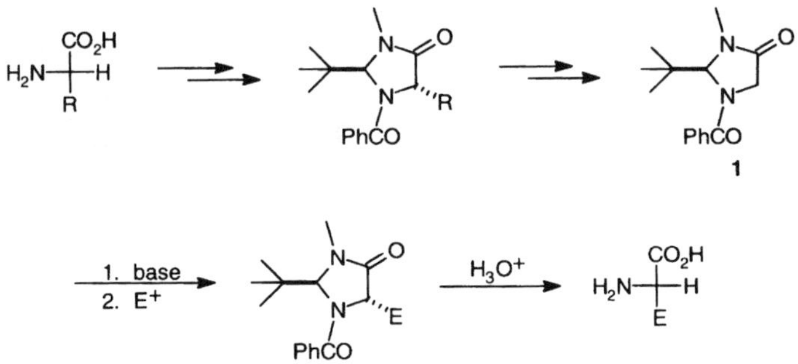

R = $CH_2OCH_2C_6H_5$ (from benzylserine) or $CH_2CH_2SCH_3$ (from methionine)

Scheme 13.2

Nevertheless, the full potential of the method was realized when both enantiomers of the starting chiral aminal were readily available. This can be accomplished by the resolution of nonbenzoylated imidazolidinone (±)-**A**, which is prepared from glycine, methylamine, and pivalaldehyde. The resolution step involved the fractional crystallization of the diastereomeric mandelate salts[7,8] (Scheme 13.3).

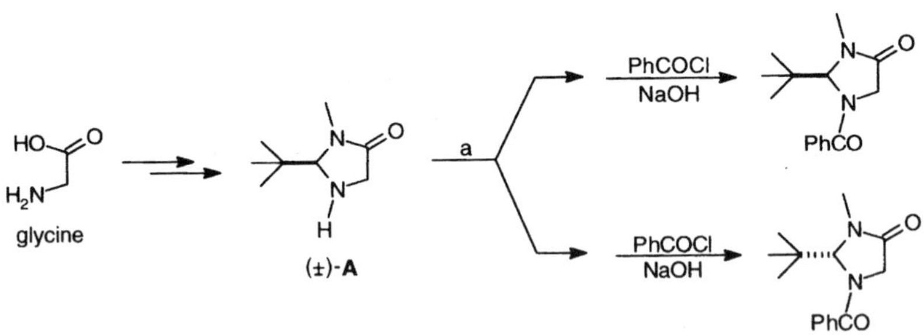

a Fractional crystallization with mandelic acid.

Scheme 13.3

Encouraged by the enormous potential of nonracemic derivatives of glycine as precursors of optically active α-amino acids (see above), we decided to explore the usefulness of chiral enolates derived in principle from β-aminopropionic acid as starting materials for the preparation of enantiopure α-substituted β-amino acids (Scheme 13.4).

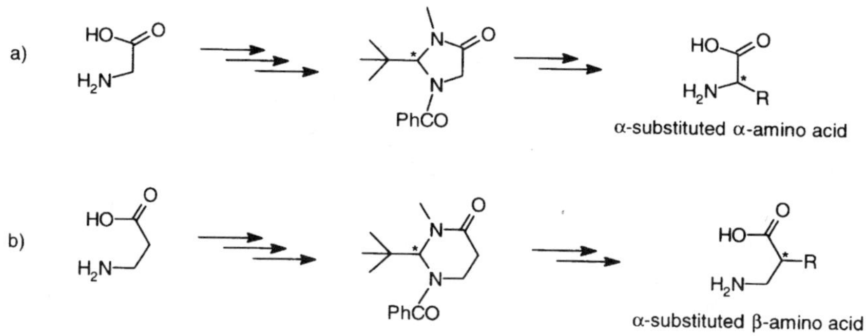

Scheme 13.4

13.2 Results and Discussion

13.2.1 Synthesis of 1-Benzoyl-2-*tert*-Butyl-3-Methylperhydropyrimidin-4-One (2)[9]

The heterocycle *rac*-**2** was prepared from β-aminopropionic acid by initial conversion of its methyl ester to the corresponding N-methylamide, which formed a Schiff base with pivalaldehyde. Finally, cyclization of the resulting imine afforded the desired heterocycle, with an overall yield of 44% (Scheme 13.5).

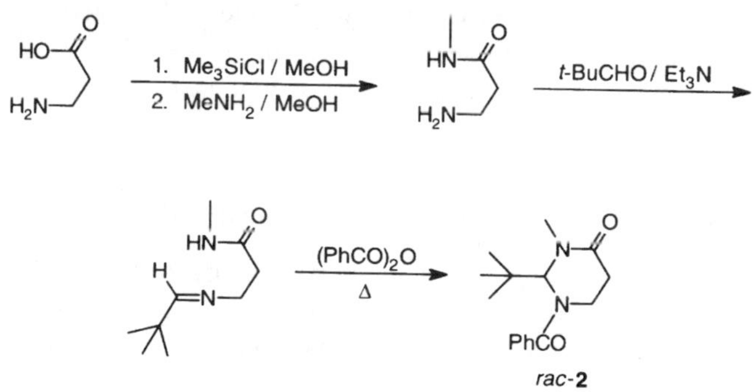

Scheme 13.5

Because the precise structure of reactive heterocycles can be important in ascertaining the factors responsible for the stereoselectivities observed,[10] we carried out an X-ray analysis with a suitable crystal of perhydropyrimidinone **2**. A view of the solid-state structure is provided in Figure 13.1. The pyrimidinone ring is rather flat and has a sofa conformation with five of the six atoms approximately in a plane, and the aminal carbon C(2) out of plane.

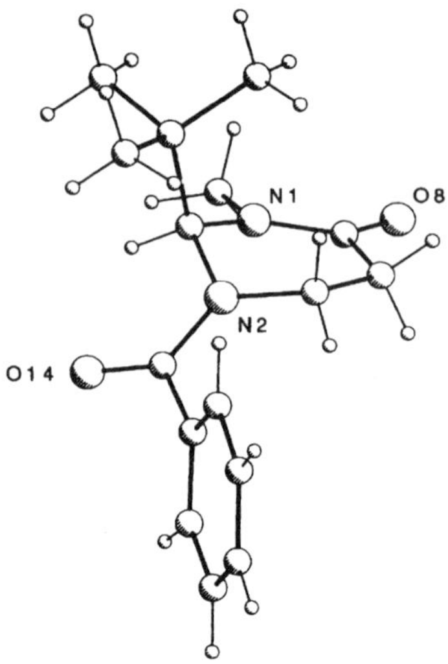

Figure 13.1. Crystal structure of *rac*-**2**, showing the axial *tert*-butyl group.

The most interesting feature of the crystal structure of **2** is, however, that the six-membered ring adopts a conformation with an *axial tert-butyl group*! The axial orientation of the bulky substituent is a classical manifestation of the so-called $A^{1,3}$ effect exerted by an amide group[9,11,12]: The energy required to rotate an amide bond is 15–20 kcal/mol, much larger than the conformational strain that is generated by the axial *tert*-butyl group. Six-membered rings bearing axial *tert*-butyl groups are rare,[13] so this finding is of general interest; nevertheless, the practical consequences are also significant: It should cause steric blocking of one of the ring faces, for example, for attack by electrophiles. This was indeed the case, as described in the following section.

13.2.2 Diastereoselectivity of Alkylation of Enolate 2-Li

The alkylation products **3**–**6** (Eq. 13.1) are formed by treatment of enolate **2**-Li, generated with lithium diisopropylamide (LDA) in THF, with halides *RX* at –75°C. High diastereoselectivity (ds > 95%) was found, as indicated by integration of the ^{13}C-NMR signals of the crude products.

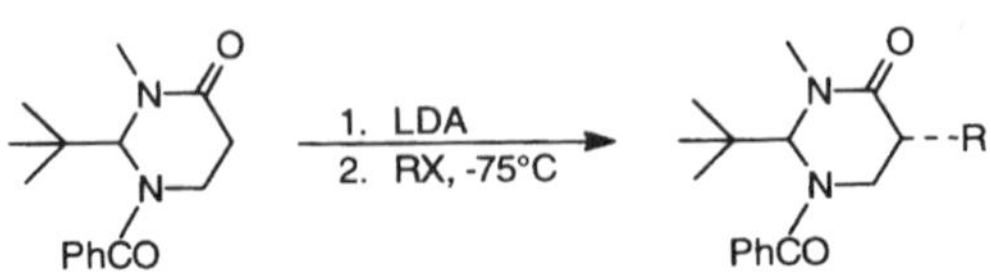

rac-**2** *rac*-**3**-**6**

(eq. 13.1)

R = CH_3, $C_6H_5CH_2$, n-C_4H_9, n-C_6H_{13}

Proton and carbon-13 NMR spectroscopy, in combination with an X-ray diffraction study of the benzylated derivative **4** ($R = C_6H_5CH_2$),[9] were used to determine that addition took place mainly from the side opposite to the *tert*-butyl group, to afford the *trans* products.

13.2.3 Hydrolysis of the Pyrimidinone Adducts to Give the α-Substituted β-Amino Acids

The final step in the overall conversion outlined in Scheme 13.4b requires hydrolysis of the heterocyclic products with cleavage of the ring and isolation of the α-substituted β-amino acid. Hydrolysis of adducts **3–6** was thus carried out under acidic conditions (which are less likely to cause epimerization at the stereogenic center). Indeed, hydrolysis of **3–6** (6*N* HCl, 160–180°C, sealed tube) proceeded in good yields (62–69% isolated) to furnish a mixture of the desired amino acid and methylammonium chloride. Final purification of the amino acids was then achieved by means of acidic Dowex 50W x 8 ion-exchange resin[9] (Eq. 13.2).

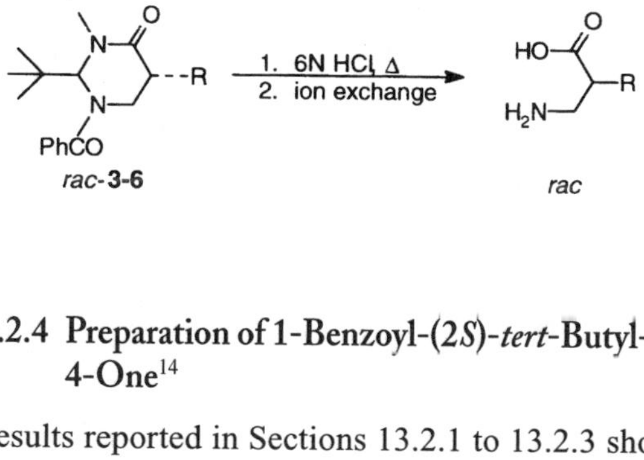

rac-**3-6** *rac* (eq. 13.2)

13.2.4 Preparation of 1-Benzoyl-(2*S*)-*tert*-Butyl-3-Methylperhydropyrimidin-4-One[14]

The results reported in Sections 13.2.1 to 13.2.3 showed the feasibility of using β-aminopropionic acid, an inexpensive and achiral amino acid, for the preparation of chiral derivative **2**, which could be alkylated with high *trans* diastereoselectivity, and then hydrolyzed to the expected α-substituted β-amino acids. A new method was now required for the efficient preparation of enantiomerically pure starting pyrimidinone **2**.[15]

(*S*)-Asparagine was condensed with pivalaldehyde according to the procedure described by Konopelski et al.[16] Benzoylation of the amine nitrogen in **7** (PhCOCl, $NaHCO_3$) yielded **8** as a 86:14 *cis*/*trans* diastereomer mixture. Slow addition of HCl resulted in the preferential precipitation of the main (*cis*) isomer in 74% yield from asparagine. Treatment of *cis*-**8** with $Pb(OAc)_4$/catalytic $Cu(OAc)_2$ (an oxidative decarboxylation[16,17]) afforded enone **9** in 60% yield, and N-methylation of **9** (*n*-BuLi, CH_3I) proceeded in 55% yield. Finally, palladium-catalyzed hydrogenation yielded (+)-**2** in 98% yield. The overall yield of (+)-**2** from (*S*)-asparagine is 25%, allowing, for example, the preparation of 7.8 g of (+)-**2** from 15-g lots of (*S*)-asparagine (Scheme 13.6).

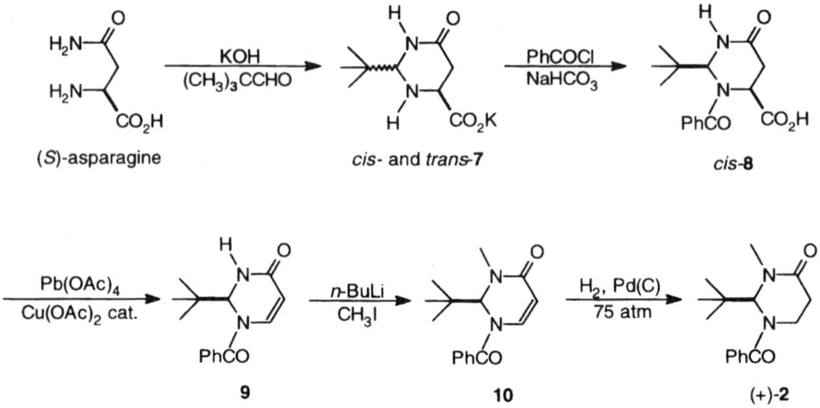

Scheme 13.6

13.2.5 Assignment of Configuration in (+)-2

It is well established that addition of **2**-Li (the enolate derived from **2**) to electrophiles takes place from the face *trans* to the *tert*-butyl group (vide supra). Thus the observation that hydrolysis of (+)-**11**, the product of methylation of (+)-**2**, affords levorotatory α-methyl-β-aminopropionic acid, allows the assignment of the absolute configuration of (+)-**2**, since the absolute configuration of (−)-**12** is known.[17,18] Scheme (13.7).

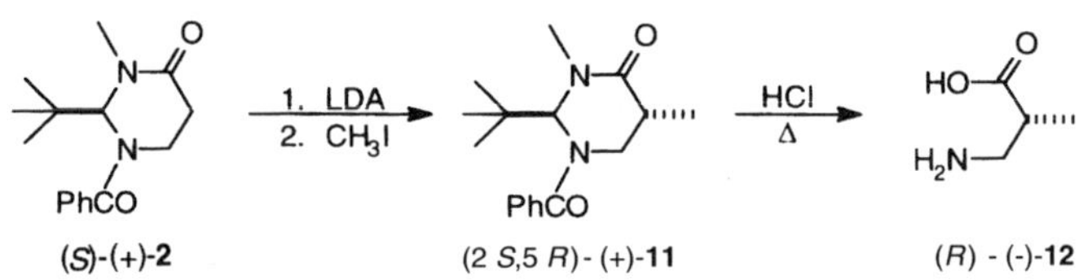

Scheme 13.7

The combination of Schemes 13.6 and 13.7 provides a classical example of the concept of "self-regeneration of a stereogenic center" (see the Introduction). (*S*)-asparagine contains the initial stereogenic center inducing the stereoselective formation of the new, temporary stereogenic center. The former stereogenic center of asparagine is then removed during decarboxylation, but it is now the aminal center of chirality that directs the stereoselective formation of the new chiral segment at the α-carbon. Finally, hydrolysis provides for removal of the temporary stereogenic center.

13.2.6 Stereoselective Alkylation of (*S*)-2

Enantiopure pyrimidinone (*S*)-**2** was then tested as a general starting material for the enantioselective synthesis of α-substituted β-aminopropionic acids. To this end,

Table 13.1 DIASTEREOSELECTIVITY OF ENOLATE (*S*)-2-Li ALKYLATIONS AND HYDROLYSIS OF THE CORRESPONDING PRODUCTS 11 AND 13–15 TO GIVE α-SUBSTITUTED β-AMINO ACIDS

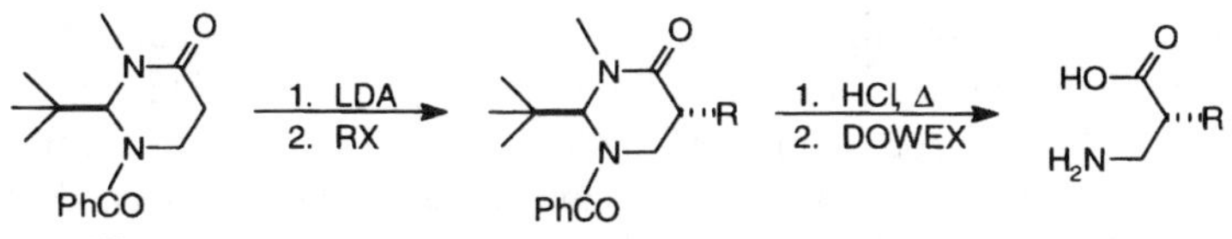

(*S*)-2 | **11, 13-15** | (*R*)-**16-9**

RX	Alkylated product	ds (%)	Yield (%)	Free amino acid	Yield (%)[a]
CH_3I	**11**	>96	76	(*R*)-**16**	80
n-C_4H_9I	**13**	95	75	(*R*)-**17**	75
n-$C_6H_{13}I$	**14**	95	76	(*R*)-**18**	70
$PhCH_2Br$	**15**	>96	77	(*R*)-**19**	80

[a] After ion-exchange chromatography and crystallization.

enolate (*S*)-**2**-Li was generated upon treatment of the heterocycle with LDA, and the electrophile was then added at –75°C to give as expected the *trans*-alkylated products in high selectivity and good yields (Table 13.1). Hydrolysis of heterocycles **11** and **13–15** (6*N* HCl, 90–100°C, sealed tube) afforded the free amino acids **14–17**, which were purified by chromatography on an ion-exchange column.[14b] (Table 13.1).

13.2.7 Epimerization of Adducts 11 and 13–15, and Hydrolysis to Give the Enantiomeric α-Alkylated β-Aminopropionic Acids

For the preparation of the enantiomeric β-amino acids (*S*)-**16–19**, one could in principle use (*R*)- instead of (*S*)-asparagine, so that the resulting pyrimidinone (*R*)-**2** may be alkylated and hydrolyzed following the same protocol described in Section 13.2.6. Nevertheless, a convenient alternative exists in the epimerization of *trans* adducts **11** and **13–15** to the *cis* diastereoisomers **20–23**, followed by acid hydrolysis to the amino acids (Scheme 13.8). It turned out that protonation (aqueous NH_4Cl) of enolates **11**-Li and **13–15**-Li also takes place with high stereoselectivity, again on the face opposite to the *tert*-butyl group.[14b]

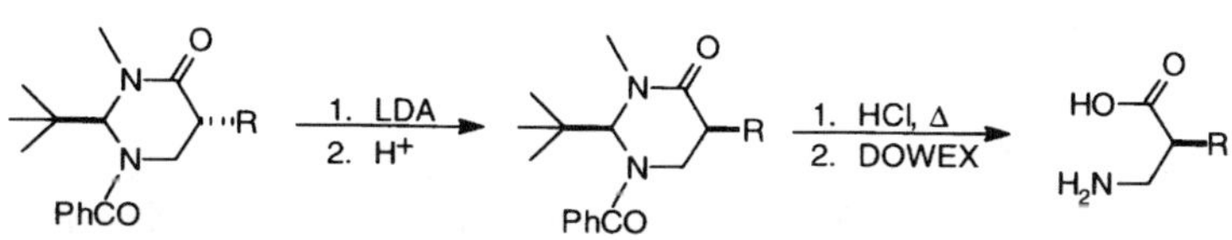

11 and **13-15** | **20-23** | (*S*)-**16-19**

Scheme 13.8

13.2.8 Preparation of the Like Stereoisomers of 2-Methyl- and 2-Benzyl-3-Aminobutanoic Acid

(*R*)- and (*S*)-3-Aminobutanoic acids, (*R*)- and (*S*)-**24**, were prepared from methyl crotonate and (*S*)-α-methylbenzylamine[19]; Michael addition of the chiral amine to the α,β-unsaturated ester furnishes a 3:2 mixture of the (3*R*,1′*S*) and (3*S*,1′*S*) isomers, which were separated by flash chromatography. Hydrogenolytic removal of the phenethyl group with concomitant ester hydrolysis gave the acid **24** (Scheme 13.9). Conversion through the ester **25** to the amide **26** was followed by Schiff base formation with pivalaldehyde. The imine **27** was then cyclized by treatment with benzoyl chloride/dimethylaminopyridine (DMAP) to give perhydropyrimidinone **28**.[20] (Scheme 13.10).

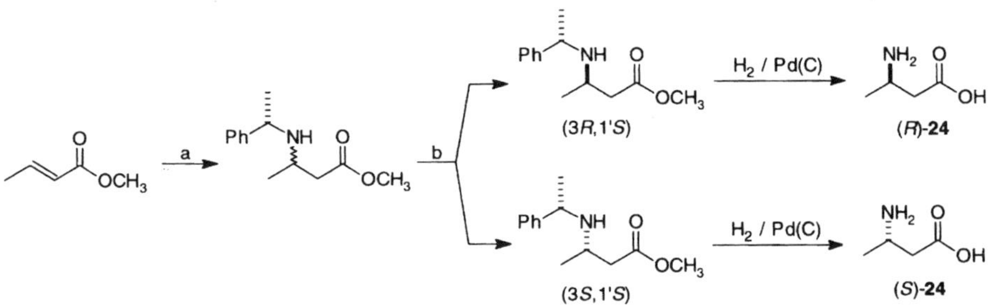

a (*S*)-α-Methylbenzylamine. b Flash column chromatography

Scheme 13.9

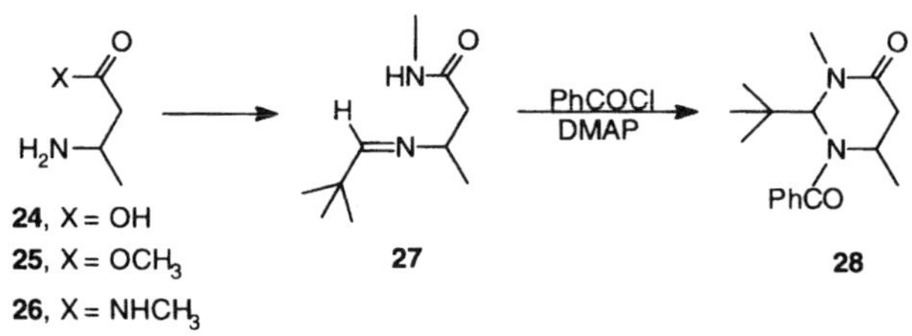

Scheme 13.10

The *cis* products (2*S*,6*R*)- and (2*R*,6*S*)-**28** predominated over the *trans* isomers (2*R*,6*R*)- and (2*S*,6*S*)-**28** in the product mixture (ca. 95:5). Chromatographic separation and recrystallization gave pure samples of all four stereoisomers. The configuration assignment was based on two X-ray crystal structure determinations of alkylation products (see below).

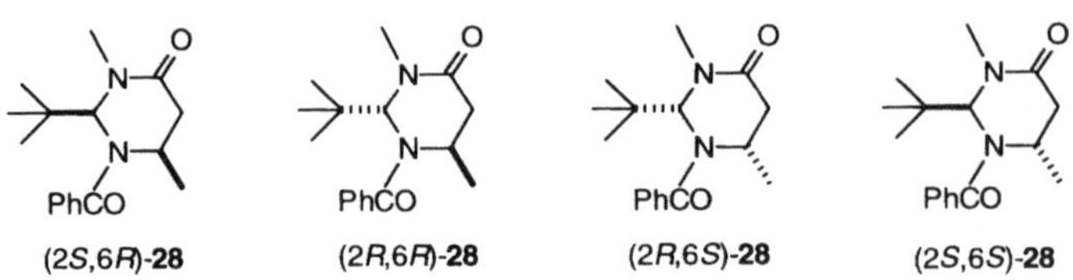

The major isomers of cyclization were deprotonated with LDA to give the lithium enolates, which were alkylated with methyl iodide or benzyl bromide to give exclusively the products from *trans* addition **29**, *ent*-**29**, **30**, and *ent*-**30** in yields of chromatographically purified samples ranging from 75 to 95%. No second diastereoisomer could be detected in the crude products by ^{1}H-NMR spectroscopy[20] (Scheme 13.11).

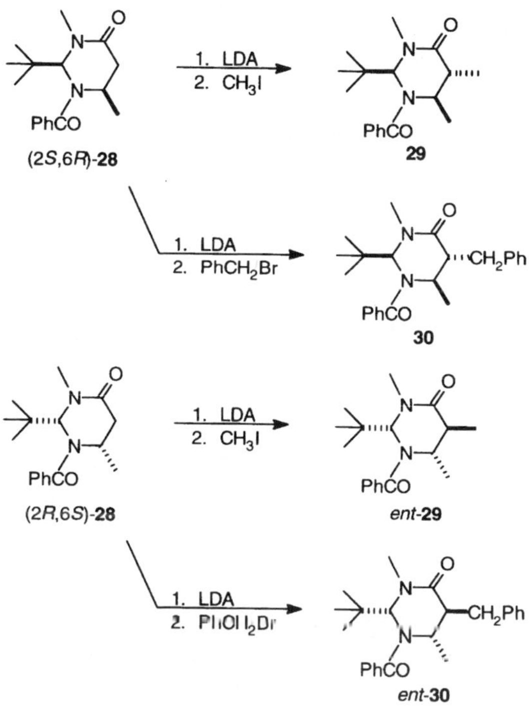

Scheme 13.11

The final step of the overall conversion of 3-aminobutanoic acid to 2-alkyl-3-aminobutanoic acid, the hydrolysis of heterocycles **29** and **30**, was achieved by heating with 6*N* HCl in a sealed tube. The free amino acids (*R*,*R*)-**31**, (*S*,*S*)-**31**, (*R*,*R*)-**32**, and (*S*,*S*)-**32** were purified by chromatography on an ion-exchange column; the yields were ca. 50% of the methylated compounds **31** and ca. 90% of the benzyl derivatives **32**.

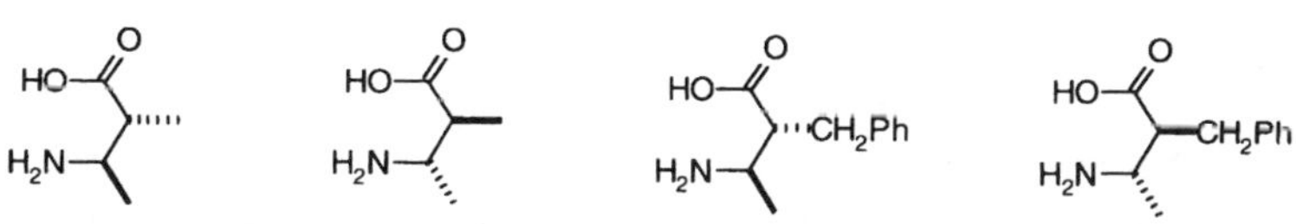

(*R*,*R*)-**31** (*S*,*S*)-**31** (*R*,*R*)-**32** (*S*,*S*)-**32**

13.2.9 Stereoselective Reaction of Chiral Pyrimidinone Enolates with Aldehydes[21]

Having established that enantiopure perhydropyrimidinones **28** can be alkylated with formation of a single diastereoisomer to provide (following hydrolysis) the important α,β-disubstituted β-amino acids **31** and **32** (Section 13.2.8),[22] it was decided to determine the diastereoselectivity of the aldol addition of enolate (2*S*,6*R*)-**28**-Li to various aldehydes *R*CHO. The creation of two stereogenic centers in this reaction (Scheme 13.12) could afford up to four diastereoisomeric products.[23]

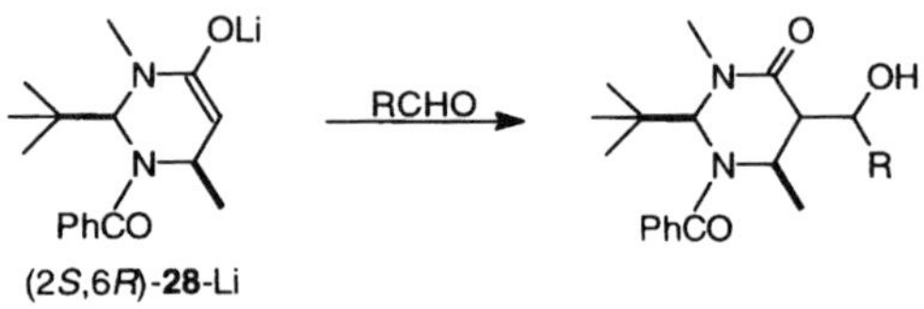

Scheme 13.12

Furthermore, the addition of enantiopure dienolate **33**-Li to aldehydes was also studied (Scheme 13.13).

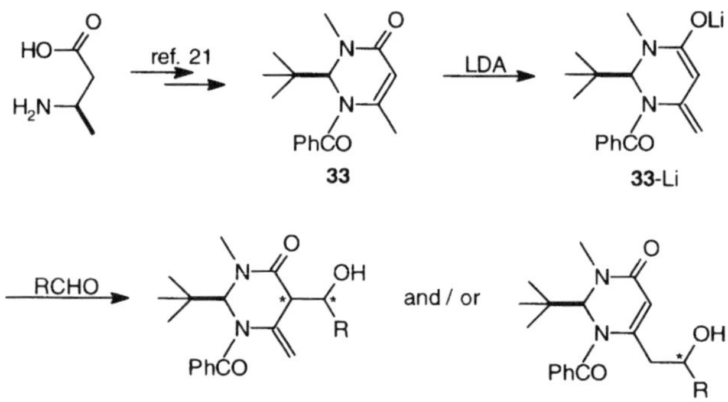

Scheme 13.13

In the event, (2*S*,6*R*)-**28**-Li (in THF at −78°C) was treated with 1.3–2.1 equivalents of aldehyde to give only two diastereoisomeric products, the ratio of which were determined from the crude reaction mixture by ^{13}C-NMR spectroscopy. The configuration was established by single-crystal X-ray structure analysis.[21] As summarized in Table 13.2, addition of (2*S*,6*R*)-**28**-Li to both aliphatic and aromatic aldehydes occurs exclusively from the enolate face opposite to the axial *tert*-butyl and the C(6) methyl groups. (See Section 13.2.8 and Ref. 20).

With regard to the relative topicity with which the trigonal faces of the enolate and the aldehyde prefer to approach each other, the configuration at (C(1′) in **34a**–**36a** indicates that it is *like* (*lk*, *Re*/*Re*).[24]

On the other hand, access to enantiopure (*S*)-**33** was achieved via radical bromination of the saturated precursor (2*S*,6*R*)-**28**, which afforded a ca. 2:1 mixture of the

Table 13.2 ALDOL ADDITIONS OF THE ENOLATE (2*S*,6*R*)-**28**-Li

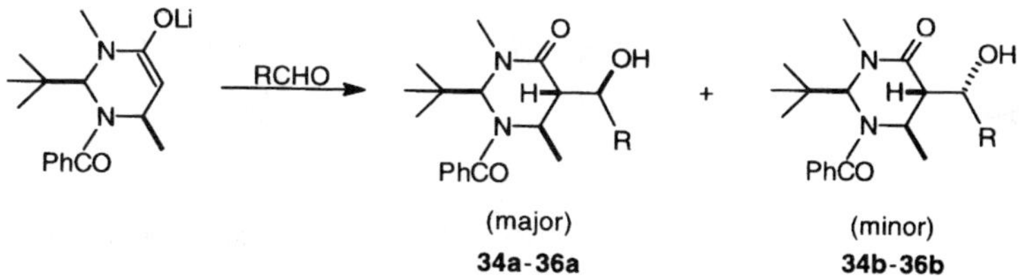

R	Products	Epimer ratio, **a**:**b**	Yield (%)[a]
Phenyl	**34**	2.2:1	57
Isopropyl	**35**	4:1	51
Methyl	**36**	3.5:1	58

[a] Referring to the major diastereoisomer after purification.

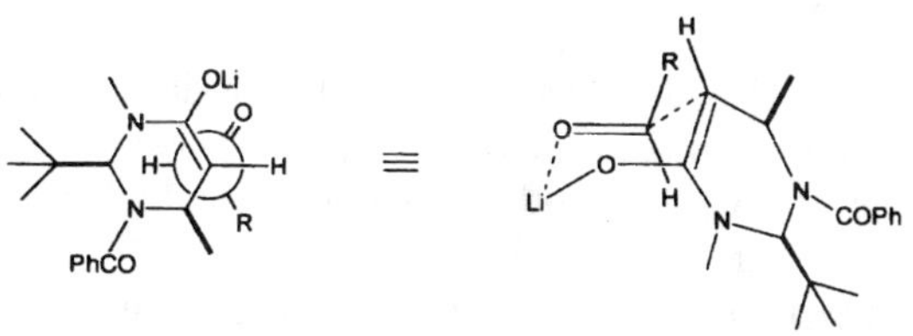

monobromide **37** and the dibromide **38**.[21] This mixture was reduced with hydrogen (palladium/C catalyst) to give the desired product (*S*)-**33** in 55–60% yield over two steps (Scheme 13.14).

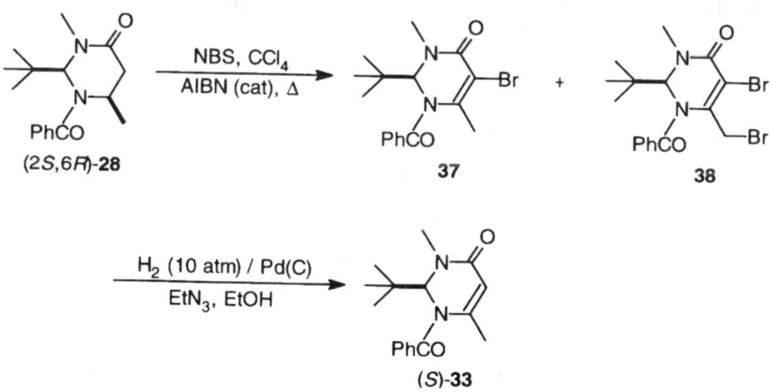

Scheme 13.14

The aldol reaction of dienolate (*S*)-**33**-Li with aldehydes can in principle proceed at the *endo*- or *exocyclic* position of the dienolate, such that up to eight regio- and diastereoisomeric products may be formed (Scheme 13.15). Furthermore, in the case of α,β-unsaturated aldehydes, addition of the enolate may proceed in a Michael fashion (see **42** and **43** in Scheme 13.15).

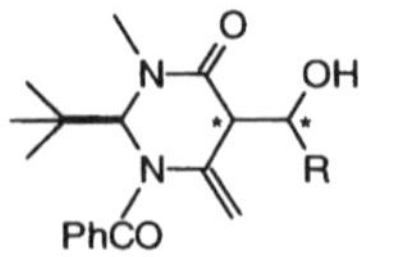

39 (four stereoisomers)

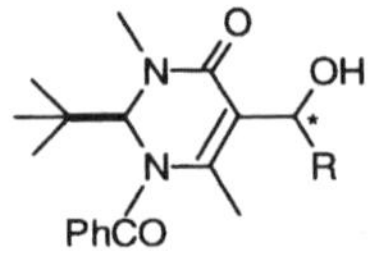

40 (two stereoisomers)

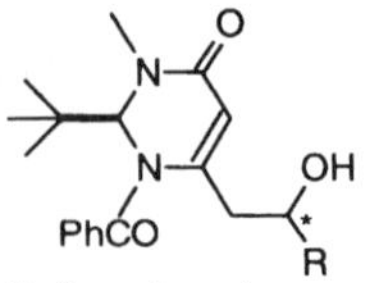

41 (two stereoisomers)

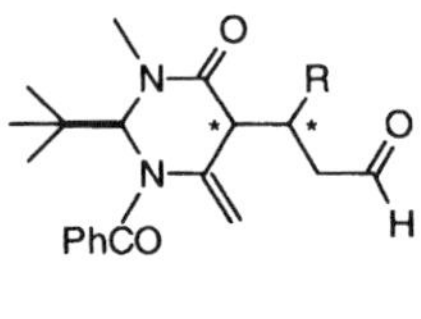

42 (four stereoisomers)

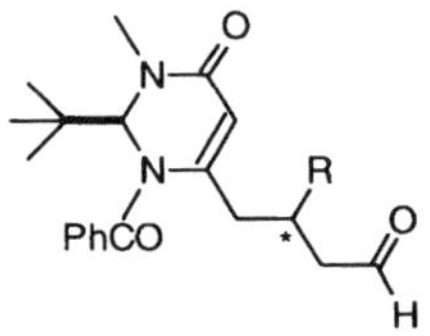

43 (two stereoisomers)

Scheme 13.15

In the actual experiment, (*S*)-**33**-Li was generated with lithium-hexamethyldisilazide (LHMDS) and addition of the aldehydes at –78°C gave the results summarized in Table 13.3. It can be seen that, with all aldehydes studied, only the products of addition at the endocyclic carbon were observed.

Only the products of *trans* addition were observed. The assignment of the configuration of both the major and minor products was secured by X-ray crystallographic analysis.[21] Thus it is concluded that the aldol additions of (*S*)-**33**-Li are completely regioselective, and proceed from the enolate face opposite to the *tert*-butyl group. Furthermore, with benzaldehyde and crotonaldehyde, the diastereoselectivity of the reaction is essentially complete, to afford the aldol products originating from a relative topicity *unlike*[24] in the coupling step.

Table 13.3 ALDOL REACTIONS OF DIENOLATE (*S*)-33-Li[21]

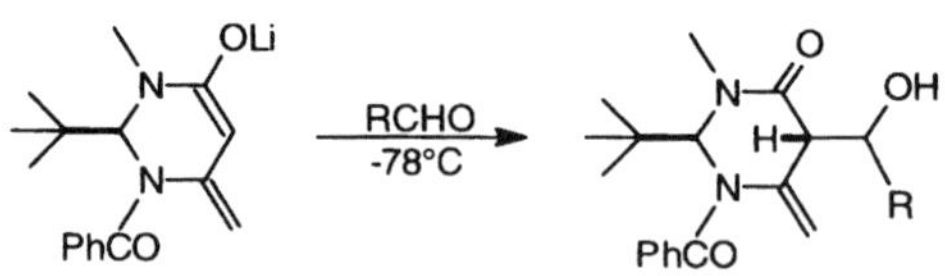

R	Products	Ratio of diastereoisomers (dr)	Yield (%)[a]
Phenyl	**44**	>50:1	82
Methyl	**45**	3:1	78
trans-CH=CHMe	**46**	>50:1	56

[a] After flash chromatography.

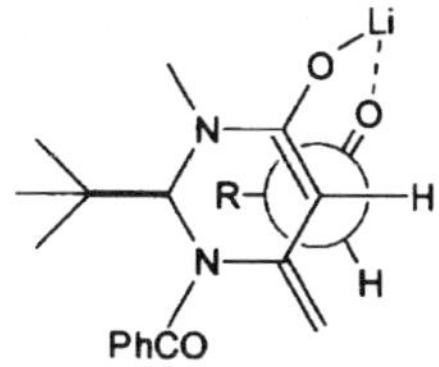

Aldol products **34–36** and **44–46** are potential precursors of enantiopure β-amino-β′-hydroxy acids (**47**) as well as of 1′-hydroxy-β-lactams (**48**), the importance of which as antibiotics is well recognized.[25] Nevertheless, preliminary observations indicate that aldol products **34–36** are hydrolyzed with anchimeric participation by the hydroxy group to give lactones **49**.[26]

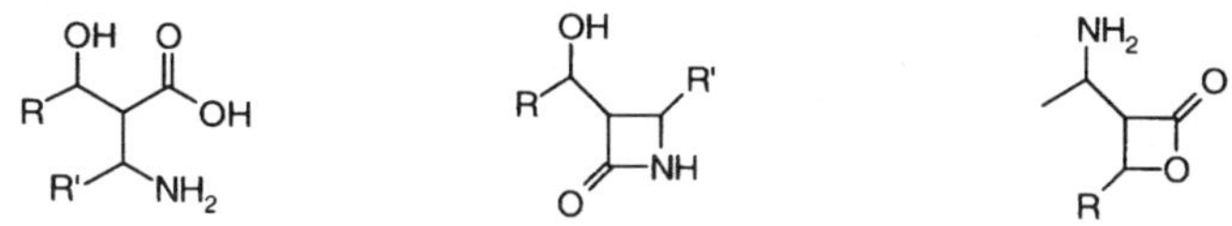

47 **48** **49**

13.2.10 1,2-Asymmetric Induction in the Alkylation of 1-Benzoyl-3,6(*S*)-Dimethylperhydropyrimidin-4-One, (*S*)-50[27]

(*S*)-3-Aminobutanoic acid [(*S*)-**24**, Scheme 13.9] was converted to the desired perhydropyrimidinone (*S*)-**50** according to the protocol outlined in Scheme 13.10, with formaldehyde instead of pivalaldehyde during Schiff base formation. The alkylation products **51** and **52** were formed by treatment of the enolate (*S*)-**50**-Li (LDA, THF) with methyl iodide and benzyl bromide at –75°C (Eq. 13.3)

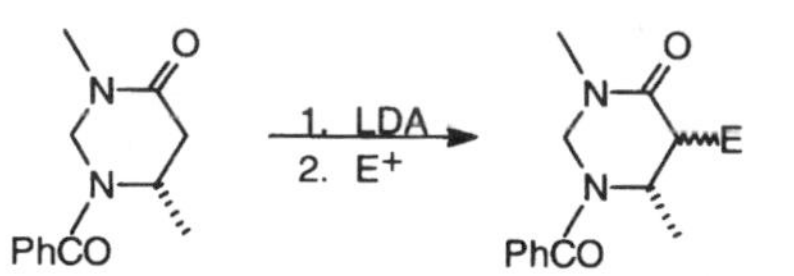

(*S*)-**50**

95 - >99% yield
51, E = CH_3
52, E = CH_2Ph

(eq. 13.3)

^{13}C-NMR analysis of the crude products indicated the formation of two diastereomers in a 4:1 ratio. Chemical correlation with known α,β-disubstituted β-amino acids confirmed that the major product corresponds to the isomer of *trans* relative configuration. Hydrolysis of heterocycles **51** and **52** was accomplished by heating to

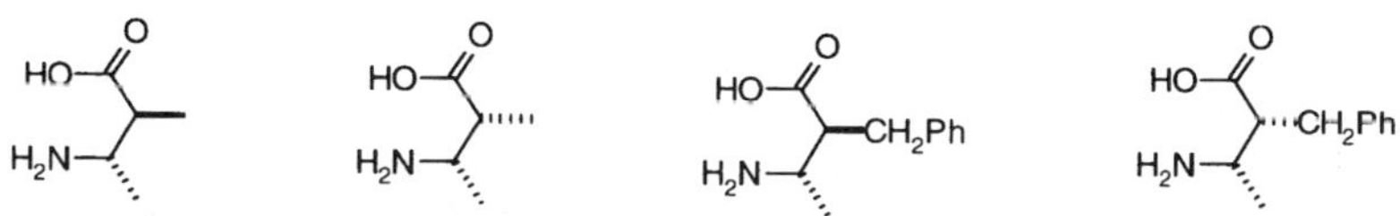

(2*S*,3*S*)-**53** (2*R*,3*S*)-**53** (2*S*,3*S*)-**54** (2*R*,3*S*)-**54**

115–120°C with 6*N* HCl in a sealed tube. The free amino acids **53** and **54** were purified by chromatography on an ion-exchange column; the yields were ca. 84% of the methyl and ca. 70% of the benzyl derivatives.

13.2.11 High 1,2-Stereoinduction in the Preparation of Enantiopure 2(*R*)-Hydroxy-3(*R*)-N-Benzoylamino-3-Phenylpropionic Acid (*Like* Stereoisomer of Taxol's Side Chain)[28]

In contrast with the moderately diastereoselective alkylation of the enolate (*S*)-**50**-Li described in the previous section, we found very high 1,2-stereoinduction in the hydroxylation of the phenyl analogue (*S*)-**55**, which allowed an efficient elaboration to the enantiopure α-hydroxy-β-amino-β-phenylpropionic acid **56**[29] (Scheme 13.16).

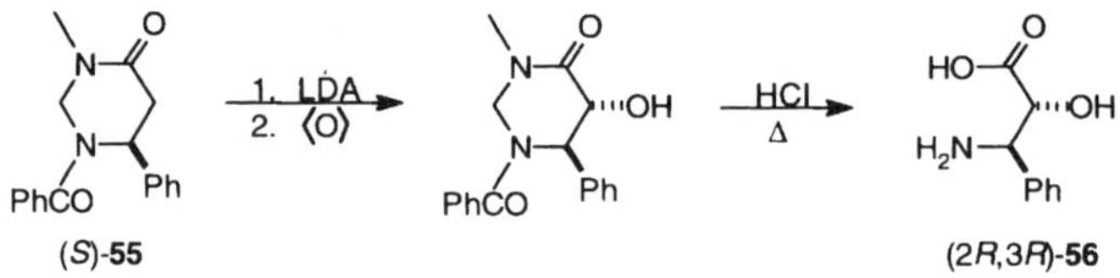

Scheme 13.16

Methyl (*S*)-3-amino-3-phenylpropionate, (*S*)-**57**, was prepared according to the procedure described by Davies et al.[30] (Scheme 13.17). Conversion of ester (*S*)-**57** to the desired perhydropyrimidinone (*S*)-**55** was carried out according to the general procedure (Scheme 13.17).

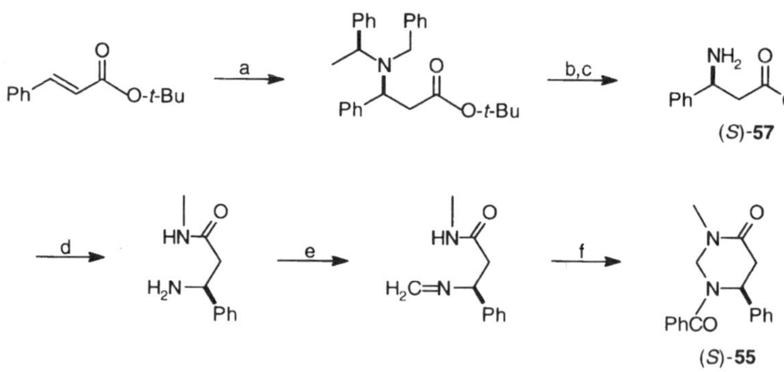

a (*R*)-Benzyl(α-methylbenzyl)amide. b H_2, Pd/C, AcOH. c $(CH_3)_3SiCl$, CH_3OH.
d CH_3NH_2 / CH_3OH. e $(CH_2O)_n$. f PhCOCl / DMAP.

Scheme 13.17

Enolization of heterocycle (*S*)-**55** with LDA was followed by hydroxylation at −78°C with (+)-(camphorsulfonyl)oxaziridine[31] to afford the *trans* product, which was then hydrolyzed by heating to 100–105°C with 6*N* HCl in a sealed tube (Scheme 13.16).

13.3 Summary

β-Aminopropionic acid was converted efficiently into the cyclic, chiral aminal *rac*-**2**. X-ray crystallographic analysis of **2** revealed the axial orientation of the *tert*-butyl group, which is responsible for high *trans* diastereoselectivity found in the reaction of enolate **2**-Li with electrophiles. The hydrolysis of the resulting adducts proceeded with 6*N* HCl to give the expected α-substituted β-amino acids in good yields. These results constituted the basis for the development of a convenient method for the preparation of enantiopure β-amino acids via perhydropyrimidinones of type **2**.

Indeed, (*S*)-asparagine was efficiently converted into enantiopure (*S*)-**2** and then into (*R*)-α-alkylated β-amino acids via diastereoselective alkylation followed by acid hydrolysis. The enantiomeric (*S*)-α-alkylated β-amino acids were obtained by epimerization of the original adducts **11** and **13**–**15**, followed by the usual hydrolysis of isomeric **20**–**23**. This work demonstrates the utility of aminal (*S*)-**2**, a chiral derivative of 3-aminopropionic acid, for the enantioselective synthesis of α-substituted β-amino acids.

α,β-Disubstituted β-amino acids were obtained in a similar fashion via the alkylation/hydrolysis protocol with enantiopure 6-substituted perhydropyrimidinones **28**, **33**, **50**, and **55**. In a relevant application, 6-phenyl derivative (*S*)-**55** was hydroxylated with complete diastereoselectivity; hydrolysis of the product gave enantiopure 2(*R*)-hydroxy-3(*R*)-N-benzoylamino-3-phenylpropionic acid, the *like* stereoisomer of Taxol's side chain.

13.4 References

1. D. Seebach, R. Naef. *Helv. Chim. Acta* **64**, 2704 (1981).
2. D. Seebach, R. Naef, G. Calderari, *Tetrahedron* **40**, 1313 (1984).
3. See also: D. Seebach, R. Imwinkelried, T. Weber, in *Modern Synthetic Methods 1986*, R. Scheffold, Ed. (Springer-Verlag, Berlin, 1986), Vol. 4.
4. For a review, see: A. R. Sting, D. Seebach, *Angew. Chem., Int. Ed. Engl.*, in preparation.
5. (a) D. Seebach, D. D. Miller, S. Müller, T. Weber, *Helv. Chim. Acta* **68**, 949 (1985). (b) T. Weber, R. Aeschimann, T. Maetzke, D. Seebach, *Helv. Chim. Acta* **69**, 1365 (1986).
6. See also: D. Seebach, E. Juaristi, D. D. Miller, C. Schickli, T. Weber, *Helv. Chim. Acta* **70**, 237 (1987).
7. (a) R. Fitzi, D. Seebach, *Angew. Chem., Int. Ed. Engl.* **25**, 345 (1986). (b) R. Fitzi, D. Seebach, *Tetrahedron* **44**, 5277 (1988).
8. An alternative method for the preparation of enantiomerically pure (*R*)- and (*S*)-**1**, which makes use of (*S*)-α-methylbenzylamine as the chiral adjuvant for the resolution step, has also been developed: (a) E. Juaristi, B. Rizo, V. Natal, J. Escalante, I. Regla, *Tetrahedron: Asymmetry* **2**, 821 (1991). (b) O. García-Barradas, E. Juaristi, *Tetrahedron* **51**, 3423 (1995).
9. E. Juaristi, D. Quintana, B. Lamatsch, D. Seebach, *J. Org. Chem.* **56**, 2553 (1991).
10. For a recent review, see: R. W. Franck, in *Conformational Behavior of Six-Membered Rings: Analysis, Dynamics, and Stereoelectronic Effects*, E. Juaristi, Ed. (VCH Publishers, New York, 1995), Chapter 5.

11. R. W. Hoffmann, *Chem. Rev.* **89**, 1841 (1989).

12. For a discussion of the consequences of $A^{1,3}$ strain on structure and reactivity of five- and six-membered N,N-, N,O-, and O,O-acetals, see: D. Seebach, B. Lamatsch, R. Amstutz, A. K. Beck, M. Dobler, M. Egli, R. Fitzi, M. Gautschi, B. Herradón, P. C. Hidber, J. J. Irwin, R. Locher, M. Maestro, T. Maetzke, A. Mouriño, E. Pfammatter, D. A. Plattner, C. Schickli, W. B. Şchweizer, P. Seiler, G. Stucky, W. Petter, J. Escalante, E. Juaristi, D. Quintana, C. Miravitlles, E. Molins, *Helv. Chim. Acta* **75**, 913 (1992).

13. (a) K. D. Hargrave, E. L. Eliel, *Tetrahedron Lett.* 1987 (1979). (b) D. J. Goldsmith, J. K. Thottahil, *J. Org. Chem.* **47**, 1382 (1982).

14. (a) Preliminary communication: E. Juaristi, D. Quintana, *Tetrahedron: Asymmetry* **3**, 723 (1992). (b) Full paper: E. Juaristi, D. Quintana, M. Balderas, E. Garcia-Pérez, *Tetrahedron: Asymmetry*, **7**, 2233 (1996).

15. Limited amounts of both enantiomers of **2** became available via preparative resolution via recycling chromatography on *chiraspher*: J. N. Kinkel, U. Gysel, D. Blaser, D. Seebach, *Helv. Chim. Acta* **74**, 1622 (1991).

16. (a) K. S. Chu, G. R. Negrete, J. P. Konopelski, F. J. Lakner, N.-T. Woo, M. M. Olmstead, *J. Am. Chem. Soc.* **114**, 1800 (1992). (b) See also: P. M. Hardy, D. J. Samworth, *J. Chem. Soc., Perkin Trans. 1* 1954 (1977), and Chapter 11 in the present volume.

17. K. Balenovic, N. Bregant, *Tetrahedron* **5**, 44 (1959).

18. Confirmation of the absolute configuration in (+)-**2** comes from the X-ray crystallographic structure of *cis*-**8**: E. Molins, unpublished.

19. (a) H. Estermann, D. Seebach, *Helv. Chim. Acta* **71**, 1824 (1988). (b) D. Seebach, H. Estermann, *Tetrahedron Lett.* **28**, 3103 (1987). (c) See also: S. G. Davies, O. Ichihara, *Tetrahedron: Asymmetry* **3**, 183 (1991).

20. E. Juaristi, J. Escalante, B. Lamatsch, D. Seebach, *J. Org. Chem.* **57**, 2396 (1992).

21. P. Murer, B. Rheiner, E. Juaristi, D. Seebach, *Heterocycles* **39**, 319 (1994).

22. Closely related heterocycles have been also exploited by Cardillo et al. in their program directed towards the synthesis of α-substituted β-amino acids: (a) R. Amoroso, G. Cardillo, C. Tomasini, *Heterocycles* **34**, 349 (1992). (b) R. Amoroso, G. Cardillo, C. Tomasini, P. Tortoreto, *J. Org. Chem.* **57**, 1082 (1992). (c) G. Cardillo, C. Tomasini, Chapter 10 in this volume, and references therein.

23. (*R*,*R*)-2-*tert*-Butyl-6-methyl-1,3-dioxan-4-one has been shown to react stereoselectively with aliphatic or aromatic aldehydes: W. Amberg, D. Seebach, *Chem. Ber.* **123**, 2413 (1990).

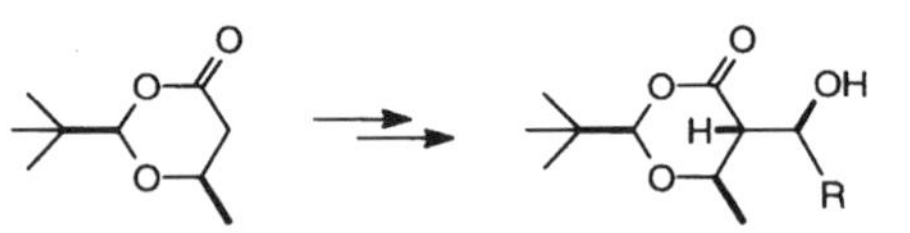

(major isomer with aliphatic aldehydes)

24. For a definition of the *like*/*unlike* stereochemical descriptors, see: (a) D. Seebach, V. Prelog, *Angew. Chem., Int. Ed. Engl.* **21**, 654 (1982). (b) E. Juaristi, *Introduction to Stereochemistry and Conformational Analysis* (Wiley, New York, 1991), pp. 52–54.

25. J. Tamariz, Chapter 2 in this volume, and references therein.

26. J. Escalante, *Doctoral thesis*, Instituto Politécnico Nacional, México, 1995.

27. E. Juaristi, J. Escalante, *J. Org. Chem.* **58**, 2282 (1993).

28. J. Escalante, E. Juaristi, *Tetrahedron Lett.* **36**, 4397 (1995).

29. Taxol derivatives equipped with either enantiomer of the *like* side chain have been found to be almost as active as Taxol itself: G. I. Georg, Z. S. Cheruvallath, R. H. Himes, M. R. Mejillano, C. T. Burke, *J. Med. Chem.* **35**, 4230 (1992). See also: G. I. Georg, Chapter 1 in this volume.

30. M. E. Bunnage, A. N. Chernega, S. G. Davies, C. J. Goodwin, *J. Chem. Soc., Perkin Trans. 1* 2373 (1994).

31. F. A. Davis, R. T. Reddy, R. E. Reddy, *J. Org. Chem.* **57**, 6387 (1992).

CHAPTER

14

The Synthesis of β-Amino Acids and Their Derivatives from β-Lactams

Claudio Palomo, Jesús M. Aizpurua, and Iñaki Ganboa

14.1 Introduction

β-Amino acids hold interest as analogues of α-amino acids. For example, the limited normal metabolism of β-amino acids suggests that many β-amino analogues of α-amino acids will be metabolically stable in vivo; enzyme inhibitors may thus have extended biological half-life when designed as β- rather than α amino acids. Similarly, incorporation of β-amino acids into peptides of pharmacological interest has, in some cases, been found advantageous in terms of biological activity, metabolic stability, or both.[1] Interest in β-amino acids also stems from their occurrence in macrocyclic peptides, antibiotics, and an array of natural products and biologically active compounds.[2] Furthermore, derivatizing appropriately substituted β-amino acids may provide the possibility to build up a wide range of heterocyclic structures, from among β-lactams are of particular relevance as they are the key structural element of the most widely employed class of antimicrobial agents, the β-lactam antibiotics.[3] Nevertheless, β-lactams may also be simply viewed as cyclized forms of β-amino acids in which the amino group and the carboxyl moiety are simultaneously protected. As a result, chemical reactions at both α- and β-positions are more readily accessible on β-lactams than on β-amino acids themselves. Furthermore, the rigidity of the four-membered ring makes these reactions often highly stereoselective. On the other hand, the regeneration of the corresponding β-amino acid–derived product from the previously elaborated β-lactam framework can be achieved by reaction with nucleophiles under appropriate acid or basic conditions. In addition, the intramolecular version of such nucleophilic reactions between the β-lactam amide and suitable substituents on the four-membered ring may provide a straightforward access to cyclic compounds containing β-amino acids and derivatives thereof.

These aspects have been reviewed on two occasions by Bose and Manhas,[4] revealing the potential utility of β-lactams for preparing valuable precursors of a wide variety of target molecules including carbohydrates, alkaloids, amino acids, and peptides. This latter subject has also been recently reviewed by Ojima, illustrating the potential of 3-amino-4-aryl β-lactams in the context of α-amino acid synthesis through the hydrogenolytic cleavage of the N_1-C_4 β-lactam bond discovered by the same author.[5] Other reviews concerning ring enlargement of these small heterocycles to large ring structures have also been reported by Wasserman[6] and Hesse.[7] Since other chapters of this book deal with the synthesis of β-amino acids, and two general reviews on this subject have already been published,[8] this account focuses only attention on the chemistry of β-lactams as building blocks of optically active substances containing β-amino acids and/or derived β-amine ketone and β-amino alcohol subunits. Emphasis will be placed on β-lactams as new acylating reagents for peptide coupling reactions.

The difficult task to class the many differently substituted β-amino acids obtained through appropriate N_1-C_2 β-lactam bond cleavage has prompted us to divide the present review into four sections. The first two sections (14.2 and 14.3) outline the presence of additional functionality or substitution besides the β-amino acid moiety and deal, respectively, with α-substituted and γ-substituted β-amino acids, covering important products like aspartic acids, hydroxyamino acids, and diamino acids (Figure 14.1). In Section 14.4 attention is drawn to the carbon nucleophile mediated N_1-C_2 β-lactam bond cleavage, leading to β-amino ketones and related compounds. Finally, large ring heterocycles containing β-amino acid fragments are discussed in Section 14.5.

The reader is referred to previous reviews for the synthesis and reactions of β-lactams.[9] In particular, the ester enolate-imine condensation,[10] the cycloaddition reaction

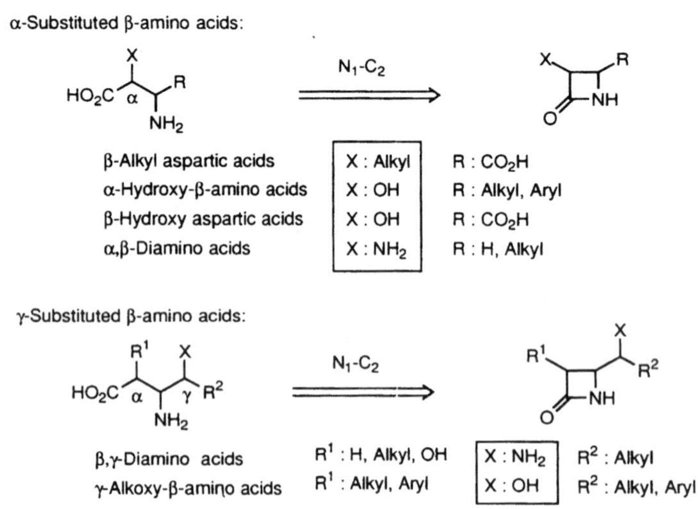

Figure 14.1

of ketenes with imines,[11] and the hydroxamate methodology developed by Miller[12] should be mentioned as the most direct approaches to the β-lactam ring. Additional strategies have also been recently reviewed.[13] Therefore, construction of β-lactams will be only described with regard to the synthesis of β-amino acids and related compounds.

14.2 α-Substituted β-Amino Acids

As indicated before, in the following section a discussion is given on the application of 3-alkyl-, 3-hydroxy-, and 3-amino β-lactams for the synthesis of α-substituted β-amino acids and/or the construction of relatively more complex molecules containing such units.

14.2.1 β-Alkyl Aspartic Acids and Derived Compounds

Taking into account the relative *trans*- or *cis*-stereochemistry of the β-lactam ring, *erythro* and *threo* β-alkyl aspartic acids can be formed. Both of them will be discussed separately, along with strategies leading to α,β-alkyl β-amino acids with desired stereochemistry at the α- and β-carbon stereogenic centers. These strategies are mainly based on the easy functionalization of the carboxy group in 4-carboxy β-lactams, providing suitable access to 3,4-dialkyl-β-lactams, which are not readily available by the hitherto established direct routes to β-lactams, that is, ester enolate–imine condensation[10] and cycloaddition reaction.[11c]

14.2.1.1 erythro β-Alkyl Aspartic Acids

Most of the investigations on the synthesis of *erythro* β-alkyl aspartic acids and derivatives have dealt with the use of the 4-benzyloxycarbonyl β-lactam **1** derived from the aspartic acid itself.[14] For example, Baldwin[15] has reported a synthesis of dealanylalahopcin **7** as precursor of the dipeptide antibiotic alahopcin **8**, involving, as the crucial step, the stereoselective alkylation of the dianion of the 4-carboxy β-lactam **2** with allyl bromide to give **3** as the only alkylation product. Another interesting feature of the approach is the incorporation of the *Boc* group at N-position of the β-lactam ring to promote ring opening by the corresponding heteronucleophile. In this way **4**, upon exposure to sodium ethanethiolate and O-benzylhydroxylamine, led to the corresponding open product, which was next transformed into **6** through the aldehyde **5** (Scheme 14.1).

Baldwin has also recently described a β-lactam approach to the aspartic acid **11** employed for the synthesis of (2*S*,3*R*)-3-carboxyprolines[16] (Scheme 14.2). Interest in these compounds arises from the development of the structure activity relationships of excitatory amino acid receptor agonists and competitive antagonists, which indicate that such substituted prolines show activity at a number of different classes of receptor sites.[17] The synthesis comprises an alkylation–lactonization of β-lactam **2** to

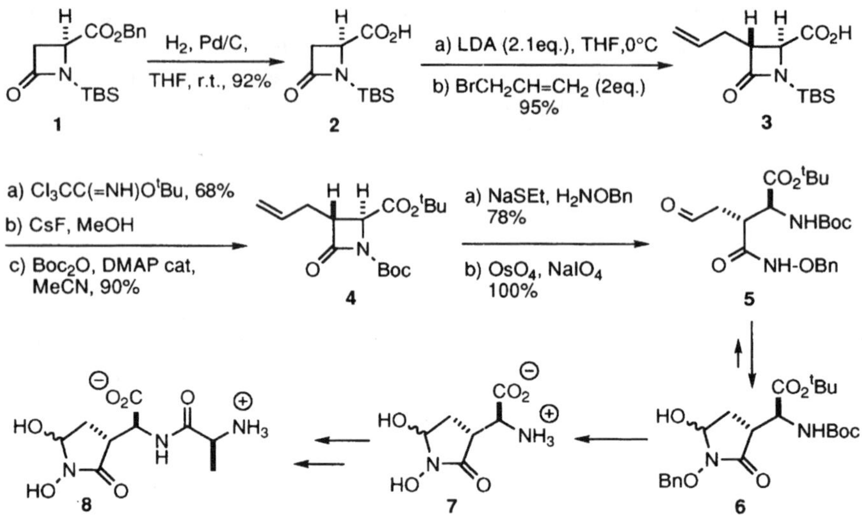

Scheme 14.1

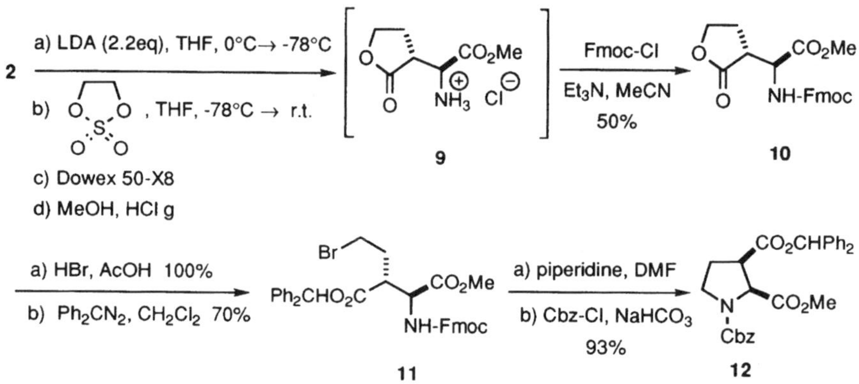

Scheme 14.2

give the 9-fluorenyloxycarbonyl derivative **10**, which is cleaved to the ω-bromo ester **11** and finally cyclized to the *cis*-proline derivative **12**.

The same group has also reported (Scheme 14.3) the synthesis of (2*S*,3*R*)-3-aminoproline **14** from **12**. The strategy is based on a Curtius rearrangement of **13** to generate the required 3-amino moiety. Compound **14** is a naturally occurring amino acid, which has been isolated from the *Morchella* genus of mushrooms and has been described in the structural elucidation of viomycidine, **15**.[18]

In general, the stereoselective alkylation of the 4-carboxy β-lactam **2** by a variety of alkyl halides has provided a useful entry to *trans*-3-alkyl-4-carboxy β-lactams of type **16**, which in turn can be transformed into (2*S*,3*R*)-3-alkyl-L-aspartic acids **17** (Scheme 14.4).[19]

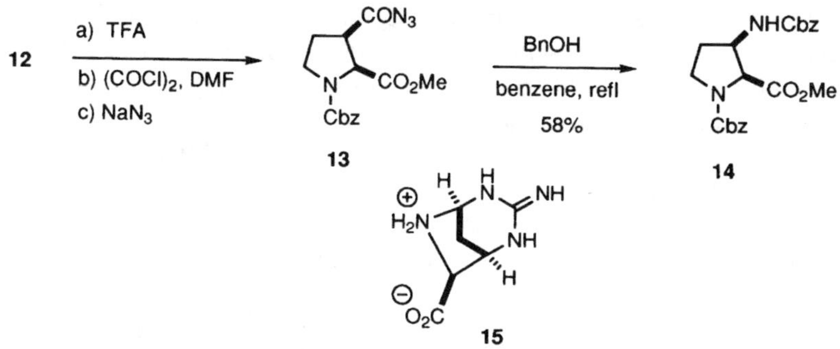

Scheme 14.3

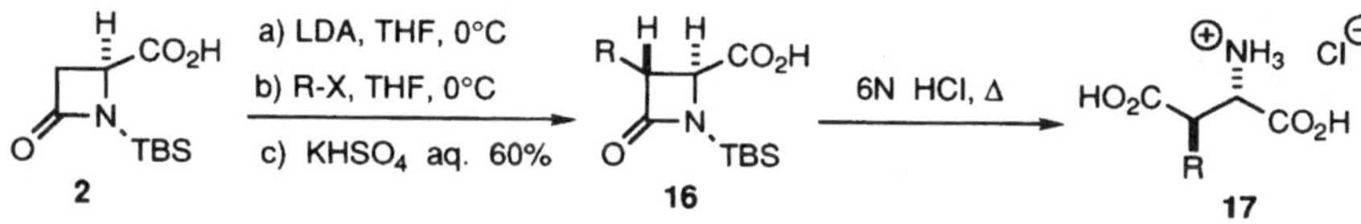

Scheme 14.4

Since the enantiomer of the (*R*)-aspartic acid-derived β-lactam **2** is also readily available, (2*R*,3*S*)-3-alkyl D-aspartic acids can also be prepared by this approach. It is interesting to note that the direct alkylation of the lithium enolate of the β-anion of aspartic acid **18** (Scheme 14.5) with reactive alkyl halides results in formation of the β-alkylated adducts **19a** and **19b** as a mixture of epimers, 5:1 and 3:1, respectively.[20] In a similar way, it has also been reported that the lithium enolate of **18** reacts with benzaldehyde or propanal to give hydroxy esters **20a** and **20b**, respectively, as an equimolar mixture of two diastereomers with the hydroxyl and β-ester groups in a *syn* relationship in both cases.[21]

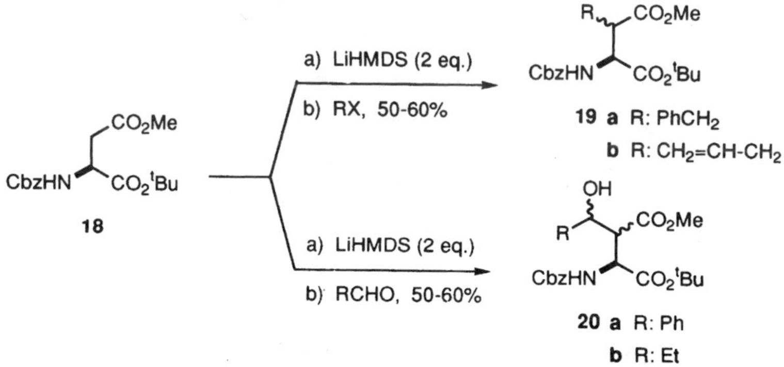

Scheme 14.5

For comparison, carrying out the aldol reaction onto the lithium enolate of the carboxy β-lactam **2**, a mixture of two epimeric carbinols **21a** and **21b** is produced without appreciable diastereoselection at the exocyclic stereogenic center.[22] However, the *trans* relative stereochemistry at the C_3 and C_4 positions of both β-lactams takes place as expected by the attack of the carbonyl electrophile from the less hindered face of the enolate anion. An improvement of this procedure takes advantage of this fact to control the new chirality at the exocyclic center of carbinol **21a** by prior acylation of the above lithium enolate and subsequent stereoselective reduction of the keto group of **22**.[23] Ring opening of these β-lactam products would provide the corresponding β-substituted aspartic acid derivatives (Scheme 14.6).

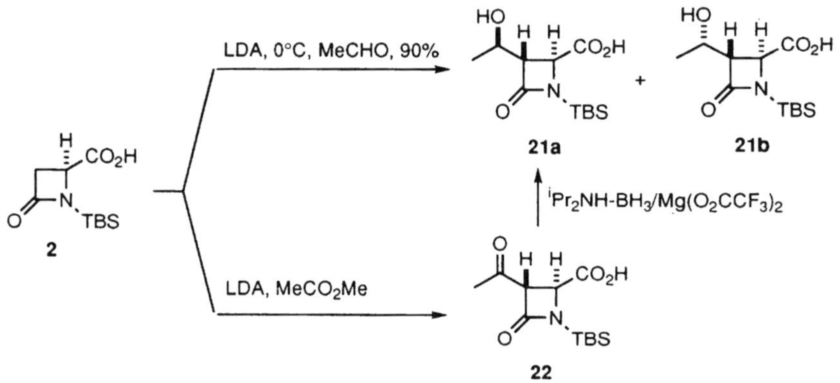

Scheme 14.6

The synthesis of the antibiotic lankacidin C **23** illustrates the above reactions for the construction of more complex molecules from suitable β-lactam frameworks. The first formal synthesis of such a compound has been reported simultaneously by Thomas[24] and Kende,[25] and the latter has recently described its total enantioselective synthesis.

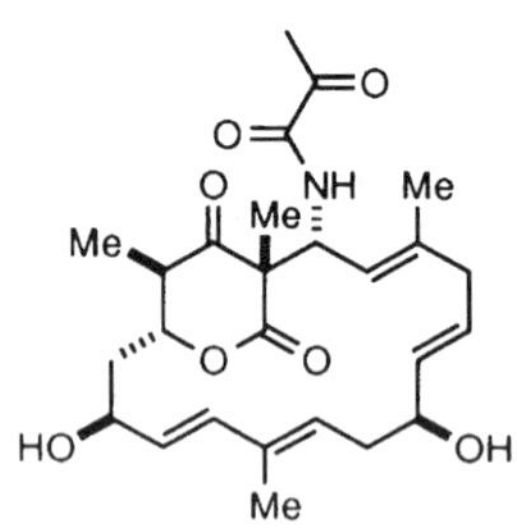

23

As reported by Thomas[24a,b] (Scheme 14.7), the benzyloxycarbonyl β-lactam **1** is transformed into **24**, followed by stereoselective methylation of the corresponding lithium enolate to give **25** in 98% yield. Subsequent aldol reaction of the lithium enolate of **25** with the aldehyde **26** and further oxidation of the resulting mixture of epimeric carbinols led to **27** as a key intermediate of **23**.

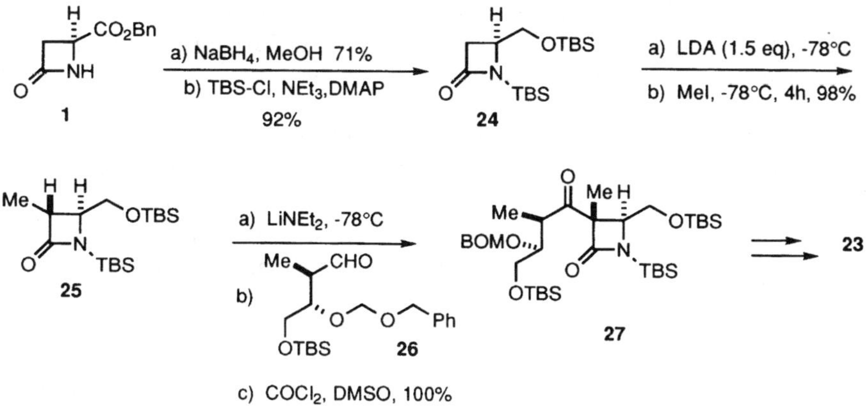

Scheme 14.7

A related approach[24c] to an analogue of **23** involves the use of the 4-formyl β-lactam **28** easily available from **1** (Scheme 14.8). Namely, Wittig reaction of **28** with ethoxycarbonylethylidene triphenylphosphorane furnished **29**. Desilylation and reduction using DIBAL gave the β-lactam alcohol **30**, which was reprotected and methylated using LDA and methyl iodide to give (3*R*,4*R*)-β-lactam **31** as the sole reaction product. Condensation of thioester **32** with the lithium enolate of β-lactam **31** was also highly stereoselective, giving rise to the (3*S*,4*R*)-β-lactam **33**.

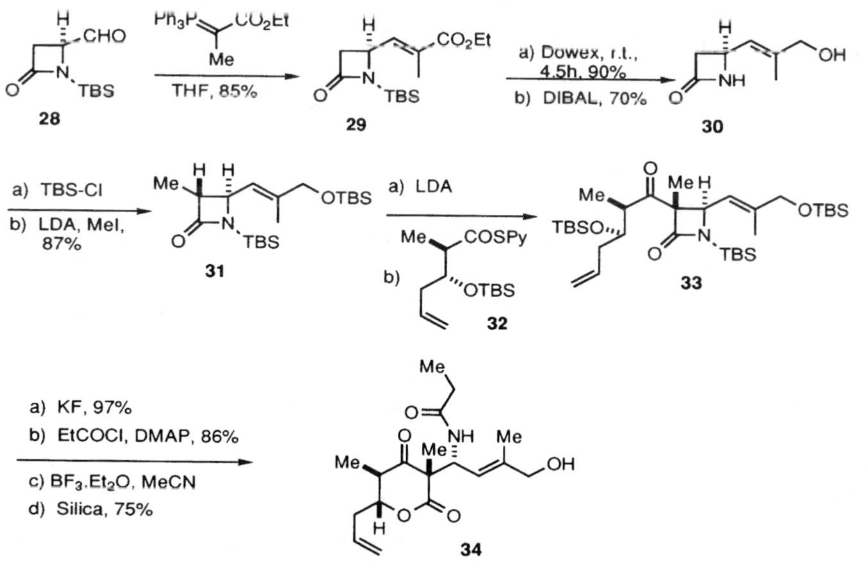

Scheme 14.8

Selective N-disilylation of **33** using KF in methanol, followed by N-acylation using propanoyl chloride, and subsequent O-desilylation using $BF_3.Et_2O$ in acetonitrile gave an imide-diol, which on further treatment with silica led to the fully functionalized lactone fragment **34** of lankacidin C.

The approach developed by Kende[25] (Scheme 14.9) involves acylation of the lithium enolate of the β-lactam **35** with **36** and subsequent stereoselective reduction of the resulting ketone by KEt_3BH to produce **37** as single carbinol. This compound was desilylated and subjected to $MeSO_3H$-catalyzed N→O acyl migration and then Im_2CO trapping to yield **38**. Hydrolysis, Dess–Martin oxidation, and PMB scission gave the stable iodoaldehyde **39**.

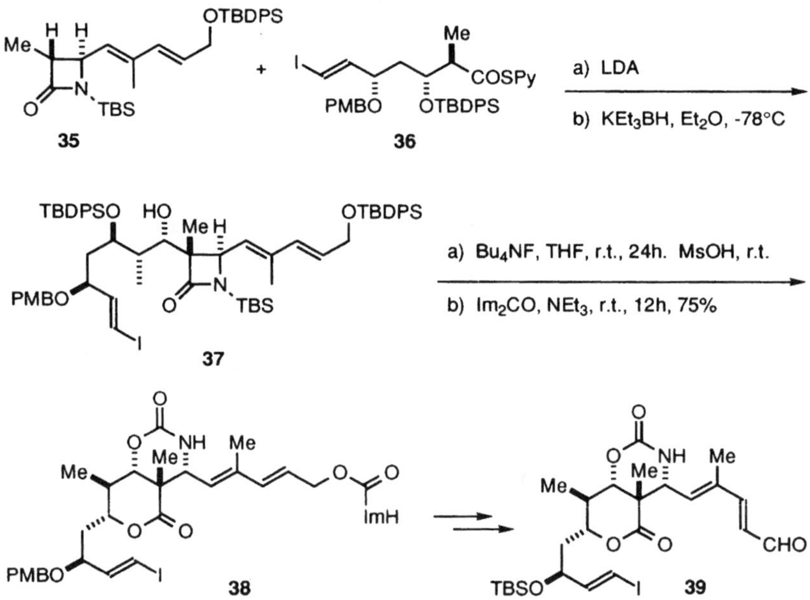

Scheme 14.9

Completion of the synthesis of **23** was accomplished according to the reactions in Scheme 14.10. Stille coupling of **39** with the stannane **40** gave the tetraene **41a**. The chloride **41b** on reaction with TMSCN, subsequent cyclization, stereoselective reduction with diborane under oxazaborole catalyst, and N-acylation led to **42**. Compound **42** was then subjected to ring opening and subsequent Dess–Martin oxidation to give **43** as a direct precursor of **23**.

The required starting β-lactam product **35** was prepared (Scheme 14.11) by reaction of the aldehyde **28** with the lithium salt of **44** followed by equilibration of the resulting crude enals to produce the (*E*)-α,β-unsaturated aldehyde **45**. A second Schlessinger–Peterson condensation with the lithium salt of **46** gave on workup the corresponding (*E,E*)-dienal, which was reduced with $LiBH_4$, O-silylated and C-methylated to give **35** in 30% yield over the six steps.

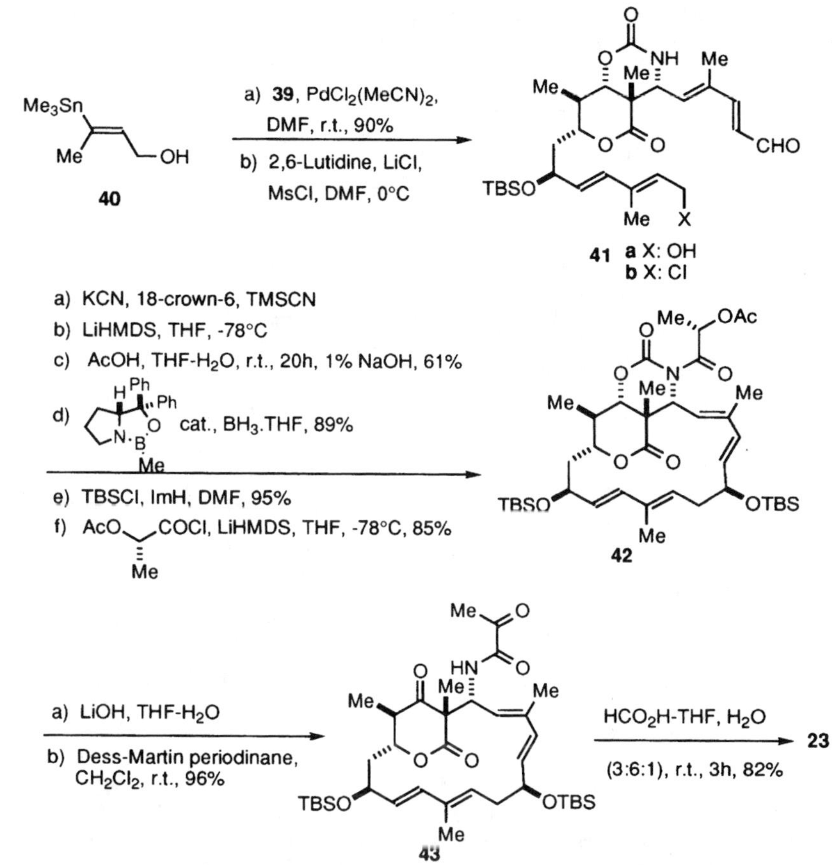

Scheme 14.10

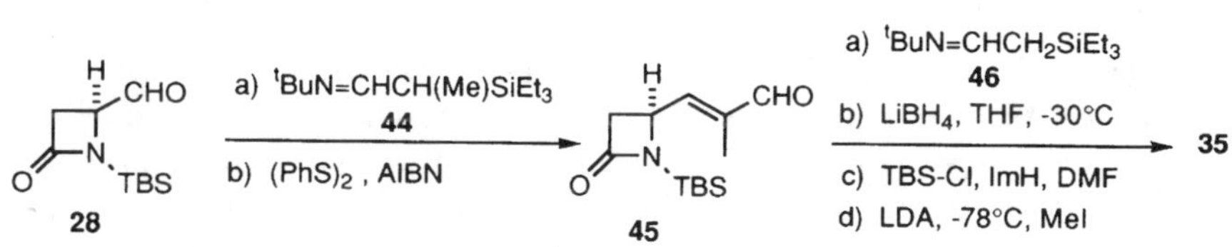

Scheme 14.11

On the basis of these precedents, work in this laboratory has shown that the aldehyde **28** could also be transformed into **47** as the cyclized form of β-substituted β-amino acids.[26] Activation of the β-lactam carbonyl by installation of the Boc group at the N-position led to **48**, which on treatment with both (*S*)-phenylalanine benzyl ester and (*S*)-leucine benzyl ester provided dipeptide products **49a** and **49b** in excellent yields. In a similar way, alkylation of the lithium enolate of **47** and subsequent ring opening of the resulting *trans*-β-lactams led to the formation of α,β-disubstituted β-amino esters, which were isolated as N-Boc derivatives **50** (Scheme 14.12).

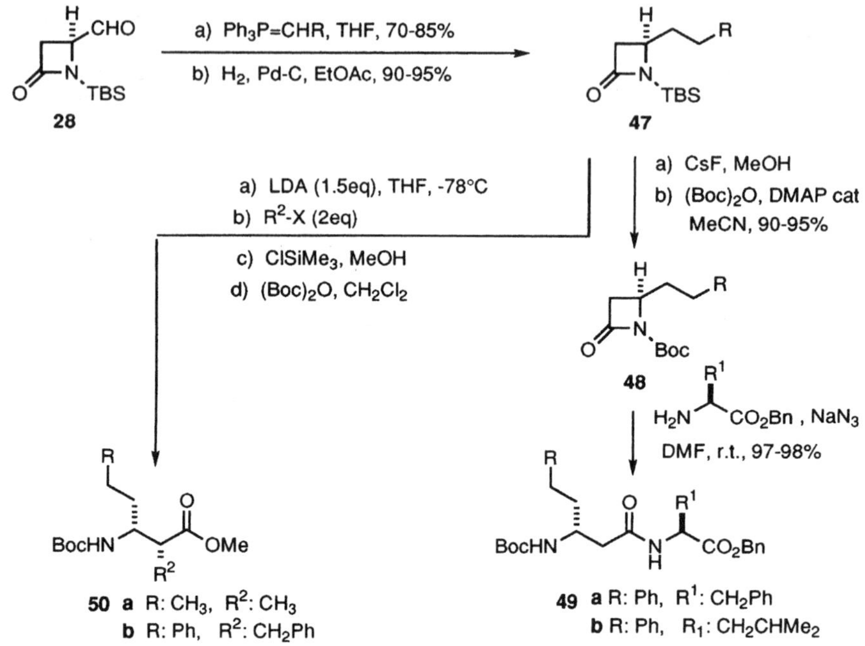

Scheme 14.12

The coupling of other α-amino esters with structurally different N-Boc-β-lactams has been studied in our laboratory to develop this tactically new approach to peptide synthesis, and a more detailed discussion will be presented in Sections 14.2.2.2 and 14.2.2.3.

14.2.1.2 threo β-Alkyl Aspartic Acids

Whilst the synthesis of *erythro* β-alkyl aspartates and *erythro* α,β-disubstituted β-amino acids can be efficiently achieved by alkylation, aldol reaction, or acylation of lithium enolates of the corresponding 3-unsubstituted β-lactams, the synthesis of *threo*-derivatives is necessarily guided by the prior construction of a suitable 4-carboxy-β-lactam with relative *cis*-stereochemistry at the C_3 and C_4 positions. This topic has been widely documented in the literature,[9,10] and the most common approach to these compounds appears to be the use of β-lactams with the styryl moiety at the C_4 position as the latent carboxyl functionality. As an example (Scheme 14.13), reaction of the lithium enolate of the Oppolzer's chiral ester **51** with the N-(p-methoxyphenyl) cinnamaldimine gave, as reported by Hart,[27] the *cis*-β-lactam **52** in 81% yield and in 92% enantiomeric excess. Subsequent N-dearylation of **52** and further N-protection afforded **53**, which was then transformed into the 4-carboxy β-lactam **54** in reasonably good yield. Although this compound was next used as a synthetic intermediate for carbapenem synthesis [(+)-PS-5], obviously it may also be considered as a β-alkyl aspartic acid precursor.

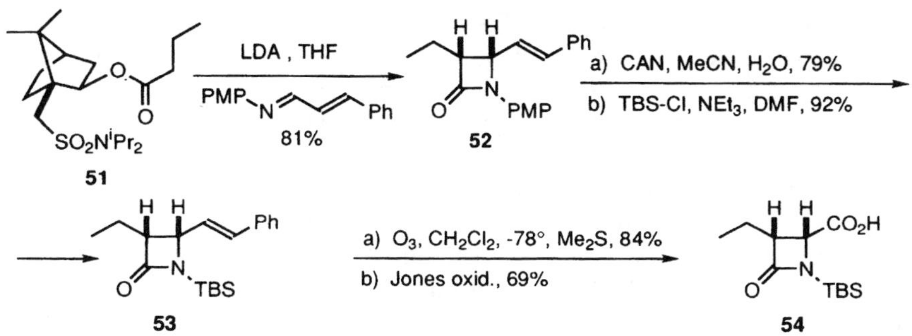

Scheme 14.13

Another illustrative example of the utility of 4-styryl β-lactams (Scheme 14.14) has also been described by Hart[28] in an access to 2,3,6-trideoxyaminohexoses, the key structural subunits of glycosidic acid polysaccharide antibiotics. Although a mixture of β-lactams **56**, **57**, and **58** was produced by condensation of the dianion of **55** with

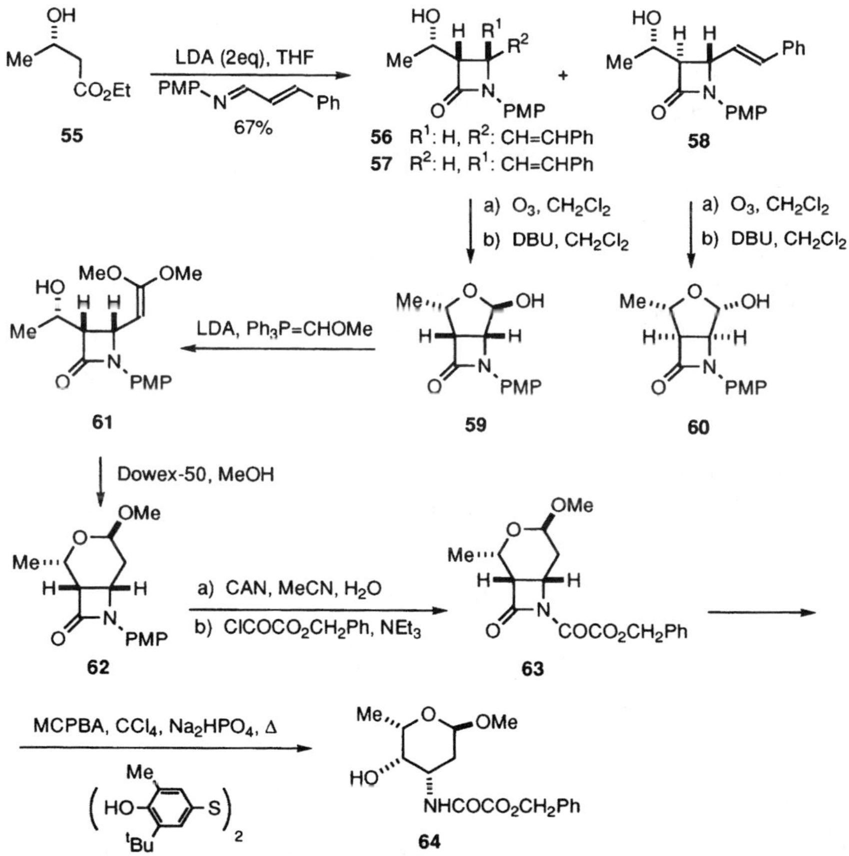

Scheme 14.14

N-(*p*-methoxyphenyl) cinnamaldimine, sequential treatment of these products with ozone, dimethylsulfide, and 1,8-diazabicyclo[5.4.0]undec-7-ene (DBU) gave a separable mixture of hemiacetals **59** and **60** in 68% and 6% isolated yields, respectively. Deprotonation of **59** with LDA and excess of (methoxymethylidine)-triphenylphosphorane afforded enol ether **61**, which was converted to the tetrahydropyran **62** upon treatment with acidic Dowex-50 in methanol. Ceric ammonium nitrate (CAN) oxidation of **62** followed by acylation with benzyl chlorooxalate of the resulting N-unsubstituted β-lactam led to the imide **63**. The authors found that warming **63** with *m*-chloroperbenzoic acid (MCPBA) in carbon tetrachloride containing Na_2HPO_4 and 3-*tert*-butyl-4-hydroxy-5-methylphenyldisulfide gave daunosamine derivative **64** in 35% yield.

On the other hand (Scheme 14.15), **62** has also been transformed into **66** through ring opening of **65** by means of DBU in refluxing methanol. Under these conditions the reaction proceeded with concomitant epimerization at the α-position of the methoxycarbonyl group. Compound **66** was then transformed into the acosamine derivative **67** under similar conditions as above.

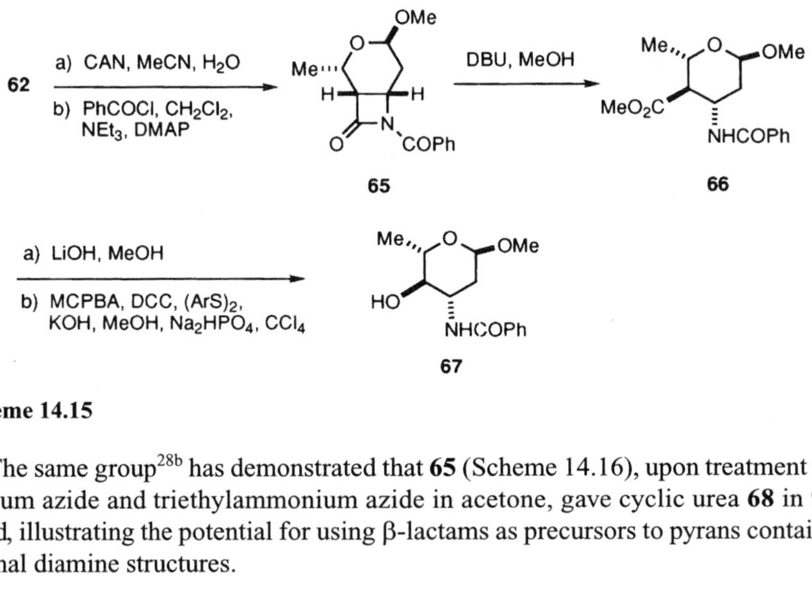

Scheme 14.15

The same group[28b] has demonstrated that **65** (Scheme 14.16), upon treatment with sodium azide and triethylammonium azide in acetone, gave cyclic urea **68** in 98% yield, illustrating the potential for using β-lactams as precursors to pyrans containing vicinal diamine structures.

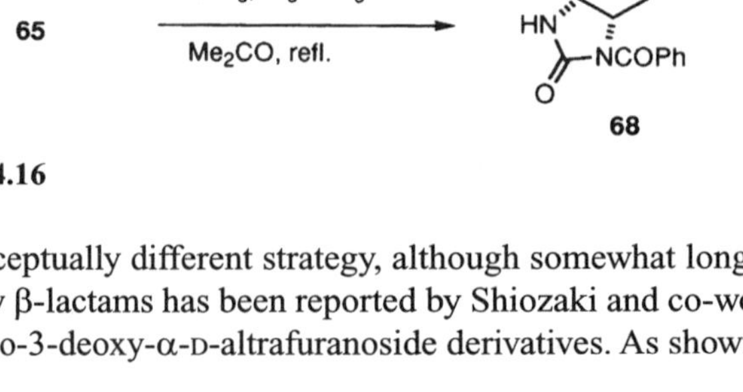

Scheme 14.16

A conceptually different strategy, although somewhat longer, for the synthesis of 4-carboxy β-lactams has been reported by Shiozaki and co-workers[29] in an approach to 2-amino-3-deoxy-α-D-altrafuranoside derivatives. As shown in Scheme 14.17, the synthesis starts from **69**, which in turn was prepared in two steps by reaction of 2,3-O-isopropylidene-D-glyceraldehyde with dimethylthetin anion in DMSO and separation of diastereomers. Such a compound was cyclized to the β-lactam **70**, which on treatment with DBU and azeotropic removal of the ethanol released during the reaction led to the lactone **71**. Compound **71** was then transformed into **72** in three steps with good overall yield. Subsequent N-deprotection was followed by N-benzoylation to give **73**. Treatment of **73** in methanol containing DBU as catalyst at reflux temperature gave the isomerized methyl ester **74**, which was then saponified, and the resulting free carboxylic acid reacted with MCPBA and DCC to give the diaryl peroxide **75**. Thermal decomposition of **75** provided the altrosamine analogue **76** along with **77** in 32% and 36% yields, respectively, for the last step.

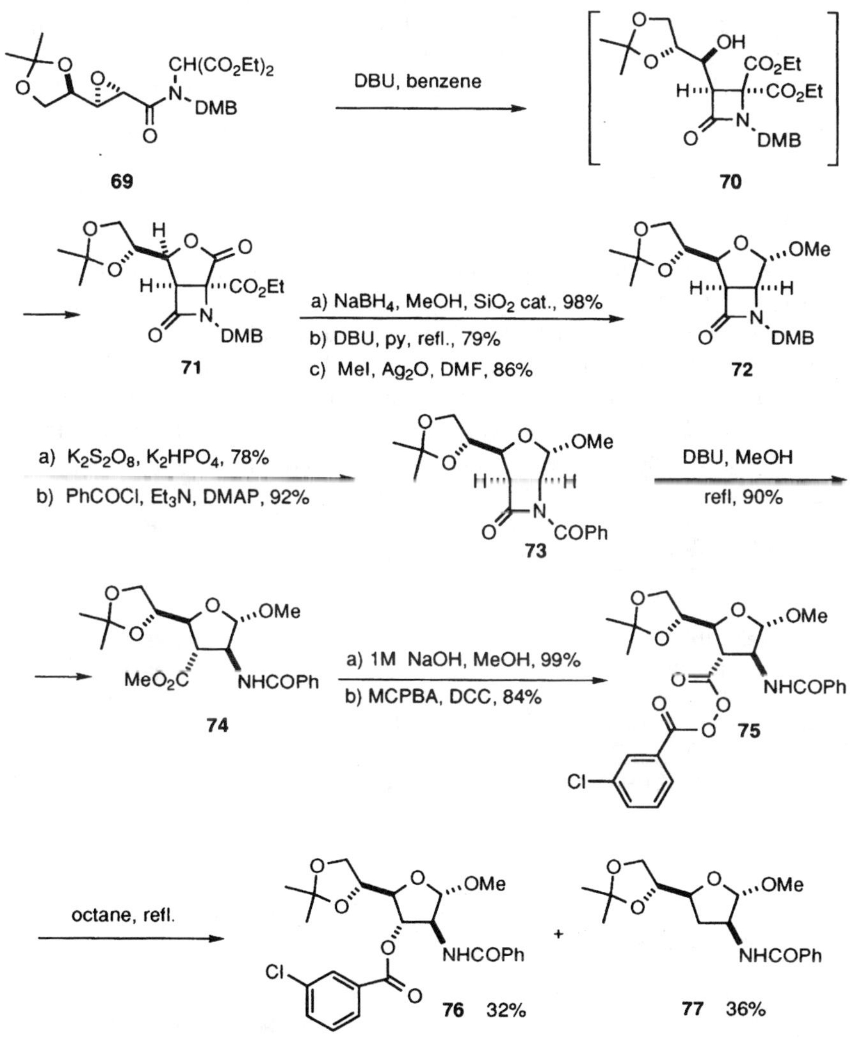

Scheme 14.17

In our own laboratory it has been found that the cycloaddition of alkanoyl chlorides to imines of glyoxylic esters provides an alternative way for the construction of *cis*-β-lactams as precursors of *threo*-β-substituted aspartic acids and derived products.[30] In particular, using β-(dimethylphenylsilyl)alkyl ketenes, it is possible: (a) to exert considerable chiral induction in the [2+2] cycloaddition, (b) to remove the silyl moiety from the reaction products, and/or (c) to transform it into a hydroxyl group with retention of configuration. For instance, *rac*-**79** (Scheme 14.18), when allowed to react with imine **78a**, afforded a mixture of *cis*-**81a**, *cis*-**83a**, and the corresponding *trans* isomers in a ratio of 38:28:34, respectively.

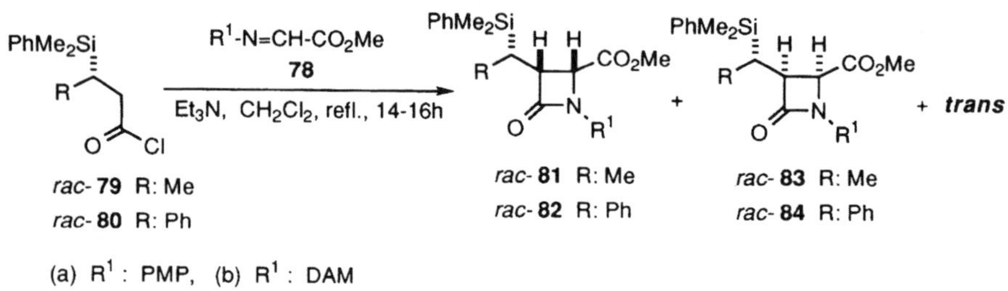

Scheme 14.18

After attempts to improve this stereochemical outcome, the best result was obtained using the imine **78b**, giving an 88% yield of the corresponding *cis*-**81b** and *cis*-**83b**, in a ratio of 66:34, without traces of the corresponding *trans* isomers. When the acid chloride *rac*-**80** was employed in such a cycloaddition reaction, *cis*-**82b** and *cis*-**84b** were obtained in 71% combined yield and, most notably, in a ratio of 91:9, respectively. Conversion of these β-lactams into β-(hydroxyalkyl)-aspartic acids was accomplished following the sequences illustrated in Scheme 14.19.

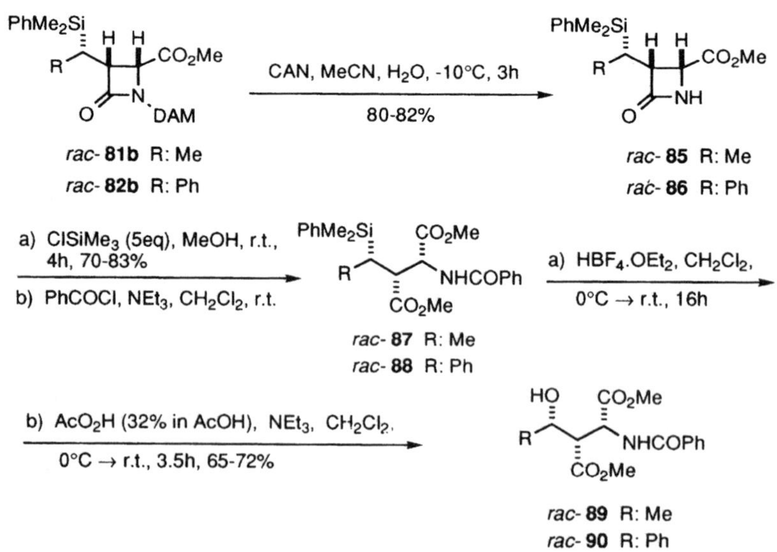

Scheme 14.19

First, **81b** and **82b** were N-deprotected to give **85** and **86**, respectively. The β-lactam ring opening was achieved by means of trimethylchlorosilane in methanol, to afford the expected dimethyl aspartates, which were isolated as the N-benzoyl derivatives **87** and **88**. Conversion of these compounds into their corresponding hydroxy derivatives was performed according to the usual Fleming protocol. On the other hand (Scheme 14.20), when this latter protocol was performed on β-lactam *rac*-**81a** followed by boron trifluoride treatment, the lactone **91** was produced in 70% yield, which is closely related to compounds **59** and **71**, employed by Hart and Shiozaki, respectively, in their approaches to amino sugars.

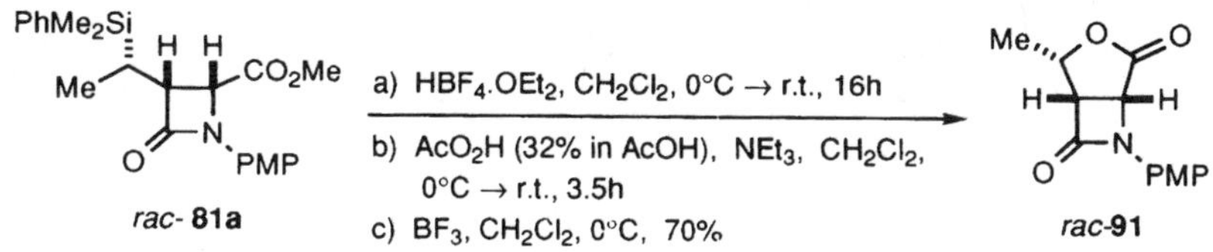

Scheme 14.20

Optically active substituted aspartic acids also became available, taking advantage of the ready access to enantiomerically pure β-silylalkanoyl chlorides. For instance (Scheme 14.21), **93** was prepared through the asymmetric 1,4-addition of "higher-order" dimethylphenylsilyl lithium cyanocuprate to Oppolzer's N-cinnamoyl sultam **92**, followed by removal of the chiral auxiliary and further carboxyl group activation by oxalyl chloride.[31]

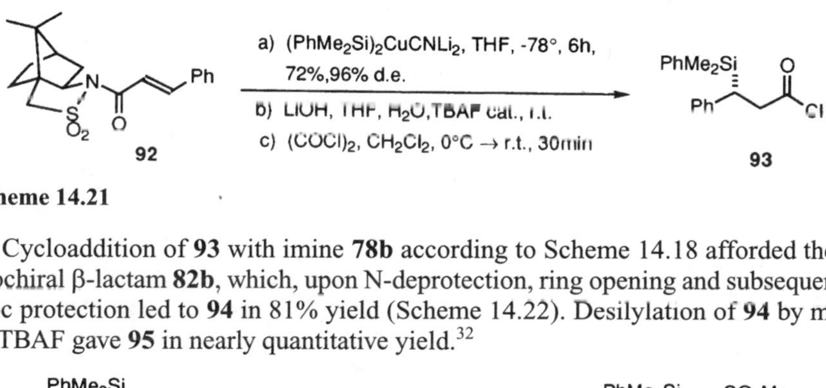

Scheme 14.21

Cycloaddition of **93** with imine **78b** according to Scheme 14.18 afforded the homochiral β-lactam **82b**, which, upon N-deprotection, ring opening and subsequent N-Boc protection led to **94** in 81% yield (Scheme 14.22). Desilylation of **94** by means of TBAF gave **95** in nearly quantitative yield.[32]

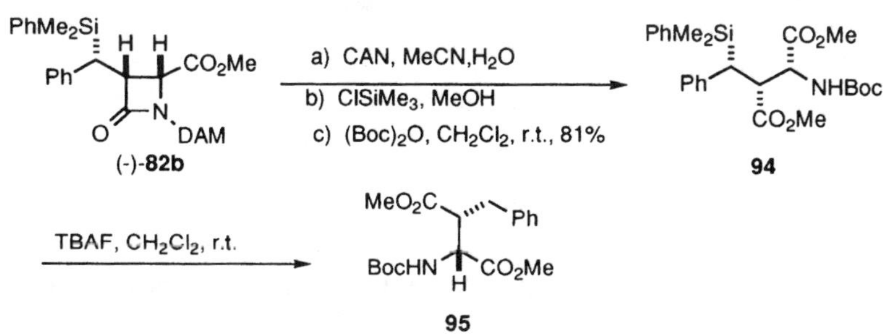

Scheme 14.22

The influence of glyoxylate imines **96** derived from some representative chiral glyoxylic esters, on the stereochemical course of the reaction (Scheme 14.23) was also examined with the aim to obtain a wider range of aspartic acid precursors. For example, when the imine **96a**, derived from menthyl glyoxylate and DAM-NH_2, was allowed to react with butanoyl chloride and triethylamine, a mixture of *cis*-diastereomers **97a** and **98a** was obtained in a 1.5:1 ratio, respectively. The replacement of DAM-NH_2 was the amino component of the imine partner by *p*-anisidine (PMP) caused formation of *cis* and *trans* isomers in a total ratio of 3:1, and virtually no diastereoselection was observed for any of the four diastereomers formed. A similar diastereoselection level was attained when imine **96b** was employed as the source of chirality, and only a negligible diastereoselection increase was observed starting from the imine **96c**, which provided **97c** and **98c** in a ratio of 1:2, respectively.

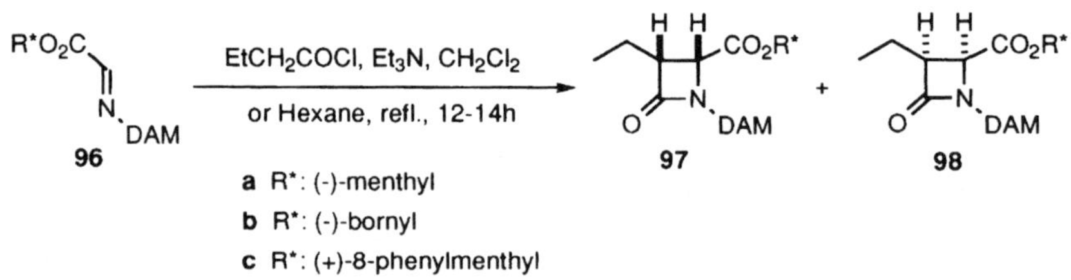

Scheme 14.23

Asymmetric induction from the amine component was also tested starting from **99** as the chiral source (Scheme 14.24), but it was unsuccessful.[30a] In this case, a mixture of both *cis*-isomers **100** and **101** in a 1:1 diastereomeric ratio was produced.

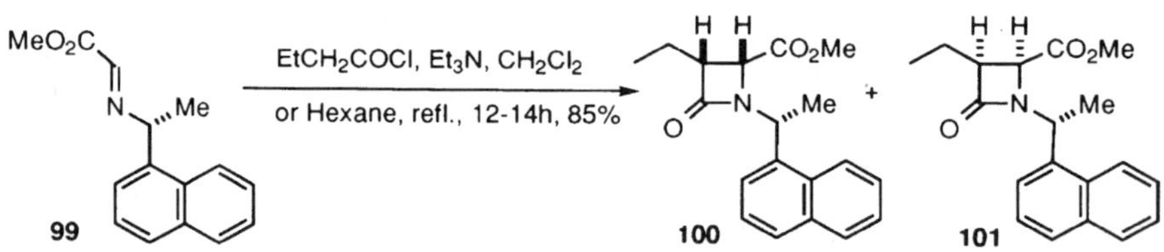

Scheme 14.24

On the other hand, the methoxycarbonyl group in these β-lactam products can be further transformed into alkyl chains via Wittig reactions (vide supra) performed on the corresponding derived 4-formyl β-lactams **103**. By that means (Scheme 14.25), it is possible to obtain 3,4-alkyl substituted β-lactams **104**, which, on the other hand, would not be accessible by both the ester enolate–imine condensation and the cycloaddition reaction owing to the ease of enolization of imines derived from aliphatic aldehydes. Subsequent ring opening of the resulting β-lactams has provided α,β-substituted β-amino acids **105**.[26]

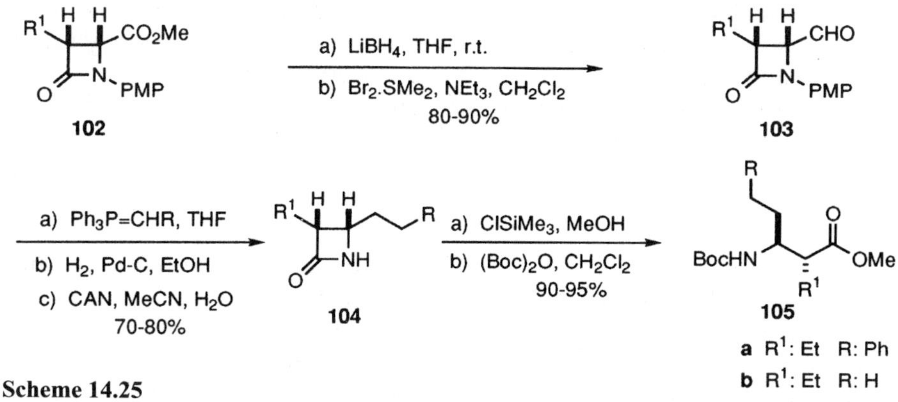

Scheme 14.25

In conclusion, it is obvious that the successful implementation of the approach based on the use of homochiral *cis*-3-alkyl-4-carboxy β-lactams requires still important improvements of the limited stereoselectivity displayed by the cycloaddition reaction or, alternatively, the development of complementary and convergent approaches to these β-amino acid precursors, that is, the metal enolate–glyoxylic ester imine condensation.[10]

14.2.2 α-Hydroxy-β-Amino Acids and Derived Compounds

α-Hydroxy β-amino acids, isoserines, constitute another important class of amino acids because of their occurrence in many biologically important compounds. Three representative examples are Taxol **106**, a highly promising anticancer product,[33] bestatin **107**, and amastatin **108**, two low-molecular-weight peptidic immunomodifiers[34] with antitumor and antimicrobial activity[35] (Figure 14.2).

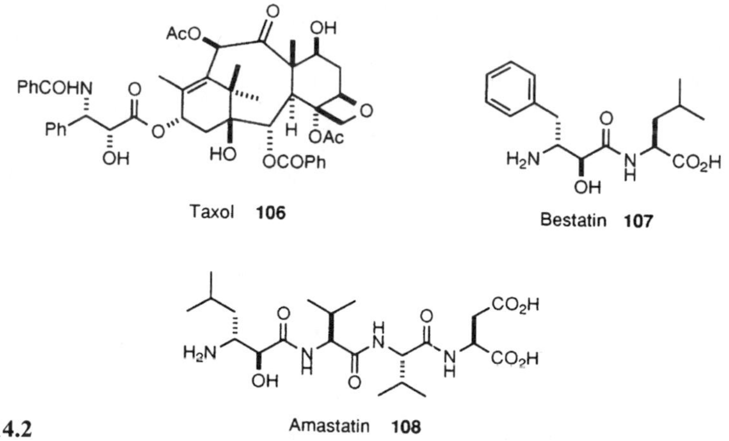

Figure 14.2

This section describes β-lactam routes to these amino acid fragments as well as the approaches to building up peptides containing α-hydroxy-β-amino acid residues. β-Aryl isoserines, β-alkyl isoserines, and β-hydroxy aspartic acids will be described separately with regard to the reported procedures for the preparation of their corresponding β-lactam precursors.

14.2.2.1 β-Aryl Isoserines

We first reported the synthesis of the side chain of Taxol from its corresponding racemic β-lactam.[36] Thus, under nonoptimized conditions (Scheme 14.26), the β-lactam *rac*-**112** was obtained in 60% yield by cycloaddition of methoxyketene, generated from methoxyacetyl chloride **109** and triethylamine, with imine **111**. Similarly, acetoxyacetyl chloride **110** in the presence of triethylamine and the imine **111** afforded the β-lactam *rac*-**113** in 56% yield. The β-lactam *rac*-**112**, upon N-dearylation and further treatment with chlorotrimethylsilane in methanol, provided the corresponding β-amino ester, which was isolated as the N-benzoyl derivative **114** in 92% yield.

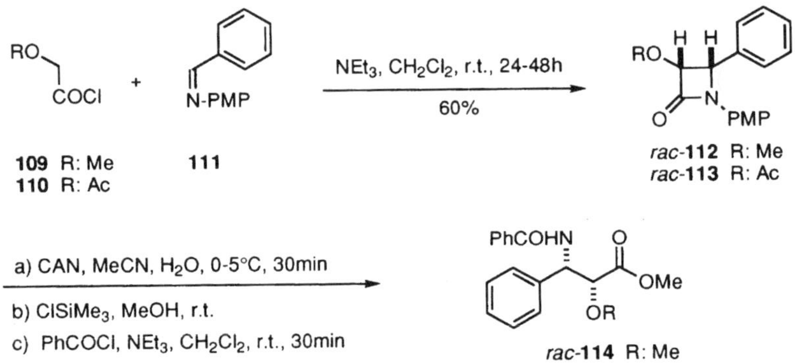

Scheme 14.26

rac-(3*R*,4*S*)-*trans* β-Lactams **112** and **113** have also been prepared by the above reaction under microwave irradiation using N-methylmorpholine as base and chlorobenzene as solvent[37] or by the slow addition of triethylamine to a refluxing solution of the acid chloride **110** and the imine **111** in toluene solution.[38] Recently, enantiomerically pure 3-hydroxy-4-phenyl β-lactam derivatives (Scheme 14.27) have been successfully prepared via enantioselective hydrolysis and transesterification of the azetidinone *rac*-**113** catalyzed by bacterial lypases.[39]

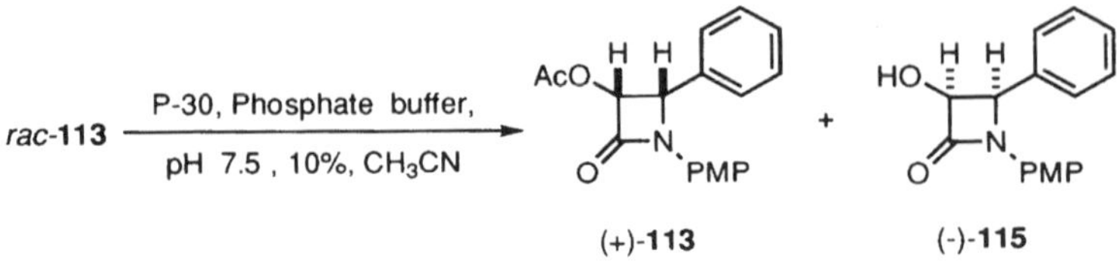

Scheme 14.27

Alternatively, chemical resolution of racemic 3-hydroxy β-lactams, via Ferrier rearrangement, has also been recently developed.[40] For example, on treatment with glucal triacetate **116** and iodine in tetrahydrofuran solution, *rac*-**115** (Scheme 14.28), afforded in 60% yield a mixture of two diastereomeric compounds **117** and **118** in a ratio of 40:60, which could be separated by chromatography over a silica gel column. Mild acid hydrolysis of **117** followed by acetylation gave pure (+)-**113**.

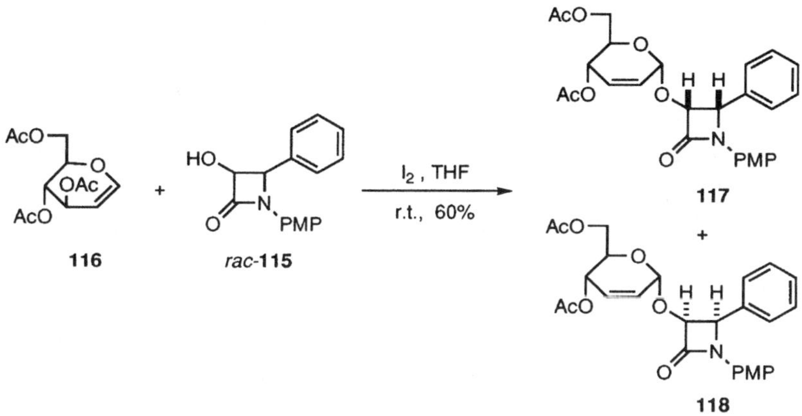

Scheme 14.28

Since the side-chain moiety N-benzoyl (2*R*,3*S*)-3-phenylisoserine is essential for the potent antitumor activity of Taxol, efforts have been made to obtain optically pure 3-acetoxy-4-aryl β-lactams through diastereoselective reactions. Georg[41] described the preparation of 3-alkoxy-β-lactams **121** and **122** by cycloaddition reaction of alkoxyketenes with imines **119** and **120** derived from galactosamino (Scheme 14.29). However, in all cases studied, a very low reaction diastereoselection was observed, typically 20–30% d.r.

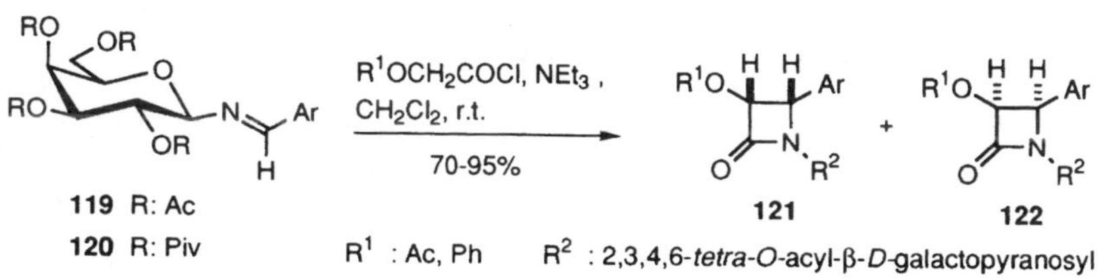

Scheme 14.29

A promising approach to the synthesis of the Taxol side chain through α-hydroxy β-lactams has recently been reported by Farina and co-workers.[42] The method (Scheme 14.30) is based on previous reports using L-threonine as the chiral source in the cycloaddition reaction.[42b,c] The authors found that whereas benzyloxyacetyl chloride reacted with the imine **123** with poor diastereoselectivity, acetoxyacetyl chloride gave the adduct **124** along with its diastereomer in a ratio of 92:8. The major diastereomer could be easily transformed into the intermediate **125** by removal of the silyl blocking group in **124** followed by the standard mesylation–elimination process and further ozonolysis of the resulting double bond. The oxalamide **125** was then directly cleaved with aqueous $NaHCO_3$ to provide β-phenylisoserine precursor **126** in 80% yield over the last three steps.

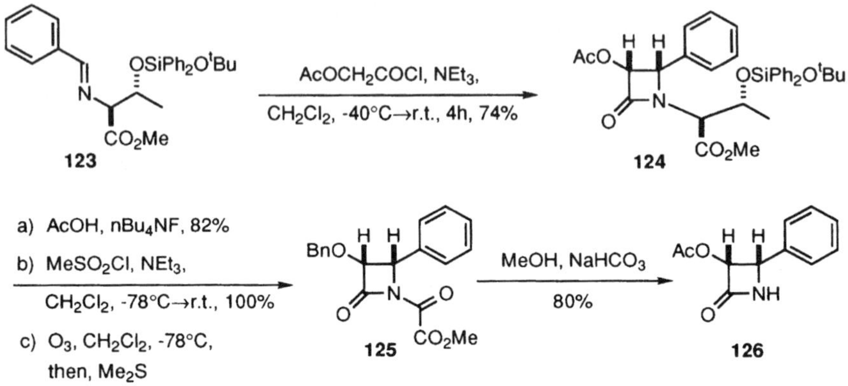

Scheme 14.30

Another approach to the Taxol side chain via β-lactams has been reported by Commerçon[43] (Scheme 14.31), employing the imine **127** derived from (*S*)-α-methylbenzylamine as the chiral source. However, a mixture of *cis*-β-lactams **128** and **129** was produced in a 25:75 diastereomeric ratio. The major isomer was then transformed into the β-phenylisoserine **130** in reasonably good yield. On the other hand, it should be mentioned that 1-(1-naphthyl)ethylimines have proved to be better chiral sources in such a cycloaddition reaction than 1-(phenyl)ethylimines,[44] albeit that the diastereoselectivities displayed were still moderate.

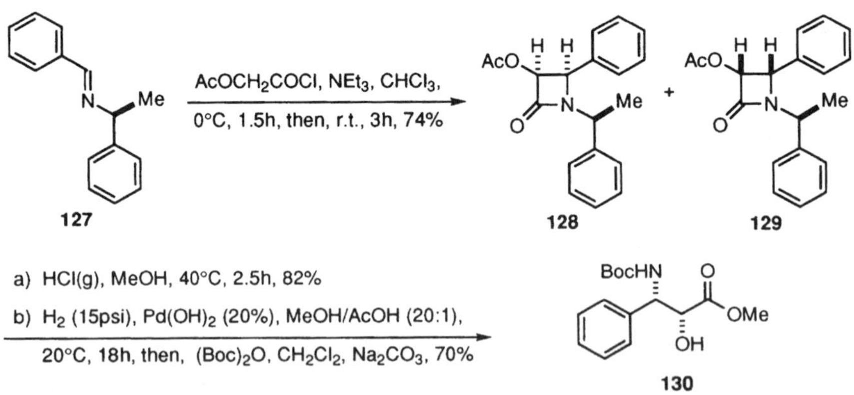

Scheme 14.31

Our own approach to optically pure α-hydroxy β-lactams was guided by the excellent diastereoselectivity observed in reactions involving Evans–Sjögren ketenes and achiral aldehyde-derived imines.[45] As illustrated in Scheme 14.32, the starting ketene was prepared from the corresponding acid chloride **131** and triethylamine and reacted with the imine **111** under the usual conditions to form the β-lactam **132** as single diastereomer. We found that when **132** was deprotonated using lithium bis(trimethylsilyl)amide in THF at –78°C followed by hydroxylation with Vedje's molybdenum peroxide reagent (MoOPH), a mixture of the starting β-lactam **132** and the amido carbinol **133** was produced together with the oxazolidinone chiral auxiliary after column chromatography. Exposure of **133** to silica gel or simple heating in THF gave **134** in 40% overall yield. The reduction of the α-keto β-lactam **134** by means of $NaBH_4$ proceeded with complete stereoselectivity to give the corresponding α-hydroxy β-lactam, which was next transformed into **126** in 70% yield over the last three steps.

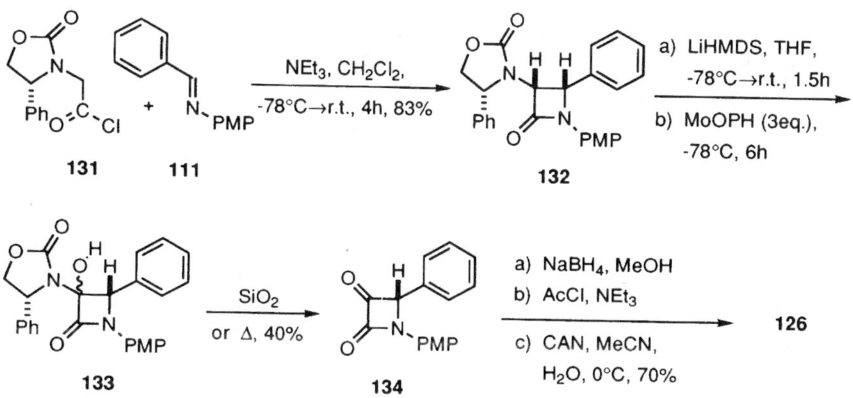

Scheme 14.32

A conceptually similar approach to the Taxol side chain was also reported by Holton[46] (Scheme 14.33). The most remarkable point is the unprecedented formation and reaction of the lithium enolate of **132** in dichloromethane as solvent, which was then trapped by N-chlorosuccinimide to form a mixture of epimeric α-chloro β-lactams **135** in greater than 95% yield. Subsequent dechlorination led to the α-keto β-lactam **134**, which was transformed into **136**. The same author[36b] found that coupling of **136** (5 equiv.) with 7-TES-baccatin III **137** proceeded at 25°C in the presence of DMAP and pyridine to give 7-TES-2′-EE-taxol in 92% yield, which was deprotected with 0.5% HCl in ethanol at 0°C to afford Taxol in ca. 90% yield.

Somewhat better results were achieved by Ojima[47] using the sodium salt of **137** generated by reacting **137** with NaH in THF. Although the reaction conditions have not been optimized, the coupling reaction could be performed employing only a slight excess of the β-lactam **136**, thus providing 7-TES-2′-EE-Taxol in 40% yield (80% yield based on **137** reacted) and recovering unreacted **137** in 37%. The same author found (Scheme 14.34) that under optimized conditions with the use of NaHMDS (1.2 eq.) as the base, the coupling of the β-lactam **138** with 7,10-diTroc-10-deacetylbaccatin III **139** (1 eq.) proceeded very smoothly in THF at –30°C to give 2′-EE-7,10-

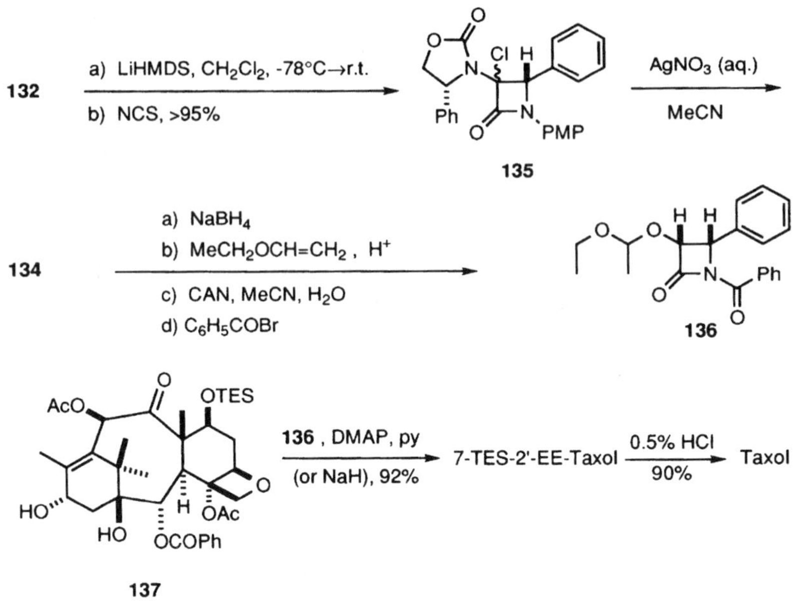

Scheme 14.33

diTroc-taxotere **140** in >92% isolated yield within 10 min, which then was converted into taxotere **141**.

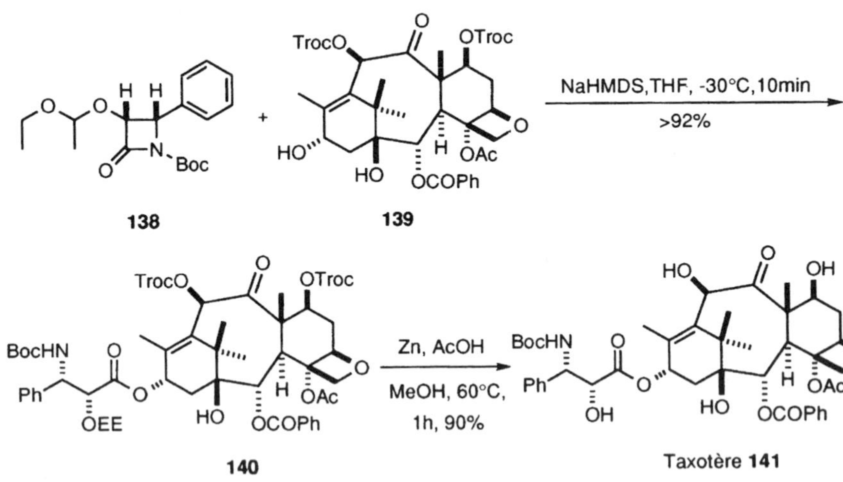

Scheme 14.34

In general, the use of optically active O-protected α-hydroxy β-lactams in combination with Holton–Ojima's coupling method seems to be one of the most effective protocols for the semisynthesis of Taxol and related analogues.[48] In addition, the required starting α-hydroxy β-lactams can also be efficiently prepared by the ester enolate–imine condensation reaction of chiral glycolates developed by Ojima and his colleagues.[47b] In their synthesis (Scheme 14.35), the use of glycolates **142** derived from Whitesell's chiral auxiliary produced the best enantiomeric excesses, the enantiomeric purity being dependent on the hydroxyl protecting group of the glycolic ester. For example, it was found that the benzyloxy group gave very low enantiomeric excesses if at all, and the use of the *tert*-butyldimethylsilyl protecting group only provided enantiomeric excesses of 76%. However, the triisoprophylsilyl ether gave an excellent enantiomeric excess of 96–98% in the formation of the corresponding 3-hydroxy-4-aryl-2-azetidinones. In addition to these findings, it is also worthy of note that aldimines of enolizable aldehydes can be successfully used for this reaction. The resulting β-lactams **143**, upon N-dearylation, provided **144**, which, in turn, can be easily transformed into α-hydroxy-β-amino acid hydrochlorides **145** or employed for subsequent coupling reactions with baccatin III derivatives. In this latter context, it should be mentioned, as pointed out by Georg,[49] that the triisopropylsilyl group, needed in the ester enolate–imine reaction to obtain high enantiomeric excesses, has to be removed prior to the coupling reaction with 7-TES-baccatin III and replaced by a sterically less demanding protecting group, such as the ethoxyethyl (EE) or the triethylsilyl (TES), otherwise the coupling product is obtained in less than 5% yield.

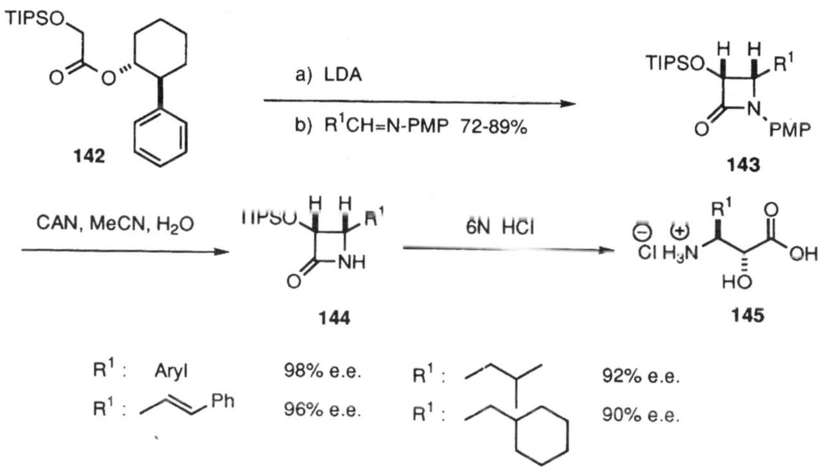

Scheme 14.35

To address this issue, Georg[50] reinvestigated the asymmetric ester enolate–imine cyclocondensation reaction in the hope of finding a chiral auxiliary that would not need the assistance of a sterically demanding protecting group at the hydroxyl group of the glycolic ester moiety to yield β-lactams in high enantiomeric excess. As shown in Scheme 14.36, the use of Oppolzer's chiral auxiliary, (–)-10-dicyclohexylsulfamoyl-D-isoborneol in the ester enolate–imine reaction fulfills this criterion, providing a route to diverse α-silyloxy β-lactams in high enantiomeric purity.[50a] For example, the reaction of the lithium enolate of **146** with previously distilled N-

(trimethylsilyl)benzaldimine **147** afforded β-lactams **148** in good chemical yields and high enantiomeric excesses. Actually, the β-lactam **148b** upon N-benzoylation could also be efficiently coupled with 7-TES-baccatin III **137** and the resulting coupling product transformed into Taxol in good overall yield.

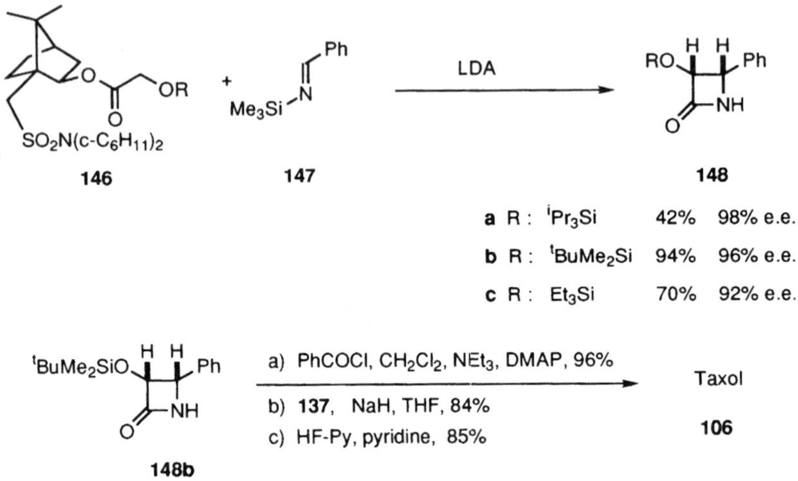

Scheme 14.36

This enolate–imine reaction has also been extended to imines **149**, derived from heteroaromatic aldehydes (Scheme 14.37) to form β-lactams **150**, which then were also incorporated into 7-TES-baccatin III. However, with the exception of the 4-(4-pyridyl)-β-lactam and 4-(2-furyl)-β-lactam, enantiomeric excesses were low.[50b]

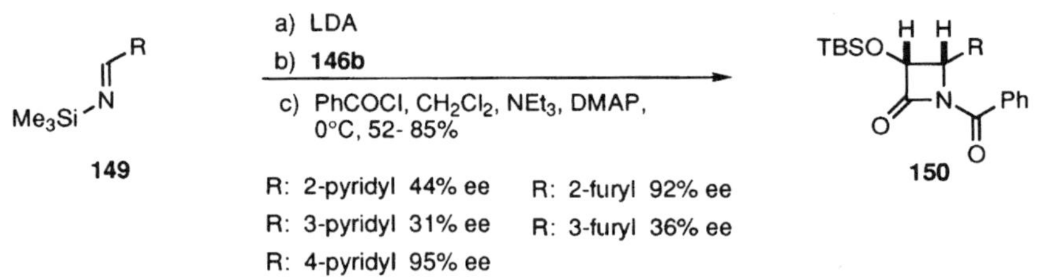

Scheme 14.37

The phenyl substituent in β-lactam **148b** has also been hydrogenated (Scheme 14.38) to the 4-cyclohexyl derivative **151**, which then was coupled through its N-benzoyl derivative **152** with 7-TES-baccatin III for biological assays.[51]

An interesting access to the Taxol side chain, via ester enolate–imine condensation, has been recently reported by Cainelli and co-workers.[52] They found that condensation of ethyl diethoxyacetate **153** (Scheme 14.39) with imine **111** in the presence of potassium *tert*-butoxide in DMF as solvent gave the β-lactam **154** in 88% yield, which upon mild acid hydrolysis led to the α-keto β-lactam **155**. Although this compound is obtained in its racemic form, the most striking feature of the approach

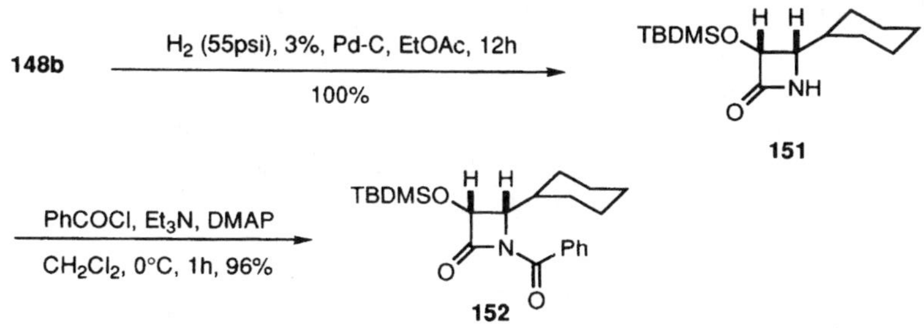

Scheme 14.38

is the use of a weak base for enolate generation, thus making this reaction very suitable for a large production scale of β-aryl isoserines.

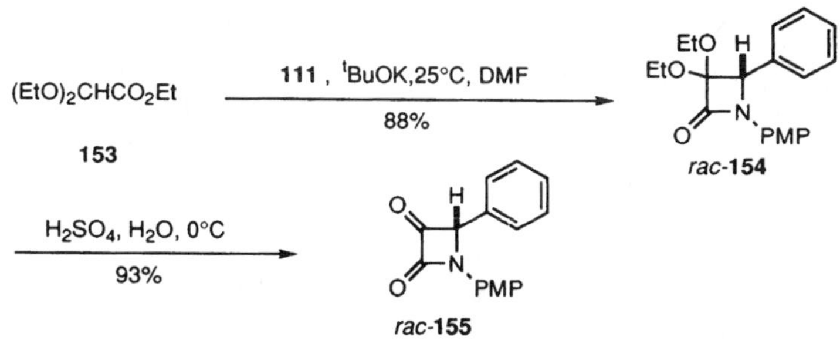

Scheme 14.39

14.2.2.2 β-Alkyl Isoserines

Because of the importance of β-alkyl isoserines as constituents of biologically significant compounds and in view of the lack of generally applicable methods for direct construction of β-lactams with alkyl side chains at the C_4 position, alternative approaches to 3-hydroxy-4-substituted β-lactams have also been developed. For example, we found that Wittig reaction of aldehyde **156** (Scheme 14.40) with various ylides afforded **157**, which on treatment with ammonium formate and Pd/C led to O-deprotection and concomitant double bond reduction, giving **158** in excellent yields.[53]

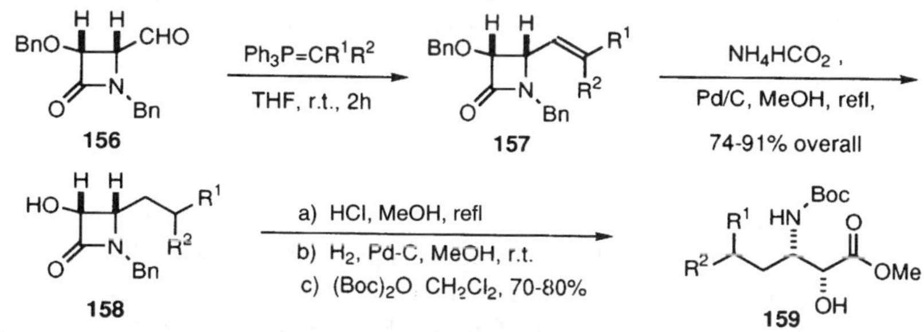

Scheme 14.40

These compounds have been opened by acid hydrolysis and transformed into the corresponding β-amino esters **159**.

Another example is shown in Scheme 14.41. Thus the Reformatsky reaction of methylbromoacetate with the aldehyde **160** followed by dehydration of the resulting intermediate led to **161** in 70% overall yield. Compound **161** was N-dearylated to the corresponding N-unsubstituted β-lactam and converted into the β-amino ester **162** in 88% yield.[36a]

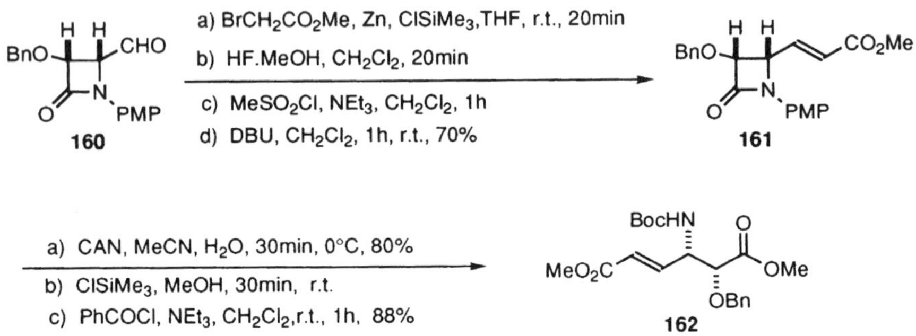

Scheme 14.41

We also examined the addition of Grignard reagents to the above 4-formyl β-lactams.[53] For example (Scheme 14.42), the addition of aryl Grignard reagents to **156** proceeded cleanly to give an equimolar mixture of carbinols, which were acylated with AcCl and triethylamine and the resulting crude products **163** subjected to simultaneous deacetoxylation and debenzylation to give **165**. Particularly noteworthy is the fact that structurally diverse 3-hydroxy-4-substituted β-lactams, as precursors of α-hydroxy β-amino acids, can be easily obtained from a single common starting material. A similar approach has also been developed by Cozzi and Cinquini[54] and then by Li and co-workers.[55]

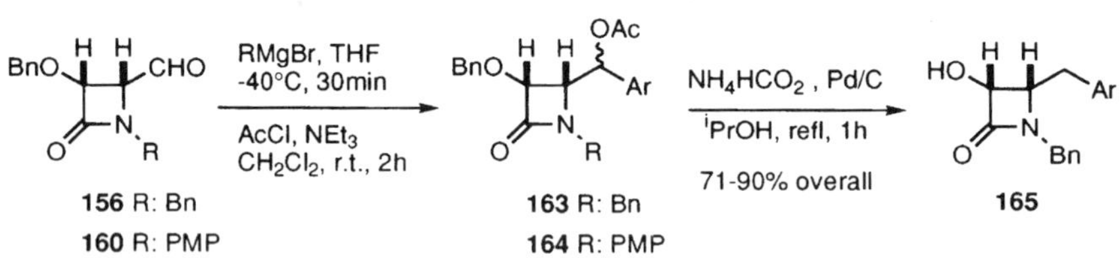

Scheme 14.42

However, when the hydrogenolysis step was performed on compounds of type **164**, amides **167** were the major products, together with the desired β-lactams **166a** (e.g., 7:1 ratio for Ar= Ph). The same results have also been observed by Bose and coworkers.[56] A partial solution to this problem was provided, however, by deoxygenation of the β-lactam **164** under modified Barton's conditions, and performing the reaction on a 2 millimole scale to give **166b** in 64% yield; otherwise, the yields were not reproducible on scaling up (Scheme 14.43).

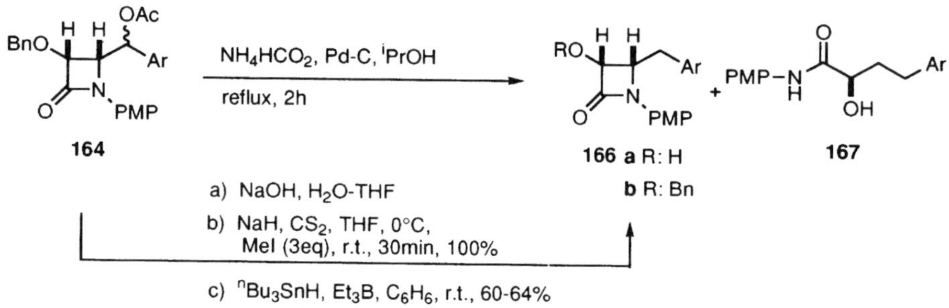

Scheme 14.43

The starting aldehydes **156** and **160** required for these reactions can easily be prepared (Scheme 14.44) in gram quantities either by the cycloaddition reaction of the glyceraldehyde-derived imines **168** and **169** with benzyloxyketene, generated "in situ" from benzyloxyacetyl chloride and triethylamine[57] or by the reaction of the titanium enolate of O-benzyl-(2-pyridylthio)glycolate with the imines **172a** and **172b**.[54] Recently, another route to these formyl β-lactams from L-tartaric acid has also been described.[58]

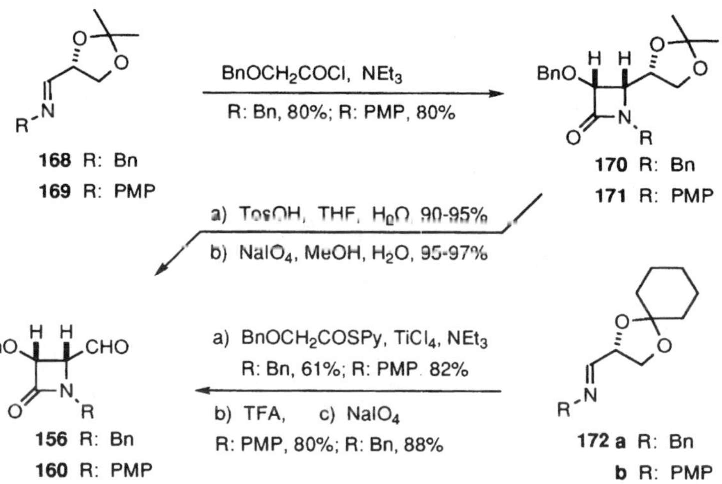

Scheme 14.44

On the other hand, while aryl Grignard reagents reacted with **156** without any diastereoselectivity, we also found that the same reaction carried out on the formyl β-lactam **160** (Scheme 14.45) afforded carbinol **173** together with its epimer **174** in a ratio 90:10.[34a,59] A complementary approach to compounds **173** and/or **175** has been described by Terashima[60] using the cycloaddition reaction of the benzyloxyketene to the imine **176** derived from (*R*)-[*tert*-butyldimethylsilyloxy]-2-phenylacetaldehyde.

This author has also reported that ring opening of the β-lactam **177** (Scheme 14.46), prepared by the cycloaddition of benzyloxyketene with the corresponding

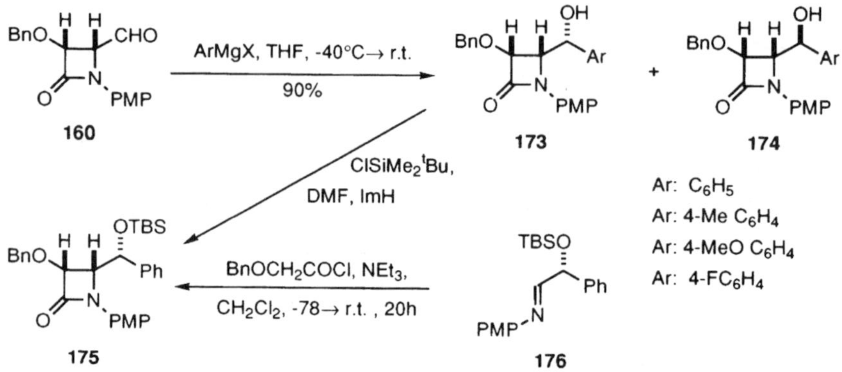

Scheme 14.45

imine according to the former procedure, by means of HCl in isopropanol, proceeded with concomitant removal of the TBS and DAM groups to afford the isopropyl ester **178**. Conversion of **178** into the 2-oxazolidinone derivative **179** was followed by hydrogenolysis over 10% Pd on charcoal and further reduction of the resulting intermediate over 5% Rh on alumina to give **180**, the key component of the renin inhibitor **181**.

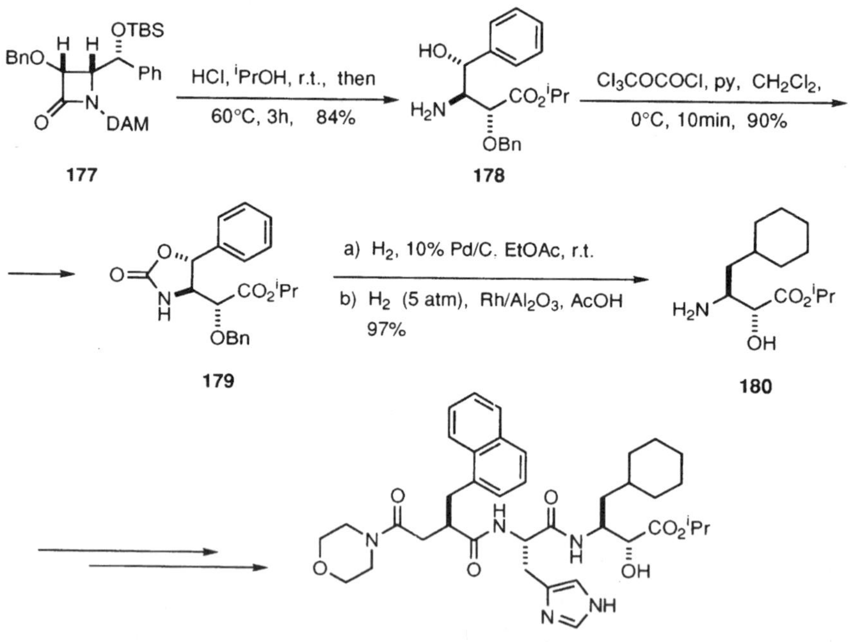

Scheme 14.46

A more direct access to isopropyl (2*R*,3*S*)-3-amino-4-cyclohexyl-2-hydroxybutyrate **180**, albeit in somewhat lower yield, has been provided by Cozzi and Cinquini[61] via condensation of the titanium enolate of O-benzyl-(2-pyridylthio)glycolate with the chiral imine **182** (Scheme 14.47). By that means, β-lactams *cis*-**183**, *cis*-**184**, and *trans*-**185** were produced in 66% yield as a 89:11 mixture of *cis* and *trans* products. The overall **183** (major *cis*)/**184** (minor *cis*)/**185** (*trans* product) was 69.4:19.6:11, respectively. The major *cis*-isomer **183** could be separated from the reaction mixture and then transformed, as above, into **180**.

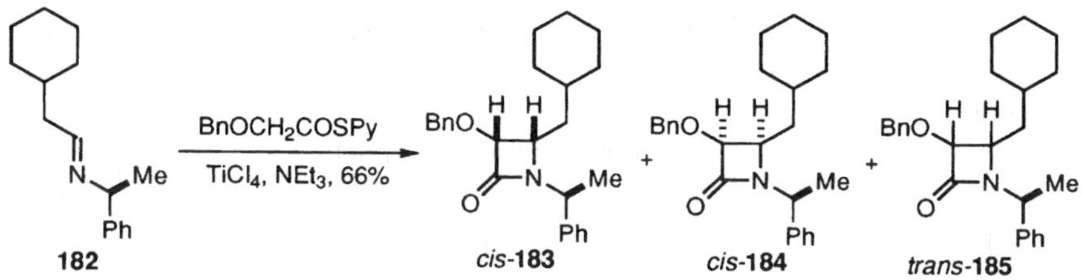

Scheme 14.47

The synthesis of the bestatin side chain **187** (Scheme 14.48) was also achieved by Terashima[60] using the same reaction protocol as above starting from the β-lactam **186**, an enantiomer of **177**. The acid **187** could be subsequently N-protected and coupled, through established techniques, with the corresponding α-amino acid to give the desired dipeptide **107**.

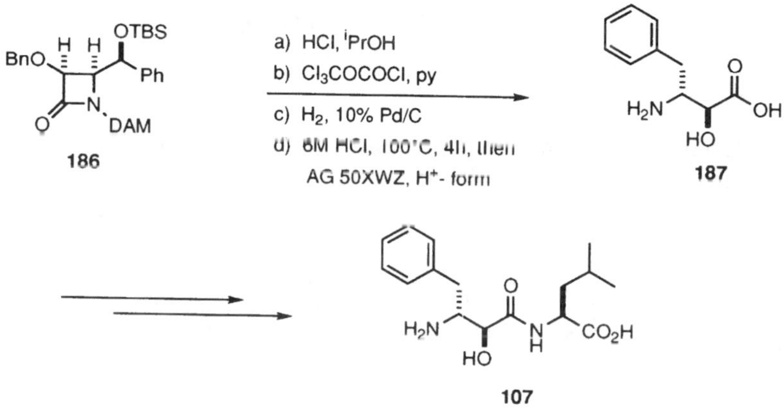

Scheme 14.48

We have also been concerned with this latter subject and focused on the direct coupling of a β-lactam framework with α-amino esters, to obtain short peptide fragments without the requirements imposed by conventional procedures.[62] Our first attempts to perform this goal starting from N-alkyl, N-aryl, and even N-unsubstituted β-lactams were completely unfruitful under conditions suitable for peptide synthesis. Efforts from our laboratory to address this issue satisfactorily have revealed, however, that N-Boc β-lactams can be efficiently employed as acylating agents for α-amino esters in the presence of NaN_3 as promoter of the coupling reaction (see also Section

14.2.1.1). This was illustrated by the synthesis of (–)-bestatin **107** from the N-Boc β-lactam **190**. As shown in Scheme 14.49, the requisite β-lactam was prepared by deoxygenation of **188** under the aforementioned modified Barton's conditions followed by N-dearylation of **189** and further exposure of the resulting crude N-unsubstituted β-lactam to di-*tert*-butyldicarbonate. The coupling reaction of **190** with *S*-leucine benzyl ester and NaN_3 proceeded to afford the dipeptide product **191** in 88% yield. Treatment of **191** with trifluoroacetic acid followed by exposure of the resulting crude free amino dipeptide to H_2 (1 atm) over 10% Pd/C gave (–)-bestatin **107** in 98% overall yield. Another example illustrating the effectiveness of NaN_3 for these coupling reactions is shown by the high yield formation of **192** under conditions that, in the absence of NaN_3, gave only a trace of the dipeptide product.

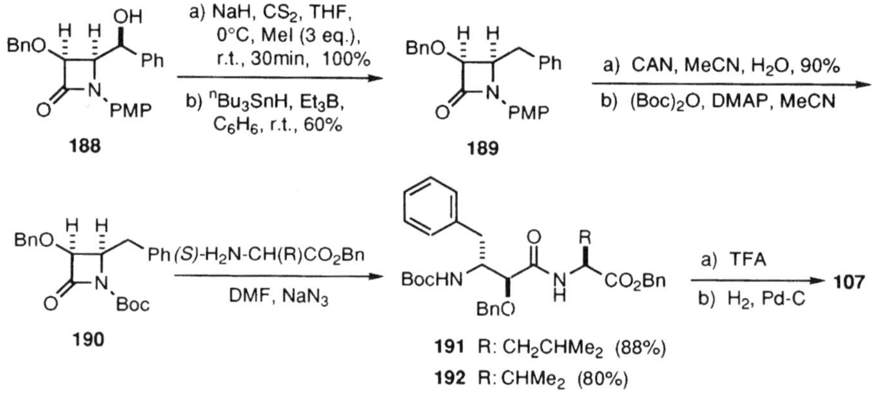

Scheme 14.49

Dipeptides **195a**, **195b**, and **195c** (Scheme 14.50), have also been prepared in good yields by coupling of the β-lactam **194** with nearly equimolar amounts of (*S*)-

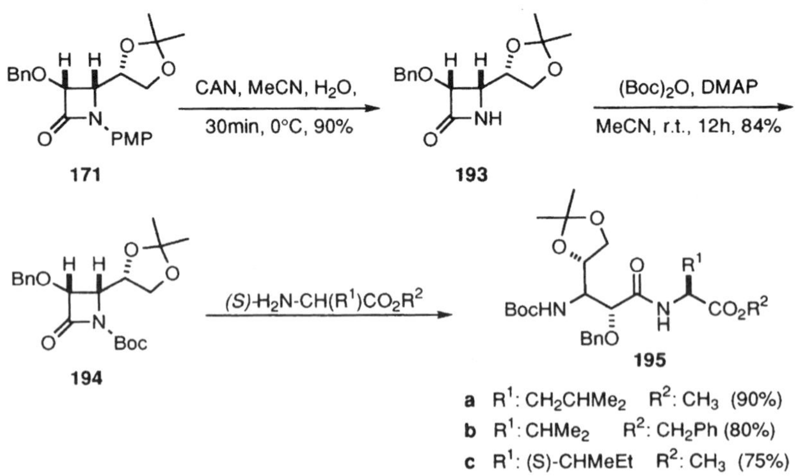

Scheme 14.50

leucine methyl ester and the corresponding β-branched α-amino esters (*S*)-valine benzyl ester and (*S*)-isoleucine methyl ester. In each case, the absence of NaN_3 in the coupling reaction gave only very poor yields of the expected dipeptide products. The starting **194** was prepared from **171** via the N-unsubstituted β-lactam **193** in good overall yield.

Simultaneously with our work, Ojima[63] has revealed that N-Boc β-lactams can be coupled with α-amino esters using the solid-phase peptide technique (Scheme 14.51). For instance, the coupling of **196** with the resin-bound glycine **197a** and the (*S*)-phenylalanine **197b** (Wang resin) at 50°C gave the corresponding resin-bound dipeptides **198a** (2 h) and **198b** (30 h) in high yields, which were treated with trifluoroacetic acid (TFA) at 25°C for 2 to give the TFA salt of dipeptides **199a** and **199b** in 72% and 78% isolated yields, respectively. The reaction of **196** with α-amino esters in solution has also been examined, but at least a twofold excess of the amino ester was required to achieve good yields (90%) of the expected dipeptide products.

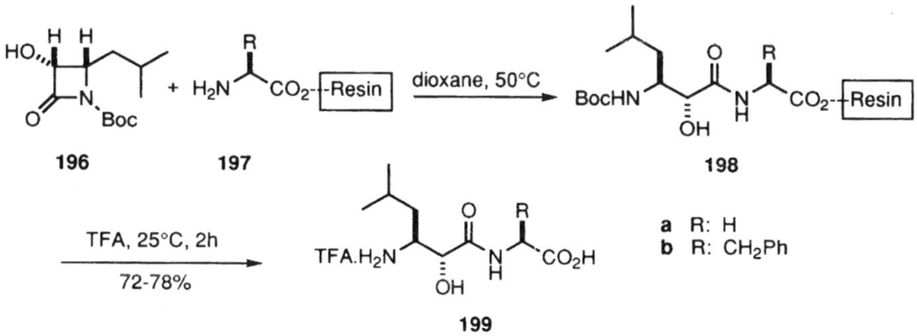

Scheme 14.51

14.2.2.3 β-Hydroxy Aspartic Acids

Interest in β-hydroxy aspartic acids stems from their occurrence in microorganisms[64] and mainly as constituents of peptides.[65] For example, the L-*erythro*-β-hydroxy aspartic acid is a constituent of several proteins involved in the blood clotting cascade, and various syntheses of this amino acid for its further incorporation into peptides have recently been reported.[66] However, a central problem in the synthesis of aspartic acid–containing peptides is aspartimide formation under acidic or basic conditions. The subsequent ring opening of the five-membered α-amino-succinimide ring by aqueous bases provides largely the wrong isomer containing a β-amide linkage, and the net effect is an α- to β-amide bond isomerization.[67]

Taking into account the results presented in the preceding section, we expected that the use of N-Boc 4-carboxy β-lactams might provide an efficient solution to this problem. The exocyclic carboxyl group could be coupled with α-amino acids under standard peptide coupling reactions, and the N_1-C_2 bond cleavage of the β-lactam ring would permit the incorporation of α-amino acids at the β-carboxyl moiety of the aspartic acid without side reactions. To illustrate this latter aspect, the synthesis of β-hydroxy aspartic acid–derived peptides was examined. The required alkoxycarbonyl β-lactams **200** and **201** needed for the study could be readily prepared in a few simple steps from **171**[68] (Scheme 14.52). Further incorporation of the Boc group in **200** provided **202**, which was allowed to react with both (*S*)-phenylalanine benzyl ester and (*S*)-valine benzyl ester in the presence of NaN_3. The reaction proceeded selectively at the imide function to give products **203** and **204** in 80% and 75% yield, respectively. In the absence of NaN_3, only a 50% conversion was observed in the reaction of **202** with (*S*)-valine benzyl ester, thus confirming the effectiveness of NaN_3 in promoting these coupling reactions.

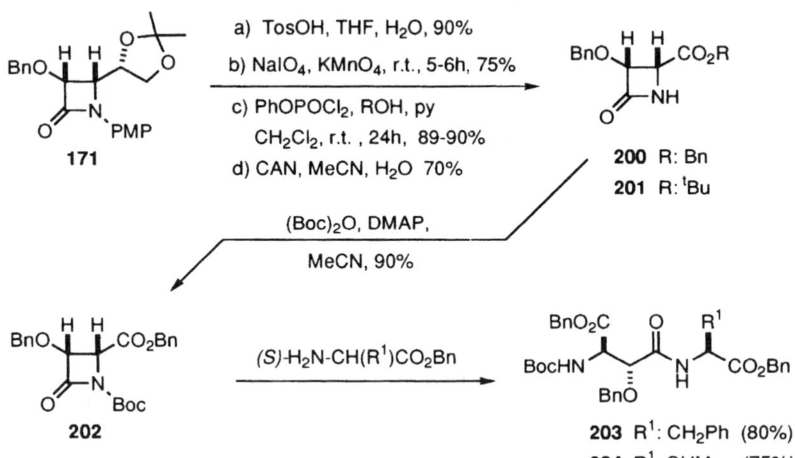

Scheme 14.52

On the other hand (Scheme 14.53), compound **201** could also be opened by means of $ClSiMe_3$ in methanol without *tert*-butyl ester deprotection. N-Boc protection, O-debenzylation, and reduction of the methoxycarbonyl group in **205** furnished the *threo*-β-hydroxymethyl (*R*)-serine **206**.[69]

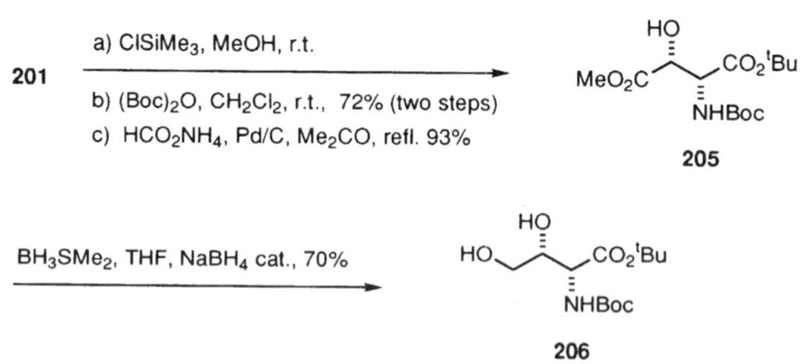

Scheme 14.53

Another example that illustrates the efficiency of this β-lactam route to peptide units is further shown in the synthesis of the tripeptide fragment A of the antibiotic lysobactin[70]

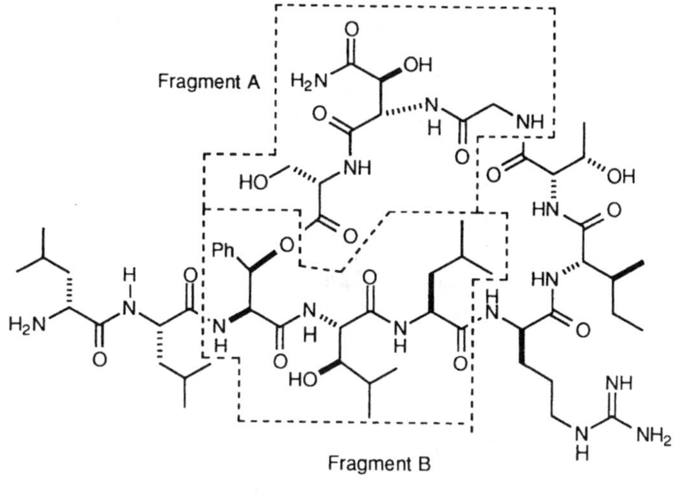

Figure 14.3.

(Figure 14.3). The mode of antibacterial action of this antibiotic has been shown to be comparable to the well-known clinically useful antibiotic vancomycin. However, lysobactin appears to be more toxic than vancomycin and, as suggested by Tymiak,[70d] the development of new semisynthetic peptides of lysobactin with improved therapeutic index becomes of considerable interest. We recently undertook a study on this subject and found that tripeptide A as well as tripeptide B of lysobactin can be efficiently obtained starting from appropriately substituted 2-azetidinone frameworks.[71]

As shown in Scheme 14.54, **207** was first activated with cyanuric fluoride and subsequently coupled with L-serine, according to Carpino's procedure to give the dipep-

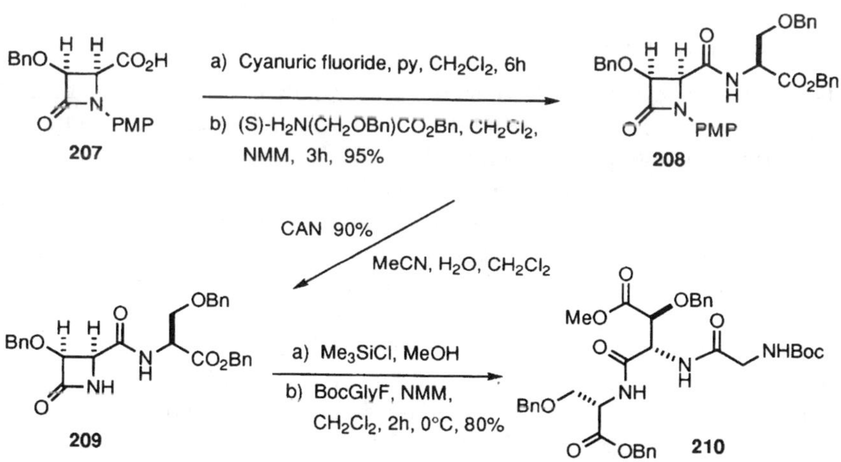

Scheme 14.54

tide product **208** in 95% overall yield. N-deprotection of **208** proceeded in the usual way with the addition of dichloromethane as covalent to furnish **209** as a pale yellow solid in 90% yield. Finally, ring opening of **209** by means of $ClSiMe_3$ in methanol led to the corresponding methyl ester, which was further acylated with BocGlyF and N-methylmorpholine (NMM) to give the tripeptide **210** in 80% yield.

Alternatively, to facilitate the β-lactam cleavage by other nucleophiles, **209** was treated with $(Boc)_2O$ and DMAP to give **211** with complete chemoselectivity (Scheme 14.55). Such a compound, on exposure to ammonia (25% aq) in DMF as solvent, cleanly furnished the dipeptide **212** in 90% isolated yield, which was then transformed into the asparagine tripeptide **213** in 70% yield.

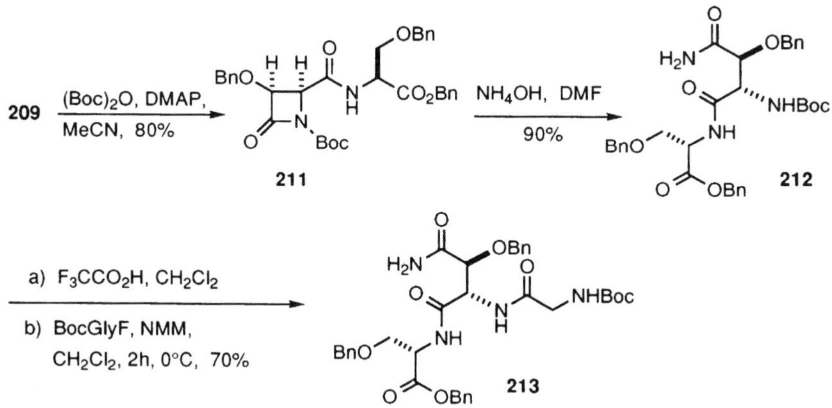

Scheme 14.55

Synthesis of tripeptide fragment B of lysobactin via another β-lactam approach will be outlined later (Section 14.3.2.1).

14.2.3 α,β-Diamino Acids

2,3-Diamino carboxylic acids are uncommon naturally occurring amino acids that have been arousing considerable interest.[72] This particular class of amino acids, when incorporated into peptides with either the β- or the α-amino group as part of the backbone, might provoke changes in polarity, additional hydrogen-bonding interactions, and basic character. α,β-Diamino acids can also serve as substitutes of α-hydroxy-β-amino acid–derived peptides to probe their structural specificity and topology.[72f] Interest in α,β-diamino acids also stems from their occurrence in cyclic peptides and biologically active substrates (Figure 14.4). For example, the first member of the series, the 2,3-diaminopropionic acid, is found in the antitubercular antibiotic capreomycin **214**[73] in the pyrimidoblamic acid **215**,[74] in the neuroexcitatory quisqualic acid **216**,[75] and in the nonproteinogenic amino acids willardiine **217**[76] and mimosine **218**.[77] β-Substituted 2,3-diamino acids are also useful synthetic intermediates for the construction of conformationally restricted peptides of type **219**.[78]

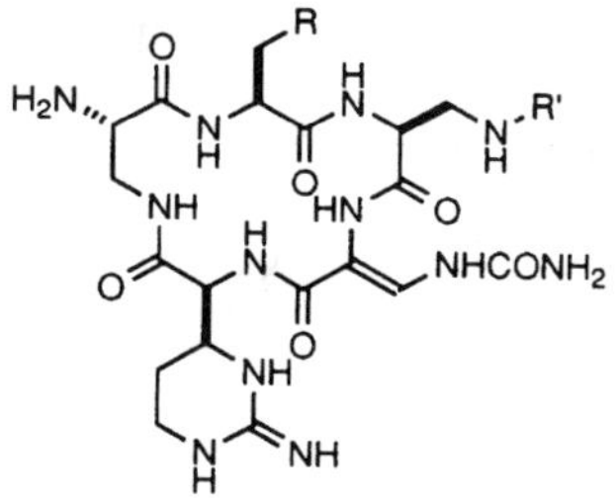

214 **a** R: OH R' : β-lysine
b R: H R' : β-lysine
c R: OH R' : H
d R: H R' : H

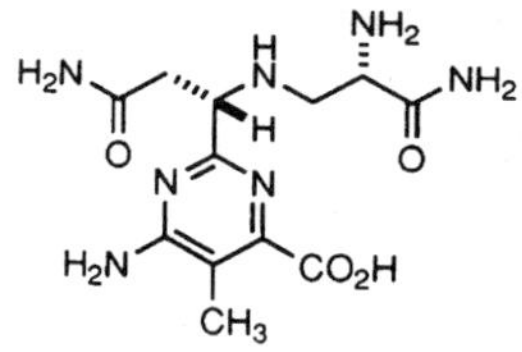

215

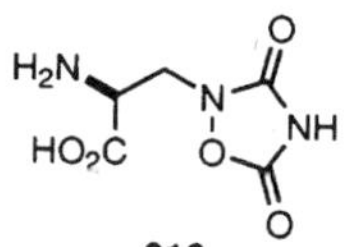

216

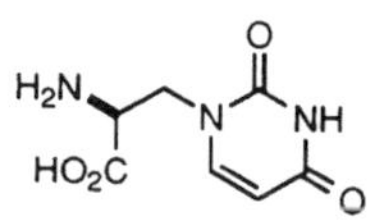

217

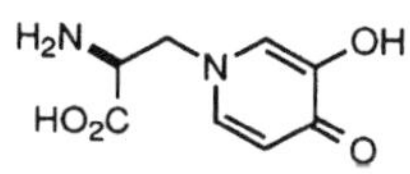

218

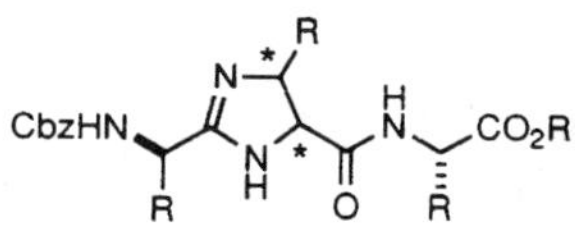

219

Figure 14.4

One of the most practical approaches to β-aminoalanine derivatives has been reported by Baldwin[79] (Scheme 14.56), involving the obtention of L-quisqualic acid **216** from the N-benzyloxy β-lactam **220** in excellent yield. For example, hydrogenolysis of **220** followed by treatment with a catalytic amount of lithium ethanethiolate gave

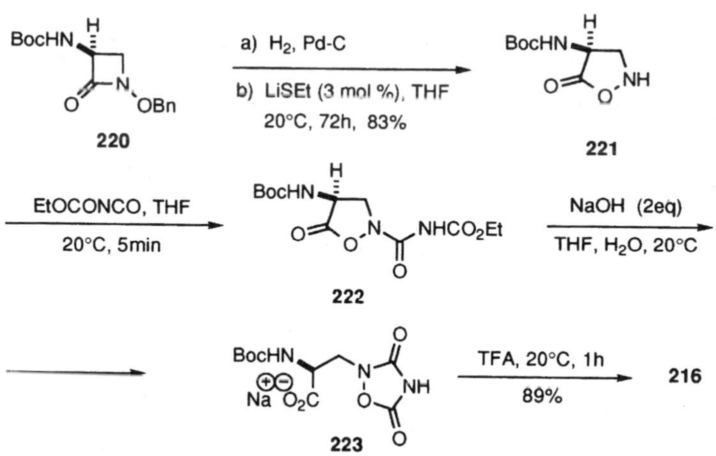

Scheme 14.56

the isooxazolidin-5-one **221** in 83% yield. Subsequent reaction with ethoxycarbonyl isocyanate furnished the urea **222**, which was immediately ring opened to the salt **223**. Upon treatment of **223** with trifluoroacetic acid, followed by ion-exchange chromatography, **216** was obtained in 89% yield.

In a similar way, Hecht[80] reported the synthesis of (2*S*)-3-amino-2-[(*tert*-butoxycarbonyl)amino]propionamide **224** (Scheme 14.57) from (2*S*)-N-(benzyloxy)-2-azetidin-one **220**. Compound **224** was next incorporated into **225** to give **226**, which was employed in the synthesis of both pyrimidoblamic acid **215**, and *epi*-**215**.

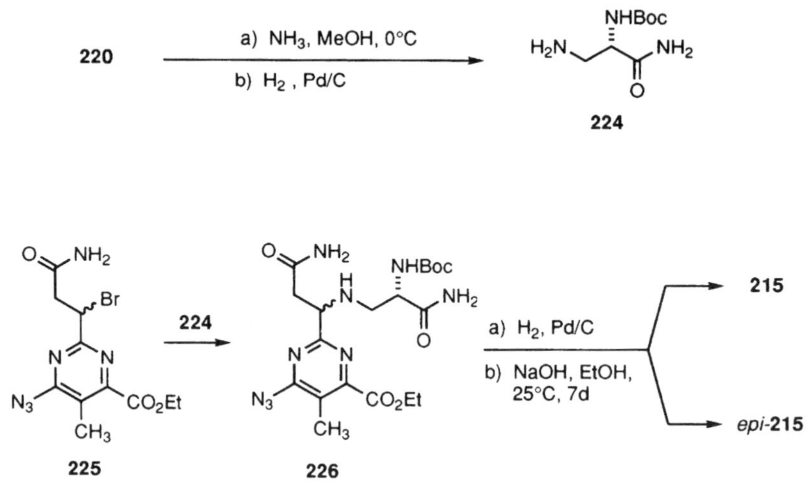

Scheme 14.57

More recently, Ojima and co-workers[81] have illustrated the use of optically active 3-amino β-lactams for the construction of β-substituted α,β-diamino acids (Scheme 14.58), via acidic ring opening. In this approach, the *cis*-β-lactam **227** could also be epimerized to its *trans*-isomer **228**, through a protection, deprotonation, protonation, and deprotection sequence. As the enantiomer of *cis*-**227** can readily be obtained, all four stereoisomers of α,β-diamino acids **229** and **230** can be synthesized by this protocol.

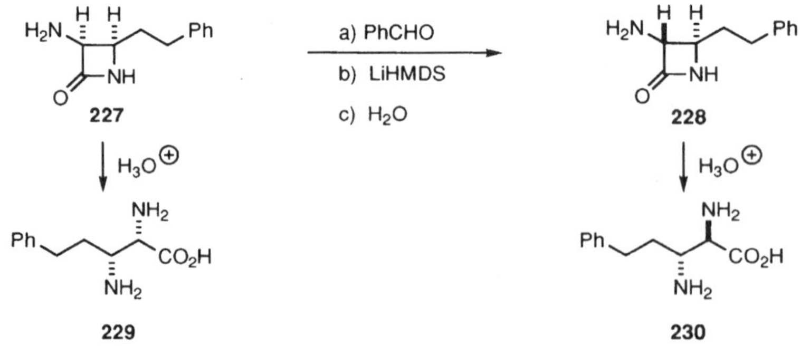

Scheme 14.58

In a similar fashion (Scheme 14.59), α-branched α,β-diamino acids **233** can be prepared from **232**, available by asymmetric alkylation of the lithium enolate of **231**. Subsequent hydrogenation, deprotection, and ring opening of the resulting β-lactam product lead to **233**. In addition, α,β-diamino alcohols **234** bearing a quaternary stereogenic center at the C_2 position can also be obtained from this approach.

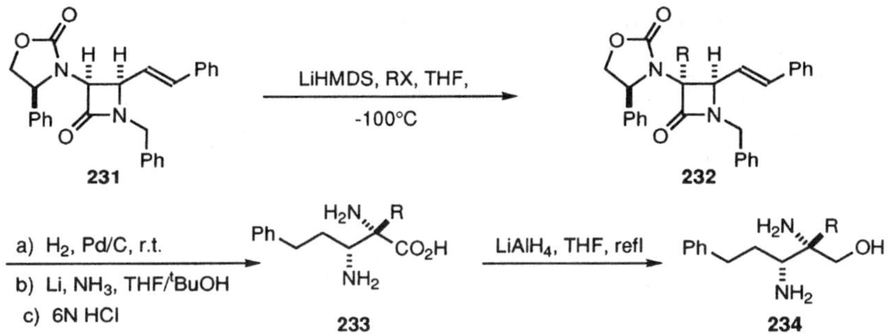

Scheme 14.59

Prior to these works, Golding[82] reported a nonstereocontrolled β-lactam route (Scheme 14.60) to the 2-piperidylglycine *rac*-**236**, an amino acid related to the antitumor agent 593A, the 3,6-bis-(5-chloro-2-piperidyl)piperazine-2,5-dione **237**. This author obtained *rac*-**236** as a 2:1 mixture of diastereomers **236a** and **236b** by heating **235** in a strongly acidic medium.

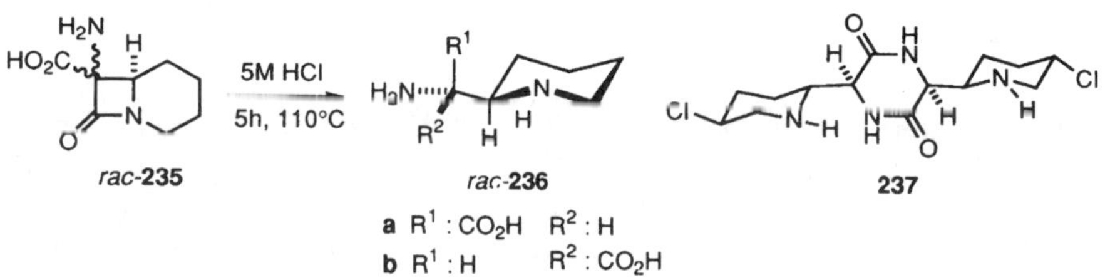

Scheme 14.60

Unfortunately, dimerization of acid *rac*-**236** to the corresponding piperazinedione failed. A solution to this problem was reported by Fukuyama[83] by building up the piperidine ring on a preformed piperazinedione. As shown in Scheme 14.61, the β-lactam *rac*-**239** required for dimerization was prepared by reaction of azidoacetyl chloride with the aldimine **238** in the presence of triethylamine. Subsequent N-dearylation, activation of the β-lactam ring by installation of the carbobenzoxy group at N_1, and reduction of the azido group led to the β-lactam **240**. Upon standing at room temperature, **240** dimerized to give a 3:2 mixture of *cis*-piperazinedione **241** and *trans*-piperazinedione **242** in 80% yield, which was chromatographically separated. Selective catalytic hydrogenation of **241** followed by cyclization led to **243**. Compound **243** was next transformed into the antibiotic 593A (±)-**237** in 62% overall yield from **243** (18% overall yield from **239**).

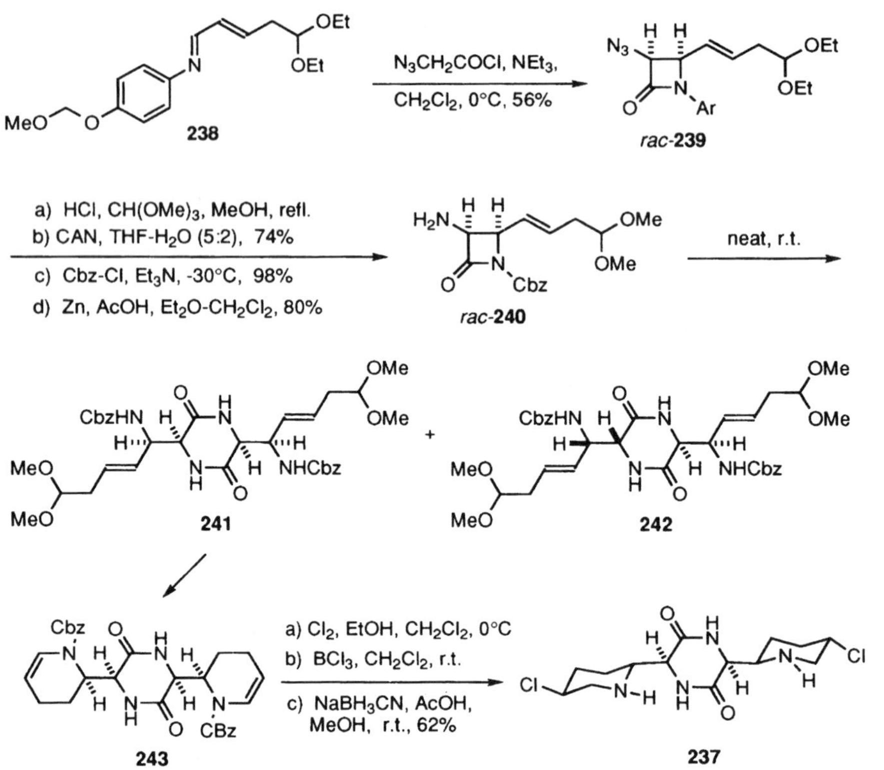

Scheme 14.61

Our laboratory has also been engaged in α,β-diamino acid chemistry with the aim to address two major issues: first, their synthesis in optically pure form, and second, the chemoselective introduction of different protective groups at either the α- or the β-amino function in order to allow their ulterior incorporation into peptide chains.

The approach we have developed[84] (Scheme 14.62) is based on the observation that the introduction of the Boc group in Cbz-β-lactams **244a** and **244b** to activate the β-lactam carbonyl towards nucleophilic attacks takes place chemoselectively at the endocyclic nitrogen atom to give products **245a** and **245b**, respectively. In the latter case, however, a small amount of the product derived from the attack to the exocyclic nitrogen atom was also obtained.

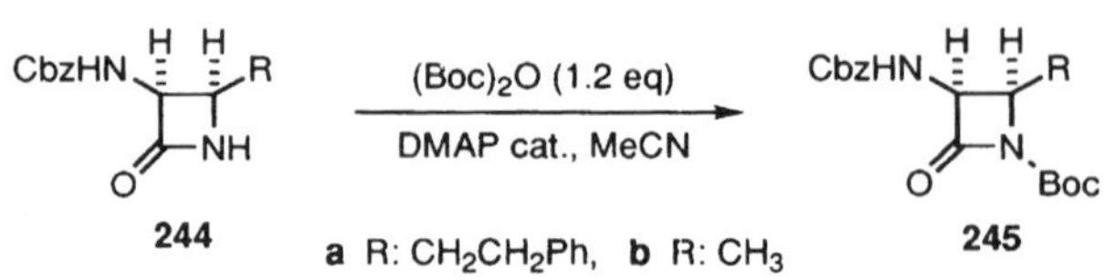

Scheme 14.62

The coupling reaction of such β-lactams with three representative alcohols (Scheme 14.63) proceeded smoothly in N,N-dimethylformamide in the presence of either NaN_3 or KCN, giving rise to the α,β-diamino esters **246**, **247**, and **248** in 80–90% yields. Similarly, acylation of *S*-phenylalanine methyl ester as well as *S*-leucine methyl ester with **245a** proceeded cleanly to give the corresponding dipeptide products **249** and **250** in excellent yields and without loss of optical integrity. The β-branched α-amino ester *S*-valine methyl ester could also be efficiently coupled with **245a** to give **251** in 75% isolated yield.

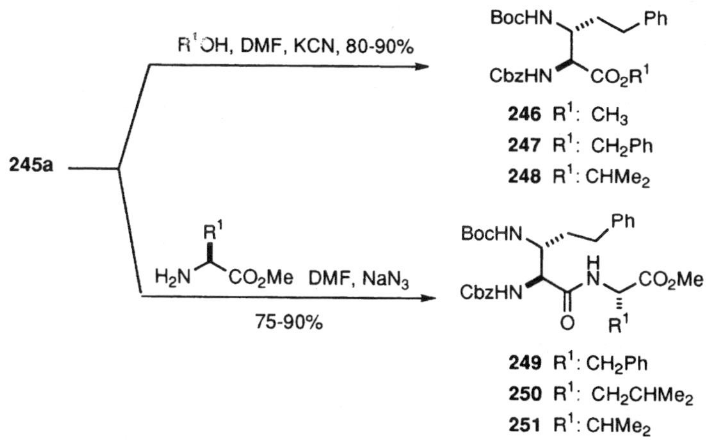

Scheme 14.63

We have also devised an alternative route to α,β-diamino acids and derived dipeptides, which consists of a regioselective Baeyer–Villiger rearrangement of α-keto β-lactams **254** as the key reaction element.[60] As Scheme 14.64 illustrates, the starting α-hydroxy β-lactams **253** were readily obtained as single diastereomers by the cycloaddition of benzyloxyketene to -Boc-α-amino imines **252** developed in our laboratory,[85] followed by O-debenzylation. At first, these α-hydroxy β-lactams were

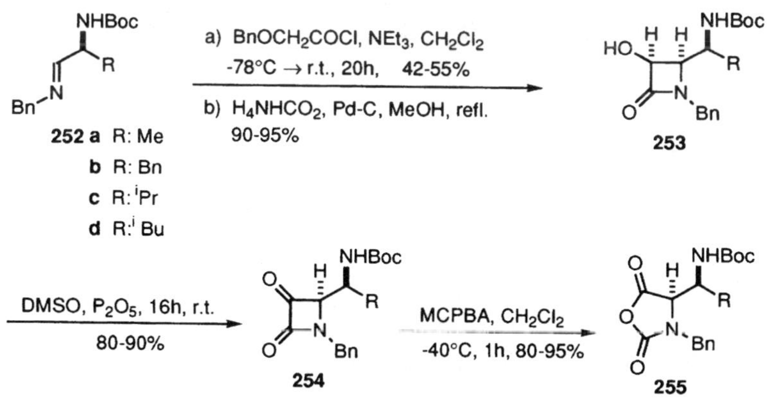

Scheme 14.64

resistant to oxidation with usual oxidizing agents, such as PDC, PCC, and Swern, but on treatment with P_2O_5 in DMSO produced the α-keto β-lactams **254** in excellent yields. These compounds on exposure to MCPBA underwent ring expansion to give exclusively α-amino acid N-carboxy anhydrides (NCAs) or Leuchs anhydrides[86] formally derived from α,β-diamino acids.

Ring opening of each NCA **255** by means of methanol (Scheme 14.65) led to the corresponding esters in good to excellent yields and without loss of optical integrity, as indicated by the ^{1}H- and ^{19}F-NMR spectra of the corresponding Mosher amides **256**.

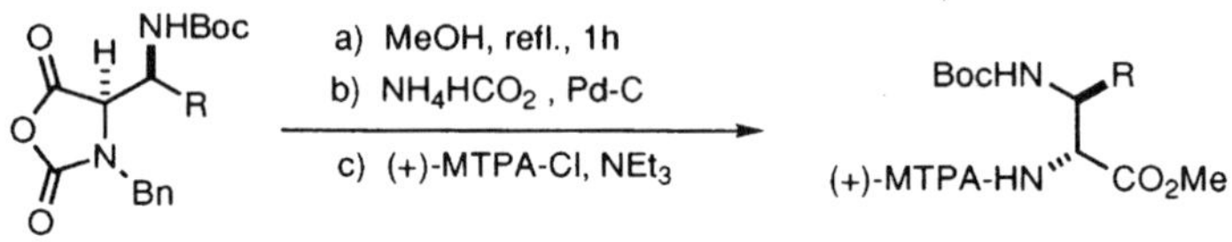

Scheme 14.65

Likewise, the α-hydroxy β-lactam **257** (Scheme 14.66), upon oxidation and Baeyer–Villiger rearrangement, gave the NCA **258**, which then was coupled with (*S*)-leucine methyl ester.[87] Subsequent N-debenzylation of the resulting dipeptide product led to the formation of the aminodeoxybestatin **259**, an AP-M inhibitor equipotent to the known bestatin.[72f]

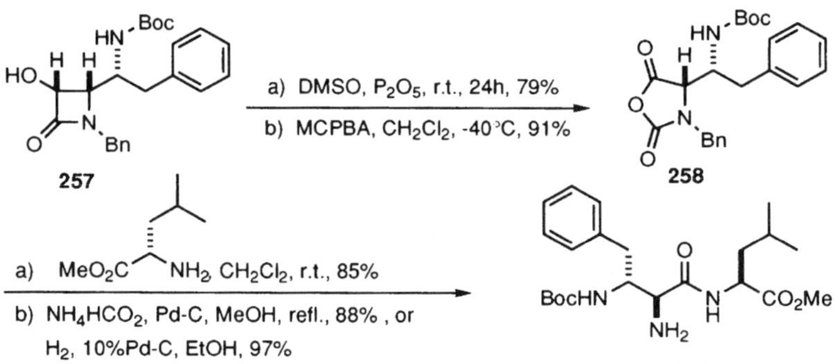

Scheme 14.66

As shown in Scheme 14.67, we have also extended this procedure to the synthesis of α,β-diaminoacids **262** bearing additional functionality[87] occurring in streptothricin-type antibiotics.[2b,c]

Concurrent with these studies, we also examined the method for the synthesis of 2-pyrrolidinylglycine and derived dipeptides as potential analogs of the 2-piperidylglycine.[88] Taking advantage of the ready availability of the prolinal imine **263** (Scheme 14.68), the α-hydroxy β-lactam **264** was prepared in good overall yield and transformed into the NCA **265**.

With NCA **265** in hand, the 2-pyrrolidinylglycine **266** was obtained in 70% yield, which, on coupling with **265** and subsequent N-debenzylation of the resulting inter-

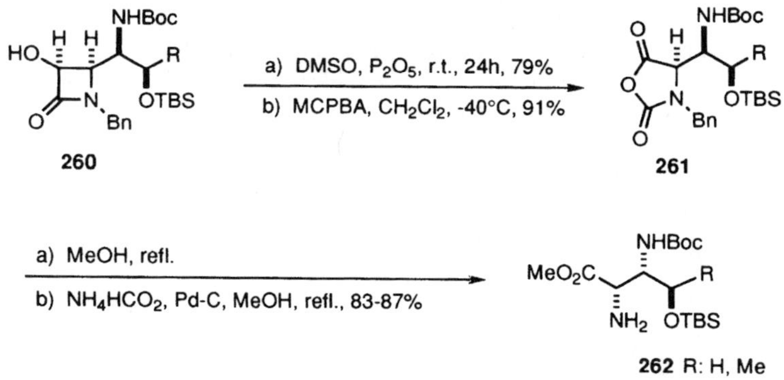

Scheme 14.67

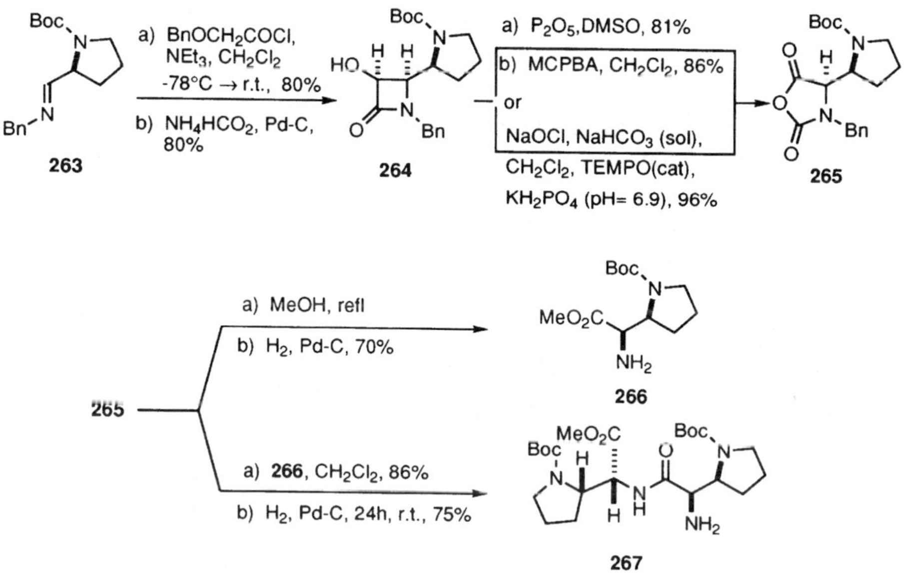

Scheme 14.68

mediate, led to the dipeptide product **267**. After completion of these studies we have found that α-hydroxy β-lactams on exposure to commercial bleach and a catalytic amount of TEMPO within about 5 min afforded any desired NCA in a one-pot operation in yields as high as 95–98%.[88]

14.3 γ-Substituted β-Amino Acids

14.3.1 β,γ-Diamino Acids and Derived Compounds

In the course of our study on synthetic applications of homochiral β-lactams derived from N-Boc α-amino imines, we anticipated that their ring opening (Scheme 14.69) should provide a route to β,γ-diamino acids and derived products.[85b,c] For example, compounds **268** and **270**, upon treatment with hydrochloric acid in refluxing ethanol, furnished pyrrolidin-2-ones **269** and **271** in excellent yields.

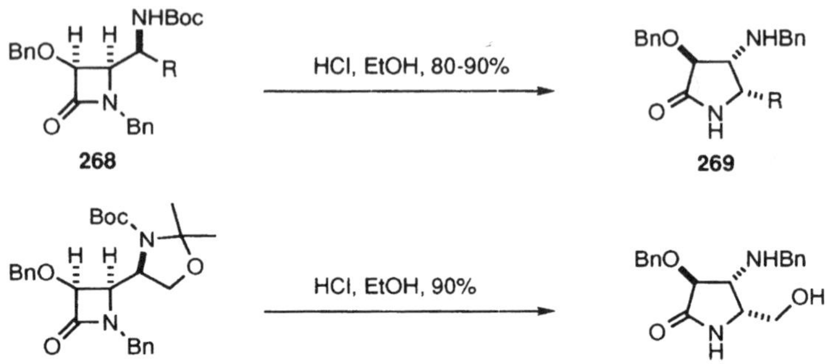

Scheme 14.69

As to be expected, extension of this rearrangement to the β-lactam **272** (Scheme 14.70) led to the bicyclic compound **273** in 80% yield. In particular, reduction of the carbonyl group in **273** gave the necine base **274** carrying the 1,2-amino alcohol subunit, which occurs, for instance, in some pyrrolizidine alkaloids, such as lolidine **275** and related systems.[89]

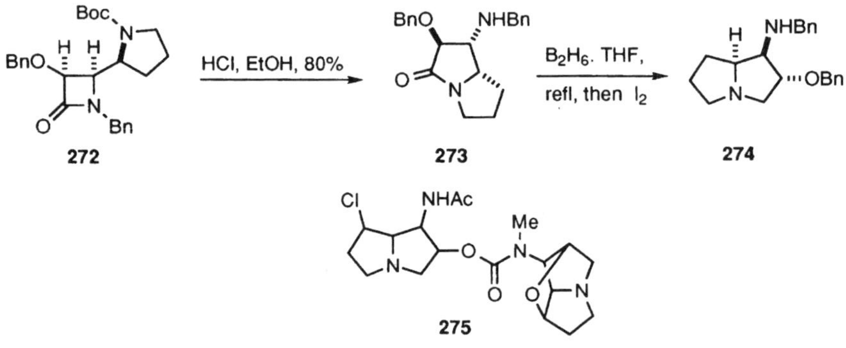

Scheme 14.70

Recently, a similar rearrangement was observed on *bis*-β-lactams **276**, generated by cycloaddition reaction of alkoxyketenes with 1,4-diaza-1,3-dienes derived from glyoxal.[90] As Scheme 14.71 illustrates, the resulting bicyclic γ-lactams **277** can be transformed either into **278** or into **279**, both of them azaanalogs of isosorbide, a powerful diuretic, and some furo[3,2-b]pyrrole derivatives, which are active as bronchodilators. The elongation of glyoxal in four carbons bearing four contiguous stereogenic centers becomes one of the most notable features of the method. The starting *bis*-β-lactams can also be prepared by cycloaddition of alkoxyketenes with imines derived from 4-formyl β-lactams.[91]

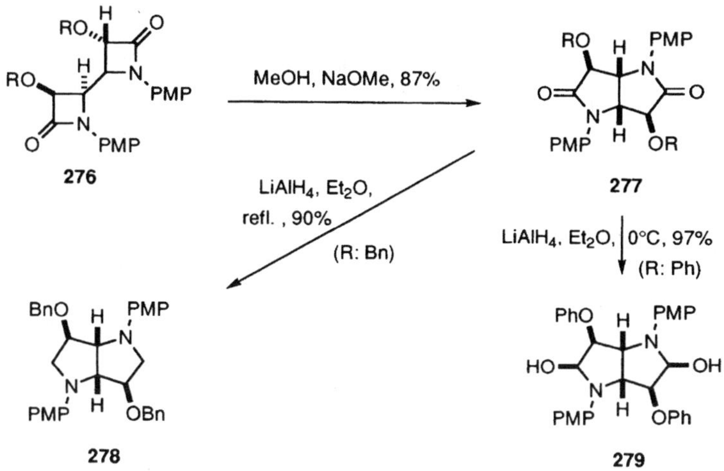

Scheme 14.71

Interest in β,γ-diamino acids also stems from the search for potent renin inhibitors. Statine (Figure 14.5), (3*S*,4*S*)-4-amino-3-hydroxy-6-methylheptanoic acid **280**, is the crucial structural unit of the potent aspartyl proteinase inhibitor, pepstatin A **281**.[92] It has been described, for example, that peptides containing aminostatine, obtained by the replacement of the (3*S*)-hydroxy group of statine by the nearly isosteric ammonium group, have similar potency to their statine congeners. Likewise, the replacement of the Leu-10 residue of renin substrate sequences with aminostatine results in novel potent inhibitors of human renin.[93]

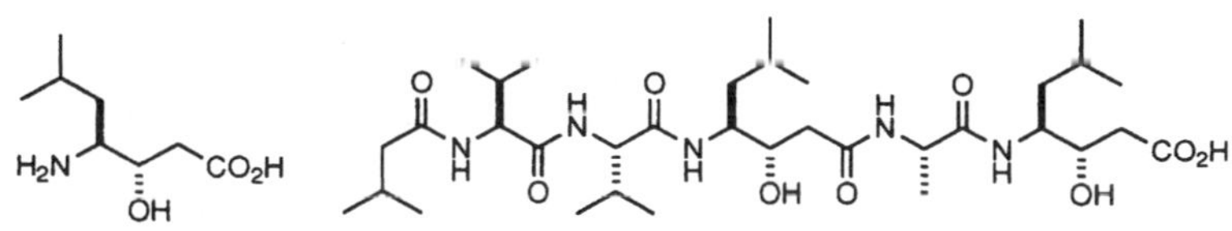

Figure 14.5

One approach to this β,γ-diamino acid has been reported by Shostarez[94] (Scheme 14.72) from the β-lactam framework **283**, prepared via the hydroxamate methodology using the *epi*-statine **282**. N-deprotection and further ring opening under basic conditions gave aminostatine, which in turn was transformed into the cyclic form **285** for structural assignments.

Renin inhibitors containing an α,α-difluoro-β-aminodeoxystatine moiety have also been described.[95] As shown in Scheme 14.73, cyclization of **286** gave the difluoro β-lactam **287**, which was then opened under basic conditions to give **288**. Incorporation of the protected difluoro β-aminodeoxystatine in an appropriate peptide sequence followed by deprotection with cerium(IV) ammonium nitrate provided **289**, a compound displaying relatively weak inhibition of human plasma renin (IC_{50} = 2800 nM).

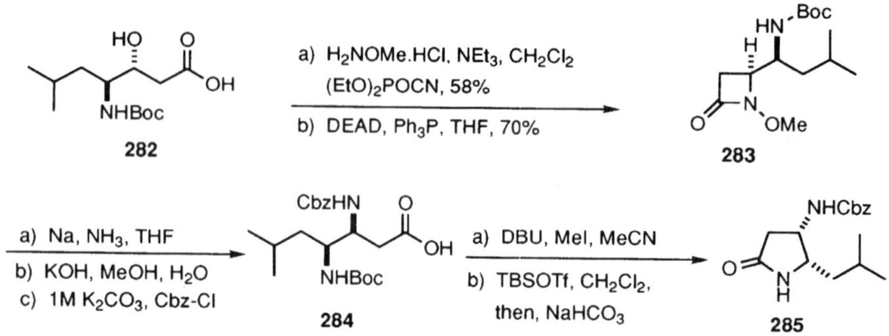

Scheme 14.72

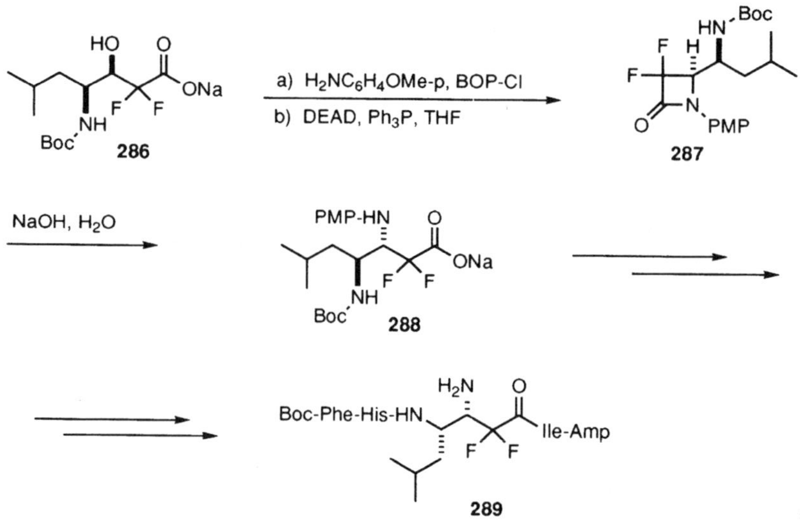

Scheme 14.73

Compound **291**, possessing the (*R*)-amino group, was a better inhibitor of human plasma renin (IC_{50} = 340 nM) and has been synthesized from **290**[95] (Scheme 14.74). The extension of this approach to other β,γ-diamino acids is, however, necessarily predicated on the availability of β-hydroxy γ-amino acids with desired configurations as the starting materials. This situation becomes more problematic when α-substituted β,γ-diamino acids, with specific absolute configurations, are required.

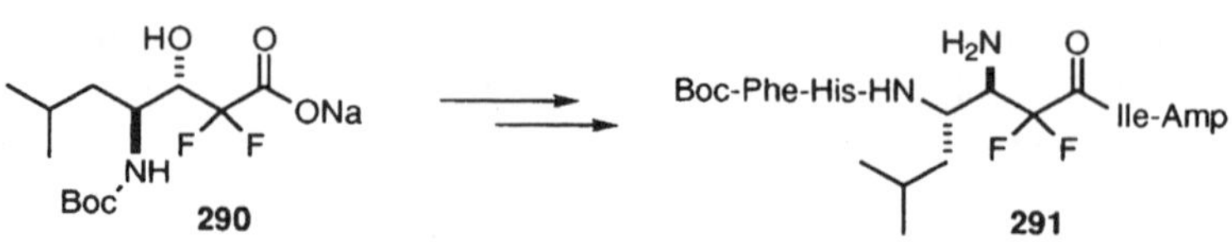

Scheme 14.74

In this context, we have developed an approach to these compounds with perfect control of the stereochemistry at the α and β positions of the amino acid as well as the corresponding derived γ-lactam products.[96] For example, the approach to γ-lactams having all the substituents in a *cis*-relationship relies on the highly stereoselective α-alkylation of lithium enolates of 3-unsubstituted β-lactams leading to *trans*-β-lactam products (see Section 14.2.1.1). As Scheme 14.75 illustrates, the 3-unsubstituted β-lactam **293**, readily obtainable from the 3-hydroxy derivative **292**,[97] after treatment with LDA and subsequent quenching with a fourfold excess of iodoalkyl electrophiles, afforded the corresponding alkylated products **294a–c** in excellent yields; each of them as a single *trans*-isomer. N-Dearylation of **294b**, for example, afforded the expected N-unsubstituted β-lactam, which upon intramolecular N_1-C_2 bond cleavage led to the corresponding γ-lactam, which was isolated as the Cbz derivative **295**.

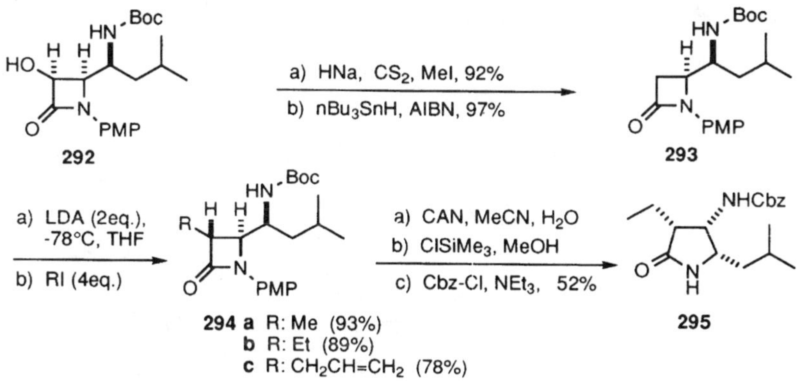

Scheme 14.75

Alternatively, we expected that deoxygenation of 3-alkyl-3-hydroxy β-lactams, which could be prepared from azetidine-2,3-diones, would generate the corresponding *cis*-isomers in a highly stereocontrolled fashion and, thereby, γ-lactams epimeric at the C_α position. The strategy, depicted in Scheme 14.76, was guided by our previous observations that tributyltin hydride desulfuration of 3-alkyl-3-arylthio β-lactams gave mainly *cis*-3-alkyl β-lactams.[98] Thus compound **296** was prepared in 90% yield by oxidation of **292** by means of DMSO in combination with P_2O_5. Subsequent Grignard addition to the carbonyl group in **296** afforded the carbinol **297** as solely the *cis* isomer. The stereoselectivity observed could be explained by the attack of the nucleophile from the less hindered side of the carbonyl group. In support of this, borohydride reduction of **296** provided the starting α-hydroxy β-lactam without traces of the *trans*-isomer. Next, compound **297** was subjected to Barton–McCombie deoxygenation to give the expected compound **298** as single diastereomer. N-Dearylation of **298** and cyclization led to the corresponding γ-lactam, which was isolated as the Cbz derivative **299** in 58% yield over the three last steps.

The examples described in this section indicate that a variety of related structures might be prepared via this approach including new γ-lactam analogs of β-lactam antibiotics of type **300**[99] or new γ-lactam bridged dipeptides of type **301**[100] (Figure 14.6).

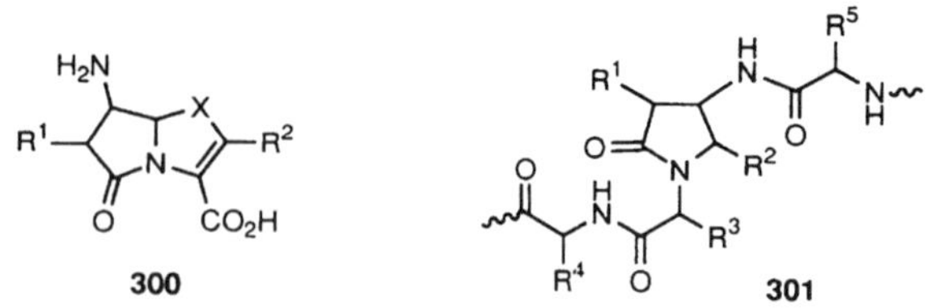

Figure 14.6

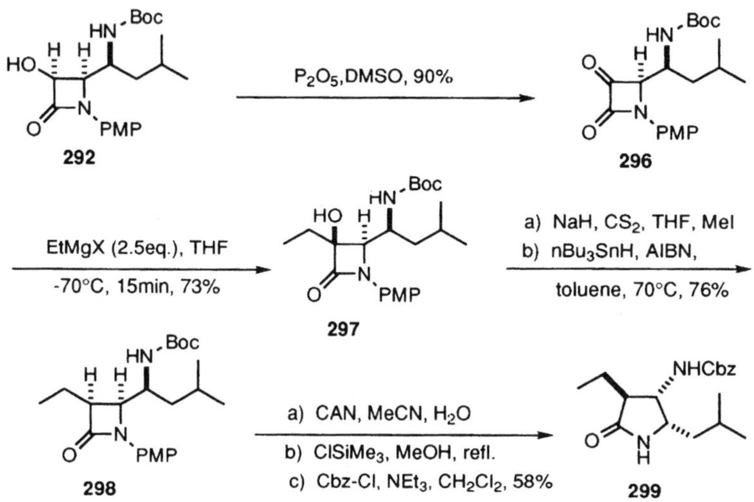

Scheme 14.76

Preliminary evidence from this laboratory[101] (Scheme 14.77) also indicates that N-Boc-β-lactams **303** might be employed as acylating agents for peptide synthesis, thus providing a practical route for the study and design of new enzyme inhibitors.[102]

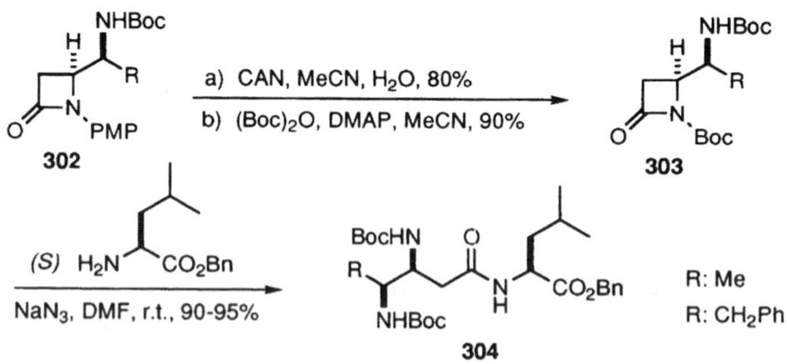

Scheme 14.77

14.3.2 γ-Alkoxy-β-Amino Acids and Derived Compounds

The former section delineates the potential use of β-lactams with α-amino alkyl groups at the C_4 position to generate target compounds containing the β,γ-diamino acid subunit. This section is related to 4-(α-alkoxyalkyl)-β-lactams and β-lactams carrying polyoxygenated side chains at the C_4 position for the construction of β-amino γ-alkoxy acids and derived products like those depicted in Figure 14.7.

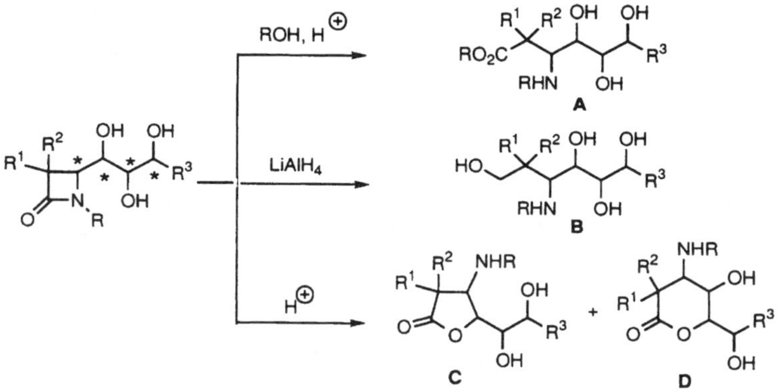

Figure 14.7

In this respect, there are at least three possible pathways to achieve this goal: first, the usual alcoholysis of O-protected polyoxygenated β-lactams to give the corresponding β-amino ester A; second, direct reductive cleavage of the N_1-C_2 bond to furnish the open chain aminopolyol B; finally, intramolecular rearrangement of these β-lactams to provide cyclic lactones C and/or D. The following examples illustrate the implementation of these pathways to the synthesis of natural products through α-substituted- and α-unsubstituted γ-alkoxy-β-amino acids.

14.3.2.1 α-Substituted γ-Alkoxy-β-Amino Acids

One of the most important contributions to this field has been reported by the group of Bose and his colleagues.[103] They found, for example, that cycloaddition reaction of methoxyketene with imine **305** led to the β-lactam **306** as single *cis*-isomer (Scheme 14.78). Hydrolytic cleavage of the β-lactam bond gave the polyoxygenated β-amino ester **307**, which on subsequent steps was easily transformed into 6-*epi*-lincosamine **308**.[104]

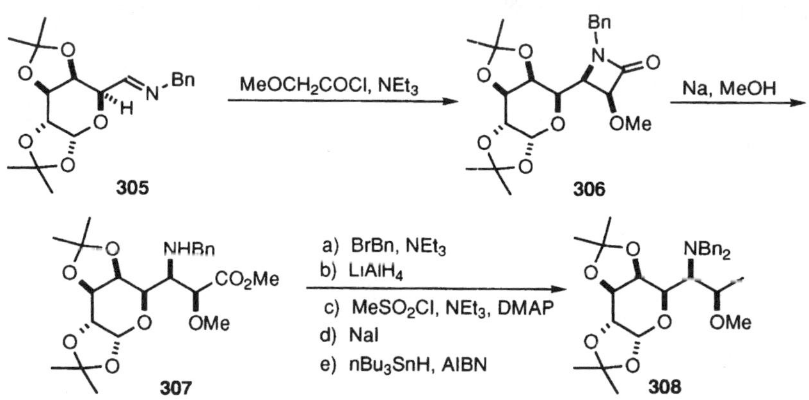

Scheme 14.78

In a similar way (Scheme 14.79), the cycloaddition reaction of benzyloxyketene with the imine **309** proceeded under complete control of the diastereoselectivity to give the β-lactam **310**. The acetonide **311**, readily obtained by the selective hydrolysis of **310**, was transformed under standard conditions into the α-hydroxy β-lactam **312**, which was then opened to the aminopolyol **313** using lithium aluminum hydride. Such a compound, upon oxidative cleavage and further exhaustive protection, gave (−)-polyoxamic acid **314**, the enantiomer of the component of the antifungal polyoxin antibiotic.[105]

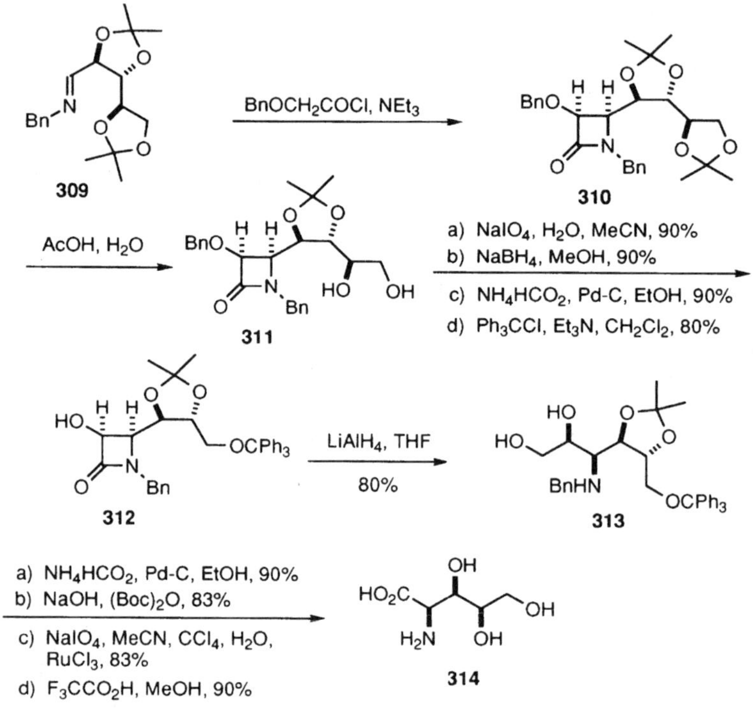

Scheme 14.79

The same group (Scheme 14.80) has found that β-lactam **315**, prepared by cycloaddition reaction of methoxyketene with the imine **168**, upon exposure to trifluo-

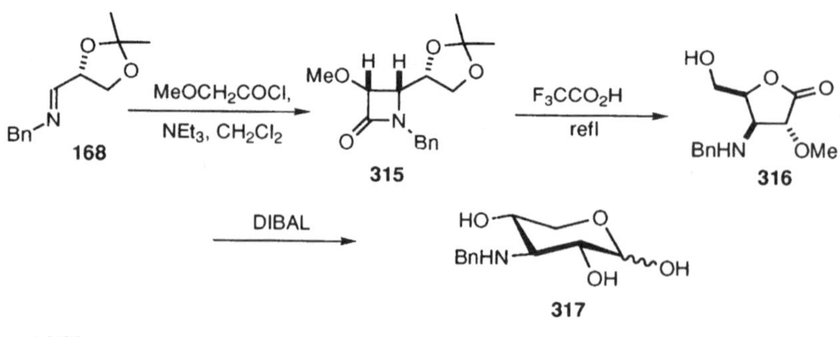

Scheme 14.80

roacetic acid, rearranges to the γ-lactone **316**, providing a way to the gentosamine amino sugar derivative **317**.[4,106]

Recently, Welch and co-workers[107] have reported a β-lactam route to 2-fluoroamino sugars, by using the cycloaddition reaction between fluoroketene and imine **169** as the access to the starting fluorinated β-lactam **318**[108] (Scheme 14.81). The target 3-amino-2,3-dideoxy-2-fluoro-D-*xylo*-furanose **320** was obtained through **319** in a similar way to that of Bose for the synthesis of gentosamine.

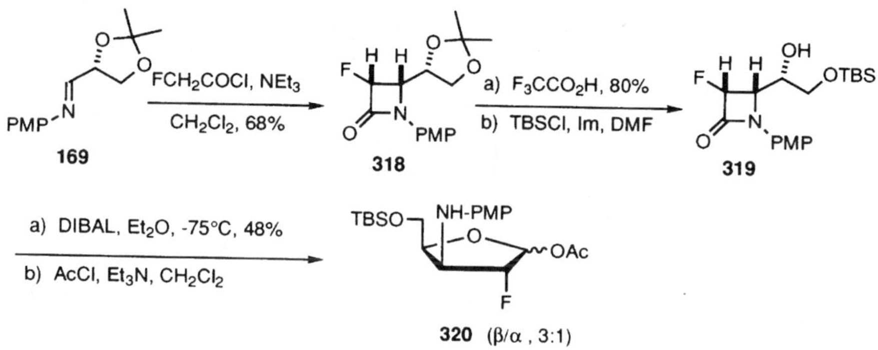

Scheme 14.81

The synthesis of the 3-amino-2,3-dideoxy-2-fluoro-D-*xylo*-pyranose derivative **323** (Scheme 14.82) by DIBAL reduction of **318** followed by acid hydrolysis of the resulting aminals **321** and **322** has also been reported by the same group.[107]

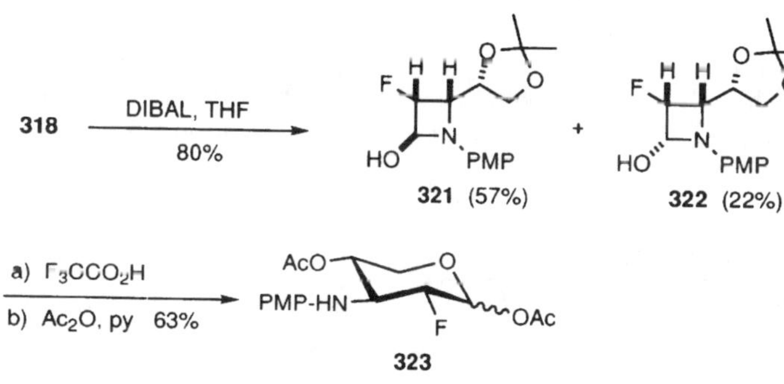

Scheme 14.82

The Reformatsky reaction between the imine **324** and ethyl difluoroiodoacetate has been reported by Kobayashi[109] as an approach to difluoroglucopiranosides **327** and **328**, via β-lactams **325** and **326**, respectively (Scheme 14.83). Unfortunately, the first step showed little diastereoselectivity (**325**/**326**:1/4.3) when compared with the former ketene–imine cycloaddition.

Our own group has also been concerned with some of these aspects, and an approach to 2-amino-1,3-polyol chains of type **329** and **330** via 3-alkoxy-4-(1-

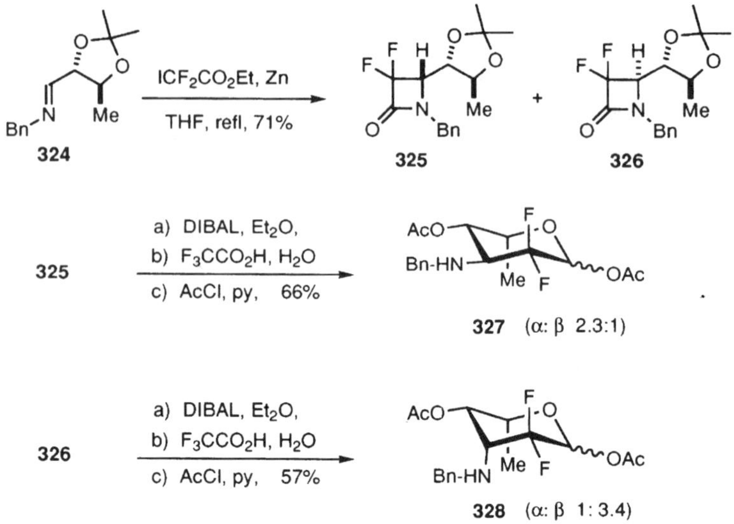

Scheme 14.83

alkoxyalkyl)-β-lactams was delineated by using an iterative cycloaddition of alkoxyketenes with α-silyloxyaldehyde–derived imines as the key reaction.[110]

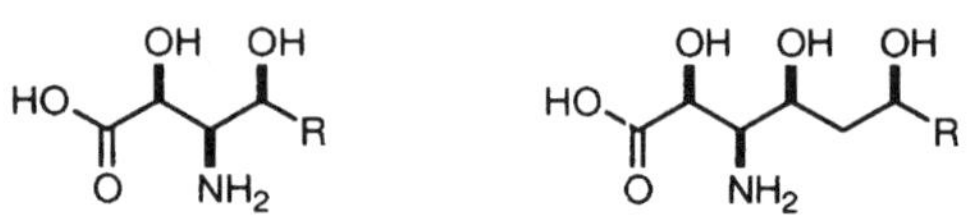

329 330

As Scheme 14.84 illustrates, the synthesis starts with the β-lactam product **332** derived from the imine **331**. Subsequent ring opening under mild acid conditions pro-

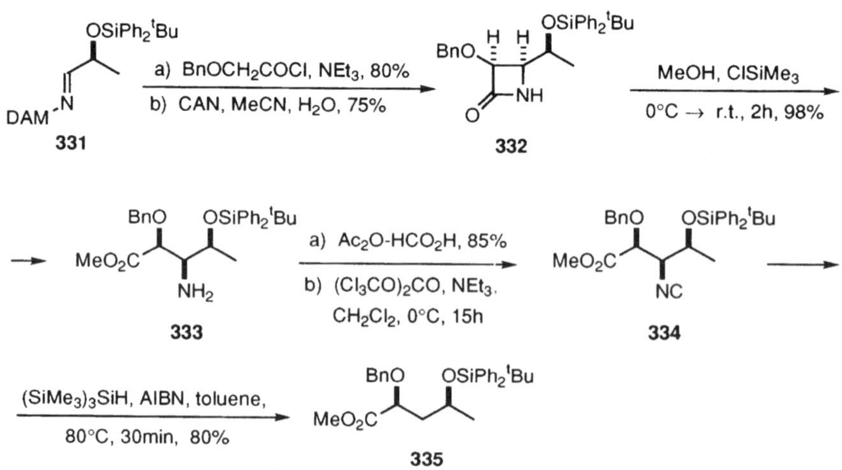

Scheme 14.84

vided the α,γ-dialkoxy β-amino ester **333** in almost quantitative yield, which was next converted into the isocyanide derivative **334**. Reduction of isocyanide **334** proved to be difficult under Barton's conditions, but using Chatgilialoglu's reagent, tris(trimethylsilyl)silane, the reduction proceeded efficiently to give the 1,3-diol **335** in 80% yield.

The second cycle of the iterative process is shown in Scheme 14.85. The hydroxy-derivative **336**, obtained by DIBAL reduction of **335**, was oxidized to the corresponding aldehyde and thus transformed into the imine **337**. Treatment of such an imine with benzyloxyacetyl chloride and triethylamine led to the β-lactam **338** in 75% yield as a single diastereomer, as judged by ^{1}H-NMR analysis of the crude compound. Subsequent N-dearylation and further ring opening provided the β-amino α,γ,ε-polyol chain **339** in 50% overall yield. Deamination of **339** followed by repetition of the above sequence of reactions would generate more extended amino polyols.

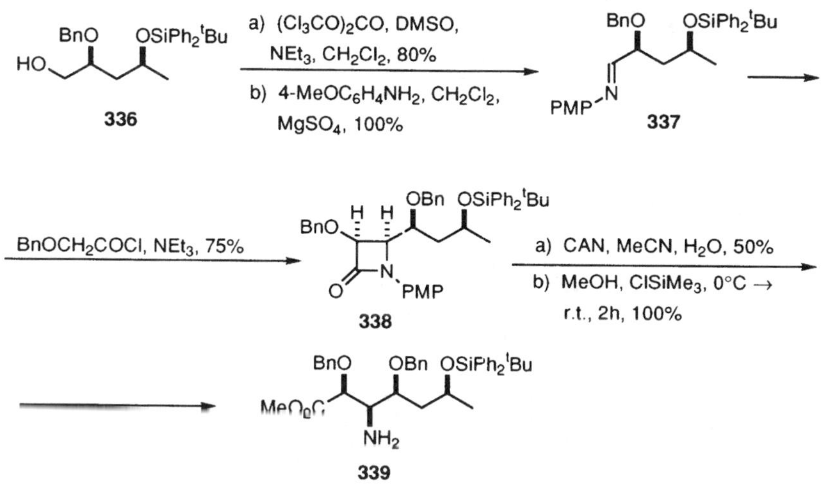

Scheme 14.85

On the other hand (Scheme 14.86), protection of the free β-amino moiety in **333** followed by reduction of the methoxycarbonyl group led to the hydroxy derivative **340**, which was further transformed into the (*R*,*S*)-threonine **341** in a similar way to

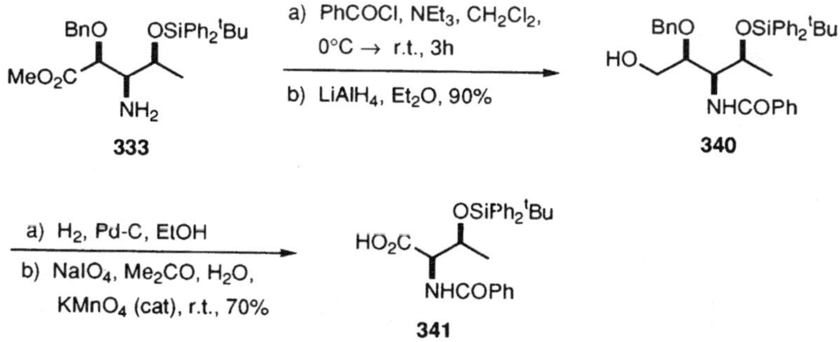

Scheme 14.86

that of Bose for polyoxamic acid synthesis (Scheme 14.79). The ready availability of α-hydroxy β-lactams greatly expands the scope of this approach to natural and unnatural α-amino β-hydroxy acids.

Concurrent with these studies and in conjunction with general interest in α-amino-β-hydroxy acid–derived peptides, we have also developed a route to these α-amino acids, which, in turn, can directly be combined with a peptide coupling reaction.[111] As shown in Scheme 14.87, ring expansion of the α-hydroxy β-lactams **342** by means of NaOCl and a catalytic amount of TEMPO leads to the formation of α-amino acid N-carboxy anhydrides **343**, formally derived from β-alkyl serines. The process occurs through a regioselective Baeyer–Villiger rearrangement of an in situ generated α-keto-β-lactam (see Section 14.2.3). The most remarkable point of the approach is the construction of α-amino acid N-carboxy anhydrides or Leuchs anhydrides from non-α-amino acid precursors.

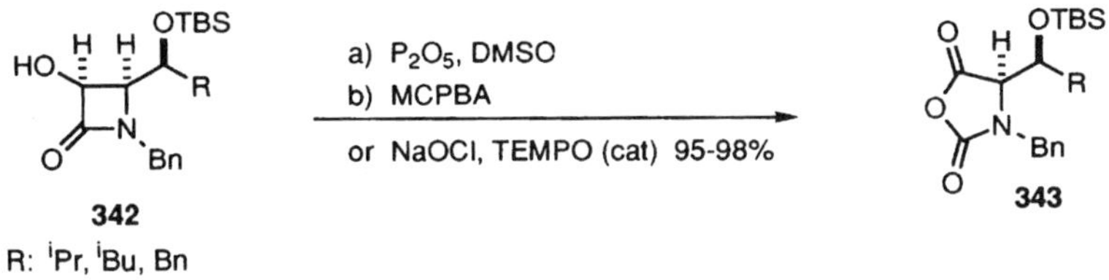

Scheme 14.87

The preparation of the tripeptide fragment B of the antibiotic lysobactin (see Section 14.2.2.3) illustrates this approach for peptide synthesis.[71]

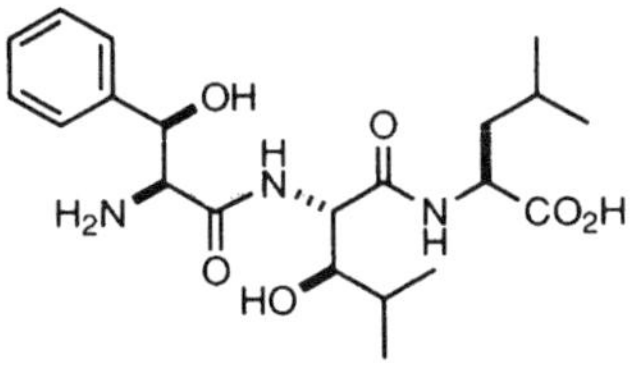

Fragment B

First, the synthesis of the NCAs **346a** and **346b**, formally derived from the corresponding nonproteinogenic amino acids *threo*-β-hydroxyleucine and *threo*-β-hydroxyphenylalanine, was accomplished (Scheme 14.88) from the respective α-hydroxy β-lactams **345a** and **345b**, readily obtained by the cycloaddition reaction of benzyloxyketene with imines **344a** and **344b**, respectively.

The NCA **346a** (Scheme 14.89) was then coupled with (*S*)-leucine methyl ester in methylene chloride as solvent, followed by N-debenzylation of the resulting coupling product to give dipeptide **348** in good overall yield. Surprisingly, the NCA **346b** was resistant to ring opening by dipeptide **348** but, with the addition of NaN_3 in DMF as solvent, coupled efficiently to give the desired tripeptide product in 70% yield, which upon N-debenzylation and simultaneous N-protection led to the N-Boc derivative **349**.

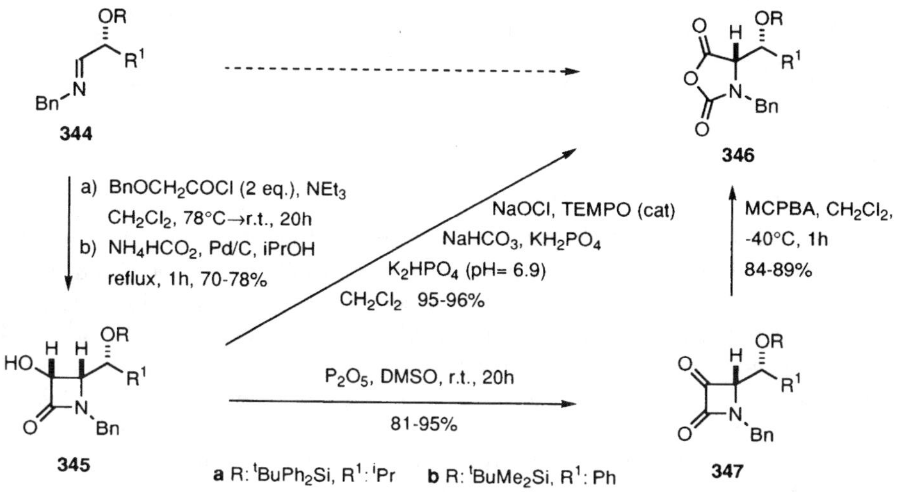

Scheme 14.88

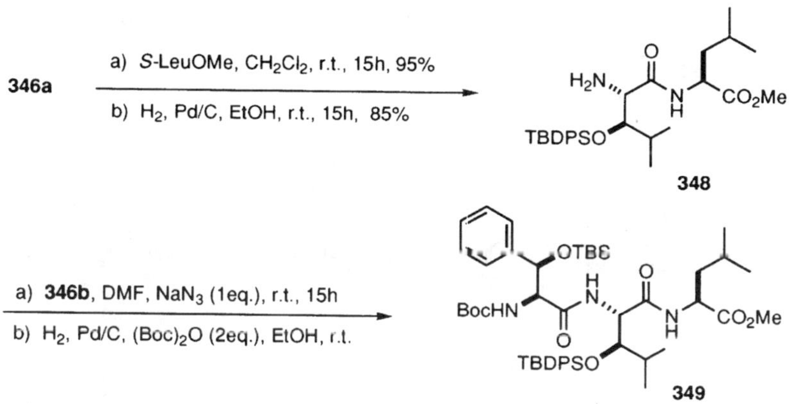

Scheme 14.89

14.3.2.2 α-Unsubstituted γ-Alkoxy-β-Amino Acids

As pointed out (Section 14.3.1), α-hydroxy β-lactams with α-aminoalkyl groups at the C_4 position can be easily transformed into the corresponding 3-unsubstituted β-lactams (*X*: NH_2) making possible to obtain new peptidomimetics by coupling their N-Boc derivatives with α-amino esters (Figure 14.8). Preliminary evidence from this laboratory[101] has also shown that deoxygenation of α-hydroxy β-lactams bearing a hydroxylated side chain at the C_4 position (*X*: OH) provides suitable intermediates for the synthesis of short peptide fragments containing α-unsubstituted γ-alkoxy-β-amino acids.

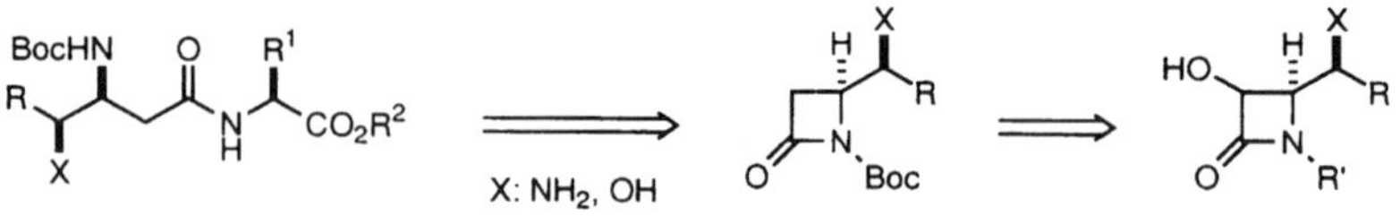

Figure 14.8

For example (Scheme 14.90), the N-Boc-β-lactam **351a**, obtained by Barton's deoxygenation of **350a**[112] followed by N-dearylation and subsequent Boc introduction, could be coupled efficiently with (*S*)-leucine benzyl ester to provide dipeptide **352a** in 90% yield. Extension of this approach to the N-Boc-3-unsubstituted β-lactam **351b** provided dipeptide **352b** containing the regioisomer of the well-known renin inhibitor statine **280** (see Section 14.3.1). Furthermore, taking advantage of the cycloaddition reaction of benzyloxyketene to sugar-aldehyde derived imines discussed in the preceding section, a wide variety of 3-unsubstituted β-lactams with hydroxylated side chains at the C_4 position would be accessible and easily converted into their corresponding β-amino acids or dipeptides containing them.

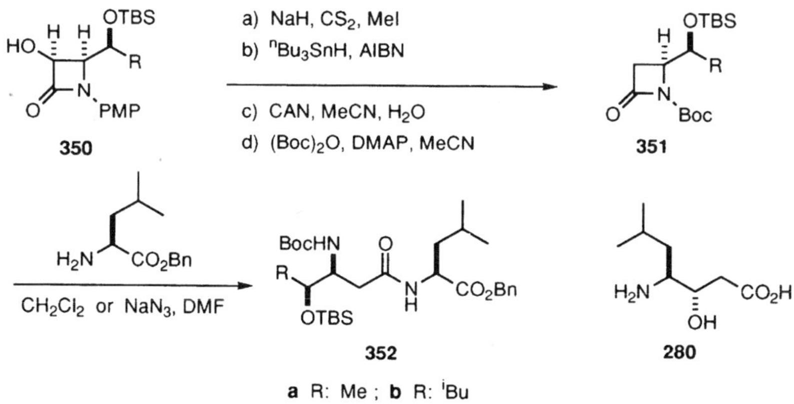

Scheme 14.90

In the above context, a tactically different approach to construct β-lactams with hydroxylated side chains at the C_4 position has also been delineated by Thomas to access the β-amino acid present in the 3,4-dihydroisocumarin AI-77-B **353**, a compound that exhibits biological activity against stress ulcers in rats, yet is noncentral suppressive, nonanticholergenic, and nonantihistaminergic.[113]

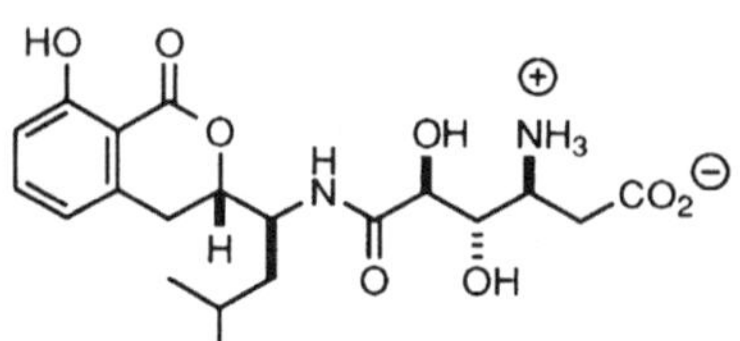

353

The approach (Scheme 14.91) is based on a stereoselective *cis*-dihydroxylation of the β-lactam **354** readily available from the aldehyde **28** (see Section 14.2.1.1). Thus condensation of **28** with benzyl bis(2,2,2-trifluoroethyl)phosphonoacetate afforded the (*Z*)-ester **354** together with a small amount (11%) of its (*E*)-isomer. Oxidation with osmium tetroxide proceeded to give a mixture of diols **355a** and **355b** in a ratio of 80:20. Diol protection and hydrogenolysis gave the acid **356**. This acid was then coupled with the amine **357**, followed by diol deprotection to give the dihydroxy-dipeptide **358**. This compound upon β-lactam ring opening under basic conditions followed by careful acidification to pH 6.5, and further demethylation of the phenolic methyl ether provided **353**.

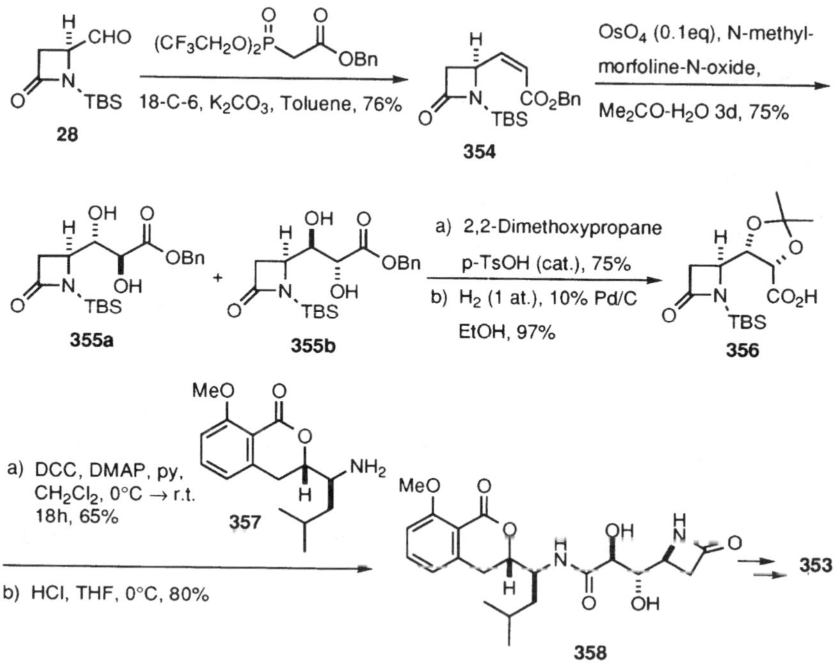

Scheme 14.91

The aldehyde **28** has also been employed by Robinson[114] in an approach to peptidyl difluoro ketones bearing an aspartic acid side chain. As shown in Scheme 14.92, **28** reacted smoothly with the Reformatsky reagent of ethyl bromodifluoroacetate to afford an epimeric mixture of carbinols in a 1:2 ratio from which the major isomer **359** could be separated by chromatography in as high as 94% enantiomeric purity. Subsequent amide formation led to **360** (73% overall from **28** in a 1:2 ratio), which was then treated with Boc-L-alanine N-hydroxysuccinimide ester to give the lactone **361**. The reaction is presumed to occur via acylation on nitrogen, attack by N-hydroxysuccinimide on the β-lactam carbonyl, and lactonization of the resulting activated γ-hydroxy ester. Finally, **361** was transformed into **362**, designed as an inhibitor of interleukin-1β converting enzyme (ICE), thus providing a route for the synthesis of other structural classes of protease inhibitors.

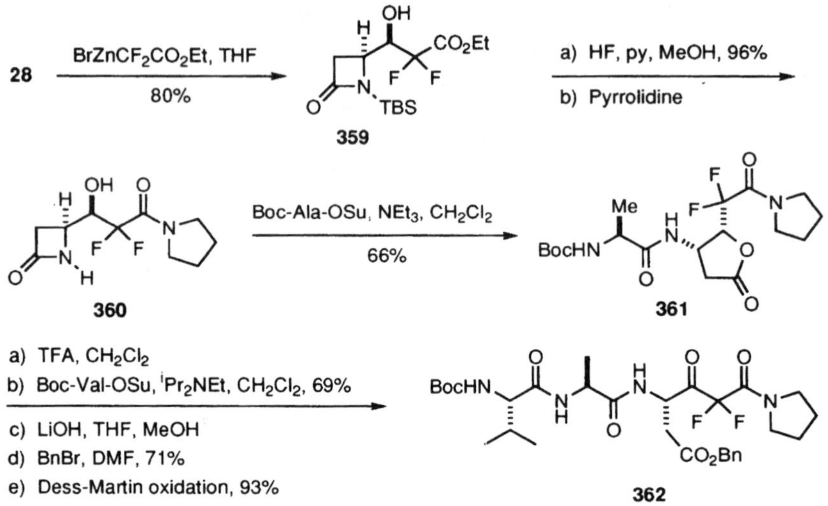

Scheme 14.92

14.4 β-Amino Ketones

According to our interest in aminopolyols and biologically active peptides, we have also devised another approach to these attractive targets. The approach (Figure 14.9) was based on the successful implementation of the N_1-C_2 β-lactam cleavage by means of heteronucleophiles discussed in the early sections. On the one hand, we expected that the ring opening of N-Boc β-lactams of type **363** by carbon nucleophiles should

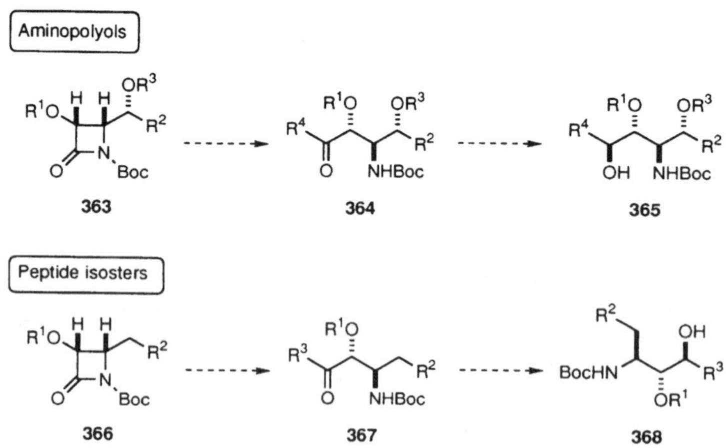

Figure 14.9

provide β-amino ketones **364**, which on diastereoselective reductions, through chelation and/or nonchelation pathways, would provide a route to more elaborated aminopolyols **365**. Furthermore, the same reaction carried out on N-Boc β-lactams **366** with alkyl substituents at the C_4 position would generate precursors of peptide isosteres and/or renin inhibitors.[102] This reaction has also been simultaneously developed by Baldwin in an approach to γ-keto α-amino acids and derived compounds. These aspects are discussed below.

14.4.1 β-Amino Ketone-Derived 1,2,3-Amino Diols

Our approach to aminopolyols[115] is illustrated in Scheme 14.93. The most significant feature is that the reaction of aryl Grignard reagents at low temperature (–40°C) with the N-Boc β-lactam **194** affords exclusively β-amino ketones **369** in 90–96% yields. In no case was overaddition observed under these conditions, even when an excess of the reagent was present in the reaction medium. On the other hand, when the reaction was performed at room temperature instead of –40°C, only tertiary alcohols were obtained. In contrast to these observations, reaction of **194** with primary alkylmagnesium halides as well as alkyllithiums at –40°C invariably led to a mixture of the corresponding β-amino ketone **369** and β-amino carbinols **370**, along with the starting β-lactam. However, the use of Gilman cuprates afforded the best results in terms of chemoselectivity, although the yields were only moderate (*R*: Me, 50%; *R*: *n*-Bu, 67%).

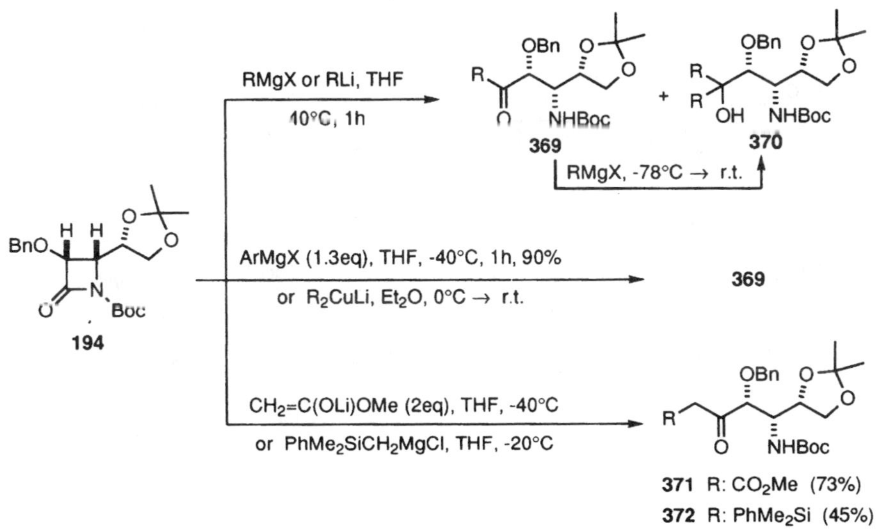

Scheme 14.93

As shown in the scheme, β-amino ketones carrying additional functionality, such as **371** and **372**, could also be prepared by this method in good yields. We also observed that compound **371** (Scheme 14.94), upon reduction with zinc borohydride, afforded a mixture of epimeric carbinols **373** without any diastereofacial discrimination. Attempts to carry out the reduction of **371** with L-selectride were unfruitful, recovering the starting material unchanged. Although no diastereoselection was observed, the mixture of carbinols **373** could be transformed into **374**, which on simple elaboration led to a mixture of the epimeric piperidines **375** and **376**.

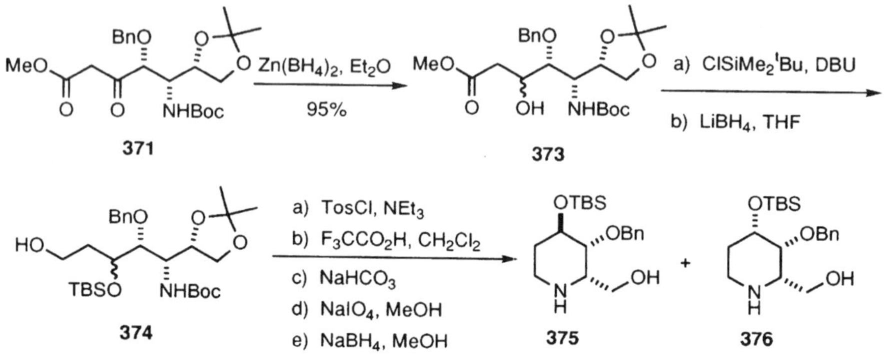

Scheme 14.94

This new approach to β-amino ketones has also been extended to the synthesis of alkyl-substituted 1,2,3-aminodiols.[101] As shown in Scheme 14.95, **377** upon treatment with isobutylmagnesium chloride afforded the β-amino ketone **378** in 70% isolated yield along with the corresponding overaddition product (10%). The reduction of the β-amino ketone **378** under chelation controlled conditions led to the corresponding

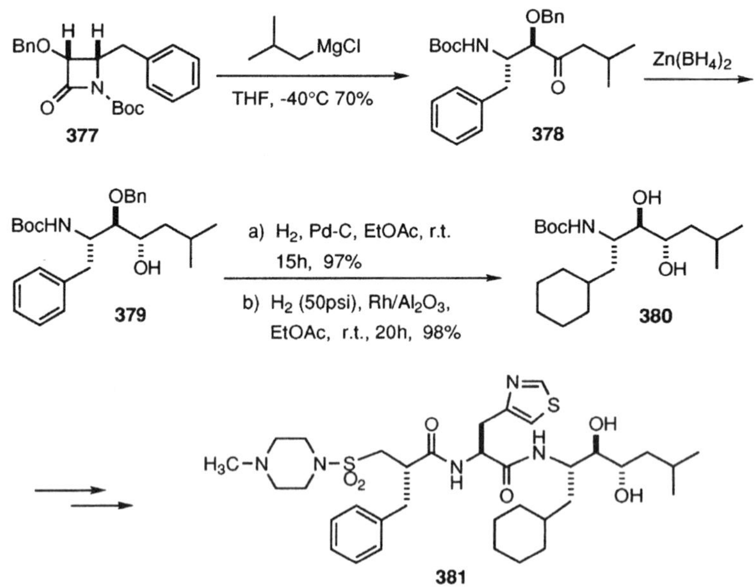

Scheme 14.95

alcohol **379** together with its epimer in a ratio of 90:10, respectively. The major isomer **379** on exposure to hydrogenolysis and further saturation of the phenyl ring afforded the 1,2,3-aminodiol **380**, a constituent of the potent inhibitor of human renin A-725-17 **381**, which is undergoing clinical evaluation as a therapeutic agent for the treatment of hypertension and congestive heart failure.[116]

Both the reaction of aryl Grignard reagents with **377** and the acylation of the lithium enolate of methyl acetate with the same compound (Scheme 14.96) proceeded very efficiently to afford the respective β-amino ketones **382** and **383** in excellent yields, making this approach suitable for other kinds of enzyme inhibitors.[101]

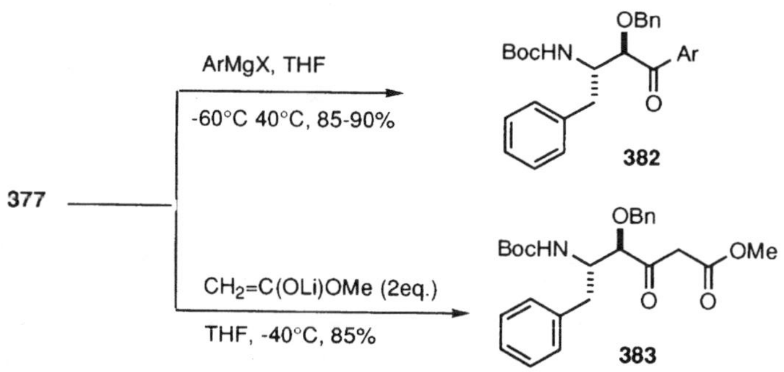

Scheme 14.96

After completion of our work, Spero and co-workers[117] have reported the same conceptual approach to 1,2,3-aminodiols. In their approach (Scheme 14.97), the β-lactam **384** obtained by known protocols[37,51,60] is subjected to treatment with isobutylmagnesium chloride to give, as expected, the β-amino ketone **385**, which on disilylation and further reduction[118] provided **380**.

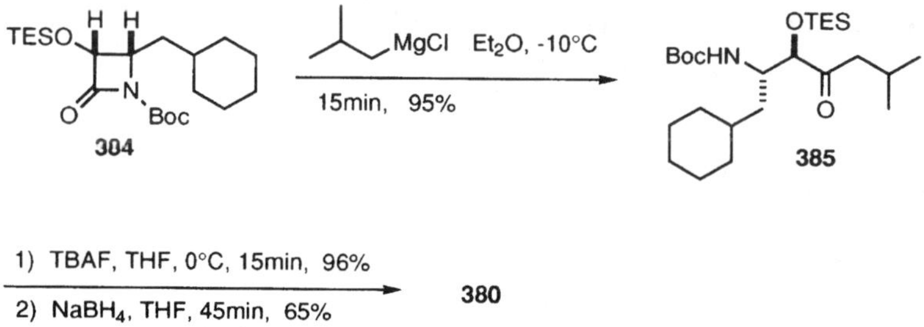

Scheme 14.97

In a closely related paper, Ojima[119] has also reported the synthesis of hydroxy (keto)ethylene dipeptide isosteres **387** and **388** through the reaction of N-Boc-β-lactams **386** with lithium enolates of ketones and esters (Scheme 14.98), respectively.

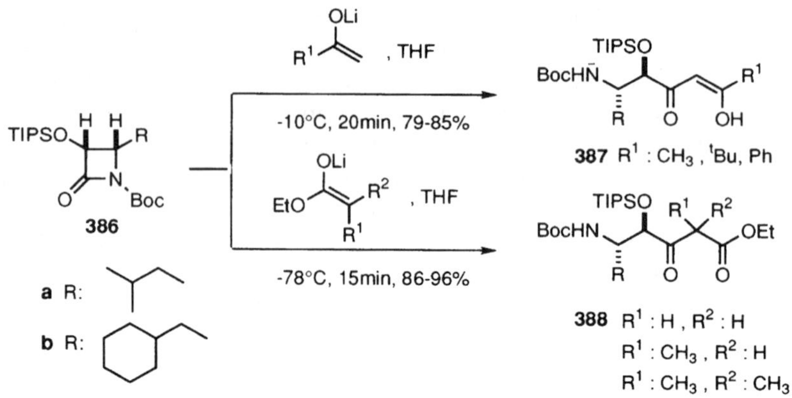

Scheme 14.98

14.4.2 β-Amino Ketone–Derived α-Amino Acids

In the course of our investigations on β-lactams as acylating agents, we have also demonstrated the utility of the reaction between N-Boc-4-carboxy β-lactams and carbon nucleophiles providing functionalized α-amino acids.[115] One example is the synthesis of the γ-hydroxy α-amino acid **394** present in the antibiotic Nikkomycin (Scheme 14.99).[120] We have found that Grignard reagents, on treatment with *rac*-**389**, react selectively with the imide function to afford exclusively the products **390** and **391**. On the other hand, while methylmagnesium bromide gave a mixture of **390a**, **391a**, and the starting product **389** in a ratio of 70:10:20, aryl Grignard reagents only gave β-amino ketones **390b** and **390c** in 80% and 72% yields, respectively. L-Selectride reduction of **390b** gave the amino lactone **392** and the hemiacetal **393** in a ratio of 62:38. This hemiacetal was oxidized to the amino lactone by using the mixed nicotinic–chromic anhydride (NDC) reagent developed in our laboratory.[121] By that means, the overall yield of **392** was 73%. This result

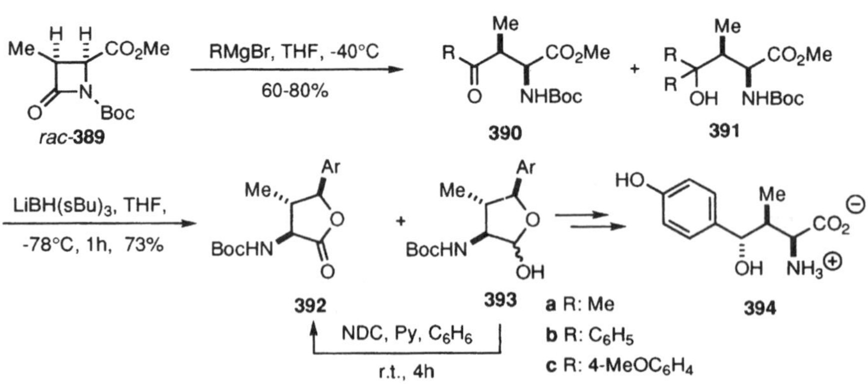

Scheme 14.99

illustrates the viability of the approach to the synthesis of β-alkyl γ-hydroxy α-amino acids in optically pure forms starting from homochiral *cis*-3-alkyl-4-alkoxycarbonyl β-lactams.

In the above context (Scheme 14.100), Baldwin and co-workers[122] have reported the synthesis of γ-keto-α-amino acids and their further conversion into targets of biological interest. In his approach, the N-Boc β-lactam **395** derived from aspartic acid was opened by means of lithiated sulfones to give **396**. Subsequent removal of the sulfone moiety afforded **397** and **398** in good yields.

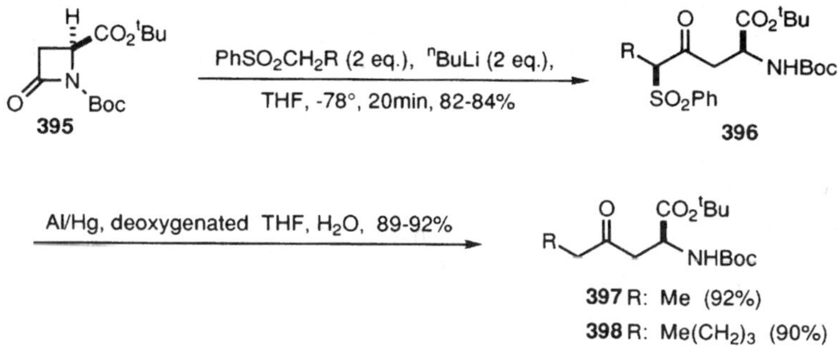

Scheme 14.100

The reaction of **395** with "higher-order cyano cuprates" was also assessed but, with limited success, as judged by the low yield attained for the expected coupling product **399** (Scheme 14.101).

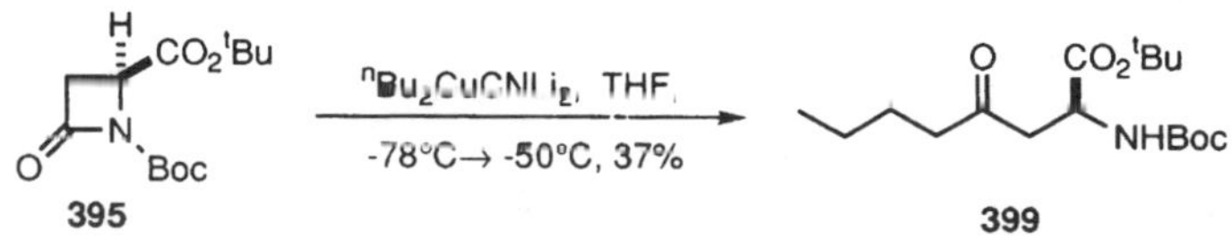

Scheme 14.101

As Scheme 14.102 illustrates, this reaction has been employed to synthesize the angiotensin converting enzyme (ACE) inhibitor WF-10129 **405**.[123] Ring opening of **400** with sulfone **402** and subsequent removal of the sulfone group followed by hydrogenolysis gave **403**, which was reacted with the triflate derived from (D)-lactic acid to give **404**. Further hydrogenolysis of **404** and DCC/HOBT mediated coupling to (L)-tyrosine *t*-butyl ester gave the corresponding dipeptide, which underwent deprotection with TFA/anisole, affording **405** as its trifluoroacetate salt.

The β-lactam ring opening also proceeds with phosphonate-stabilized carbon nucleophiles to afford β-keto-δ-phosphonato α-amino acids **406** (Scheme 14.103).

The same group[124] has also reported the reaction of **395** with trimethylsulfoxonium ylide to give **407**, which on transition metal catalyzed decomposition produces

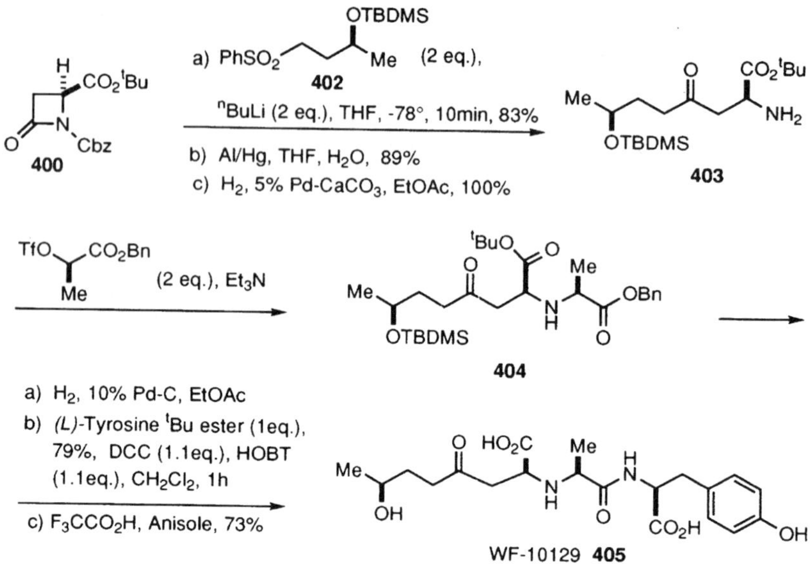

Scheme 14.102

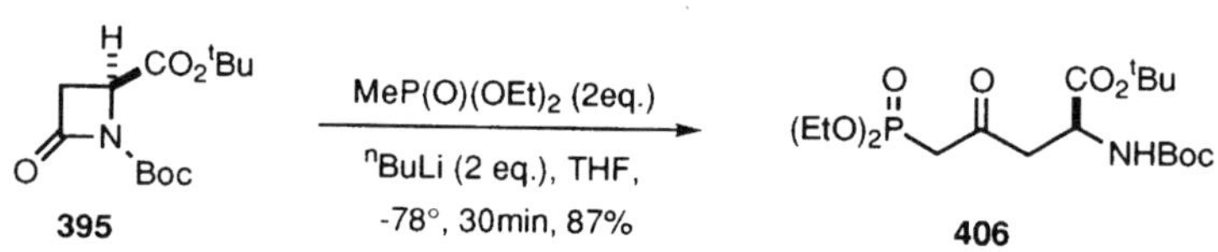

Scheme 14.103

the 4-oxoproline derivative **408** via an active carbenoid intermediate (Scheme 14.104).

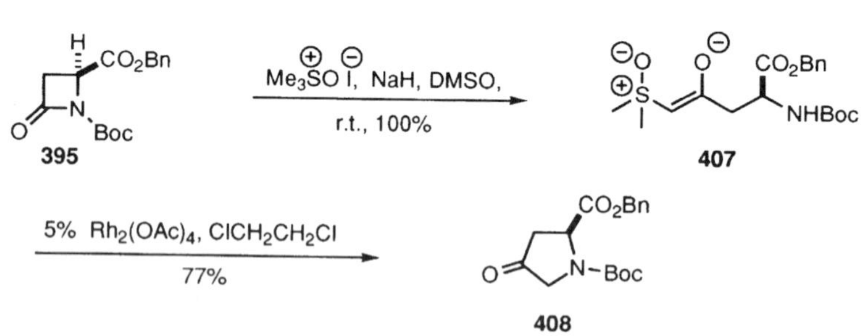

Scheme 14.104

The utility of 4-oxoproline derivatives for building up 4-aryl kainoid acids has been illustrated by Baldwin.[125] The kainoids exhibit a powerful neuroexcitatory activity,[126] and it has been reported that introduction of an aryl substituent at C_4 of the pyrrolidine ring provokes a strong potentiation of its activity.[127] As Scheme 14.105 illustrates, the approach starts from the enamine **410**, derived from the oxoproline **409** obtained as above. First, **410** was alkylated with *tert*-butyl bromoacetate to give **411** in fair yield. Subsequent Grignard addition followed by cleavage of the *tert*-butyl esters was accompanied by concomitant lactonization to give **412**. Finally, this compound was transformed into the kainoid hydrochloride **413**.

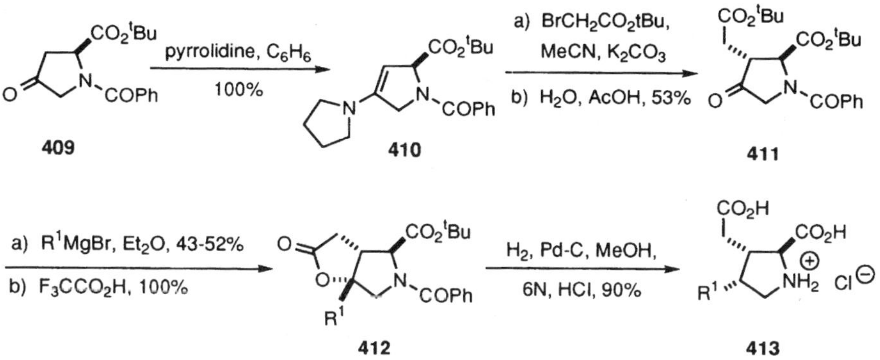

Scheme 14.105

14.5 β-Lactam-Derived Large Ring Heterocycles: Cyclopeptides, Alkaloids, and Cyclams

The preceding section delineated the utility of the intermolecular N_1-C_2 β-lactam bond cleavage provoked by nucleophiles, leading to compounds containing β-amino acid residues. In a similar way, several works have documented the intramolecular versions of such nucleophilic reactions between the β-lactam amide and suitable substituents on the four-membered ring, resulting in a cycloexpansion of the β-lactam nucleus to either medium or large ring heterocycles. Some medium rings have already been discussed in this chapter, and others, like pyrrolidines, pyrrolizidines, piperidines, morpholines, and so on have been reviewed elsewhere.[4] On the other hand, of the more representative examples of large ring heterocycles in the literature, cyclopeptidomimetics,[128] some spermine backbone containing alkaloids,[6,7] and cyclams have been chosen for discussion.

For instance, Greene and co-workers[129] have reported the first low-molecular-mass immunoglobulin mimetic **418** developed on the basis of an X-ray structure analysis of the antigen–antibody complex (Scheme 14.106). Cyclopeptide **418** is resistant toward proteases and imitates the binding and functional properties of the native antibody.

The synthesis of **418** is based on an intramolecular aminolysis of the β-lactam ring in **417**. This latter compound was prepared by stereoselective alkylation of **414** (obtained from (*R*)-aspartic acid; see Section 14.2.1.1) with 1-bromo-3-butene and subsequent oxidative cleavage of the double bond to give the carboxylic acid **415**, which next was coupled with **416**. The resulting peptide product **417**, on reductive depro-

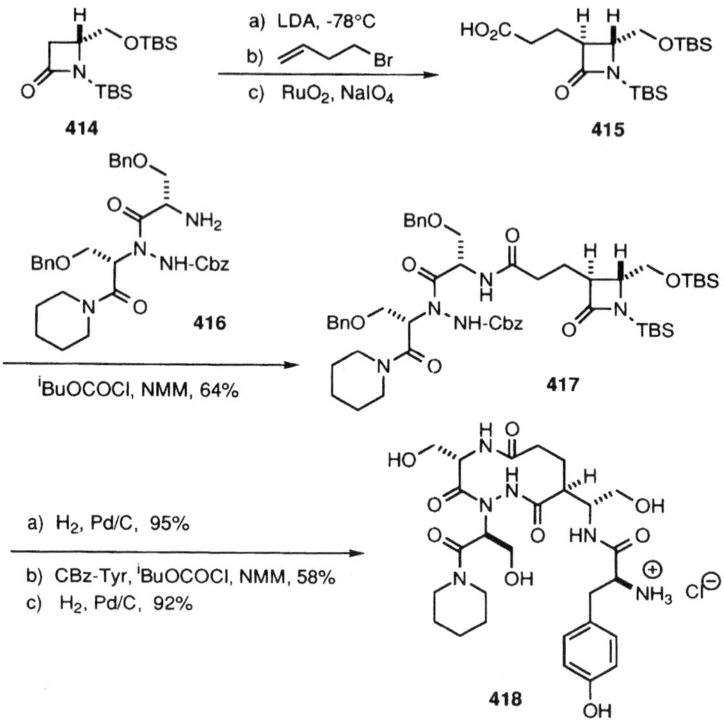

Scheme 14.106

tection of the hydrazino group, rapidly cyclized to a ten-membered ring compound, which was next coupled with N-carbobenzoxytyrosine to give **418**.

Another example is compound **425**, which imitates the thrombin-bound structure of the fibrin peptide A and is an inhibitor of this protease. The first step of this synthesis (Scheme 14.107) involved opening **419** to the seven-membered ring compound **420**. Subsequent acylation with **421** to form **422** was followed by an intramolecular

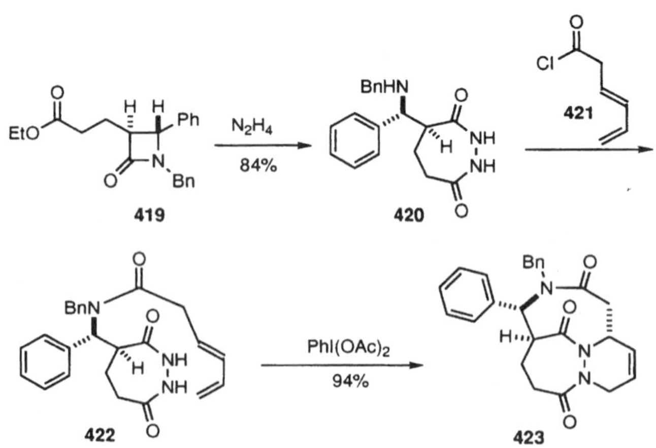

Scheme 14.107 *(continued)*

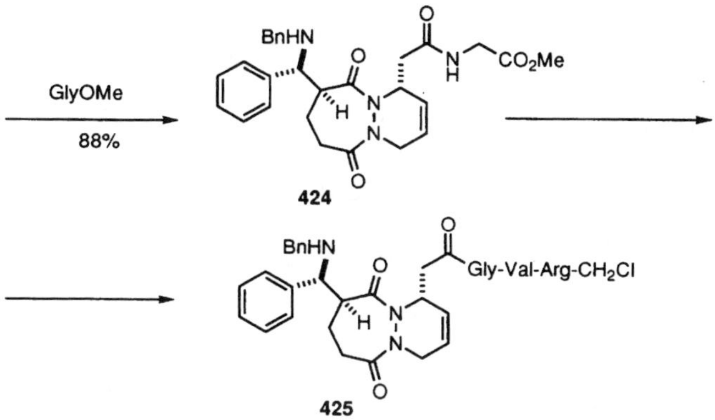

Scheme 14.107 (*continued*)

oxidative cycloaddition to provide tricyclic lactam **423**. Nucleophilic ring opening proceeded readily with glycine methyl ester to afford lactam **424** in 88% yield, which was then transformed into **425** using solid-phase protocols.[130]

The same group has also recently reported the synthesis of the nonpeptide β-turn mimetic **430** of enkephalin.[131] The approach (Scheme 14.108) involved the β-lactam **426** as the starting material, which was transformed into **427** via mixed anhydride coupling with benzyl glycinate and subsequent hydrogenolysis. A further coupling reaction of **427** with **428** provided the peptide product **429** ready for intramolecular cyclization to **430**.

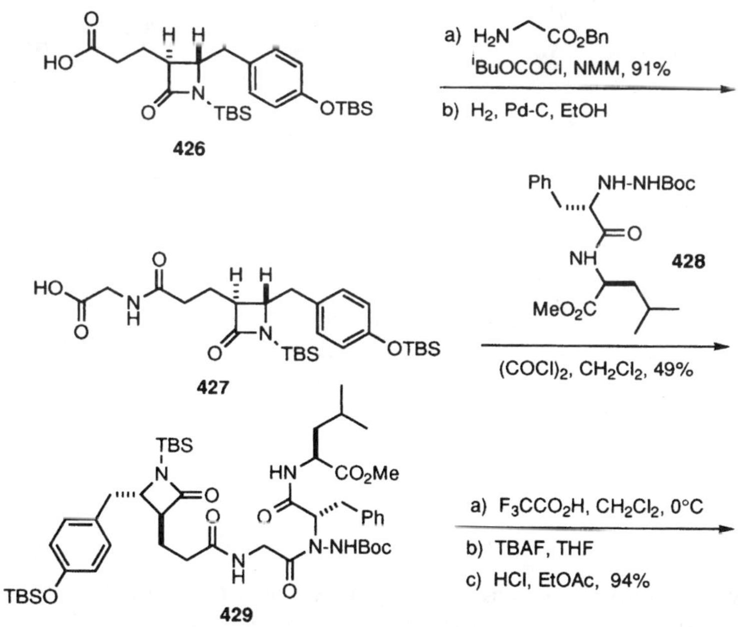

Scheme 14.108

(*continued*)

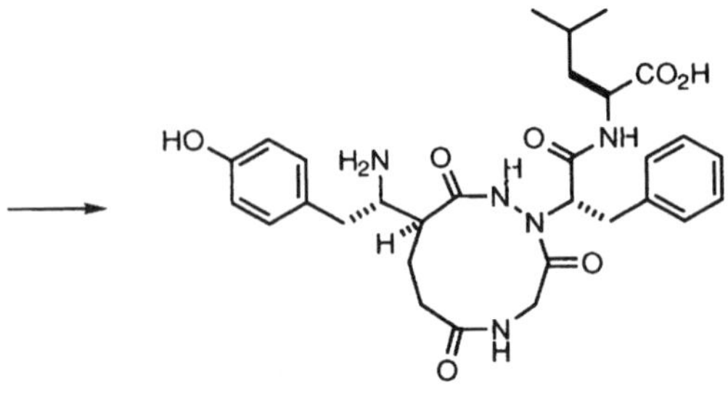

430

Scheme 14.108 (*continued*)

Boger et al.[132] have also described (see Scheme 14.109) the synthesis of the β-amino acid–containing macrocyclic compound **434**, a structural component of deoxybouvardin lacking the 14-membered N-methylcycloisodityrosine subunit. The intended 4-methoxycarbonyl-1-methyl-3-(N-methylamino)-β-lactam **432** was prepared according to the Evans methodology[45a] and coupled with the tetrapeptide **431** to provide **433**. Sequential methyl ester hydrolysis and Cbz hydrogenolysis followed by subjection of the resulting amino acid to diphenyl phosphorazidate promoted macrocyclization in 42%. It is apparent from these contributions that the current availability of β-lactams should greatly extend the scope of this strategy to diverse conformationally constrained reverse-turn mimetics.

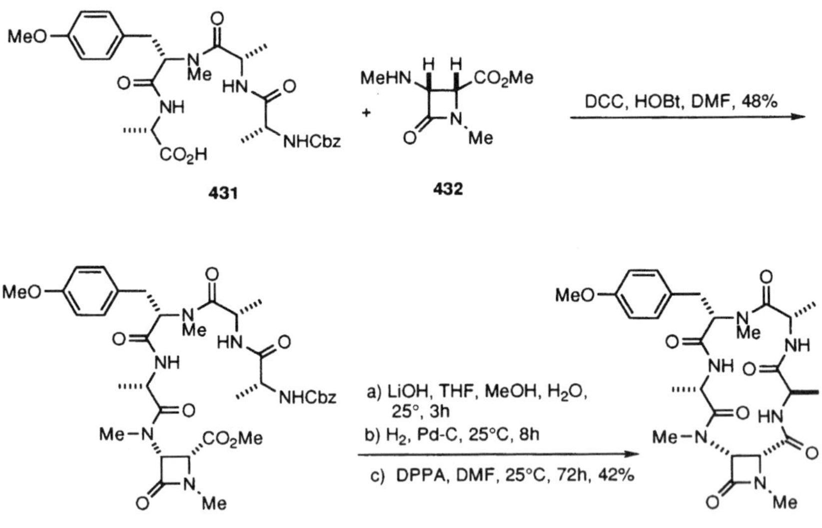

433 434

Scheme 14.109

In a related context, β-amino acid fragments containing large heterocycles also occur in some alkaloids, and approaches to these compounds using β-lactams as key intermediates have long been known since the pioneering contributions of Wasserman[6] and Hesse.[7] Recently, Crombie and Jones[133] have described the internal rearrangements on N-(ω-haloalkyl) β-lactams as a valuable method for the total synthesis of several homalium alkaloids containing a spermine structural backbone. These include (*S,S*)-(−)-homaline **435**, *rac*-hopromine **436**, *rac*-hoprominol **437**, and *rac*-hopromalinol **438** (Scheme 14.110). As an example, the homochiral β-lactam **440**, prepared from **439**, experiences a transamidative ring expansion in liquid ammonia on prolonged standing to afford the macrocyclic β-amino amide **441**, which is converted into (*S,S*)-(−)-homaline **435**.

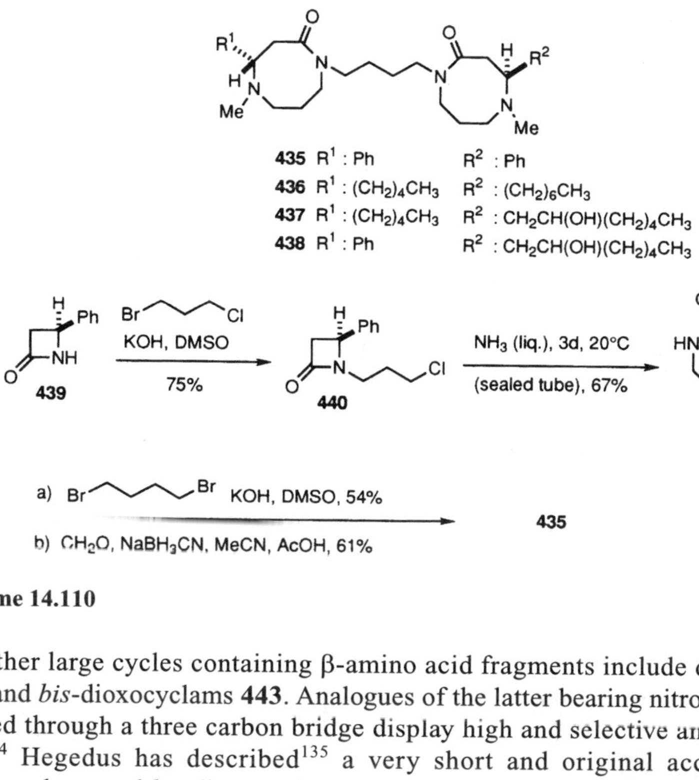

Scheme 14.110

Other large cycles containing β-amino acid fragments include dioxocyclams **442** and *bis*-dioxocyclams **443**. Analogues of the latter bearing nitrogen atoms attached through a three carbon bridge display high and selective anti-HIV activity.[134] Hegedus has described[135] a very short and original access to these dioxocyclams and *bis*-dioxocyclams using either optically active or racemic 1-N-Boc-azapenams **442** as key intermediates. The approach (Scheme 14.111) involves the photolysis of chromium alkoxycarbene complexes **444** with N-Boc-imidazolines **445** to give the expected azapenams **446** in a fully stereoselective way. Remarkably, changing the acidity of the hydrogenolysis medium from camphorsulfonic acid to triethylamine results in a dramatic difference in chemical behavior, leading in the first case to seven-membered hexahydrodiazepinones **447** and to fourteen-membered dioxocyclams **442** in the second. An extension of this methodology has been reported by the same author to prepare the five-atom system bridged [-O(CH_2)$_3$O-] **443** from *bis*-chromium alkoxycarbene complexes **448**.[135b]

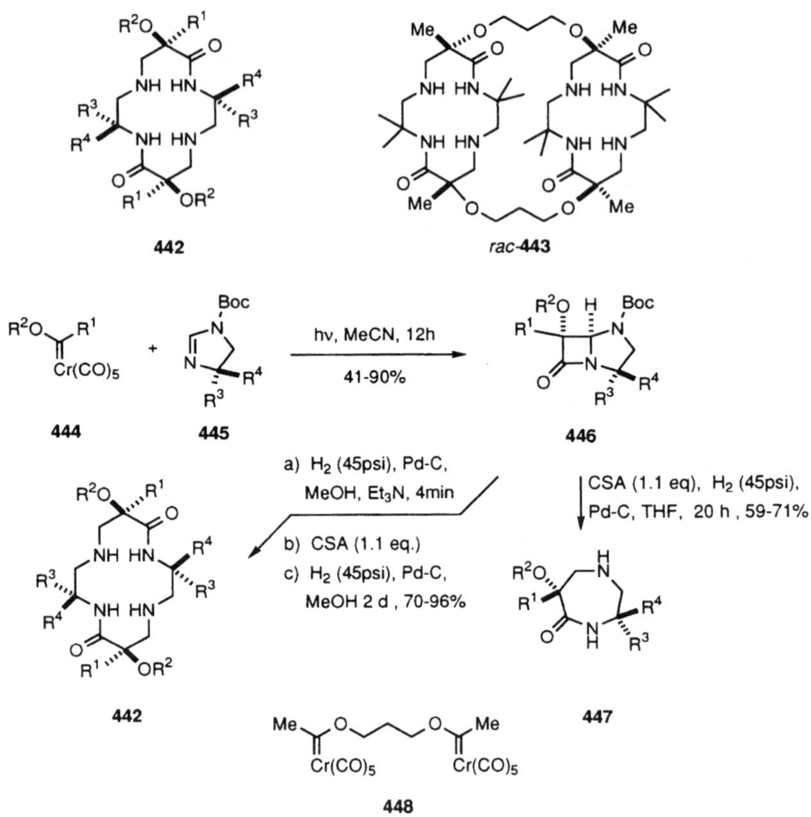

Scheme 14.111

14.6 Miscellaneous β-Amino Acids

This section deals with a few methods that have appeared in the literature describing the use of β-lactams as precursors of isotopically labeled (deuterated) and phosphonated β-amino acids.

β-Deuterated racemic β-amino amides of type **452** and **453** have been prepared[136] (Scheme 14.112), using a $LiAlD_4$-mediated reduction of the *cis*-4-amino-β-lactams **450** and **451** obtained by phenyl isocyanate cycloaddition to enamine **449**. The method showed itself to be very limited in scope owing to the low stereoselectivities achieved during the β-lactam formation and the deuteration steps.

Much more convenient in terms of stereocontrol was the approach of Basak[137] (Scheme 14.113), although limited to the preparation of α-unsubstituted β-deutero-β-alanine **458** and its enantiomer. The method is based on the use of (+)-cinchonine or (–)-cinchonine to prepare homochiral *trans*-3-trimethylsilyl-4-thiophenyl β-lactam **455** (or its enantiomer), which then reacts with tributyltin deuteride in a completely stereoselective fashion.

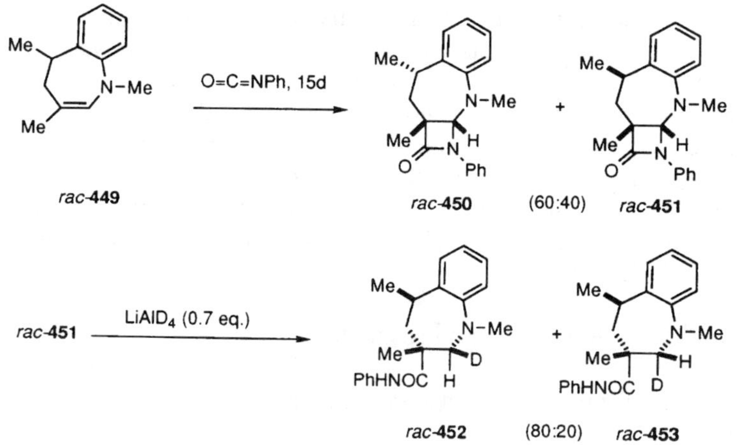

Scheme 14.112

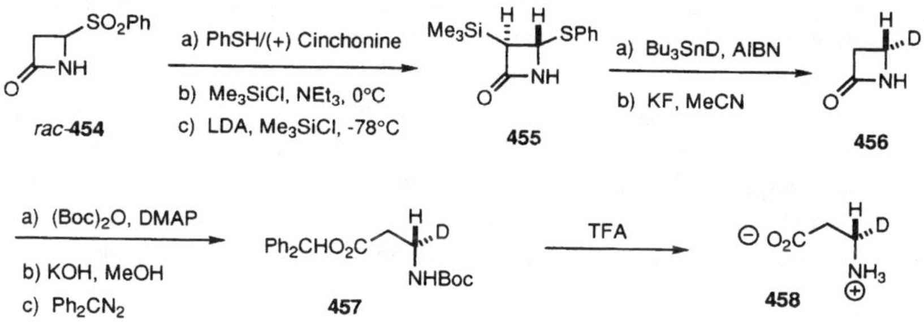

Scheme 14.113

Finally, Campbell and co-workers described (Scheme 14.114) a novel entry to racemic β-phosphonic and β-phosphinic β-amino acid analogues of aspartic acid[138] through 4-phosphono- and 4-phosphino β-lactams **460** by means of an Arbusov-like reaction of phosphites and phosphonites onto 4-acetoxyazetidin-2-one **459**. Ring opening of **460** in acidic medium afforded α-phosphono aspartic acid **461**, which on

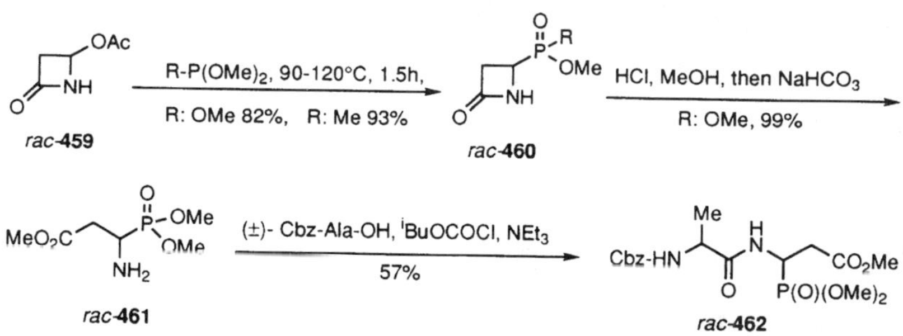

Scheme 14.114

reaction with α-amino acids by using the mixed anhydride coupling method led to dipeptides like **462**. Recently, this approach has been employed by Vo-Quang and coworkers[139] as an access to potential enzyme inhibitors.

In general, the former approach should also provide other kind of β-substituted β-amino acids, owing to the easy displacement of the acetoxy group by means of a variety of nucleophiles such as silylenol ethers, allylstannanes, silanes, Grignard reagents, cuprates, and so on.[140] Therefore, the development of these reactions in an enantioselective fashion would provide a novel entry to β-amino acids and derivatives.

14.7 Concluding Remarks and Prospects

It is evident from the results presented in this chapter that appropriately substituted β-lactam frameworks constitute effective tools for the construction of a wide variety of structurally different products. Within this context, the β-lactam coupling reactions with α-amino acids and the β-lactam ring opening with carbon nucleophiles, although at their initial stages, will probably find further applications for the development of new bioactive targets and heterocyclic compounds of biological interest. In some way, owing to the reliable and straightforward methods now available for β-lactam ring opening under racemization-free conditions, the synthesis of many enantiomerically pure β-amino acid derivatives may be considered a routine task, whenever the adequate β-lactams could be prepared. Therefore, it is reasonable to expect in the near future research to procure asymmetric azetidin-2-ones bearing any substitution pattern by short and direct approaches, particularly in a single synthetic step.

14.8 Acknowledgments

We thank the following organizations for financial support of our research in β-lactam chemistry: Ministerio de Educación y Ciencia (Spanish Government); Eusko Jaurlaritza (Basque Government), Euskal Herriko Unibertsitatea—Universidad del Pais Vasco (Basque Country University). Special thanks to Professor M. Oiarbide for helpful discussions of the manuscript. All collaborators cited within the references are also acknowledged.

14.9 Abbreviations

Ac	Acetyl
aq	Aqueous
AIBN	2,2′-Azobisisobutyronitrile
Ar	Aryl
Bn	Benzyl
Boc	*tert*-Butoxycarbonyl
BOM	Benzyloxymethyl

BOP-Cl	N,N-bis(2-Oxo-3-oxazolidinyl)phosphinic chloride
Bu	Butyl
t-Bu	*tert*-Butyl
CAN	Ceric(IV) ammonium nitrate
Cbz	Benzyloxycarbonyl
CSA	Camphorsulfonic acid
d	Day
DAM	Di-(*p*-anisyl)methyl
DBU	1,8-diazabicyclo[5.4.0]undec-7-ene
DCC	Dicyclohexylcarbodiimide
DEAD	Diethyl azodicarboxylate
DIBAL	Diisobutylaluminum hydride
DMAP	4-Dimethylaminopyridine
DMB	2,4-Dimethoxybenzyl
DMF	N,N-Dimethylformamide
DMS	Dimethyl sulfide
DMSO	Dimethylsulfoxide
EE	Ethoxyethyl
equiv.	Equivalent
Et	Ethyl
EtOAc	Ethyl acetate
Fmoc	Fluorenylmethoxycarbonyl
HMPA	Hexamethylphosphoric triamide
HOBT	1-Hydroxybenzotriazole
HPLC	High-performance liquid chromatography
Im	Imidazolyl
LDA	Lithium diisopropylamide
LiHMDS	Lithium hexamethyldisilylamide
MCPBA	*m*-Chloroperbenzoic acid
Me	Methyl
Ms	Methanesufonyl (Mesyl)
MsOH	Methanesulfonic acid
NMM	N-Methylmorpholine
MoOPH	Oxodiperoxymolybdenum(pyridine)hexamethylphosphoramide
MTPA	Methoxytrifluoromethylphenylacetic acid (Mosher acid)
NCS	N-Chlorosuccinimide
Ph	Phenyl
Piv	Pivaloyl (trimethylacetyl)
PMB	*p*-Methoxybencyl
PMP	*p*-Methoxyphenyl
PNB	*p*-Nitrobenzyl
i-Pr	Isopropyl
Py	Pyridyl or pyridine
TBAF	Tetrabutylammonium fluoride
TBDPS	*tert*-Butyldiphenylsilyl
TBS	*tert*-Butyldimethylsilyl
TEA	Triethylamine
TEMPO	2,2,6,6-Tetramethyl-1-piperidinyloxy, free radical

TES	Triethylsilyl
TFA	Trifluoroacetic acid
THF	Tetrahydrofuran
TMS	Trimethylsilyl
Troc	Trichloroethoxycarbonyl

14.10 References

1. O. W. Griffith, *Ann. Rev. Biochem.* **55**, 855 (1986).

2. (a) C. N. C. Drey, in *The Chemistry and Biochemistry of the Amino Acids*, B. Weinstein, Ed. (Dekker, New York, 1976), Vol. 4, p. 241. (b) *Amino Acids, Peptides and Proteins. Specialist Periodical Reports, Chem. Soc.*; London, 1971, *3*, 1. (c) I. Wagner, H. Musso, *Angew. Chem. Int. Ed. Engl.* **22**, 816 (1983). (d) *Chemistry and Biochemistry of the Amino Acids*, G. C. Barrett, Ed. (Chapman and Hall, New York, 1985).

3. For some reviews, see: (a) *Chemistry and Biology of β-Lactam Antibiotics*; R. B. Morin, M. Gorman, Eds. (Academic, New York, 1982), Vols. 1–3. (b) *Recent Advances in the Chemistry of β-Lactam Antibiotics*, A. G. Brown, S. M. Roberts, Eds. (The Royal Society of Chemistry, Burlington House: London, 1984). (c) W. Dürckheimer, J. Blumbach, R. Latrell, K. H. Sheunemann, *Angew. Chem. Int. Ed. Engl.* **24**, 180 (1985). (d) *The Chemistry of β-Lactams*, M. I. Page, Ed. (Chapman and Hall, London, 1992). (e) R. Southgate, C. Branch, S. Coulton, E. Hunt, in *Recent Progress in the Chemical Synthesis of Antibiotics and Related Microbial Products*, G. Lukacs, Ed. (Springer-Verlag, Berlin, 1993), Vol. 2, p. 621.

4. (a) M. S. Manhas, S. G. Amin, A. K. Bose, *Heterocycles* **5**, 669 (1976). (b) M. S. Manhas, D. R. Wagle, J. Chiang, A. K. Bose, *Heterocycles* **27**, 1755 (1988).

5. I. Ojima, in *The Organic Chemistry of β-Lactams*, G. I. Georg, Ed. (VCH, New York, 1992), Chapter 4, p. 197. For another β-lactam route to α-amino acids, see: P.-J. Coleon, L. S. Hegedus, *J. Org. Chem.* **58**, 5918 (1993).

6. H. H. Wasserman, *Aldrichimica Acta* **20**, 63 (1987).

7. M. Hesse, in *Ring Enlargement in Organic Chemistry* (VCH, Weinheim, 1991).

8. (a) E. Juaristi, D. Quintana, J. Escalante, *Aldrichimica Acta* **27**, 3 (1994). (b) D. C. Cole, *Tetrahedron* **50**, 9517 (1994).

9. (a) R. Labia, C. Morin, *J. Antibiotics* **37**, 1103 (1984). (b) A. G. M. Barret, M. A. Sturgess, *Tetrahedron* **44**, 561 (1988). (c) J. Backes, in *Houben-Weyl, Methoden der Organischen Chemie*, E. Muller, O. Bayer, Eds. (Band E16B. Thieme, Stuttgart, 1991), p. 31. (d) L. Ghosez, J. Marchand-Brynaert, in *Comprehensive Organic Synthesis*, Vol. 5, B. M. Trost, I. Fleming, Eds. (Pergamon, Oxford, 1991), p. 85. (e) F. Van der Steen, G. Van Koten, *Tetrahedron* **47**, 7503 (1991). For reactions of natural and synthetic β-lactams, see: (f) A. K. Mukerjee, A. K. Singh, *Synthesis* 547 (1975). *Idem.*, *Tetrahedron* **34**, 1731 (1978). (g) H. Wild, in *The Organic Chemistry of β-Lactams*, G. I. Georg, Ed. (VCH, New York, 1992), p. 44.

10. (a) D. J. Hart, D. C. Ha, *Chem. Rev.* **89**, 1447 (1989). (b) M. J. Brown, *Heterocycles* **29**, 2225 (1989). (c) G. I. Georg, in *Studies in Natural Product Chemistry*, A. Ur Rahman, Ed. (Elsevier, Amsterdam, 1989), Vol. 4, p. 431.

11. (a) R. D. G. Cooper, B. W. Daugherty, D. B. Boyd, *Pure Appl. Chem.* **59**, 485 (1987). (b) R. C. Thomas, in *Recent Progress in the Chemical Synthesis of Antibiotics*, G. Lukacs, M. Ohno, Eds. (Springer-Verlag, Berlin, 1990), Vol. 1, p. 533. (c) G. I. Georg, V. T. Ravikumar, in *The Organic Chemistry of β-Lactams*, G. I. Georg, Ed. (VCH, New York, 1992), p. 295. (d) S. Dumas, L. S. Hegedus, *J. Org. Chem.* **59**, 4967 (1994).

12. M. J. Miller, *Acc. Chem. Res.* **19**, 49 (1986).

13. R. J. Ternansky, J. M. Morin, Jr., in *The Organic Chemistry of β-Lactams*, G. I. Georg, Ed. (VCH, New York, 1992), p. 257.

14. G. M. Coppola, H. F. Schuster, in *Asymmetric Synthesis, Construction of Chiral Molecules Using Amino Acids* (John Wiley, New York, 1987), p. 204.

15. (a) J. E. Baldwin, R. M. Adlington, D. J. Birch, *J. Chem. Soc; Chem. Commun.* 256 (1985). (b) J. E. Baldwin, R. M. Adlington, D. W. Gollins, C. J. Schofield, *Tetrahedron* **46**, 4733 (1990).

16. (a) J. E. Baldwin, R. M. Adlington, D. W. Gollins, C. R. A. Godfrey, *Tetrahedron* **51**, 5169 (1995). For related work, see: (b) F. Cavagna, A. Linkies, H. Pietsch, D. Reuschling, *Angew. Chem. Int. Ed. Engl.* **19**, 129 (1980). (c) D. R. Wagle, M. G. Monteleone, L. Krishnan, M. S. Manhas, A. K. Bose, *J. Chem. Soc; Chem. Commun.* 915 (1989). See also Ref. 4(b).

17. (a) J. C. Watkins, H. J. Olverman, *Trends Neurosci* **10**, 265 (1987). (b) J. C. Watkins, P. Krogsgaard-Larsen, T. Honoré, *Trends Pharmacol. Sci.* **11**, 25 (1990).

18. (a) C. Gallina, V. Koch, A. Romeo, *Tetrahedron Lett.* 3055 (1969). (b) C. Gallina, C. Marta, C. Colombo, A. Romeo, *Tetrahedron* **27**, 4681 (1971).

19. S. Hanessian, K. Sumi, B. Vanasse, *Synlett.* 33 (1992).

20. (a) J. E. Baldwin, M. G. Moloney, M. North, *Tetrahedron* **45**, 6309 (1989). See also: (b) R. Cotton, A. N. C. Johnstone, M. North, *Tetrahedron* **51**, 8525 (1995).

21. J. E. Baldwin, M. G. Moloney, M. North, *Tetrahedron* **45**, 6319 (1989).

22. P. J. Reider, E. J. J. Grabowskii, *Tetrahedron Lett.* 2293 (1982).

23. T. N. Salzmann, R. W. Ratcliffe, B. G. Christensen, F. A. Bouffard, *J. Am. Chem. Soc.* **102**, 6161 (1980).

24. (a) E. J. Thomas, A. C. Williams, *J. Chem. Soc. Chem. Commun.* 992 (1987). (b) E. J. Thomas, A. C. Williams, *J. Chem. Soc. Perkin Trans 1*, 351 (1995). (c) J. M. Roe, E. J. Thomas, *Synlett* 727 (1990).

25. A. S. Kende, K. Koch, G. Dorey, I. Kaldor, K. Liu, *J. Am. Chem. Soc.* **115**, 9842 (1993). *Idem. J. Am. Chem. Soc.* **117**, 8258 (1995).

26. C. Palomo, J. M. Aizpurua, J. J. Gracenea (unpublished results).

27. D. J. Hart, C.-S. Lee, *J. Am. Chem. Soc.* **108**, 6054 (1986).

28. (a) D. J. Hart, D. C. Ha, *Tetrahedron Lett.* **28**, 4489 (1987). (b) J. C. Galluci, D. C. Ha, D. J. Hart, *Tetrahedron* **45**, 1283 (1989). For another β-lactam route to 2,3-dideoxy-3-aminopentoses and 2,3-dideoxy-3-C-methyl-3-aminopentoses, see: (c) F. M. Hauser, S. R. Ellenberger, R. P. Rhee, *J. Org. Chem.* **52**, 5041 (1987).

29. (a) M. Shiozaki, N. Ishida, *Chem. Lett.* 1403 (1987). (b) M. Shiozaki, N. Ishida, S. Sato, *Bull. Chem. Soc. Jpn.* **62**, 3950 (1989).

30. (a) C. Palomo, J. M. Ontoria, J. M. Odriozola, J. M. Aizpurua, I. Ganboa, *J. Chem. Soc.; Chem. Commun.* 248 (1990). (b) (a) C. Palomo, J. M. Aizpurua, R. Urchegui, M. Iturburu, *J. Org. Chem.* **57**, 1571 (1992).

31. (a) C. Palomo, J. M. Aizpurua, M. Iturburu, R. Urchegui, *J. Org. Chem.* **59**, 240 (1994). (b) I. Fleming, N. D. Kindon, *J. Chem. Soc. Perkin Trans 1* 303 (1995).

32. C. Palomo, J. M. Aizpurua, J. M. Ontoria, M. Iturburu, *Tetrahedron Lett.* **33**, 4823 (1992).

33. (a) E. K. Rowinsky, L. A. Casenave, R. C. Donehower, *J. Natl. Cancer Inst.* **82**, 1247 (1990). (b) R. K.-Y. Zee- Cheng, C. C. Cheng, *Drugs of the Future* **11**, 45 (1986). For recent reviews of the chemistry of Taxol, see: (c) D. G. I. Kingston, *Pharm. Ther. 52*, 1 (1991). (d) M. Suffiness, G. A.

Cordell, *The Alkaloids. Chemistry and Pharmacology*, in *The Alkaloids*, A. Brossi, Ed. (Academic, New York, 1985), Vol. 25, p. 3. (e) K. C. Nicolau, W.-M. Dai, R. K. Guy, *Angew. Chem. Int. Ed. Engl.* **33**, 15 (1994). (f) A. N. Boa, P. R. Jenkins, N. J. Lawrence, *Contemporary Organic Synthesis* **1**, 47 (1994).

34. (a) H. Umezawa, T. Aoyagi, H. Suda, M. Hamada, T. Takeuchi, *J. Antibiot.* **29**, 97 (1976). (b) H. Umezawa, in *Small Molecular Immunomodifiers of Microbial Origin. Fundamental and Clinical Studies of Bestatin*, H. Umezawa, Ed. (Pergamon, Oxford, 1981). (c) H. Umezawa, *Drug Exptl. Clin. Res.* **10**, 519 (1984). (d) K. Miura, T. Sawa, T. Takeuchi, H. Umezawa, *J. Antibiot.* **34**, 734 (1986).

35. H. Blomgren, J. Wasserman, *Canc. Lett.* **11**, 303 (1981).

36. (a) C. Palomo, A. Arrieta, F. P. Cossio, J. M. Aizpurua, A. Mielgo, N. Aurrekoetxea, *Tetrahedron Lett.* **31**, 6429 (1990). (b) R. A. Holton, *Eur. Pat. Appl.* EP400,971, 1990; *Chem. Abstr.* **114**, 164568q (1990).

37. A. K. Bose, B. K. Banik, M. S. Manhas, *Tetrahedron Lett.* **36**, 213 (1995).

38. M. Endo, R. Droghini, *Bio. Med. Chem. Lett.* **3**, 2483 (1993).

39. R. Brieva, J. Z. Crich, C. J. Sih, *J. Org. Chem.* **58**, 1068 (1993).

40. B. K. Banik, M. S. Manhas, A. K. Bose, *J. Org. Chem.* **59**, 4714 (1994).

41. G. I. Georg, E. Akün, P. M. Mashava, M. Milstead, H. Ping, Z.-J. Wu, D. Vander Velde, *Tetrahedron Lett.* **33**, 2111 (1992).

42. (a) V. Farina, S. I. Hauck, D. G. Walker, *Synlett* 761 (1992). (b) S. M. Tenneson, B. Bellau, *Can. J. Chem.* **58**, 1605 (1980). (c) A. K. Bose, M. S. Manhas, J. M. Van der Veen, S. S. Bari, D. R. Wagle, *Tetrahedron* **48**, 4831 (1992).

43. J. D. Bouzart, A. Commerçon, *Tetrahedron Lett.* **34**, 6049 (1993).

44. G. I. Georg, Z. Wu, *Tetrahedron Lett.* **35**, 381 (1994).

45. (a) D. A. Evans, E. B. Sjögren, *Tetrahedron Lett.* **26**, 3783 (1985). *Idem, ibid.* **26**, 3787 (1985). (b) C. Palomo, J. M. Aizpurua, J. I. Miranda, A. Mielgo, J. M. Odriozola, *Tetrahedron Lett.* **34**, 6325 (1993).

46. R. A. Holton, J. H. Liu, *Bio. Med. Chem. Lett.* **3**, 2475 (1993).

47. (a) I. Ojima, C. M. Sun, M. Zucco, Y. H. Park, O. Duclos, S. Kiuduk, *Tetrahedron Lett.* **34**, 4149 (1993). (b) I . Ojima, I. Habus, M. Zhao, M. Zucco, Y. H. Park, C. M. Sun, T. Brigaud, *Tetrahedron* **48**, 6985 (1992).

48. (a) I. Ojima, M. Zucco, O. Duclos, S. D. Kuduk, C. M. Sun, Y. H. Park, *Bio. Med. Chem. Lett.* **3**, 2479 (1993). (b) G. I. Georg, T. C. Boge, Z. S. Cheruvallath, G. C. B. Harriman, M. Hepperle, H. Park, R. H. Himes, *Bio. Med. Chem. Lett.* **4**, 1825 (1994). (c) D. Dasgupta, H. Park, G. C. B. Harriman, G. I. Georg, R. H. Himes, *J. Med. Chem.* **37**, 2976 (1994).

49. G. I. Georg, Z. S. Cheruvallath, R. H. Himes, M. R. Mejillano, C. T. Burke, *J. Med. Chem.* **35**, 4230 (1992).

50. (a) G. I. Georg, Z. S. Cheruvallath, G. C. B. Harriman, M. Hepperle, H. Park, *Bio. Med. Chem. Lett.* **3**, 2467 (1993). (b) G. I. Georg, G. C. B. Harriman, M. Hepperle, R. H. Himes, *Bio. Med. Chem. Lett.* **4**, 1381 (1994).

51. (a) I. Ojima, O. Duclos, M. Zucco, M.-C. Bissery, C. Combeau, P. Vrignaud, J. F. Riou, F. Lavelle, *J. Med. Chem.* **37**, 2602 (1994). (b) T. C. Boge, R. H. Himes, D. G. Vander Velde, G. I. Georg, *J. Med. Chem.* **37**, 3337 (1994).

52. G. Cainelli, M. Panunzio, D. Giacomini, B. DiSomone, R. Camerini, *Synthesis* 805 (1994).

53. C. Palomo, J. M. Aizpurua, I. Ganboa, E. Maneiero, B. Odriozola, *J. Chem. Soc. Chem. Commun.* 1505 (1994).

54. R. Annunziata, M. Benaglia, M. Cinquini, F. Cozzi, F. Ponzini, *J. Org. Chem.* **58**, 4746 (1993).

55. L. Li, S. A. Thomas, L. L. Klein, C. M. Yeung, C. J. Maring, D. J. Grampovnik, P. A. Lartey, J. J. Plattner, *J. Med. Chem.* **37**, 2655 (1994).

56. A. K. Bose, B. K. Banik, K. J. Barakat, M. S. Manhas, *Synlett* 575 (1993).

57. (a) D. R. Wagle, G. Garai, J. Chiang, M. G. Monteleone, B. E. Kurys, T. W. Strohmeyer, V. R. Hedge, M. S. Manhas, A. K. Bose, *J. Org. Chem.* **53**, 4227 (1988). (b) B. K. Banik, M. S. Manhas, Z. Kaluza, K. J. Barakat, A. K. Bose, *Tetrahedron Lett.* **33**, 3603 (1992). For the synthesis of enantiomeric β-lactams, see: (c) C. Hubschwerlen, G. Schmid, *Helv. Chim. Acta* **66**, 2206 (1983).

58. M. Jayaraman, A. R. A. S. Deshmukh, B. M. Bhawal, *J. Org. Chem.* **59**, 932 (1994).

59. C. Palomo, J. M. Aizpurua, I. Ganboa, F. Carreaux, C. Cuevas, E. Maneiro, J. M. Ontoria, *J. Org. Chem.* **59**, 3123 (1994).

60. (a) Y. Kobayashi, Y. Takemoto, Y. Ito, S. Terashima, *Tetrahedron Lett.* **31**, 3031 (1990). (b) T. Matsumoto, Y. Kobayashi, Y. Takemoto, H. Kamijo, Y. Harada, Y. Ito, S. Terashima, *Tetrahedron* **48**, 1853 (1992).

61. R. Annunziata, M. Benaglia, M. Cinquini, F. Cozzi, L. Raimondi, *Tetrahedron* **50**, 9471 (1990).

62. C. Palomo, J. M. Aizpurua, C. Cuevas, *J. Chem. Soc. Chem. Commun.* 1957 (1994).

63. I. Ojima, C. M. Sun, Y. H. Park, *J. Org. Chem.* **59**, 1249 (1994).

64. (a) H. L. Kornberg, J. G. Morris, *Biochem. J.* **95**, 577 (1965). (b) T. Ishiyama, T. Furuta, M. Takai, Y. Okimoto, S. Aizawa, A. Shiman, H. Yonehara, *J. Antibiot.* **23**, 821 (1975).

65. (a) T. Wieland, *Helv. Chim. Acta* **44**, 919 (1961). (b) W. A. Bulen, J. R. LeCompte, *Biochem. Biophys Res. Commun.* **9**, 523 (1962). (c) M. Teintze, M. B. Hossain, C. L. Barnes, J. Leong, D. Van der Helm, *Biochemistry* **20**, 446 (1981). (d) K. Sugawara, K.-I. Numata, M. Konishi, H. Kawaguchi, *J. Antibiotics* **37**, 958 (1984). (e) S. Matsunaga, N. Fusetani, *J. Org. Chem.* **60**, 1177 (1995). (f) J. F. Okonya, T. Kolasa, M. S. Miller, *J. Org. Chem.* **60**, 1932 (1995).

66. (a) T. G. Hansson, J. O. Kihlberg, *J. Org. Chem.* **51**, 4490 (1986). (b) R. Wagner, J. W. Tilley, *J. Org. Chem.* **55**, 6289 (1990). (c) R. Wagner, J. W. Tilley, K. Lovey, *Synthesis* 785 (1990).

67. J. P. Tam, M. W. Riemen, R. B. Merrifield, *Peptide Research* **1**, 6 (1988).

68. For other methods to prepared 3-alkoxy-4-carboxy β-lactams, see: (a) T. Kawabata, T. Minami, T. Hiyama, *J. Org. Chem.* **57**, 1864 (1992). (b) D. H. R. Barton, J. Cleophax, A. Gateau-Olersker, S. D. Gero, C. Tachdjian, *Tetrahedron* **49**, 8381 (1993).

69. C. Palomo, F. Cabre, J. M. Ontoria, *Tetrahedron Lett.* **33**, 4819 (1992).

70. (a) J. O'Sullivan, J. E. McCullough, A. A. Tymiak, D. R. Kirsch, W. H. Trejo, P. A. Principe, *J. Antibiot.* **41**, 1740 (1988). (b) T. Kato, H. Hinoo, Y. Tervi, J. Kikuchi, J. Shoji, *J. Antibiot.* **41**, 719 (1988). (c) T. Kato, *J. Antibiot.* **42**, C-2 (1989). (d) A. A. Tymiak, T. J. McCormick, S. E. Unger, . *Org. Chem.* **54**, 1149 (1989).

71. C. Palomo, J. M. Aizpurua, I. Ganboa, B. Odriozola, E. Maneiro, J. I. Miranda, R. Urchegui, *J. Chem. Soc. Chem. Commun.* 161 (1996).

72. (a) P. J. Dunn, R. Haner, H. Rapoport, *J. Org. Chem.* **55**, 5017 (1990). (b) Y. Nakamura, C. Shin, *Chem. Lett.* 49 (1992). (c) D. Seebach, A. Studer, E. Pfammatter, H. Widmer, *Helv. Chim. Acta* **77** 2035 (1994). (d) D. F. Rane, V. M. Girijavallabhan, A. K. Gauguly, R. E. Pike, A. K. Saksena, A. T. McPhail, *Tetrahedron Lett.* **34**, 3201 (1993). (e) W. Hartwig, L. Mittendorf, *Synthesis* 939 (1991). (f)

R. Herranz, S. Vinuesa, J. Castro-Pichel, C. Perez, M. T. Garcia Lopez, *J. Chem. Soc. Perkin Trans 1* 1825 (1992).

73. (a) M. Wang, S. J. Gould, *J. Org. Chem.* **58**, 5176 (1993). (b) S. Nomoto, T. Teshima, T. Wakamiya, T. Shiba, *Tetrahedron* **34** 921 (1978).

74. (a) Y. Umezawa, H. Morishima, S. Saito, T. Takita, H. Umezawa, S. Kobayashi, M. Otsuka, M. Narita, M. Ohno, *J. Am. Chem. Soc.* **102**, 6630 (1980). (b) H. Arai, W. K. Hagmann, H. Suguna, S. M. Hecht, *J. Am. Chem. Soc.* **102**, 6631 (1980).

75. (a) T. Takemoto, T. Nakajima, S. Arihara, K. Koike, *Yakugaku Zasshi* **95**, 326 (1975). (b) J. F. Flippen, R. D. Gilardi, *Acta Crystallogr. Sec. B.* **32**, 951 (1976).

76. J. H. Dwar, G. Shaw, *J. Chem. Soc.* 583 (1962).

77. I. Murakoshi, F. Ikegami, Y. Hinuma, Y. Hanma, *Phytochemistry* **23**, 1905 (1984).

78. (a) I. H. Gilbert, D. C. Rees, A. K. Crockett, R. C. F. J ones, *Tetrahedron* **51**, 6315 (1995). (b) R. C. F. Jones, A. K. Crockett, D. C. Rees, I. H. Gilbert, Tetrahedron Asymmetry **5**, 1661 (1994). (c) I. Gilbert, D. C. Rees, R. S. Richardson, *Tetrahedron Lett.* **32**, 2277 (1991). (d) R. C. F. Jones, G. J. Ward, *Tetrahedron Lett.* **29**, 3853 (1988). For a review, see: (e) R. M. J. Liskamp, *Recl. Trav. Chim. Pays Bas* **113**, 1 (1994).

79. J. E. Baldwin, R. M. Adlington, D. J. Birch, *J. Chem. Soc. Chem. Commun.* 256 (1985).

80. Y. Aoyagi, M. S. Chorghade, A. A. Padmapriya, H. Suguna, S. M. Hecht, *J. Org. Chem.* **55**, 6291 (1990).

81. I. Ojima, Y. Pei, *Tetrahedron Lett.* **31**, 977 (1990).

82. B. T. Golding, A. J. J. Smith, *J. Chem. Soc. Chem. Commun.* 702 (1980).

83. T. Fukuyama, R. K. Frank, C. F. Jewell, Jr., *J. Am. Chem. Soc.* **102**, 2122 (1980).

84. (a) C. Palomo, J. M. Aizpurua, C. Cuevas, A. Mielgo, R. Galarza, *Tetrahedron Lett.* **36**, 9027 (1995). (b) C. Palomo, J. M. Aizpurua, A. Mielgo, R. Galarza, *J. Chem. Soc. Chem. Commun.* **633** (1996).

85. (a) C. Palomo, F. P. Cossio, C. Cuevas, B. Lecea, A. Mielgo, P. Roman, A. Luque, M. Martinez-Ripoll, *J. Am. Chem. Soc.* **114**, 9360 (1992). (b) C. Palomo, F. P. Cossio, C. Cuevas, J. M. Ontoria, J. M. Odriozola, S. Munt, *Bull. Soc. Chim. Belg.* **101**, 541 (1992). (c) C. Palomo, J. M. Aizpurua, *Trends in Organic Chemistry* **4**, 637 (1993).

86. (a) H. Leuchs, *Ber. Dtsch. Chem. Ber. Ges.* **39**, 857 (1906). For a review, see: (b) H. R. Kricheldorf, *α-Amino Acid N-Carboxy-Anhydride and Related Heterocycles* (Springer-Verlag, Berlin, 1987).

87. C. Palomo, J. M. Aizpurua, F. Cabre, C. Cuevas, S. Munt, J. M. Odriozola, *Tetrahedron Lett.* **35**, 2725 (1994).

88. C. Palomo, J. M. Aizpurua, C. Cuevas, R. Urchegui, A. Linden, *J. Org. Chem.* **61**, 4400 (1996).

89. C. Palomo, J. M. Aizpurua, C. Cuevas, R. Urchegui, P. Roman, A. Lugue, M. Martinez-Ripoll, *Anales de Quimica Int. Ed.* **92**, 134 (1996).

90. B. Alcaide, Y. Martin-Cantalejo, J. Perez-Castells, J. Rodriguez-Lopez, M. A. Sierra, *J. Org. Chem.* **57**, 5921 (1992).

91. A. K. Bose, J. F. Womelsdorf, L. Krishnan, Z. Urbanczyk-Lipkowska, D. C. Shelly, M. S. Manhas, *Tetrahedron* **47**, 5379 (1991).

92. (a) H. Umezawa, T. Aoyagi, H. Morishima, M. Matsuzaki, M. Hamada, T. Takeuchi, *J. Antibiotics* **23**, 259 (1970).

93. R. J. Arrowsmith, K. Carter, J. G. Dann, D. E. Davies, C. J. Harris, J. A. Morton, P. Lister, J. A. Robinson, D. J. Williams, *J. Chem. Soc. Chem. Commun.* 755 (1986).

94. H. J. Schostarez, *J. Org. Chem.* **53**, 3628 (1988).

95. (a) S. Thaisrivongs, D. T. Pals, W. M. Kati, S. R. Turner, L. M. Thomasco, W. Watt, *J. Med. Chem.* **29**, 2080 (1986). (b) S. Thaisrivongs, H. J. Schostarez, D. J. Pal, S. R. Turner, *J. Med. Chem.* **30**, 1837 (1987).

96. C. Palomo, F. P. Cossio, C. Cuevas, J. M. Odriozola, J. M. Ontoria, *Tetrahedron Lett.* **33**, 4827 (1992).

97. For a review on α-unsubstituted β-lactams, see: A. K. Bose, M. S. Manhas, A. Mathur, D. R. Wagle, in *Recent Progress in the Chemical Synthesis of Antibiotics and Related Microbial Products*, Vol. 2, G. Lukacs, Ed. (Springer-Verlag, Berlin, 1993), p. 551.

98. C. Palomo, F. P. Cossio, J. M. Odriozola, M. Oiarbide, J. M. Ontoria, *J. Org. Chem.* **56**, 4418 (1991).

99. J. E. Baldwin, G. P. Lynch, J. Pitlik, *J. Antibiotics* **44**, 1 (1991).

100. N. J. Ede, I. O. Rae, M. T. W. Hearn, *Tetrahedron Lett.* **31**, 6071 (1990). *Idem*, *Aust. J. Chem.* **44**, 891 (1991).

101. Presented at the IXth European Symposium on Organic Chemistry, Warszawa, Poland, June 18–23, 1995. C. Palomo, I. Ganboa, J. M. Aizpurua, Diastereoselective Construction of the 1,2-Aminoalcohol, 1,2-Diamine and 1,2-Diol Subunits via [2+2] Cycloaddition Reactions.

102. For recent reviews on enzyme inhibitors, see: (a) J. R. Luly, A. K. L. Fung, J. J. Pattner, P. Marcotte, N. Ba-Maung, J. L. Soderquist, H. Stein, in *Peptides: Chemistry and Biology*, G. R. Marshall, Ed. (ESCOM Science Publishers, Leiden, The Netherlands, 1988), p. 487. (b) A. Giannis, T. Kolter, *Angew. Chem. Int. Ed. Engl.* **32**, 1244 (1993). (c) D. Dorsch, P. Raddatz, C. J. Schmitges, K. Van der Helm, F. Rippmann, *Kontakte* **2**, 48 (1993). (d) J. Gante, *Angew. Chem. Int. Ed. Engl.* **33**, 1699 (1994).

103. M. S. Manhas, D. R. Wagle, C. Garai, T. Strohmeyer, J. Chiang, A. K. Bose, Abstracts of papers 193rd ACS National Meeting, Denver, Colorado, April, 1987, ORGN 215. (b) M. S. Manhas, D. R. Wagle, A. K. Bose, paper presented at the 15th IUPAC International Symposium on the Chemistry of Natural Products, The Hague, The Netherlands, August, 1986. See also Ref. 4.

104. A. K. Bose, C. Mathur, D. R. Wagle, R. Nagui, M. S. Manhas, *Heterocycles* **39**, 491 (1994).

105. B. K. Banik, M. S. Manhas, A. K. Bose, *J. Org. Chem.* **58**, 307 (1993).

106. For two closely related intramolecular β-lactam lactonizations, see: (a) B. Alcaide, A. Gomez, J. Plumet, J. Rodriguez-Lopez, *Tetrahedron* **45**, 2751 (1989). (b) F. Farouz, M. J. Miller, *Tetrahedron Lett.* **32**, 3305 (1991).

107. A. Koichi, J. C. O'Toole, J. T. Welch, *Bio. Med. Chem. Lett.* **3**, 2457 (1993).

108. For other routes to fluorinated azetidin-2-ones, see: (a) J. T. Welch, K. Araki, R. Kawecki, J. A. Wichtowski, *J. Org. Chem.* **58**, 2454 (1993). (b) R. Kawecki, J. T. Welch, *Tetrahedron Lett.* **34**, 3087 (1993). (c) A. Abouabdellah, J. T. Welch, *Tetrahedron Asymmetry* **5**, 1005 (1994). For a review, see: (d) O. A. Mascaretti, C. E. Boscheti, G. O. Danelon, in *Organofluorine Compounds in Medicinal Chemistry and Biomedical Applications*, R. Filler, Ed. (Elsevier, Amsterdam, 1993).

109. (a) T. Taguchi, O. Kitagawa, Y. Suda, S. Ohkawa, A. Hashimoto, Y. Iitaka, Y. Kobayashi, *Tetrahedron Lett.* **29**, 5291 (1988). See also: (b) J. E. Baldwin, G. P. Lynch, C. J. Schofreld, *J. Chem. Soc. Chem. Commun.* 736 (1991).

110. C. Palomo, J. M. Aizpurua, R. Urchegui, J. M. Garcia, *J. Org. Chem.* **58**, 1646 (1993).

111. C. Palomo, J. M. Aizpurua, F. Cabré, J. M. García, J. M. Odriozola, *Tetrahedron Lett.* **35**, 2721 (1994).

112. C. Palomo, F. P. Cossio, J. M. Ontoria, J. M. Odriozola, *Tetrahedron Lett.* **32**, 3105 (1991).

113. S. D. Broady, J. E. Rexhausen, E. J. Thomas, *J. Chem. Soc. Chem. Commun.* 708 (1991).

114. R. P. Robinson, K. M. Donahue, *J. Org. Chem.* **57**, 7309 (1992).

115. C. Palomo, J. M. Aizpurua, J. M. García, M. Iturburu, J. M. Odriozola, *J. Org. Chem.* **59**, 5184 (1994).

116. (a) H. D. Kleinert, S. H. Rosenberg, W. R. Baker, H. H. Stein, V. Klinghofer, J. Barlow, K. Spina, J. Polakowski, P. Kovar, J. Cohen, J. Denissen, *Science* **257**, 1940 (1992). (b) W. R. Baker, S. L. Condon, *J. Org. Chem.* **58**, 3277 (1993).

117. D. M. Spero, S. Kapadia, V. Farina, *Tetrahedron Lett.* **36**, 4543 (1995).

118. J. R. Luly, C.-N. Hsiao, N. BaMaung, J. J. Plattner, *J. Org. Chem.* **53**, 6109 (1988).

119. (a) C. M. Sun, E. W. Ng, I. Ojima, Novel Coupling Methods for the Synthesis of Peptide Mimetics Bearing Norstatine Residue and Their Analogues, 208th American Chemical Society National Meeting, Washington, D.C.; August 21–25, 1994; Abstracts ORGN 314. *Tetrahedron Lett.* **36**, 4547 (1994).

120. (a) V. Dähn, H. Hagenmaier, H. Höhne, W. A. König, G. Wolf, H. Zähner, *Arch. Microbiol.* **107**, 249 (1976). (b) K. Kobinata, M. Uramoto, M. Nishii, H. Kusakabe, G. Nokamura, K. Isono, *Agric. Biol. Chem.* **44**, 1709 (1980). (c) K. Isono, S. Suzuki, *Heterocycles* **13**, 333 (1979).

121. F. P. Cossio, M. C. Lopez, C. Palomo, *Tetrahedron* **43**, 3963 (1987).

122. J. E. Baldwin, R. M. Adlington, C. R. A. Godfrey, D. W. Collins, M. L. Smith, A. T. Russell, *Synlett* 51 (1993).

123. J. E. Baldwin, R. M. Adlington, A. T. Rusell, M. L. Smith, *Tetrahedron* **51**, 4733 (1995).

124. J. E. Baldwin, R. M. Adlington, C. R. A. Godfrey, D. W. Gollins, J. G. Vaughan, *Chem. Soc. Chem. Commun.* 1434 (1993).

125. J. E. Baldwin, H. Rudolph, *Tetrahedron Lett.* **35**, 6163 (1994).

126. E. G. McGeer, *Kainic Acid as a Tool in Neurobiology*, J. W. Olney, D. L. McGeer, Eds. (Raven Press, New York, 1978).

127. (a) K. Hashimoto, M. Horikawa, H. Shirahama, *Tetrahedron Lett.* **31**, 7047 (1990). (b) K. Hashimoto, H. Shirahama, *Tetrahedron Lett.* **32**, 2625 (1991).

128. M. Kahn, *Synlett* 821 (1993).

129. (a) H. U. Saragovi, D. Fitzpatrick, H. Raktabur, M. Nakanishi, M. Kahn, M. I. Greene, *Science* **253**, 792 (1991). (b) W. V. Williams, T. Kieber-Emmons, J. VonFeldt, M. I. Greene, D. B. Weiner, *J. Biol. Chem.* **266**, 5182 (1991).

130. M. Kahn, S. Bertenshaw, *Tetrahedron Lett.* **30**, 2317 (1989).

131. B. Gardner, H. Nakanishi, M. Kahn, *Tetrahedron* **49**, 3433 (1993).

132. D. L. Boger, J. B. Myers, Jr., *J. Org. Chem.* **56**, 5385 (1991).

133. (a) M. J. Begley, L. Crombie, D. Haigh, R. C. F. Jones, S. Osborne, R. A. B. Webster, *J. Chem. Soc. Chem. Commun.* 2027 (1993). (b) L. Crombie, D. Haigh, R. C. F. Jones, R. Mat-Zin, *Ibid.* 2047 (1993). (c) *Idem., Ibid.* 2055 (1993).

134. E. De Clerq, N. Yamamoto, R. Pauwels, M. Baba, D. Schols, H. Nakashima, J. Balzarini, Z. Debyser, B. A. Murrer, D. Schwartz, D. Thornton, G. Bridger, S. Fricker, G. Henson, M. Abrams, D. Picker, *Proc. Natl. Acad. Sci. U.S.A.* **89**, 5286 (1992).

135. (a) C. Betschart, L. S. Hegedus, *J. Am. Chem. Soc.* **114**, 5010 (1992). (b) S. Dumas, E. Lastra, L. S. Hegedus, *J. Am. Chem. Soc.* **117**, 3368 (1995).

136. V. L. Eifler-Lima, P. Uriac, J. Huet, S. Sinbandhit, L. Toupet, *J. Chem. Res. (S)* 300 (1994).

137. A. Basak, *Synth. Commun.* **23**, 1985 (1993).

138. M. M. Campbell, N. I. Carruthers, S. J. Mickel, *Tetrahedron* **38**, 2513 (1982).

139. A. M. Chollet-Gravey, L. Vo-Quang, Y. Vo-Quang, F. Le Goffic, *Synth. Commun.* **21**, 1847 (1991).

140. For comprehensive reviews, see: (a) D. E. Davies, R. C. Storr, in *Comprehensive Heterocyclic Chemistry*, W. Lwowski, Ed. (Pergamon, New York, 1984), Vol. 7, p. 237. (b) S. Mickel, *Aldrichimica Acta* **18**, 95 (1985). (c) G. A. Veinberg, E. Lukevics, *Heterocycles* **38**, 2309 (1994).

CHAPTER

15

Enantioselective Synthesis of β-Amino Acids Using the Nicholas Reaction

Peter A. Jacobi and Wanjun Zheng

15.1 Introduction

The Nicholas reaction was discovered in 1977 and provides an efficient means for effecting the nucleophilic displacement of propargyl alcohols and related materials.[1] This transformation takes advantage of the fact that cobalt complexes of general structure **2** greatly facilitate the heterolytic cleavage of adjacent alcohols or ethers, which upon HBF_4[1b] or Lewis acid[1e] catalysis afford cobalt-stabilized carbocations of type **3** (Scheme 15.1a). Capture of these carbocations with nucleophiles such as electron-rich aromatics,[1a] ketone or ester enolates,[1d,e] and enol ethers[1c] then affords the products of nucleophilic displacement **4**, which may conveniently cleaved to the parent acetylenes **5** under a variety of mild oxidative conditions.[1c] In contrast to more traditional methodology involving direct displacement of propargyl tosylates and

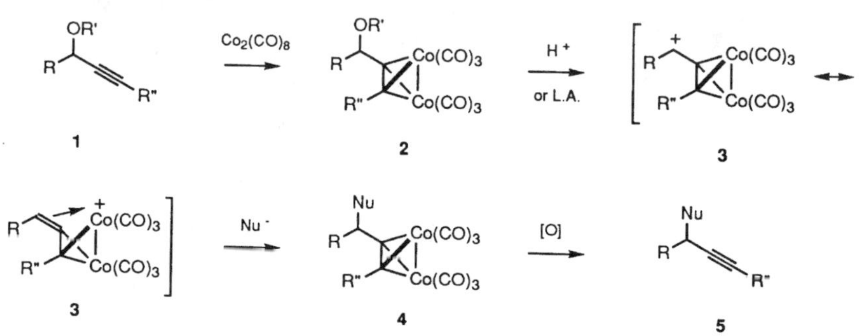

Scheme 15.1a

halides,[1f] this approach avoids complications arising from the formation of allenic, elimination, and other byproducts. Since propargyl alcohols are readily derived by addition of acetylides to carbonyl compounds, the overall transformation constitutes a flexible carbonyl-to-geminal dialkyl transposition.

Among other uses, the Nicholas reaction has been widely employed in cyclization strategies leading to fused-ring cyclopentenones.[1g] For example, in combination with the Pauson–Khand protocol, this methodology provides a novel means for the synthesis of complex ring systems of type **8** related to epoxydictymene (Scheme 15.1b).[1c] In 1987, Schreiber et al. considerably expanded the scope of the Nicholas reaction with their finding that alkylations can be carried out with both excellent diastereo- *and* enantioselectivity with homochiral nucleophiles.[1h] Thus alkylation of Evans' enolate **9** with cobalt complex (±)-**10** resulted in an 80% yield of a 12:1 mixture of *syn*-adduct **11s** together with the corresponding *anti*-isomer **11a** (Scheme 15.2). This selectivity was elegantly rationalized by postulating a novel double stereodifferentiating process, in which cations **12** and **13** interconvert via enantiomerization at a rate that is fast relative to alkylation (kinetic resolution). Reaction of **9** with the "matched" *syn*-cation **13** would then be expected to occur at a faster rate than with the corresponding "mismatched" pair **9**–**12**.[1h] In principle, this observation provides the

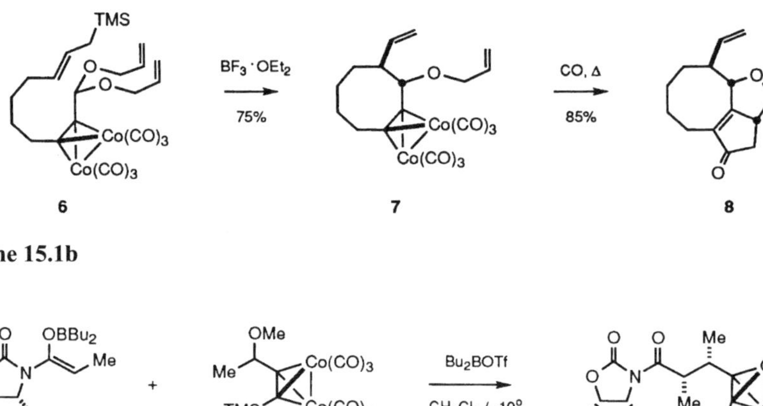

Scheme 15.1b

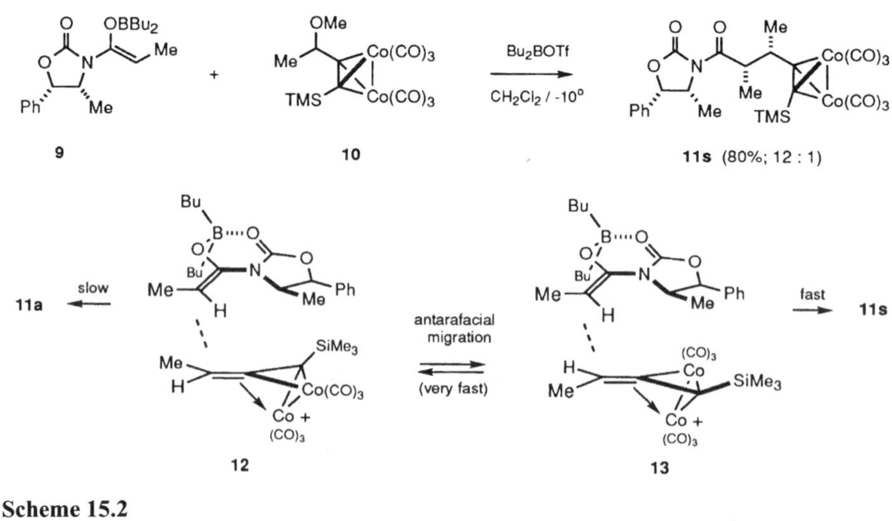

Scheme 15.2

basis for a general synthesis of homochiral acetylenic acid derivatives that might find widespread utility in organic synthesis. However, until the present studies little was known about the effect of chiral substituents on such transformations.[2a,b]

We have been interested in the possibility that acetylenic acid derivatives of general structure **17** might be prepared in homochiral form by extension of the Nicholas–Evans–Schreiber methodology (Scheme 15.3). By way of summary, it appeared likely that condensation of chiral aldehydes **14** with acetylides **15** would afford propargyl alcohol derivatives **16**, which upon Nicholas reaction with an appropriate chiral enolate, followed by decomplexation and hydrolysis, should provide **17** with complete control over both relative and absolute stereochemistry (*).

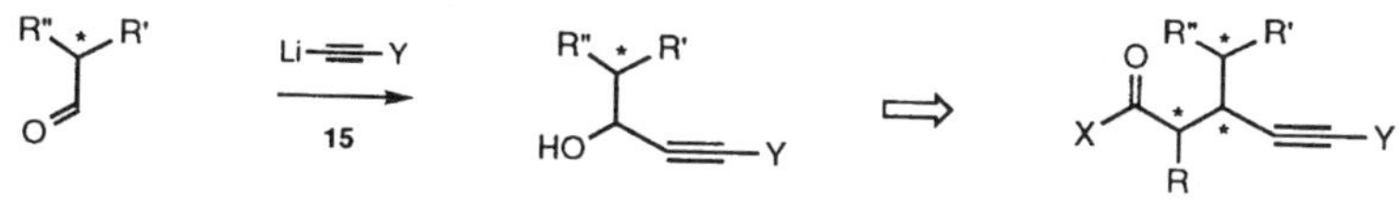

14 15 16 17

Scheme 15.3

Chiral fragments related to **17** are found widely distributed in nature, occurring in some compounds as nearly intact structural units, and in others in easily derivable form. An example of the first case is provided by acetylenic amides of type **18**, which are attractive intermediates for the synthesis of biologically important tetrapyrroles such as phytochrome (**19**), phycocyanin (**20**), and phycoerythrin (**21**) (Scheme 15.4).[2c–h] An example of the second case is afforded by acetylenic acids of general structure **25**, which should be readily available by condensation of cobalt complexes **23** with the Evans enolate **22** (Scheme 13.5). In principle, acids of type **25** might be efficiently employed in the synthesis of β-lactam antibiotics of the carbapenem class.[3] For example, Curtius rearrangement of **25d** (*R′* = *R*-OBn),[1a] followed by oxidative cleavage of the acetylenic bond,[3e,5] would afford amino acid derivatives **26d** of a type that have been efficiently converted to thienamycin (**27**) and imipenem (**28**) (Scheme 15.6).[3a–f,l] In analogous fashion, **25c** (*R′* = *S*-OBn) would provide access to members of the olivanic acid class of antibiotics, such as MM-22381 (**29**), which differ from the thienamycins primarily in having the *S*-configuration at C_8.[3h] Finally, acetylenic acids **25a** (*R′* = H) and **25b** (*R′* = Me) are viable precursors to the important des-hydroxy derivatives PS-5 (**30**) and PS-6 (**31**), respectively.[3a–c,i,j] The considerable synthetic potential of this methodology derives from its highly convergent nature, and the fact that both relative and absolute stereochemistry at all stereogenic centers can be varied at will.

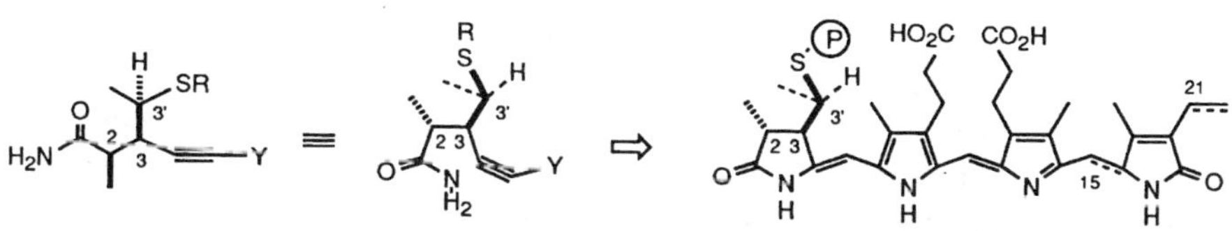

18 18 **19** (Δ-15, Δ-21); **20** (Δ-15); **21** (Δ-21)

Scheme 15.4

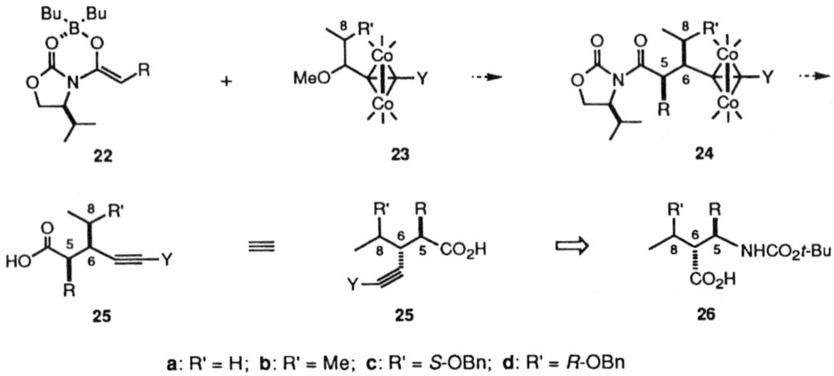

Scheme 15.5

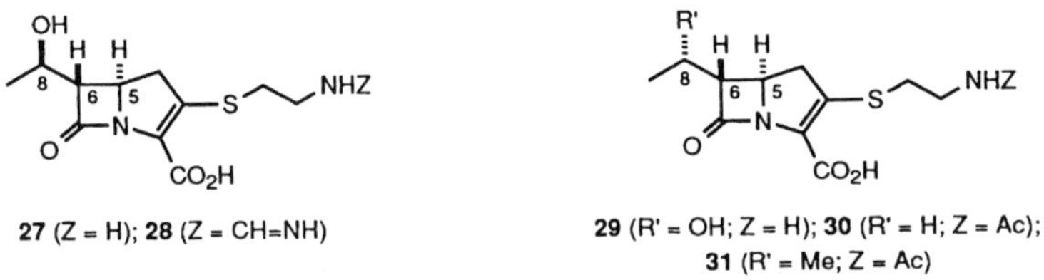

Scheme 15.6

15.2 Model Studies with Simple Nicholas Substrates (*R* = Me)

15.2.1 Synthesis of Homochiral Acetylenic Acids

Our initial efforts in this area were directed toward the preparation of acetylenic acids **36a–c**, in order to explore the generality of the Nicholas–Schreiber methodology (Scheme 15.7). Cobalt derivatives **33a–c** and *ent*-**33a–c** (*ent* = mirror image of structure shown) were readily obtained by condensation of lithiotrimethylsilylacetylene with the appropriate aldehydes *R′*CHMeCHO (*R′* = H, Me, *S*-OBn, *R*-OBn),[2] followed by in situ methylation (DMS), and complexation of the resulting methylpropargyl ethers with $Co_2(CO)_8$.[1] Imide enolate **32** was prepared following the general procedure of Schreiber et al.,[1h] employing 1.0 equivalents each of $(i\text{-}C_3H_7)_2NEt$ and Bu_2BOTf at 0°C in CH_2Cl_2. The resulting enolate solutions were then cooled to –78°C, treated with an additional 0.5–1.0 equiv. of Bu_2BOTf, followed by an equimolar quantity of **33** (based on excess Bu_2BOTf), and warmed to 0°C to afford the desired adducts **35** after oxidative cleavage with ceric ammonium nitrate (CAN) (Table 15.1). In identical fashion, imide enolate *ent*-**32** afforded the enantiomeric adducts *ent*-**35** upon condensation with cobalt complexes *ent*-**33** (Scheme 15.8).

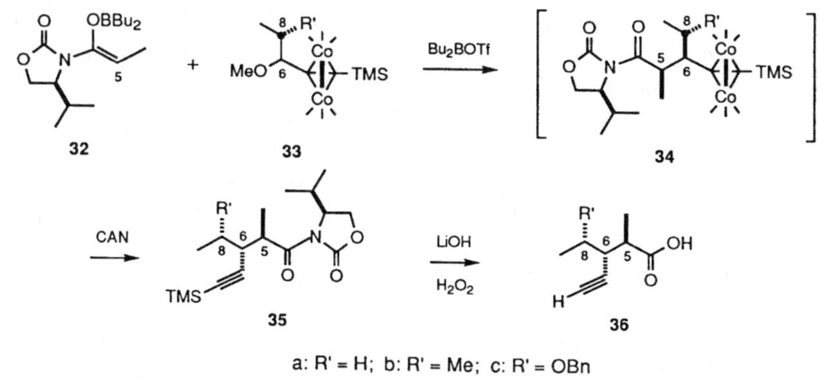

Scheme 15.7

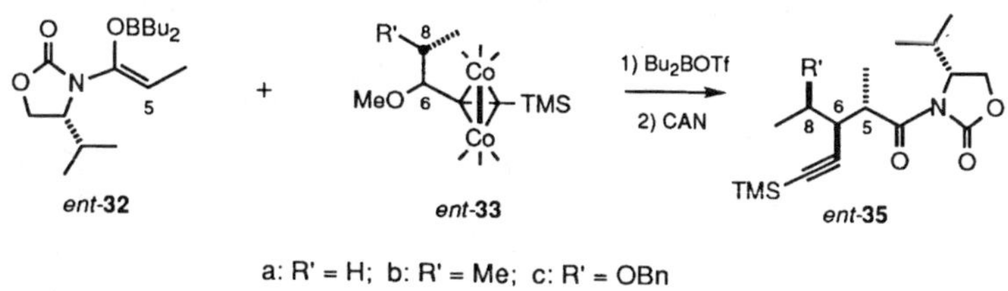

Scheme 15.8

In general, yields for this reaction were excellent (85–98%) utilizing a ratio of **32**:**33** = 2:1 (entries 1 and 4, Table 15.1), and only slightly less satisfactory (75–85%) employing a ratio of **32**:**33** = 1:1 (entries 2 and 5). Not surprisingly, however, **35b** (R' = Me) was obtained in considerably lower yield (19%, entry 3), presumably due to steric hindrance and competing elimination reactions in the stabilized carbocation derived from **33b**.[1] Diastereo- and enantioselectivities were also generally excellent, with *syn:anti* ratios of >98:2 employing chiral enolates **32** and *ent*-**32** with achiral cobalt complexes **33a,b** (en-

Table 15.1 SUMMARY DATA FOR COMPOUNDS 35–36

Entry	Compd.	Yield[a]	$[\alpha]^{25}_D$ (c)[d]	Compd.	Yield[a]	$[\alpha]^{25}_D$ (c)[d]
1	**35a**	94[b]	+28.7 (12.4)	**36a**	95	+11.7 (27.8)
2	*ent*-**35a**	81[c]	–26.0 (30.6)	*ent*-**36a**	91	–11.5 (60.7)
3	**35b**	19[b]	+15.9 (4.5)	**36b**	34	–1.3 (45.8)
4	**35c**	93[b]	–33.1 (6.4)	**36c**	85	–31.6 (7.8)
5	*ent*-**35c**	75[c]	+37.4 (14.7)	*ent*-**36c**	77	+30.2 (19.5)

[a] Average yield for several runs.

[b] Yield employing 2 equiv. **32**

[c] Yield employing 1 equiv. *ent*-**32**.

[d] Measured in MeOH (c = mg/ml).

tries 1–3; slightly lower *syn*-selectivity was observed employing chiral enolates **9** and *ent*-**9**; cf. Scheme 15.2). Equally impressive ratios (>98:2) were obtained with the "matched" chiral substrates **32** + **33c** → **35c**, and *ent*-**32** + *ent*-**33c** → *ent*-**35c** (entries 4 and 5; the case of "mismatched" substrates will be discussed below). These results are in full accord with the transition-state model proposed by Schreiber et al. (vide supra).[1h]

Once in hand, oxazolidinones **35** were readily converted to the corresponding acetylenic acids **36** by hydrolysis with excess lithium hydroperoxide,[6] which effected concomitant cleavage of the TMS group (cf. Scheme 15.7). In analogous fashion, oxazolidinones *ent*-**35** gave the enantiomeric acids *ent*-**36**. As indicated (Table 15.1), yields for this step were excellent (80–98%), except for the special case where R' = Me (entry 3). In this example, steric hindrance once again had a deleterious effect on reactivity. Finally, 5*R*,6*R*-diastereoselectivity of >98:2 was also realized upon condensation of chiral cobalt complex **33c** with *achiral* enolate **37** (Scheme 15.9), as judged by conversion of the derived adduct **38c** to the identical homochiral acetylenic acid **36c** obtained employing chiral enolate **32** (cf. Scheme 15.7).[7] This result provides some indication of the powerful directing influence chiral substituents can exert on the Nicholas reaction.

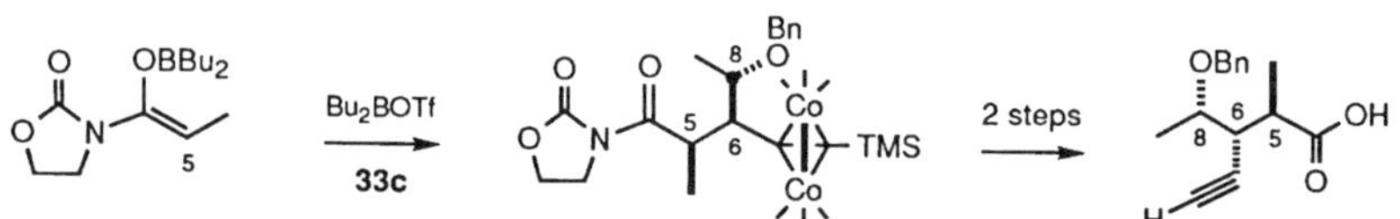

37 38c 36c

Scheme 15.9

15.2.2 Conversion of Acetylenic Acids to Amino Acids

We next studied the conversion of acetylenic acids **36** to homochiral amino acid derivatives **41** (Scheme 15.10). As indicated, Curtius rearrangement of carboxylic acids **36** to afford carbamates **39** was conveniently carried out with diphenylphosphoryl azide (DPPA),[4a] followed by HCl-catalyzed capture of the intermediate isocyanate (not isolated) with *tert*-butanol.[4b] Similar results were obtained with the enantiomeric acids *ent*-**36**. Within the limits of detection, this step occurred with complete retention of stereochemistry, as determined by NMR analysis and comparison of the specific rotations for **39a,c** and *ent*-**39a,c** (Table 15.2). However, we experienced some initial difficulties in effecting the oxidative cleavage of acetylenic carbamates **39**/*ent*-**39** to the desired amino acid derivatives **41**/*ent*-**41**. For **39a,b** and *ent*-**39a**, this transformation was best accomplished with $KMnO_4/NaIO_4$,[3e] which afforded ~50:50 mixtures of the corresponding acids **41a,b** and *ent*-**41a** together with the N-formyl derivatives **40a,b** and *ent*-**40a**. We believe that formyl derivatives **40** were derived by intramolecular formylation involving oxidation intermediates of type **I** (Scheme 15.11), since **41** itself was unreactive to formic acid. In any event, these mixtures were usually not separated, but rather were directly hydrolyzed (KOH) to afford pure **41a,b** and *ent*-**41a** in >90% overall yield (entries 1–3, Table 15.2). With **39c** and *ent*-**39c** (R' = OBN), however, $KMnO_4/NaIO_4$ caused extensive decomposition due to oxidation of the benzyl protecting group to produce benzoic acid. This difficulty was eventually circumvented with the discovery that $OsO_4/NaIO_4$ provided the desired chemoselectivity,[5] leading exclusively to the N-formyl derivatives **40c** and *ent*-**40c**. As with **40a,b** and *ent*-**40a** described above, the latter materials were then readily cleaved with KOH to afford the desired carbamates **41c** and *ent*-**41c** (entries 4, 5). The utility of these amino acid derivatives for the synthesis of β-lactams was then convincingly demonstrated by their facile conversion to amino acids **42a–c** and *ent*-**42a–c**, and subsequently to lactams **43a–c** and *ent*-**43a**–c following standard literature procedures (see Scheme 15.10).

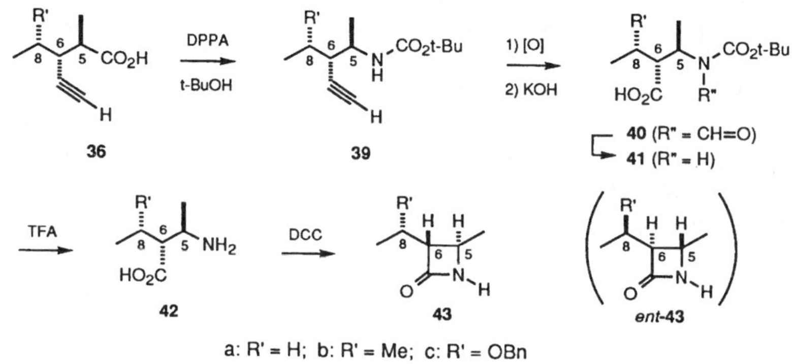

Scheme 15.10

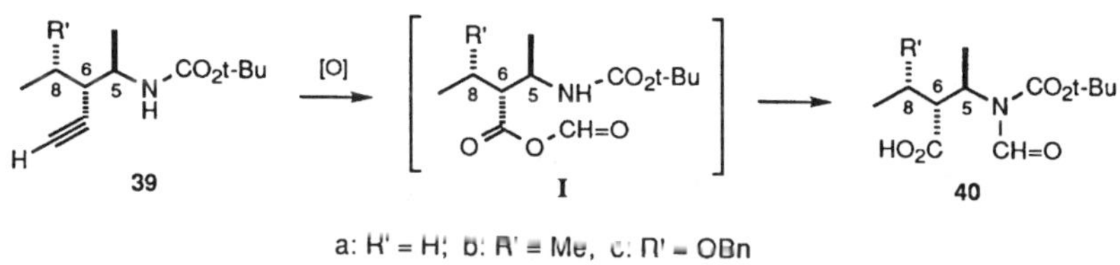

Scheme 15.11

Table 15.2 SUMMARY DATA FOR COMPOUNDS 39–43

Entry	Compd.	Yield[a]	$[\alpha]^{25}_D$ (c)[b]	Compd.	Yield[a]	$[\alpha]^{25}_D$ (c)[b]
1	**39a**	82	+61.1 (24.8)	**41a**	93	+12.7 (2.7)
2	*ent*-**39a**	74	−64.0 (20.1)	*ent*-**41a**	92	−13.8 (7.2)
3	**39b**	62	+55.2 (20.4)	**41b**	97	+23.8 (15.1)
4	**39c**	83	+34.9 (19.8)	**41c**	64	+62.4 (16.3)
5	*ent*-**39c**	82	−34.4 (4.7)	*ent*-**41c**	68	−64.7 (12.8)

Entry	Compd.	Yield[a]	$[\alpha]^{25}_D$ (c)[c]
1	**43a**	68	+19.8 (10.3)
2	*ent*-**43a**	78	−18.3 (19.5)
3	**43b**	82	+11.3 (23.0)
4	**43c**	73	+35.5 (19.8)
5	*ent*-**43c**	79	−36.5 (10.5)

[a] Average yield for several runs.

[b] Measured in MeOH (*c* = mg/ml).

[c] Measured in CH_2Cl_2 (*c* = mg/ml).

15.3 Formal Synthesis of PS-5

β-Lactam **43a** (cf. Scheme 15.10) bears a close structural resemblance to the known alcohol derivative **44** (5*R*,6*R* stereochemistry), which has previously been employed in the synthesis of the potent antibiotic PS-5 (**30**) (Scheme 15.12).[8a] It appeared likely that precursors of type **44** might be conveniently prepared using methodology analogous to that described in Schemes 15.7 and 15.10. In order to explore this possibility, it was first necessary to prepare the chiral oxazolidinone derivative **50**, which was derived in excellent overall yield beginning with 1,4-butanediol (**45**) (Scheme 15.13). Thus selective monoprotection of **45** with TBDPSCl gave a 90% yield of the corresponding alcohol **46**, which upon oxidation with PDC in DMF afforded carboxylic acid derivative **47** in 75% yield. Treatment of **47** with oxalyl chloride in benzene then provided the acid chloride **48** (99%), which was directly converted to the desired acyl derivative **50** by condensation with the lithium anion derived from oxazolidinone **49** (89% yield from **47**).[9]

Scheme 15.12

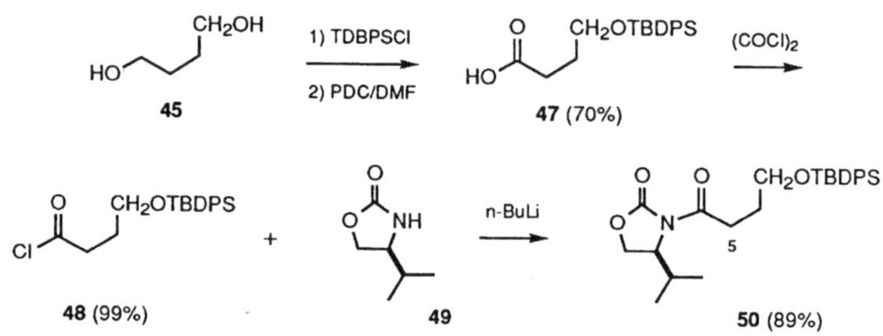

Scheme 15.13

Once in hand, oxazolidinone **50** was readily converted to the corresponding boron enolate **51** (vide supra), which underwent smooth Nicholas condensation with cobalt complex **33a** to give *syn*-adduct **52** in 88% overall yield after decomplexation (Scheme 15.14; >98:2 *syn*-selectivity). Surprisingly, however, hydrolysis of **52** under the standard conditions (LiOOH, 3:1 THF/H_2O) afforded complex mixtures of products,[6] consisting in part of *endo*-ring opened alcohol **53**, together with lesser amounts of the desired carboxylic acid **54** derived by *exo*-hydrolysis (conditions A). This unexpected reaction pathway might be due to complexation of Li cation between the *exo*-carbonyl group and the -OTBDPS functionality (cf. **II**, Scheme 15.15), since related alkyl derivatives (OTBDPS = Me) underwent normal hydrolysis (vide infra). In any event, addition of DMF to the hydrolysis reaction completely reversed the regioselectivity (3:3:1 DMF/THF/H_2O), and afforded an 81% yield of the desired acetylenic acid **54** with no trace of *endo*-ring opened product **53** (conditions B).

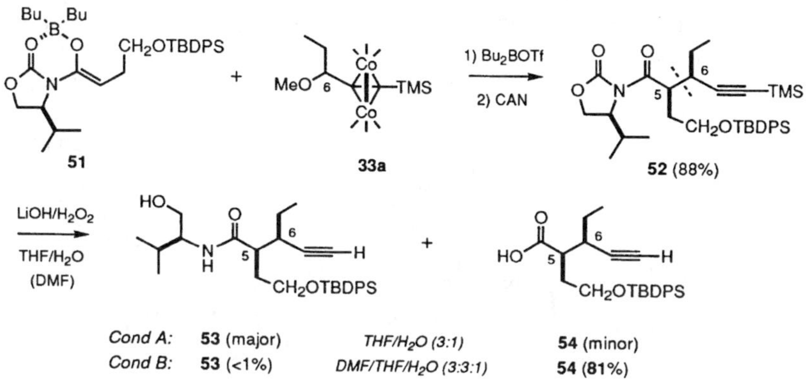

Scheme 15.14

We believe that DMF functions by solvating chelated structures of types **II** and **III** (Scheme 15.15). This result is in qualitative agreement with the suggestion of Evans et al. that the rate-determining step in cleavage by HOO^- is *collapse* of the initially formed tetrahedral intermediate.[6] In the case of *endo*-attack, collapse of **III** to **53** is presumably accelerated by chelation of the type illustrated in **III**, which renders the oxazolidinone C-N bond relatively labile. In contrast, collapse of intermediate **IV**, derived by *exo*-attack, should exhibit little rate enhancement. As predicted on the basis of this hypothesis, DMF had no effect on the regioselectivity of hydrolysis by HO^-, where the rate-determining step is *formation* of the tetrahedral intermediate (steric control.[6] In both cases (with and without added DMF) hydrolytic cleavage was predominantly endocyclic.

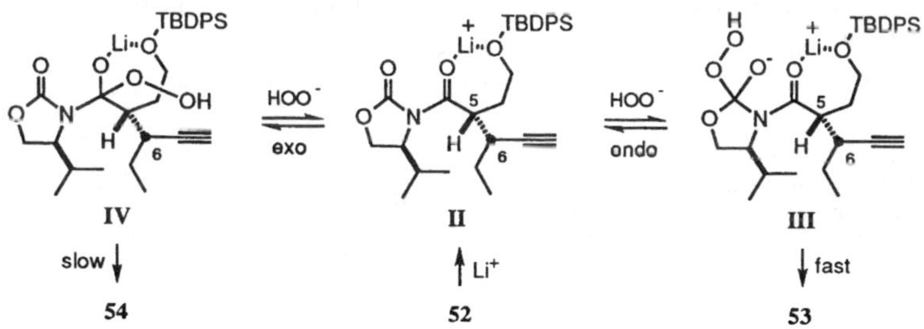

Scheme 15.15

The remaining steps necessary for the conversion of **54** to the PS-5 (**30**) precursor **44** were then accomplished in a straightforward fashion, in exact analogy to our model studies described in Scheme 15.10. Thus Curtius rearrangement of **54** with DPPA afforded an 83% yield of the carbamate derivative **55**, which upon oxidative cleavage with $KMnO_4$/$NaIO_4$ gave the desired protected amino acid **56** in 91% yield

(Scheme 15.16). This last compound, upon acid-catalyzed deprotection, then gave a quantitative yield of the β-amino acid **57**, which in turn afforded 91% of the target β-lactam **44** upon cyclization with DCC. The material thus obtained had identical chemical and physical properties as those reported in the literature,[8a] and was prepared as a single enantiomer.

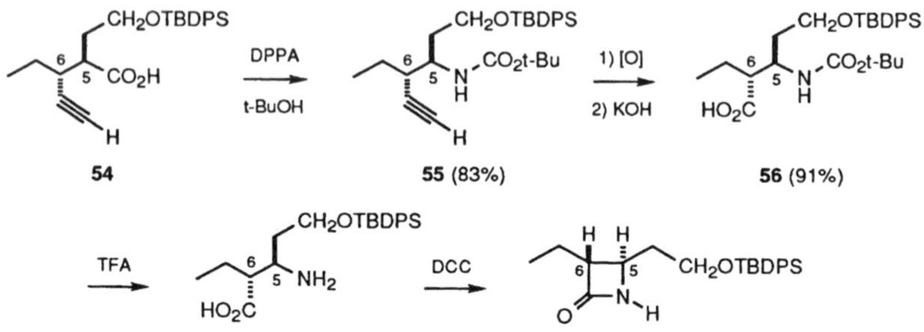

54 **55** (83%) **56** (91%) **57** (100%) **44** (91%)

Scheme 15.16

15.4 Formal Synthesis of Thienamycin

15.4.1 Model Studies Related to MM-22381 and Thienamycin

Our studies with PS-5 (**30**) were readily extended to a synthesis of the potential olivanic acid precursor **58** (Scheme 15.17). Amino acid **58** has the 5*R*,6*S*,8*S*-stereochemistry characteristic of MM-22381 (**29**) (cf. Scheme 15.6), and it served as a useful probe of stereochemical control in the Nicholas reaction. As in the case with oxazolidinone **32** (cf. Scheme 15.7), oxazolidinone **51** condensed in a "matched" fashion with cobalt complex **33c**, affording an 80% yield of adduct **52b** with >98:2 *syn*-selectivity. Adduct **52b** was then converted without complication to the desired amino acid **58**, employing the conventional three-step sequence of hydrolysis (DMF/THF/H_2O),[6] followed by Curtius rearrangement,[4] and oxidative cleavage of the

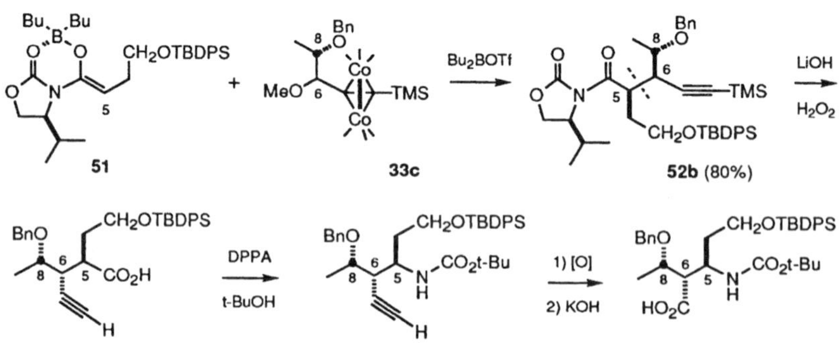

51 **33c** **52b** (80%) **54b** (66%) **55b** (71%) **58** (88%)

Scheme 15.17

acetylenic triple bond.[5] In these examples, the term *matched* refers to those condensations in which the chirality of the oxazolidinone precursors reinforces the directing effect of the C-8 stereogenic center in cobalt complexes **33**. As previously noted (Scheme 15.9), chiral cobalt complex **33c** combines even with *achiral* oxazolidinone **37** to provide adduct **38c** with excellent diastereo- and enantioselectivity.

Based upon these results, it seemed likely that a similar approach could be applied to the synthesis of amino acids having the 5*R*,6*S*,8*R*-configuration found in thienamycin (**27**) and imipenem (**28**) (cf. Scheme 15.6).[3] In principle, this substitution pattern was available by "mismatched" condensation of oxazolidinone **51** with cobalt complex *ent*-**33c**. This combination would ultimately lead to the known thienamycin precursor **59** if transition-state interactions were dominated by chiral auxiliary **51** (Scheme 15.18); cf. also Scheme 15.17). However, our initial model studies in this direction were not encouraging. Thus all attempts at the condensation of boron enolate **32** with *ent*-**33c** provided only complex mixtures of products, which contained at least three isomeric adducts **60** in a ratio of ~7:3:1 (29% combined yield, Scheme 15.18). This result stands in marked contrast to that obtained in the "matched" condensation of oxazolidinone **32** with **33c** (cf. Scheme 15.7), which gave adduct **35c** in 93% yield with >98:2 *syn*-selectivity.

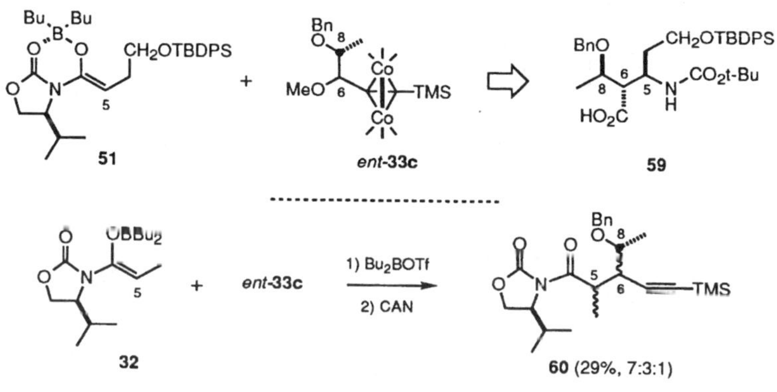

Scheme 15.18

In contrast to the case with **32**, oxazolidinone *ent*-**9** underwent clean "mismatched" condensation with *ent*-**33c** to provide a ~13:1 mixture of two isomeric acetylenic acid derivatives (Scheme 15.19). These were subsequently identified as *syn*-adduct **61** and *anti*-isomer **62**. Interestingly, however, the major isomer proved to be the *anti*-adduct **62**, as demonstrated by chemical correlation (it was impossible to distinguish between **61** and **62** based upon spectral data alone). Thus **62** was readily converted to the carboxylic acid **63**,[6] which upon Curtius rearrangement (81%),[4] followed by oxidative cleavage (68%),[5] afforded amino acid derivative **65** in exact analogy to our earlier studies with **41** (cf. Scheme 15.10).

Upon cyclization of **65** to β-lactam **67** the *cis*-relationship between H_5 and H_6 was immediately apparent from their relatively large coupling constant ($J_{5,6}$ = 6.0 Hz) (Scheme 15.20). For *trans*-β-lactams this coupling constant is typically <3 HZ.[8b] The

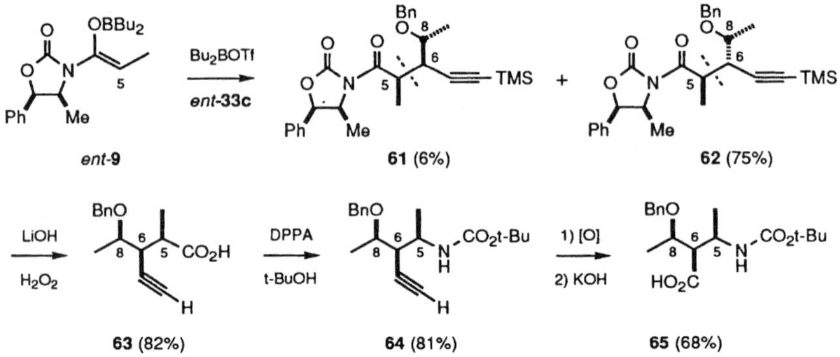

Scheme 15.19

remaining question pertaining to absolute stereochemistry at C_5-C_6 was then resolved by epimerization studies. As expected, **67** was readily epimerized to the desired *trans*-isomer **68** ($J_{5,6}$ = 1.8 Hz),[10] which proved to be identical to the material obtained directly from the minor *syn*-adduct **61**. By way of comparison, if **67** had been of opposite absolute configuration at C_5-C_6 (i.e., **69**), epimerization would have afforded the previously synthesized β-lactam *ent*-**43c** (cf. Scheme 15.10). These results provide further testimony to the powerful directing influence chiral substituents can exert on the Nicholas reaction (see also Scheme 15.9).

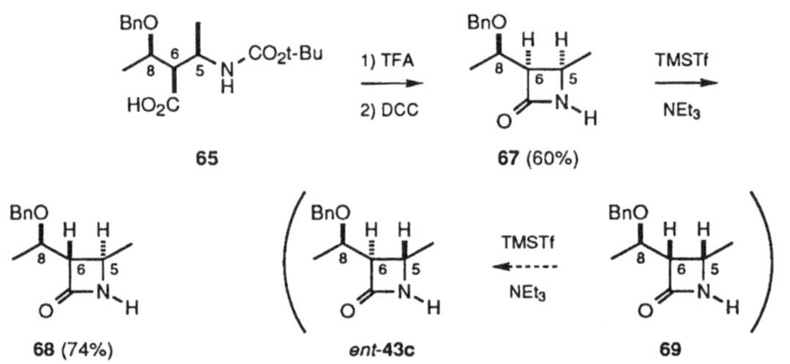

Scheme 15.20

At present we cannot offer a detailed rationale for this difference in reaction pathway between "matched" and "mismatched" condensations (to the best of our knowledge, this is the first example of an *anti*-selective Nicholas reaction). However, with further study we hope to develop a transition-state model that will permit logical predictions regarding the stereochemical outcome of Nicholas reactions with chiral substrates.

15.4.2 Synthesis of the Key Thienamycin Precursor 77

Finally, these model studies were readily extended to a formal total synthesis of thienamycin (**27**).[3] Thus condensation of chiral enolate **71** with *ent*-**33c** provided a 79% yield of the Nicholas adduct **72**, which was obtained with ~17:1 *anti*-selectivity (Scheme 15.21). As in the case with **52** (cf. Scheme 15.14), hydrolysis of **72** under standard conditions (LiOOH, 3:1 THF/H_2O) afforded a complex mixture of products,[6] from which *endo*-ring opened product **73** could be isolated in 30% yield (conditions A). No trace of the desired carboxylic acid **74** derived from *exo*-nucleophilic attack could be detected. Once again, however, addition of DMF to the hydrolysis reaction completely reversed the regioselectivity (3:3:1 DMF/THF/H_2O), and afforded a 74% yield of the desired acetylenic acid **74** together with only traces of **73** (conditions B).

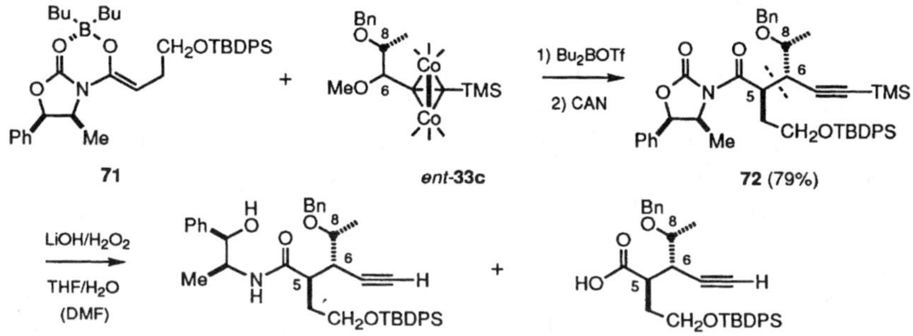

Cond A:	**73** (30%)	*THF/H2O (3:1)*	**74** (<1%)
Cond B:	**73** (<1%)	*DMF/THF/H2O (3:3:1)*	**74** (74%)

Scheme 15.21

As described above for **63** (Scheme 15.19), **74** was then converted in two steps to the homochiral amino acid derivative **59**, which upon deprotection and cyclization with DCC afforded the *cis*-β-lactam **76** in 56% yield (Scheme 15.22). Finally, epimerization of **76** according to the procedure of Nakai et al. afforded the known thienamycin precursor **77**,[9,5b] which had identical spectral data as those reported by Grieco et al. for the racemic material ($[\alpha]^{25}_{D}$ −1.9°).[3e]

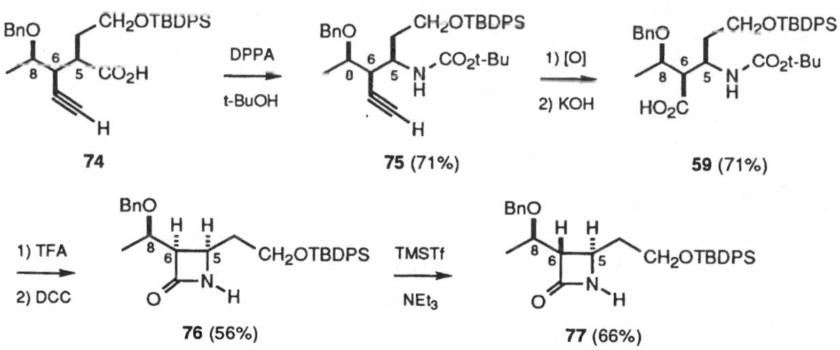

Scheme 15.22

15.5 Summary

The control of relative and absolute stereochemistry at three and more contiguous centers is an important problem in organic synthesis. In certain cases we believe that the Nicholas reaction could provide an attractive alternative to more traditional methodology in this area. This will be especially true for molecules containing carboxylate, alkene, amino, and related functionalities that are easily derived from acetylenic acid derivatives of type **17**. As one example of the utility of this process, we have developed unequivocal syntheses of two of the most important members of the carbapenem class of antibiotics. Extension of this methodology to the synthesis of other biologically important molecules is currently under investigation.

15.6 Acknowledgments

It is a pleasure for me to acknowledge the dedicated undergraduate, graduate, and postdoctoral associates whose efforts made this work possible: Ms. Karine Chung, Mr. Frederic Rupprecht, Mr. Chaowen Zheng, Dr. Shaun Murphree, and in particular, Dr. Wanjun Zheng. The continuing support of this program by the National Science Foundation is also gratefully acknowledged.

15.7 References

1. (a) R. F. Lockwood, K. M. Nicholas, *Tetrahedron Lett.* **18**, 4163 (1977). (b) K. M. Nicholas, M. O. Nestle, D. Deyferth, *Transition Metal Organometallics*, Halper, Ed. (Academic Press, New York, 1978), Vol. 2, p. 1. (c) S. L. Schreiber, T. Sammakia, W. E. Crowe, *J. Am. Chem. Soc.* **108**, 3128 (1986). (d) K. M. Nicholas, M. Mulvaney, M. Bayer, *J. Am. Chem. Soc.* **102**, 2508 (1980). (e) H. D. Hodes, K. M. Nicholas, *Tetrahedron Lett.* **19**, 4349 (1978). (f) A. F. Bramwell, L. Crombie, M. H. Knight, *Chem. Ind. (London)* 1265 (1965) and references cited therein. (g) M. Saha, B. Bogby, K. M. Nicholas, *Tetrahedron Lett.* **27**, 915 (1986). (h) S. L. Schreiber, M. T. Klimas, T. Sammakia, *J. Am. Chem. Soc.* **109**, 5749 (1987).
2. (a) P. A. Jacobi, W. Zheng, *Tetrahedron Lett.* **34**, 2581 (1993). (b) P. A. Jacobi, W. Zheng, *Tetrahedron Lett.* **34**, 2585 (1993). (c) P. A. Jacobi, S. Rajeswari, *Tetrahedron Lett.* **33**, 6231 (1992). (d) *Ibid.* **33**, 6235 (1992). (e) P. A. Jacobi, R. W. DeSimone, *Tetrahedron Lett.* **33**, 6239 (1992). (f) P. A. Jacobi, H. L. Brielmann, S. I. Hauck, *Tetrahedron Lett.* **36**, 1193 (1995). (g) P. A. Jacobi, J. Guo, W. Zheng, *Tetrahedron Lett.* **36**, 1197 (1995). (h) P. A. Jacobi, J. Guo, *Tetrahedron Lett.* **36**, 2717 (1995).
3. For recent reviews on the synthesis of thienamycin (**27**) and related materials, see (a) T. Nagahara, T. Kametani, *Heterocycles* **25**, 729 (1987). (b) G. I. Georg, in *Studies in Natural Product Chemistry*, A-ur Rahman, Ed. (Elsevier Science, Amsterdam, 1989), Vol. 4. (c) J. H. Bateson, in *Progress in Heterocyclic Chemistry*, H. Suschitzky, E. F. V. Scrier, Ed. (Pergamon Press, Oxford, 1991), Vol. 3. See also, (d) S. Hanessian, D. Desilets, Y. L. Bennani, *J. Org. Chem.* **55**, 3098) (1990) and references cited therein. (e) P. A. Grieco, D. L. Flynn, R. E. Zelle, *J. Am. Chem. Soc.* **106**, 6414 (1984). (f) D. G. Melillo, R. J. Cvetovich, K. M. Ryan, M. Sletzinger, *J. Org. Chem.* **51**, 1498 (1986). (g) D. A. Evans, E. B. Sjogren, *Tetrahedron Lett.* **27**, 4961 (1986). (h) D. F. Corbett, S. Coulton, R. Southgate, *J. Chem. Soc., Perkin Trans. 1* 3011 (1982), and references cited therein. (i) D. A. Evans, E. B. Sjogren, *Tetrahedron Lett.* **27**, 3119 (1986). (j) K. Yamamoto, T. Yoshioka, Y. Kato, N. Shibamoto, K. Okamura, Y. Shimauchi, T. Ishikura, *J. Antibiot.* **33**, 796 (1980). (k) T. Shono, N. Kise, F. Sanda, S. Ohi, K. Yoshioka, *Tetrahedron Lett.* **30**, 1253 (1989). (l) D. G. Melillo, R. J. Cvetovich, K. M. Ryan, M. Sletzinger, *J. Org. Chem.* **51**, 1498 (1986), and references cited therein.
4. (a) T. Shioiri, K. Ninomiya, S. Yamada, *J. Am. Chem. Soc.* **94**, 6203 (1972). (b) A. Benalil, P. Roby, B. Carboni, M. Vaultier, *Synthesis* 787 (1991).
5. (a) R. Pappo, D. S. Allen,Jr., R. U. Lemieux, W. S. Johnson, *J. Org. Chem.* **21**, 478 (1956). (b) R. M. Moriarty, R. Penmasta, A. K. Awasthi, I. Prakash, *J. Org. Chem.* **53**, 6164 (1988), and references cited therein.
6. D. A. Evans, T. C. Britton, J. A. Ellman, *Tetrahedron Lett.* **28**, 6141 (1987).
7. We are grateful to Ms. Gayle Schulte, of Yale University, for carrying out an X-ray analysis of acetylenic acid **36c**.
8. (a) D. Tanner, P. Somfai, *Tetrahedron* **44**, 619 (1988). See also, (b) T. Shono, N. Kise, F. Sanda, S. Ohi, K. Yoshioka, *Tetrahedron Lett.* **30**, 1253 (1989).
9. D. A. Evans, J. Bartroli, T. L. Shih, *J. Am. Chem. Soc.* **103**, 2127 (1981).
10. T. Chiba, T. Nakia, *Tetrahedron Lett.* **26**, 4647 (1985).

CHAPTER

16

Asymmetric Synthesis of β-Amino Acids from Catalytic Sharpless Epoxidation

Miquel A. Pericàs, Antoni Riera, and Albert Moyano

16.1 Enantioselective Synthesis of *anti* N-Boc 3-Amino-1,2-Diols

Since the introduction by Sharpless and co-workers of the asymmetric epoxidation of allylic alcohols,[1] chiral epoxyalcohols **1** have become a most useful building block for the stereocontrolled synthesis of complex molecules.[2]

One of the usual ways for the integration of **1** into synthetic sequences involves the titanium(IV)-mediated regioselective ring opening by nucleophiles, a process also disclosed by Sharpless.[3] In this way, two or three adjacent chiral centers can be constructed in a stereocontrolled manner by operationally simple methodology (Scheme 16.1).

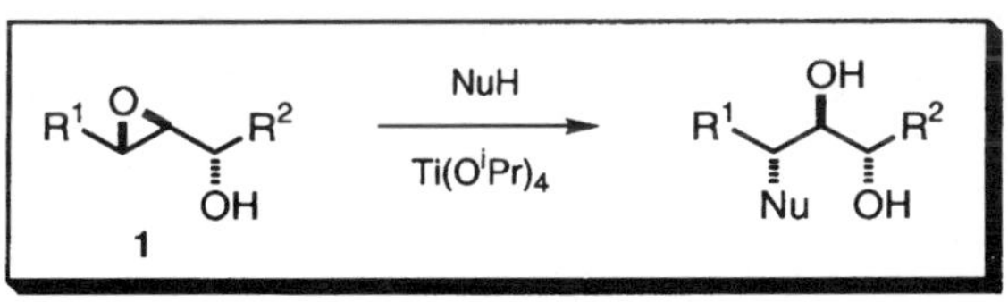

Scheme 16.1

The 3-amino-1,2-diol array, which could conveniently arise from the reaction of **1** with primary amines able to experience a subsequent cleavage leaving a free amino group, is perhaps the most attractive one among those directly derivable from the reaction. We have recently shown[4] that benzhydrylamine, as well as other primary benzylic amines, performs a completely stereoselective and highly regioselective ring opening of a variety of epoxy alcohols in the presence of titanium tetraisopropoxide, thus providing an easy entry to the aforementioned 3-amino-1,2-diols (**2**) (Scheme 16.2).

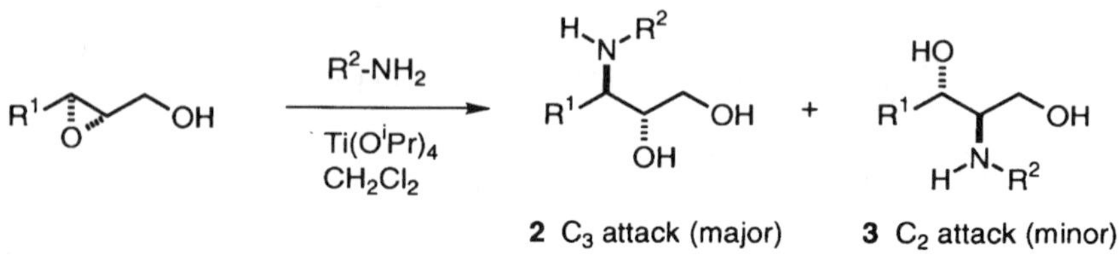

Scheme 16.2

Some of the prepared derivatives **2** are shown in Figure 16.1 along with key details of their syntheses. Good regiochemical control is observed in all cases, rather independently of the nature of R^1. Moreover, the minor regioisomer **3** can be easily removed, so that regiochemically pure **2** can be secured in all instances. The only *syn* derivative in Figure 16.1, **2c**, is obtained in significantly lower yield than the corresponding *anti* diastereomer, **2b**. Given that the *syn* derivatives arise from the nucleophilic ring opening of *cis* epoxyalcohols, which, in turn, come from the rather stereochemically inefficient epoxidation of (*Z*)-allyl alcohols, it is strongly indicated that the *syn* derivatives, when required, be prepared by modification of the *anti* ones.

The N-benzhydryl substituent, which both controls the regioselectivity of the ring opening process and facilitates the purification of the reaction products, can be then changed to the more convenient *tert*-butoxycarbonyl (Boc) or benzyloxycarbonyl (Cbz) ones through the high-yield sequences shown in Scheme 16.3. According to that, benzhydrylamine behaves in the ring opening process as a safe synthetic equiv-

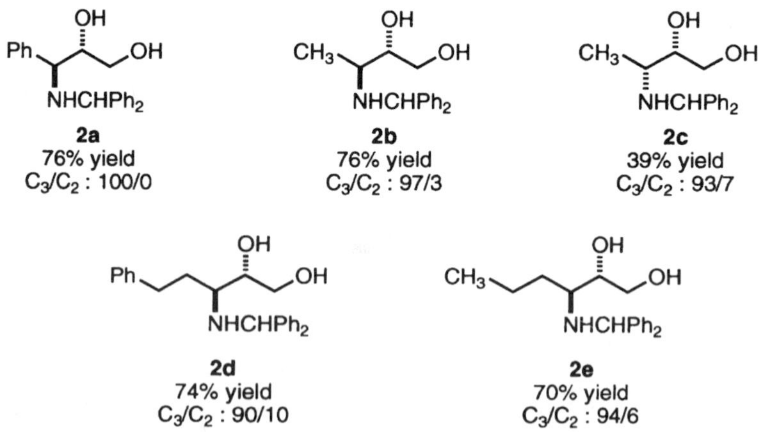

Figure 16.1. N-Benzhydryl-3-amino-1,2-diols by nucleophilic ring opening of epoxyalcohols.

alent of ammonia, allowing avoidance for the same purpose of the potentially risky azide anion.[5,6]

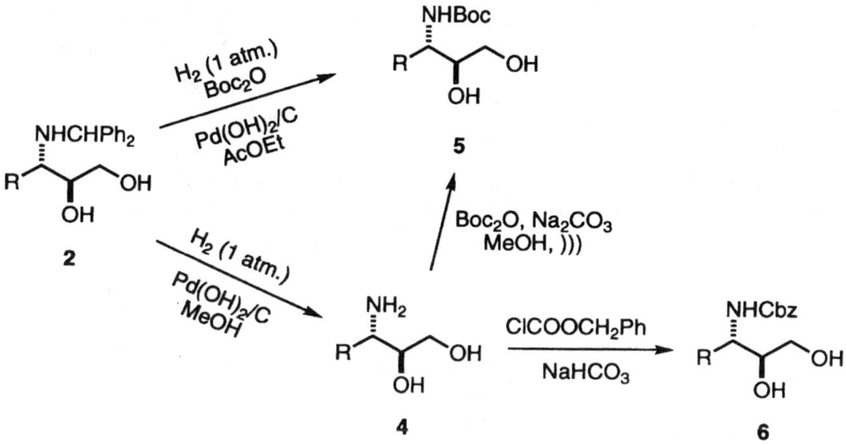

Scheme 16.3

The N-Boc derivatives **5** tend to be highly crystalline, allowing in many instances an easy improvement of the enantiomeric purity by simple crystallization. The particular arrangement of functional groups in these molecules, with a completely stereodefined chiral center bearing a protected amino group β to a primary alcohol, converts them in highly suitable precursors for the synthesis of β-amino acid derivatives. Moreover, the presence of the secondary hydroxy group increases its synthetic potential, offering opportunity for the preparation of the highly interesting α-hydroxy-β-amino acids (Scheme 16.4).

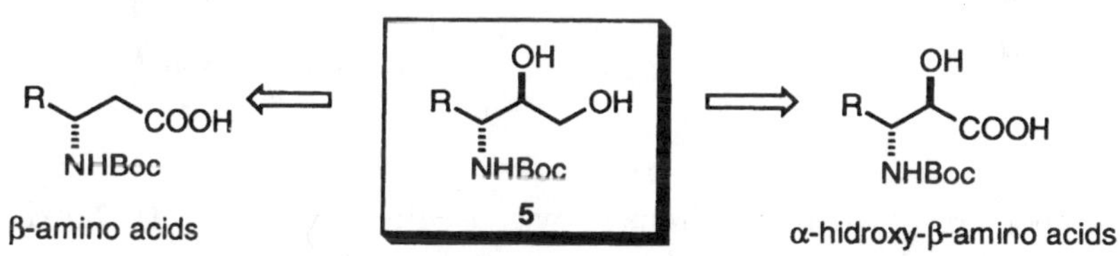

Scheme 16.4

16.2 Enantioselective Synthesis of N-Boc β-Amino Acids[7]

The synthetic strategy selected for the conversion of N-Boc aminodiols **5** into the target β-amino acids involved the intermediate formation of allyl amines, through a 1,2-diol deoxygenation protocol, followed by a hydroboration–oxidation sequence. Enantiopure allyl amines are interesting on their own,[8] so that the selected methodology offered clear advantages over the apparently more direct, but in practice more problematic, selective deoxygenation of the secondary alcohol followed by oxidation of the primary one.

For the 1,2-diol deoxygenation, a variety of reported procedures were attempted with disappointing results. After much experimentation, it was finally found that the Corey–Hopkins protocol[9] afforded N-Boc allyl amines **7** in good yields (Scheme 16.5). Although beyond the scope of this review, it is worth noting that allyl amines **7** are convenient precursors of N-Boc amino acid esters **8**[10] through a Marshall ozonolysis sequence.[11] Chiral HPLC analysis[12] of **8** simultaneously proves that both the Corey–Hopkins protocol and the ozonolysis take place with conservation of the stereochemical integrity.

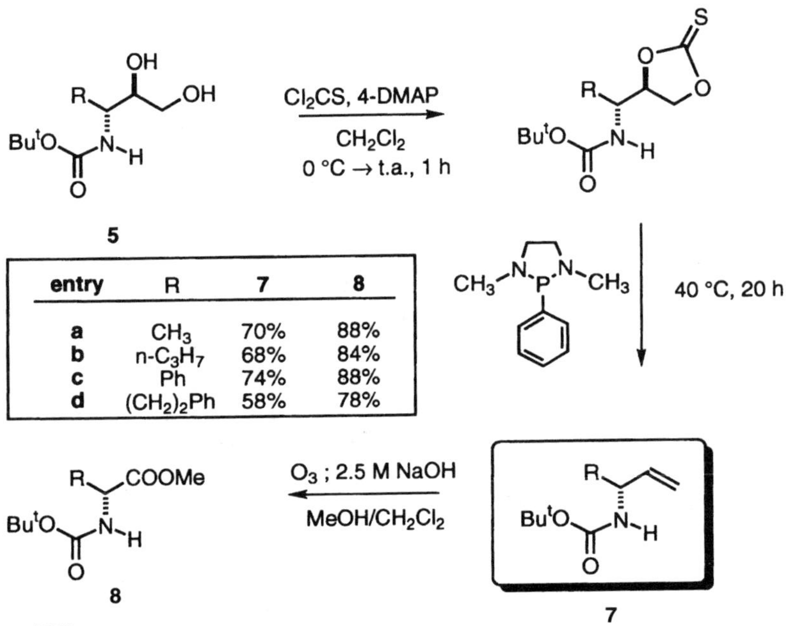

entry	R	7	8
a	CH_3	70%	88%
b	n-C_3H_7	68%	84%
c	Ph	74%	88%
d	$(CH_2)_2Ph$	58%	78%

Scheme 16.5

Hydroboration of N-Boc allylamines **7** with 9-BBN[13] followed by H_2O_2 oxidation afforded the N-Boc 3-aminoalcohols **9**, which could be oxidized to the corresponding N-Boc β-amino acids **10** with PDC in DMF[14] in satisfactory overall yields (Scheme 16.6). In order to facilitate isolation, the crude β-amino acids can be converted into the corresponding methyl esters by treatment with diazomethane.

The ultimate source of chirality in our procedure is the catalytic Sharpless epoxidation.[1c,d] According to that, both enantiomeric series of the target β-amino acids are equally available by simply selecting the tartrate ester used in the epoxidation step. Moreover, since a large number of epoxyalcohols are available in high enantiomeric purity by the use of the Sharpless epoxidation, the present methodology[7] promises a broad applicability. On the other hand, enantioenriched allylamines **7** are also an actively pursued class of materials.[8] Their preparation from aminodiols **5** opens a straightforward route to this interesting class of compounds.

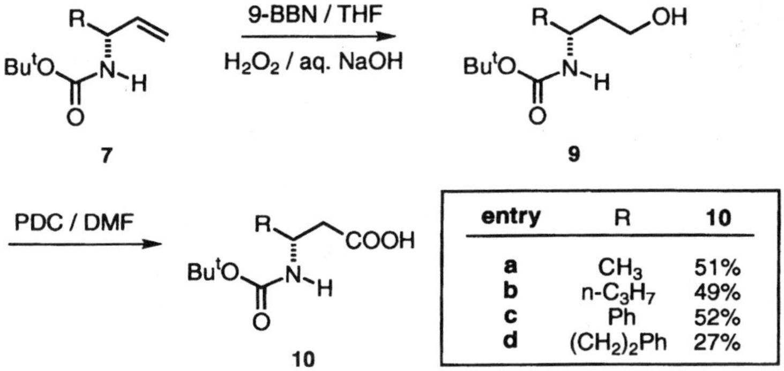

entry	R	10
a	CH_3	51%
b	n-C_3H_7	49%
c	Ph	52%
d	$(CH_2)_2Ph$	27%

Scheme 16.6

16.3 Development of Synthetic Methodology for the Preparation of α-Hydroxy-β-Amino Acids from N-Boc-3-Amino-1,2-Diols[15,16]

α-Hydroxy-β-amino acids are well-recognized key components for a variety of protease inhibitors[17] and other new-generation pharmaceutics.[18,19] Examples of interesting bioactivity are known between both the *syn* and the *anti* stereoisomers, but for a given activity precise absolute and relative stereochemistries are required.[20] Among the *anti* stereoisomers, for instance, the benzyl ester **11** (Figure 16.2) is an inhibitor of LTA_4 hydrolase with anti-inflammatory activity,[21] whereas allophenylnorstatine (**12**) is the key component of the kinostatins,[22] which are inhibitors of the HIV protease. Among the *syn* stereoisomers, cyclohexylnorstatine (**13**) is a component of renin inhibitors,[23] and the N-acyl phenylisoserine derivatives **14** and **15** are the side chain of the potent antitumor agents Taxol and Taxotere®, respectively.[18]

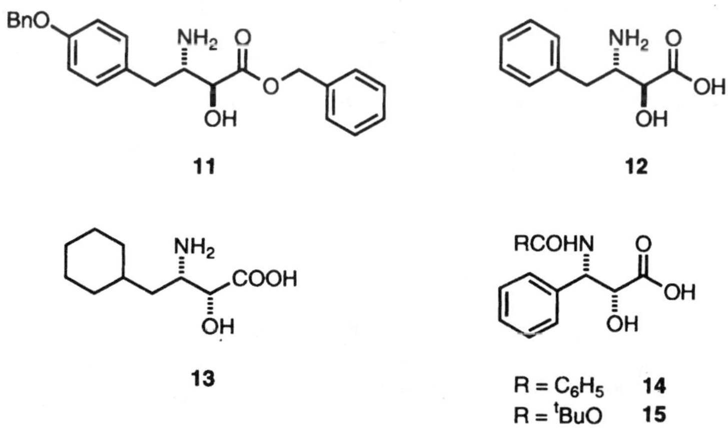

Figure 16.2. Some pharmacologically interesting α-hydroxy-β-amino acids.

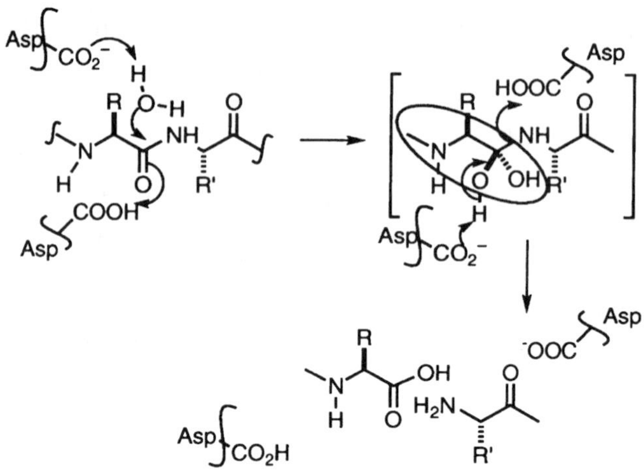

Figure 16.3. Mechanism of hydrolysis of proteins by an aspartyl protease. The region of the tetrahedral intermediate isosteric with an α-hydroxy-β-amino acid is shown encircled.

In many instances, the bioactivity of α-hydroxy-β-amino acids is expressed in the inhibition of aspartyl proteases (renin, HIV protease, etc.). This mode of action has been related to the ability of these amino acids to mimic the tetrahedral intermediate in the hydrolysis of proteins mediated by the aspartyl proteases, which results in a blocking of the enzymes (Figure 16.3).[17,24]

Our strategy for the conversion of protected aminodiols **5** or **6** into *anti* α-hydroxy-β-amino acids was a priori very simple, only requiring orthogonal protection of the 1,2-diol system followed by selective deprotection of the primary hydroxyl group and oxidation (Figure 16.4).

To develop the synthetic procedure, we used the readily available *anti* 3-aminobutan-1,2-diol derivatives **5a** and **6a**.[4,6a]

For the orthogonal protection of the 1,2-diol system, the formation of a *p*-methoxybenzaldehyde acetal was selected. In fact, it has been reported[25] that similar acetals can be regioselectively cleaved with DIBAL-H, leaving a secondary *p*-methoxybenzyl ether and a free primary alcohol ready for oxidation. In our case, the

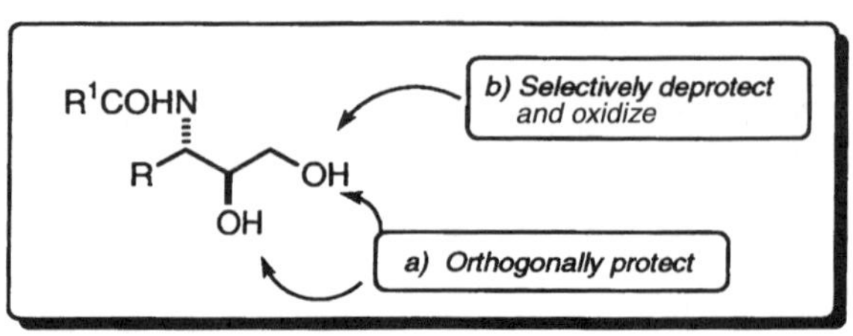

Figure 16.4. Synthetic strategy for *anti* α-hydroxy-β-amino acids.

preparation of the acetals **16/17** could be conveniently performed by standard methodology (Scheme 16.7). However, the cleavage with DIBAL-H took place with the opposite regioselectivity,[25b] affording in high yield the secondary alcohols **18/19**. This unexpected result forced a modification in our initial strategy. Thus **18/19** were converted by treatment with 2,2-dimethoxypropane[26] into the 2,2-dimethyloxazolidiones **20/21**, which still permit selective deprotection of the primary aryloxy group by oxidation. This was performed with DDQ,[27] leading to the primary alcohols **22/23**. Final oxidation with PDC in DMF[14] led to the fully protected *anti* 3-amino-2-hydroxy-butyrates **24/25**, which were best isolated as their methyl esters **26/27**.

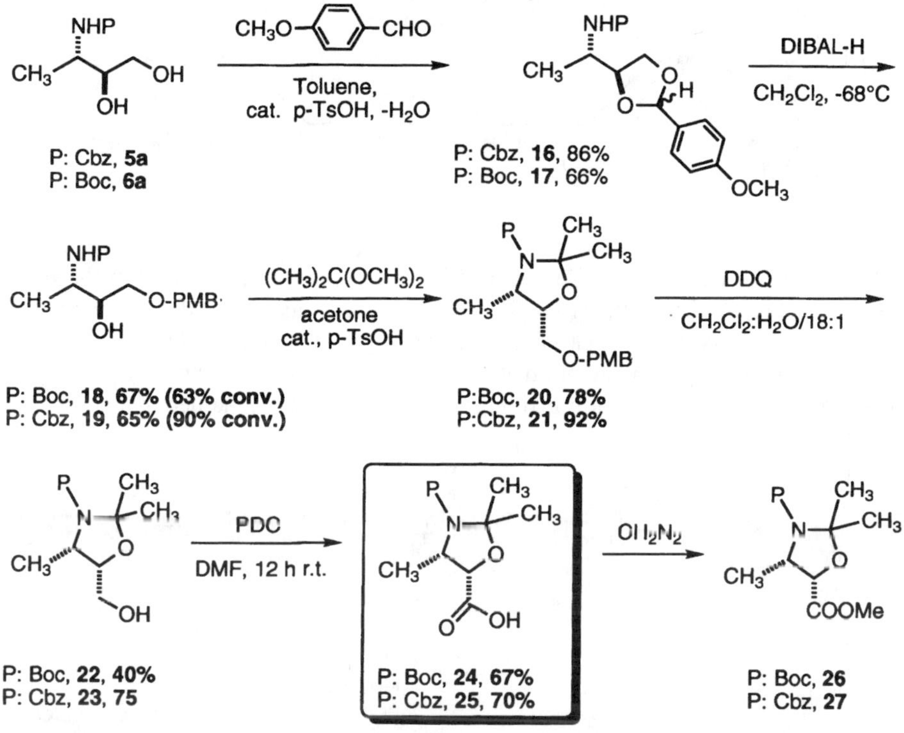

Scheme 16.7

Intermediates **18/19** were subjected to preliminary evaluation as precursors of the diastereomeric *syn* α-hydroxy-β-amino acids (Scheme 16.8).

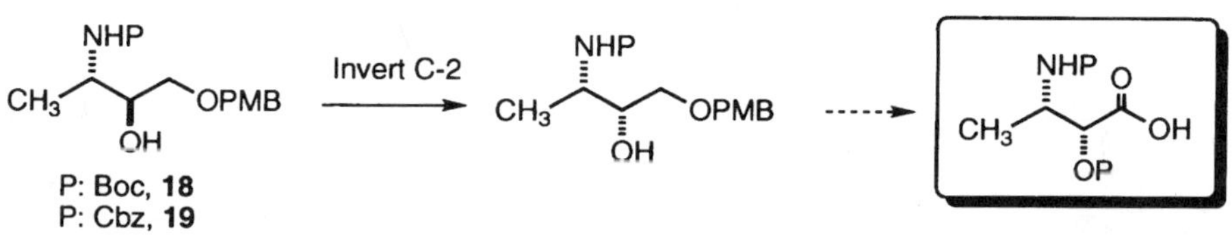

Scheme 16.8

An attempted Mitsunobu reaction on **19** using *p*-nitrobenzoic acid as a nucleophile[28] did not proceed at all. As a less direct alternative, an oxidation–reduction sequence was also studied (Scheme 16.9). The oxidation of **18/19** with PDC in DMF[14] cleanly provided (*S*)-3-alkoxycarbonylamino-1-*p*-methoxybenzyloxy-2-butanones **28/29**. The reduction of these ketones with DIBAL-H at low temperature took place with very high yields and excellent diastereoselectivities. Unfortunately, however, the observed stereoselectivity was opposite to the desired one, the process leading back to the *anti* series.

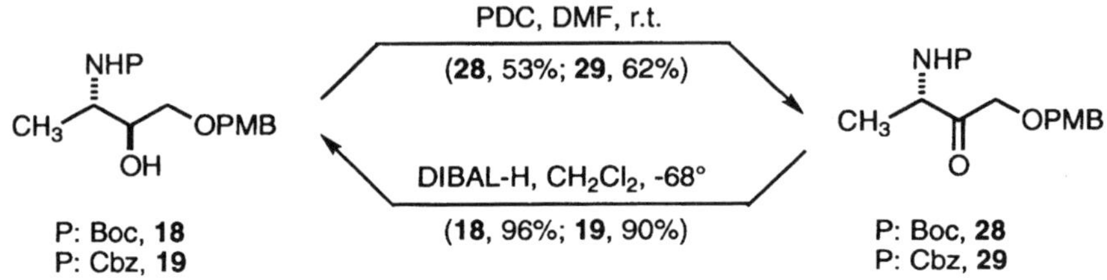

Scheme 16.9

In order to simplify our original synthetic scheme, making it suitable at the same time for the preparation of the *syn* α-hydroxy-β-amino acids, it was decided to protect the primary hydroxy group in **6a** as a silyl ether, hoping that this would not impede the necessary inversion of configuration at C-2. It is worth noting that an orthogonal protection scheme of the different functional groups would be maintained in this way.

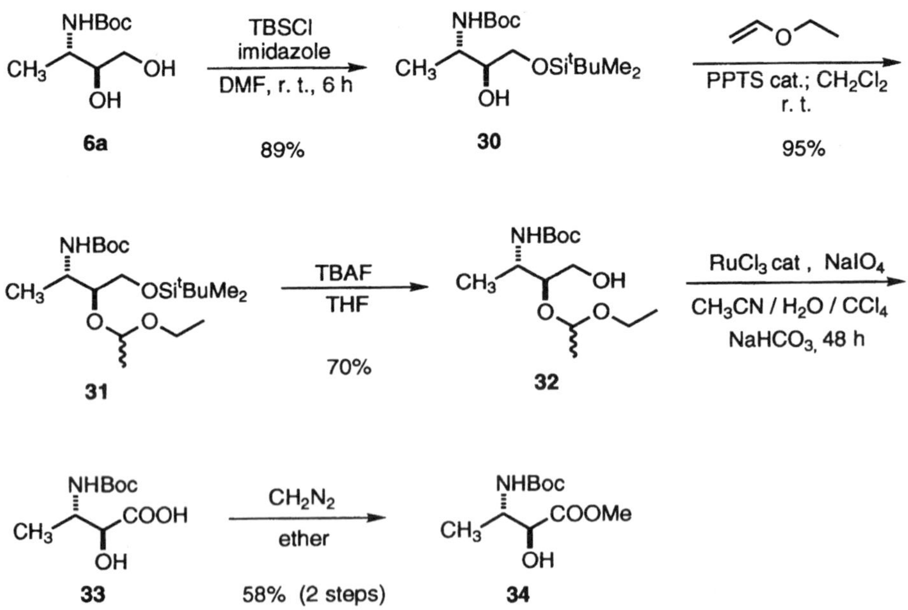

Scheme 16.10

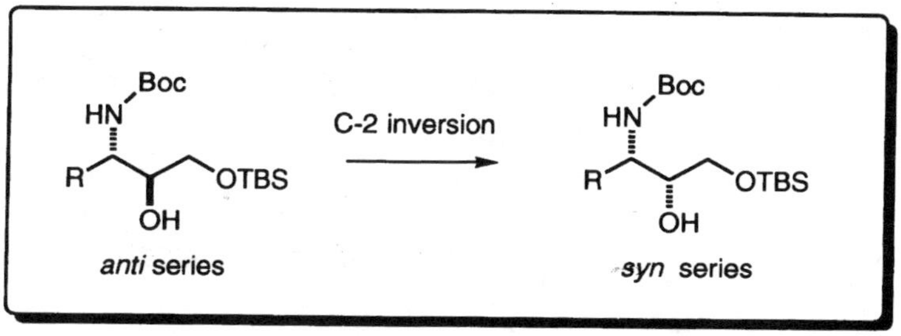

Figure 16.5. From *anti* to *syn* α-hydroxy-β-amino acid precursors.

Thus silylation of **6a** with *t*-butyldimethylsilyl chloride[29] (Scheme 16.10) cleanly afforded the primary silyl ether **30**, which, by subsequent acid-catalyzed treatment with excess ethyl vinyl ether in dichloromethane,[30] provided access to the fully protected derivative **31** in very high yield as a mixture of diastereomers at the acetal carbon.

Selective deprotection of **31** with 1 *M* tetrabutylammonium fluoride in THF led to the primary alcohol **32**, which was converted into the target α-hydroxy-β-amino acid by oxidation with $RuCl_3/NaIO_4$ in $CH_3CN/CCl_4/H_2O$.[31] Deprotection of the secondary alcohol occurred during the subsequent acidic workup, and, to facilitate isolation and characterization, the free acid **33** was converted into its methyl ester **34** by treatment with diazomethane. Both **33** and **34** were obtained in diastereomerically pure form according to high-resolution NMR.

As already mentioned, the early intermediate **30** and similar species can provide a good opportunity for accessing the *syn* stereoisomers, provided that an inversion of configuration can be performed at C-2 (Figure 16.5).

Starting from **30**, inversion at C-2 was performed by Mitsunobu methodology. With benzoic acid as the nucleophile[32] low conversions were recorded. To circumvent this difficulty, *p*-nitrobenzoic acid was used instead,[28] and the *syn p*-nitrobenzoate **35** was obtained in high yield (Scheme 16.11). Conversion of **35** into the target *syn* α-hydroxy-β-amino acid was initially attempted by deprotection of the primary alcohol

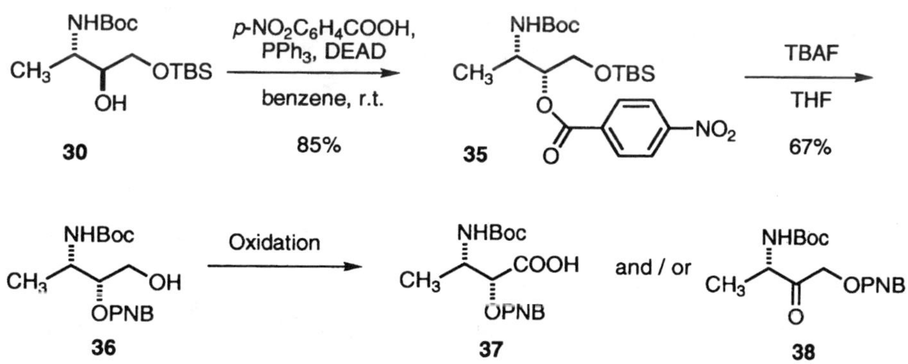

Scheme 16.11

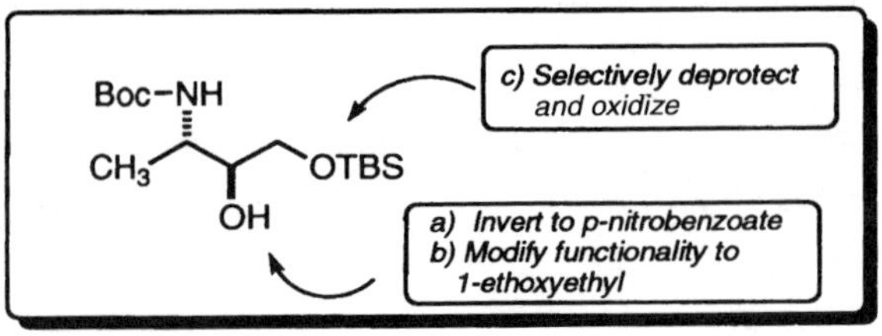

Figure 16.6. Synthetic strategy for *syn* α-hydroxy-β-amino acids.

with tetrabutylammonium fluoride in THF, leading to **36**, and subsequent oxidation. Although the target acid **37** could be isolated in low yield, under all the experimental conditions tested the major product was the ketone **38**, arising from a transacylation–oxidation sequence.

This result clearly indicated that the functionality at C-2 of **35** had to be modified, given the incompatibility between the *p*-nitrobenzoate with the oxidizing conditions required for the completion of the synthesis. Our ultimate strategy for the *syn* α-hydroxy-β-amino acids was then established as shown in Figure 16.6.

Removal of the *p*-nitrobenzoate ester from **35** by saponification was accompanied by significant epimerization at C-2. This was also the case for reductive removal with a variety of metal hydrides. Fortunately, the reduction with DIBAL-H at low temperature takes place with complete preservation of the stereochemical integrity, leading to **39** in high yield (Scheme 16.12).

The conversion of **39** into the target *syn* α-hydroxy-β-amino **42** acid is completely parallel to that of **30** into **34**, which we have already discussed, and is summarized in

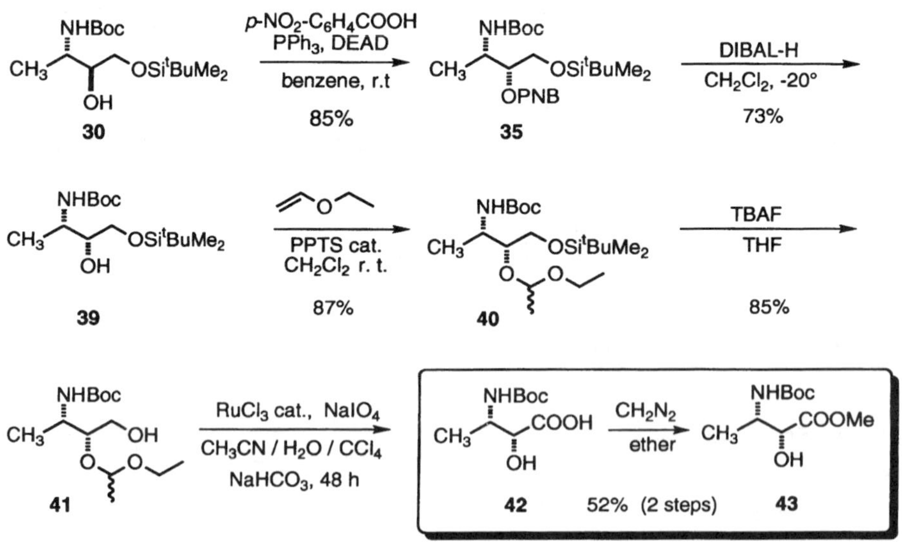

Scheme 16.12

Scheme 16.12. Also in this case, the oxidation of the primary alcohol **41** takes place with concomitant deprotection of the secondary alcohol, and isolation of the final product is facilitated by conversion into its methyl ester **43**. From the TBS derivative **30**, the synthesis of the diastereomerically pure (2*R*,3*S*)-3-*tert*-butoxycarbonylamino-2-hydroxybutyric acid **42** involves only five steps and takes place in more than 25% overall yield.

The broad scope of the methodologies presented here for the diastereo- and enantioselective synthesis of α-hydroxy-β-amino acids[15,16] is demonstrated by the syntheses of cyclohexylnorstatine and of the Taxotere® side chain described in the following sections.

16.4 Enantioselective Synthesis of Cyclohexylnorstatine[33]

(2*R*,3*S*)-3-Amino-4-cyclohexyl-2-hydroxybutanoic acid **44**, also known as cyclohexylnorstatine and nor-C-statine, is an important component of several peptides with inhibitory properties against renin.[34] The development of enantioselective syntheses of **44** is thus a matter of interest, and several approaches to this molecule have been reported in recent times.[35]

Our methodology for the enantioselective synthesis of *syn* α-hydroxy-β-amino acids[16] is perfectly suited for the preparation of **44**; the corresponding retrosynthetic analysis leads to the N-Boc aminodiol **45** as the key intermediate and to the readily available[36a] allyl alcohol **46** as the ultimate starting material (Figure 16.7).

Putting ideas to work, the allyl alcohol **46** was submitted to catalytic Sharpless epoxidation[1c] with D-(–)-DIPT, to afford epoxyalcohol **47** in good yield and 90% enantiomeric excess (Scheme 16.13). Regioselective ring opening of this epoxide was first performed with benzhydrylamine in the presence of $Ti(O\text{-}i\text{-}Pr)_4$,[4] leading to aminodiol **48**. Even better results were obtained when the ring opening was performed with $Ti(O\text{-}i\text{-}Pr)_2(N_3)_2$,[37] the corresponding azidodiol **49** being isolated in quantitative yield. Both **48** and **49** could be converted into the key N-Boc aminodiol intermediate **45** by hydrogenolysis in the presence of Boc_2O.[6a,36b]

Compound **45** is highly crystalline and offers a good opportunity for enantioenrichment. A single recrystallization from hexane–ether upgrades its enantiomeric excess up to *ca.* 95%.

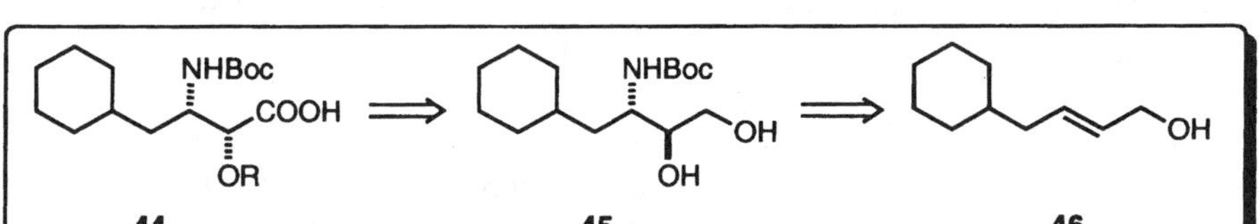

Figure 16.7. Retrosynthetic analysis of cyclohexylnorstatine.

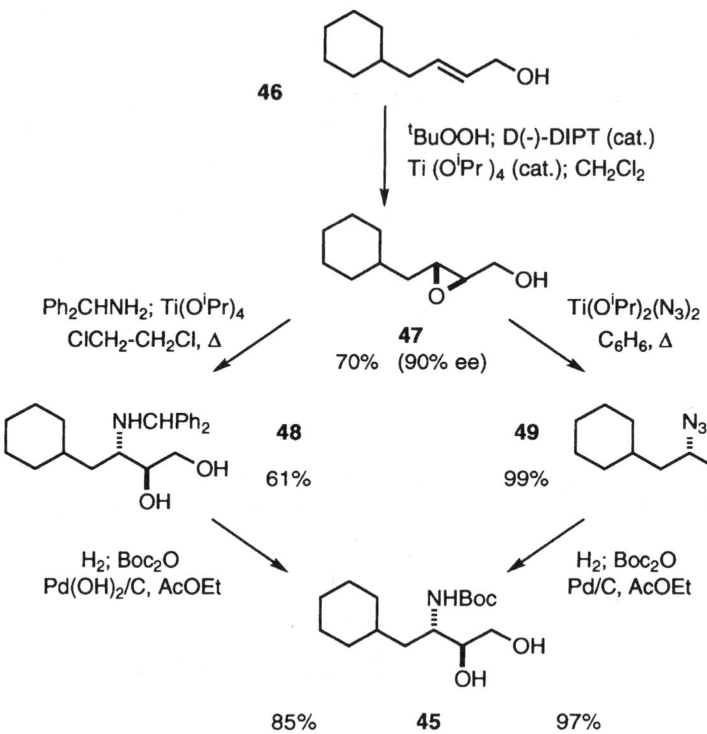

Scheme 16.13

Conversion of **45** into the target amino acid **44** (*R*=1-ethoxyethyl) is performed by application of the general methodology described in the preceding section (Scheme 16.14).

In the present case, careful control of pH at the final oxidation stage allowed the isolation of the desired amino acid **44** in fully protected form (cf. Section 16.3). The synthesis involves eight steps from the epoxyalcohol **47**, and the overall yield is 32%. According to HPLC analysis on a Chiralcel® ODR column, the enantiomeric excess of **44** is 94%.

16.5 Enantioselective Synthesis of the Ready-for-Coupling Taxotere® Side Chain[16]

Taxotere® (**55**) is a highly promising cancer chemotherapeutic agent that has very recently been approved for clinical use in the United States. This semisynthetic drug is closely related to Taxol (**56**), a cytotoxic principle isolated from the bark of *Taxus brevifolia.*[18] Taxotere® is prepared from 10-deacetylbaccatin III (**57**), a ready available natural product,[38] by integration of the phenylisoserine side chain (Figure 16.8). This explains the enormous interest on this α-hydroxy-β-amino acid and the plethora of synthetic approaches reported in recent years to the Taxol and Taxotere® side chains.[39]

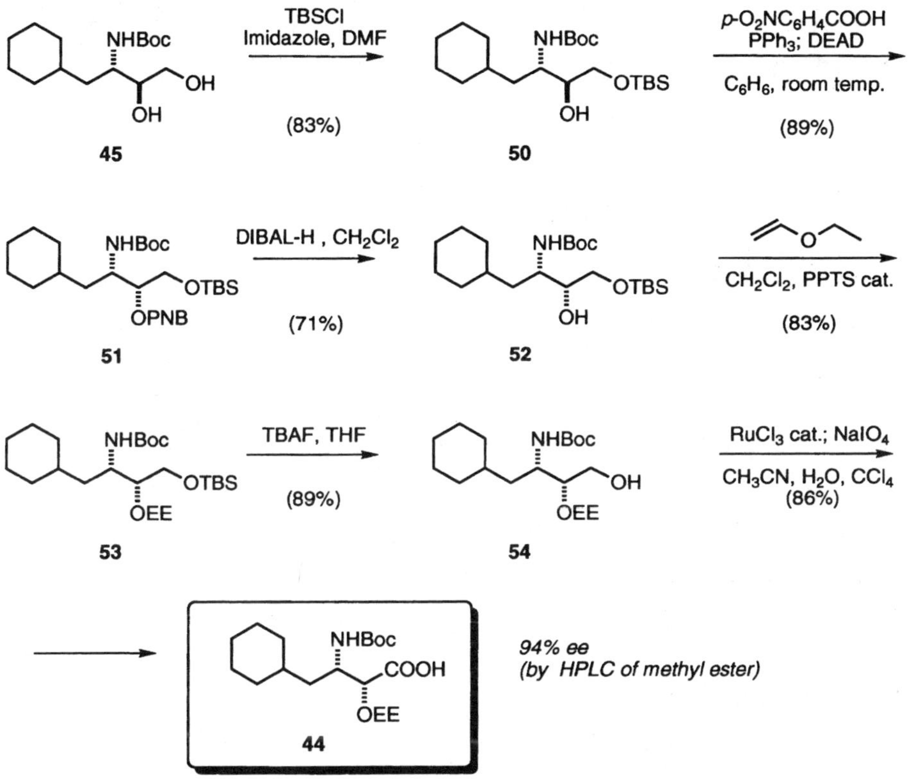

Scheme 16.14

In the Potier–Greene semisynthesis of Taxotere®, the 1-ethoxyethyl derivative **58** is directly coupled with 10-deacetylbaccatin III.[40] Since our methodology directly leads to *syn* α-hydroxy-β-amino acids with this protection scheme, was decided to apply it to a synthesis of **58**.

Enantiomerically pure (2*S*,3*S*)-N-Boc-3-amino-3-phenylpropane-1,2-diol **59**, readily secured from (*E*)-3-phenyl-1-propanol via Sharpless epoxidation and subsequent regioselective ring opening,[6a] could be selectively converted to the mono TBS derivative **60** in high yield (Scheme 16.15).

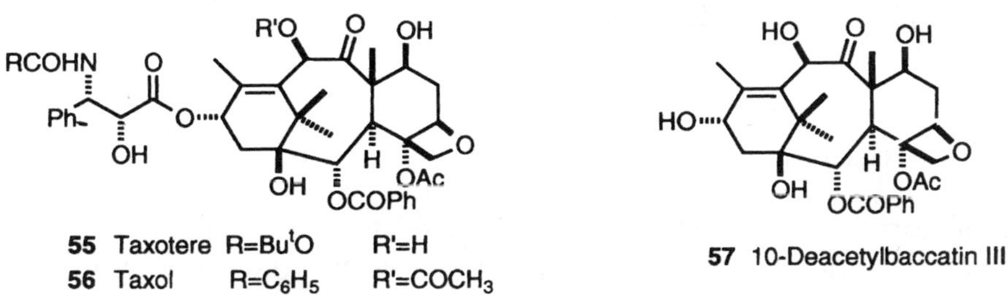

Figure 16.8. Structures of Taxotere®, Taxol, and 10-deacetylbaccatin III.

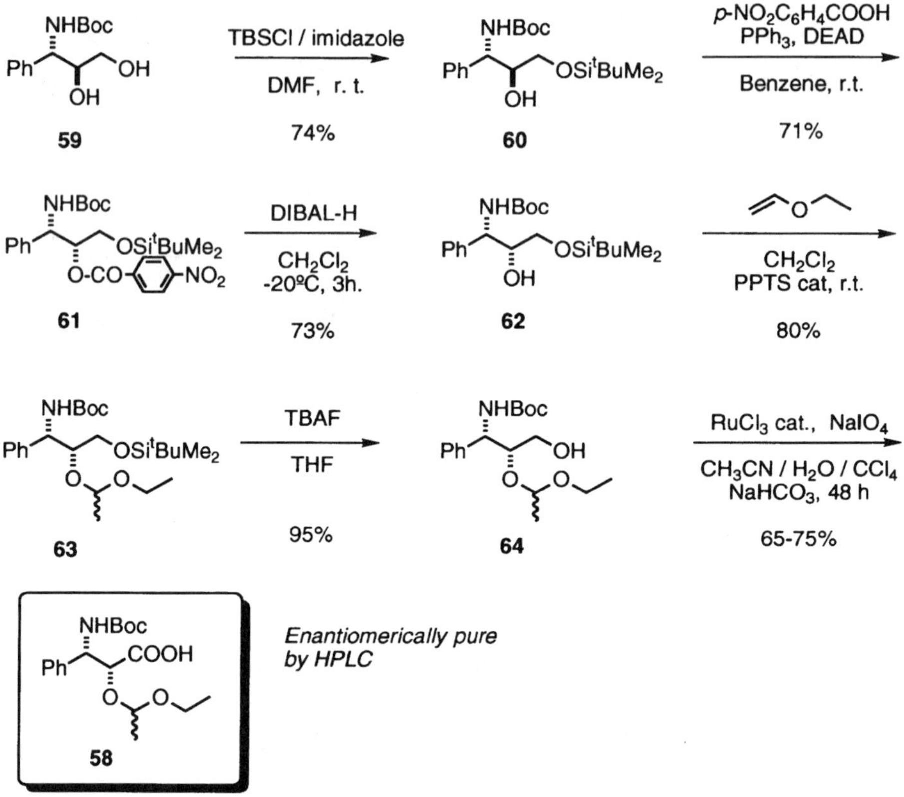

Scheme 16.15

The more critical steps in the sequence, Mitsunobu inversion with *p*-nitrobenzoic acid as the nucleophile leading to **61** and the subsequent reduction with DIBAL-H to **62** under the previously optimized conditions, occurred with complete control of stereochemistry at C-2. This final manipulation of protecting groups and oxidation took place in a satisfactory manner, affording **58** in high yield. According to HPLC analysis on a Chiralcel® ODR column, **58** is enantiomerically pure. Its synthesis can be accomplished in five steps and 20–22% overall yield from aminodiol **59**. The Taxotere® side chain has been obtained in both diastereomerically and enantiomerically pure form and can be directly coupled with protected derivatives of 10-deacetylbaccatin III without any additional functional group manipulation.[40]

16.6 Conclusions

In summary, we have developed a flexible, stereodivergent approach to β-amino acids and to *syn* and *anti* α-hydroxy-β-amino acids in enantioenriched or enan-

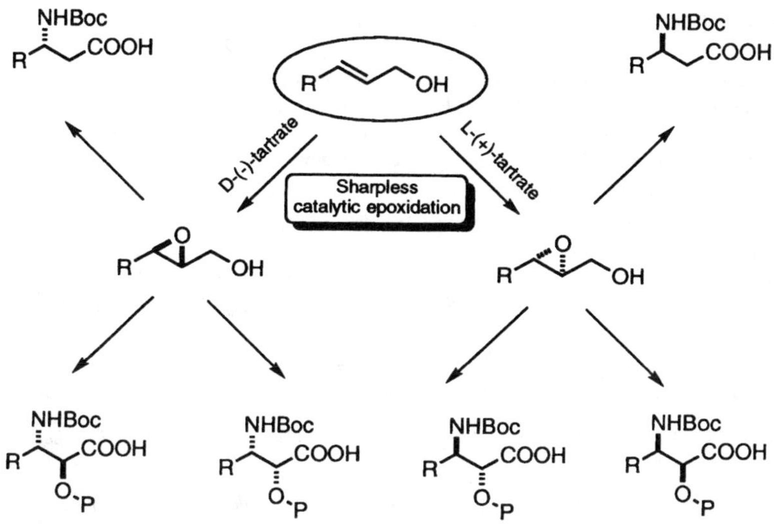

Figure 16.9. Stereodivergent approach to β-amino acids and to α-hydroxy-β-amino acids from (*E*)-allyl alcohols.

tiomerically pure form from *anti* N-Boc-3-amino-1,2-diols. These precursors are readily available through Sharpless catalytic epoxidation of (*E*) allyl alcohols, subsequent regioselective opening with benzhydrylamine, and change of the amino protecting group. Selection of the tartrate ester (D or L) employed in the epoxidation step and use of the methodologies presented here provide completely stereocontrolled access to any of the two enantiomeric simple β-amino acids or to any of the four possible α-hydroxy-β-amino acid stereoisomers from a single (*E*)-allyl alcohol (Figure 16.9).

In the present methodology, two ammonia synthons can be used at will, whereas azide can be more suitable in some instances for small-scale work, benzhydrylamine is a more convenient alternative when scaleup is considered. Given the broad scope of the Sharpless epoxidation, it is presumed that the procedures presented here can provide access to a great variety of enantiomerically pure β-amino acids and diastereo- and enantiomerically pure α-hydroxy-β-amino acids. Work towards this end is in progress in our laboratories.

16.7 Acknowledgements

The work described here could not have been possible without the efforts of our dedicated coworkers, Montse Alcón, Dr. Marc Canas, Dr. Patricia Castejón, Mireia Pastó, and Dr. Marta Poch. Financial support from CIRIT–CICYT and from DGICYT is also gratefully acknowledged.

16.8 References

1. (a) T. Katsuki, K. B. Sharpless, *J. Am. Chem. Soc.* **102**, 5974 (1980). (b) M. G. Finn, K. B. Sharpless, in *Asymmetric Synthesis*, J. D. Morrison, Ed. (Academic Press, New York, 1985), Vol. 5, Chap. 8, p. 247. (c) Y. Gao, R. M. Hanson, J. M. Klunder, S. Y. Ko, H. Masamune, K. B. Sharpless, *J. Am. Chem. Soc.* **109**, 5765 (1987). (d) R. A. Johnson, K. B. Sharpless, in *Catalytic Asymmetric Synthesis*, I. Ojima, Ed. (VCH, New York, 1993), Chap. 4. 1, p. 103.

2. (a) A. Pfenninger, *Synthesis*, 89 (1986). (b) B. E. Rossiter, in *Asymmetric Synthesis*, J. D. Morrison, Ed. (Academic Press, New York, 1985), Vol. 5, Chap. 8, p. 193. (c) K. B. Sharpless, S. S. Woodard, M. G. Finn, *Pure & Appl. Chem.* **55**, 1823 (1983). (d) K. B. Sharpless, C. H. Behrens, T. Katsuki, A. W. M. Lee, V. S. Martin, M. Takatani, S. M. Viti, F. J. Walker, S. S. Woodard, *Pure & Appl. Chem.* **55**, 589 (1983).

3. (a) M. Caron, K. B. Sharpless, *J. Org. Chem.* **50**, 1557 (1985). (b) S. Y. Ko, K. B. Sharpless, *J. Org. Chem.* **51**, 5413 (1986).

4. M. Canas, M. Poch, X. Verdaguer, A. Moyano, M. A. Pericàs, A. Riera, *Tetrahedron Lett.* **32**, 6931 (1991).

5. M. Poch, X. Verdaguer, A. Moyano, M. A. Pericàs, A. Riera, *Tetrahedron Lett.* **32**, 6935 (1991).

6. (a) M. Poch, M. Alcón, A. Moyano, M. A. Pericàs, A. Riera, *Tetrahedron Lett.* **34**, 7781 (1993). (b) P. Castejón, A. Moyano, M. A. Pericàs, A. Riera, *Synthetic Commun.* **24**, 1231 (1994).

7. M. Alcón, M. Canas, M. Poch, A. Moyano, M. A. Pericàs, A. Riera, *Tetrahedron Lett.* **35**, 1589 (1994).

8. Some representative enantioselective syntheses of allylamines: (a) R. B. Grossman, W. M. Davis, S. L. Buchwald, *J. Am. Chem. Soc.* **113**, 2321 (1991). (b) D. Enders, J. Schankat, *Helv. Chim. Acta* **76**, 402 (1993). (c). Z. Y. Wei, E. E. Knaus, *J. Org. Chem.* **58**, 1586 (1993).

9. E. J. Corey, P. B. Hopkins, *Tetrahedron Lett.* **23**, 1979 (1982).

10. M. Alcón, A. Moyano, A. Riera, M. A. Pericàs, unpublished results.

11. J. A. Marshall, A. W. Garofalo, *J. Org. Chem.* **58**, 3675 (1993).

12. Chiralcel® ODR column, 0.5 *M* aq. $NaClO_4$/MeOH as eluent.

13. (a) H. C. Brown, E. F. Knights, C. G. Scouten, *J. Am. Chem. Soc.* **96**, 7765 (1974). (b) K. Burgess, M. J. Ohlmeyer, *Tetrahedron Lett.* **30**, 5857 (1989).

14. E. J. Corey, G. Schmidt, *Tetrahedron Lett.* **20**, 399 (1979).

15. M. Pastó, A. Moyano, M. A. Pericàs, A. Riera, *Tetrahedron: Asymmetry*, **6**, 2329 (1995).

16. M. Pastó, A. Moyano, M. A. Pericàs, A. Riera, *Tetrahedron: Asymmetry*, **7**, 243 (1996).

17. (a) S. Thaisrivongs, *Ann. Reports Med. Chem.* **29**, 133 (1994). (b) A. E. P. Adang, P. H. H. Heunkens, J. T. M. Linders, H. C. J. Ottenheim, C. J. van Staveren, *Recl. Trav. Chim. Pays-Bas* **113**, 63 (1994). (c) J. Gante, *Angew. Chem. Int. Ed. Engl.* **33**, 1699 (1994).

18. K. C. Nicolaou, W.-M. Dai, R. K. Guy, *Angew. Chem. Int. Edn. Engl.* **33**, 15 (1994).

19. S.-S. Moon, J. L. Chen, R. E. Moore, G. M. L. Patterson, *J. Org. Chem.* **57**, 1097 (1992).

20. In most protease inhibitors the relative configuration of the α-hydroxy-β-amino acid fragment is *syn*. *Cf.*: K. Iizuka, T. Kamijo, H. Harada, K. Akahane, T. Kubota, H. Umeyama, T. Ishida, Y. Kiso, *J. Med. Chem.* **33**, 2707 (1990). R. Nishizawa, T. Saino, T. Takita, H. Suda, T. Aoyagl, H. Umezawa, *J. Med. Chem.* **20**, 510 (1977).

21. W. Yuan, C.-H. Wong, J. Z. Haeggström, A. Wetterholm, B. Samuelsson, *J. Am. Chem. Soc.* **114**, 6552 (1992).

22. M. E. Bunnage, S. G. Davies, C. J. Goodwin, *Synlett*, 731 (1993).

23. (a) K. Iiuzuka, T. Kamijo, H. Harada, K. Akahane, T. Kubota, H. Umeyama, Y. Kiso, *J. Chem. Soc. Chem. Commun.*, 1678 (1989). (b) T. Matsumoto, Y. Kobayashi, Y. Takemoto, Y. Ito, T. Kamijo, H. Harada, S. Terashima, *Tetrahedron Lett.* **29**, 4175 (1990).

24. D. C. Chatfield, B. R. Brooks, *J. Am. Chem. Soc.* **117**, 5561 (1995), and references cited therein.

25. (a) A. B. Smith, K. J. Hale, L. M. Laakso, K. Chen, A. Riera, *Tetrahedron Lett.* **30**, 6963 (1991). (b) S. Takano, M. Akiyama, S. Sato, K. Ogasawara, *Chemistry Lett.*, 1593 (1983).

26. U. Schmidt, R. Meyer, V. Leitenberger, A. Lieberknecht, H. Grisser, *J. Chem. Soc., Chem. Commun.*, 275 (1991).

27. Y. Oikawa, T. Yoshioka, O. Yonemitsu, *Tetrahedron Lett.* **23**, 885 (1982).

28. S. F. Martin, J. A. Dodge, *Tetrahedron Lett.* **32**, 3017 (1991).

29. E. J. Corey, A. Venkateswarlu, *J. Am. Chem. Soc.* **94**, 6190 (1972).

30. N. Miyashita, A. Yoshikoshi, P. A. Grieco, *J. Org. Chem.* **42**, 3772 (1977).

31. P. H. J. Carlsen, T. Katsuki, V. S. Martin, K. B. Sharpless, *J. Org. Chem.* **46**, 3936 (1981).

32. (a) O. Mitsunobu, *Synthesis*, 1 (1981). (b) P. L. Robinson, C. N. Barry, S. W. Bass, S. E. Jarvis, S. A. Evans, Jr., *J. Org. Chem.* **48**, 5396 (1983).

33. M. Pastó, P. Castejón, A. Moyano, M. A. Pericàs, A. Riera, *J. Org. Chem.* **61**, 6033 (1996).

34. D. V. Patel, K. Rielly-Gauvin, D. E. Ryono, C. A. Free, S. A. Smith, E. W. Petrillo, Jr., *J. Med. Chem.* **36**, 2431 (1993).

35. (a) T. Matsumoto, Y. Kobayashi, Y. Takemoto, Y. Ito, T. Kamijo, H. Harada, S. Terashima, *Tetrahedron Lett.* **29**, 4175 (1990). (b) T. Inokuchi, S. Tanigawa, M. Kanazaki, S. Torii, *Synlett*, 707 (1991). (c) R. W. Dugger, J. L. Ralbovsky, D. Bryant, J. Commander, S. S. Massett, N. A. Sage, J. R. Selvidio, *Tetrahedron Lett.* **33**, 6763 (1992). (d) K. Hattori, H. Yamamoto, *Tetrahedron* **50**, 2785 (1994).

36. (a) This alcohol was obtained in three steps (Swern oxidation, 99%; Wittig reaction with ethyl triphenylphosphoranylacetate, 92%; DIBAL-H reduction, 96%) from commercial 2-cyclohexylethanol. (b) P. Castejón, M. Pastó, A. Moyano, M. A. Pericàs, A. Riera, *Tetrahedron Lett.* **36**, 3019 (1995).

37. (a) M. Caron, B. Carlier, K. B. Sharpless, *J. Org. Chem.* **46**, 3936 (1981). (b) R. Choukroun, D. Gervais, *J. Chem. Soc., Dalton Trans.*, 1800 (1980).

38. G. Chauvière, D. Guénard, F. Picot, V. Sénilh, P. Potier, *C. R. Séances Acad. Sci. Paris (Série 2)* **293**, 501 (1981).

39. Some representative enantioselective synthesis of the Taxol and/or Taxotere® side chain: (a) J.-N. Denis, A. M. Kanazawa, A. E. Greene, *Tetrahedron Lett.* **35**, 105 (1994). (b) A. M. Kanazawa, A. E. Greene, *J. Org. Chem.* **59**, 1238 (1994). (c) A. M. Kanazawa, A. E. Greene, *J. Chem. Soc. Chem. Commun.*, 2591 (1994). (d) A. M. Kanazawa, J. N. Denis, A. Correa, A. E. Greene, *J. Org. Chem.* **58**, 255 (1993). (e) I. Ojima, I. Habus, M. Zhao, *J. Org. Chem.* **56**, 1681 (1991). (f) D. Guénard, F. Guéritte-Voegelein, P. Potier, *Acc. Chem. Res.* **26**, 160 (1993). (g) A. Commerçon, D. Bézard, F. Bernard, J. D. Bourzart, *Tetrahedron Lett.* **33**, 5185 (1992). (h) A. Commerçon, J. D. Bourzart, *Tetrahedron Lett.* **34**, 6049 (1993). (i) F. A. Davis, R. T. Reddy, R. E. Reddy, *J. Org. Chem.* **57**, 6387 (1992). (j) L. Deng, E. N. Jacobsen, *J. Org. Chem.* **57**, 4320 (1992). (k) M. E. Bunnage, S. G. Davies, C. J. Goodwin, *J. Chem. Soc., Perkin Trans I*, 1375 (1993). (l) K. Hattori, M. Miyata, H. Yamamoto, *J. Am. Chem. Soc.* **115**, 1151 (1993). (m) R. Brieva, J. Z. Crich, C. J. Şih, *J. Org. Chem.* **58**, 1068 (1993). (n) C. Palomo, J. M. Aizpurúa, J. I. Miranda, A. Mielgo, J. M. Odriozola, *Tetrahedron Lett.* **34**, 6325 (1993). (o) Z.-M. Wang, H. C. Kolb, K. B. Sharpless, *J. Org. Chem.* **59**, 5104 (1994). (p) J. Kearns, M. M. Kayser, *Tetrahedron Lett.* **35**, 2845 (1994). (q) A. M. P. Koskinen, E. K. Karvinen, J. P. Siirilä, *J. Chem. Soc., Chem. Commun.*, 21 (1994). (r) C. Bonini, G. Righi, *J. Chem. Soc., Chem. Commun.*, 2767 (1994).

40. J.-N. Denis, A. E. Greene, D. Guénard, F. Guéritte-Voegelein, L. Mangatal, P. Potier, *J. Am. Chem. Soc.* **110**, 5917 (1988).

CHAPTER

17

Stereoselective Ring Opening of Chiral Oxazolidines: A Way to Enantiopure α- and β-Amino Acids

Rafael Pedrosa

17.1 Introduction

Chiral 1,3-oxazolidines, obtained by condensation of an aldehyde with a suitable 1,2-aminoalcohol, have been widely used in asymmetric synthesis for the past 10 years. The synthetic potential of the heterocycle has been mainly exploited in two different ways: as a chiral masking of α,β-unsaturated or functionalized aldehyde, or as a chiral template in nucleophilic ring opening transformations.

Whereas functionalization of a double bond in a 2-alkenyloxazolidine opens the way to different types of compounds, including amino acids, and has been previously reviewed,[1] 2-acyl substituted oxazolidines served as chiral templates in the preparation of protected 2-hydroxyalkanals,[2] chiral 1,2-diols,[3] or diastereomerically pure cyclohexanone derivatives.[4] Husson and colleagues[5] have extensively used 2-cyano-6-phenyloxazolopiperidine and pyrrolidine homolog as chiral template in the asymmetric synthesis of alkaloids.

17.2 Synthesis of Chiral 1,3-Oxazolidines

The preparation of the requisite chiral oxazolidines is achieved by condensation of aldehydes with the corresponding chiral 1,2-aminoalcohol. Generally the reaction is carried out in refluxing benzene with azeotropic removal of water, although it occurs in different solvents, in very mild conditions at room temperature. The common chiral 1,2-aminoalcohols used are ephedrine,[6,7] norephedrine,[8,9] pseudoephedrine,[10] valinol,[11] and phenylglycinol[12] or their derivatives.

There are two remarkable aspects in this reaction: on one hand the nature of the products depending on the substitution at the nitrogen atom at the starting aminoalcohol; on the other, the stereochemistry of the cyclization product.

In fact, it has been demonstrated[10,13] that condensation of N-unsubstituted aminoalcohols **1** with aldehydes leads to an equilibrium mixture of *E*-configured hydroxy imines **2** and the two epimers of 1,3-oxazolidines **3**. The ratio of tautomers and epimers depends on the solvent, the substituents, the stereochemistry at the aminoalcohol,[10] and the structure of the aldehyde, but the hydroxy imines are always the major tautomer (83:17 to 93:7) present in the mixture. As an example, in Scheme 17.1, are collected the data for (*R*)-phenylglycinol and different aromatic aldehydes.[14]

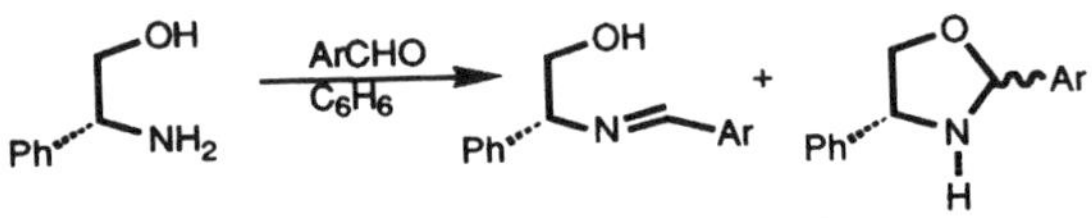

1 **2** **3**

Ar	Yield (%)	Ratio **2**:**3**
Phenyl	78	83:17
4-Methoxyphenyl	72	93:7
2-Furyl	88	94:6
2-Thienyl	98	94:6
3-Thienyl	92	93:7

Scheme 17.1

1,3-Oxazolidines are formed as a single tautomer by condensation of carbonyl compounds or their derivatives with N-substituted 1,2-aminoalcohols. In this case, a mixture of two different diastereomers at C-2 are also formed in different ratios depending on the reaction time and the solvent used in the reaction, but a constant preference (85:15 to 99:1) for the *cis* isomer is observed. The stereochemistry of the major isomers **5** has been assigned on the basis of X-ray structural analysis for a number of these compounds[7,9,12] (Scheme 17.2 summarizes some selected data for different oxazolidines).

This general stereochemical behavior in the cyclization process is explained by accepting that the *cis* isomer is favored by kinetic as well as thermodynamic conditions.[9,15] However condensation of methylhemiacetal of ethyl glyoxylate with (*R*)-N-benzyl-2-phenylglycinol in refluxing CH_2Cl_2 for 2 h led to a mixture (83:17) of 2-ethoxycarbonyl-1,3-oxazolidine, where the major component was the *trans* isomer **8**, as determined by X-ray analysis (Scheme 17.3).[16]

On the other hand, deprotonation of **8** with the lithium salt of 2,2,6,6-tetramethyl piperidine in dry THF at –78°C for 1 h and quenched by deuterium oxide yields the 2-deuteriated *cis* isomer **10** as a single compound.[17]

The *trans* isomers are also obtained as major diastereomers when 2-methoxy-1,3-oxazolidines derived from ephedrine or phenylglycinol are allowed to react with allyltrimethylsilane in the presence of zinc chloride etherate or a catalytic amount of

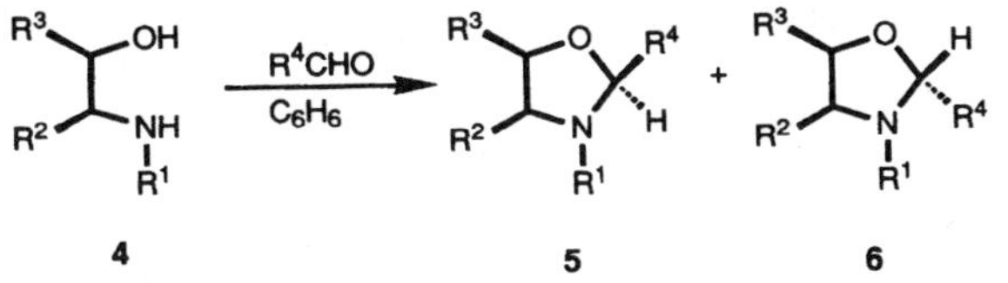

R^1	*R^2*	*R^3*	*R^4*	Ratio **5**:**6**
CBZ	Me	Ph	$CH_2CH{=}CHCO_2Me$	94:6
Ts	Me	Ph	$CH_2CH{=}CHCO_2Me$	93:7
CBZ	Me	Ph	$CH_2CH{=}CHCO_2Bn$	85:15
Me	Ph	H	Ph	97:3
Me	Ph	H	3-Thienyl	99:1
Me	Ph	H	Ph	>95:5
Bn	Ph	H	Me	88:12

Scheme 17.2

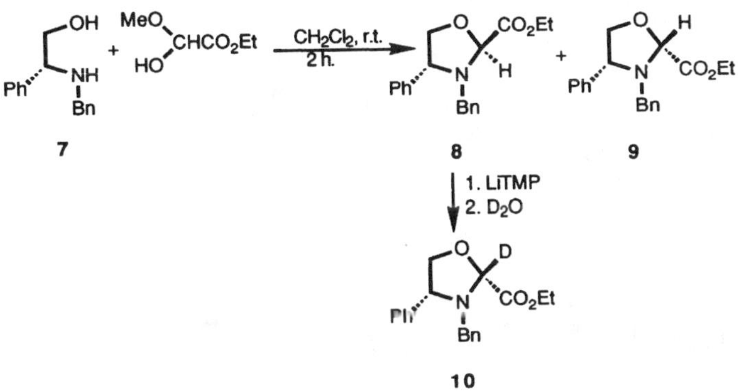

Scheme 17.3

trimethylsilyl triflate. The product[12] is formed by kinetic control because it epimerizes completely to the *cis* diastereomer **13**, when treated with TMSOTf at room temperature (Scheme 17.4).[18]

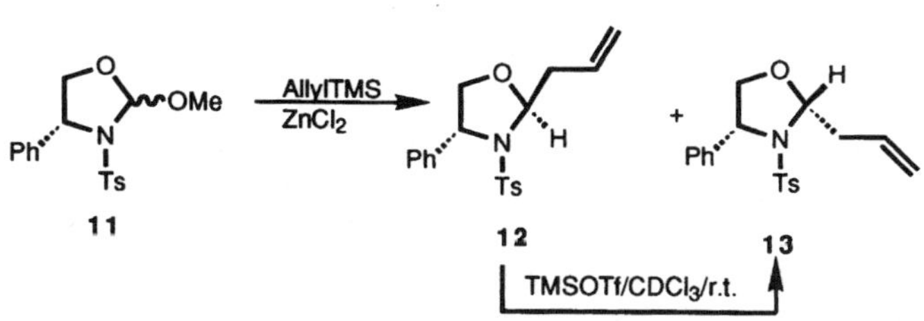

Scheme 17.4

17.3 Stereoselective Ring Opening of Chiral 1,3-Oxazolidines

17.3.1 Stereoselective Synthesis of Amines

The stereoselective ring opening of chiral oxazolidines by Grignard reagents was first reported by Neelakantan[19] more than 20 years ago, and later widely exploited as a way to enantiopure amines.

Thus the reaction of 2-alkyl-substituted 6-oxazolopiperidine **14** with alkylmagnesium halides constitutes one of the key steps in the asymmetric synthesis of (+)- and (–)-dihydropinidine[20] and some other alkaloids (Scheme 17.5).

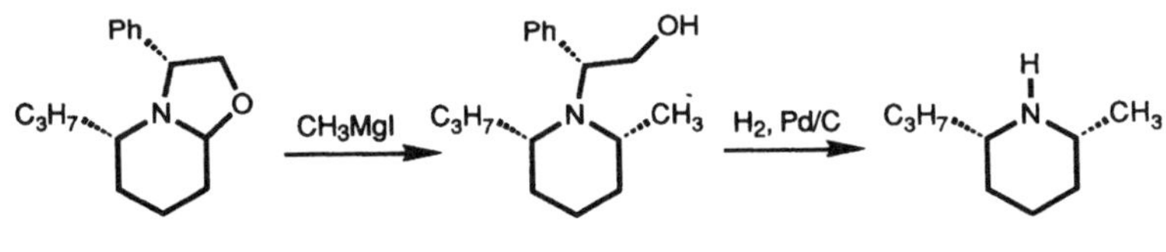

Scheme 17.5

Very recently, and without apparent notice of the previous work, this method has been adapted by Higashiyama[21] to the asymmetric synthesis of 2-aryl- and 2,6-diaryl-substituted pyrrolidines (Scheme 17.6).

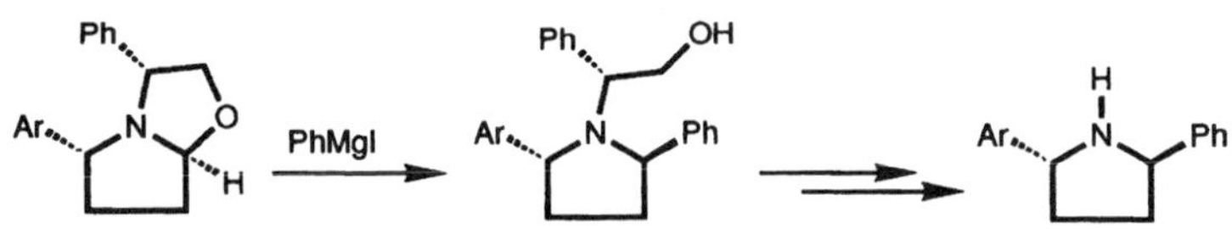

Scheme 17.6

It is worth noting that the *trans* geometry in the final pyrrolidine compared with the *cis* in the piperidine derivative is due to the different stereochemistry at the starting condensed bicyclic 1,3-oxazolidine, and not to a different stereochemical approach of the organometallic in the ring opening process.

However, a more systematic study on the characteristics of this reaction has been made by Takahashi and co-workers.[12,22] These authors have shown that the stereochemistry at the new created stereocenter in the open products depends, not only on the configuration at the 1,2-aminoalcohol moiety, but also on the organometallic used. The Grignard or organolithium derivatives[22a] led to the same major stereoisomer, whereas the minor product in these reactions was obtained as major component when the opening was carried out with organotitanium reagents.[22d] Moreover, the stereoselectivity of the reaction is increased with increasing bulkiness of the N-substituent[12] (Scheme 17.7).

Hydroxylamines, accompanied by small amount of tautomeric 1,3-oxazolidines, obtained by condensation of chiral N-unsubstituted 1,2-aminoalcohols with alde-

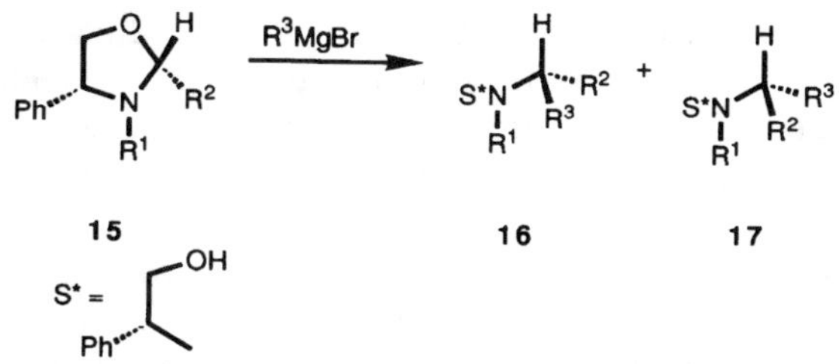

15	R^1	R^2	R^3	Ratio **16**:**17**
a	Me	p-Br-C_6H_4	Me	66:34
b	Me	Ph	Me	66:34
c	Bn	Ph	Me	84:16
d	*i*-Pr	Ph	Me	97:3
e	Me	Me	Ph	78:22
f	Bn	Me	Ph	77:23
g	*i*-Pr	Me	Ph	88:11

Scheme 17.7

hydes, have also been used as the starting material in a related synthesis of enantiopure amines and aminodiols.[23]

To this end, allylcerium chloride,[24] alkylmagnesium chlorides,[14,25] methyllithium,[14,26] vinylmagnesium bromide,[23] or phenylcerium dichloride[23] have been used as nucleophiles. In general, the chiral discrimination in the addition reaction is very good, and, in contrast with 1,3-oxazolidines derived from N-substituted (*R*)-phenylglycinol, the isomeric hydroxyimines derived from the unsubstituted aminoalcohol lead to final amines with contrary stereochemistry at the stereocenter (Scheme 17.8)

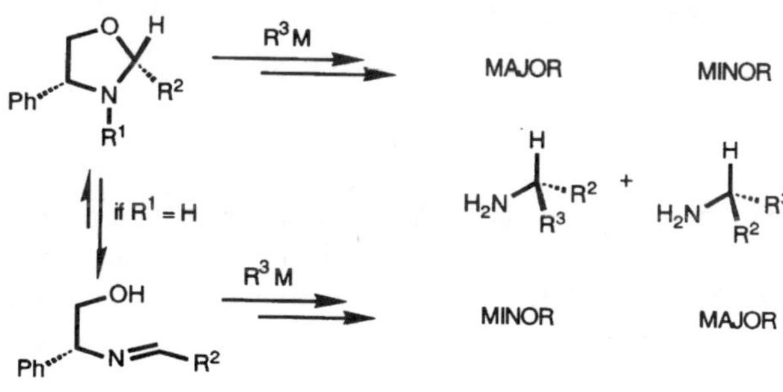

Scheme 17.8

17.3.2 Stereoselective Synthesis of α-Amino Acids

Chiral 1,3-oxazolidines have been used as starting templates in the stereoselective synthesis of α-amino acids in two ways: as chiral glycine or related synthon, or as starting structures in an asymmetric Strecker-type reaction.

As previously mentioned, condensation of (*R*)-N-benzyl-2-phenylglycinol with the methyl hemiacetal of ethyl glyoxilate in refluxing CH_2Cl_2 and molecular sieves for 2 h gave a mixture (83:17) of epimeric 2-ethoxycarbonyl-1,3-oxazolidines. After column chromatography the major (2*S*,4*R*)-isomer **8** was obtained as a white solid in 66% total yield.

The stereoselective introduction of the chain into **8** was carried out by reaction with 4 equiv. of dialkylzinc reagents prepared from alkylmagnesium iodides and zinc chloride in diethyl ether. The reaction occurs in mild conditions, 0°C for 2–3 h, leading to a mixture of stereoisomers **18** and **19** in good diastereomeric excess.[16] (Scheme 17.9)

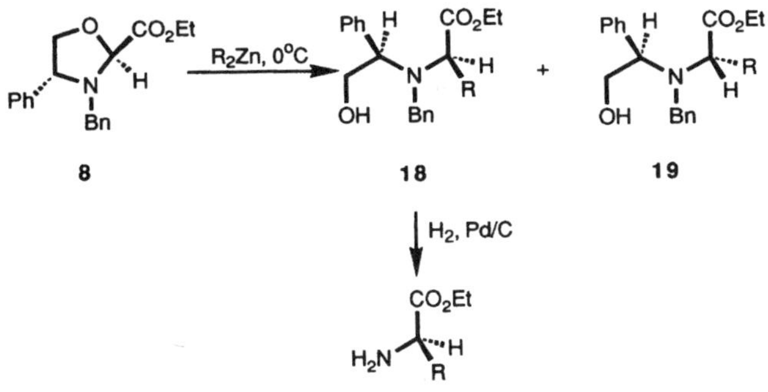

R	Ratio **18**:**19**	**29** Yield (ee, conf.)
Me	86:14	45 (>98, *R*)
Et	87:13	42 (>98, *R*)
Pr	88:12	47 (>98, *R*)
i-Pr	97:3	56 (>98, *R*)
n-Bu	92:8	54 (>98, *R*)
i-Bu	89:11	47 (>98, *R*)

Scheme 17.9

Moreover, the major stereoisomers **18** were obtained as pure compounds by flash chromatography, and then transformed into the enantiomerically pure (*R*)-amino esters in one step by hydrogenolytic *bis*-debenzylation at room temperature.

Taking into account the stereochemistry of the starting 1,3-oxazolidine and the final products, and that the presence of a magnesium salt is necessary to the reaction, it was assumed that the opening of the heterocycle occurs from the "nitrogen side" in a chelated ion pair **21** (Scheme 17.10).

We have used the epimeric *cis* 2-deuteriated-1,3-oxazolidine **10** as the starting compound in the diastereoselective synthesis of (*S*)-α-deuterated α-amino acids **22** according to a related protocol[10] (Scheme 17.11). However, the behavior of **10** towards dialkylzines is quite different from that shown by **8**; it is necessary to use boron trifluoride etherate as a Lewis acid to get the transformation into the opening products, probably because the reaction occurs via an open iminium ion.

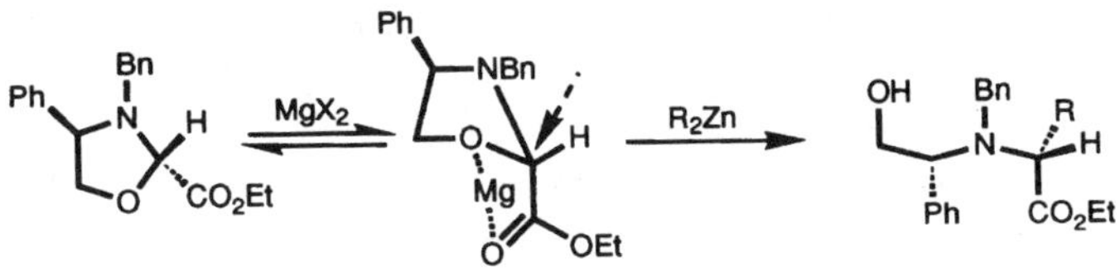

8 21 20

Scheme 17.10

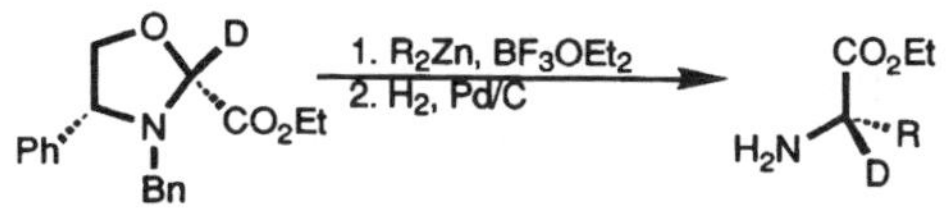

10 22

Scheme 17.11

In a related work, Pridgen et al.[27] has recently reported the enantioselective synthesis of aldehydes from N-unsubstituted chiral 1,3-oxazolidines. In this case cyclic acetals **23** derived from 1,4-dimethyl-2-4-pentanediol of 2-formyl-1,3-oxazolidine have been used as the starting material in the addition of magnesium derivatives. Grignard addition to **23** in toluene at –45°C gave the aminoalcohols **24** in excellent chemical yields and diastereomeric excesses ranging from 90 to 98%. The diastereometric discrimination diminished to 60–75% yield when *tert*-butyllithium or mixtures of this reagent and cerium trichloride or ethylmagnesium chloride and cerium trichloride were used as nucleophiles.

The conversion of **24** into the final chiral α-aminoaldehydes is described for the ethyl derivative (*R* = Et) as a three-step protocol, including elimination of the chiral appendage by hydrogenolysis with palladium–ammonium formate, transformation into N-Cbz protected aminoacetal **25**, followed by transacetalation and subsequent hydrolysis in aqueous DMSO (Scheme 17.12). The starting compound

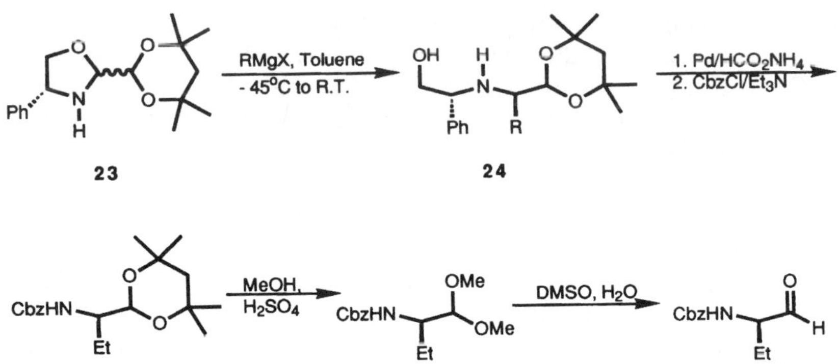

23 24

25

Scheme 17.12

derived from *tert*-leucinol slightly increased the selectivity with respect to the phenylglycinol derivative.

The diastereoselective incorporation of a cyanide (Strecker-type reaction) into a chiral 1,3-oxazolidine was first reported by Husson.[28] The reaction is not directed to the synthesis of α-amino acids, but it constitutes a key step in the asymmetric synthesis of (+)-solenopsin A. The 2-methyl substituted 6-oxazolopiperidine **26** was reacted with trimethylsilyl cyanide in the presence of a catalytic amount of $ZnBr_2$ in refluxing CH_2Cl_2, leading to the α-amino nitrile **27** in nearly quantitative yield as a mixture (7:3) of epimers (Scheme 17.13).

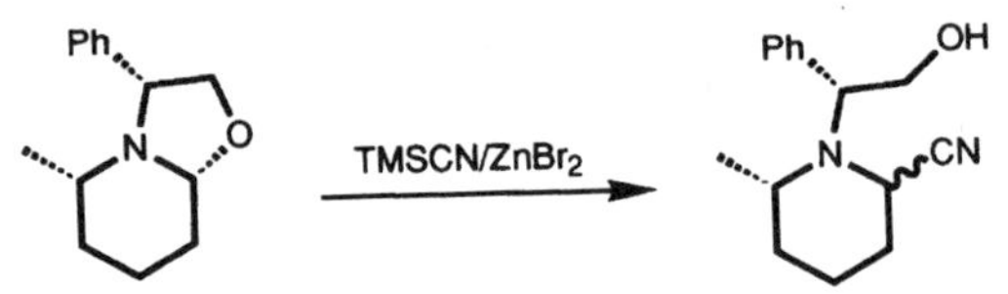

26 **27** (7:3 mixture)

Scheme 17.13

α-Amino nitriles, later converted into α-amino acids or esters, have been also prepared[29] by addition of trimethylsilyl cyanide to the tautomeric mixture of hydroxyimines and N-unsubstituted 1,3-oxazolidines prepared from (*R*)-phenylglycinol and aldehydes.

The degree of diastereoselectivity depends on the structure of the aldehyde, and varies from phenylacetaldehyde (8% de) to *p*-methoxybenzaldehyde (80% de). Otherwise, the stereochemical discrimination also depends on the solvent used. The best results are obtained in chloroform, and the worst in THF, although in this solvent the de increases by addition of anhydrous magnesium chloride before adding TMSCN.

It is worthy of note that the major stereoisomers obtained in these processes have the *S* configuration at the new stereocenter, and contrary to those obtained by addition of organometallics to the same imines (Scheme 17.14).

A better stereochemical discrimination was observed when diethylaluminum cyanide was used as nucleophile. Treatment of (2*R*,4*R*)-2-alkylsubstituted-1,3-oxazolidines **31** with Nagata's reagent at –80°C in toluene yields chiral α-aminonitriles with de ranging from 54 to 86%.[30]

This reaction constitutes a complementary method to the above described because the major diastereomer obtained has configuration *R* at the newly created stereocenter. Attempts to debenzylate α-aminonitriles **32** failed, but they were easily transformed into α-amino esters **33** by treatment with saturated hydrogen chloride ethanol, and then with water. Debenzylation (H_2, Pd/C, room temperature, 9 atm.) led to (*R*)-α-amino esters **34** without loss of their stereochemical integrity (Scheme 17.15).

The presence of an additional donor heteroatom at the alkyl substituent at C-2 substantially varies the stereochemical behavior of these substrates.[31] Thus the reaction of 1,3-oxazolidine **35**, derived from phenoxyacetaldehyde, when submitted to the de-

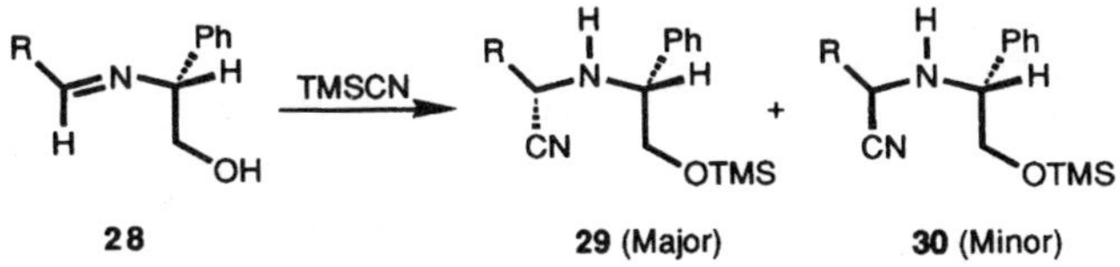

R	Ratio **29**:**30**
Ph	82:18
4-Me-C_6H_4	85:15
4-MeO-C_6H_4	90:10
$C_6H_6CH_2$	54:46
i-Pr	84:16
t-Bu	88:12

Scheme 17.14

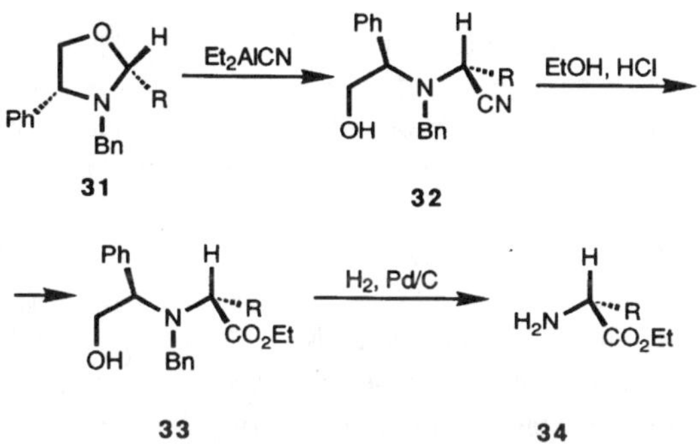

R	**32** d.e.
Me	58
Et	86
i-Pr	86
n-Bu	54
i-Bu	84
$PhCH_2CH_2$	62

Scheme 17.15

scribed reaction conditions, led to a mixture of epimeric amino nitriles **36** and **37** in a 7:3 ratio, whereas the oxazolidine **38** derived from 3-benzyloxypropanal yielded an equimolar mixture of diastereomers **39** and **40** (Scheme 17.16).

On the other hand, oxozolidinium salts, obtained by reaction of chiral 1,3-oxazolidines with methyl iodide, react with sodium cyanide in DMSO, leading to chiral α-cyanoethers, later transformed into enantiopure α-hydroxy acids. This process changes not only the stereochemistry but also the regiochemistry of the reaction.[32]

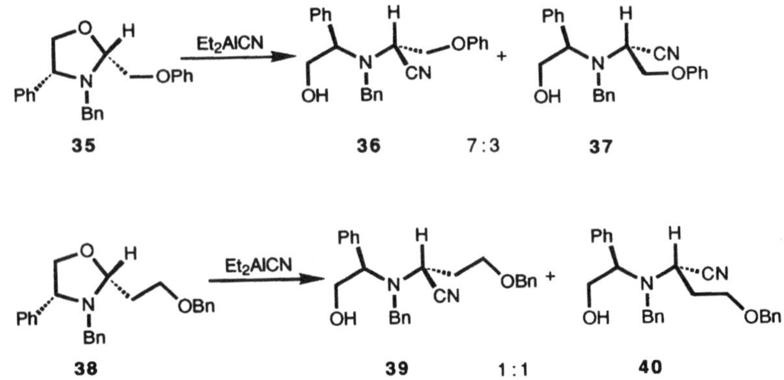

Scheme 17.16

17.3.3 Synthesis of β-Amino Acids

Based on previous work[33] showing the regioselective ring opening of N-benzyl-1,3-oxazines by Reformatsky reagents to give β-amino ester derivatives with excellent chemical yields, we extended our method to the stereoselective synthesis of β-amino carboxylates from chiral 1,3-oxazolidines **31** derived from (*R*)-N-benzyl-2-phenylglycinol.[34]

2-Alkyl substituted 1,3-oxazolidines readily react with the zinc derivative of ethyl bromoacetate in mild conditions (Et_2O or CH_2Cl_2, 0°C), affording the open products **41** with moderate to very good diastereoselectivities, depending on the bulkiness of the substituent at C-2. Moreover, the change of the solvent in the reaction has only a minor effect on both the chemical yields and the diastereomeric excess (Scheme 17.17).

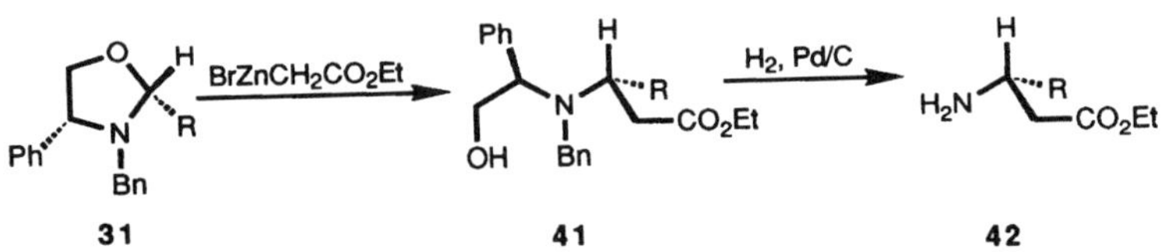

R	**41** (dr)	**42** (ee, Conf.)
Me	80:20	(60, *R*)
Et	86:14	(72, *R*)
n-Pr	89:11	(78, *R*)
Bu	96:4	(92, *R*)
i-Bu	97:3	(86, *R*)
$C_6H_5CH_2CH_2$	96:4	(92, *R*)
i-Pr	69:31	(38, *R*)

Scheme 17.17

An exception to this behavior is the 2-isopropyl derivative. This compound does not react with Reformatsky reagent at 0°C, and it is necessary to heat the reaction mixture to 35°C for 3 h. In these conditions a mixture (69:31) of epimeric perhydro oxazepinones was obtained. Otherwise, compounds **41** were easily *bis*-debenzylated without loss of their stereochemical integrity into a mixture of ethyl β-amino carboxylates **42** by stirring for 15–24 h at room temperature under H_2 atmosphere in the presence of 10% palladium on carbon as catalyst.

The stereochemistry of the final β-amino esters was proved to be *R*, indicating that the Reformatsky reagent preferentially enters by the "nitrogen side" of the heterocycle following an S_N2-type reaction.

Pridgen et al.[35] have also reported a highly diastereoselective preparation of β-aryl substituted β-amino esters by reaction for a long period of time (72–264 h) of a mixture of hydroxyimines and N-unsubstituted 1,3-oxazolidines **43** with the stable tributylstannylacetate, in the presence of Lewis acids (Scheme 17.18).

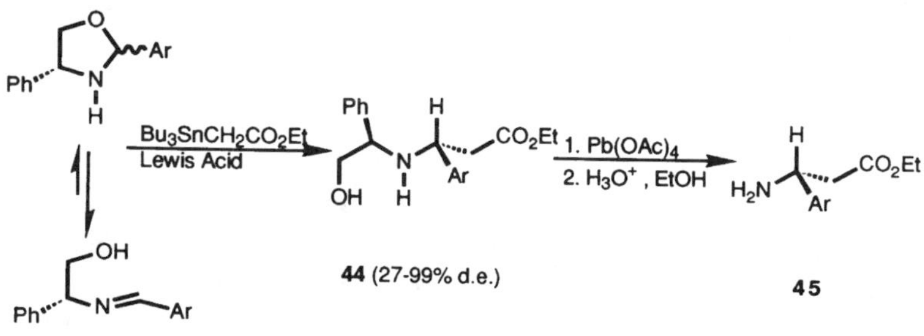

43	Ar	L. A.	44 de
a	Ph	$ZnCl_2$	96
a	Ph	BF_3OEt_2	28
a	Ph	$TiCl_4$	40
a	Ph	$Sn(OTf)_2$	70
a	Ph	$ZnCl_2/BF_3OEt_2$	92
b	4-Me-C_6H_4	$ZnCl_2/BF_3OEt_2$	92
c	4-Br-C_6H_4	$ZnCl_2$	96
c	4-Br-C_6H_4	BF_3OEt_2	27
c	4-Br-C_6H_4	$ZnCl_2/BF_3OEt_2$	91
d	1-naphthyl	$ZnCl_2/BF_3OEt_2$	99
e	cyclohexyl	$ZnCl_2/BF_3OEt_2$	94
f	*i*-propyl	$ZnCl_2/BF_3OEt_2$	96

Scheme 17.18

Among the different Lewis acids assayed, the best stereochemical results correspond to $ZnCl_2$, but the chemical yield was poor (39%); on the contrary, boron trifluoride etherate as catalyst on the same substrate accelerated the rate of addition (82% chemical yield), albeit at the expenses of stereoselectivity (28% de).

A combination of a half-equivalent each of $ZnCl_2$ and BF_3OEt_2 represents the catalyst of choice to obtain an excellent diastereofacial selectivity (92–99% de) and chemical yields (33–71%). These results led the authors to propose that in the transition state, the $ZnCl_2$ acts as chelator for the hydroxyimine, and BF_3OEt_2 complexes the ester, acting as activator. The nucleophilic addition occurs by the *Re* face of the imino moiety, opposite the (*R*)-phenyl group of the auxiliary, and leading to *S*-configured β-amino ester (Scheme 17.19).

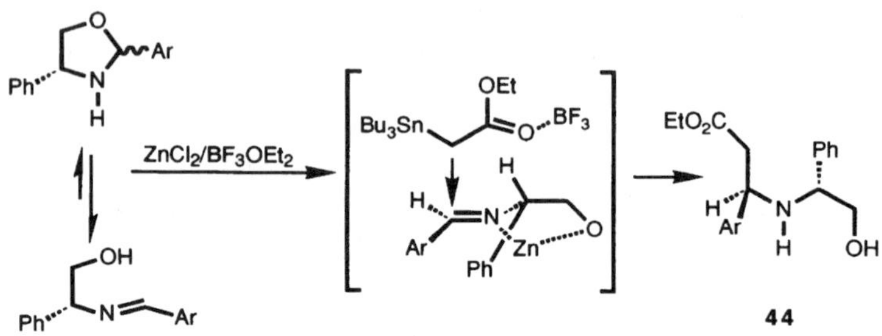

44

43

Scheme 17.19

In this respect, the addition of the stannyl derivative is complementary to the ring opening of 1,3-oxazolidines by Reformatsky reagents[34]; the stereochemistry of the final products are opposite, starting from the same chiral adjuvant. Otherwise, the transformation of products **44** in to the final chiral β-amino esters was achieved by oxidative cleavage of the auxiliary with $Pb(OAc)_4$ in 73–86% yields, and without apparent racemization.[25]

Although authors claim that the secondary amino function is responsible for the great selectivity, it is most probable that, owing to the chemical constitution of the mixture, a good discrimination is only achieved in the case of a free hydroxy group present in the imine.[14] In this respect, the transformation of the OH in the methyl ether derivative highly decreased the facial discrimination in related additions to chiral hydroxyimines.[29]

Very recently, Royer, Husson, et al.[36] have proposed a very interesting approach to β-lactams from 6-cyano-2-oxazolopiperidine. After transformation of the starting compound into the *n*-propyl derivative **14**, the key step in the synthesis is the diastereoselective ring opening of the oxazolidine by reaction with the TBS-enol ether of ethyl acetate in acetonitrile at –78°C in the presence of BF_3OEt_2.

The β-amino ester **46** is obtained as a 9:1 mixture of epimers in favor of the *cis* isomer, which after chromatographic separation, hydrogenolysis of the chiral appendage, and hydrolysis led to the β-amino acid **47**. The cyclization to the β-lactam **48** was achieved, in 22% yield, using Mukaiyama's reagent [37] (Scheme 17.20).

This approach was extended to the synthesis of carbacephams bearing a carboxylic substituent at the C-2 position. The commercially available 6-cyano-2-oxazolopiperidine was transformed into an epimeric mixture (7:3) of the esters **49** by acidic ethanolysis. Addition of TBS-enol ether of benzyl acetate to the expected iminium ion, performed by treatment of **49** with BF_3OEt_2, led to the *cis* open product **50** in 75% yield. Debenzylation by hydrogenolysis and cyclization yielded the final ethoxycarbonyl derivative **51** (Scheme 17.21).

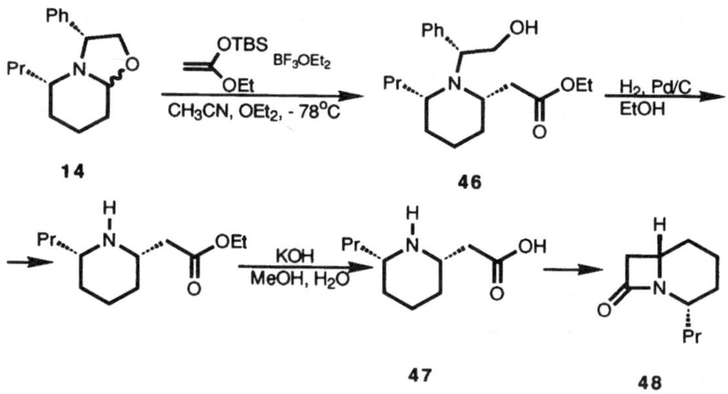

Scheme 17.20

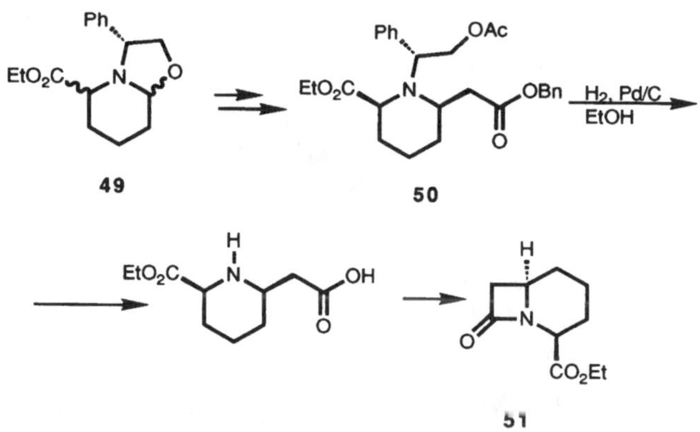

Scheme 17.21

17.4 Concluding Remarks

Highly diastereoselective carbon–carbon bond forming reactions have been developed by opening chiral 1,3-oxazolidines. The development of functionalized nucleophiles and efficient reaction conditions that allow the control of diastereoselectivity will offer attractive opportunities and further applications in the years to come.

17.5 Acknowledgments

It is a great pleasure to thank my co-workers, whose dedication, skill, and creativity have made this review possible. The Spanish DGICYT (Projects PB89-0356 and PB92-0262) is also acknowledged for funding the research summarized here.

17.6 References

1. C. Scolastico, *Pure and Appl. Chem.* **60**, 1689 (1988).
2. K. C. Frieboes, T. Harder, D. Aubert, C. Strahringer, M. Bolte, D. Hoppe. *Synlett*, 921 (1993).
3. C. Agami, F. Couty, C. Lequesne, *Tetrahedron Lett.* **35**, 3309 (1994).
4. O. Prien, H. Hoffmann, K. Conde-Frieboes, T. Krettek, B. Berger, K. Wagner, M. Bolte, D. Hoppe, *Synthesis*, 1313 (1994), and references therein.
5. See, for instance: S. Arseniyadis, P. Q. Huang, D. Piveteau, H.-P. Husson, *Tetrahedron* **44**, 2457 (1988).
6. H. A. Beckett, G. R. Jones, *Tetrahedron* **33**, 3313 (1977).
7. G. Just, P. Potvin, P. Uggowitzer, P. Bird, *J. Org. Chem.* **48**, 2923 (1983).
8. L. Guerrier, J. Royer, D. S. Grierson, H.-P. Husson, *J. Am. Chem. Soc.* **105**, 7754 (1983).
9. A. Bernardi, S. Cardani, T. Pilati, G. Poli, C. Scolastico, R. Villa, *J. Org. Chem.* **53**, 1600 (1988).
10. F. Fülöp, G. Bernáth, J. Mattinen, K. Pihlaja, *Tetrahedron* **45**, 4317 (1989).
11. H. Takahashi, Y. Suzuki, T. Kametani, *Heterocycles* **20**, 607 (1983).
12. H. Takahashi, B. C. Hsieh, K. Higashiyama, *Chem. Pharm. Bull.* **38**, 2429 (1990).
13. M. B. Eleveld, H. Hogeveen, E. P. Schudde, *J. Org. Chem.* **51**, 3635 (1986).
14. K. Higashiyama, H. Inoue, H. Takahashi, *Tetrahedron Lett.* **33**, 235 (1992).
15. C. Agami, T. Rizk, *Tetrahedron* **41**, 537 (1985).
16. C. Andrés, A. González, R. Pedrosa, A. Pérez-Encabo, S. García-Granda, M. A. Salvadó, F. Gómez-Beltrán, *Tetrahedron Lett.* **33**, 4743 (1992).
17. J. M. Andrés, R. Pedrosa, A. Pérez-Encabo, unpublished results.
18. K. Conde-Frieboes, D. Hoppe, *Synlett*, 99 (1990).
19. L. Neelakantan, *J. Org. Chem.* **36**, 2253 (1971). L. Neelakantan, *ibid.* **32**, 2256 (1972).
20. L. Guerrier, J. Royer, D. S. Grierson, H.-P. Husson, *J. Am. Chem. Soc.* **105**, 7754 (1983).
21. K. Higashiyama, H. Inoue, H. Takahashi, *Tetrahedron* **50**, 1083 (1994).
22. (a) H. Takahashi, Y. Suzuki, T. Kametani, *Heterocycles* **20**, 607 (1983). (b) H. Takahashi, Y. Chida, T. Suzuki, H. Onishi, S. Yanaura, *Chem. Pharm. Bull.* **32**, 2714 (1984). (c) H. Takahashi, Y. Chida, K. Higashiyama, H. Onishi, *Chem. Pharm. Bull.* **33**, 4662 (1985). (d) H. Takahashi, H. Niwa, K. Higashiyama, *Heterocycles* **27**, 2099 (1988). (e) H. Takahashi, Y. Chida, T. Yoshii, T. Suzuki, S. Yanaura, *Chem. Pharm. Bull.* **34**, 2071 (1986).
23. M. A. Scialdone, A. I. Meyers, *Tetrahedron Lett.* **35**, 7533 (1994).
24. M.-J. Wu, L. N. Pridgen, *Synlett,* 636 (1990).
25. M.-J. Wu, L. N. Pridgen, *J. Org. Chem.* **56**, 1340 (1991).
26. K. Higashiyama, H. Inoue, T. Yamaguchi, H. Takahashi, *J. Chem. Soc., Perkin Trans 1*, 111 (1995).
27. K. R. Muralidharan, M. K. Mokhallalati, L. N. Pridgen, *Tetrahedron Lett.* **35**, 7489 (1994).
28. D. S. Grierson, J. Royer, L. Guerrier, H.-P. Husson, *J. Org. Chem.* **51**, 4475 (1986).
29. T. K. Chakraborty, G. Reddy, K. A. Hussain, *Tetrahedron Lett.* **32**, 7597 (1991).
30. C. Andrés, A. Maestro, R. Pedrosa, A. Pérez-Encabo, M. Vicente, *Synlett*, 45 (1992).

31. C. Andrés, J. M. Iglesias, J. Nieto, R. Pedrosa, unpublished results.

32. C. Andrés, M. Delgado, R. Pedrosa, R. Rodríguez, *Tetrahedron Lett.* **34**, 8325 (1993).

33. A. Alberola, M. A. Alvarez, C. Andrés, A. González, R. Pedrosa, *Synthesis*, 1057 (1990).

34. C. Andrés, A. González, R. Pedrosa, A. Pérez-Encabo, *Tetrahedron Lett.* **33**, 2895 (1992).

35. M. K. Mokhallalati, M.-J. Wu, L. N. Pridgen, *Tetrahedron Lett.* **34**, 47 (1993).

36. J. F. Berrien, M. A. Billion, H. P. Husson, J. Royer, *J. Org. Chem.* **60**, 2922 (1995).

37. H. Huang, N. Iwasawa, T. Mukaiyama, *Chem. Lett.*, 1465 (1984).

CHAPTER

18

Carbohydrates as Chiral Templates in the Stereoselective Synthesis of β-Amino Acids

Horst Kunz, Markus Weymann, and Andreas Burgard*

18.1 Introduction

The stereoselective formation of chiral amino acids is of particular interest in the field of asymmetric synthesis. Until the end of the 1980s, numerous methods for the enantioselective preparation of α-amino acids[1] had been developed, whereas only a few isolated reports on asymmetric syntheses of β-amino acids were published.[2] β Amino acids are valuable precursors for the syntheses of β-lactams.[3] The increasing interest in chiral β-amino acids and α-hydroxy-β-amino acids mainly arises from the important roles they play in a variety of biologically active molecules. They are found in peptides, such as in the enzyme inhibitor and antitumor reagent bestatin,[4] or in more complex structures such as Taxol,[5] a promising lead compound for cancer therapy. The demand for new efficient synthetic methods for the stereoselective syntheses of chiral β-amino acids has initiated intensive research activities in this field.[6] In our concept, we use carbohydrate-derived auxiliaries for enantio- and diastereoselective syntheses of chiral β-amino acids, as is outlined in this chapter.

18.2 Carbohydrates in Stereoselective Syntheses

Carbohydrates and related natural polyhydroxy compounds have received increasing interest as starting materials in ex-chiral pool synthesis of chiral drugs and natural compounds during the past decades.[7] Bearing up to five stereogenic centers, common monosaccharides include the most inexpensive natural chiral compounds. In contrast to their application in ex-chiral pool syntheses and despite the stereochemical information they contain, the utility of carbohydrates as chiral auxiliaries in asymmetric syntheses was hardly recognized for a long time. Until the mid-1980s only a few examples of applications in this sense had been described.[8] Carbohydrates were considered to be too complex, and due to the presence of up to five very similar hydroxyl groups, "overfunctionalized" to be successfully applicable as stereodifferentiating tools in stereoselection processes. Meanwhile, carbohydrate-derived auxiliaries have been successfully applied in numerous stereoselective transformations. They were used to attain high asymmetric induction in processes such as [4+2]- and [2+2]-cycloadditions, cyclopropanations, Michael addition, or Strecker and Ugi syntheses of α-amino acids.[9]

Our concept of using carbohydrate-derived auxiliaries in stereoselective syntheses originally arose from mechanistic rationalizations of side reactions in the synthesis of biologically interesting glycopeptides.[10] The base-catalyzed β-elimination of the carbohydrate moiety from glycosyl serine and glycosyl threonine derivatives was traced back to the complexing abilities of polyfunctional carbohydrates. The use of these complexing properties together with the stereochemical information carbohydrates contain on one molecular unit appeared to be a promising principle in stereoselective syntheses.[11]

18.3 Glycosylamines as Chiral Templates

Glycosylamines, in particular, the 2,3,4,6-tetra-O-pivaloyl-β-D-galactopyranosylamine **1**,[12] are chiral cyclic amino acetals, concomitantly providing high stereoselectivities and good yields in a variety of asymmetric transformations. Cleavage of the target compound from the carbohydrate template is readily accomplished under mild conditions. Galactosylamine **1** is easily obtained in large scale in a four-step synthesis starting from commercially available penta-O-acetyl-galactose **2** (Scheme 18.1). Transformation of **2** with trimethylsilyl azide under Lewis acid catalysis[13] furnishes β-configurated azide **3** in high yield. After removal of the O-acetyl groups and subsequent reaction with pivaloyl chloride/pyridine to give O-pivaloylated galactosyl azide **4**, the synthesis of galactosyl amine **1** is accomplished by hydrogenation of the azide **4** using Raney nickel as the catalyst.[14]

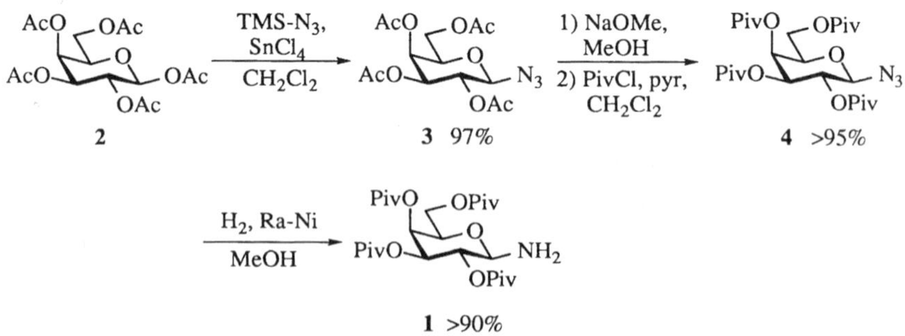

Scheme 18.1

Due to steric, stereoelectronic, and complexing effects, aldimines of glycosylamine **1** proved to be very effective in asymmetric Strecker[12,15] and Ugi[14,16] reactions to furnish α-amino acids, which were obtained in high yield, in excellent diastereomeric excess, and finally in enantiomerically pure form. These results encouraged us to focus on stereoselective constructions of chiral nitrogen containing heterocycles. Starting from galactosylamine **1**, the Schiff bases **5** were synthesized by reaction of **1** with aromatic aldehydes in isopropanol in the presence of catalytic amounts of acetic acid. The aldimines of aliphatic aldehydes were prepared in pentane in the presence of 4 Å molecular sieves at room temperature.[17] Because of the electron-withdrawing amino acetal structure, the aldimines **5** are less electron-rich than common Schiff bases. Therefore, they undergo Lewis acid–promoted aza-Diels–Alder reactions with isoprene and 2,3-dimethyl-butadiene at temperatures of 0 up to 20°C to give the corresponding N-glycosylated piperidine derivatives in good yield and a diastereoselectivity up to ratios of 10:1.[18]

It appeared reasonable that reaction of **5** with electron-rich dienes, as, for example, the Danishefsky diene 1-methoxy-3-trimethylsiloxy-butadiene[19] **6** (Scheme 18.2, Table 18.1), can be conducted at lower temperatures, providing higher stereoselectivities. In fact, the diastereomeric ratios achieved in these syntheses ranged from 20 to 50:1.[20] Mechanistic studies revealed that the reaction does not proceed via a concerted cycloaddition pathway. Instead, an initial Mannich reaction of the silyl enolether moiety with the aldimine leads to Mannich base **8**, which in a second step cyclizes via a Michael-type addition. A subsequent condensation forms the N-glycosylated piperidinone **7**.

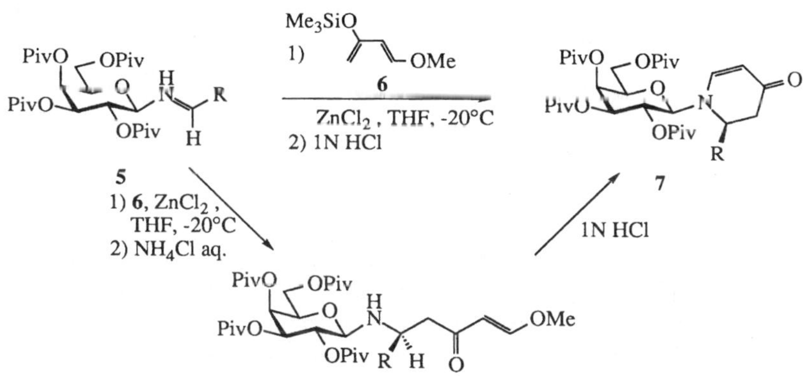

Scheme 18.2

The assignment of the absolute configuration of the chiral dehydropiperidinones **7** was provided by transformation of **7d** (*R* = 3-pyridyl) into the known alkaloid (*S*)-anabasin[21] in four steps. The same reaction sequence, applied to the Schiff base **5** of butyraldehyde, gave the propyl-substituted piperidinone **7a**, which after reduction of a double bond and carbonyl group and acidolytic release from the carbohydrate furnished the piperidine alkaloid coniine as its hydrochloride. According to the optical rotation value of the obtained coniine hydrochloride and its comparison

Table 18.1 DIASTEREOSELECTIVE REACTION OF N-GALACTOSYL IMINES **5** WITH DANISHEFSKY'S DIENE **6** ACCORDING TO SCHEME 18.2

Product	*R*	Yield (%)[a] **7/8**	Yield (%)[b] **7**	(*S*):(*R*)[c]
7a/8a	*n*-Prop	90/–	96	2.5:97.5[d]
7b/8b	4-Cl-Ph	–/82	95	98:2
8c/8c	2-Furyl	60/36	95	96:4
7d/8d	3-Pyr	–/57	90	96:4

[a] Hydrolysis with aq. NH_4Cl.

[b] Hydrolysis with 1*N* HCl.

[c] Analytical HPLC of the crude products.

[d] Reversal of selectivity due to changed priority of the residue *R*.

with the data reported in the literature,[22] the absolute configuration of **7a** was assigned (*S*), which would imply a reversal of diastereoselectivity relative to the anabasin synthesis.

Contradictory results concerning the absolute configuration and the corresponding optical rotation values of coniine hydrochloride described in the literature[22,23] inspired further investigations. Recently, we performed an X-ray analysis of **7a**, which revealed that the absolute configuration at C-2 of alkyl-substituted piperidinones **7** is (*R*).

The efficient diastereoselectivity of the initial Mannich reaction of aldimines **5** in the formation of the nitrogen heterocycles was considered a promising principle for the asymmetric synthesis of β-amino acids, starting from ester enolate equivalents. The realization of this concept will be dealt with in the following chapters.

18.4 Asymmetric Mannich Reactions

The nucleophilic addition of carbanions to Schiff bases provides a general access to β-amino acids. Before we had published the asymmetric Mannich reaction of N-glycosyl imines with silyl ketene acetals,[24] chiral β-amino acids had been obtained, as a rule, in ex-chiral pool syntheses from enolates of chiral carbocyclic esters.[25] Schiff bases of α-amino acids have been applied in β-lactam synthesis, mainly resulting in products from which the stereodifferentiating N-substituent is not easily removable or has to be cleaved off with destruction of the auxiliary.[26] A stereoselective β-amino acid synthesis was achieved by reaction of silyl ketene acetals of ephedrine-containing esters with aldimines.[27] More recently, stereoselective additions of organometallic reagents to aldimines and subsequent transformations to β-amino acids, such as by oxidative degradations, have been reported,[28] as well as other stereoselective additions of enolates to imines.[29]

Stimulated by the outcome of diastereoselective Mannich reactions of Schiff bases **5** of 2,3,4,6-tetra-O-pivaloyl-β-D-galactopyranosylamine **1** with vinologous silyl ketene acetal **6** (Danishefsky's diene), the addition of 1-methoxy-2-methyl-1-

trimethylsiloxy-propen **9** was examined (Scheme 18.3). The reactions proceed at low temperature and afford the N-glycosylated β-amino acid esters **10** in high yield and diasteroselectivity within 24 h (Table 18.2). Even at elevated temperatures, the transformation proceeds with remarkable selectivity (entries 2 and 4, Table 18.2). The β-amino acid esters **11** are released from the carbohydrate in almost quantitative yield by treatment of **10** with HCl in methanol (Scheme 18.4). The carbohydrate template **12** can be recovered in yields of more than 90%.

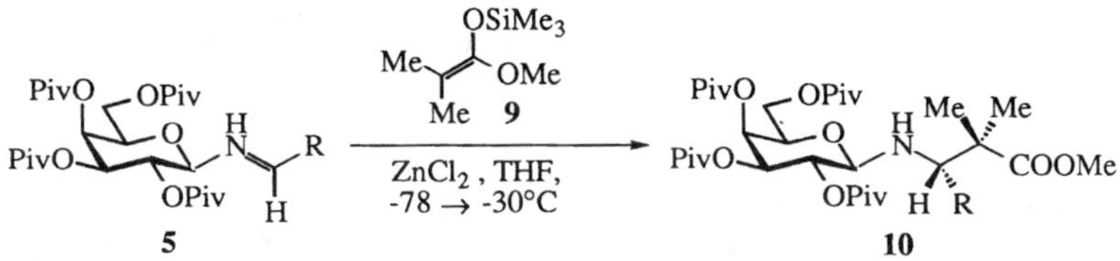

Scheme 18.3

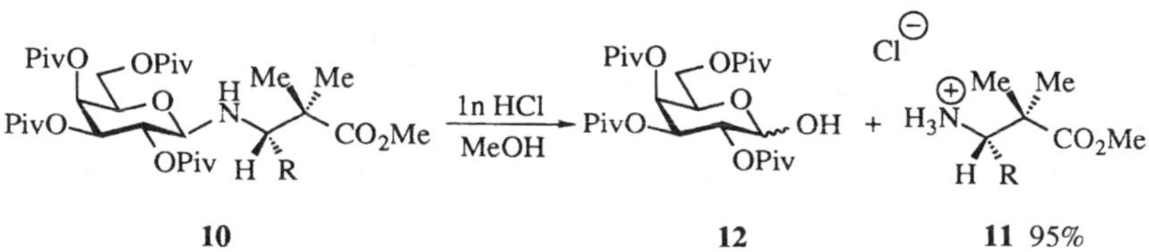

Scheme 18.4

The absolute configuration of the major diastereomers was assigned by detachment of β-amino acid ester **11a** and subsequent conversion to the known amino acid **13** (Scheme 18.5). Comparison of the optical rotation value of **13** with that quoted for the (*R*) compound[30] suggested that **13** has (*S*) configuration and was obtained in high enantiomeric purity. However, the stereoselectivity of this β-amino acid synthesis seems to contradict the recently confirmed stereochemical course of the addition of Danishefsky's diene **6** to aldimines **5**. Unfortunately, attempts to crystallize either glycosylated amino acid esters **10** or target compounds **11** and **13** were not

Table 18.2 DIASTEREOSELECTIVE ADDITION OF SILYL KETENE ACETAL 9 TO N-GALACTOSYL IMINES 5 ACCORDING TO SCHEME 18.3

Entry	Ester	*R*	T (°C)	*t* (h)	Yield (%)	(*S*):(*R*)[a]
1	**10a**	Ph	–30	24	89	150:1
2	**10a**	Ph	+30	1	90	70:1
3	**10b**	3-Cl-Ph	–30	24	89	200:1
4	**10c**	2-Naph	+30	72	90	10:1
5	**10c**	*n*-Prop	–30	24	83	250:1

[a] Analytical HPLC of the crude products (diode array detection; RP 18 columns; solvents acetonitrile and water).

successful so far. Therefore, the stereochemistry of the formed β-amino acids **10** remains uncertain.

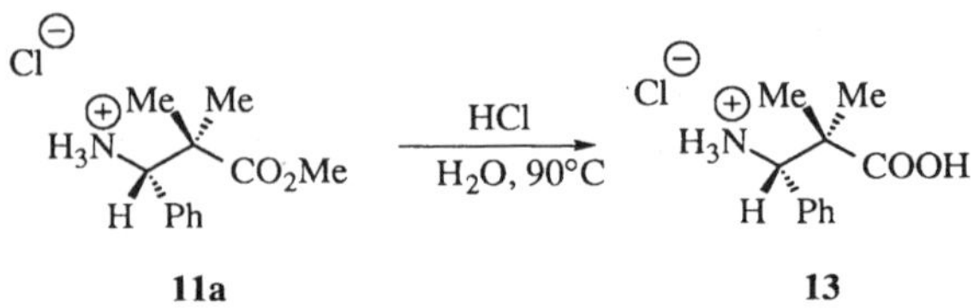

Scheme 18.5

In a series of experiments, the addition of prochiral bissilyl ketene acetals **14** to aldimines **5** was investigated (Scheme 18.6).[31] N-Glycosylated β-amino acids **15** bearing two newly formed chiral centers are obtained within 48 h. The ratio of diastereomers was determined by ^{1}H-NMR of the crude products. The reactions proceed with high stereoselectivity and give only two out of four possible diastereomers (Table 18.3). To assure a complete turnover of the imine, at least one equivalent of $ZnCl_2$ is required. Lewis acid in excess effected neither the stereoselectivity nor the yield. Catalysts such as $SnCl_4$, $AlCl_3$, or $BF_3{\cdot}OEt_2$, besides others, only caused anomerization or decomposition of starting material **5**. In general, aromatic imines **5** are applicable, independent of their substitution pattern. Only the Schiff base derived from benzaldehyde showed a lower reactivity and required a higher temperature (entries 1 and 7, Table 18.3). Detachment from the chiral template under mild conditions affords β-amino acids **16** and the carbohydrate **12** in yields of more than 90% (Scheme 18.7).

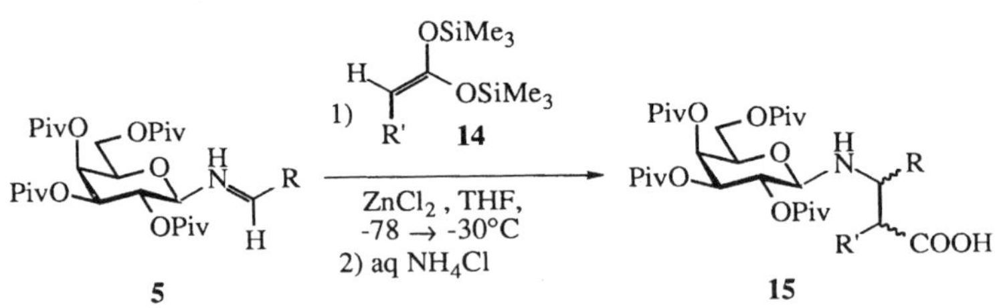

Scheme 18.6

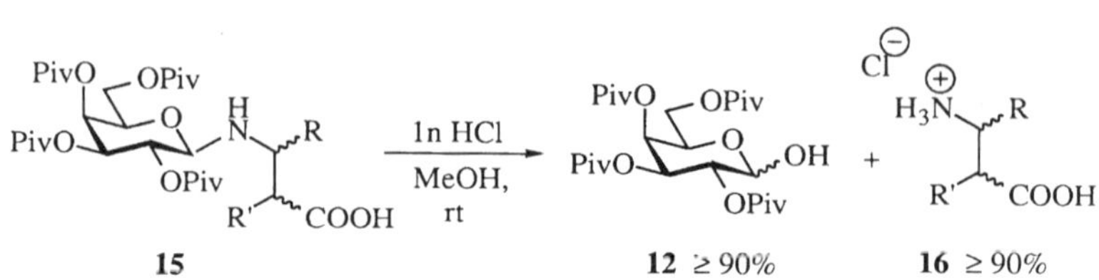

Scheme 18.7

Table 18.3 STEREOSELECTIVE SYNTHESIS OF N-GLYCOSYLATED β-AMINO ACIDS 15 ACCORDING TO SCHEME 18.6

Entry	Acid	*R*	*R′*	*T* (°C)	*t* (h)	Yield (%)[a]	de (ratio)[b]
1	**15a**	Ph	Me	0	6d	45	≈20:1
2	**15b**	4-Cl-Ph	Me	–30	48	90	>20:1
3	**15c**	3-Cl-Ph	Me	–30	48	94	>20:1
4	**15d**	4-Me-Ph	Me	–30	48	90	>20:1
5	**15e**	4-Cl-Ph	Et	–30	48	90	≈10:1
6	**15f**	3-Cl-Ph	Et	–30	48	90	≈20:1
7	**15g**	Ph	Ph	rt	12	65	≈ 8:1
8	**15h**	4-Cl-Ph	Ph	–30	72	94	≈18:1
9	**15i**	3-Cl-Ph	Ph	–30	72	95	≈20:1
10	**15j**	*n*-Pentyl	Ph	–30	96	72	≈15:1

[a] Referred to crude products.

[b] Determined by ^{1}H-NMR of the crude products.

In order to clarify the stereochemical course of the addition of prochiral bissilyl ketene acetals **14** to N-galactosyl aldimines **5**, β-amino acid **16g** was subjected to reduction of its carboxy function to yield amino propanol **17**, of which all four diastereomers have been described in the literature (Scheme 18.8). Comparison of the reported NMR data[32] with those obtained for **17** proved the *anti* configuration of **17**. The diastereomeric *syn* compounds were not detectable. According to the optical rotation value of **17** and those reported for the amino propanol,[33] the absolute configuration of the major diastereomer of **17** was assigned as *anti*-(2*R*,3*S*).

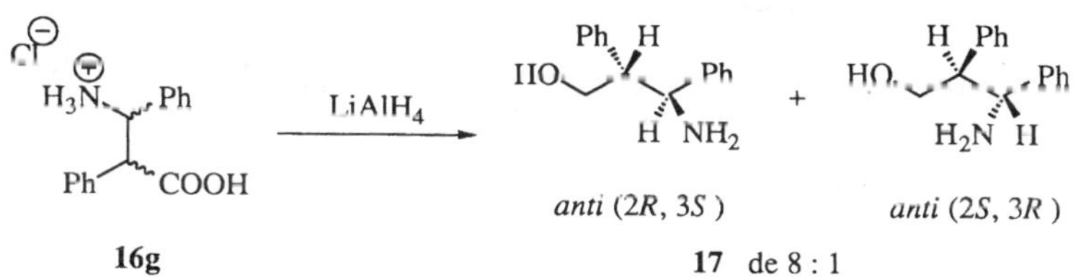

Scheme 18.8

The preferred conformation of aldimines **5**,[16b] supported by NOE measurements, suggest the *Si* side as being susceptible to nucleophilic attack. The *Re* side obviously is strongly shielded by the sterically demanding pivaloyl group at the 2-position of the carbohydrate template.

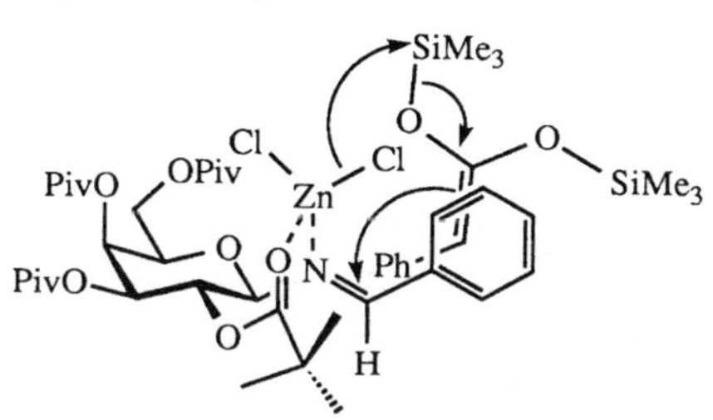

Coordination of the Lewis acid $ZnCl_2$[15b,16b] does not affect the conformation of Schiff base **5**. Logically, sterically demanding nucleophiles should approach the imine from the less hindered *Si* side (shown above). According to the exclusive formation of *anti*-(*erythro*) products, the *cis* silyloxy group of ketene acetal **14** should preferable interact with the coordinatively fixed $ZnCl_2$.

The stereoselective formation of diastereomers and enantiomers with reversed stereochemical configuration can be achieved by two different methods. In a few cases, the modification of the catalyst–solvent system may cause a completely different pathway of stereoselective reactions, as was demonstrated in asymmetric Strecker syntheses on carbohydrate templates.[34] A more general strategy includes the use of enantiomeric auxiliaries or catalysts. Being formally enantiomeric to 2,3,4,6-tetra-O-pivaloyl-β-D-galactopyranosylamine **1**, the glycosylamine derived from α-D-arabinose **18** (Scheme 18.9) was revealed to be very efficient in the synthesis of α-amino acids via the asymmetric Ugi reaction.[14,35]

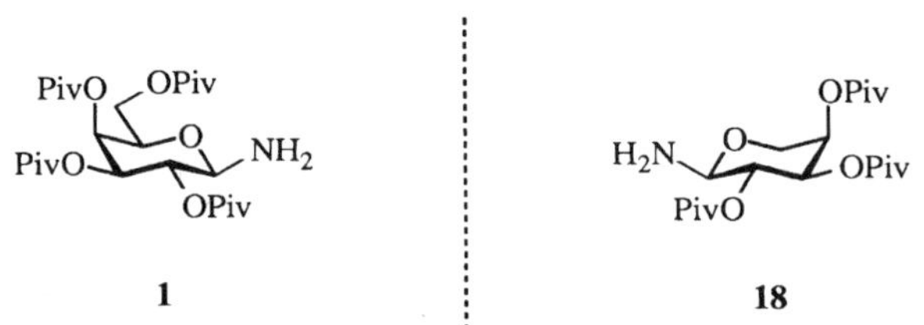

Scheme 18.9

The Lewis acid–induced reaction of arabinosylamine-derived aldimines **19**[14] with prochiral silyl ketene acetal **14** furnishes the N-glycosylated β-amino acids **20** in high yield and high diastereomeric excess (Scheme 18.10, Table 18.4).[36] Therefore, it is concluded that the nucleophilic addition of **14** to **19** proceeds via a transition state formally enantiomeric to that depicted previously.

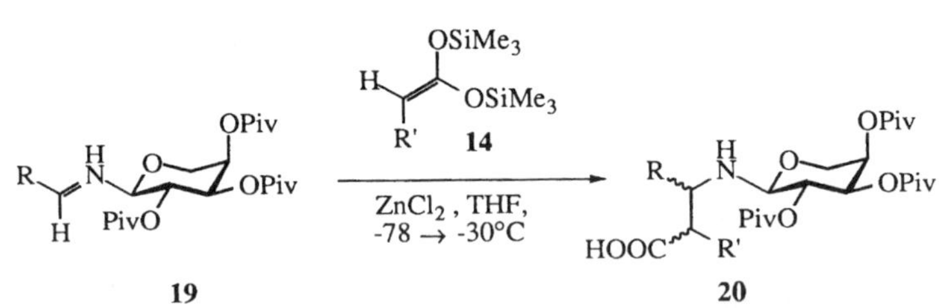

Scheme 18.10

The addition of ester enolates to chiral imines provides another strategy for stereoselective syntheses of β-amino acids.[37] However, reactions of this type with imines carrying removable auxiliaries have not been described in the literature. Encouraged by the successful nucleophilic additions of silyl ketene acetals to N-glycosyl imines, we investigated the addition of lithium ester enolates to these imines.

Table 18.4 DIASTEREOSELECTIVE ADDITION OF SILYL KETENE ACETAL 14 TO N-ARABINOSYL AMINES 19 ACCORDING TO SCHEME 18.9

Entry	Acid	*R*	*R′*	*T* (°C)	*t* (h)	Yield (%)[a]	de (ratio)[b]
1	**20a**	3-Cl-Ph	Me	–30	48	86	≈16:1
2	**20b**	4-F-Ph	Me	–30	48	85	≈14:1
3	**20c**	4-Me-Ph	Me	–30	48	87	≈15:1
4	**20d**	4-Cl-Ph	Et	–30	48	94	≈12:1

[a] Referred to crude products.

[b] Determined by ^{1}H-NMR of the crude products.

Unfortunately, the application of N-glycosyl imines **5** to transformations with lithium ester enolates proved to be limited. Enolates derived from esters like *tert*-butyl propionate show only low reactivity (turnover < 30%). The more stabilized lithium enolate of *tert*-butyl phenylacetate **21** affords the desired N-glycosylated β-amino acid esters **22** in moderate yield (Scheme 18.11).[31] To determine the diastereo-

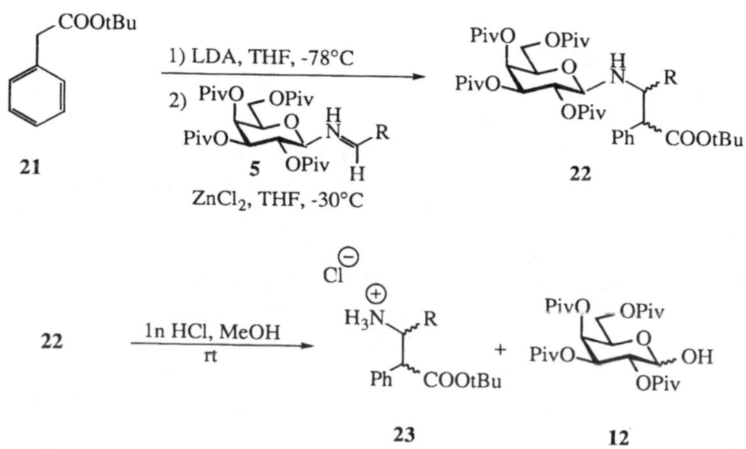

Scheme 18.11

meric ratios, crude esters **22** are subsequently hydrolyzed to give **23** as the hydrochloride. The diastereomeric ratios of the isolated hydrochlorides **23** is rather low (Table 18.5). However, the stereochemical construction of these products is surprising and interesting. Reduction of 3-amino-2,3-diphenyl-*tert*-butylester hydrochloride **23a** gave known amino alcohols of type **17**.

Comparison of their NMR and optical rotation data with those reported in the literature[32,33] revealed that the major diastereomers **17**, in contrast to the products from addition of silyl ketene acetals **14** to aldimines **5**, have the *syn*-(*threo*) configuration (Scheme 18.12). The formation of the corresponding *anti*-diastereomers could not be observed.

Table 18.5 SYNTHESIS OF β-AMINO ACID ESTERS **24** ACCORDING TO SCHEME 18.10

Entry	Ester	*R*	Yield (%)[a]	de (ratio)[b]
1	**23a**	Ph	54	2.7:1
2	**23b**	4-F-Ph	46	2.0:1
3	**23c**	4-Me-Ph	58	1.7:1

[a] Overall yield.
[b] Determined using Mosher's reagent.

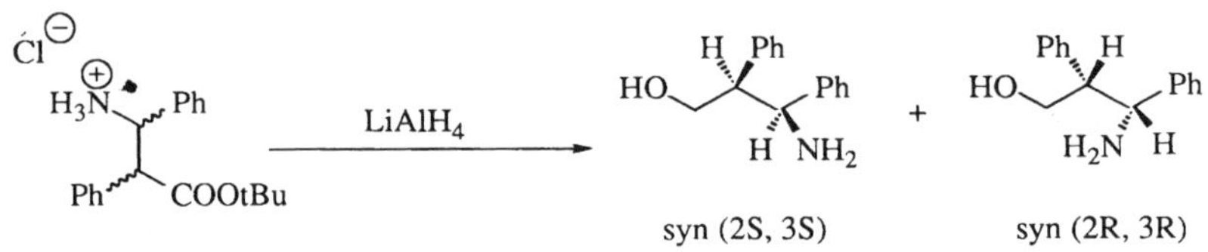

23a

17 de: 2.7 : 1

Scheme 18.12

Differences and similarities between the reactions of N-galactosyl imines **5** with bissilyl ketene acetals **14** on the one hand and lithium ester enolates on the other presumably arise from complexation effects and depend on the structure of the enolate. In both reactions, at least one equivalent of Lewis acid is required. In the case of the lithium ester enolate addition, complexes of the aldimines **5** with $ZnCl_2$ have to be preformed, otherwise no reaction occurs. In the major diastereomers of the Mannich adducts **17** of both transformations, the absolute configuration at the amino-substituted carbon is (*S*). Therefore, in both reactions, the aldimines **5** obviously are attacked from the *Si* side of the C-N double bond. For the addition of lithium ester enolates **21** to Schiff bases **5** (shown below), it is assumed that the O-*tert*-butyl group prefers the *anti*-position within the (*Z*)-ester enolate and, thus, could be responsible for the inverse diastereofacial attack of the enolate carbon.

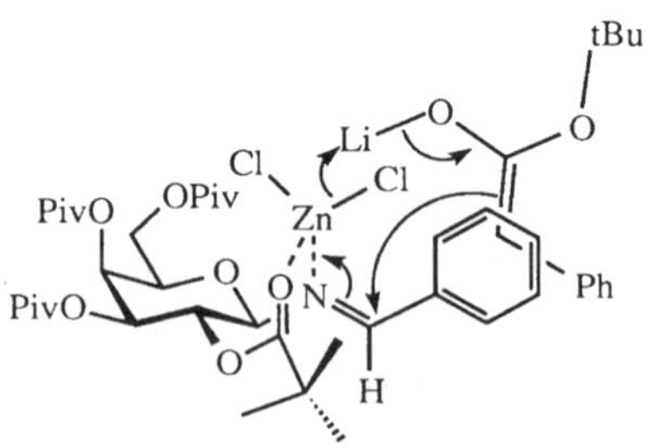

Of course, these pictures only illustrate hypotheses that may rationalize the surprisingly different stereochemical outcome of the Mannich reactions of ester enolates and bisilyl ketene acetals.

18.5 Asymmetric Syntheses of Homoallylamines

Chiral homoallylamines are versatile precursors for a variety of functionalized products. They can easily be transformed into 1,3-amino alcohols, 1-amino 3,4-epoxides, and β-amino acids. Only a few examples of asymmetric additions of allylic organometallic reagents to aldimines have been reported in the literature.[38] In contrast, numerous allylic additions to aldehydes were described.[39] In particular, the aza analogue to the Sakurai reaction,[40] that is, the addition of allylsilanes to imines, was unknown, when we investigated the Lewis acid–induced reaction of allylsilanes and allylstannanes to the Schiff bases of galactopranosylamine **5** (Scheme 18.13).[41] The N-galactosyl-1-aryl-homoallylamines **24** are obtained by reaction of **5** with allyltrimethylsilane in THF in the presence of 2.2 equiv. of $SnCl_4$ at temperatures between 0 and 25°C. The homoallylamines are formed with high diastereoselectivity (Table 18.6). A variety of Lewis acids has been tested. Besides $SnCl_4$ only $BF_3 \cdot OEt_2$ was useful, but less effective, in initiating the reaction. Other promoters, such as $ZnCl_2$, $SnCl_4$/LiBr, and TBAF, caused anomerization of the imine **5**.

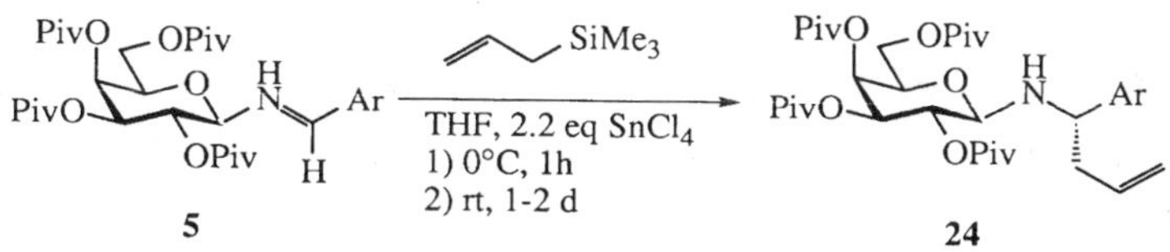

Scheme 18.13

Increased concentration of Lewis acid (2 equiv.) sometimes resulted in higher yields, but slightly decreased stereoselectivity. Reactions with the more nucleophilic allyl tributylstannane afforded, for example, amine **24a** in a higher yield (68%), but reduced diastereomeric ratio (10:1). (See Table 18.7.)

The less stable aldimines derived from galactosylamine **1** and aliphatic aldehydes undergo anomerization rather than reaction with trimethylsilane at 0°C.

Table 18.6 STEREOSELECTIVE SYNTHESIS OF N-GALACTOSYL (*S*)-HOMOALLYLAMINES 24 ACCORDING TO SCHEME 18.12

Entry	Amine	Ar	Yield (%)	(*S*):(*R*)[a]
1	**24a**	4-Cl-Ph	49	22:1
2	**24b**	2-Cl-Ph	82	27:1
3	**24c**	Ph	65	14:1
4	**24d**	4-Me-Ph	55	22:1
5	**24e**	3-Pyridyl	28	11:1
6	**24f**	2-Naphthyl	49	16:1
7	**24g**	CH=CH-Ph	70	15:1

[a] Determined by HPLC of the crude products.

Table 18.7 ADDITION OF ALLYL TRIBUTYLSTANNANE TO N-GALACTOSYL IMINES 5 OF ALIPHATIC ALDEHYDES

Entry	Amine	*R*	Yield (%)	dr (β)[a]	dr (α)[a]	β:α[a]
1	**24h**	*n*-C_3H_7	32	>24:1	>24:1	4:1
2	**24i**	*n*-C_9H_{19}	37	8:1	2:1	3.5:1

[a] Determined by ^{1}H-NMR.

In order to clarify the stereochemical course of the addition of allyl trimethylsilane to N-galactosyl imines **5**, N-glycosyl homoallylamine **24c** was converted to a known β-amino acid. After quantitative detachment from the carbohydrate template, the α-phenyl homoallylamine **25** is subjected to N-protection, oxidation of the double bond, and subsequent removal of the N-Boc group to yield β-phenyl-β-alanine **27**, a precursor of Winterstein's acid,[42] in an overall yield of 58% (Scheme 18.14). Comparison of the optical rotation value with that reported for the (*R*) compound in the literature[43] proves the (*S*)-configuration of **27** for the obtained β-phenyl-β-alanine.

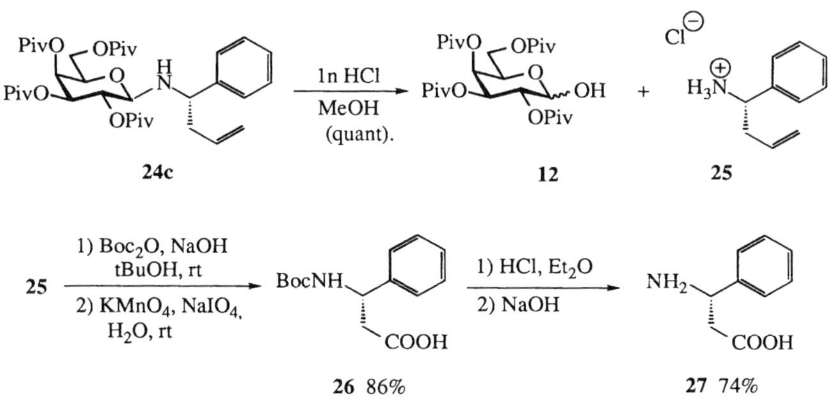

Scheme 18.14

The observed stereoselectivity of the Lewis acid–initiated allylic addition to N-galactosyl aldimines **5** can be rationalized in analogy to the addition of silyl ketene acetals. Octahedral coordination of tin in the Lewis acid–aldimine complex (shown below) results in an effective shielding of the *Re* side of the C-N double bond. Therefore, the allylsilan attacks from the *Si* side.

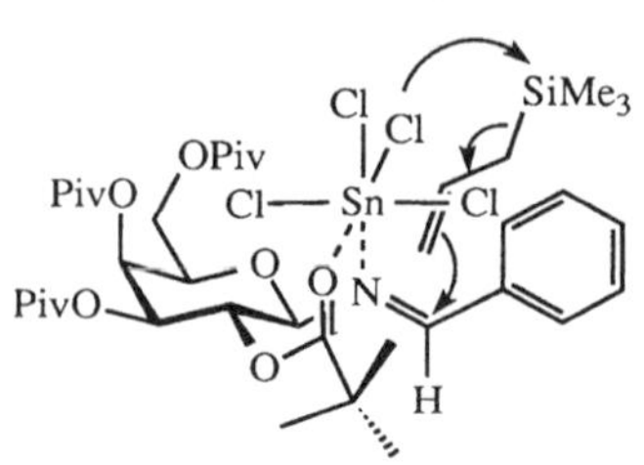

The Lewis acid $SnCl_4$ induces both activation of the imine double bond and eliciting of the latent nucleophilicity of the allylsilane by interaction of a chlorine ligand with the silicon. Since the aza-analogous Sakurai reaction of the N-galactosyl imines **5** resulted in an efficient synthesis of (*S*)-configured homoallylamines **25** (*R*=aryl), further investigations focused on the stereoselective preparation of the corresponding (*R*) compounds (*R*=aryl). Unexpectedly, the aldimines **19** derived from the α-D-arabinosylamine **18** did not react either with allyl trimethylsilan or with allyl tributylstannane, although N-arabinosyl imines **19** had given excellent results, for example, in the stereoselective synthesis of α-amino acids via an asymmetric Ugi reaction.[34] Another carbohydrate template, formally enantiomeric to β-D-galactose, the β-L-fucose, was applied in order to achieve access to (*R*)-β-amino acids. The required fucosylamine **28** (Scheme 18.15) is available by a reaction sequence analogous to that developed for galactosylamine **1**.

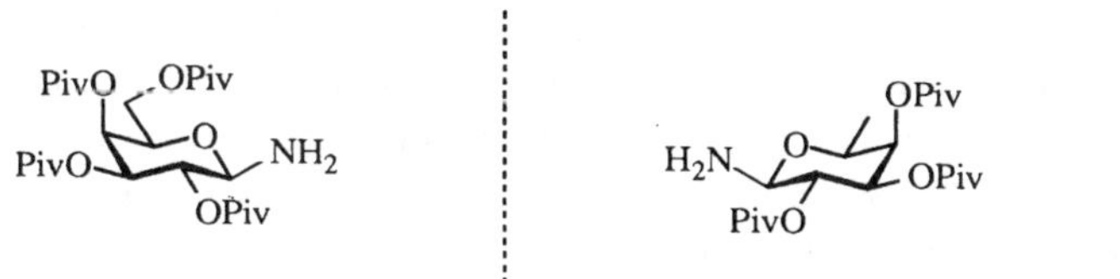

1 **28**

Scheme 18.15

The corresponding β-L-fucosyl imines **29**[14] turned out to be very efficient substrates for the addition of allyl trimethylsilane in the presence of $SnCl_4$ as the catalyst (Scheme 18.16).[44]

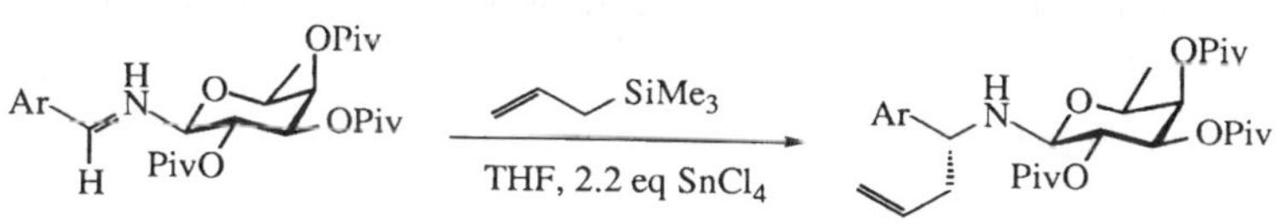

29 **30**

Scheme 18.16

The (*R*)-configured homoallylamines **30** are obtained in moderate yield but high diastereoselectivity (Table 18.8). Performance of the reaction at lower temperature causes increasing selectivity, but decreasing yields (entries 4 and 7, Table 18.8). Also for this series, the N-fucosyl imines derived from aliphatic aldehydes do not react with allyl trimethylsilane/$SnCl_4$. Again, the use of allyl tributylstannane affords the corresponding homoallylamines, although yields and stereoselectivity are lower than in the galactosyl series. It is interesting to note that the N-fucosyl (*R*)-homoallylamines **30** showed a marked tendency to crystallize. A high enrichment of the major diastereomer is often achieved by single recrystallization.

Conversion of the obtained N-fucosyl homoallylamines to the corresponding β-amino acids is carried out in strict analogy to the transformation of the N-galactosyl derivatives (Scheme 18.14).

Table 18.8 STEREOSELECTIVE SYNTHESIS OF N-FUCOSYL (*R*)-HOMOALLYLAMINES 30 ACCORDING TO SCHEME 18.15

Entry	Amine	*Ar*	*T* (°C)	Time	Yield (%)	(*R*):(*S*)[a]
1	**30a**	Ph	(1) 0, (2) rt	1) 2h, 2) rd	44	23:1
2	**30b**	4-Cl-Ph	(1) 0, (2) rt	1) 2h, 2) 3d	86	8:1
3	**30c**	2-Cl-Ph	(1) 0, (2) rt	1) 2h, 2) 3d	69	3.3:1
4	**30c**	2-Cl-Ph	(1) –10, (2) rt	1) 6h, 2) 2.5d	59	>100:1
5	**30d**	4-NO_2-Ph	(1) 0, (2) rt	1) 1h, 2) 2d	68	20:1
6	**30e**	4-NC-Ph	(1) 0, (2) rt	1) 1h, 2) 2d	56	13:1
7	**30e**	4-NC-Ph	(1) –10, (2) rt	1) 26h, 2) 15h	42	28:1
8	**30f**	4-Me-Ph	(1) 0, (2) rt	1) 1h, 2) 36h	43	49:1

[a] Determined by HPLC of the crude products.

In conclusion, imines of N-glycosylamines allow efficient stereoselective synthesis of β-amino acids of both enantiomeric configurations via two different synthetic methods: The aza-analogous Sakurai reaction delivers chiral homoallylamines that can be converted to enantiomerically pure β-amino acids. The asymmetric Mannich reaction of N-glycosyl imines with silyl ketene acetals or lithium ester enolates yields β-amino acid derivatives in high diastereoselectivity. With prochiral carbanion equivalents either the *erythro* or the *threo* compounds can be selectively synthesized depending upon the choice of the silyl ketene acetal or the lithium enolate as the nucleophilic reagent. Furthermore, glycosylamines are readily available, inexpensive chiral templates. The detachment of the β-amino acid from the carbohydrate auxiliaries is achieved under mild conditions in almost quantitative yield. Taking advantage of these properties, glycosylamines can efficiently be used as chiral templates not only in asymmetric synthesis of β-amino acids, but also in other types of stereoselective transformations.

18.6 References

1. For review, see: M. J. O'Donnell, Ed., *Tetrahedron* **44** (1988).
2. H. Estermann, D. Seebach, *Helv. Chim. Acta* **71**, 1824 (1988), and references cited therein.
3. N. Mayachi, M. Shibasaki, *J. Org. Chem.* **55**, 1975 (1990); S. Kim, P. H. Lee, T. A. Lee, *J. Chem. Soc., Chem. Commun.*, 1242 (1988); D. Tanner, P. Somfai, *Tetrahedron* **44**, 613 (1988); T. Kunieda, T. Nagamatsu, T. Kiguchi, M. Hirobe, *Tetrahedron Lett.* **29**, 2203 (1988); H. Huang, N. Iwasawa, T. Mukaiyama, *Chem. Lett.*, 1465 (1984).
4. H. Umezawa, Ed., *Small Molecular Immunomodifiers of Microbial Origin. Fundamental and Clinical Studies of Bestatin* (Pergamon Press, Oxford, 1981); D. Guénard, F. Guéritte-Voegelein, P. Potier, *Acc. Chem. Res.* **26**, 160 (1993).
5. N. Katayama, Y. Nozaki, S. Tsubotani, M. Kondo, S. Harada, H. Ono, *J. Antibiot.* **43**, 10 (1990).
6. Recent reviews: D. Enders, *Acros Org. Acta* **1**, 37 (1995); E. Juaristi, D. Quintana, J. Escalante, *Aldrichimica Acta* **27**, 3 (1994); D. C. Cole, *Tetrahedron* **50**, 9517 (1994); C. W. Jefford, Z.-H. Lu, J. B. Wang, *Pure Appl. Chem.* **66**, 2075 (1994).

7. For review, see: S. Hanessian, *Total Syntheses of Natural Products: The Chiron Approach* (Pergamon Press: Oxford, U.K., 1983).

8. S. R. Landor, B. J. Miller, A. R. Tatchell, *J. Chem. Soc. C*, 2280 (1966); S. Brandlänge, S. Josephson, L. Mörch, S. Vallén, *Acta Chem. Scand. B* **35**, 1296 (1981); C. H. Heathcock, C. T. White, J. J. Morrison, D. Van Derveer, *J. Org. Chem.* **46**, 1296 (1981); S. David, J. Eustache, A. Lubineau, *J. Chem. Soc., Perkin Trans I*, 2274 (1973); R. C. Gupta, A. M. Slavin, R. J. Stoodley, D. J. Williams, *J. Chem. Soc., Chem. Commun.*, 668 (1986); A. Vasella, R. Voeffray, *Helv. Chim. Acta* **65**, 1953 (1982); H. Felber, G. Kresze, R. Prewo, A. Vasella, *Helv. Chim. Acta* **69**, 1137 (1986).

9. For review, see: H. Kunz, K. Rück, *Angew. Chem.* **105**, 355 (1993); *Angew. Chem. Int. Ed. Engl.* **32**, 336 (1993).

10. H. Kunz, *Angew. Chem. Int. Ed. Engl.* **26**, 294 (1987).

11. H. Kunz, in *Selectivities in Lewis Acid-Promoted Reactions*, D. Schinzer, Ed. (Kluwer, Amsterdam, 1989), p. 189.

12. H. Kunz, W. Sager, W. Pfrengle, M. Decker, DE-A 3 624 376 A1, 1986: *Chem. Abstr.* **111**, 39 830 (1988).

13. H. Paulsen, Z. Györgydeak, M. Friedmann, *Chem. Ber.* **107**, 1590 (1974).

14. H. Kunz, W. Pfrengle, K. Rück, W. Sager, *Synthesis*, 1039 (1991).

15. H. Kunz, W. Sager, *Angew. Chem.* **99**, 595 (1987); *Angew. Chem. Int. Ed. Engl.* **26**, 557 (1987); H. Kunz, W. Sager, D. Schanzenbach, M. Decker, *Liebigs Ann. Chem.*, 649 (1991).

16. H. Kunz, W. Pfrengle, *J. Am. Chem. Soc.* **110**, 651 (1988); H. Kunz, W. Pfrengle, *Tetrahedron* **44**, 5487 (1988).

17. S. Laschat, H. Kunz, *Synthesis*, 90 (1992).

18. W. Pfrengle, H. Kunz, *J. Org. Chem.* **54**, 4261 (1989).

19. S. Danishefsky, *J. Am. Chem. Soc.* **96**, 7807 (1974).

20. H. Kunz, W. Pfrengle, *Angew. Chem.* **101**, 1041 (1989), *Angew. Chem. Int. Ed. Engl.* **28**, 1067 (1989)

21. E. Späth, F. Kesztler, *Ber. Dtsch. Chem. Ges.* **70**, 704 (1937).

22. W. H. Tallent, E. C. Horning, *J. Am. Chem. Soc.* **78**, 4467 (1956); see also: Beilstein E III / IV, Vol. 20, p. 1611.

23. D. Enders, J. Tiebes, *Liebigs Ann. Chem.*, 173 (1993) and references therein.

24. H. Kunz, D. Schanzenbach, *Angew. Chem.* **101**, 1042 (1989); *Angew. Chem. Int. Ed. Engl.* **28**, 1068 (1989).

25. M. Hatanaka, H. Nitta, *Tetrahedron Lett.* **28**, 69 (1987); M. Hatanaka, *Tetrahedron Lett.* **28**, 83 (1987).

26. I. Ojima, S. Inaba, *Tetrahedron Lett.* **21**, 2081 (1980), and references cited therein.

27. C. Gennari, J. Venturini, G. Gislon, G. Schimperna, *Tetrahedron Lett.* **28**, 227 (1987), and references cited therein.

28. D. Enders, M. Klatt, R. Funk, *Synlett*, 226 (1993); see also C. Boga, D. Savoia, U.-R. Achille, *Tetrahedron: Asymmetry* **1**, 291 (1990); T. Matsamuto, Y. Kobayashi, Y. Takemoto, Y. Ito, T. Kamijo, H. Harada, S. Terashima, *Tetrahedron Lett.* **31**, 4175 (1990); Y. Ukaji, J. Watai, T. Sumi, T. Fujisawa, *Chem. Lett.* 1555 (1991).

29. L. S. Liebeskind, M. E. Welker, R. W. Fengl, *J. Am. Chem. Soc.* **108**, 6328 (1986); T. Yamada, H. Suzuki, T. Mukaiyama, *Chem. Lett.*, 1013 (1986); T. Shono, N. Kise, F. Sanda, S. Ohi, K. Yoshioka, *Tetrahedron Lett.* **30**, 1253 (1989); I. Ojima, I. Habus, M. Zhao, *J. Org. Chem.* **56**, 1681 (1991); F. A.

Davis, R. T. Reddy, R. E. Reddy, *J. Org. Chem.* **57**, 6387 (1992); R. Annuziata, M. Cinquini, F. Cozzi, P. G. Cozzi, *Tetrahedron Lett.* **33**, 1113 (1992).

30. M. Furukawa, T. Okawara, H. Noguchi, Y. Terawaki, *Heterocycles* **6**, 1323 (1977).

31. H. Kunz, D. Schanzenbach, A. Burgard, unpublished; D. Schanzenbach, *Dissertation* Universität Mainz (1992); A. Burgard, *Dissertation*, Universität Mainz (1995).

32. G. Fodor, R. E. Reavill, J. Stefanovsky, B. Kurtev, H. J. Bernstein, *Tetrahedron* **22**, 235 (1966).

33. G. Fodor, R. E. Reavill, J. Stefanovsky, B. Kurtev, *Chem. Ber.* **98**, 705 (1965); N. D. Berova, B. J. Kurtev, *Tetrahedron* **25**, 2301 (1969).

34. H. Kunz, W. Sager, W. Pfrengle, D. Schanzenbach, *Tetrahedron Lett.* **29**, 4397 (1988).

35. H. Kunz, W. Pfrengle, W. Sager, *Tetrahedron Lett.* **30**, 4109 (1989).

36. H. Kunz, D. Schanzenbach, unpublished results; D. Schanzenbach, *Dissertation* Universität Mainz, 1992.

37. Review: M. J. Brown, *Heterocycles* **29**, 2225 (1989).

38. Y. Yamamoto, T. Komatsu, K. Maruyama, *J. Am. Chem. Soc.* **106**, 5031 (1984); Y. Yamamoto, T. Komatsu, K. Maruyama, *J. Chem. Soc., Chem. Commun.*, 814 (1985); Y. Yamamoto, S. Nishii, K. Maruyama, T. Komatsu, W. Ito, *J. Am. Chem. Soc.* **108**, 7778 (1986); Y. Yamamoto, *Acc. Chem. Res.* **20**, 243 (1987); D. H. Hua, S. W. Miao, J. S. Chen, S. Iguchi, *J. Org. Chem.* **56**, 4 (1991).

39. R. W. Hoffmann, *Angew. Chem. Int. Ed. Engl.* **21**, 555 (1982); D. Hoppe, *Angew. Chem. Int. Ed. Engl.* **23**, 932 (1984); Y. Yamamoto, *Acc. Chem. Res.* **20**, 243 (1987) and references therein.

40. H. Sakurai, *Synlett*, 1 (1989).

41. S. Laschat, H. Kunz, *Synlett*, 51 (1990); S. Laschat, H. Kunz, *J. Org. Chem.* **56**, 5883 (1991).

42. E. Graf, H. Boeddeker, *Liebigs Ann. Chem.* **613**, 111 (1958).

43. T. Kaseda, T. Kikuchi, C. Kibayashi, *Tetrahedron Lett.* **30**, 4539 (1989).

44. S. Laschat, H. Kunz, *Synlett*, 629 (1990).

CHAPTER

19

Efficient Stereoselective Preparation of the Taxol Phenylisoserine Side Chain

Ari M. P. Koskinen and Esko K. Karvinen

19.1 Introduction

Taxol[1] is considered the most promising new anticancer drug lead since the discovery of *Vinca* alkaloids, colchicine, and tetracyclines. Taxol is also an approved drug in many countries for the treatment of a variety of forms of cancer (ovarian and breast, typically, but experimentally implicated in most solid tumors).[2] Structure–activity studies on Taxol analogues have suggested that the C-13 side chain is necessary for activity, and its stereochemistry is also of crucial importance. Some variations are allowed in the phenyl rings of the phenylisoserine and the benzoate group.

Initially the availability of Taxol was considered to become the bottleneck of Taxol-based therapies. Although Taxol can be isolated in low yields from the bark of the Pacific yew tree (*Taxus brevifolia*), its wide use in chemotherapy is hindered by limited availability.[3] Therefore, new semisynthetic routes via more the readily available 10-deacetyl baccatin III (10-DAB-III), from the renewable leaves of *Taxus baccata*, an abundant ornamental yew growing in Asia and Europe, and properly protected 3-phenylisoserine have been developed (Scheme 19.1).[4] The initial lack of

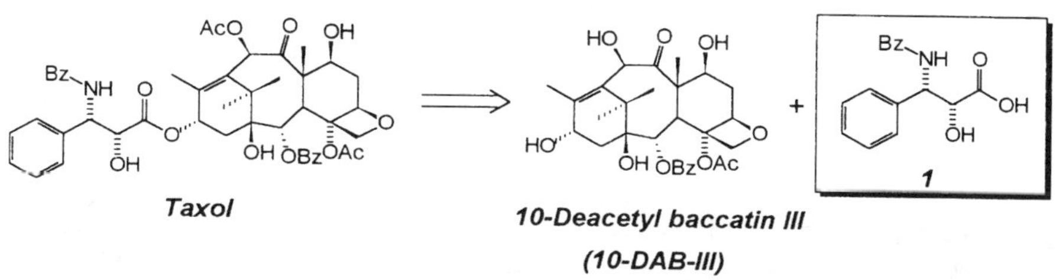

Scheme 19.1

material, together with the need to develop more readily bioavailable and less toxic analogues, has spurred a vast activity in total synthesis of Taxol, which has thus far been achieved by three groups, with a number of others on their way.[5]

Because of the ready availability of the tetracyclic Taxol nucleus, a large amount of effort has been expended in designing a number of routes to synthesize suitable precursors for coupling with the 10-DAB III derivatives. The semisynthetic process utilizes 7-O-triethylsilyl baccatin III, a derivative of the natural product 10-DAB III, together with a simple precursor for the phenylisoserine side chain **1**.

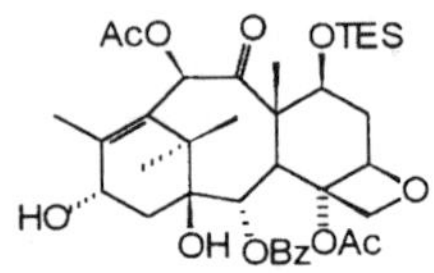

This chapter will initially review the general routes described in the literature, then give a short account of the state of our own studies, and finally some comparisons and observations of general nature will be made. Enzymatic appróaches and more detailed accounts of various strategies will be found in other chapters in this volume.

19.2 General Routes in the Literature

Practically all the general strategies for the synthesis of the phenylisoserine side-chain derivatives fall into two conceptually distinct categories: formation of the central C_2-C_3 bond or functionalization of a pre-existing double bond in a cinnamic acid derivative (Scheme 19.2).

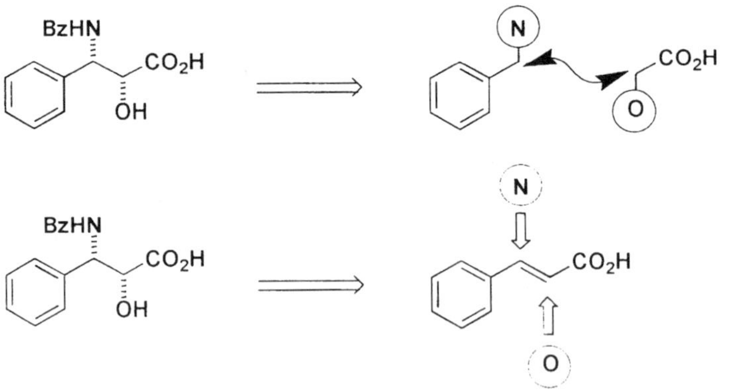

Scheme 19.2

The strategies based on the C-C bond formation have been widely utilized, and these approaches typically rely on the Staudinger reaction, proceeding via a β-lactam intermediate (Scheme 19.3).[6] It is noteworthy that the suitably protected β-lactam derivatives have been directly used in the coupling with 10-DAB-III derivatives.[7]

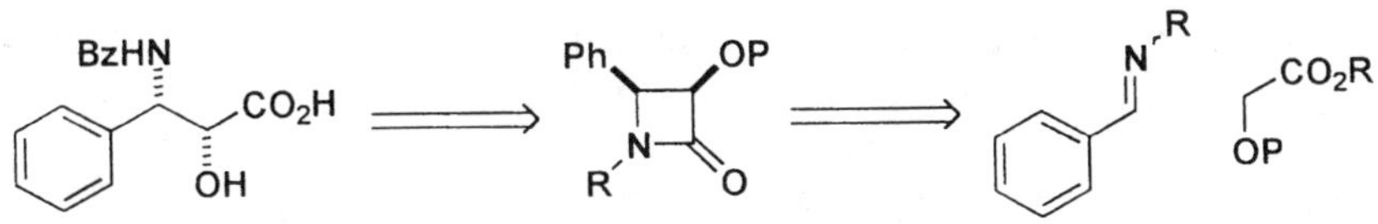

Scheme 19.3

Several attempts have been made to improve the enantio- and diastereoselectivities in the Staudinger reaction, and typically these problems can be solved by judicious choice of the reacting partners. The high diastereoselectivity can be exemplified by the procedure developed by Holton and Liu (Scheme 19.4),[6g] which utilizes the Evans-type oxazolidine as the chiral auxiliary in the nucleophile part of the molecule.

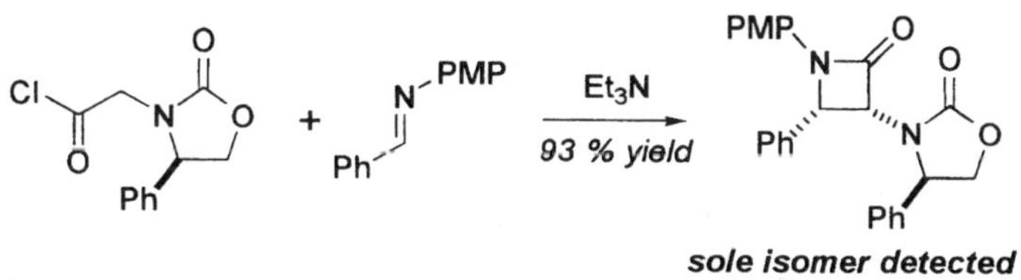

Scheme 19.4

A number of Staudinger-related approaches have appeared.[6] They also involve the coupling of an imine with a glyoxylic acid derivative, but avoid the formation of the β-lactam. An illustrative example is provided in Scheme 19.5.[8c,d] The camphor-sultam derivative is reacted with a benzoylimine to provide the desired diastereomer in 68% yield. Treatment with DDQ oxidizes the PMB-ether to the cyclic PMP-hemi-aminal, which can be coupled with 10-DAB-III after hydrolysis of the sultam.

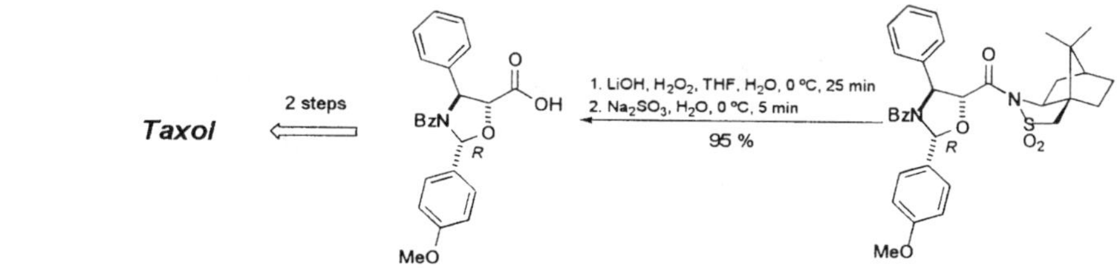

Scheme 19.5

A few approaches have been based on the Darzens condensation, which produces the cinnamate epoxide.[9] Further considerations of this intermediate will be discussed shortly, but we must first make a couple of general observations.

The strategies based on the C-C bond formation require the input of the chiral information in a stoichiometric amount via the use of a chiral auxiliary either in the nucleophilic or the electrophilic partner. Such relayed asymmetric induction processes are suboptimal, if one considers atom and/or enantiomeric economy of the overall schemes. The only exception to this is the Corey diazaborolidine–catalyzed Darzens condensation (Scheme 19.6).[9a]

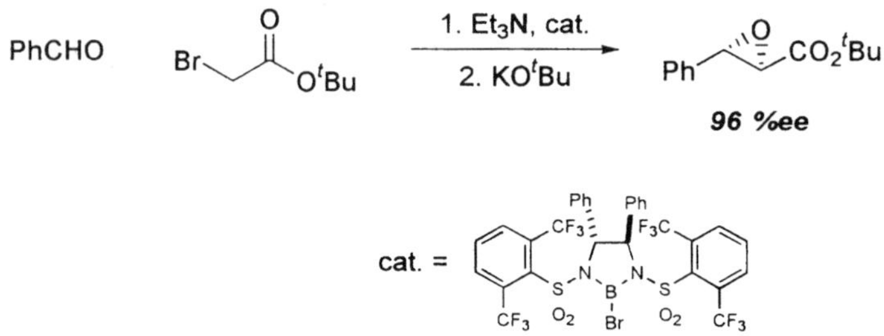

Scheme 19.6

The strategies based on the functionalization of an existing C-C double bond are based on three approaches: direct addition of a nitrogen nucleophile followed by electrophilic oxygen capture of the ensuing ester enolate,[10] methods using epoxidation followed by epoxide opening,[11] and asymmetric dihydroxylation[12,13] (Scheme 19.7).

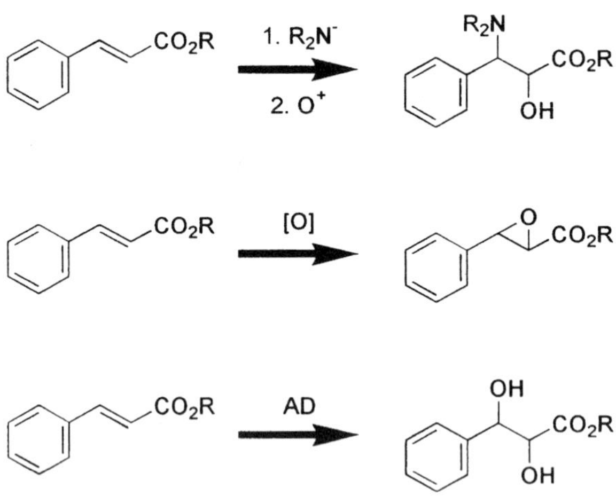

Scheme 19.7

The first one of these approaches was developed by Davies, and involved a highly successful addition of a chiral lithium amide onto *t*-butyl cinnamate.[10] The ester enolate anion was then trapped with (+)-(camphorsulfonyl)oxaziridine (Scheme 19.8). Cleavage of the protecting groups and epimerization of the hydroxyl-bearing atom then completed the sequence in high overall yield.

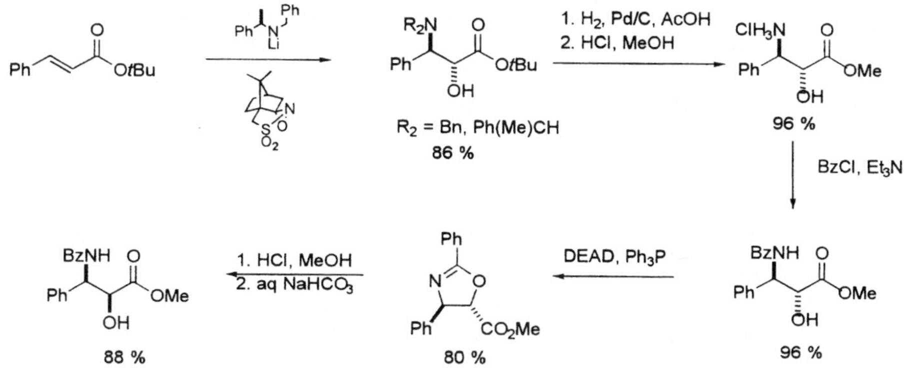

Scheme 19.8

The latter two approaches in Scheme 19.7 involve the utilization of the highly powerful catalytic technologies developed in connection with the Sharpless asymmetric epoxidation, the Katsuki–Jacobsen epoxidation, and the Sharpless asymmetric dihydroxylation (AD). The utility of these catalytic methods is represented by the Sharpless approach in Scheme 19.9.[12c] In principle, catalytic asymmetric synthesis methods will avoid the problems associated with the poor efficiency of the chiral information transfer usually inherent in both internal and relayed asymmetric transformations.

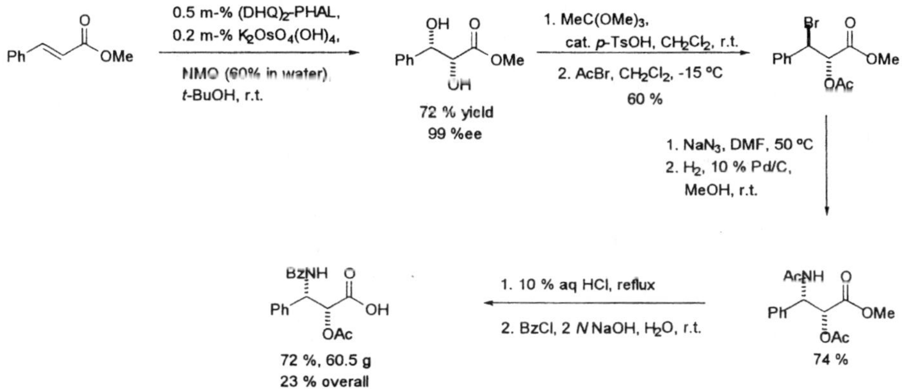

Scheme 19.9

19.3 Route via Asymmetric Dihydroxylation

In our previous report on the synthesis of Taxol and Taxotere side chains, we described a route via asymmetric dihydroxylation of both *cis*- and *trans*-methyl cinnamates.[12b] Whereas the route starting from *cis*-cinnamate was plagued by the fact that the enantiomeric purity was fatally low (7% ee), the route starting with *trans*-cinnamate led to relatively tedious separations due to both regioisomeric and diastereomeric mixtures. Facing these difficulties, we have pursued a practical solution to avoid any experimental obstacles to scaling up the reaction sequence. In this chapter, we report a simple and inexpensive synthesis of the Taxol phenylisoserine side chain starting from *trans*-methyl cinnamate.

Asymmetric dihydroxylation with $(DHQD)_2$-PHAL (AD-mix-β)[14] of *trans*-methyl cinnamate gave the diol **2** in 67% yield (Scheme 19.10).[15] The diol was then allowed to react with thionyl chloride in CCl_4 to give a diastereomeric mixture of cyclic sulfites, which were oxidized immediately to the corresponding sulfate **3** using a catalytic amount of $RuCl_3$ with sodium periodate as the stoichiometric oxidant.[16]

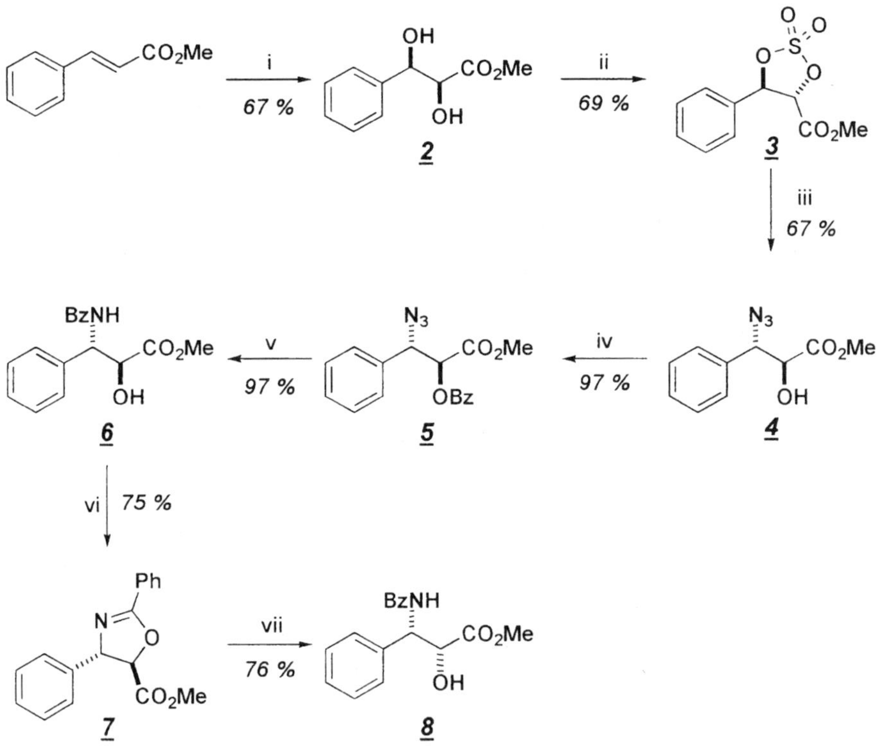

REAGENTS: (i) AD-mix β; (ii) 1.$SOCl_2$, CCl_4; 2. $RuCl_3$, $NaIO_4$, MeCN, CCl_4; (iii) NaN_3, Me_2CO, H_2O; (iv) BzCl, DMAP, CH_2Cl_2; (v) H_2, Pd/C, MeOH; (vi) $SOCl_2$, $CHCl_3$; (vii) HCl, MeOH.

Scheme 19.10

Treatment of the sulfate **3** with sodium azide in acetone–water[16a] for 3 h at room temperature gave a regioisomeric mixture of azido alcohols with the desired isomer **4** predominating. Although clearly not optimal, the mixture could be simply purified by flash chromatography to give pure **4** in 67% yield, in contrast to our previous results, where the diastereomeric mixture of azido alcohols could only be separated by HPLC.[12b] Benzoylation of the azido alcohol with benzoyl chloride in dichloromethane in the presence of DMAP[13f] gave the benzoate **5** in 97% yield. Reduction of the azide in methanol with concomitant acyl migration[11a] from oxygen to nitrogen gave the *anti* (2*S*,3*S*) 3-phenyl isoserine methyl ester **6** in 97% yield (27% overall yield from **1**). This diastereomeric phenylisoserine has recently been observed, in a cyclically protected form, to afford Taxol precursors with the *syn* (2*R*,3*S*) side chain on esterification with baccatin III derivatives (ROH) (Scheme 19.11).[17]

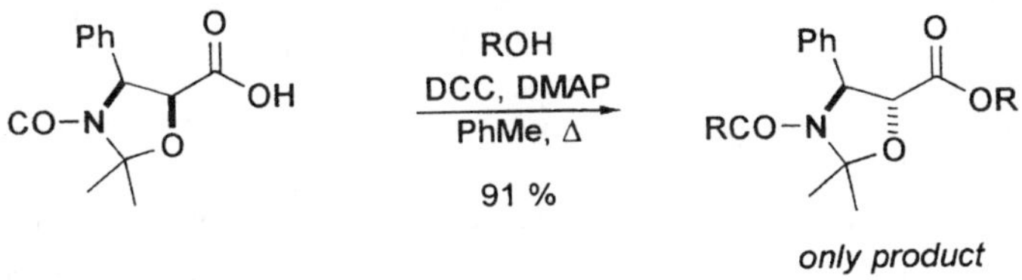

Scheme 19.11

Commerçon et al. have shown that the cyclic protection of the Taxotere side chain ethyl ester with methoxypropene gave 99% yield and the saponification of the protected ester led quantitatively to acid.[18] Thus the same protected Taxol side chain could be obtained in the acid form in 26% overall yield from **1**. Alternatively, inversion at C-2 could be effected[13f] by treatment with thionyl chloride in chloroform to give the oxazoline **7** followed by hydrolysis with aqueous HCl in methanol to give the Taxol side chain methyl ester **8**.

19.4 Asymmetric *syn*-Oxyamination

Ideally, the most direct route to the phenylisoserine side chain could be direct *syn*-oxyamination of cinnamic acid esters. The problems in this approach have mainly resided in the cleavage of the nitrogen protecting groups,[19] but recent developments have enabled a simple one-step oxyamination for the Taxol side chain (Scheme 19.12).[20]

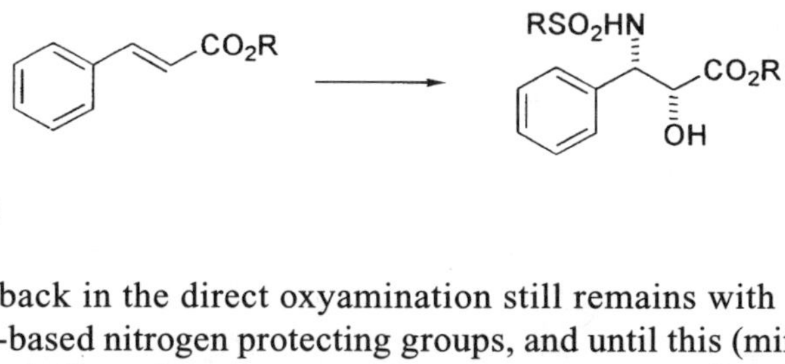

Scheme 19.12

The drawback in the direct oxyamination still remains with the necessity for sulfonamide-based nitrogen protecting groups, and until this (minor) problem can be solved, we believe that our approach represents a very direct and competitive synthesis of the Taxol side chain from inexpensive starting materials and high enantioselectivity.

19.5 Acknowledgments

This work was made possible by a generous grant from the Technology Development Centre (TEKES, Finland). Helpful discussions with Professor K. Barry Sharpless, The Scripps Research Institute, and technical assistance by Jussi Siirilä, M.Sc., are gratefully acknowledged.

19.6 References

1. M. C. Wani, H. L. Taylor, M. E. Wall, P. Coggon, A. T. McPhail, *J. Am. Chem. Soc.* **93**, 2325 (1971).

2. The FDA has recently approved Taxol for the treatment of patients with ovarian cancer, whose first-line or subsequent chemotherapy has failed; see: *Pharmaceutical Business News*, January 8, 1993, 7, *Chem. Ind.* 41 (1993).

3. S. Borman, *Chem. Eng. News*, Sept. 2, 11 (1991).

4. (a) J. N. Denis, A. E. Greene, D. Guenard, F. Gueritte-Voegelain, L. Mangatanal, P. Potier, *J. Am. Chem. Soc.* **110**, 5917 (1988). (b) S. Borman, *Chem. Eng. News* Oct. 12, 30 (1992).

5. (a) K. C. Nicolaou, Z. Zang, J. J. Liu, H. Ueno, P. G. Nantermet, R. K. Guy, C. F. Claiborne, J. Renaud, E. A. Couladouros, K. Paulvannan, E. J. Sorensen, *Nature* **367**, 630 (1994). (b) K. C. Nicolaou, P. G. Nantermet, H. Ueno, R. K. Guy, E. A. Coulandouros, E. J. Sorensen, *J. Am. Chem. Soc.* **117**, 624 (1995). (c) K. C. Nicolaou, J. J. Liu, H. Yang, H. Ueno, E. J. Sorensen, C. F. Claiborne, R. K. Guy, C. K. Hwang, M. Nakada, P. G. Nantermet, *ibid.* **117**, 634 (1995). (d) K. C. Nicolaou, Z. Yang, J. J. Liu, P. G. Nantermet, C. F. Claiborne, J. Renaud, R. K. Guy, K. Shibayama, *ibid.* **117**, 645 (1995). (e) K. C. Nicolaou, H. Ueno, J. J. Liu, P. G. Nantermet, Z. Yang, J. Renaud, K. Paulvannan, R. Chadha, *ibid.* **117**, 653 (1995). (f) R. A. Holton, C. Somoza, H. B. Kim, F. Liang, R. J. Biediger, D. Boatman, M. Shindo, C. C. Smith, S. Kim, H. Nadizadeh, Y. Suzuki, C. Tao, P. Vu, S. Tang, P. Zhang, K. K. Murthi, L. S. Gentile, J. H. Liu, *J. Am. Chem. Soc.* **116**, 1597 (1994). (g) R. A. Holton, H. B. Kim, C. Somoza, F. Liang, R. J. Biediger, D. Boatman, M. Shindo, C. C. Smith, S. Kim, H. Nadizadeh, Y. Suzyki, C. Tao, P. Vu, S. Tang, P. Zhang, K. K. Murthi, L. S. Gentile, J. H. Liu, *ibid.* **116**, 1599 (1994). (h) J. Masters, J. T. Link, L. T. Snyder, W. B. Young, S. J. Danishefsky, *Angew. Chem., Int. Ed. Engl.* **34**, 1723 (1995).

6. (a) C. Palomo, A. Arrieta, F. P. Cossio, J. M. Aizpurua, A. Mielgo, N. Aurrekoetxea, *Tetrahedron Lett.* **31**, 6429 (1990). (b) I. Ojima, I. Habus, M. Zhao, *J. Org. Chem.* **56**, 1681 (1991). (c) G. I. Georg, P. M. Mashava, E. Akgün, M. W. Milstead, *Tetrahedron Lett.* **32**, 3151 (1991). (d) V. Farina, S. I. Hauck, D. G. Walker, *Synlett*, 761 (1992). (e) G. I. Georg, E. Akgün, P. M. Mashava, M. Milstead, H. Ping, Z. Wu, D. V. Velde, *Tetrahedron Lett.* **33**, 2111 (1992). (f) G. I. Georg, Z. S. Cheruvallath, G. C. B. Harriman, M. Hepperle, H. Rark, *Bioorg. Med. Chem. Lett.* **3**, 2467 (1993). (g) R. A. Holton, J. H. Liu, *ibid.* **3**, 2475 (1993). (h) M. Endo, R. Droghini, *ibid.* **3**, 2483 (1993). (i) J. D. Bourzat, A. Commercon, *Tetrahedron Lett.* **34**, 6049 (1993). (j) G. Cainelli, M. Panunzio, D. Giacomini, B. Di Simone, R. Camerini, *Synthesis*, 805 (1994).

7. R. A. Holton, *U.S. Patent* 5,015,744 (1991); R. A. Holton, PCT Patent Application WO 93/06079 (1993).

8. (a) K. Hattori, M. Miyata, H. Yamamoto, *J. Am. Chem. Soc.* **115**, 1151 (1993). (b) C. S. Swindell, M. Tao, *J. Org. Chem.* **58**, 5889 (1993). (c) A. M. Kanazawa, J.-N. Denis, A. E. Greene, *ibid.* **59**, 1238 (1994). (d) A. M. Kanazawa, J.-N. Denis, A. E. Greene, *J. Chem. Soc., Chem. Commun.*, 2591 (1994).

9. (a) E. J. Corey, S. Choi, *Tetrahedron Lett.* **32**, 2857 (1991). (b) A. Commercon, D. Bézard, F. Bernard, J. D. Bourzat, *ibid.* **33**, 5185 (1992). (c) R. P. Srivastava, J. K. Zjawiony, J. R. Peterson, J. D. McChesney, *Tetrahedron: Asymmetry* **5**, 1683 (1994).

10. M. E. Bunnage, S. G. Davies, C. J. Goodwin, *J. Chem. Soc. Perkin Trans. I*, 1375 (1993).

11. (a) J.-N. Denis, A. E. Greene, A. A. Serra, M.-J. Luche, *J. Org. Chem.* **51**, 46 (1986). (b) J. Legters, L. Thijs, B. Zwanenburg, *Tetrahedron Lett.* **30**, 4881 (1989). (c) L. Deng, E. Jacobsen, *J. Org. Chem.* **57**, 4320 (1992). (d) C. Bonini, G. Righi, *J. Chem. Soc., Chem. Commun.*, 2767 (1994).

12. (a) J.-N. Denis, A. Correai, A. E. Greene, *J. Org. Chem.* **55**, 1957 (1990). (b) A. M. P. Koskinen, E. K. Karvinen, J. P. Siirilä, *J. Chem. Soc., Chem. Commun.*, 21 (1994). (c) Z.-M. Wang, H. C. Kolb, K. B. Sharpless, *J. Org. Chem.* **59**, 5104 (1994).

13. For other nonenzymatic syntheses, see: (a) H. Hönig, P. Seufer-Wasserthal, H. Weber, *Tetrahedron* **46**, 3841 (1990). (b) J.-N. Denis, A. Correa, A. E. Greene, *J. Org. Chem.* **56**, 6939 (1991). (c) F. A. Davis, R. T. Reddy, R. E. Reddy, *ibid.* **57**, 6387 (1992). (d) C. Mukai, I. J. Kim, M. Hanaoka, *Tetrahedron: Asymmetry* **3**, 1007 (1992). (e) A. M. Kanazawa, A. Correa, J.-N. Denis, M.-J. Luche, A. E. Greene, *J. Org. Chem.* **58**, 255 (1993). (f) D.-M. Gou, Y.-C. Liu, C.-S. Chen, *J. Org. Chem.* , 1287 (1993). (g) C. Mukai, I. J. Kim, E. Furu, M. Hanaoka, *Tetrahedron* **49**, 8323 (1993). (h) R. N. Patel, A. Banerjee, J. M. Howell, C. G. McNamee, D. Brozozowski, D. Mirfakhrae, V. Nanduri, J. K. Thottathil, L. J. Szarka, *Tetrahedron: Asymmetry* **4**, 2069 (1993). (i) F. A. Davis, R. E. Reddy, *ibid.* **5**, 955 (1994). (j) J. Kearns, M. M. Kayser, *Tetrahedron Lett.* **35**, 2845 (1994).

14. (a) K. B. Sharpless, W. Amberg, Y. L. Bennani, G. A. Crispino, J. Hartung, K.-S. Jeong, H.-L. Kwong, K. Morikawa, Z.-M. Wang, D. Xu, X.-L. Zhang, *J. Org. Chem.* **57**, 2768 (1992) and references therein. (b) For a review, see B. B. Lohray, *Tetrahedron: Asymmetry* **3**, 1317 (1992). (c) H. C. Kolb, M. S. VanNieuwenhze, K. B. Sharpless, *Chem. Rev.* **94**, 2483 (1994).

15. Yields are given for purified and chromatographically homogenous products. All substances exhibited satisfactory analytical and spectral data. Selected physical data for compounds: **2**: $[\alpha]_D$ - 9.1 (*c* 1.5, CH_3OH), 98% ee by chiral phase GLC (β-cyclodextrin column, isothermal, 150°C); **3**: $[\alpha]_D$ + 39 (*c* 1.4, CH_2Cl_2); **4**: $[\alpha]_D$ + 119 (*c* 2.1, $CHCl_3$), mp 43–44°C {Lit.,[13a] $[\alpha]_D$ + 125.0 (*c* 2, CH_2Cl_2)]; **5**: $[\alpha]_D$ + 97 (*c* 0.78, CH_3OH); **6**: $[\alpha]_D$ + 8.8 (*c* 1.0, CH_3OH), mp 157–158°C {Lit.,[13c] $[\alpha]_D$ + 8.7 (*c* 1.03, CH_3OH), mp 158–159°C}; **7**: $[\alpha]_D$ + 17 (*c* 1.0, $CHCl_3$) {Lit.,[13f] $[\alpha]_D$ + 13 (*c* 1, $CHCl_3$)}; **8**: $[\alpha]_D$ – 43 (*c* 0.33, CH_3OH) {Lit.,[13f] $[\alpha]_D$ – 49.6 (*c* 1, CH_3OH)}.

16. (a) Y. Gao, K. B. Sharpless, *J. Org. Chem.* **110**, 7538 (1988). (b) B. M. Kim, K. B. Sharpless, *Tetrahedron Lett.* **30**, 655 (1989). (c) M. S. Berrige, M. P. Franceschini, E. Rosenfeld, T. J. Tewson, *J. Org. Chem.* **55**, 1211 (1990). (d) For a review on cyclic sulfites and cyclic sulfates, see: B. B. Lohray, *Synthesis*, 1035 (1992).

17. J.-N. Denis, A. M. Kanazawa, A. E. Greene, *Tetrahedron Lett.* **35**, 105 (1994).

18. A. Commerçon, D. Bezard, F. Bernard, J. D. Bourzat, *Tetrahedron Lett.* **33**, 5185 (1992).

19. T. Fukuyama, C.-K. Jow, M. Cheun, *Tetrahedron Lett.* **36**, 6373 (1995).

20. K. B. Sharpless, Nobel Symposium on *Catalytic Asymmetric Synthesis*, September 3–7, 1995 at Tammsvik, Bro, Sweden.

CHAPTER

20

Chemoenzymatic Synthesis of the Side Chain of Taxol

Charles J. Sih

20.1 Introduction

In the late 1950s the National Cancer Institute (NCI) initiated a bioassay-directed screening program of plant extracts using tumor model systems in vivo and tumor cell lines in vitro. From these studies, it was shown that the stem bark extracts of the Pacific yew tree, *Taxus brevifolia*, displayed cytotoxicity in the KB (epidermoid carcinoma cell line) assay and also activity against carcinoma in rats and leukemia in mice. Subsequently, Wall and his collaborators[1] fractionated the active extract, and in 1969 they isolated Taxol, the most active constituent of the extract. The structure of Taxol (**1**) was elucidated and published in 1971. Methanolysis of Taxol led to the formation of 10-deacetylbaccatin III (**2a**) and the methyl ester of (2*R*,3*S*)-N-benzoyl-3-phenylisoserine (**4b**). The structure and the stereochemistry of Taxol were deduced by the X-ray analysis of a derivative of 10-deacetyl baccatin III (**2a**) and a derivative of N-benzoyl-3-phenylisoserine methyl ester (**4b**).[1]

Taxol (**1**) was shown to be a potent inhibitor of cell division with a unique mechanism of action. Other natural antitumor agents such as vinblastine, colchicine, and maytansine interact with tubulin by preventing the assembly of tubulin into microtubules.[2] In contrast, Taxol acts as a promoter of tubulin assembly as well as an inhibitor of the disassembly process. Today Taxol has been approved by the FDA for use in the clinical treatment of ovarian and breast cancers.[3]

For many years, the source of Taxol was the bark of the slow-growing Western yew, *Taxus brevifolia*, found in the forests in the Pacific northwest. Procurement of the bark and isolation of Taxol results in a low yield of drug and death of the tree.[4] Although two total syntheses of Taxol have been reported,[5] neither one is suited for industrial utilization.

To overcome the Taxol supply problem, the French group headed by Potier turned to a semisynthetic approach.[6] They discovered high concentrations of 10-deacetylbaccatin III[7] (**2a**) (1 g/kg) in the needles (renewable) of *Taxus baccata* and were successful in transforming it into Taxol or the analog Taxotere (**3**) via coupling to an appropriate derivative of N-benzoyl-(2*R*,3*S*)-3-phenylisoserine (**4a**).[6] Alternatively, Taxol may also be obtained by the direct condensation of suitably protected derivatives of 10-deacetylbaccatin III and (3*R*,4*S*)-3-hydroxy-4-aryl-β-lactam (**5**).[8]

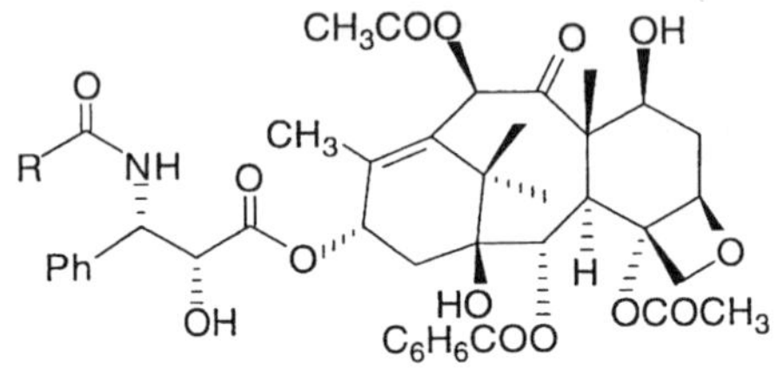

1 R = Ph
3 R = $OC(CH_3)_3$

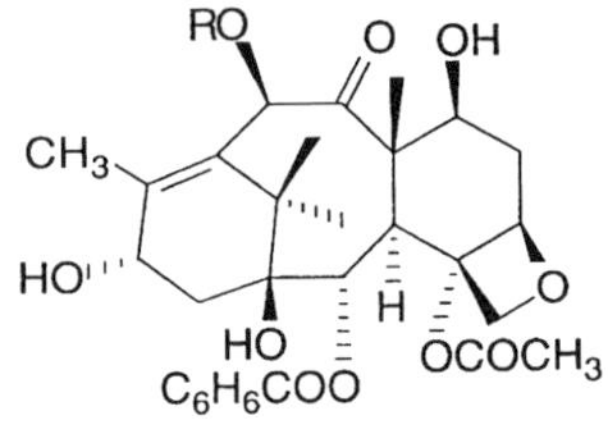

2a R = H
2b R = CH_3CO

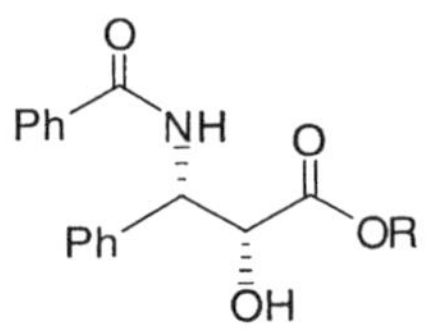

4a R = H
4b R = CH_3

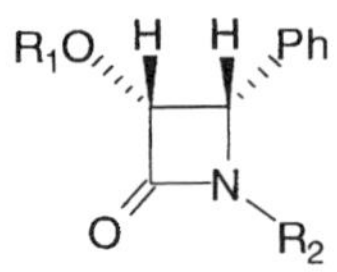

5 $R_1 = CH_3CH(C_2H_5O)$
$R_2 = PhCO$

Since the N-benzoyl-(2*R*,3*S*)-3-phenylisoserine moiety is essential for the potent antitumor activity of Taxol,[9] many groups have been interested in the development of practical syntheses of the phenylisoserine side chain of Taxol.[10]

20.2 Biocatalytic Methods

The application of biocatalytic methods to organic synthesis has become a well-defined area of research and several books,[11] and review articles[12] are now available.

In recent years, there has been an increasing interest in the use of biocatalytic methods to produce optically active compounds. In particular, hydrolytic enzymes such as the microbial lipases have been extensively used for the kinetic resolution of racemic alcohols and carboxylic esters.[12] Their commercial accessibility and relative stability make them an attractive class of biocatalysts for effecting industrial-scale kinetic resolutions. This chapter describes several different biocatalytic strategies used for the preparation of the Taxol side chain in its enantiomerically pure form.

Enantioselective biocatalysis has been increasingly used for the optical resolution of many different classes of molecules such as amino acids, lactones, esters, and hydroxy acids. Azido alcohols are suitable precursors of amino alcohols, and it was shown that enzymatic hydrolyses of the butyl esters proceeded with acceptable conversion rates and with high degrees of stereoselectivities.[13]

20.3 Suitable Ester Substrates for Lipase Hydrolysis

Honig and co-workers[14] first reported the chemoenzymatic synthesis of 3-phenylisoserine derivatives. They synthesized the substrates, *syn*- and *anti*-3-azido-2-butanoyloxy-3-phenylpropionic esters (**6**) and (**7**) and subjected to kinetic resolution using several microbial lipases.

(±)-6 (±)-7

(a) *R* = H
(b) *R* = acetyl
(c) *R* = butyryl
(d) *R* = chloracetyl
(e) *R* = isobutyryl
(f) *R* = cyclohexylcarboxyl
(g) *R* = octanoyl

To find a suitable acyl group to be hydrolyzed, a series of esters of (+)-methyl-*anti*-3-azido-2-hydroxy-3-phenyl propionate (**7**) was prepared and first examined with lipases of *Candida rugosa* and *Pseudomonas cepacia* (P). As a rule, the acetate (**7b**) and butyrate (**7c**) was cleaved at useful rates with high degrees of enantioselectivity; the chloracetate (**7d**) was cleaved rapidly, but chemical hydrolysis also occurred during the reaction, which could account for the decrease in the optical purity of the product; the isobutyrate (**7e**) and cyclohexylcarboxylate (**7f**) were hydrolyzed slowly with good enantioselectivity, whereas the octanoate (**7g**) was hydrolyzed at an acceptable rate, but the ee of the product was low. Both lipases preferentially cleaved the 2*S*,3*R* enantiomer. The most suitable ester was the butyl ester (**7c**), which was hydrolyzed by the *Pseudomonas cepacia* lipase in 40% yield in 1 h with an *E* value of 100; *Candida rugosa* lipase gave only modest *E* values on all the esters examined. Attempts to achieve enantioselective esterification of (+)-methyl-*anti*-3-azido-2-hydroxy-3-phenylpropionate (**7a**) with acetic anhydride, and *P. cepacia* lipase in benzene afforded a lower conversion rate with product of lower ee. Similar attempts using racemic phenylisoserinamide as substrate and vinyl acetate as acyl donor likewise gave unsatisfactory results.

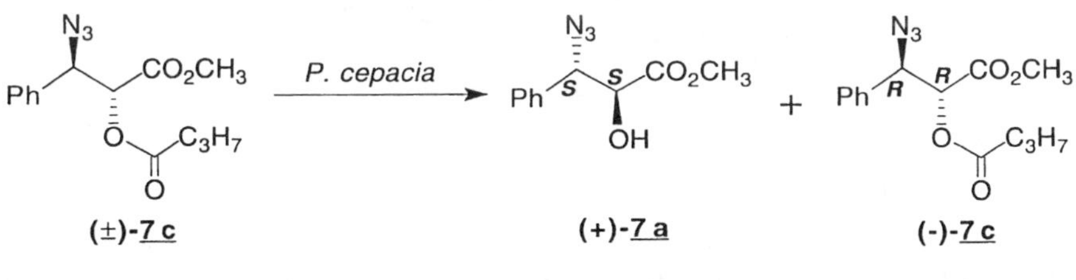

(±)-7c (+)-7a (-)-7c

As far as the hydrolysis of (±)-ethyl-*syn*-3-azido-2-hydroxy-3-phenylpropionate butyl ester (**6c**) is concerned, the most suitable enzyme for this transformation was found again to be *P. capacia*, which preferentially cleaved the 2*S*,3*R* enantiomer, but the reaction rate was found to be considerably slower than **7c**. In

68 h, approximately 35% of the butyl ester of **6c** was hydrolyzed with an *E* value of 100.

In contrast, the *Candida rugosa* lipase exhibited good reaction rate towards **6c**, but the enantioselectivity was considerably lower.

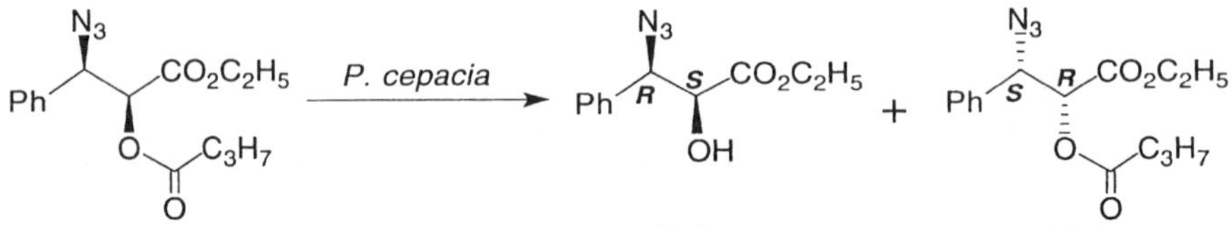

(±)-6 c (-)-6 a ee = 98% (+)-6 c ee = 98%

Chen and co-workers[15] devised a practical chemoenzymatic synthesis of the Taxol C-13 side chain using enzymatically prepared enantiomerically pure *trans*-β-phenyl-glycidic esters as starting materials for the synthesis of N-benzoyl-(2*R*,3*S*)-3-phenylisoserine (**4g**). They were able to achieve the lipase-mediated enantioselective transesterification of (±)-methyl trans-β-phenyl-glycidate (**8**) in organic media by circumventing several obstacles along the way.

The racemic methyl ester (**8**) was first exposed to different lipases in hexane with *n*-butanol as the acyl donor. The most suitable enzyme to catalyze this enantioselective transesterification was the lipase of *Mucor miehei* (Amano, MAP-10). However, the enantioselectivity of this acyl transfer reaction was complicated by the intrinsic thermodynamic equilibrium. That is, the overall selectivity gradually diminished as the reaction proceeded close to 50% conversion. Under the prescribed conditions, the kinetic and thermodynamic parameters were estimated to be E=33 and K=0.11. Two strategies were used to minimize the extent of reverse catalysis. By selecting a sterically hindered alcohol as the acyl acceptor and increasing the alcohol concentration, the backward reaction became virtually negligible. For example, when the methyl ester (±)-**8** was exposed to the *Mucor miehei* lipase in hexane-isobutyl alcohol (1:1) at 36% and 45% conversion, the enantiomeric excess (ee) values for the acylated product were found to be 97% and 95%, respectively, and the corresponding remaining substrate fractions were 56% and 77%, respectively.

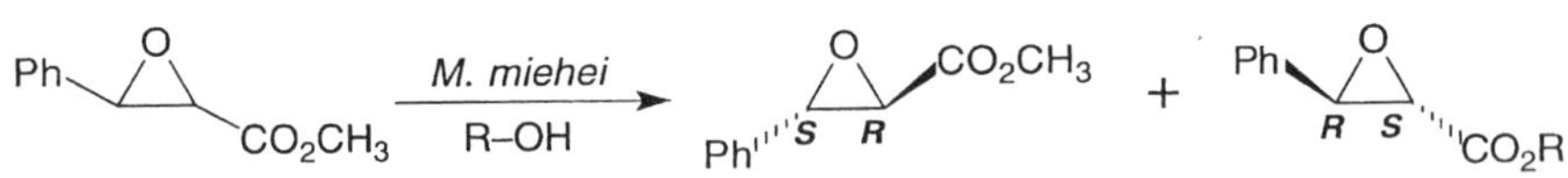

(±)-8 (-)-8 (+)-9 R = isobutyl

Thus, on a 20 g scale conversion, the isolated yield of the product as the isobutyl ester, (+)-**9**, was 43% (11.3 g), and the recovered methyl ester, (–)-**8**, was 42% (8.5 g); the optical purity of both fractions were determined to be 95% ee. The advantages of this enzymatic process are that the product and substrate could be readily isolated by fractional distillation and the crude enzyme retained 85% of its original activity after one cycle and could be easily recovered from the mixture by filtration. Moreover, the lipase is relatively inexpensive ($150/kg), and operational costs could be further reduced by recycling of the biocatalyst.

The chemical sequence used for the transformations of 2*R*,3*S* and 2*S*,3*R* isomers of *trans*-β-phenylglycidate, (–)-**8** and (+)-**9**, into N-benzoyl-(2*R*,3*S*)-3-phenylisoserine (**4a**) is outlined in Scheme 20.1.

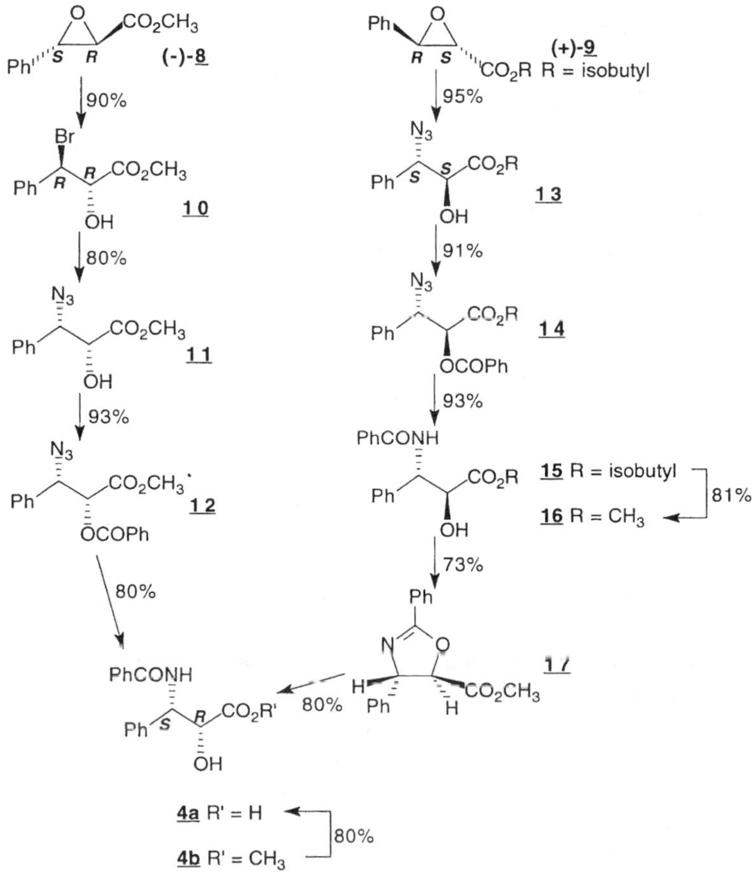

Scheme 20.1

Regioselective cleavage of the (2*R*,3*S*)-epoxide (**8**) with diethylamine hydrobromide and diethylaluminum chloride afforded the (2*R*,3*R*)-hydroxy bromide (**10**) in 90% yield. SN^2 Azide displacement of the bromide (**10**) proceeded in 80% yield. The hydroxy azide (**11**) was then benzoylated followed by hydrogenation and then recrystallized to give enantiomerically pure methyl ester (**4b**), which was then hydrolyzed with K_2CO_3/CH_3OH-H_2O to give the C-13 side chain (**4a**) in 80% yield. The overall yield from the epoxide (**8**) was 40%.

The 2*S*,3*R* epoxide (**9**) was refluxed with sodium azide and ammonium chloride in acetone–water to generate the 2*S*,3*S*-hydroxy azide (**13**), which was subjected to benzoylation and then hydrogenated to give the key intermediate (**15**), which was converted into its methyl ester (**16**). This *anti* isomer was treated with thionyl chloride to give the oxazoline intermediate (**17**), resulting in an inversion of the C-2 configuration.

Subsequent hydrolysis with aqueous 1*N* HCl in methanol give enantiomerically pure *syn* isomer (**4b**) with an overall yield of 38% from the 2*S*,3*R*-epoxide (**9**). Alkaline hydrolysis yields the target side chain (**4a**).

This chemoenzymatic approach provides an efficient route to the Taxol C-13 side chain, for both isomers secured from the enzymatic resolution were efficiently utilized in the process.

Although Shibatani[16] of Tanabe reported a method for preparing optically active 3-phenylglycidic acid esters using a variety of microorganisms, the only example given was for the stereoselective hydrolysis of racemic methyl 3-(4-methoxyphenyl)glycidate (**18**). The microbe, *Serratia marcescens* ATCC 27117, catalyzed the enantioselective hydrolysis to yield (2*R*,3*S*)-3-(4-methoxyphenyl)glycidate (**19**) in 40% yield and 100% optical purity. Although this conversion may be useful for the preparation of analogs of the Taxol side chain, unfortunately, no experimental information was given for the preparation of the requisite optically active *trans*-β-phenylglycidate, **8** or **9**.

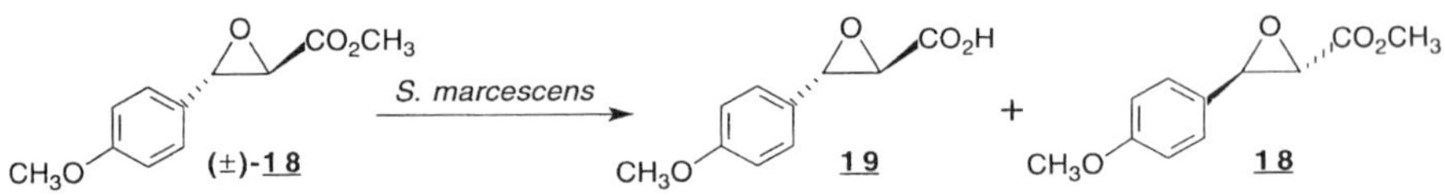

Similarly, Servi and co-workers[17] reported the enzymatic preparation of (2*R*,3*S*)-phenylglycidic acid esters in connection with their investigations on their chemoenzymatic synthesis of (2*S*,3*S*)-diltiazem, an enantiomerically pure drug with calcium antagonist activity. They found the same enantioselective transformation was catalyzed by the *Candida cylindracea* lipase in cyclohexane–water mixture to furnish **19** in 96% ee. However, it is not known whether this lipase could also catalyze the enantioselective hydrolysis of (±)-**8**, since this lipase was not as enantioselective in catalyzing the transesterification of (±)-**8** as compared to *Mucor miehei*.

Since enantiomerically pure 3*R*-hydroxy-4*S*-phenyl-β-lactam derivatives[10c] are key chiral intermediates to **4a**, or they may be coupled directly to **2b** to form Taxol.[8] For this reason, the enzymatic kinetic resolutions of 1-substituted 3-acetoxy-4-phenyl-β-lactam derivatives were investigated,[18] and the results are summarized in Scheme 20.2.

The most suitable lipases for these transformations were catalyzed by the lipases of *Pseudomonas cepacia* (p-30) or *Pseudomonas sp.* (AK), which afforded good reaction rates and high enantioselectivity. For water-insoluble substrates in aqueous media, the reaction rate was markedly improved by the addition of a cosolvent such as acetonitrile.

One unique feature of the lipases is their remarkable stability in organic solvents such as *tert*-butyl ether (*t*-BuOMe), even at 50°C, provided only a small quantity of water is present (<0.4%). The conformational flexibility of the enzyme becomes more

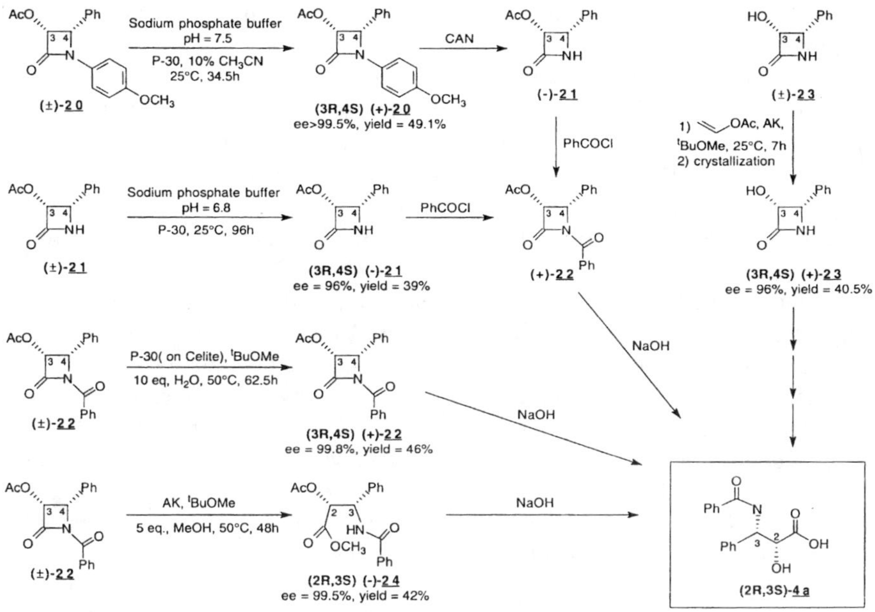

Scheme 20.2

restricted when very small quantities of water are present. This provides the investigator with additional versatility to allow one to conduct enantioselective hydrolytic reactions in organic solvents. Hence, the enantioselectivity of the reaction may be improved by systematically manipulating the organic solvent media.

It is worthy of note that the reaction pathway was markedly influenced by the nucleophile in the medium. When water was used as the nucleophile, only the cleavage of the C-3 acetoxy ester of (±)-**22** was noted. On the other hand, when methanol was substituted for water, methanolysis of the β-lactam ring was observed by the two *Pseudomonas* lipases p-30 and AK. Interestingly, the reaction rate and enantioselectivity were considerably better than the penicillinases, enzymes that are known to cleave the β-lactam ring of penicillins. This observation showed clearly the different reaction pathway of hydrolysis versus methanolysis. Moreover, lipases P-30 and AK preferentially hydrolyzed the acetoxy ester of the (3*S*,4*R*)-enantiomer of **22**. In contrast, under solvolytic conditions, these enzymes preferentially cleaved the β-lactam ring of the (3*R*,4*S*)-enantiomer of **22**, presumably via the formation of an acyl–enzyme intermediate. Hence these investigators substituted baccatin III (**2b**) for methanol as the nucleophile in the reaction mixture, hoping that these lipases would transfer the phenylisoserine moiety directly onto the C-13 hydroxyl group of **2b**.

However, no coupling product was detected. It is possible that the size of the acyl acceptor (**2b**) may be too large to be accommodated optimally by the enzymes, because of the lack of conformational mobility under a dehydrated environment.

The Squibb group[19] was also successful in achieving the enantioselective hydrolysis of (±)-**21** using a wide variety of lipases, including P-30 form the Amano company.

Finally, the Bristol–Myers Squibb group[20] developed a microbial asymmetric reduction process for the preparation of the phenylisoserine side chain of Taxol. A variety of microorganisms were found to catalyze the stereoreduction of the keto ester (**25**) to the desired 2*R*,3*S*-ester, **26**, which included organisms from the following genera: *Nocardia*, *Candida*, *Rhodococcus*, *Mortierella*, *Saccharomyces*, and *Hansenula*. The reaction yields varied from 31 to 86%, and the optical purity of **26** ranged from 26 to 99% ee.

In principle, the reduction of the keto ester, **25**, could result in this formation of four possible diastereomeric alcohols as shown. However, reaction conditions were found under which predominantly the desired 2*R*,3*S*-isomer, **26a**, was formed in the microbial reduction. Since the proton next to the carbonyl is sufficiently acidic to undergo enolization under the reaction conditions, all of **25** could be converted into **26** via this second-order-asymmetric transformation, which is an advantage over resolution methods. The most suitable microorganisms were found to be *Hansenula polymorpha* and *Hansenula fabianii*, which effectively reduced **25** to the desired *syn* diastereomer **26a**. Reaction yields of >80% and optical purity of >98% were obtained. About 10–20% of the *anti* diastereomers **26c**, **26d** were also produced but were readily separated from **26a**.

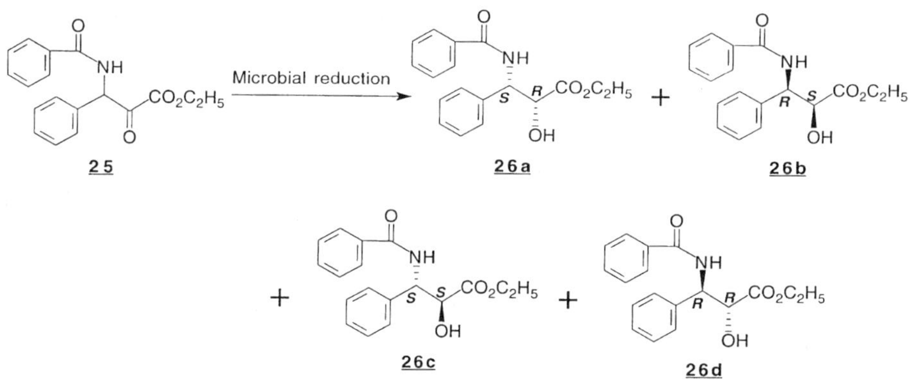

20.4 Conclusion

Several different chemoenzymatic methods are now available for the preparation of the side chain of Taxol, **4b**. Each approach has its merits in terms of yields and convenience of operation. It is likely that one or more of these procedures will be utilized for the industrial semisynthesis of Taxol and Taxotere. No doubt, new improved syntheses will be developed in the future. An especially challenging problem is the enzymatic attachment of the phenylisoserine moiety directly onto **2b**.

20.5 Acknowledgment

This work was supported by National Institutes of Health grant GM 331449.

20.6 References

1. M. C. Wani, H. L. Taylor, M. E. Wall, P. Coggon, A. T. McPhail, *J. Am. Chem. Soc.* **93**, 2325 (1971).
2. P. B. Schiff, J. Fant, S. B. Horwitz, *Nature* **277**, 665 (1979).
3. E. K. Rowinsky, R. C. Donehower, *Pharmacol. Ther.* **52**, 35 (1991).
4. (a) E. K. Rowinski, L. A. Casenave, R. C. Donehower, *J. Nat'l. Cancer Inst.* **82**, 1247 (1990). (b) E. J. Blume, *J. Nat'l. Cancer Inst.* **81**, 1122 (1989).
5. (a) R. A. Holton, H. B. Kim, C. Somoza, F. Liang, R. J. Biediger, P. D. Boatman, M. Shindo, C. C. Smith, S. Kim, et al., *J. Am. Chem. Soc.* **116**, 1599 (1994). (b) K. C. Nicolaou, Z. Yang, J. J. Liu, H. Ueno, P. G. Nantermet, R. K. Guy, C. F. Claiborne, J. Renaud, E. R. Couladouros, K. Paulvannan, et al., *Nature* **367**, 593 (1994).
6. (a) J. N. Denis, A. E. Greene, D. Guénard, F. Guéritte-Voegelin, L. Mangatal, P. Potier, *J. Am. Chem. Soc.* **110**, 5917 (1988). (b) L. Mangatal, M. T. Adeline, D. Guénand, F. Guéritte-Voegelin, P. Potier, *Tetrahedron* **45**, 4177 (1989).
7. G. Chauvière, D. Guénard, F. Picot, V. Sénilh, P. Potier, *C. R. Seances Acad. Sci. Ser.* **293**, 501 (1981).
8. R. A. Holton, Eur. Pat. Appl. 0400-971 A2 (1990).
9. D. Guénard, R. Guéritte-Voegelein, P. Potier, *Acc. Chem. Res.* **26**, 160 (1993).
10. (a) J. N. Denis, A. E. Greene, A. A. Sera, M. J. Luche, *J. Org. Chem.* **51**, 46 (1986). (b) J. N. Denis, A. Correa, A. E. Greene, *ibid.* **55**, 1957 (1990). (c) I. Ojima, H. Habas, M. Zhao, G. I. Georg, L. R. Jayasinghe, *ibid.* **56**, 1681 (1991). (d) J. N. Denis, A. Correa, A. E. Greene, *ibid.* **56**, 6939 (1991).
11. (a) K. Drauz, H. Waldmann, Eds., *Enzyme Catalysis in Organic Synthesis: A Comprehensive Handbook* (VCH Publishers, New York, 1995). (b) C. H. Wong, G. M. Whitesides, *Enzymes in Synthetic Organic Chemistry*, Tetrahedron Organic Chemistry Series, Vol. 12 (Elsevier Science Publishers, New York, 1994).
12. (a) E. Santaniello, P. Ferraboschi, P. Grisenti, A. Manzocchi, *Chem. Rev.* **92**, 1071 (1992). (b) J. A. Jongejan, J. B. A. Vantol, J. A. Daine, *Chemistry Today* **12**, 15 (1994). (c) W. Boland, C. Fröbl, M. Lorenz, *Synthesis*, 1049 (1991).
13. (a) K. Faber, H. Honig, P. Seufer-Wasserthal, *Tetrahedron Lett.* **29**, 1903 (1988). (b) H. Honig, P. Seufer-Wasserthal, F. Fülüp, *J. Chem. Soc., Perkin Trans. I*, 2341 (1989). (c) E. Foelsche, A. Hickel, H. Hönig, P. Seufer-Wasserthal, *J. Org. Chem.* **55**, 1749 (1990).
14. H. Hönig, P. Seufer-Wasserthal, H. Weber, *Tetrahedron* **46**, 3841 (1990).
15. D. M. Gou, Y. C. Liu, C. S. Chen, *J. Org. Chem.* **48**, 1287 (1993).
16. T. Shibatani, Eur. Pat. Appl. 0362-556 A1 (1990).
17. A. Gentile, C. Giordano, C. Fuganti, L. Ghirotto, S. Servi, *J. Org. Chem.* **57**, 6635 (1992).
18. R. Brieva, J. Z. Crich, C. J. Sih, *J. Org. Chem.* **58**, 1068 (1993).
19. R. N. Patel, A. Banerjee, R. Y. Ko, J. M. Howell, W. S. Li, F. T. Comezoglu, R. A. Partyka, L. Szarka, *Biotechnol. Appl. Biochem.* **20**, 23 (1994).
20. R. N. Patel, A. Banerjee, J. M. Howell, C. G. McNamee, D. Brozozowski, D. Mirfakhraea, V. Nanduri, J. K. Thottathil, L. J. Szarka, *Tetrahedron Asymmetry* **4**, 2069 (1993).

CHAPTER

21

Biocatalytic Entry to Enantiomerically Pure β-Amino Acids

Vadim A. Soloshonok

21.1 Introduction

Chiral chemistry of living systems, which are made up of chiral nonracemic components, causes apparent discrimination between enantiomers of a chiral compound, leading to the manifestation of differences in their biological activity as a macroscopic phenomenon of a long chain of substrate–receptor diastereomeric interactions.[1] Examples in which the biological behavior of chiral compounds dramatically depends on stereochemistry can be drawn from many classes of natural and manmade products, such as terpenes, pheromones, drugs, analgesics, food additives, perfumes, and crop protection agents.[1–3] Now it is finally acknowledged that two enantiomers of a chiral compound should be recognized as different substances,[4] one of which (eutomer[5]) can display desired biological properties whereas the other (distomer) could possess opposite biological properties, be less effective or inactive, or, at worst, toxic.[6] As a consequence of the general awareness of the relevance of chirality to biological activity, the development of EPC synthesis[7] is currently a topic of fundamental importance[8] for the life science fields concerned with the design, investigation, and production of compounds for biological purposes.

The chemistry and biology of β-amino acids are an integral part of the expanding area of research directed towards nonproteinogenic natural and manmade amino acids, which are, apart from their own usually high biological activity,[9] of paramount importance in the synthesis of complex compounds such as peptides with

*New address: National Industrial Research Institute of Nagoya, Hirate-cho 1-1, Kita-ku, Nagoya, Japan.

designed conformational properties and biological functions.[10] In particular, most β-amino acids themselves exhibit powerful antibacterial properties and are key constituents of many naturally occurring peptides, terpenes, alkaloids, macrolides, and β-lactam antibiotics.[9,11] In general, β-amino acids are recognized as promising biologically active compounds of high pharmaceutical and medicinal interest. Consequently, in recent years considerable efforts have been expended to devise stereocontrolled methods for their synthesis. The overwhelming majority of synthetic methodologies described in the literature for preferential formation of a single enantiomer of a targeted chiral β-amino acid deal with purely chemical means, using the innate chirality of naturally occurring organic molecules (manipulation of the chiral pool), stoichiometric, and catalytic asymmetric syntheses; these are discussed in detail in the corresponding chapters of this book. By contrast, biotransformations, as an approach to enantiomerically pure β-amino acids, remain, to date, virtually unexplored. Obviously, it is largely due to the fact that β-amino acids turned out to be challenging targets for biocatalytic techniques. For instance, the enzymes, commonly used for the resolution of α-amino acids, such as aminoacyclases,[12] amino peptidases,[13] or lipases,[14] do not resolve β-amino acids or show low activity and stereoselectivity of biocatalytic process. Thus the huge positive potential of biotransformations achieved in the area of α-amino acids[12,13,15,16] cannot be directly applied to the β-amino acid family. Nevertheless, the lure of biocatalytic preparation of enantiopure β-amino acid, especially on a large scale, has stimulated research activity in this area.

Synthetic methods that make use of the biocatalytic reaction to achieve preparation of enantiomerically pure β-amino acids, the relevance of β-amino acid chemical structure to the stereochemical outcome of the enzymatic process, the synthetic merits and limitations of a biocatalytic approach to stereochemically defined β-amino acids, form the subject of the present chapter. This review is not intended to be comprehensive; however, we have attempted to summarize all that has been reported so far on the subject. Material presented in the chapter has been organized according to the structure of β-amino acids available by biocatalytic methods but not to the nature of enzyme or microorganism used, as usually adopted for reviews on enzymatic subjects. The β-amino acids reviewed in this chapter were divided into four main groups of general structures **1**–**4**. Each group possesses some characteristic features, such as number of stereogenic centers and functionalities, which determine feasibility of biocatalytic method to be used for prepared of corresponding enantiomerically pure β-amino acids. Furthermore, we believe that this presentation would be more convenient for understanding the perspectives of biocatalytic entry to enantiopure β-amino acids and familiar to readers of this book.

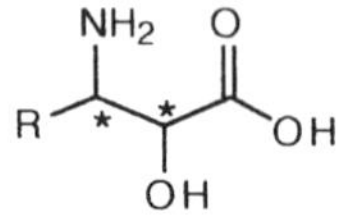

1

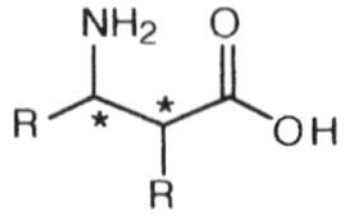

2

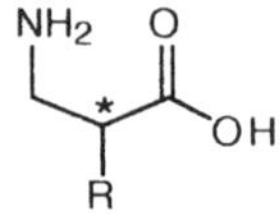

3

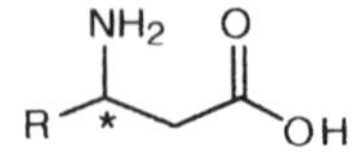

4

21.2 β-Amino-α-Hydroxy Acids (Isoserines)

The recent surge of research in the area of the Taxol-related family of naturally occurring diterpenes has generated a great deal of attention towards β-amino-α-hydroxy acids[17] and particularly to the β-phenylisoserine. Without this compounds, [(2*R*,3*S*)-N-acyl-3-phenylisoserine moieties **9** and **10**], the strong anticancer activity of Taxol **7** and Taxotere® **8** is not expressed.[18] The evolution of **6** and **7** into highly significant antitumor drugs[18b,19] has focused interest on their semisynthetic preparation via attachment of the A-ring side chain (compounds **9**, **10**) to naturally derived taxanes **5**, **6**, which are more readily available than Taxol **7** and Taxotere® **8**.[18] Among the methods reported so far on the synthesis of β-phenylisoserine,[18] the biocatalytic approach seems to be able to offer the most practical access to β-phenylisoserine and its analogs in quantity.

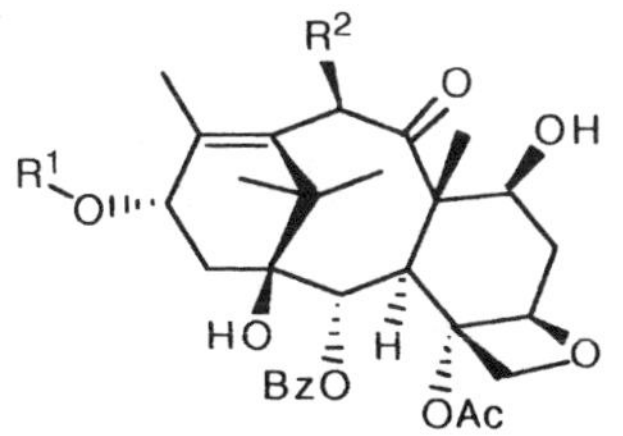

5: 10-Deacetylbaccatin III, R^1 = H, R^2 = OH;
6: Baccatin III, R^1 = H, R^2 = OAc;
7: Taxol, R^1 = **8**, R^2 = OAc
8: Taxotere®, R^1 = **9**, R^2 = OH

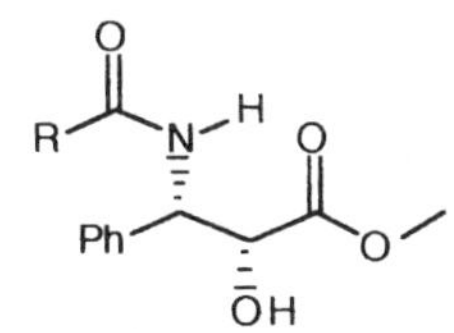

(2*R*,3*S*)-*N*-acyl-3-phenylisoserine moiety, C-13 ester side chain of **7** and **8**

9: R = Ph;
10: R = *t*-BuO

21.2.1 Biocatalytic Reductions

Enzymatic reduction of the carbonyl group is a well-established synthetically useful generalized method for the preparation of various enantiomerically pure alcohols of both (*S*) and (*R*) absolute configuration.[13,16] Kearns and Kayser[20] have demonstrated recently an application of yeast-catalyzed reductions of α-keto esters for preparation of (2*R*,3*S*) and (2*R*,3*R*)-β-phenylisoserines (Scheme 21.1). Commercial (*S*)-phenylglycine (**11**), without protection of an amino group, was readily converted to methyl (*S*)-amino-2-keto-3-phenylpropionate (**14**), which was reduced with fermenting baker's yeast to (2*R*,3*S*)-3-phenylisoserine methyl ester (**15**) in 72% chemical yield. Examination of the crude reaction product by 250 MHz ^{1}H-NMR has revealed that only a single diastereomer was produced by biocatalytic transformation. Rapid chromatography over a short column of silica gel furnished diastereo- and enantiomerically pure amino ester **15**, which was converted to N-benzoyl derivative **18**, for further attachment to baccatin III (**6**), through O-protected compounds **16** or directly by the Schotten–Baumann reaction.

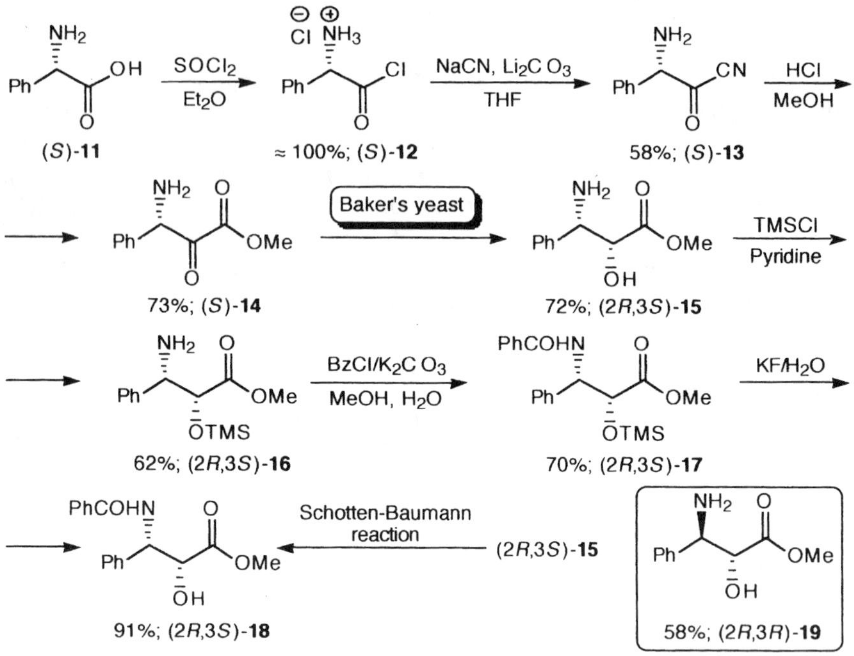

Scheme 21.1

The diastereomeric (2*R*,3*R*)-3-phenylisoserine **19** was prepared by the same procedure starting with (*R*)-phenylglycine. The yield of the yeast-catalyzed reduction in this case was lower; however, the isolated product (2*R*,3*R*)-**19** was diastereomerically pure.

In another biocatalytic protocol, developed by Patel et al.,[21] phenylglycine (**11**) was first benzoylated to N-benzoyl derivative **20** and then converted to racemic α-keto ester **22** through enol ester **21** (Scheme 21.2). An appealing feature of this method is that under the conditions reported certain growing microorganisms preferentially reduced the (*S*) enantiomer of keto ester **22** to the isoserine derivative (2*R*,3*S*)-**23**, while (*R*) isomer underwent racemization, thus allowing one to achieve complete transformation of starting material **22** to the desired product (2*R*,3*S*)-**23**. Screening of the series of microorganisms for asymmetric reduction of **22** (see Table 21.1) has revealed that specimens from the genus *Nocardia* (entries 1–3), *Candida* (entry 4), *Rhodococcus* (entries 5,6), *Mortierella* (entry 7), *Saccharomyces* (entry 8), and *Hansenula* (entries 9,10) are capable of producing the isoserine derivative (2*R*,3*S*)-**23** in high enantiomeric purity. *H. fabianii* SC 13894 (entry 9) and *H. polymorpha* SC 13865 were most effective in giving rise to the target compound (2*R*,3*S*)-**23** in >80% chemical yield and >95% optical purity. Asymmetric reduction of **22** to **23** on a preparative scale was carried out in a 5 L fermentor using cell suspensions of *H. fabianii* SC 13894 and *H. polymorpha* SC13865 or corresponding cell extracts. From one experiment (*H. polymorpha* SC 13865), 5.2 g (65% overall yield) of (2*R*,3*S*)-N-benzoyl-3-phenylisoserine ethyl ester (**23**) was isolated in 99.6% optical purity (by chiral HPLC analysis).

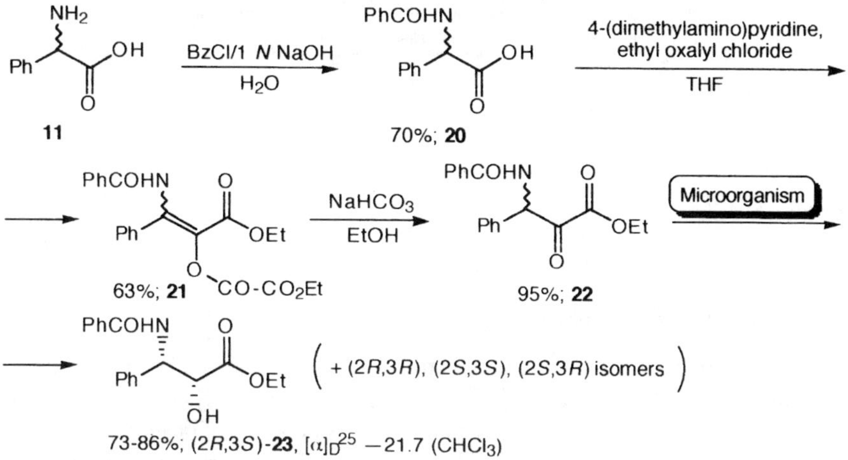

Scheme 21.2

The same group also reported asymmetric reduction of α-keto ester **22** to (2*R*,3*R*)-N-benzoyl-3-phenylisoserine ethyl ester (**24**) Scheme 21.3) by the purified β-hydroxy butyrate dehydrogenase (Sigma) in the presence of NADH. Purification of the crude reaction product by flash column chromatography and recrystallization from acetonitrile furnished enantiomerically pure (2*R*,3*R*)-**24** with 70% chemical yield. The original stereoselectivity of this biocatalytic process is not determined.[21] Remarkably, in sharp contrast to the microorganism reduction discussed above, β-hydroxy butyrate dehydrogenase preferentially reduced the (*R*)-enantiomer of α-keto ester **22**, while the stereochemical result of the asymmetric reduction step was the same for both biocatalytic reactions, the (2*R*) absolute configuration of the products **23** and **24**.

Table 21.1 BIOCATALYTIC REDUCTION OF ETHYL 3-(N-BENZOYL)AMINO-2-KETO-3-PHENYLPROPIONATE

Entry	Microorganism	Yield of **23**	Optical purity of (2*R*,3*S*)-**23**[a]
1	*Nocardia restricta* SC 13107	75	88
2	*Nocardia salmonicolor* SC 6310	45	99
3	*Nocardia globerula* ATCC 21505	36	92
4	*Candida guilliermondi* ATCC 20318	31	95
5	*Rhodococcus sp.* ATCC 29675	53	91
6	*Rhodococcus erythropolis* ATCC 4277	39	96
7	*Mortierellarammanianna* ATCC 38191	35	97
8	*Saccharomyces cerevisiae* ATCC 24702	35	94
9	*Hansenula fabianii* SC 13894	85	95
10	*Hansenula polymorpha* SC 13865	80	99

[a] Determined by HPLC analysis.

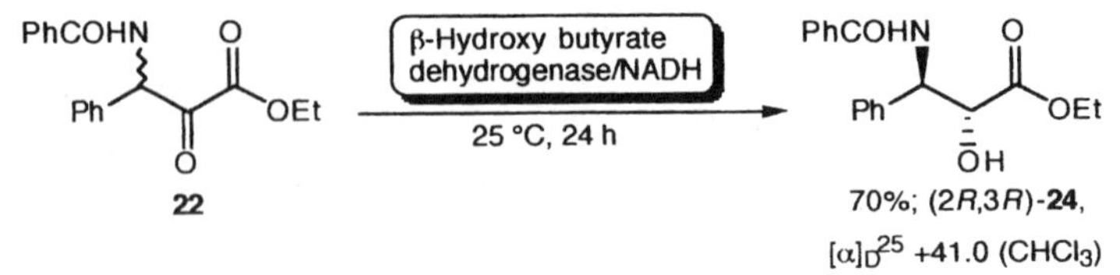

Scheme 21.3

21.2.2 Biocatalytic Resolutions

The presence of α-hydroxy functionality in the isoserine skeleton makes it possible to access the enantiomers of this class of β-amino acids by lipase-catalyzed kinetic resolution methods, which have been widely employed for the resolutions of racemic secondary alcohols.[16,23] Commercial availability and relative stability in both aqueous and organic media, usually high stereochemical specificity and broad substrate specificity of the lipases, make them attractive catalysts for effecting large-scale resolutions. Thus Kunesch et al.[23] described a convenient preparation of both enantiomers of isoserine via biocatalytic transesterification with *Candida cylindracea* lipase (CCL) in organic medium (Scheme 21.4). This five-step procedure starts from the readily available ethyl glyoxylate (**25**), which was converted to racemic isoserine (**27**) with 85% overall yield through cyanohydrin **26**. Acetylation of amino acid **27** on amino and hydroxy groups gave the diacetyl derivative **28** subjected next to the key biocatalytic step. Asymmetric transesterification of **28** by CCL was conducted in the presence of *n*-butanol (≈ 1 equiv.) in

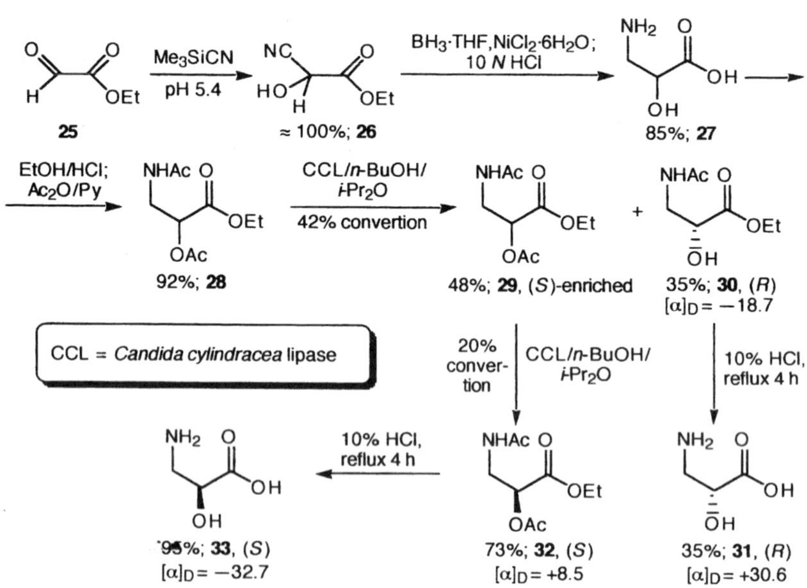

Scheme 21.4

diisopropyl ester solution. To obtain both enantiomers (*R*)-**31** and (*S*)-**33** of isoserine,[23b] the enzymatic reaction was quenched after 42% conversion of **28**. This gave (*S*)-enriched diacetate **29** and (*R*)-alcohol **30**, which were easily separated by column chromatography. Acidic hydrolysis of **30** yielded (*R*)-isoserine (**31**) with 94% ee (calculated from the optical rotation). The unreacted diacetate **29** was further submitted to enzymatic transesterification stopped after 20% conversion. Hydrolysis of diacetate **32** gave enantiomerically pure (*S*)-isoserine (**33**). With the aim to prepare only (*S*)-enantiomer **33**,[23a] the enzymatic reaction was stopped after 52% conversion of **28** that provided diacetate **29** with 43% chemical yield and 90% ee. Hydrolysis of **29** to (*S*)-**33** followed by crystallization of crude **33** furnished enantiopure (*S*)-isoserine (**33**) in 33% overall yield.

This methodology seems to be easily generalized for resolution of other amino acids containing a β-aminoalcohol functionality associated with a carboxylic group, which, for instance, was demonstrated by preparation of both enantiomers of γ-amino-β-hydroxybutanoic acid.[24]

Systematic study of lipase-catalyzed kinetic resolution of β-lactams, as precursors to the C-13 Taxol side chain [(2*R*,3*S*)-3-phenylisoserine], was presented recently by Sih et al.[25] Four different compounds **34**, **37**, **40** (Scheme 21.5), and **45** (Scheme 21.6)

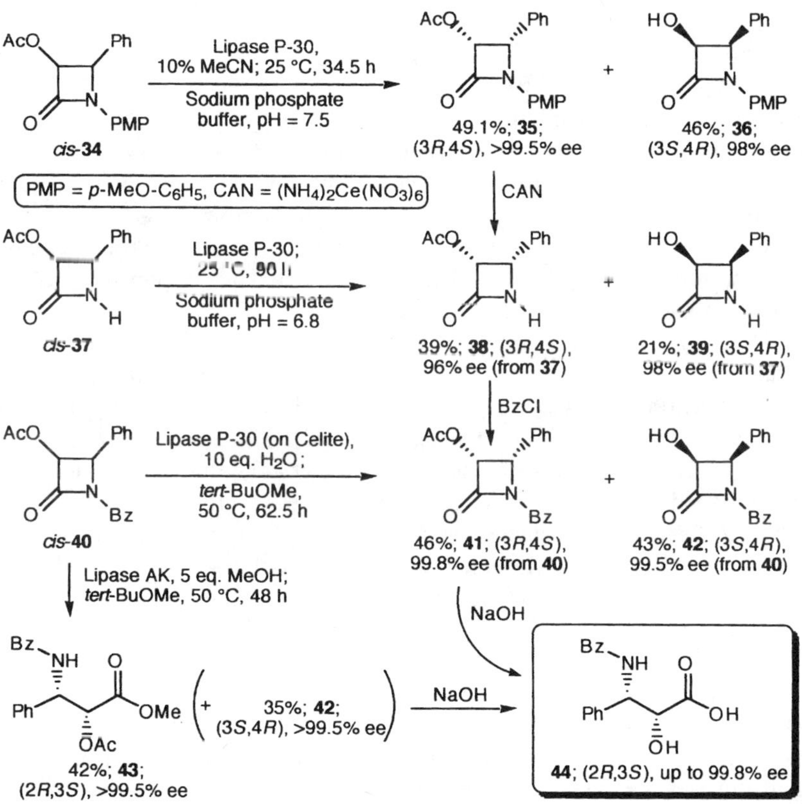

Scheme 21.5

were selected as substrates in view of their ready availability[26] in the *cis*-diastereomerically pure state and their potential utility in the Taxol synthesis. *Pseudomonas cepacia* lipase P-30 was found to be superior for enantioselective cleavage of the ester group at C-3 of racemic *cis*-β-lactam **34** (Scheme 21.5). Thus subjection of substrate *cis*-**34** to the action of lipase P-30 in the aqueous solution, containing 10% of acetonitrile as a co-solvent, for 34.5 h at 25°C, gave (3*R*,4*S*) O-acetyl derivative **35** and deacetylated product (3*S*,4*R*)-**36** in both excellent yield and enantiomeric purity. It was pointed out that the presence of co-solvent is crucial for the observed rate and (3*S*,4*R*)-enantioselectivity of the enzymatic reaction.

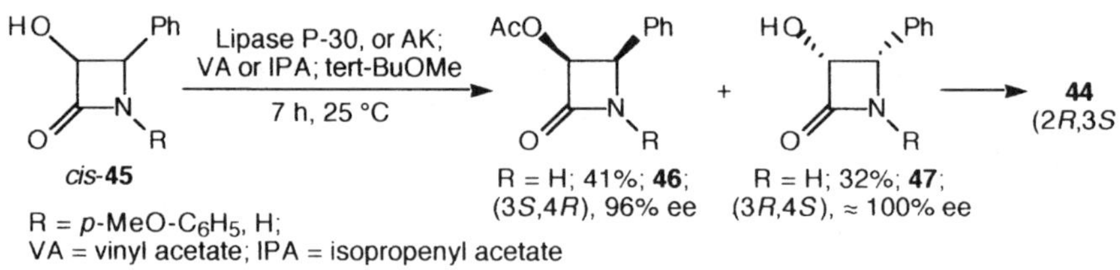

Scheme 21.6

In contrast to this, the most synthetically interesting result on P-30 lipase-catalyzed resolution of N-unsubstituted β-lactam *cis*-**37** was obtained without addition of a co-solvent. However, the rate of biocatalytic hydrolysis and yields of resolved compounds (3*R*,4*S*)-**38** and (3*S*,4*R*)-**39** were lower than that of *cis*-**34** resolution process. The presence of an N-benzoyl group in the *cis*-β-lactam **40** caused a very low water solubility and an easy β-lactam ring cleavage in pH 7.5 phosphate buffer. Thus to achieve preparative resolution of *cis*-**40** immobilized P-30 enzyme on Celite was used in an organic solvent. *tert*-Butyl methyl ether containing 10 molar equivalents of water was found to be the most suitable medium for biocatalytic resolution of *cis*-**40** giving compounds (3*R*,4*S*)-**41** and (3*S*,4*R*)-**42** of excellent enantiomeric purity (>99% ee) in 46 and 43% chemical yield, respectively. The economic efficiency of this resolution procedure was shown by the feasibility of recycling the immobilized lipase three times without sizable loss of the activity and enantioselectivity of enzyme. It is also noteworthy that the stereochemical preference of the P-30 lipase-catalyzed hydrolysis was the same in aqueous and organic media; (3*S*,4*R*)-enantiomer was the fast-reacting isomer in both cases. Substitution of water with methanol, as the nucleophile for lipase-catalyzed resolution of *cis*-β-lactam **40** in *tert*-butyl methyl ether, led to a rather surprising result. Thus the presence of methanol in the reaction medium caused involvement of both reactive sites of β-lactam **40** in the enzymatic reaction giving rise to (2*R*,3*S*)-3-phenylisoserine derivative **43** as a dominant product, along with expected deacetylated (3*S*,4*R*)-β-lactam **42**. The highest yields of the resolved compounds (2*R*,3*S*)-**43** (42%) and (3*S*,4*R*)-**42** (35%) were achieved using *Pseudomonas* lipase AK. Remarkably, both enzyme-catalyzed reactions were highly stereoselective and opposite in stereochemical preference. Normally, lipases are unsuitable for cleaving amide bonds due to very slow reaction rates and, as was pointed out by Sih et al., this result demonstrates an exciting possibility of expanding the range of substrates for lipase-catalyzed resolutions. All enzymatically prepared compounds **35**, **38**, and **41**, of (3*R*,4*S*) absolute configuration, and

(2*R*,3*S*)-**43** of high diastereo- and enantiomeric purity, were transformed by the established chemical procedures to the target (2*R*,3*S*)-N-benzoyl-3-phenylisoserine (**44**).

Lipase-catalyzed acetylation of racemic *cis*-**45**, as an alternative approach to (2*R*,3*S*)-**44**, was also studied. The transesterification reaction was carried out in *tert*-butyl methyl ether in the presence of lipase P-30 or AK using isopropenyl acetate (IPA) or vinyl acetate (VA) as the acyl donor. The reaction rate of lipase P-30–catalyzed transesterification of *cis*-**45** ($R = p$-MeO-C_6H_5) was slow; however, enantioselectivity of the process approached was shown to be very high (>99% ee). The most synthetically useful result was obtained using N-unsubstituted substrate *cis*-**45** ($R = H$) and lipase AK. After only 7 h, acylated product (3*S*,4*R*)-**46** (96% ee) and hydroxy derivative (3*R*,4*S*)-**47** (≈ 100% ee) were isolated in 41 and 32% chemical yield, respectively. It is worth noting that the faster-reacting enantiomer was the same for both enzyme-catalyzed hydrolytic and transesterification reactions. This promising biocatalytic approach is covered in a separate chapter of the present book in greater detail.

Concerning other approaches that make use of lipase-catalyzed resolution to produce diastereo- and enantiomerically pure Taxol side chain, N-benzoyl-(2*R*,3*S*)-3-phenyl isoserine (**44**), the resolutions of 3-azido-2-(butanoyloxy)-3-phenylpropionic esters[27a] and *trans*-β-phenyl-glycidic esters[27b] are worth mentioning. In these methods the structure of the target β-amino acid was formed by chemical transformations of enzymatically resolved species and thus are out of the scope of the present chapter.

Finally, the amino group of isoserines can also be involved in a biocatalytic resolution process. This possibility was demonstrated by penicillin acylase–catalyzed preparation of (2*S*,3*S*)-**50** and (2*R*,3*R*)-3-phenylisoserine (**52**) (Scheme 21.7).[28] Racemic N-phenylacetyl derivative **49**, prepared by phenylacetylation of commercially available diastereomerically pure β-amino acid (2*R**,3*)-**48**, was exposed to penicillin acylase (EC 3.5.1.11) from *Escherichia coli* ATCC 9637 in aqueous medium for 9 h at 23°C. Smooth enzymatic reaction gave free amino acid (2*S*,3*S*)-**50** and N-phenylacetyl derivative (2*R*,3*R*)-**51** in appreciable chemical yields and high enantiomeric purity (by chiral ligand-exchanged HPLC analysis), which was converted to amino acid (2*R*,3*R*)-**52**. Penicillin acylase was shown to be a unique enzyme for resolution of β-amino acids. Its application for the preparation of enantiomerically pure β-amino acids will be discussed in the following sections.

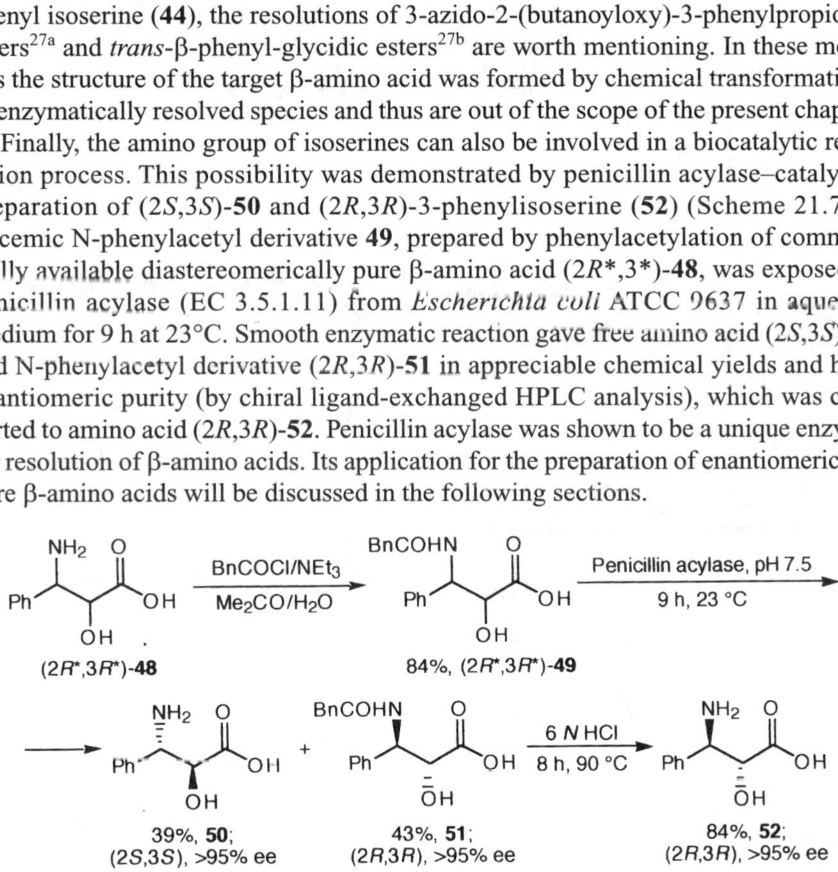

Scheme 21.7

21.3 α,β-Disubstituted β-Amino Acids

α,β-Disubstituted β-amino acids are components of a number of biologically interesting natural products such as the antifungal agents Majusculamide C, 57-Normajusculamide C, and the antitumor depsipeptides Dolastatins D, 11, and 12.[29] These disubstituted β-amino acids are not easily prepared in a stereochemically pure state, however.

The feasibility of a biocatalytic resolution approach to individual stereoisomers of this class of β-amino acids has been demonstrated by the preparation of all four possible enantiomers of 2-methyl-3-amino-4,4,4-trifluorobutanoic acid **57**, **58**, **60**, **61** via penicillin acylase–catalyzed resolution (Scheme 21.8).[30] Diastereomerically pure N-phenylacetyl derivative (2*R**,3*S**)-**54**, synthesized by the [1,3]-proton shift reaction starting with β-keto ester **53** and subsequent N-phenylacetylation, was exposed in aqueous medium at pH 7.5 to penicillin acylase (PA) (EC 3.5.1.11) from *Escherichia coli* for 9 h at room temperature. The enzymatic reaction proceeded in a highly stereoselective way, releasing free amino acid (2*S*,3*R*)-**57** and retaining (2*R*,3*S*)-N-phenylacetyl derivative **56**. Due to the great difference in physical properties between

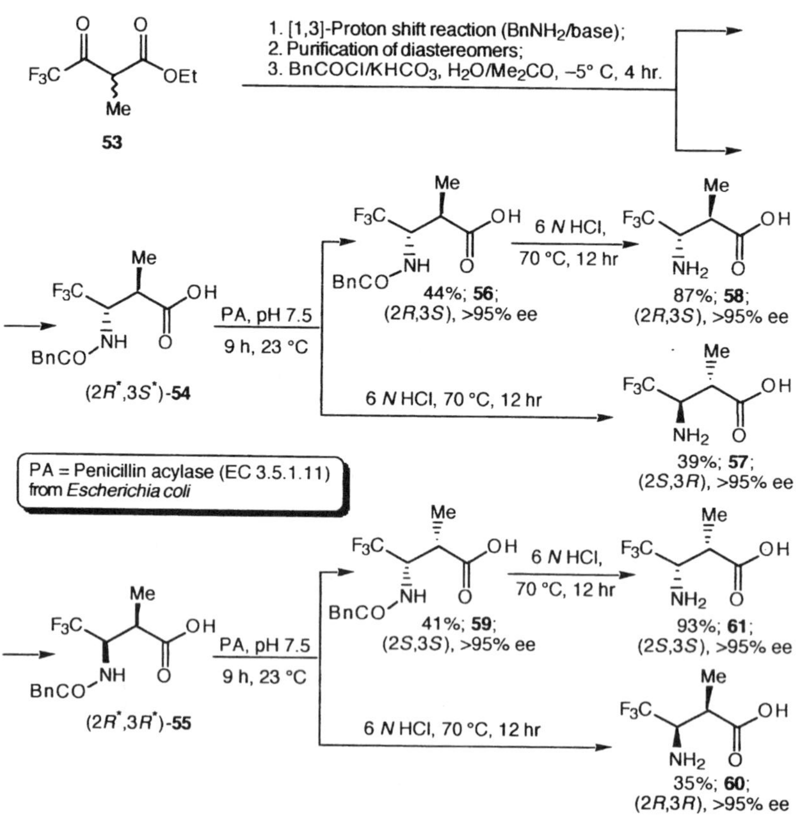

Scheme 21.8

resolved compounds **56** and **57**, they were easily separated and isolated in 44 and 39% chemical yield, respectively. (2*R*,3*S*)-N-Phenylacetyl derivative **56** was hydrolyzed under acidic conditions to give (2*R*,3*S*)-enantiomer **58** in high yield. The enantiomeric purity of amino acids (2*S*,3*R*)-**57** and (2*R*,3*S*)-**58** was shown to be >95% ee (chiral HPLC analysis). A pair of (2*R*,3*R*)-**60** and (2*S*,3*S*)-**61** enantiomers were prepared by the same biocatalytic procedure starting with (2*R**,3*R**)-diastereomer **55**.[30]

This penicillin acylase–catalyzed resolution could be generalized for the preparation of other α-alkyl-β-perfluoroalkyl β-amino acids and their fluorine-free analogs as well.[28]

21.4 α-Monosubstituted β-Amino Acids

α-Monosubstituted β-amino acids **3** are of interest as analogs of natural β-alanine and, being methylene homologues of proteinogenic amino acids, might be valuable building blocks in the study of natural peptides.[31]

21.4.1 Biocatalytic Michael Addition Reactions

The ability of hydrolytic enzymes for asymmetric catalysis of the Michael addition reaction has been explored by Kitazume et al. for the preparation of optically active α-trifluoromethyl-β-alanines **63** (Scheme 21.9).[32] Thus *Candida cylindracea* lipase (CCL) was shown to assist the asymmetric addition of diethylamine to 2-(trifluoromethyl)propenoic acid (**62**) (*R* = H) in aqueous buffer solution (pH 8) at 40–41°C. In 92 h corresponding N,N-diethyl-α-trifluoromethyl-β-alanine **63** (*R* = H) was isolated with 47% chemical yield and moderate (71% ee) enantiomeric purity (entry 1). Under the same reaction conditions pig liver esterase, as a catalyst, gave the desired amino acid **63** (*R* = H) in both lower chemical yield and enantiomeric excess (entry 2). A similar process, but conducted in an organic medium (benzene) with ethyl 2-(trifluoromethyl)propenate as a Michael acceptor, allowed increased chemical yields of corresponding adducts **63** (entries 3–6) without a sizable decrease in enantioselectivity of lipase-catalyzed addition reaction. The absolute configuration of products **63** was not determined.

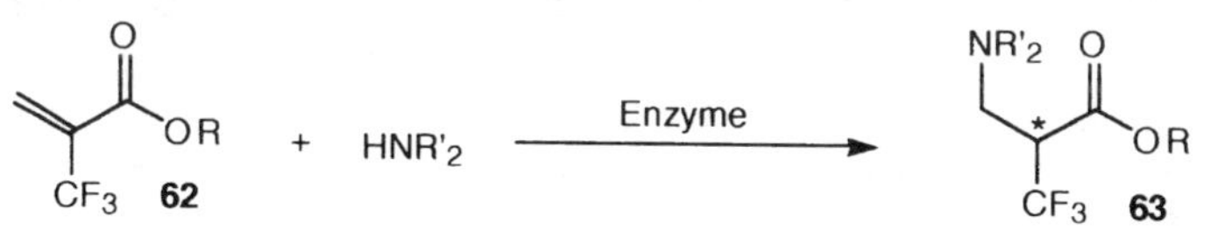

Entry	Enzyme	R	R'	**63**, Yield, %	**63**, % ee
1	*Candida cylindracea*	H	Et	47	71
2	Pig liver esterase	H	Et	39	67
3	*Candida cylindracea*-C(O)(CF$_3$)C=CH$_2$	Et	Et	82	35
4	Lipase PL 679 (*Alcaligenes sp* 679)	Et	Et	82	65
5	Lipase AL 865 (*Achromobacter* 865)	Et	Et	69	62
6	Lipase PL 679 (*Alcaligenes sp* 679)	Et	Bu	68	67

Scheme 21.9

21.4.2 Biocatalytic Resolution

Enzymatic resolution of α-substituted β-amino acids **3** is virtually unexplored. An attempt to prepare enantiomers of α-methyl-β-alanine (**65**) by benzylpenicillinacylase (PA) from *Escherichia coli* ATCC 9637[33] was described by Rossi et al.[34] Incubation of N-phenylacetyl derivative **64** in aqueous solution (pH 7.0–7.5) of PA at 37°C for 1.2 h (50% conversion of starting **64**) gave (*S*)-amino acid **65** with 37% chemical yield and only 7.2% ee (Scheme 21.10). Such a low stereodifferentiation of PA between enantiomers of **64** could be reasonably ascribed to the fact that the N-phenylacetylamino group in **64** is not directly bonded to the stereogenic center.

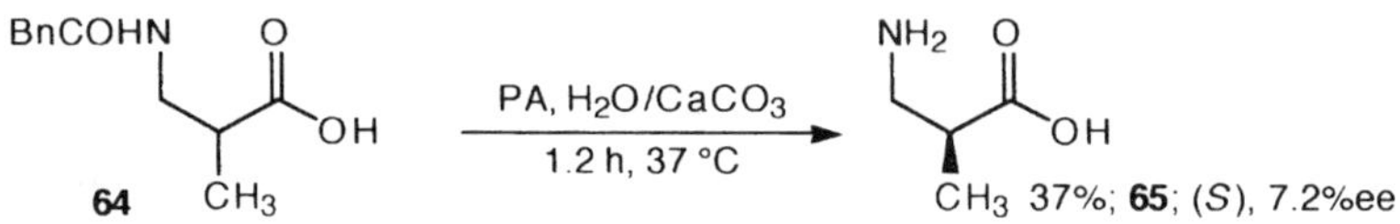

Scheme 21.10

21.5 β-Substituted β-Amino Acids

β-Substituted β-amino acids **4** are the most abundant and well-studied species of β-amino acids. The directions of their proved and potential biological activity and applications are amply discussed in other chapters of this book. Some of these amino acids are currently required in multigram quantities,[35] which greatly stimulated the development of potent and economical routes to these β-amino acids in an enantiomerically pure state. The possibilities offered by a biocatalytic approach for the preparation of enantiomerically pure β-amino acids **4**, particularly on a large scale, are discussed below.

21.5.1 Biocatalytic Michael Addition Reactions

Rao et al.[36] reported biocatalytic asymmetric synthesis of β-aryl-β-amino esters **69** of moderate enantiomeric purity by the Michael addition of various amines to the C-C double bond of β-aryl-α,β-unsaturated esters **66** in the presence of Baker's yeast and β-cyclodextrin (Scheme 21.11). The aqueous solution of the inclusion complex between an ester **66** and β-cyclodextrin (1:1) was added to the buffer solution (pH 7.2) of corresponding amine and Baker's yeast (*Saccharomyces cerevisiae*). The resulting mixture was incubated at 37°C for 48 h and then worked out. The summary of the biocatalytic process is shown in Scheme 21.11. Thus the structure of the starting ester was not influential for the yield of β-aryl-β-amino esters **66** but sizably affected the enantioselectivity of the reaction. The highest percent ee (72.5%) for **67** was obtained with ethyl cinnamate (entries 1,2), while its derivative-bearing electron donor methoxy group gave corresponding product **67** of poor enantiomeric purity (entry 4). It is interesting to note that the Baker's yeast–catalyzed addition reaction conducted without β-cyclodextrin gave products **67** in much lower enantiomeric purity, however with the same sign of optical rotation. The effect of β-cyclodextrin on the stereochemical outcome of this biocatalytic reaction was ascribed by the authors[36] to the fixed geometry of the ester **66** in its complex with β-cyclodextrin, which is incapable of catalyzing the addition reactions alone. The absolute configuration of the amino esters **67** was not determined.

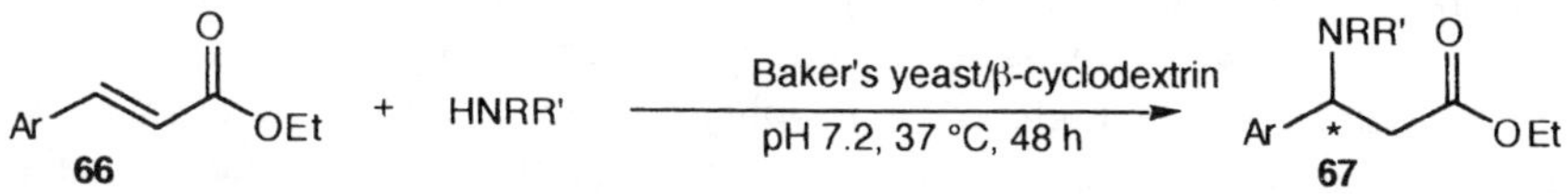

Entry	Ar	Amine	**67**, Yield, %	**67**, % ee
1	C_6H_5	$NHCH_2C_6H_5$	70	72.5
2	C_6H_5	$NH(CH_2)_2C_6H_5$	42	67.2
3	4-Me-C_6H_4	$NHCH_2C_6H_5$	72	47.2
4	4-MeO-C_6H_4	$NHCH_2C_6H_5$	71	23.1

Scheme 21.11

Application of purified enzymes to achieve asymmetric addition of amines to the ethyl *trans*-2-(trifluoromethyl)propenate **68** was described by Kitazume and Murata (Scheme 21.12). N,N-Dialkyl-substituted β-trifluoromethyl-β-alanines **69** were prepared with an ee up to 77% by lipase-catalyzed addition of dialkylamine to **68** in benzene at 40–41°C. The authors[32c] also reported that the use of (+) or (–)-menthyl *trans*-2-(trifluoromethyl)propenate as Michael acceptor instead of **68** can significantly increase diastereoselectivity of lipase-catalyzed addition reactions. The absolute configuration of the products **69** was not reported.

N,N-Dialkyl β-amino acids available by these biocatalytic Michael addition reactions are quite rare; however, they might be biologically interesting compounds, and some representatives of this type of β-amino acids are found in nature.[37]

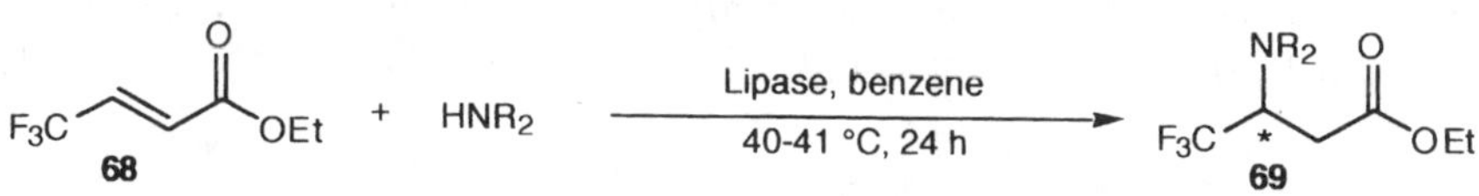

Entry	Enzyme	R	**69**, Yield, %	**69**, % ee
1	Lipase PL 679 (*Alcaligenes sp* 679)	Et	53	65
2	Lipase AL 865 (*Achromobacter* 865)	Et	62	77
3	Lipase PL 679 (*Alcaligenes sp* 679)	Bu	55	49

Scheme 21.12

21.5.2 Biocatalytic Resolution

The structure of β-substituted β-amino acids **4** offers two potential ways for their biocatalytic resolution involving amino and carboxy groups, respectively. Since the carboxy function is located one methylene away from the stereogenic center, one could expect low enantioselectivity of enzyme-catalyzed hydrolytic reactions on this group. Nevertheless, there are a few examples of successful resolutions of β-substituted β-amino acids **4** via biocatalytic hydrolysis of esters derived from **4**.

One of the earliest enzymatic resolutions of β-substituted β-amino acids **4** was reported by Cohen et al.[38a,b] on the preparation of (*S*)-β-phenyl-β-acetamidopropionic acid **71** by α-chymotrypsin (Scheme 21.13). The low solubility of ethyl β-phenyl-β-acetamidopropionate **70** in water (pH 7.8) and very slow reaction rate necessitated the use of a rather large amount of the enzyme to achieve 50% conversion of racemic **70** in 90 h at 27°C. However, the stereoselectivity of α-chymotrypsin–catalyzed hydrolysis was shown to be very high, and (*S*)-β-phenyl-β-acetamidopropionic acid **71** was isolated in enantiomerically pure form. It is worth noting that the stereochemical preference of α-chymotrypsin revealed in this reaction is opposite to that normally observed in the α-chymotrypsin–catalyzed hydrolyses of α-amino acid derivatives. The authors[38a,b] suggested that both the slow reaction rate and reversed stereochemical outcome could be caused by distortion in the enzyme-adopted abnormal substrate possessing an acetamido group one methylene away from the hydrolyzing ester function.

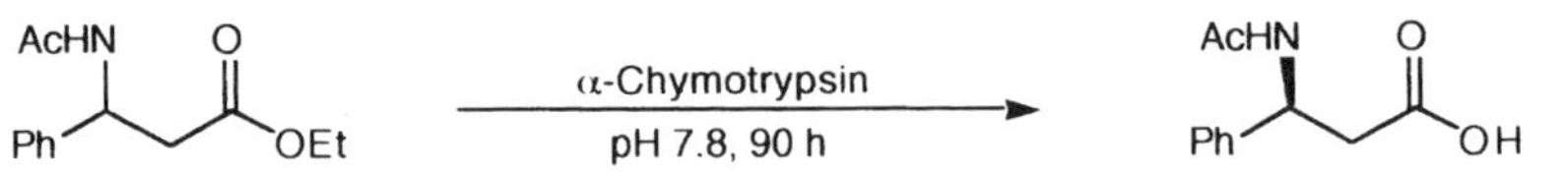

70 28%; **71**; (*S*), ≈ 100% ee

Scheme 21.13

An efficient access to the chiral nonracemic compounds is the enantioselective generation of a stereogenic center from a prochiral starting material. The use of a biocatalytic kinetic resolution procedure for this purpose is particularly beneficial, since in this case the theoretical yield of final product is not limited to 50%. In this line, the enzymatic transformation of *meso*-diethyl β-aminoglutarate derivatives **72** to highly synthetically useful chiral monoesters **73, 74** have received a great deal of attention (Scheme 21.14). Application of α-chymotrypsin for enantioselective hydrolysis of *meso*-diethyl β-acetamidoglutarate (**72**) (*R* = Et, *R′* = Ac, Scheme 21.13, entry 1) to

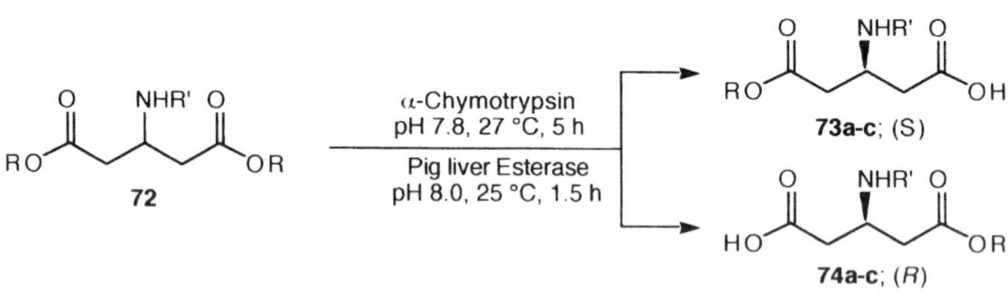

Entry	Enzyme	R	R'	Yield, %	Config.	% ee
1	α-Chymotrypsin	Et	Ac	57 (**74a**)	(*R*)	≈ 100
2	Pig liver Esterase	Me	H	94 (**74b**)	(*R*)	42
3	Pig liver Esterase	Me	Ac	81 (**74c**)	(*R*)	93
4	Pig liver Esterase	Me	*tert*-BuOCO	50 (**73a**)	(*S*)	93
5	Pig liver Esterase	Me	MeCH=CHCO	60 (**73b**)	(*S*)	≈ 100
6	Pig liver Esterase	Me	$PhCH_2OCO$	93 (**73c**)	(*S*)	>96

Scheme 21.14

half-ester (*R*)-**74a** was described Cohen et al.[38c] Diethyl ester **72** (entry 1), lacking the structural features of typical substrates for α-chymotrypsin, was hydrolyzed by this enzyme at 27°C, pH 7.8, with very slowly reaction rate. Upon reaching (5 h) 83% conversion of starting **72** (entry 1), the enzymatic reaction was quenched and the solution worked up to give enantiomerically pure half-ester (*R*)-**74a** in 57% yield. Similarly to the previously discussed ethyl β-phenyl-β-acetamidopropionate **70** resolution by α-chymotrypsin, a large amount of the enzyme (equal weights of α-chymotrypsin and substrate) was used to provide a convenient rate of diethyl ester **72** (entry 1) hydrolysis to (*R*)-**74a**. In view of the fact that α-chymotrypsin is very fastidious about the structure of potential substrates, this enzyme seems to be of limited use for the resolution of β-amino acids **4**. For example, β-methyl-β-alanine derivative, *rac*-ethyl β-acetamidobutyrate, was not hydrolyzed at all by α-chymotrypsin.[38a]

Application of pig liver esterase (PLE) for this biotransformation, disclosed by Ohno et al.[39] unambiguously is more synthetically useful. In contrast to α-chymotrypsin, PLE was found to hydrolyze the diethyl ester **72** (entry 2) under the similar reaction conditions (pH 8.0, 25°C) much more efficiently (1.5 h) to give the half-ester (*R*)-**74b** in 94% yield. However, the enantiomeric purity of isolated (*R*)-**74b** was shown to be poor, about 40% ee. The authors[39a] revealed that a low % ee of (*R*)-**74b** was caused by the nonenzymatic partial hydrolysis of **72** (entry 2) under the reaction conditions used. Protection of the amino function was thought to be essential to avoid undesirable chemical hydrolysis of substrate; therefore, PLE-catalyzed hydrolysis was studied using N-benzyloxycarbonyl diester **72** (entry 6), readily prepared from amino diester **72** (*R* = Me, *R′* = H) in high chemical yield. Incubation of N-benzyloxycarbonyl derivative **72** (entry 6) with PLE under the same reaction conditions furnished corresponding half-ester **73c** in 93% yield. Surprisingly, the product **73c** proved to be the (*S*) monoester and was formed in excellent enantiomeric purity (>96% ee). Further experiments have revealed that long-chain and bulky acyl substituents on the amino group of starting diester **72** (entries 4–6) provided preferential PLE-catalyzed cleavage of pro-(*R*) methyl ester of **72**, giving the corresponding half-ester **73a–c** of (*S*) absolute configuration, while the N-acetyl derivative **72** (entry 3) was hydrolyzed by PLE to give the (*R*) monoester **74c**. Thus the pattern of enantioselectivity of PLE-catalyzed hydrolysis of diesters **72** can be effectively controlled by careful selection of the N-protecting group, providing ready excess to both enantiomers (*S*)-**73** and (*R*)-**74** of high optical purity. The synthetic merit of this method was further enhanced by selection of certain microorganisms, capable of catalyzing enantioselective hydrolysis of **72**, in order to substitute PLE, which is quite expansive and impractical on a very large scale. Some strains, from amongst 500 species of microorganisms tested by Ohno et al.,[40] have shown a synthetically useful rate and enantioselectivity of N-benzyloxycarbonyl diester **72** hydrolysis to (*S*)-**73c**. For instance, *Achromobacter Lyticus* IFO 12725 and IFO 12726, *Gluconobacter dioxyacetonicus* IFO 3271, and *Flavobacterium lutescens* IFO 3084 and IFO 3085 produced monoester (*S*)-**73c** in up to 98% ee. The practical importance of this biocatalytic approach was demonstrated by the synthetic versatility of half-esters (*S*)-**73** and (*R*)-**74** in the total synthesis of various carbapenem antibiotics and other chiral organic compounds of biological interest.[41]

One more example of the PLE-catalyzed kinetic resolution of β-amino acids has been described by Estermann and Seebach (Scheme 21.15).[14] Incubation of *rac*-methyl β-benzamidobutyrate **75** with a large amount of PLE or PLAMP at pH 7 for 41 h (61% conversion of **75**) gave resolved species **76** and **77** in both moderate chemical yield and enantiomeric purity. The authors[14] also reported unsuccessful attempts to resolve **75** by porcine pancreatic lipase, *Candida cylindracea* lipase, α-chymotrypsin, and acylase **I**.

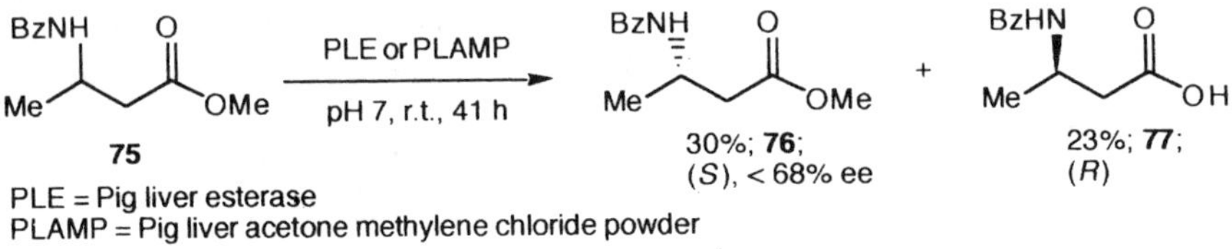

Scheme 21.15

Obviously, a more efficient biocatalytic resolution approach to enantiomerically pure β-amino acids in general, and **4** in particular, would involve an amino functionality, directly bonded to the stereogenic center, in an enzyme-catalyzed reaction. However, most amidases,[42] commonly used for the resolution of α-amino acids,[16] are highly directed towards the structure of amino acid being resolved, and thus β-amino acids are not the substrates for these enzymes.[12–16] A fortunate exception to the trend is penicillin acylase (PA) (EC 3.5.1.11) from *Escherichia coli*, the successful application of which for the preparation of optically pure 3-phenylisoserines **50**, **52**, and 2-methyl-3-amino-4,4,4-trifluorobutanoic acids **57**, **58**, **60**, **61** was already discussed above (Schemes 21.7 and 21.8, respectively). In view of generalized feasibility of PA for the kinetic resolution of various β-amino acids, this enzyme deserves some special comments. PA has found industrial application in the production of 6-aminopenicillanic acid, 7-aminodeacetoxycephalosporanic acid, a β-lactam intermediate in the synthesis of loracarbef, and also for the preparation of a variety of semisynthetic β-lactam antibiotics.[43] Previous investigations of PA substrate specificity have disclosed an exceptionally high affinity of this enzyme to the derivatives of phenylacetic acid.[44] PA is capable of catalyzing hydrolytic cleavage of the phenylacetyl group from amines,[45a,b] α-,[44,45a,d–f] and γ-amino carboxylic acids,[45g] α-amino alkylphosphonic acids,[45h] with moderate-to-excellent stereochemical discrimination between corresponding enantiomers. Consequently, in sharp contrast to other amidases, the α-disposition of the hydrolyzing acylamido group to the carboxy function is not responsible for the high hydrolytic activity of PA.

Thus a series of β-alkyl and β-fluoroalkyl-β-alanines **80** and **82** were prepared in enantiomerically pure form via PA-catalyzed hydrolysis of corresponding N-phenylacetyl derivatives **79** (Scheme 21.16).[34,46] Compounds **79**, synthesized in high chemical yield (83–99%) from readily available free amino acids **78**, were incubated with a truly catalytic amount of PA (concentration 5×10^{-7}) in aqueous medium at pH 7.5, room temperature. The rate of the enzymatic hydrolysis of long-chain derivatives **79d,i,j** was slightly lower than that of short-chain compounds; however, the nature of β-substituent *R* has no effect on the stereochemical preference of PA-catalyzed hydrolysis. In all cases (*S*)-β-alkyl and (*R*)-β-fluoroalkyl-β-alanines **80**,[47] released by the enzyme, and their antipodes **82**, prepared by acidic hydrolysis of enzymatically unreactive N-phenylacetyl derivatives **81**, were obtained in high enantiomeric purity (>95% ee, by chiral HPLC). Kinetic studies into the reaction have shown high enantioselectivity of the PA-alanine was hydrolyzed by PA in 500 times faster than corresponding (*S*)-enantiomer, that is, $(k_{cat}/K_M)_R/(k_{cat}/K_M)_S = 500$.

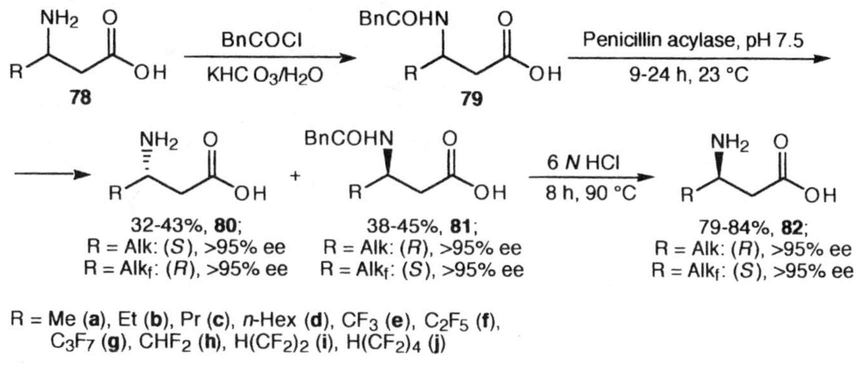

Scheme 21.16

An analogous biocatalytic procedure with application of PA was recently developed also for the resolution of an aromatic series of β-amino acids (Scheme 21.17).[34,46a,48] Starting racemic β-aryl-β-alanines **83** were readily available by Rodionow reaction, which consists of an interaction between an aromatic aldehyde with malonic acid in the presence of ammonium acetate, which serves at the same time as a base and as a source of amino group. Upon formation, the desired β-amino acid precipitates from solution, which allows the final product to be isolated in 51–74% chemical yields simply by filtration of the reaction mixture. Phenylacetylation of amino acids **83** required modification of the standard Schotten–Baumann conditions with application of triethylamine in place of $KHCO_3$. Enzymatic resolution of N-phenylacetyl derivatives **84** were carried out under the usual reaction conditions at pH 7.5, room temperature, using 10^{-7} *M* PA. Separation of the resultant amino acids **85** and unconverted N-phenylacetyl derivatives **86** was easily achieved with cation-exchange resin using distilled water to isolate **86** and 0.1 *N* NH_4OH to isolate free amino acids **85**. The enzymatically unconverted N-phenylacetyl derivatives **86** were hydrolyzed by 6 *N* HCl to give amino acids **87**. According to chiral HPLC analysis (see below), both the biocatalytically released, and those obtained by acidic hydrolysis, amino acids **85** and **87**, respectively, after a single recrystallization from water–ethanol were enantiomerically pure (>95% ee). Regardless of the nature of substituent on the phenyl ring, corresponding (*R*)-enantiomers were the fast-reacting isomers in all cases studied. Experiments on the catalytic constants of the PA-catalyzed hydrolyses of (*R*)- and (*S*)-N-phenylacetyl derivatives of β-aryl-β-amino acids have revealed a great difference in the rates of

(*R*)- and (*S*)-enantiomer hydrolysis. Thus enantioselectivity of the enzymatic process, expressed as the ratio of the bimolecular rate constants for the hydrolysis of (*R*)- and (*S*)-enantiomers, exceeds 10,000.

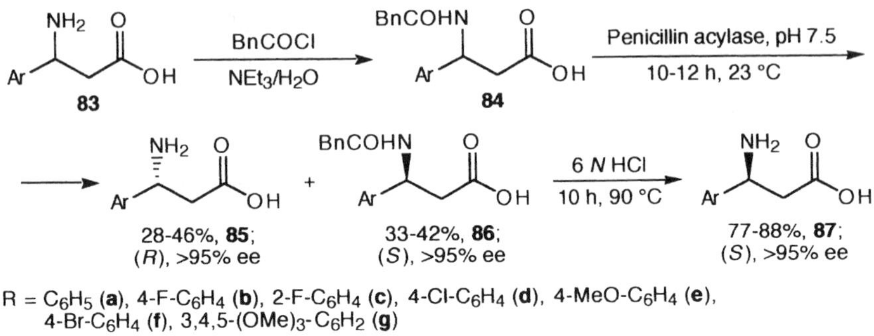

Scheme 21.17

21.6 Conclusions

There is no single "best" approach towards enantiomerically pure β-amino acids. Each method itself possesses its own advantages and limitations. Despite the power and flexibility of synthetic asymmetric approaches, in some cases, a biocatalytic approach would be the method of choice to prepare enantiomers of the target β-amino acid. The use of microorganisms and/or isolated, partially purified enzymes for the preparation of a desired product, holds several important advantages over chemical transformations. In contrast to chemical reactions, enzymatic reactions are, with some exceptions, carried out in aqueous solutions under very mild conditions with minimum protection of functional groups and easily reproducible within a simple chemistry laboratory without the requirement of special equipment. Many of the biotransformations can be easily generalized and conducted on a multigram or even kilogram scale. Finally, the cost of commercially available microorganisms and enzymes is normally quite low, and the whole biocatalytic procedure would be economically appealing. Among the disadvantages of the biocatalytic approach, the limit of 50% theoretical yield for the resolution process (if recycling of undesired enantiomer is impossible) and the usual stereochemical inflexibility of biocatalytic asymmetric transformations should be pointed out. Thus the biocatalytic method could be preparatively useful only provided three criteria are met. First, the starting educt must be cheap and easily prepared. Second, the biocatalytic step should be highly enantioselective and/or the enantiomeric purity enhancement of the products should be simple. And finally, there should be a convenient method for the complete separation of the resolved/synthesized species. Some of the methods discussed in this chapter meet these criteria and would be attractive for large-scale preparation of enantiomerically pure β-amino acids. The growing importance of enantiopure β-amino acids in general, and the practical needs in large quantity of some of them, should stimulate further achievements in this area.

21.7 Acknowledgments

We would like to thank Professors Margaret M. Kayser, Ramesh N. Patel, Nicole Kunesch, Takashi, Yamazaki, and Tomoya Kitazume for providing reprints of their papers.

21.8 References

1. (a) *Chirality in Drug Design and Synthesis*, C. Brown, Ed. (Academic Press, New York, 1990). (b) *Chirality and Biological Activity*, H. Frank, B. Holmsted and B. Testa, Eds. (Liss, New York, 1991).

2. (a) W. H. Kemp, *Chirality* **1**, 2 (1989). (b) S. Borman, *Chem. Eng. News*, July 9, 9 (1990). (c) Stinson, S. C. *Chem. Eng. News*, September 28, 46 (1992).

3. For instance, (*S*)-limonene has a lemon odor while the (*R*)-antipode has an orange one; the less potent (*R*)-enantiomer of commercial anesthetic isoflurane has a typical ether odor, while the more potent (*S*)-enantiomer has no odor at all; (*R*)-asparagine tastes sweet, whereas the (*S*)-enantiomer has a bitter taste; (*S,S*)-ethambutol is a tuberculostatic agent and the (*R,R*)-antipode causes blindness; (*S*)-propranolol is a β-blocker while the (*R*)-isomer is a contraceptive; and the most sad example in this line is thalidomide, the (*R*)-isomer of which serves as a powerful tranquilizer while its (*S*)-antipode has been shown to cause ill effects in fetal development (teratogenic activity).

4. In FDA guidelines stereoisomer(s) of biologically active compound may be considered as impurities and, consequently, data on safety and efficacy have to be produced for each diastereo- and/or enantiomer. *Guideline for Submitting Supporting Documentation in Drug Applications for the Manufacture of Drug Substances*, Office of Drug Evaluation and Research, Food and Drug Administration, Washington, DC, 3 (1987).

5. The terms *eutomer* and *distomer* (the good and the bad isomers) were suggested to classify active (desired activity) and inactive stereoisomers; E. J. Ariens, *Eur. J. Clin. Pharmacol.* **26**, 663 (1984).

6. D. Enders, *Chemica Scripta* **25**, 139 (1985).

7. The abbreviation "EPC" was suggested by Seebach as a generic term for the preparation of enantiomerically pure compounds; D. Seebach, E. Hungerbuhler, *Mod. Synth. Methods 2*, 91 (1980).

8. For a general review see: D. Seebach, *Angew. Chem. Int. Ed. Engl.* **29**, 1320 (1990).

9. (a) C. N. Drey, in *Chemistry and Biochemistry of the Amino Acids*; B. Weinstein, Ed. (Dekker, New York, 1976), Vol. 4, p. 241. (b) C. N. Drey, in *Chemistry and Biochemistry of the Amino Acids*; G. C. Barett, Ed. (Chapman and Hall, New York, 1985). (c) A. F. Spatola, in *Chemistry and Biochemistry of the Acids, Peptides and Proteins*, B. Weinstein, Ed. (Marcel Dekker, New York, 1983), Vol. 7, p. 331. (d) O. W. Griffith, *Ann. Rev. Biochem.* **55**, 855 (1986).

10. (a) *Peptide Chemistry: Design and Synthesis of Peptides, Conformational Analysis and Biological Functions*. Tetrahedron-Symposia-in-Print **31**, V. J. Hruby and R. Schwyzer, Eds., *Tetrahedron* **44**, 661 (1988). For recent publications see: (b) G. Li, M. A. Jarosinski, V. Hruby, *Tetrahedron Lett.* **34**, 2561 (1993). (c) A. Giannis, T. Kolter, *Angew. Chem. Int. Ed. Engl.* **32**, 1244 (1993). (d) K. Paulvannan, J. R. Stille, *Tetrahedron Lett.* **34**, 8197 (1993). (e) J. A. Robl, M. P. Cimarusti, L. M. Simpkins, H. N. Weller, Y. Y. Pan, M. Malley, J. D. DiMarco, *J. Am. Chem. Soc.* **116**, 2348 (1994).

11. (a) T. N. Salzmann, R. W. Ratcliffe, B. G. Christensen, F. A. Bouffard, *J. Amer. Chem. Soc.* **102**, 6161 (1980). (b) K. Okano, T. Izawa, M. Ohno, *Tetrahedron Lett.* **24**, 217 (1983). (c) H. Huang, N. Iwasawa, T. Mukaiyama, *Chem. Lett.* 1465 (1984). (d) S. Kim, S. B. Chang, P. H. Lee, *Tetrahedron Lett.* **28**, 2735 (1987). (e) S. Kim, P. H. Lee, T. A. Lee, *J. Chem. Soc. Chem. Commun.* 1242 (1988). (f) T. Kunieda, T. Nagamatsu, T. Higuchi, M. Hirobe, *Tetrahedron Lett.* **29**, 2203 (1988). (g) D. Tanner,

P. Somfai, *Tetrahedron* **44**, 613 (1988). (h) J. M. Waisvisz, M. G. van der Hoeven, B. te Nijenhuis, *J. Am. Chem. Soc.* **79**, 4524 (1957), (i) G. L. Helms, R. E. Moore, W. P. Niemczura, G. M. L. Patterson, K. B. Tomer, M. L. Gross, *J. Org. Chem.* **53**, 1298 (1988). (j) S. M. Hecht, *Acc. Chem. Res.* **19**, 383 (1986).

12. H. K. Chenault, Jeffrey Dahmer, G. Whitesides, *J. Am. Chem. Soc.* **111**, 6354 (1989).

13. E. M. Meijer, W. H. Y. Boesten, H. E. Schoemaker, J. A. M. van Balken, in *Biocatalysis in Organic Synthesis*; J. Tramper, H. C. van der Plas, P. Links, Eds. (Amsterdam, Elsevier, 1985), p. 135.

14. H. Estermann, D. Seebach, *Helv. Chim. Acta* **71**, 1824 (1988).

15. (a) *Kirk-Othmer Encyclopedia of Chemical Technology*, 3rd ed. (Wiley , New York, 1978), Vol. 2, pp. 376–410. (b) *Chemical and Process Technology Encyclopedia*, D. M. Considine, Ed. (McGraw-Hill, New York, 1974), pp. 93–106. (c) *Biotechnology of Amino Acids Production*, K. Aida, I. Chibata, K. Nakayama, K. Takinami, H. Yamada, Eds. (Kodansha, Tokyo, 1986). (d) K. Yonaha, K. Soda, in *Advances in Biochemical Engineering Biotechnology*, A. Fiechter, Ed. (Springer-Verlag, Berlin, 1986), Vol. 33, pp. 95–130.

16. H. G. Davies, R. H . Green, D. R. Kelly, S. M. Roberts, *Biotransformations in Preparative Organic Chemistry* (Academic Press, London, 1989).

17. (2*S*,3*R*)-3,7-Diamino-2-hydrohyheptanoic acid (a), (2*S*,3*S*) (b) and (2*S*,3*R*)-3-amino-2-hydroxy-4-phenylbutyric acid (c), (2*S*,3*R*)-3-amino-2-hydroxydecanoic acid (d), and (*S*)-isoserine (e) are particularly noteworthy specimens of the class of β-amino-α-hydroxy acids. For their synthesis and bioactivity see: (a) R. Herranz, S. Vinuesa, C. Perez, M. T. Garcia-Lopez, E. Lopez, M. L. de Ceballos, J. Del Rio, *J. Med. Chem.* **35**, 889 (1992). (b) M. E. Bunnage, S. G. Davies, C. J. Goodwin, *Synlett*, 731 (1993). (c) B. H. Norman, M. L. Morris, *Tetrahedron Lett.* **33**, 6803 (1992). (d) M. E. Bunnage, A. J. Burke, S. G. Davies, C. J. Goodwin, *Tetrahedron: Asymmetry* **5**, 203 (1994). (e) A. Solladie-Cavallo, N. Khiar, *Tetrahedron Lett.* **29**, 2189 (1988).

18. (a) *Total and Semi-Synthetic Approaches to Taxol*. Tetrahedron-Symposia-in-Print **48**, J. D. Winkler, Ed., *Tetrahedron* **48**, 6953 (1992). (b) K. C. Nicolaou, W.-M. Dai, R. K. Guy, *Angew. Chem. Int. Ed. Engl.* **33**, 15 (1994).

19. A. Commercon, D. Bezard, F. Bernard, J. D. Bourzat, *Tetrahedron Lett.* **33**, 5185 (1992).

20. (a) J. Kearns, M. M. Kayzer, *Tetrahedron Lett.* **35**, 2845 (1994). (b) M. M. Kayser, personal communication.

21. R. N. Patel, A. Banerjee, J. M. Howell, C. G. McNamee, D. Brozozowski, D. Mirfakhrae, V. Nanduri, J. K. Thottathil, L. J. Szarka, *Tetrahedron: Asymmetry* **4**, 2069 (1993).

22. (a) W. Boland, C. Frobl, M. Lorenz, *Synthesis* 1049 (1991). (b) C. J. Sih, S. H. Wu, *Top. Stereochem.* **19**, 63 (1989).

23. (a) Y. Lu, C. Miet, N. Kunesch, J. Poisson, *Tetrahedron: Asymmetry* **2**, 871 (1991). (b) Y. Lu, C. Miet, N. Kunesch, J. Poisson, *Tetrahedron: Asymmetry* **4**, 893 (1993).

24. Y. Lu, C. Miet, N. Kunesch, J. Poisson, *Tetrahedron: Asymmetry* **1**, 707 (1990).

25. R. Brieva, J. Z. Crich, C. J. Sih, *J. Org. Chem.* **58**, 1068 (1993).

26. C. Palomo, A. Arrieta, F. P. Cossio, J. M. Aizparua, A. Meilgo, N. Aurrekoetxea, *Tetrahedron Lett.* **31**, 6429 (1990).

27. (a) H. Honig, P. Seufer-Wasserthal, H. Weber, *Tetrahedron* **46**, 3841 (1990). (b) D.-M. Gou, Y.-C. Liu, C.-S. Chen, *J. Org. Chem.* **58**, 1287 (1993).

28. V. A. Soloshonok, unpublished data.

29. For instance, (1*R*,2*S*)-2-aminocyclopentane-1-carboxylic acid (cispentacin) was shown to possess potent in vivo antifungal activity (a); (2*R*,3*R*)-2-methyl-3-aminobutanoic acid is a constituent of cyto-

toxic cyclic depsipeptide, Dolastatin D (b); (2*S*,3*R*)-2-methyl-3-amino-penatnoic acid was found in Majusculamide C (c); 57-Norajusculamide C (d); Dolastatins 11 and 12 (e). (a) K. Kawabata, Y. Inamoto, K. Sakane, T. Iwamoto, S. Hashimoto, *J. Antibiotics* **43**, 513 (1990). (b) H. Sone, T. Nemoto, H. Ishiwata, M. Ojika, K. Yamada, *Tetrahedron Lett.* **34**, 8449 (1993). (c) D. C. Carter, R. E. Moore, J. S. Mynderse, W. P. Niemczura, J. S. Todd, *J. Org. Chem.* **49**, 236 (1984). (d) J. S. Mynderse, A. H. Hunt, R. E. Moore, *Journal of Natural Products* **51**, 1299 (1988). (e) G. R. Pettit, Y. Kamano, H. Kizu, C. Dufresne, C. L. Herald, R. J. Bontems, J. M. Schmidt, F. E. Boettner, R. A. Nieman, *Heterocycles* **28**, 553 (1989).

30. V. A. Soloshonok, A. G. Kirilenko, N. A. Fokina, V. P. Kukhar, S. V. Galushko, V. K. Svedas, G. Resnati, *Tetrahedron: Asymmetry* **5**, 1225 (1994).

31. See, for instance, B. J. Mavunkel, Z. Lu, D. J. Kyle, *Tetrahedron Lett.* **34**, 2255 (1993).

32. (a) T. Kitazume, T. Ikeye, K. Murata, *J. Chem. Soc. Chem. Com.* 1331 (1986). (b) T. Kitazume, K. Murata, *J. Fluor. Chem.* **36**, 339 (1987). (c) T. Kitazume, K. Murata, *J. Fluor. Chem.* **39**, 75 (1988).

33. In Ref. 34 the authors used the term *benzylpenicillinacylase*. Actually, it is the same enzyme that is referred to as penicillinacylase (a) or penicillin acylase (b) in more recent literature. (a) C. Fuganti, P. Grasselli, S. Servi, A. Lazzarini, P. Casati, *J. Chem. Soc. Chem. Commun.* 538 (1987). (b) A. L. Margolin, *Tetrahedron Lett.* **34**, 1239 (1993).

34. D. Rossi, G. Lucente, A. Romeo, *Experientia* **33**, 1557 (1977).

35. For instance, (*S*)-3-amino-3-(3-pyridine)propanoic acid is a constituent part of (*S*)-ethyl β-[[4-[[4-(aminoiminomethyl)phenyl]amino]-1,4-dioxobutyl]amino]-3-pyridinepropanoate, a peptidomimetic for the Arg-Gly-Asp-Phe sequence of fibrinogen, is an orally active antiplatelet agent that may be useful in preventive therapy for myocardial infarction and unstable angina. J. G. Rico, R. J. Lindmark, T. E. Rogers, P. R. Bovy, *J. Org. Chem.* **58**, 7948 (1993).

36. K. R. Rao, Y. V. D. Nageswar, H. M. S. Kumar, *Tetrahedron Lett.* **32**, 6611 (1991).

37. For instance, Winterstein's acid, (*S*)-3-dimethylamino-3-phenyl propionic acid. E. Graf, H. Boeddeker, *Ann. Chem.* **613**, 111 (1958).

38. (a) S. G. Cohen, Y. Sprinzak, E. Khedouri, *J. Am. Chem. Soc.* **83**, 4225 (1961). (b) S. G. Cohen, S. Y. Weinstein, *J. Am. Chem. Soc.* **86**, 725 (1964). (c) S. G. Cohen, E. Khedouri, *J. Am. Chem. Soc.* **83**, 1093 (1961).

39. (a) M. Ohno, S. Kobayashi, T. Limori, Yi-Fong Wang, T. Izawa, *J. Am. Chem. Soc.* **103**, 2405 (1981). (b) M. Ohno, S. Kobayashi, K. Adachi, in *Enzymes as Catalysts in Organic Synthesis,* M. P. Schneider Ed., p. 123, and references therein. D. Reidel, Dordrecht, Holland, 1986. (c) K. Adachi, S. Kobayashi, M. Ohno, *Chimia* **40**, 311 (1986).

40. H. Kotani, Y. Kuze, S. Uchida, T. Miyabe, T. Limori, K. Okano, S. Kobayshi, M. Ohno, *Agric. Biol. Chem.* **47**, 1363 (1983).

41. (a) K. Okano, T. Izawa, M. Ohno, *Tetrahedron Lett.* **24**, 217 (1983). (b) K. Okano, Y. Kyotani, H. Ishihama, S. Kobayashi, M. Ohno, *J. Am. Chem. Soc.* **105**, 7186 (1983). (c) T. Iimori, Y. Takahashi, T. Izawa, S. Kobayashi, M. Ohno, *J. Am. Chem. Soc.* **105**, 1659 (1983). (d) Yi-Fong Wang, T. Izawa, S. Kobayashi, M. Ohno, *J. Am. Chem. Soc.* **104**, 6465 (1982).

42. *Amidases* is the term that encompasses all enzymes capable of catalyzing either cleavage or formation of amides, such as acylases, acyl transferases, carboxypeptidases, and proteases.

43. (a) D. L. Regan, P. Dunnil, M. D. Lilly, *Biotechnol. Bioeng.* **16**, 333 (1974). (b) B. J. Abbot, *Adv. Appl. Microbiol.* **20**, 203 (1976). (c) V. K. Svedas, A. L. Margolin, I. V. Berezin, *Enzyme Microb. Technol.* **2**, 138 (1980).

44. (a) M. Cole, *Nature* **203**, 519 (1964). (b) A. L. Margolin, V. K. Svedas, I. V. Berezin, *Biochim. Biophys. Acta* **616**, 283 (1980).

45. (a) M. Cole, *Biochem. J.* **115**, 733 (1969). (b) M. Cole, *Biochem. J.* **115**, 741 (1969). (c) A. Czentirmai, *Acta Microbiol. Acad. Sci. Hung.* **12**, 395 (1965/1966). (d) D. Rossi, A. Calcagni, *Experientia* **41**, 35 (1985). (e) W. Kaufmann, K. Bauer, *Nature*, **203**, 520 (1964). (f) G. Lucente, A. Romeo, D. Rossi, *Experientia* **21**, 317 (1965). (g) A. L. Margolin, *Tetrahedron Lett.* **34**, 1239 (1993). (h) V. A. Solodenko, T. N. Kasheva, V. P. Kukhar, E. V. Kozlova, D. A. Mironenko, V. K. Svedas, *Tetrahedron* **47**, 3989 (1991).

46. (a) V. A. Soloshonok, V. K. Svedas, V. P. Kukhar, A. G. Kirilenko, A. V. Rybakova, V. A. Solodenko, N. A. Fokina, O. V. Kogut, I. Yu. Galaev, E. V. Kozlova, I. P. Shishkina, S. V. Galushko, *Synlett*, 339 (1993). (b) V. P. Kukhar, V. A. Soloshonok, V. K. Svedas, N. V. Kotik, I. Y. Galaev, A. G. Kirilenko, E. V. Kozlova, *Bioorgan. Khim.* **19**, 474 (1993). (c) V. A. Soloshonok, A. G. Kirilenko, N. A. Fokina, I. P. Shishkina, S. V. Galushko, V. P. Kukhar, V. K. Svedas, E. V. Kozlova, *Tetrahedron: Asymmetry* **5**, 1119 (1994).

47. The (*R*)-configuration of β-fluoroalkyl-β-alanines, a consequence of the Cahn–Ingold–Prelog priority, is stereochemically equivalent to the (*S*)-configuration in the hydrocarbon analogs; R. S. Cahn, C. Ingold, V. Prelog, *Angew. Chem. Int. Ed. Engl.* **5**, 385 (1966).

48. V. A. Soloshonok, N. A. Fokina, A. V. Rybakova, I. P. Shishkina, S. V. Galushko, A. E. Sorochinsky, V. P. Kukhar, M. V. Savchenko, V. K. Svedas, *Tetrahedron: Asymmetry* **6**, 1601 (1995).

CHAPTER

22

Enantioselective Preparation of Deuterated β-Alanines

Amit Basak

22.1 Introduction

Isotopically labeled molecules are important synthetic targets, as they reveal valuable information regarding the mechanism of enzymatic processes.[1] Especially attractive are the chiral isotopically labeled amino acids, which have been used to obtain first-hand knowledge about the stereochemistry of such processes.[2] α-Amino acids are biosynthetic precursors to many natural products, including various antibiotics, and have generated considerable interest in bioorganic chemists.[3] β-Amino acids, although comparatively less abundant, are not lagging behind. Studies have revealed their interesting biological properties[4] and their role in natural product biosynthesis.[5] The simplest member in the β-series is 3-amino propanoic acid or β-alanine, which is a constituent of various natural products. It is the product of catabolism of the RNA base uracil[6] and is also formed via a lyase-mediated addition of ammonia to acrylyl CoA.[7] Thus isotopically labeled β-alanines are logical synthetic targets. They have been used to determine the stereochemical outcome and mechanism of various enzymatic processes as well as for the synthesis of important biosynthetic intermediates.[8]

22.2 Biosynthetic Probes

As stated earlier, in order to determine the stereochemistry of a reaction at a prostereogenic center, the substrate needs to be chirally isotopically labeled at the reaction center. The usual practice is to introduce a deuterium in a chiral fashion at the prostereogenic center and study its fate during enzymatic transformation by high-field NMR and mass spectrometry. Quite a large number of enzyme-catalyzed reactions have been studied this way. The conversion of proclavaminic acid to clavaminic acid catalyzed by clavaminate synthase (CS) is shown as an example (Scheme 22.1). The loss of 4′-(*S*) hydrogen during the reaction is demonstrated by the use of 4′-(*R*) and 4′-(*S*)-deuterated proclavaminic acids.[9]

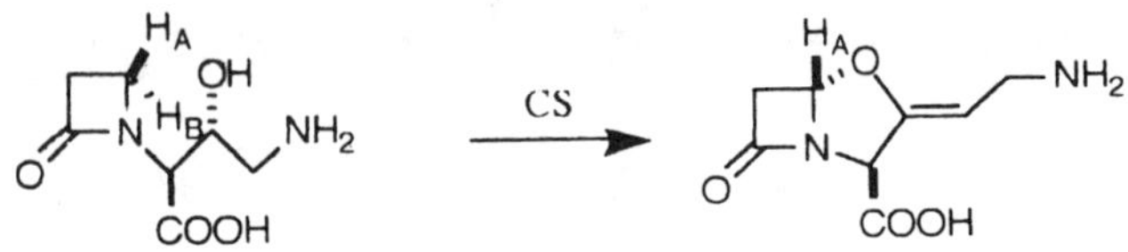

4′-(*R*) Proclavaminic acid, $H_A={}^2H$, $H_B={}^1H$ → [5-^{2}H] clavaminic acid $H_A={}^2H$

4′-(*S*) Proclavaminic acid, $H_A={}^1H$, $H_B={}^2H$ → clavaminic acid $H_A={}^1H$

Scheme 22.1

22.3 Problems of Synthesis

The major challenge associated with the preparation of enantiomerically pure amino acids is the introduction of the isotope, a small deuterium or hydrogen in a highly stereoselective manner. The possibility of scrambling at different stages of the synthesis needs to be taken care of. Evaluation of the degree of enantiomeric excess, the extent of deuteration, and the determination of configuration are the other problems that need to be addressed. The synthesis of chiral isotopomers of β-alanines, which include C2 or C3 monodeuterated and C2, C3 dideuterated isomers (Structures **1a**–**4b**), is no exception in that respect.

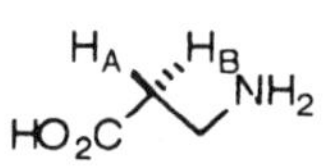

1a, $H_A={}^1H$, $H_B={}^2H$
1b, $H_A={}^2H$, $H_B={}^1H$

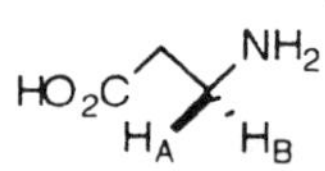

2a, $H_A={}^1H$, $H_B={}^2H$
2b, $H_A={}^2H$, $H_B={}^1H$

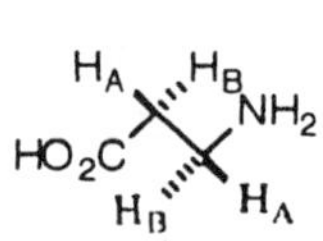

3a, $H_A={}^1H$, $H_B={}^2H$
3b, $H_A={}^2H$, $H_B={}^1H$

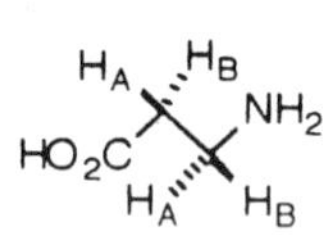

4a, $H_A={}^1H$, $H_B={}^2H$
4b, $H_A={}^2H$, $H_B={}^1H$

22.4 Types of Synthesis

Quite a few enantioselective syntheses of deuterated β-alanines have been reported. The methods developed so far can be classified into the following three categories.

22.4.1 Biocatalytic Approach

The biocatalysts are mainly enzymes whose potentials as asymmetric catalysts for the synthesis of chiral molecules have long been recognized and exploited in many cases.[10] This extraordinary power of enzymes has been applied in the synthesis of chiral deuterated β-alanines.

22.4.2 Chiral Pool as a Building Block

The chiral pool[11] consists of optically active natural products that are relatively inexpensive, readily available, and hence attractive starting materials for the enantiospecific synthesis of deuterated β-alanines and related derivatives.

22.4.3 Asymmetric Synthesis Involving Self-Regeneration of Stereogenic Centers

This method, first developed by Seebach and co-workers,[12] starts with an enantiomerically pure material with a particular stereogenic center. A temporary auxiliary stereogenic center is then created in one definite configuration at an adjacent center. The original stereogenic center is then destroyed (converted to a trigonal center or a readily invertible radical) and then regenerated. The temporary auxiliary induces stereoselectivity during regeneration. The process is outlined in Structures **5–10**.

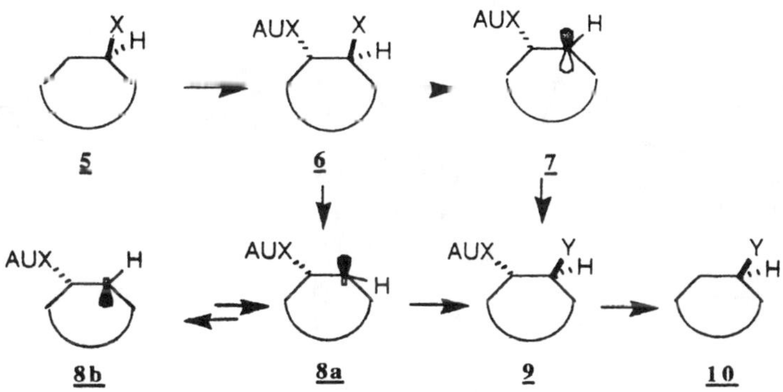

This type of methodology has been used in the synthesis of the chiral deuterated β-alanines by Basak.[13]

22.5 Actual Synthesis of Chirally Deuterated β-Alanines

Young and co-workers[14] did pioneering work in the area of chirally deuterated amino acids including β-alanine. They have synthesized all four stereoisomers of C-2 and C-3 substituted monodeuterated β-alanines and one isomer of chiral dideuterated β-alanine. In all their methods enzymes played a crucial role. These are described in this section.

22.5.1 Synthesis Involving L-Aspartase

The key step in the synthesis[15] as shown in Scheme 22.2 is the stereospecific addition of ammonia across the double bond in fumaric acid by the commercially available enzyme L-aspartase.[16] The assignment of absolute stereochemistry of the process indicated that NH_2 and H, respectively, add in a *Si-Re* fashion to the double bond.[17] Thus the product of the reaction was (2*S*,3*R*)-[3-$^2H_1$] aspartic acid (**13**) when the addition was carried out in ^2H_2O. On the other hand, incubation of [2,3-$^2H_2$]-fumaric acid (**12**) in ^1H_2O with L-aspartase afforded the (2*S*,3*S*)-[2,3-$^2H_2$] aspartic acid (**14**). These labeled aspartic acids were then converted [18] to the corresponding β-methyl esters (**15**, H_B=^2H) and (**16**, H_A=^2H), which upon decarboxylation and subsequent hydrolysis produced samples of β-alanine, stereospecifically labeled at C-2. Incidentally, [2,3-$^2H_2$] fumaric acid (**12**) was synthesized[19] from acetylene dicarboxylate using triphenyl phosphine and ^2H_2O (Scheme 22.2).

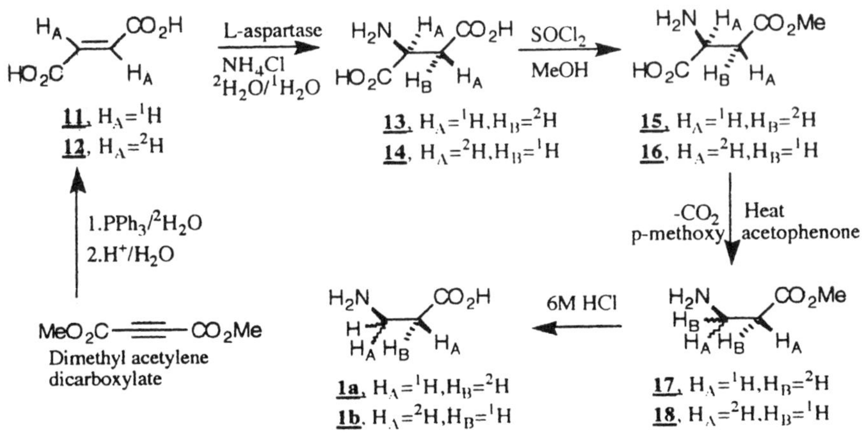

Scheme 22.2

22.5.2 Synthesis Involving Aminotransferase

For the enantioselective synthesis of β-alanines stereospecifically labeled at C-3 (Scheme 22.3), Gani and Young[20,21] used another commercially available enzyme L-alanine aminotransferase (E.C.2.6.1.2),[22] which is known to exchange the 2-*pro*-(*R*) hydrogen of glycine. Thus incubation of glycine with this enzyme in ^2H_2O produced (2*R*)[2-$^2H_1$] glycine,[21] while (2*S*)[2-$^2H_1$] glycine[22] was obtained from an exchange of [2,2-$^2H_2$] glycine with ^1H_2O. The diazoketones **25** and **26** were then obtained via protection as trifluoroacetamide[23] followed by conversion to the acid chlorides and subsequent treatment with diazomethane. The labeled diazoketones on Wolf rearrangement[24] by photolysis in methanol yielded the methyl esters **27** and **28**. These were then hydrolyzed directly into the corresponding [3-$^2H_1$] β-alanines **2a** and **2b**. As expected, the Wolf rearrangement occurred with retention of stereochemistry at the migrating center.[25]

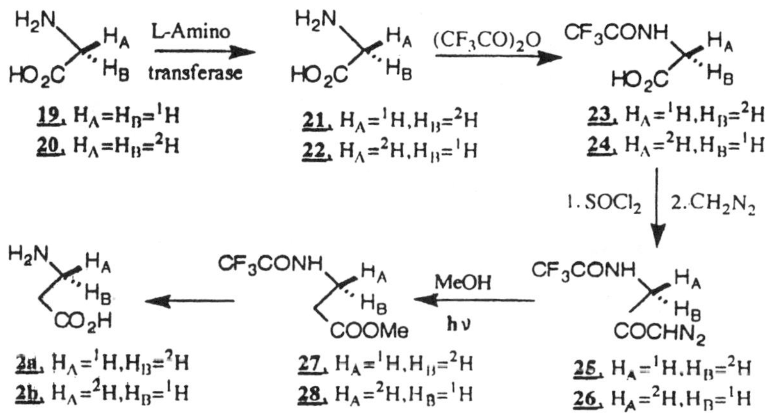

Scheme 22.3

22.5.3 Synthesis Involving Dihydrouracil Dehydrogenase

In another pioneering work Young et al.,[21] while establishing the stereochemistry of catabolism of RNA base uracil, prepared 2(*R*)-[$^2H_1$] and 3(*R*)-[$^2H_1$] β-alanines **1a** and **2a**, respectively. 5,6-Dihydrouracil was obtained when uracil was incubated with dihydrouracil dehydrogenase.[26] Subsequent hydrolysis followed by decarboxylation yielded β-alanine. Degradation of uracil is a key metabolic process, and it is interesting to note this enzyme has decreased activity in cancer cells.[27] From a mechanistic point of view the enzyme works in a highly stereospecific manner with an overall *anti* addition of hydrogens from the *Si* faces of both the prostereogenic faces at C-5 and C-6. Thus starting from [5-^2H] and (6-^2H] uracils separately and doing the incubations in ^1H_2O, stereospecifically labeled samples of β-alanine **1a** and **2a** have been prepared (Scheme 22.4).

Using the same enzyme, Young et al.[21] synthesized the (2*S*,3*S*)[2,3-$^2H_2$] β-alanine (**3b**) from uracil, following the same sequence as they did for the preparation of monodeuterated analogues. In this case, incubation was carried out in ^2H_2O, so that the two deuteriums were delivered from the *Si* face to lead ultimately to **3b** (Scheme 22.5). However, due to the severe isotope effect in ^2H_2O, the yield was poor.

22.5.4 Synthesis Involving Decarboxylase

3(*R*)-[3-^2H] β-alanine (**2a**) was also prepared from L-histidine (Scheme 22.6) by Santaniello and co-workers,[28] again exploiting the architecture of enzymes. L-Histidine decarboxylase–catalyzed decarboxylation of L-histidine in ^2H_2O took place in a stereospecific manner with complete retention of configuration. Ruthenium

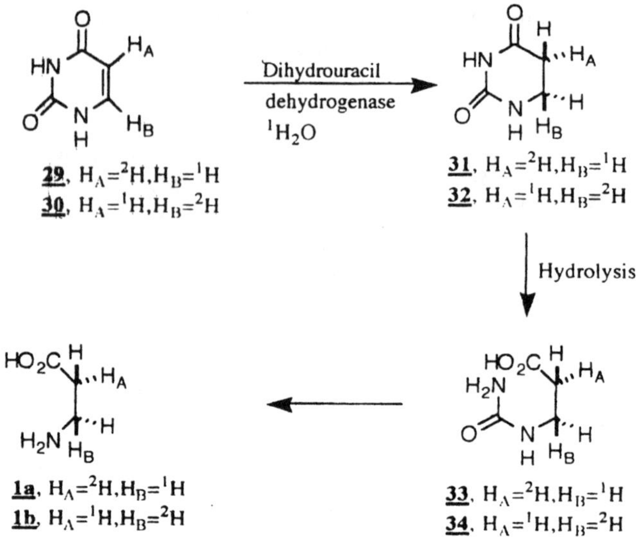

Scheme 22.4

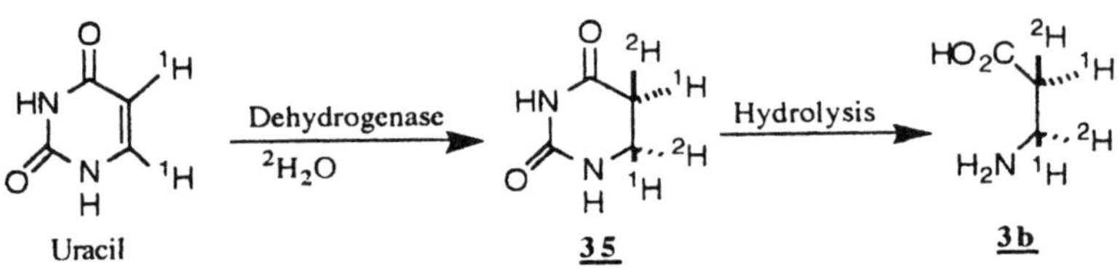

Scheme 22.5

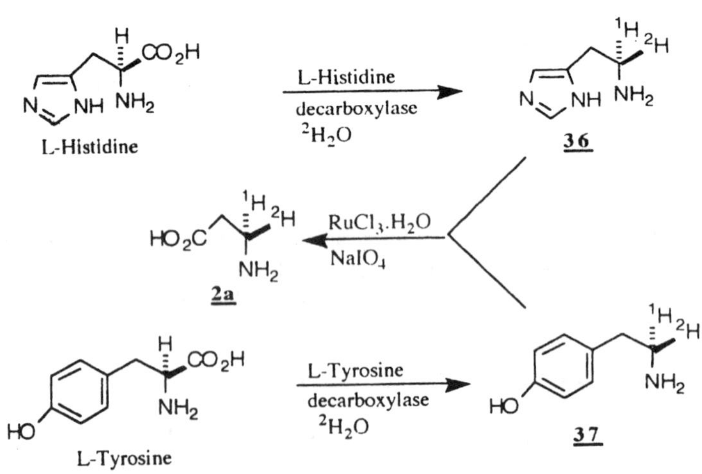

Scheme 22.6

tetroxide–mediated oxidation of the resulting amine **36** caused degradation of the imidazole ring and furnished 3(*R*)-deuterated β-alanine **2a**. The extent of deuteration was ~ 90%. **2a** was also prepared from L-tyrosine via the same sequence, this time employing L-tyrosine decarboxylase.

22.5.5 Chiral Pool Approach

The inexpensive natural product serving as the chiral pool was Mandelic acid, available in both *R*-(–) and *S*-(+) forms. Elsenbaumer and Mosher[29] prepared enantiomerically pure *R*-(+)-2-phenylethanol-2-^{2}H (**42**) starting from *R*-(–)-mandelic acid. The key steps in the synthesis (Scheme 22.7) were two stereospecific processes, namely, the S_N2 ring opening of the 1,3-dioxolane intermediate **40** upon treatment with trimethyl silyl chloride and the reduction of (*S*)-2-chloro-2-phenylethyl acetate (**41**) by lithium aluminum deuteride. Thus the overall sequence proceeded with retention of configuration via double inversion at the stereogenic center.

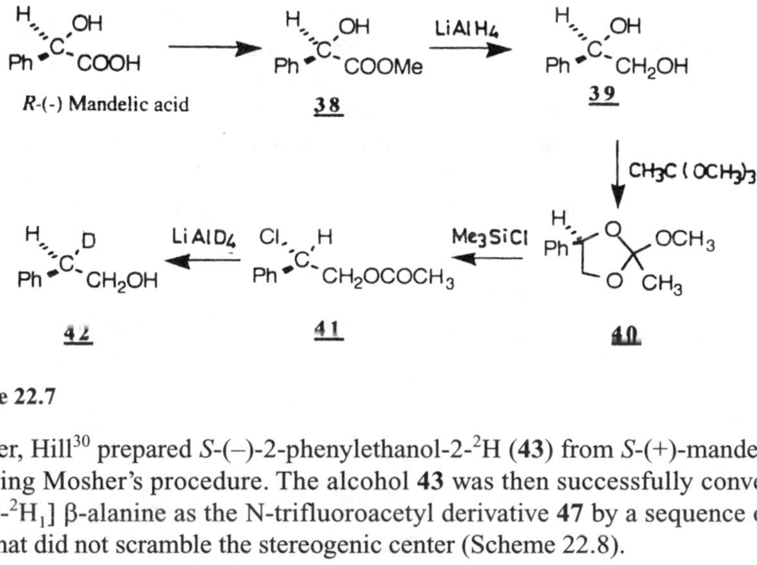

Scheme 22.7

Later, Hill[30] prepared *S*-(–)-2-phenylethanol-2-^{2}H (**43**) from *S*-(+)-mandelic acid following Mosher's procedure. The alcohol **43** was then successfully converted to (2*R*)[2-$^2H_1$] β-alanine as the N-trifluoroacetyl derivative **47** by a sequence of reactions that did not scramble the stereogenic center (Scheme 22.8).

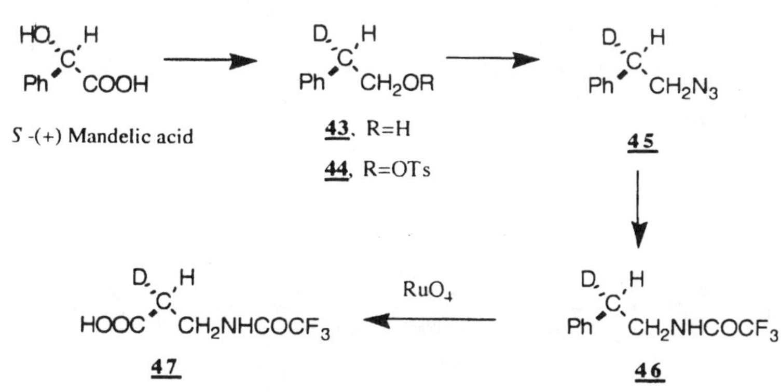

Scheme 22.8

22.5.6 Asymmetric Synthesis

Many chirally deuterated molecules have been made by asymmetric nucleophilic reduction of deuterated aldehydes by alpine borane.[31] The resulting chiral isotopically labeled alcohols can then be elaborated into various target molecules. Asymmetric protonation or deuteration[32] at a carbanion is another possible route to chirally isotopically labeled molecules. A third possibility exists where a hydrogen or deuterium is delivered onto a carbon radical in a stereospecific manner.[33] Apparently, since the radicals are extremely reactive, such types of processes are expected to have low stereoselectivity. However, with slow generation of radicals at a low concentration and differentiating between the two facial approaches of the donor agent on steric grounds, one may achieve a high degree of selectivity. Recently several stereoselective reactions based on radicals have appeared in the literature.[34] Basak[13] has synthesized 3-deuterated β-alanines based on a stereoselective radical reaction. The key step is the highly stereoselective reduction of both enantiomers of *trans*-3-trimethylsilyl-4-thiophenyl-2-azetidinones **49** and **54** by tributyl tin deuteride.[35] The deuterium was delivered to the radical generated at C-4 mainly from the face opposite that containing the trimethylsilyl group. The extent of stereoselectivity was around 18:1. The azetidinone derivatives **49** and **54** were prepared[36] in both enantiomeric forms starting from 4(*R*)- and 4(*S*)-thiophenyl azetidinones **48** and **53**. Removal[37] of the TMS group after reduction, followed by N-protection, hydrolysis, and carboxy protection, furnished the N-*t*-Boc-β-alanine benzhydryl esters **52** and **57**. Deprotection with trifluoroacetic acid finally afforded the target deuterated molecules. The entire synthesis is shown in Scheme 22.9. Thus the TMS group really played the role

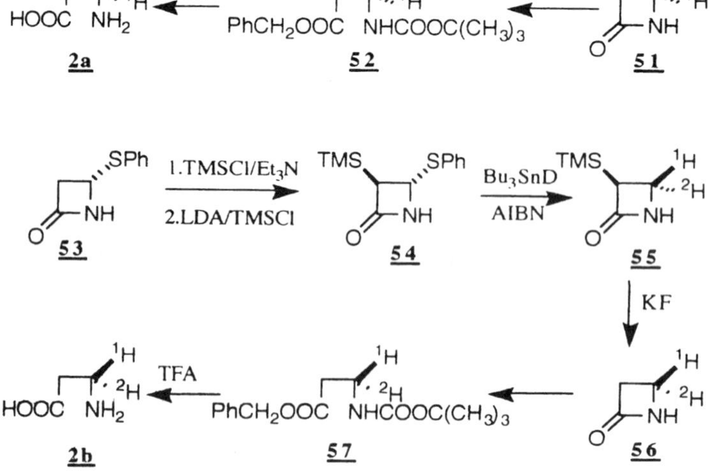

Scheme 22.9

of a bulky proton.[38] Using an even bulkier *t*-butyldimethylsilyl group, the extent of stereoselectivity could be increased to 24:1.[39]

Access to the 3*S*[3-^{2}H] β-alanine could also be achieved by the reduction of *cis*-3-trimethylsilyl-4-thiophenyl azetidinone **58**, obtained during silylation as a minor component. However, in this case, the reduction proceeded very slowly, presumably because of the steric hindrance offered by the TMS group to the tributyl tin radical approaching for the initial abstraction of the thiophenyl group (Scheme 22.10).

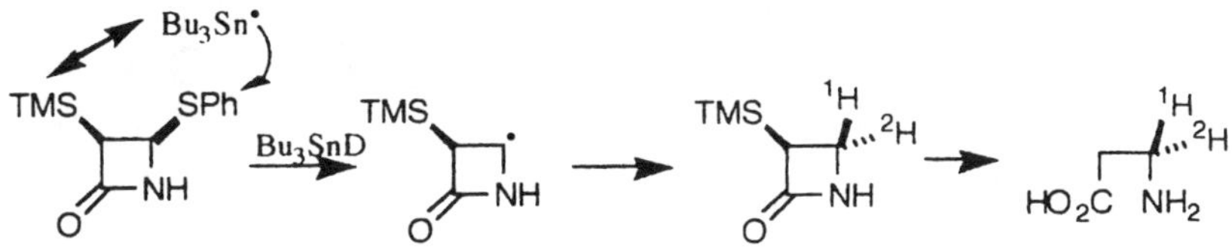

58 59 55 2b

Scheme 22.10

It is interesting to note that the method was based upon the principle of self-regeneration of chirality. The chirality at C-4 was exploited to generate chirality at C-3. When the radical was formed at C-4, the chirality of C-3 helped in the regeneration of stereogenicity at the original center by controlling the direction of delivery of deuterium.

22.5.7 Configurational Assignment

The assignment of stereochemistry for the chirally deuterated β-alanines were mainly done by high-field ^{1}H-NMR spectroscopy. It was only in the synthesis of 3(*R*)-[$^2H_1$] β-alanine by Santaniello et al.[28] that the stereochemistry was assigned on the basis of optical rotation of the N-phthaloyl methyl ester **60**. In β-alanine the C-2 or C-3 hydrogens are enantiotopic and have the same chemical shift. Young et al.[20] converted β-alanine into the camphanyl derivative **61**, which resolved the signals for C-3 hydrogens so that assignments could be made. Sections of the 360 MHz ^{1}H-NMR spectra of the camphanyl derivatives of unlabeled and stereospecifically C3-deuterated β-alanines are shown in Figure 22.1. Obviously, the 3-deuterated β-alanines were prepared with good optical purity. Assignments for the C-2 hydrogens were difficult as the signals were ill-resolved. The chirality at this center was ultimately assessed[40] by conversion (Scheme 22.11) to the 1(*R*)- and 1(*S*)-[1-$^2H_1$] ethanolamines **70** and **71**, which were then converted to N,O-dicamphanyl derivatives (for NMR, see Figure 22.2).

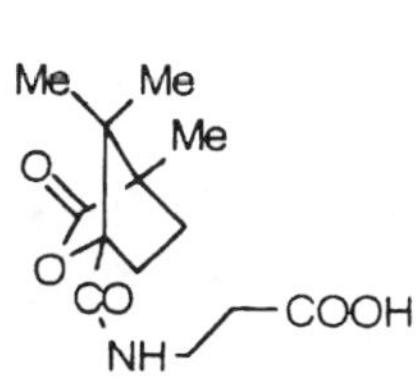

60

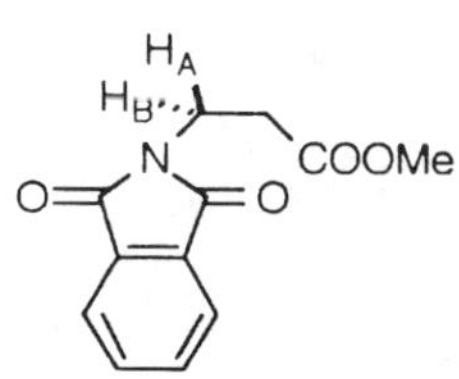

61, H_A=^{1}H, H_B=^{2}H

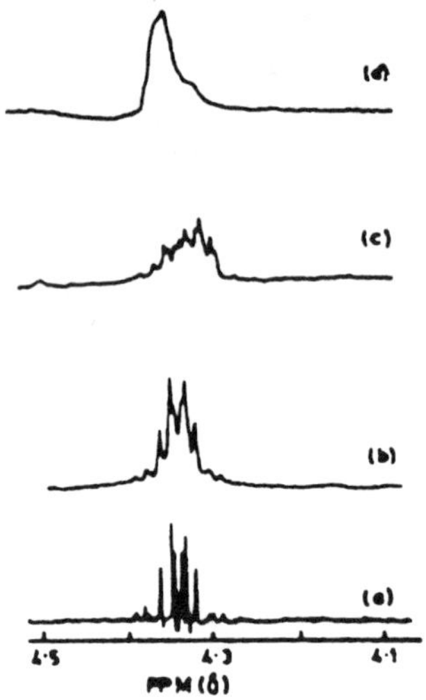

Figure 22.1. 360 MHz ^{1}H-NMR spectra ($CDCl_3$) of β-alanine C-3 protons in (a) the amide (**60**); (b) synthetic 3(*R*)-[3-$^2H_1$]-(**60**) derived from 3(*R*)-[3-$^2H_1$-] alanine; (c) synthetic 3(*S*)-[3-$^2H_1$]-(**60**) derived from 3(*S*)-[3-$^2H_1$] alanine.

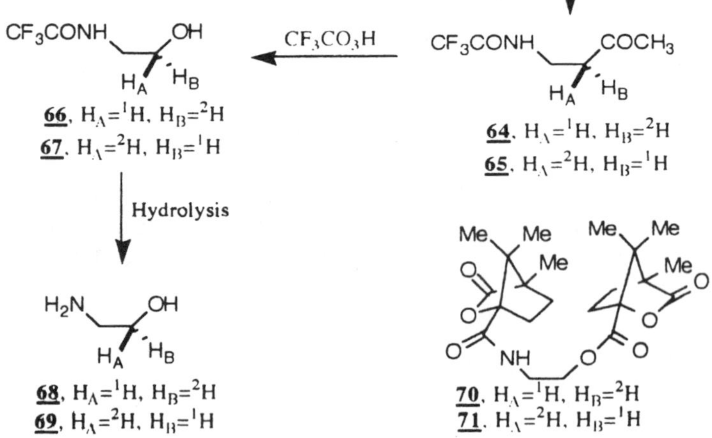

Scheme 22.11

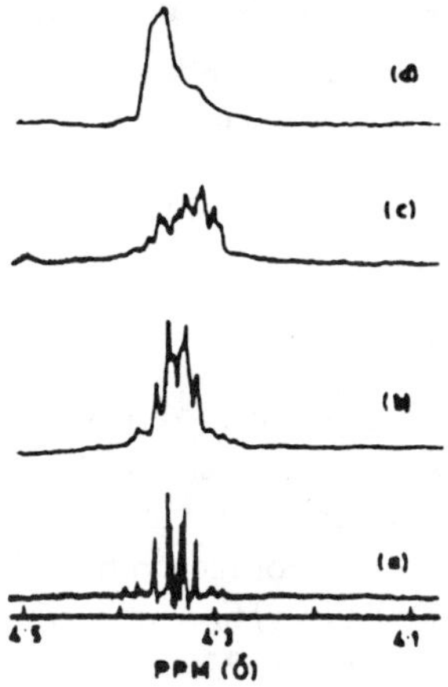

Figure 22.2. 360 MHz ^{1}H-NMR spectra ($CDCl_3$) of N,O-dicamphanyl-2-aminoethanol in (a) unlabeled compound with line narrowing; (b) unlabeled compound without line narrowing; (c) compound obtained from 2(*R*)-[2-$^{2}H_1$]-β-alanine; (d) compound obtained from (2*S*,3*RS*)-[2,3-$^{2}H_2$]-β-alanine.

In the synthesis of 3(*R*) and 3(*S*) β-alanine by Basak,[13] the chirality of the precursor trimethylsilyl-4-deutero azetidinone was also established by ^{1}H-NMR (Figure 22.3). In all the steps leading to the β-alanine, the deuterium content measured by mass spectrometry) remained the same proving there was no loss of stereochemical integrity.

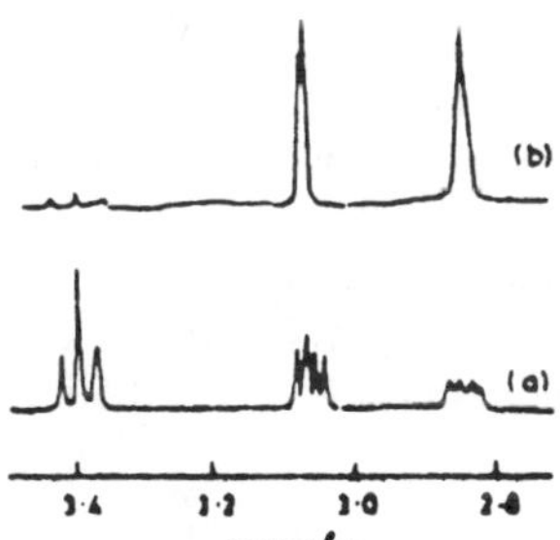

Figure 22.3. 400 MHz ^{1}H-NMR spectra ($CDCl_3$) of 3-trimethylsilyl-2-azetidinone of (a) unlabeled compound; (b) 4(*R*)-[4-^{2}H] isomer.

22.6 Enantioselective Synthesis of Other Deuterated β-Amino Acids

Apart from β-alanine, other β-amino acids are also known either as naturally occurring or as constituents of biologically important natural products.[41–46] Chirally deuterated forms of these amino acids are also invaluable entities for revealing their role in natural product biosynthesis. Significant progress has already been made towards achieving this goal. Like the deuterated β-alanines, other similar isotopically labeled β-amino acids have mostly been prepared by enzyme-mediated transformations. Few chemical methods are also known; recently Sewald et al.[47] have published a general chemical approach to the asymmetric synthesis of these target molecules. All these are described in this section.

22.6.1 Biocatalytic Approach

During the investigation of the biosynthesis of β-lysine, Aberhart et al.[48] synthesized the di-N-phthaloyl derivatives of (2*R*,3*S*)[2,3-$^2H_2$] β-lysine (**74**) and (2*S*)[2-$^2H_1$] β-lysine (**75**) starting from 3-dideuterated and unlabeled β-lysine, respectively, using the enzyme lysine-2,3-aminomutase. For the synthesis of **75**, the incubation was carried out in the presence of ^2H_2O. The reaction proceeded with inversion of configuration of C-2 and C-3 of α-lysine (Scheme 22.12).

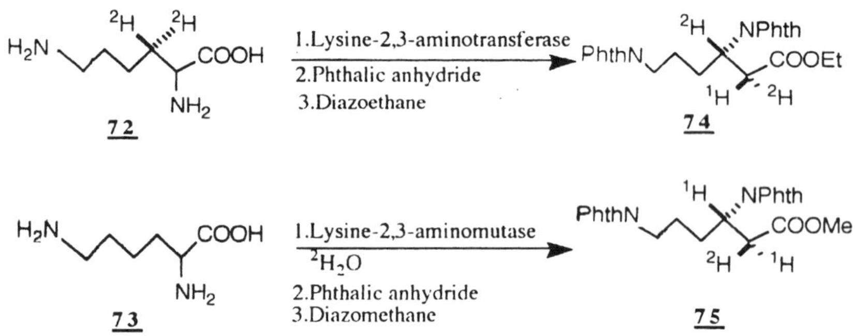

Scheme 22.12

N-Methyl β-arginine is another β-amino acid that is a constituent of the antibiotic Blasticidin.[49] While investigating the stereochemistry of the transformation of arginine to the β-isomer by arginine-2,3-aminomutase, Gould et al.[49] prepared the stereospecifically C-3 deuterated N-methyl β-arginine isolated in the form of Blasticidin (**77**) rather than the free amino acid (Scheme 22.13).

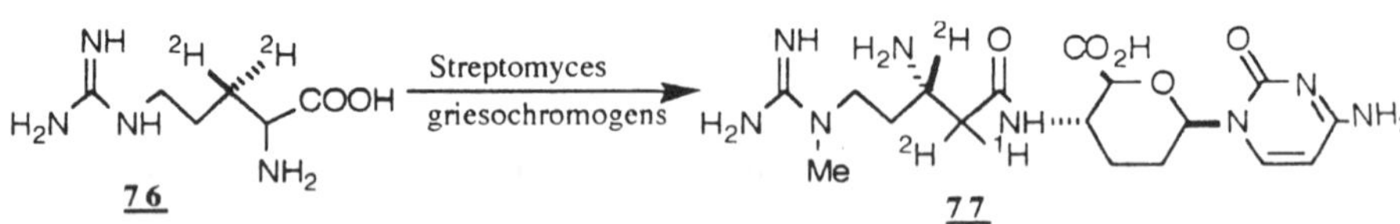

Scheme 22.13

22.6.2 Chemical Approach

Stereospecifically labeled β-amino acids are excellent precursors for labeled β-lactams. Based on this, Young et al.[50] synthesized (2*S*,3*R*) and (2*S*,3*S*) isomers of 3-amino-2-*p*-tolylsulfonamidopropanoic acids (**81** and **82**), which may be regarded as derivatives of α-amino β-amino acids. The starting materials are the specifically labeled aspartic acids **13** and **14**. The key step in the synthetic scheme is the Hofmann rearrangement of labeled asparagines **79** and **80** that occurred with retention of configuration at the migrating center (Scheme 22.14).

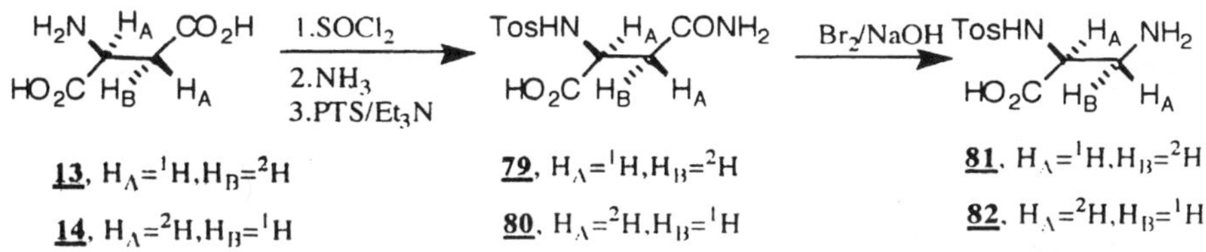

Scheme 22.14

Aberhart et al. have also reported the synthesis of racemic (2*R*,3*R*) and (2*S*,3*S*)[2,3-$^2H_2$] β-lysine as the di-N-phthaloyl derivative **87**. Stereospecific reduction of the enamine **85** with deuterium gas using Wilkinson catalyst gave racemic **86**, which on heating at 200°C was converted to the desired racemic **87** (Scheme 22.15).

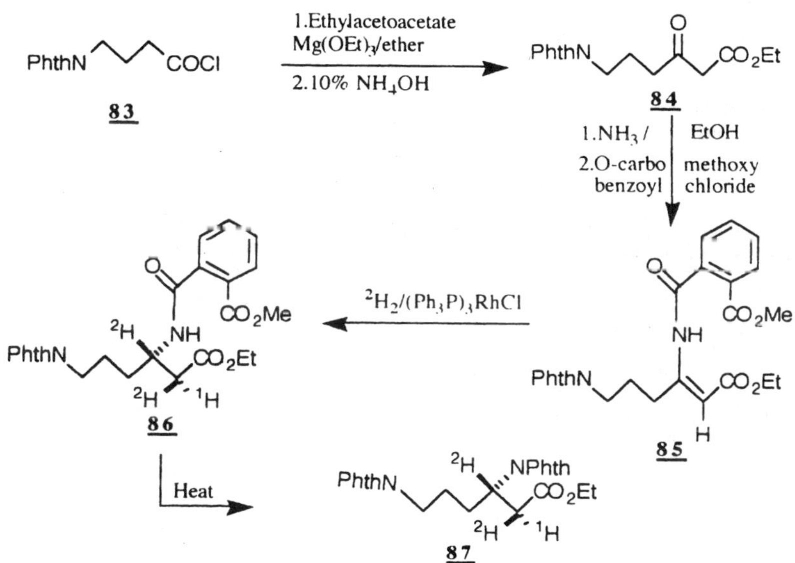

Scheme 22.15

Recently, Sewald et al.[47] have developed an extremely general method for asymmetric synthesis of α-deuterated β-amino acids via conjugate addition of the cuprate derived from N-(*R*)-1-phenylethyltrimethylsilylamine (**88**) to an α,β-unsaturated ester. The intermediate enolate **89** was then quenched with ^2H_2O. The degree of deuteration was > 95%, while the distereoselectivity was > 90%. The high stereoselectivity was explained on the basis of a coordinative stabilization of the enolate by the oxaphilic silicon substituent, which strongly favored the *Re*-face attack of the deuteron (Scheme 22.16).

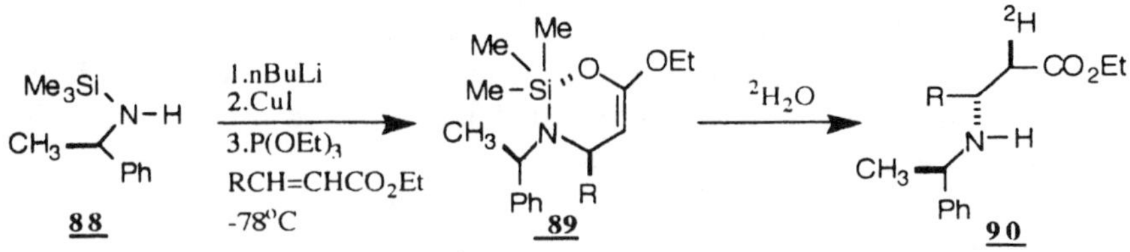

Scheme 22.16

22.7 Acknowledgments

The author thanks DST, India, for financially supporting the work carried out in his laboratory.

22.8 References

1. C. Walsh, *Enzymatic Reaction Mechanisms* (W. H. Freeman: New York, 1979), p. 108.
2. R. K. Hill, in *Biorganic Chemistry*, Chapter 5, Vol. II, E. E. Van Tamelen, Ed. (Academic Press, New York, 1978).
3. J. E. Baldwin, E. P. Abraham, *Nat. Pdt. Rep.* **5**, 129 (1988); C. A. Townsend, A. B. Thies, A. S. Neese, E. B. Barrabee, D. Poland, *J. Am. Chem. Soc.* **107**, 4760 (1985); S. J. Field, D. W. Young, *J. Chem. Soc. Perkin Trans I*, 2387 (1983); D. Gani, D. W. Young, *J. Chem. Soc. Perkin Trans. 1*, 2393 (1983); P. C. Prabhakaran, T. N. Woo, P. S. Yorgey, S. J. Gould, *J. Am. Chem. Soc.* **110**, 5785 (1988).
4. H. R. Crumpler, C. E. Dent, H. Harris, R. G. Westall, *Nature* **167**, 307 (1951); S. Kondo, S. Shibahara, S. Takahasi, K. Maeda, H. Umesawa, H. Ohno, *J. Antibiotics Ser A* **29**, 97 (1976).
5. S. M. Hecht, *Acc. Chem. Res.* **19**, 383 (1986).
6. G. D. Vogels, C. Vander Drift, *Bacteriol. Rev.* **40**, 403 (1976).
7. P. R. Vagelos, J. M. Earl, E. R. Stradman, *J. Biol. Chem.* **234**, 490 (1959).
8. W. J. Krol, A. Basak, S. P. Salowe, C. A. Townsend, *J. Am. Chem. Soc.* **111**, 7625 (1989).
9. A. Basak, S. P. Salowe, C. A. Townsend, *J. Am. Chem. Soc.* **112**, 1654 (1990).
10. E. Santaniello, P. Ferraboschi, P. Gresenti, A. Manzocchi, *Chem. Rev.* **92**, 1071 (1992).
11. J. Crossby, *Tetrahedron* **47**, 4789 (1991).
12. D. Seebach, R. Naef, *Helv. Chim. Acta* **64**, 2704 (1981): D. Seebach, R. Naef, G. Calderari, *Tetrahedron* **40**, 1313 (1984).
13. A. Basak, *Synth. Commun.* **23**, 1985 (1993).
14. B. S. Axelsson, K. J. O'Toole, P. A. Spencer, D. W. Young, *J. Chem. Soc. Perkin Trans. 1*, 807 (1994).
15. D. Gani, D. W. Young, *J. Chem. Soc. Chem. Commun.*, 576 (1983).
16. S. P. Colourick, N. O. Kaplan, Eds., *Methods in Enzymology*, Vol. II, p. 386 (Academic Press, New York, 1955).
17. D. W. Young, in *Isotopes in Organic Chemistry*, E. Buncel, C. C. Lee, Eds. (Elsevier, Amsterdam, 1987).

18. D. Gani, D. W. Young, D. M. Carr, J. P. Poyser, I. H. Sadler, *J. Chem. Soc. Perkin Trans. 1*, 2811 (1983).

19. E. M. Richards, J. C. Tebby, R. S. Ward, D. H. Williams, *J. Chem. Soc. (C)*, 1542 (1969).

20. D. Gani, O. C. Wallis, D. W. Young, *Eur. J. Biochem.* **136**, 303 (1983).

21. D. Gani, D. W. Young, *J. Chem. Soc. Perkin Trans. 1*, 1355 (1985).

22. P. Besmer, D. Arigoni, *Chimia* **22**, 494 (1968).

23. F. Weygand, P. Klinke, I. Eigen, *Chem. Ber.* **90**, 1896 (1957).

24. P. A. Smith, in *Molecular Rearrangements*, E. P. De Mayo, Ed. (Wiley Interscience, New York, 1967), p. 528.

25. S. J. Field, D. W. Young, *J. Chem. Soc. Chem. Commun.*, 1163 (1979).

26. T. Shiotani, G. Weber, *J. Biol. Chem.* **256**, 219 (1981); S. J. Field, D. W. Young, *J. Chem. Soc. Perkin Trans. 1*, 2387 (1983).

27. S. W. Queener, G. Weber, *Cancer Res.* **31**, 1004 (1971).

28. E. Santaniello, A. Manzocchi, P. A. Biondi, *J. Chem. Soc. Perkin Trans. 1*, 307 (1981).

29. R. L. Elsenbaumer, H. S. Mosher, *J. Org. Chem.* **44**, 600 (1979).

30. R. K. Hill, S. R. Prakash, T. M. Zydowsky, *J. Org. Chem.* **49**, 1666 (1984).

31. M. M. Midland, A. Tramontano, S. A. Zderic, *J. Am. Chem. Soc.* **99**, 5211 (1977).

32. K. Fuzi, K. Taneka, H. Miyamoto, *Tetrahedron: Asymmetry* **4**, 247 (1993).

33. J. A. Aimetti, E. S. Hamanaka, D. A. Johnson, M. S. Kellog, *Tetrahedron Lett.* **20**, 4631 (1979).

34. T. V. Rajan Babu, *Acc. Chem. Res.* **24**, 139 (1991); N. A. Porter, J. D. Bruhnke, W. X. Wu, I. J. Rosenstein, R. A. Breyer, *J. Am. Chem. Soc.* **113**, 7788 (1991).

35. M. S. Manhas, M. S. Khajvi, S. S. Bari, A. K. Bose, *Tetrahedron Lett.* **24**, 2323 (1983).

36. M. Shibasaki, A. Nishida, S. Ikegami, *J. Chem. Soc. Chem. Commun.*, 1324 (1982).

37. H. Fritz, P. Sutter, C. D. Weis, *J. Org. Chem.* **51**, 558 (1986).

38. J. R. Hwu, J. M. Wetzel, *J. Org. Chem.* **50**, 3946 (1985); J. R. Hwu, N. Wang, *Chem. Rev.* **89**, 1599 (1989).

39. A. Basak, unpublished work.

40. D. Gani, D. W. Young, *J. Chem. Soc. Chem. Commun.*, 867 (1982).

41. W. J. McGahren, G. O. Mortam, M. P. Kunstmann, G. A. Ellestad, *J. Org. Chem.* **42**, 1282 (1977). P. Prabhakaran, N. T. Woo, P. S. Yorgey, S. J. Gould, *Tetrahedron Lett.* **27**, 3815 (1986).

42. T. C. Stadtman, *Adv. Enzmol. Relat. Areas Mol. Biol.* **38**, 413 (1970); M. Ohsugi, J. Kahn, C. Hensley, S. Chew, H. A. Barker, *J. Biol. Chem.* **256**, 7642 (1981); S. J. Gould, K. J. Marbinkus, C. H. Tann, *J. Am. Chem. Soc.* **103**, 2871 (1981).

43. E. Graf, H. Boedekker, *Arch. Pharm. (Weinheim, Ger.)* **289**, 364 (1956).

44. Z. Kurylo-Borowska, T. Abramsky, *Biochim. Biophys. Acta* **264**, 1 (1972).

45. J. M. Poston, *Phytochemistry* **17**, 401 (1978); I. Freer, G. Pedrocchi-Fantoni, D. J. Picken, K. J. Overton, *J. Chem. Soc. Chem. Commun.*, 80 (1981).

46. R. E. Summons, *Phytochemistry* **20**, 1125 (1981).

47. N. Sewald, K. D. Hiller, B. Helmreich, *Liebigs Ann.*, 925 (1995).

48. D. J. Aberhart, S. J. Gould, H-J. Lin, T. K. Thiruvengadam, B. H. Weiller, *J. Am. Chem. Soc.* **105**, 5461 (1983).

49. P. C. Prabhakaran, N. T. Woo, P. S. Yorgey, S. J. Gould, *J. Am. Chem. Soc.* **110**, 5785 (1988).

50. D. G. Gani, D. W. Young, D. M. Carr, J. P. Poyser, I. H. Sadler, *J. Chem. Soc. Perkin Trans. 1*, 2811 (1983).

31. W. L. F. Armarego, B. A. Milloy, W. Pendergast, *J. Chem. Soc., Perkin Trans. 1*, 2229 (1976).

Index